THE EVOLUTION OF GEOTECH - 25 YEARS OF INNOVATION

The Evolution of Geotech - 25 Years of Innovation

Editors

Reginald E. Hammah

Rocscience Africa, Accra, Ghana

Thamer E. Yacoub

Rocscience Inc, Toronto, Canada

Alison McQuillan

Rocscience Australia, Southport, Australia

John Curran

Rocscience Inc, Toronto, Canada

CRC Press
Taylor & Francis Group
Boca Raton London New York Leiden

CRC Press is an imprint of the
Taylor & Francis Group, an **informa** business
A BALKEMA BOOK

CRC Press/Balkema is an imprint of the Taylor & Francis Group, an informa business

Typeset by Integra Software Services Pvt. Ltd., Pondicherry, India

Library of Congress Cataloging-in-Publication Data

A catalog record has been requested for this book

Published by: CRC Press/Balkema
Schipholweg 107C, 2316 XC Leiden, The Netherlands
e-mail: enquiries@taylorandfrancis.com
www.routledge.com – www.taylorandfrancis.com

ISBN: 978-1-032-03645-8 (Hbk)
ISBN: 978-1-032-03646-5 (Pbk)
ISBN: 978-1-003-18833-9 (eBook)
DOI: 10.1201/9781003188339

Table of contents

Technical sessions
Session 1 - Rock mass characterization

Session 2 - Tunnels & caverns I

Session 3 - Stability analysis of mining slopes I

Session 4 - Probabilistic slope stability analysis I

Session 5 - New technologies in geotechnical engineering

Session 6 - Surface excavations

Session 7 - 3D slope stability analysis

Session 15 - Spatial variability in probabilistic slope stability analysis

Session 16 - Numerical analysis of underground excavations

Session 17 - Stability analysis of jointed rock slopes

Session 18 - Limit equilibrium slope stability analysis

The Evolution of Geotech - 25 Years of Innovation – Hammah et al. (Eds)

Preface

Dear Authors and Respected Readers,

With deep satisfaction, we write this Foreword to the Proceedings of the first-ever Rocscience International Conference (RIC) held online on April 20 and 21, 2021. The conference theme, "The Evolution of Geotech: 25 Years of Innovation," was an invitation to bring the sharpest brains in geotechnical engineering – academics, researchers and practitioners – together to discuss progress and exchange ideas as part of celebrating 25 years of Rocscience's existence. We believe that the intellectually stimulating discussions from these interactions will go a long way in helping advance our field.

The day before the Conference, Rocscience awarded the 2021 Lifetime Achievement Medal to Dr. Evert Hoek, a leading international rock mechanics expert who has had an enormous impact on our field as a researcher, practitioner, educator, and mentor. Dr. Evert Hoek gave an address that outlined his journey over the past 50 years. You can access this address at https://www.rocscience.com/learning/hoeks-corner.

RIC 2021 was attended by more than 500 participants from all continents. It received over 100 abstract submissions, which were carefully reviewed. Seventy papers were accepted for presentation at the Conference. There were 18 oral sessions, with each session covering a specific topic. The papers are published in these proceedings.

In addition to the contributed papers, Prof. Will Bawden, Prof. Kemal Onder Cetin, Prof. KK Phoon, and Prof. Mark Diederichs delivered four invited keynote papers at the Conference. These keynotes discussed practical ways to maximize the benefits of existing and emerging empirical approaches and powerful numerical modelling and data analysis tools. These keynotes are also included in the proceedings.

The Conference also hosted two panel discussions aimed at helping us look at marrying geotechnical modelling to instrumentation and monitoring efforts through calibration. The first panel discussed the geotechnical industry's efforts to improve the predictive capabilities of modelling tools through feedback from monitoring and instrumentation. Presenters on the second panel shared their thoughts on the progress, promise and hiccups of numerical modelling. They provided their views on using our modelling tools to facilitate understanding rather than hinder critical analysis. The position papers of the panellists are included in these proceedings.

The Conference also featured a special session to recognize the contributions of Dr. John Curran, the founder of Rocscience, to geomechanics numerical research and modelling. Participants glimpsed John's dedication, hard work and an unremitting devotion to the things he wanted to see happen (borrowing from the words of Frank Lloyd Wright).

We want to thank the Organizing Committee, Technical Advisors and Reviewers for making the daunting task of organizing a major international conference in COVID times appear simple. We thank all keynote speakers, panellists, authors and Conference participants for their contributions.

We hope that the Conference Proceedings will be an excellent reference for geotechnical engineering. We look forward to them kindling new ideas and enhancing the practical use of the tools of our trade in our everyday work. We are confident that what we learn from these Proceedings will inspire us all to contribute to scaling the challenges we presently face.

Dr. Reginald E. Hammah
Technical Chair, RIC 2021

Dr. Thamer E. Yacoub
General Chair, RIC 2021

Dr. Alison McQuillan
Co-editor, RIC 2021

Dr. John Curran
Co-editor, RIC 2021

The Evolution of Geotech - 25 Years of Innovation – Hammah et al. (Eds)

Organization

ORGANIZING COMMITTEE

Dr. Thamer Yacoub, *Conference General Chair, Rocscience Inc.*
Dr. Reginald Hammah, *Conference Technical Chair, Rocscience Africa*
Saiyona Ghosh, *Marketing & Customer Service Coordinator, Rocscience Inc.*
Calvin Heinrichs, *Platform & Conference Logistics Coordinator, Rocscience Inc.*
Ruth Obeng-King, *Administration & Conference Coordinator, Rocscience*

AWARD RECIPIENTS
Dr. Evert Hoek, *Lifetime Achievement Medal*
Dr. John Curran, *Award for Outstanding Contributions to Geomechanics Numerical Research & Modelling*

PANEL DISCUSSIONS
Panel 1: From Monitoring and Calibration to Predictive Modelling
Dr. Alison McQuillan, *Moderator, Rocscience Australia, Australia*
Neil Bar, *Panellist, Gecko Geotechnics, Australia*
Davide Colombo, *Panellist, TRE-ALTAMIRA, Italy*
Niccolo Coli, *Panellist, Hexagon Geosystems, Italy*
Dr. Will Bawden, *Panellist, University of Toronto & Mine Design Technology, Canada*

Panel 2: Is Numerical Modelling a Solution or a Problem?
Dr. Reginald Hammah, *Moderator, Rocscience Africa, Ghana*
Dr. Joe Carvalho, *Panellist, Golder Associates, Canada*
Dr. Thomas Dick Stacey, *Panellist, University of Witwatersrand, South Africa*
Dr. Brent Corkum, *Panellist, Rocscience Inc., Canada*
Thiago Bretas, *Panellist, BVP Geotecnica e Hidrotecnia, Brazil*

INVITED KEYNOTES
Dr. Will Bawden, *University of Toronto & Mine Design Technology, Canada*
Dr. Kemal Onder Cetin, *Middle East Technical University, Turkey*
Dr. Kok-Kwang Phoon, *Singapore University of Technology and Design, Singapore*
Dr. Mark Diederichs, *Queen's University, Canada*

The Evolution of Geotech - 25 Years of Innovation – Hammah et al. (Eds)
 ISBN 978-1-032-03645-8

Technical advisors, reviewers and session chairs

Derek Martin, *University of Alberta, Canada*
Manoj Verman, *Rocscience, India*
Ana Maria Parra, *Carlos H. Parra & Asociados, Colombia*
Mason Ghafghazi, *University of Toronto, Canada*
Denis Kalumba, *University of Cape Town, South Africa*
Anil Yunatci, *Geodestek, Turkey*
Sebastian Lobo-Guerrero, *American Geotechnical & Environmental Services, US*
Erik Loehr, *University of Missouri, US*
Richard Bathurst, *Royal Military College of Canada, Canada*
David Oliveira, *Jacobs, Australia*
Navid Jafari, *Louisiana State University, US*
Sina Javankhoshdel, *Rocscience, Canada*
Sergio Fontoura , *Pontifícia Universidade Católica do Paraná, Brazil*
Antonio Samaniego, *SVS Ingenieros, Peru*
Francisco Suarez, *Pontificia Universidad Católica de Chile, Chile*
Phil De Graaf, *Anglo American, Australia*
Neil Bar, *Gecko Geotechnics, Australia*
Niccolo Coli, *Hexagon Geosystems, Italy*
Will Bawden, *University of Toronto & Mine Design Technologies, Canada*
Thiago Bretas, *BVP Geotecnica e Hidrotecnia, Brazil*
Mark Diederichs, *Queen's University, Canada*
Azadeh Riahi, *Beyond Limits, US*
Ian Williams, *Rocscience, UK*
Tolga Bilge, *Geodestek, Turkey*
Joe Carvalho, *Golder Associates, Canada*
Ashraf Mohd, *Universiti Sains Malaysia, Malaysia*
Trevor Carter, *TGC Geosolutions, Canada*

Keynote lectures

The Evolution of Geotech - 25 Years of Innovation – Hammah et al. (Eds)

Reflections on 40+ years of rock engineering practice in mining

Will F. Bawden
University of Toronto and MDT Engineering

ABSTRACT: I have been teaching and researching rock engineering and practicing as a consultant since 1972. I have spent most of that time in underground mining. I have seen significant changes, both evolutionary and radical, in the sector. In this paper, I reflect on some of these changes in mining geomechanical practice. I will also comment on some of the positive and occasionally less positive impacts of these changes. One constant throughout my career is that geomechanical designs will always be fraught with uncertainty. As a result, we will discuss geomechanical design as being, in its essence, a risk management exercise. I will argue that, as impressive as modern numerical analysis techniques are, they do not on their own mitigate design risk; under worst-case scenarios, they make things worse. Empirical and simple models must be used as a check. The paper intends to provoke healthy skepticism in geomechanics practitioners, designers and managers regarding all the aspects of the geomechanical analysis and design process. A failure to recognize the risks in these processes can have catastrophic effects.

1 INTRODUCTION

Over the past 20 – 30 years, great advances have been made in instrumentation, empirical methods and numerical modelling in mining geomechanics. Instrumentation had a head start over numerical modelling. (This was simply due to the lack, until recently, of adequate computational power.) The earlier days' instrumentation was primitive but still useful.

Charles Terzaghi (1929) presented one of the most important instrumentation pieces available then and today – a well-trained professional's eyes and brain - in a paper entitled, 'Effect of Minor Geologic Details on the Safety of Dams'! One of the earliest papers on instrumentation in mining was 'Instrumentation of room and pillar workings in a copper mine of the Copper Range Company, White Pine Michigan' by Jack Parker and James J Scott (1964). In my own experience in the mid 1980's, late-stage mining at the Norita mine in northwest Quebec was successfully completed under moderate seismic conditions with the help of output from early extensometer and stress cell data, complemented by an early experimental microseismic system.

During the 1980s and '90s developments in empirical design methods (e. g., caving chart, stability graph method, and pillar stability chart) were developed and became widely used in the mining sector. They continue to enjoy wide application today.

An article (Clarke, 1973) in South Africa described some of the earliest attempts at numerical modelling in mining rock mechanics. In 1964, Ryder et al (1964) and Ortlepp et al (1964) demonstrated that elastic theory could be applied to stress-strain problems in deep hard rock mining. Based on these findings, Salamon et al (1964) developed an electrolyte analogue simulator to study the convergence of tabular excavations, and this work was further described by Wilson and More-O'Ferrall (1970). Because the set-up of the electrical analogue simulator

DOI: 10.1201/9781003188339-1

was difficult, a computer program, MINSIM was developed to tackle tabular mining excavation modelling.

During that same period, other researchers (Deist and Oravecz, 1968, Plewman et al, 1969, Deist et al, 1972 and Kulhawy, 1973) examined the application of the Finite Element Method (FEM) to underground rock mechanics. Duncan & Goodman (1968) and Stacey (1969) applied finite element analysis to jointed rock slopes. Cundall (1971) applied discontinuum analysis to blocky rock masses.

Today, due to the unprecedented explosion in computational power and associated software, numerical modelling capabilities have surpassed instrumentation. In some practitioners' belief, modelling should replace empirical design methods.

In this paper, I will argue that eliminating empirical methods is not wise. This position does not mean to diminish the importance of numerical modelling advances. Instead, it aims to shine light on some unanticipated consequences of the advances that can and do occur; as with many rapidly advancing technologies, the impacts of powerful models are not all positive.

2 HISTORICAL PERSPECTIVE

Although mining has existed since antiquity, metal usage increased exponentially and continues to grow following the industrial revolution. Canada's diverse geology (and geography) resulted in a strong mining sector and recognition as one of the top mining countries. Until the mid-20th century, most underground mines were relatively shallow and dominated by narrow vein cut and fill and shrink stope mining methods. Larger orebodies were mined using timber set (square set) stoping methods. These mining methods all relied on handheld equipment and rail haulage, meaning that the mine development openings by today's standards were generally small (≤ 3 x 3 m). Even with these small openings, rock fall-related accidents and fatalities were common.

Around the 1960's the development of the first steel mechanical rock bolts, along with mobile rubber-tired underground equipment, led to much larger scale underground mines. For example, development openings of 5 x 5 m or larger cross-section and bulk underground stoping methods (transverse and longitudinal longhole stopes, sublevel caving, etc.) began to appear. These larger mine openings naturally resulted in increased rock-related mine accidents. Simultaneously, more mines in Canada started going deeper, and rockbursting, a long-term mining hazard in deep South African gold mines, became more severe, particularly in Ontario. Figure 1 shows reported rock-bursts and fatalities in Ontario mines from 1928 to 1990 (After Hedley, 1992). Although improved technology reduced fatalities, between the 70's and early 80's, there were nevertheless three or more provincial enquiries into mine safety in Ontario alone. Similar issues were occurring in other jurisdictions.

Increasing pressure from the public and government regulators forced companies to recognize that mine safety was becoming a central issue in the new and increasingly crucial paradigm of the 'social license to operate' and was beginning to impact mine permitting. It emerged that geotechnical problems were the driving force in many areas of mine safety. Significant research funding (both corporate and governmental) has since flowed to this area. This has resulted in impressive advances in new and improved ground support elements and practices, empirical design methods, instrumentation, mine design practices and numerical modelling software. The exponential growth is continuing and is widespread. Falling prices of significant computational power have been central to many of these technological advances.

The highly three-dimensional nature of mines, combined with computing power limitations, initially held back the adoption of numerical modelling. By the early 1980's Boundary Element and Displacement Discontinuity (DD) models made 2D elastic modelling practical. DD software also made pseudo3D simulation of tabular orebodies feasible, although it could not include 3D infrastructure. Finite Difference (FD) models (e. g. FLAC, UDEC, etc.) allowed limited nonlinear and Discrete Element modelling, although in only 2D initially.

By the turn of the century, 3D elastic modelling became widely available in software packages such as MAP3D and EXAMINE3D. 3D nonlinear versions of FD software also appeared. With the relentless increase in computational power in the last ten years, full 3D nonlinear mine-wide modelling has become commonplace, at least for the largest mining companies and specialist rock engineering consultants.

3 GEOMECHANICAL MINE DESIGN

Geomechanical mine design relies on basic geotechnical data collection followed by empirical, semi-empirical and numerical analyses and design techniques. I will discuss essential elements of these methods and their limitations in the following sections.

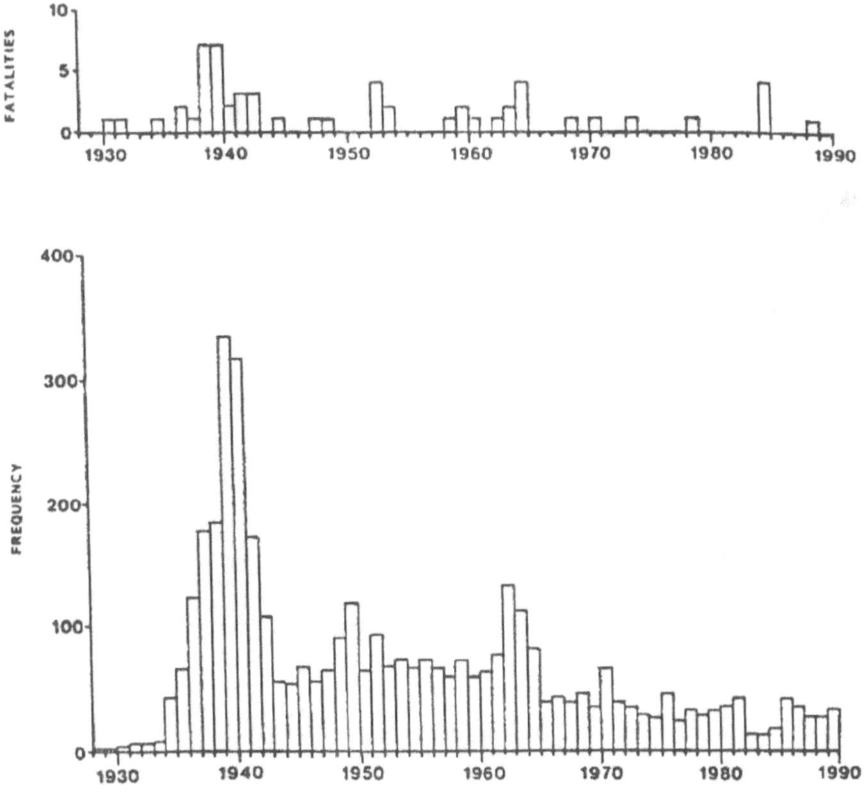

Figure 1. Reported rockbursts and fatalities in Ontario Mines1928 to 1990 [After Hedley 1992).

4 THE DISCONNECT WITH MODELLING

In this author's experience, there often appears to be a disconnect between end-users of numerical modelling software and what their models are actually meant to deliver. End-users make statements like, "I just want the answer" or "Just tell me what the input is." To me, such statements indicate a fundamental lack of understanding of what a model and its basic purpose are. Most participants in this conference fully understand that models are not reality but are rather highly simplified approximations of complex physical problems.

A model needs to do the following:

1. Represent problem geometry as accurately as possible
2. Apply appropriate boundary conditions
3. Use best estimates of the far-field stress regime, and
4. Apply the most suitable material constitutive laws.

As will be discussed below, each of the above factors carries uncertainty that must be accounted for when interpreting and using modelling outputs.

5 UNCERTAINTY IN NUMERICAL MODELS

Problem geometry is generally well defined and has reasonably high confidence. Boundary conditions can also generally be applied with acceptable confidence. On the other hand, the fractured nature of hard rock masses, combined with highly variable geological regimes at most mines, means that field-scale virgin stress levels (far-field stresses) are highly uncertain.

In most mining projects, only a few, if any, direct in-situ stress measurements are conducted. The highest calibre in-situ stress measurement techniques are expensive and time-consuming to conduct. Less costly, indirect stress measurement techniques (such as acoustic emission stress estimates) produce higher uncertainty results.

Figure 2 shows in-situ stress measurements conducted in the shaft of the Atomic Energy of Canada's Underground Research Laboratory (URL) (Martin 1990). The URL was constructed in a largely unfractured, homogenous isotropic Granitic Pluton (i.e., in a geological domain where the far-field stress regime might be assumed to be reasonably consistent) in Manitoba, Canada. Figure 3a shows the dispersion of stress values measured on given horizons in the URL shaft, even in such a homogenous rock mass. Even in one of the most homogeneous environments, significant departures from expected field stresses occurred. Figure 2b shows the strong influence fault zones exerted on the local stress field.

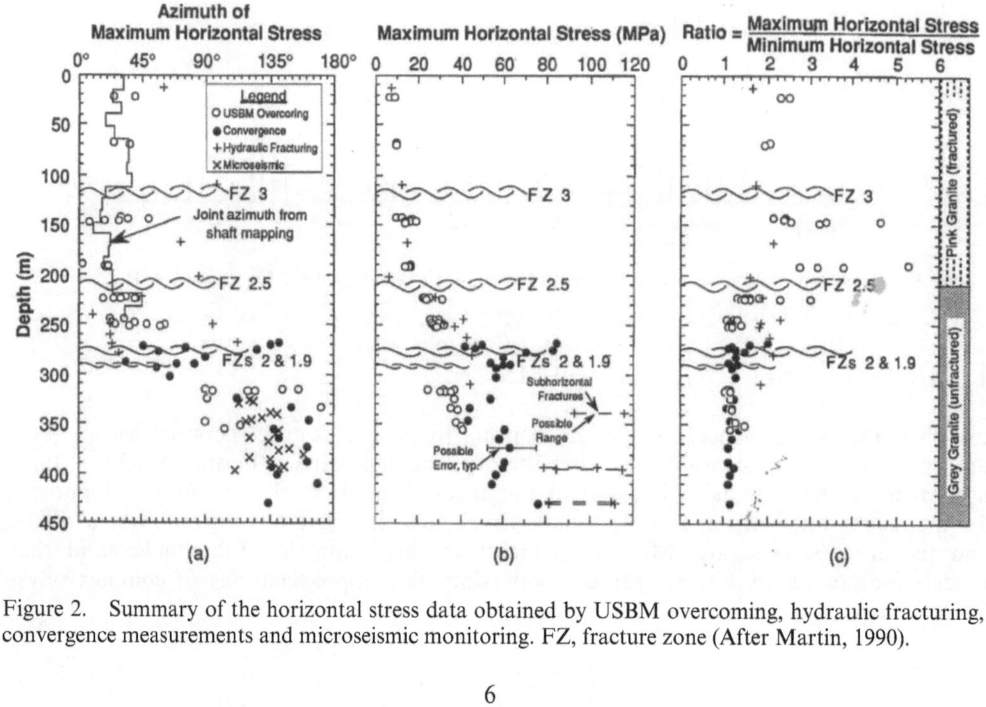

Figure 2. Summary of the horizontal stress data obtained by USBM overcoming, hydraulic fracturing, convergence measurements and microseismic monitoring. FZ, fracture zone (After Martin, 1990).

In highly structured rock masses, far greater variability is expected in stress measurements results. All this means is that, even under optimal conditions, far-field stress inputs possess a high degree of uncertainty.

Material constitutive laws are mathematical relationships that relate stress to strain or deformation. These relationships are generally derived from laboratory experiment data. The constitutive laws for most engineered materials (such as steel and aluminum) are known with a high degree of confidence. For geological materials (soils and rocks), which are inherently more heterogeneous and anisotropic, the situation is different.

Constitutive relationships for intact hard rock were initially based on elasticity theory (Jaeger & Cook, 1969). At its most complex, elastic theory for fully anisotropic materials requires 36 elastic (Young's Modulus and Poisson's ratio) constants. With symmetry assumptions, the number can be significantly reduced (for example, to two for homogeneous materials). Intact rock testing at many mine sites is limited. Defining the elastic constants described above, except for the simplest of conditions, in geomaterials is challenging. There is significant uncertainty associated with laboratory and PLT testing.

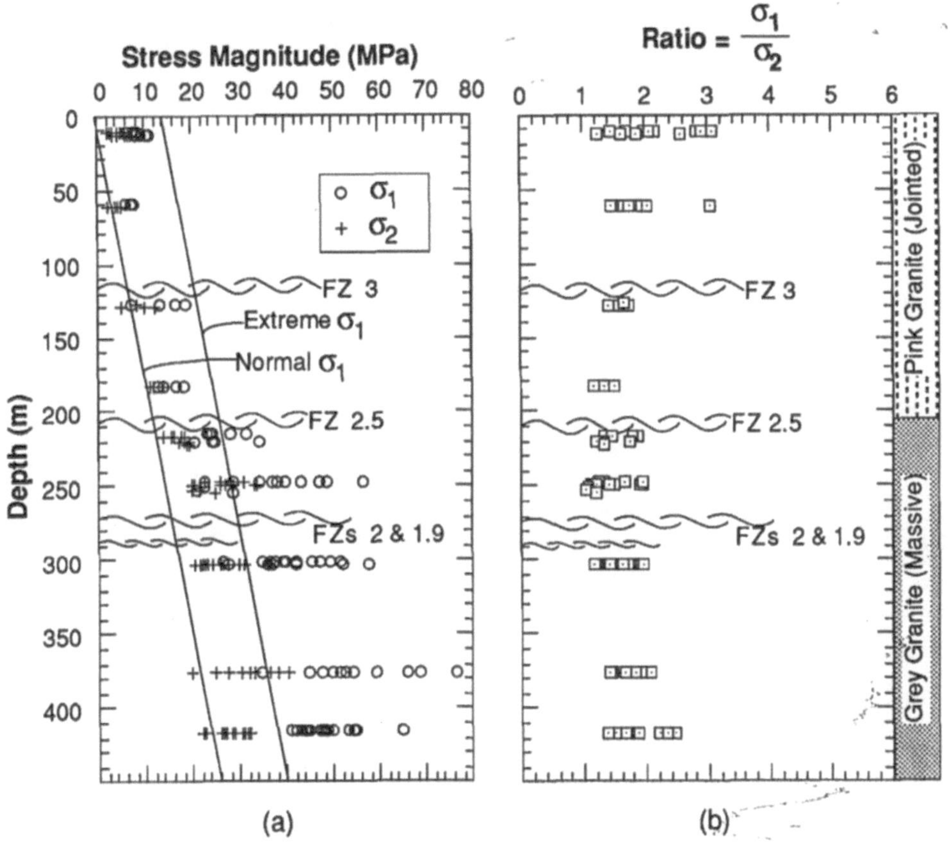

Figure 3. Summary of σ1 and σ2 measured from the URL shaft and the σ1/σ2 ratio. The extreme and normal lines are from Herget's (1986) database for the Canadian Shield. (After Martin, 1990).

Field scale strength, deformability and anisotropy of fractured rock masses are controlled by intact rock properties and natural occurring discontinuities (joints, faults, shears, bedding, foliation, etc.). For this reason, numerical models of field-scale rock masses utilize semi-empirical constitutive methods, usually the nonlinear Generalized Hoek-Brown (GHB) or linear Mohr-Cou- lomb relationships. The fractured nature of rock masses in these schemes is quantified using rock mass classification techniques (Figure 4). There is significant uncertainty with rock mass classification schemes, which can be seriously exacerbated by inexperience. These uncertainties are transferred into modelling and compound the others already discussed.

Thus far, the discussion assumes continuum behaviour in which fracture properties are 'smeared' into constitutive models. New fractures begin to form in hard rock masses as induced stresses increase, and at yield, the rock mass's load-bearing capacity begins to reduce. Figure 5 shows this strain-softening behaviour, which represents another level of complexity. Strain-softening means estimation of another set of constitutive parameters - the post-peak parameters – is required for truly nonlinear numerical models.

Unfortunately, there are currently no theoretical or empirical methods to determine these parameters a priori. Their estimation relies on the experience of the modeller (Bawden, 2010). This imposes another set of uncertainties on numerical modelling and its outputs.

Recently a novel approach has been applied to the nonlinear constitutive problem outlined above. It uses a 3D discrete element model to create a 'synthetic rock mass.' Numerical experiments are then conducted to evaluate the nonlinear constitutive parameters (Pierce et al, 2007). The approach still suffers from the inherent uncertainty and variability of the basic geological/geotechnical input parameters. As well, it is limited in the practical level of detail that can be incorporated.

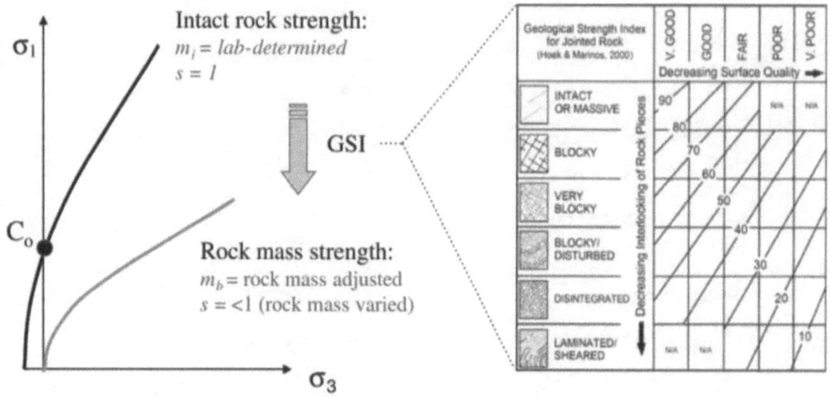

Figure 4. Scaling of Hoek-Brown failure envelope for intact rock to that for rock mass strength. See Marinos et al (2005) for full details on use of the GSI chart (After Eberhart 2012).

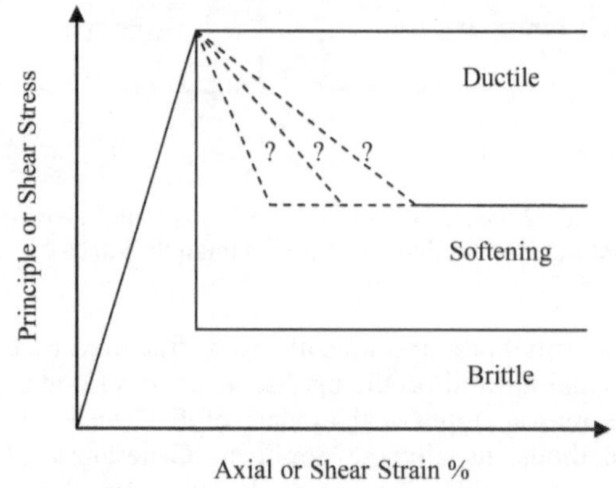

Figure 5. Post-peak rock mass behaviour models.

6 WHAT SHOULD WE DO ABOUT UNCERTAINTY IN GEOMECHANICAL DESIGN MODELS?

To answer the question of how to handle uncertainty in geomechanical models, one must begin from the fundamental understanding that all geotechnical parameters naturally vary. At the design stage, we must collect sufficient data to define statistical distributions for each parameter. We must also undertake sensitivity studies to determine how calculated variances impact design risk. Finally, the selected model needs to be calibrated against field data to validate and adjust the input parameter assumptions.

7 NUMERICAL MINE MODEL CALIBRATION

Model calibration can be qualitative or quantitative and generally requires a combination of direct observation and instrumentation. The use of instrumentation in mines has increased dramatically over the past 30 years, particularly in the last ten years. Kalenchuk (2019) presents an excellent discussion of numerical model calibration in the mining context.

Geomechanical instrumentation for underground mining generally differs from that used in open pits. In this paper, I focus on underground mine instrumentation. Underground mining instrumentation includes mine-wide microseismic systems and laser scanning surveys. It also includes location-specific instruments such as extensometers, closure meters, and stress cells.

Valuable mine-wide data, which can be used in the calibration process, also include observations from geotechnical mapping, rock mass classification and damage mapping. Plotted on 3D mine development renderings, this data can be directly correlated to numerical predictions. Comparison of seismic and aseismic locations to predicted rock mass yielded and elastic zones, for example, allows good semi-quantitative calibration of an assumed rock mass failure criterion. Measurement of mine induced stress changes and deformations (from stress cells, MPBX and closure, as shown in Figure 6) can provide high-quality quantitative model calibration. However, it does not on its own give the shape of the post-peak curve. Damage mapping can be compared to predicted yield extents to provide qualitative calibration of models over large mine areas. Such calibration efforts increase confidence in forward model predictions, and help engineers refine and reduce uncertainty in model inputs. To the author's knowledge, calibration is not taught in most mining or geotechnical engineering schools.

Calibration efforts are not problem-free, however. The effectiveness of model yield calibration using microseismic data, for example, depends on factors such as the 3D sensor coverage of the seismic system in the area of interest, and the type of sensor used. The value from using

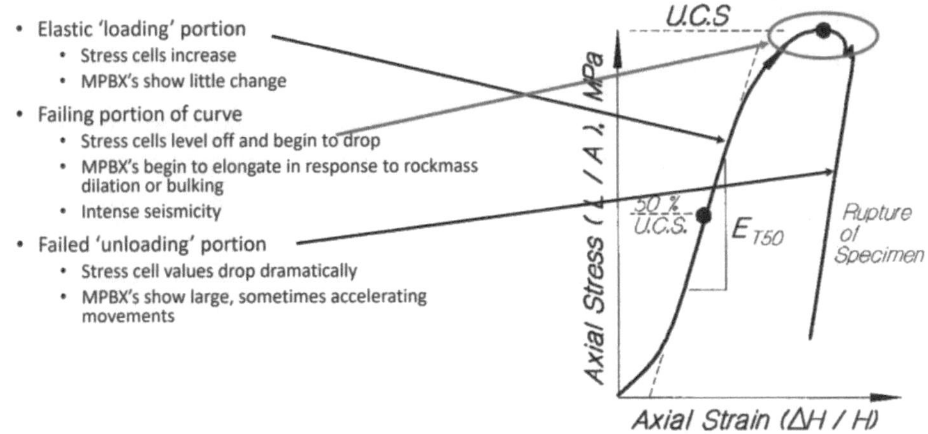

Figure 6. Interpretation of combined stress cell and MPBX data.

location- specific conventional instrumentation depends on having (i) sufficient instruments in the area of concern and (ii) recording the instrumentation data regularly (e.g., daily or weekly) to generate a database for calibration. Figure 7 provides an example of instrumentation data unsuitable for model calibration due to significant data reading gaps.

We will examine an example from the 1990s and early 2000s on one of the most intensely instrumented mines, the Willams Mine. The mine is in the Hemlo mining camp of northeastern Ontario, Canada. From 1999, an isolated sill pillar area experienced a series of major seismic events (Bawden and Jones, 2007). Later studies (Crowder et al, 2006 and Coulson et al., 2007) investigated the use of this instrumentation database to back-analyze constitutive parameters. Figure 8 shows an attempted 'calibration' to one instrument, while Figure 9

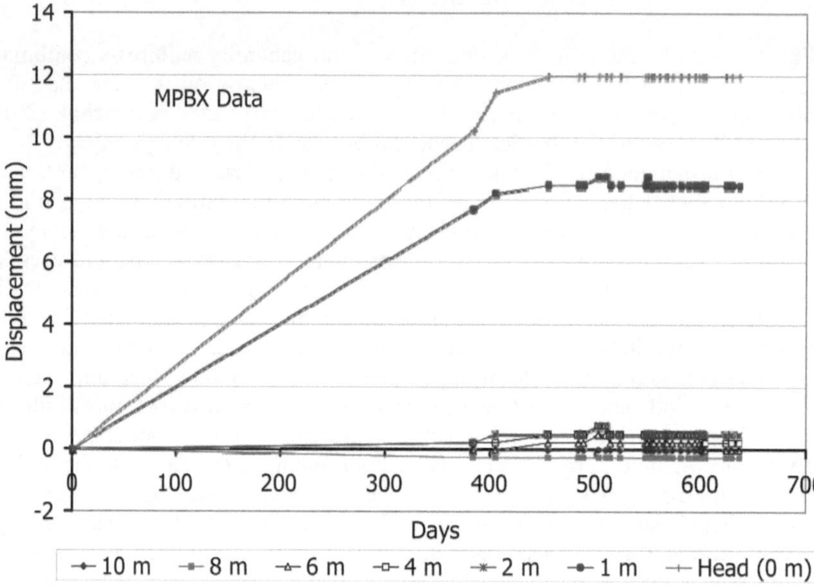

Figure 7. Example of instrumentation data with a large time gap between readings.

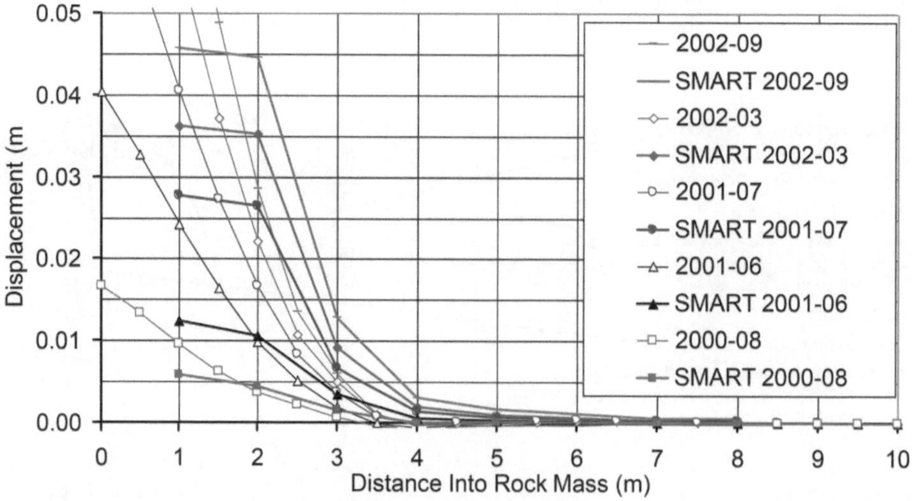

Figure 8. Attempted back analysis calibration to one instrument – solid symbols are field data – open symbols are numerical back analysis. (After Crowder et al, 2006).

indicates the exercise's outcome. The calibration, while imperfect, was the best that could be achieved with available models and data. This illustrates the type of challenges that should be anticipated in attempting a 'quantitative calibration'.

The studies showed that, of the 95 instruments deployed in the area, only 10 had good enough data quality due to infrequent data reading (Figure 7). At the time, the instrumentation had to be read manually, and the data subsequently entered by hand into the database, leading to the data gaps that rendered the calibration ineffective. For comparison, Figure 10 shows examples of good quality instrumentation data.

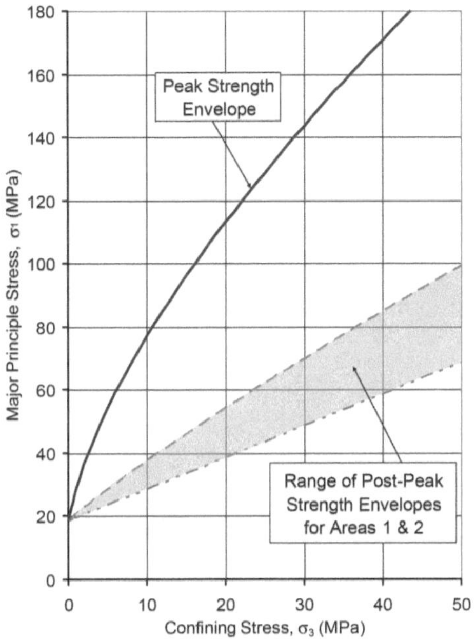

Figure 9. Generalized Hoek-Brown peak strength envelope, and a tight band representing the post-peak strength envelopes that produced the best back analysis results in Areas 1 & 2 (after Coulson et al., 2007).

Today, modern underground wireless telecommunication systems allow instrumentation data to be collected and transmitted to surface data rooms in near real-time. This eliminates the data collection barrier from the calibration process. However, calibration still requires that sufficient instruments be installed early in the mining cycle to provide appropriate baseline information.

A non- or weakly calibrated 3D nonlinear model can be very dangerous. Why? Because end users (engineers, operators and management) may blindly believe it!!

Even with 'calibrated' models, forward modelling results must be regarded with healthy skepticism until engineers can demonstrate that those results are fit for service. Modelling outcomes must be validated and pass the 'smell test,' i. e., do experienced engineers and operations personnel believe that they appear 'reasonable?'

The resources devoted to calibration should be proportional to the risk inherent in a design.

8 SO WHERE IS THE DISCONNECT WITH MODELLING?

As discussed earlier, the advent of the PC, subsequent explosion in affordable computing power, and ever more powerful software have brought us great advantages. HOWEVER, they have also created some new problems. From my perspective, critical issues that have emerged are as follow:

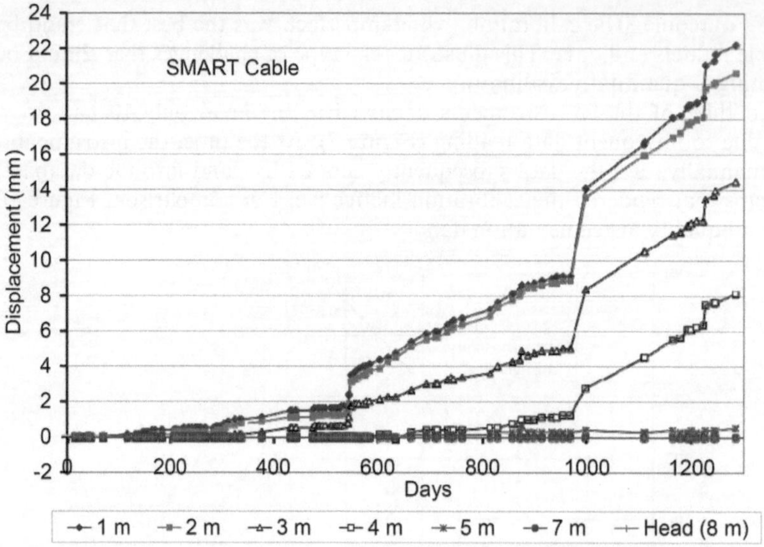

(a). An example of good quality instrumentation data from Williams Mine.

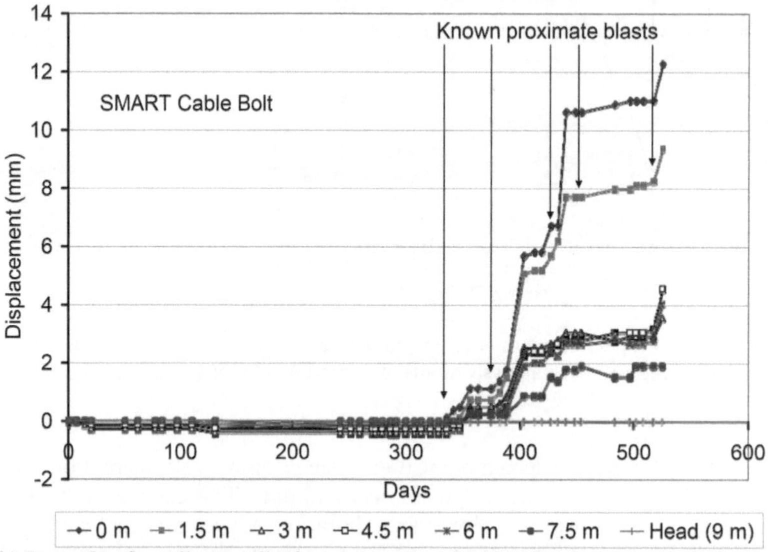

(b) Example of excellent quality instrumentation data. Notice the response of the instrument to the arrows, which indicate timing corresponding to proximate blasting.

Figure 10. High-quality instrumentation data (after Crowder et al., 2006).

1. An increasing reliance on modelling as being 'the answer'
2. A failure of many end-users and management to understand the real purpose and power of numerical modelling, which at its most basic is to
   a. Investigate potential failure modes
   b. Test sensitivity to variations in input parameters, and
   c. Assess risk under various assumed conditions (e. g., best-case and worst-case)
3. An increasing reluctance of some engineers to 'go underground and observe actual ground behaviour,' which is critical for model calibration; many instead favour office computer-modelling work

4. A fundamental lack of appreciation at almost all levels, but often most acute with upper management, of
   a. The fundamental uncertainties inherent in all geomechanical numerical models, but particularly in complex, nonlinear models
   b. The fact that complex nonlinear models do not provide 'unique solutions,' and thus their 'calibrated' models can produce results that significantly depart from real behaviour and result in substantial unanticipated risks
   c. The recognition that modelling complexity must match the available quality and quantity of data
   d. A limited understanding of basic rock engineering principles and applied geomechanics aspects such as empirical design, instrumentation and model calibration, generally due to inadequate training (both university programs and on the job)
   e. The overconfidence of senior managers, and some practitioners, on modelling results because of the 'impressive graphics' produced by modern software; they equate nice pictures to convincing geomechanics results and entirely lack appreciation for simple, but useful, empirical models founded on real-world cases
   f. A general lack of appreciation, even within the rock engineering community, of the limitations of model calibration. This situation is particularly acute with the use of highly complex nonlinear models, and the significant challenges in achieving high confidence calibrated models. It is exacerbated by the generally nonexistent to poor understanding of the difference between 'qualitative,' 'semi-quantitative' and 'quantitative' calibration (as described in some detail by Kalenchuk, 2019)
   g. A lack of appreciation of the criticality of the different kinds of instrumentation and the calibration process (how to use instrumentation measurements to validate and calibrate numerical and empirical design models).
5. An awareness that models do not, and cannot, provide 'the answer.' It is the job of the engineering and management team to determine 'the best answer' for a problem based on
   a. Assessing all design analyses and deciding what is most suitable for the specific problem
   b. Communicating the corporate appetite for risk that the engineers can translate into required factors of safety or probabilities of failure (factor of safety, sometimes known as a factor of ignorance, is meant to counter the many uncertainties inherent in design and analysis)

As an example of senior management and engineers' overconfidence, we will examine what happened at an underground mine in Canada. A novel mining method with opening dimensions well outside experience was proposed for the mine, which then engaged a consulting firm to assist with the geomechanical design. Using 3D mine-wide nonlinear numerical modelling, the design engineers concluded that the new excavation span would be stable without backfill. Unfortunately, the design did not work in practice, resulting in significant surface displacements and damage. Consequently, the mining company suffered substantial economic and reputational damage.

A forensic review concluded that the original geological model was inadequate, with significant structural details missing or misinterpreted, and contributed directly to the failure. It showed that the use of the simple empirical stability graph method would have revealed this. It would have raised red flags and initiated a thorough investigation before implementation. These investigations would likely have uncovered the deficiencies in the geological model and forced critical evaluation of the differences between the numerical modelling and empirical predictions. The case illustrates that not only geomechanical numerical models bring inherent uncertainty and risk to design. Once the revised geological model was introduced, the numerical and empirical model results aligned.

The example shows that skepticism is not restricted to geomechanical models alone. It must be extended to basic geological models, which are derived from spatially limited data and involve significant interpretation and interpolation. In fact, geology underlies all uncertainty in the mine design process.

9 WHAT ABOUT INSTRUMENTATION?

Numerical modelling is not the only problem area of geomechanics limited by results interpretation ability and use. Interpretation of basic instrumentation data and instrument layout design equally confound many people. This is partially due to instrumentation being viewed as a 'boring' topic that many mining and geotechnical schools do not typically teach. However, the subject is critical, and, at a risk management level, the industry owes a duty of care to provide proper training to personnel working in these areas.

10 WHAT ABOUT EMPIRICAL DESIGN?

Early in my career Dr. Hoek cautioned me not to undervalue empirical methods. Although empirical models are less sophisticated than numerical models, their simplicity and grounding offer advantages. Within certain limitations, discussed later, empirical and numerical results should broadly agree (for example, whether stopes are stable or unstable). However, we must exercise healthy skepticism with empirical models if we use them outside the primary database limits. As shown in Figure 11, use of empirical design charts must be restricted to areas that have sufficient direct data coverage to justify usage.

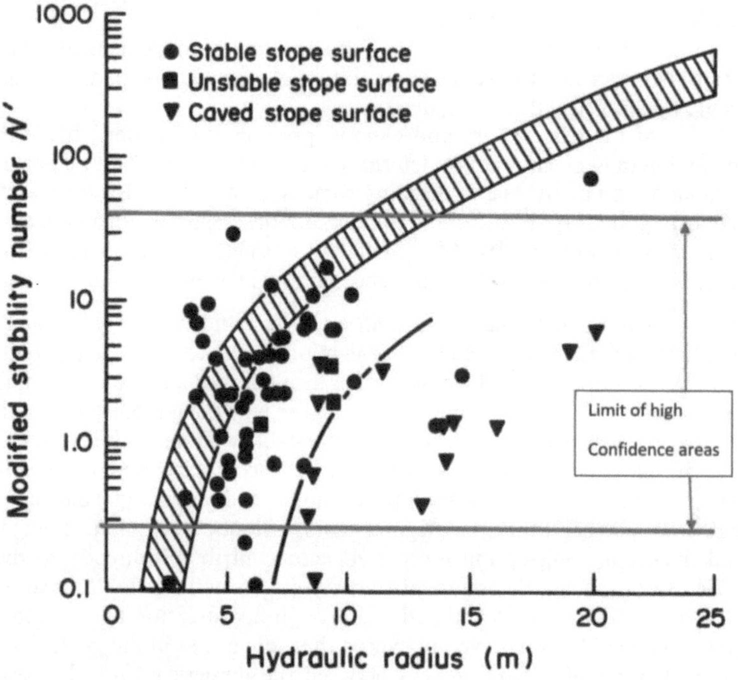

Figure 11. Stability Graph showing Potvin (1988) original database.

11 GEOMECHANICAL MINE DESIGN AS A PROCESS

As noted in the section title, geomechanical design is a PROCESS. It is not a formula or a simple one-off modelling effort. The design process is thoroughly discussed in Bieniawski (1992) and succinctly summarized in Stacey's (2009) 'Wheel of Design' (shown in Figure 12).

Geomechanical mine design should follow Einstein's maxim that "problems should be made as simple as possible but no simpler." To do so, design teams should begin with simple models (such as linear elastic or empirical models) to capture expected behaviour. They should move to more complex analyses only when they can justify such analyses with more and better data, and improved understanding.

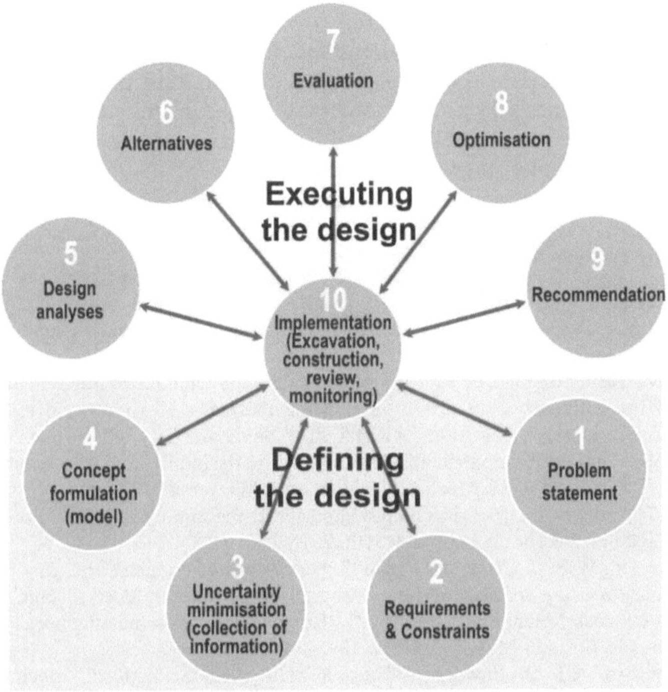

Figure 12. Wheel of Design (After Stacey, 2009).

12 FACTOR OF SAFETY VS PROBABILITY OF FAILURE

Classical engineering design uses a factor of safety (FoS) to attempt to account for various uncertainties. An alternate design approach is to utilize probabilistic analysis that calculates a probability of failure (PoF). Probabilistic analysis uses the same computational equations as deterministic analysis but requires that appropriate statistical distributions be specified for input parameters. An excellent discussion on this topic is provided in Hoek et al (1992). Many modern modelling software (such as Rocscience's RS2 and SWEDGE) incorporate these techniques. In this author's experience, using a combination of deterministic and probabilistic techniques can be beneficial. The challenge with probabilistic analysis lies in acquiring sufficient good quality data to define the required statistical distributions for input parameters.

At the end of the day, however, a competent, and therefore ultimately responsible, person or team must make decisions. All the discussion in this paper is simply meant to generate healthy skepticism around these important issues and raise awareness among decision-makers.

13 CONCLUSIONS

The advent of the PC, subsequent explosion in computing power, and increasingly powerful sofware have brought us significant advantages. However, they have also led to unanticipated consequences.

In this author's opinion, the common lack of clarity on the uncertainties and the risks inherent in a geomechanical analysis is one major issue. For nonlinear numerical models, such risks can be significantly amplified and, left unrecognized, result in serious consequences. Some specialist geomechanics practitioners fail to be fully transparent about this in their models.

Management and engineers must understand that models do not provide 'the answer.' They must be more skeptical and critical of all modelling outputs. Remember that although modern computer-generated graphics may appear impressive, they can still be wrong.

Finally, we should all remember the advice of Einstein and Dr. Hoek! Keep things as simple as possible and do not discount the value of empiricism.

REFERENCES

Bawden, W.F. & Jones, S. 2007. Ground support design and performance under strong rockburst conditions. In Proceedings, Challenges in Deep and High Stress Mining, Australian Center for Geomechanics, pp 327–336.

Bawden, W.F. 2010. Thoughts on Quantitative Field Scale Characterization on Post-failure Rock Mass Conditions and the Influence on Underground Mine Design – Keynote Address. In Proceedings, Amer- ican Rock Mechanics Symposium, ARMA 2010, Salt Lake City Utah, June 20 – 30.

Bieniawski, Z.T. 1992. Invited Paper: Principles of engineering design for rock mechanics, Rock Mechanics, Proc. 33rd US Symp. on Rock Mech., Ed. Tillerson & Wawersik, Balkema, pp 1031–1040.

Clarke, I.E. 1973. The role of the rock mechanics engineer in the mining industry, Published in Articles: EE Publishers, Articles: Vector on February 11th, 2016.

Coulson, A.L., Bawden, W.F. & Crowder, J.J. 2007. Estimation of the peak and post-peak behaviour of fractured rock masses using spatial and temporal analysis of mine induced seismicity, in proceedings of 1st Canada – US rock Mechanics Symposium - Rock Mechanics – meeting society's challenges and demands, Vancouver, Canada May 27 – 31, pp 1395–1404.

Crowder, J.J., Coulson, A.L. & Bawden W.F. 2006. The Field Scale Rock Mechanics Laboratory: Estima- tion of Post Peak Parameters and Behaviour of Fractured Rock Masses, 41st US Symposium on Rock Mechanics, Golden Colorado June 17- 21.

Cundall P.A. 1971. A computer model for simulating progressive large scale movements in blocky rock systems, Proc. ISRM Symposium, Nancy, France, Vol.1 Paper No.II-8.

Deist, F.H. & Oravecz, K.I. 1968. The finite element method and its application to problems in rock mechanics, research report, Chamber of Mines of South Africa No. 60/68.

Deist, F.H., Georgiadis, E. & Moris, J.P.E. 1972. Computer applications in rock mechanics, African Insti- tute of Mining and Metallurgy, vol. 72, pp. 265–272.

Duncan, J.M. & Goodman, R.E. (1968). Finite element analyses of slopes in jointed rock. Contract Report S-68-3. US Army Engineers Waterways Experiment Station, Vicksburg, Miss.

Eberhart, E. 2012. The Hoek-Brown Failure Criterion. Rock Mech Rock Eng 45:981–988

Hedley, D.G.F. 1992. Rockburst Handbook for Ontario Hardrock Mines. CANMET Special Report SP92-1E

Hoek, E., Kaiser, P.K. & Bawden, W.F. 1995. Design of Support for Underground Hard Rock Mines, A. A. Balkema.

Jaeger, J.C. & Cook N.G.W. 1969. Fundamentals of rock mechanics. METHUEN and CO LTD, London.

Kalenchuk, K. 2019. Mitigating a fatal flaw in modern geomechanics: understanding uncertainty, applying model calibration, and defying the hubris in numerical modelling, CGS 2019 Colloquium at the 72nd Canadian Geotechnical Conference, St. John's, NL.

Kulhawy, F.H. 1973. Analysis of Underground Openings in Rock by Finite Element Methods. Report for Advanced Research Projects Agency, USBM.

Martin, C.D. (1990) Characterizing in Situ domains at the AECL Underground Research Laboratory. CAN. GEOTECH. J. VOL. 27

Parker, J. & Scott, J.J. 1964. Instrumentation of room and pillar workings in a copper mine of the Copper Range Company, White Pine Michigan. 6th US Symposium on Rock Mechanics (USRMS), 28-30 Oc- tober, Rolla Missouri
Pierce, M., Cundall, P. & Potyondy, D. 2007. A synthetic rock mass model for jointed rock. In Frock Me- chanics: Meeting Society's Challenges and Demands. Proceedings of the 1st Canada-US Rock Mechan- ics Symposium, Vancouver, Canada.
Plewman, R.P., Deist, F.H. & Ortlepp, W. 1969. The development and application of a digital computer method for the solution of strata control problems, African Institute of Mining and Metallurgy, vol. 70, pp. 33–44.
Potvin, Y. 1988. Empirical open stope design in Canada. Ph.D. thesis, Dept. Mining and Mineral Processing, University of British Columbia.
Salamon, M.D.G., Ryder, J.A. & Ortle, W.D. 1964. An analogue solution for determining the elastic response of strata surrounding tabular mining excavations. African Institute of Mining and Metallurgy vol. 64, pp. 115–137.
Stacey, T.R. 1969. Application of the finite element method in the field of rock mechanics with particular reference to slope stability, S. Afr. Mech. Eng., v 19, Dec 1969, pp 131–134.
Stacey, T.R. 2009. Design – a Strategic Issue, Jl S. Afr. Inst. Min. Metall., Vol 109, No 3, pp 157–162.
Terzaghi, K. 1929. Effect of minor geologic details on the safety of dams. In "From Theory to Practice: Selections of writings from Karl Terzaghi" p119–133. John Wiley and Sons (1967)
Wilson, J.W. & More-O'Ferrall, R.C. 1970. Application of the electric resistance analogue to mining oper- ations. Journal of the Southern African Institute of Mining and Metallurgy, vol. 70, pp. 369–378.

The Evolution of Geotech - 25 Years of Innovation – Hammah et al. (Eds)

Recent advances in seismic soil liquefaction engineering

K.O. Cetin
Middle East Technical University, Turkey

H.T. Bilge
GeoDestek Geoengineering & Consultancy Services, Turkey

ABSTRACT: Over the last two decades, seismic soil liquefaction engineering has evolved into a field of its own right, as engineering assessment and mitigation of liquefaction hazard is increasingly well addressed in both research and practice. Assessment of liquefaction potential and consequent strength and strain response of soils are considered as the initial steps of engineering evaluation of seismic soil liquefaction, and these topics have naturally drawn significant research interest. Within in the confines of this manuscript, current state of practice on these issues will be reviewed, critically examined, and some recommendations will be listed for practitioners.

1 INTRODUCTION

Assessment of cyclic response of soils has been a major concern of geotechnical earthquake engineering since the very early days of the profession. Engineering treatment of liquefaction-induced problems evolved initially after two devastating earthquakes of 1964 (Niigata, Japan and Great Alaska, USA), at which seismically-induced soil liquefaction was listed as one of the prime causes of structural failures. While pioneer efforts focused on triggering of liquefaction in clean sands, as earthquakes have continued to provide more lessons and data, researchers have become increasingly aware of the problems associated with silt and clay mixtures along with the strength and strain/deformation response of liquefied soils.

Today, "soil liquefaction engineering" is emerging as a rapidly progressing field of practice which involves a number of assessment steps. Figure 1 schematically illustrates these key steps as introduced by Seed et al. (2001), and the discussion layout of this manuscript will also be structured to follow the same footprints by focusing on the first 3 steps of this assessment scheme.

1. Assessment of liquefaction potential and triggering.

2. Assessment of cyclic strength response of soils.

3. Assessment of seismic deformation response of soils.

4. Assessments of the consequences of these deformations.

5. Implementation of engineered mitigation, if necessary.

Figure 1. Key elements of seismic soil liquefaction engineering.

DOI: 10.1201/9781003188339-2

Among seismic soil liquefaction engineering steps, the first one involves the assessment of the presence of potentially liquefiable soils, and if their existence is confirmed then, triggering of soil liquefaction. This first step has drawn the highest level of research interest. Despite the level of some controversy, it can still be concluded as the most developed assessment stage in liquefaction engineering assessments, and will be discussed next.

2 ASSESSMENT OF LIQUEFACTION POTENTIAL AND TRIGGERING

2.1 *Potentially liquefiable soils*

There has long been a consensus in the literature that "clean" sandy soils, with limited fines, are potentially vulnerable to cyclically-induced liquefaction; however, there has been significant controversy and confusion regarding the liquefaction potential of silty soils (silt – clay mixtures), as well as gravelly soils.

The cyclic behavior of coarser soils (gravels and coarser) is not very different than that of "sandy" soils. There are now a number of well-documented field cases of liquefaction of coarse, gravelly soils (Ishihara, 1985; Harder, 1988; Yegian et al., 1994; Hatanaka et al., 1997; Chen et al., 2018). As discussed in Seed et al. (2001), these soils do differ in behavior from sandy soils in two ways: (1) they can be much more pervious, and consequently generated pore pressures can be dissipated quite rapidly, and (2) due to their larger particle masses, the gravelly soils are seldom deposited "gently" so they are not commonly encountered in loose state as compared to sandy soils. However, it should be noted that the apparent drainage advantages of these soils can be eliminated, if i) they are surrounded and encapsulated by less pervious finer materials, (ii) drainage is internally impeded by the presence of finer soils in the void spaces between the coarser particles, or (iii) the layer is too thick, which in turn increase the distance over which drainage must occur (rapidly) during an earthquake. In these cases, the coarser soils should be considered to be potentially liquefiable and be assessed for liquefaction triggering hazard. This naturally requires the estimation of in-situ density state (or the penetration resistance), for which the Becker penetration test still continues to be one of the practical tools, despite some of its limitations, along with shear wave velocity measurements.

Contrary to the consensus on liquefaction potential of clean sands, the susceptibility of silt and clay mixtures to liquefaction has been one of the most controversial and widely discussed issues. As previously stated, in the early days of the profession, plastic silt and clay mixtures were considered to be resistant to cyclic loading, and consistently, most research was focused on cyclic response of saturated sandy soils. However, in the following years, especially after fine-grained soil site failure case histories of 1975 Haicheng and 1979 Tangshan earthquakes from China (Wang, 1979), increasing number of research studies focused on understanding their cyclic response. On the basis of Wang's (1979) database and conclusions, a set of criteria to assess liquefaction potential of soils with fines (widely referred to as Chinese Criteria) was proposed by Seed and Idriss (1982). These criteria had been used widely with slight modifications (Andrews and Martin, 2000). More recently, ground failure case histories compiled after 1989 Loma Prieta, 1994 Northridge, 1999 Adapazari and Chi-Chi, 2010 Darfield, and 2011 Christchurch earthquakes have refreshed research attention on assessing cyclic mobility response of fine-grained soils. Case histories from these earthquakes highlighted that low plasticity silt and clay mixtures may significantly strain soften, which may in turn cause significant damage to overlying structural systems. As an alternative to Chinese Criteria, Seed et al. (2003), Bray and Sancio (2006) and Boulanger and Idriss (2006) proposed new susceptibility criteria based on field observations and laboratory test results. Before discussion of these methods, it is important to note that assessing susceptibility of soils to liquefaction requires a potentially liquefiable soil definition, which ideally should be independent of the intensity and duration of earthquake loading. This is a difficult to achieve requirement and is listed as one of the common drawbacks of existing susceptibility criteria. Hence, in practice, for most cases, liquefaction susceptibility (potential) assessments are -unfortunately- combined with liquefaction triggering.

Seed et al. (2003) recommended their susceptibility criteria inspired from the case histories and results of cyclic tests performed on "undisturbed" fine-grained soils documented after

1999 Adapazari and Chi-Chi earthquakes. As presented in Figure 2, Seed et al. (2003) used liquid limit (LL), plasticity index (PI) and water content (w) to assess liquefaction susceptibility of soils. Fine-grained soils with PI≤12 and LL≤37 are concluded to be potentially liquefiable, if the natural water content is wetter than 80 % of their liquid limit. Bray and Sancio (2006) developed alternative criteria based on cyclic test results performed on undisturbed fine grained soil samples retrieved from Adapazari province of Sakarya, Turkey. As presented in Figure 3, Bray and Sancio adopted the PI and w/LL ratio as the input parameters. Fine-grained soils with PI≤12 are judged to be potentially liquefiable, if their natural water content is wetter than 85 % of their liquid limit. Also, it should be noted that unlike most of available methods to assess liquefaction susceptibility of fine-grained soils, Bray and Sancio (2006) provided a complete documentation of their database (i.e., tested specimens and also test conditions), which establish the basis of their recommendation. As clearly revealed by the adopted cyclic stress ratio levels and consolidation stress histories of soil samples, the intent of these criteria seems to assess liquefaction potential (better to refer to it as triggering) response of Adapazari soils, specifically subjected to 1999 Kocaeli Earthquake (M_w=7.5) shaking.

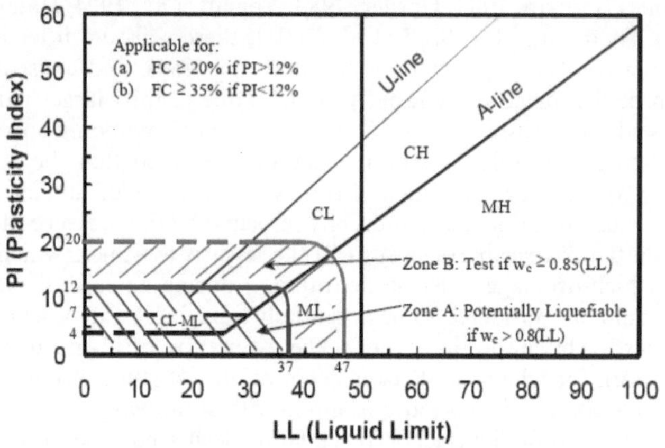

Figure 2. Criteria for liquefaction susceptibility of fine-grained soils by Seed et al. (2003) (after Seed et al., 2003).

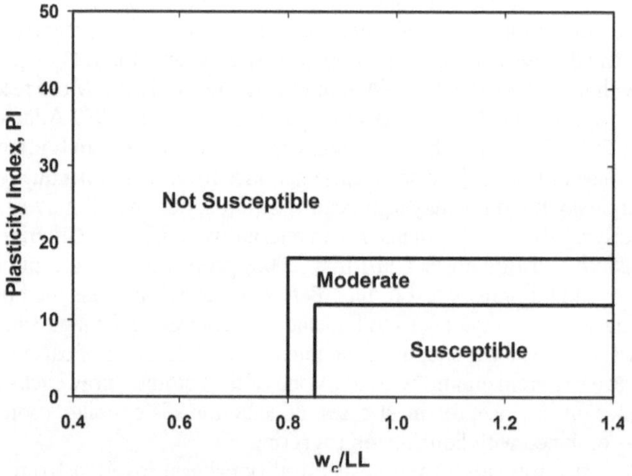

Figure 3. Criteria for liquefaction susceptibility of fine-grained soils by Bray and Sancio (2006).

Boulanger and Idriss (2006) proposed an alternative criterion based on the results of cyclic laboratory tests and the authors' engineering judgment. They categorize the cyclic response of fine-grained soils as "sand-like" and "clay-like" where soils that behave "sand-like" are judged to be potentially liquefiable and they have substantially lower values of cyclic resistance ratio (CRR) compared to those classified as to behave "clay-like". The input parameter of this criterion is PI as presented in Figure 4. Fine-grained soils with PI >7% are judged to exhibit significantly "larger" cyclic resistance.

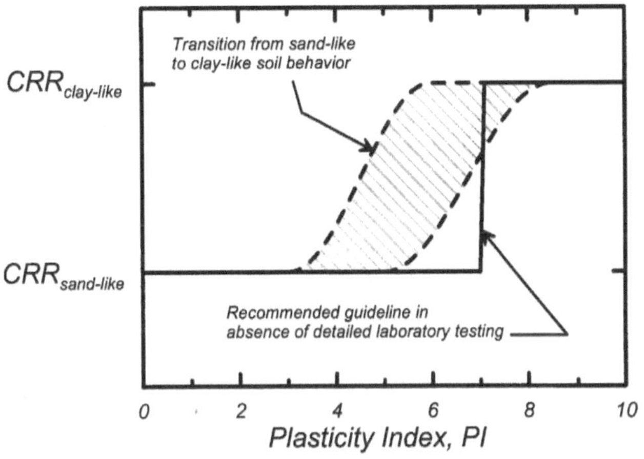

Figure 4. Criteria to differentiate sand- and clay-like soil responses proposed by Boulanger and Idriss (2006) (after Boulanger and Idriss, 2006).

Cetin and Bilge (2014) proposed an alternative susceptibility criterion based on cyclic triaxial tests performed on a wide range of high quality "undisturbed" fine-grained soil specimens. As opposed to excess pore water pressure (r_u) or maximum shear strain (γ_{max}) threshold, occurrence of contraction-dilation cycles (aka banana loops), was used as the screening evidence for liquefaction triggering. Fine-grained soils with PI values in excess of 30 are identified as "non-liquefiable" but with susceptibility to "cyclic mobility". Similarly, fine grained soils satisfying the following condition are classified as potentially "liquefiable":

$$LI 0.578 \cdot \ln(PI) - 0.940 \qquad (1)$$

where LI = liquidity index; and PI = plasticity index.

The proposed criterion along with the test database is presented in Figure 5, and a complete documentation is available in Bilge (2010).

As discussed in detail by Cetin and Bilge (2014), currently available susceptibility criteria were observed to suffer from one or more of the following issue:

i. Ideally separate assessments of a) identifying potentially liquefiable soils and b) liquefaction triggering, were combined into a single assessment. When soil layers (in the field) or samples (in the laboratory) liquefied under a unique combination of CSR and number of equivalent loading cycle, N (or moment magnitude of the earthquake), they were misleadingly labeled as"potentially liquefiable" rather than "liquefied" at the selected CSR and N combination. These types of combined assessment procedures produce some-what biased classifications of potentially liquefiable soils.

ii. Judging liquefaction susceptibility of a soil layer or a sample through a unique combination of CSR and number of equivalent loading cycle (or moment magnitude of the earthquake) requires clear definition for liquefaction triggering. Unfortunately, there exist multiple and mostly conflicting strain, pore pressure or field performance based definitions, some of which may not be even clearly documented in the source documents.

iii. Liquefaction triggering manifestations and their extent are not unique in the field (sand boils, extensive settlements, lateral spreading etc.). There is no single liquefaction definition (exceedance of threshold r_u or γ_{max} levels) for laboratory-based evaluations either. The success rate of the existing assessment methodologies for identifying liquefiable soils depend strongly on the adopted threshold levels.

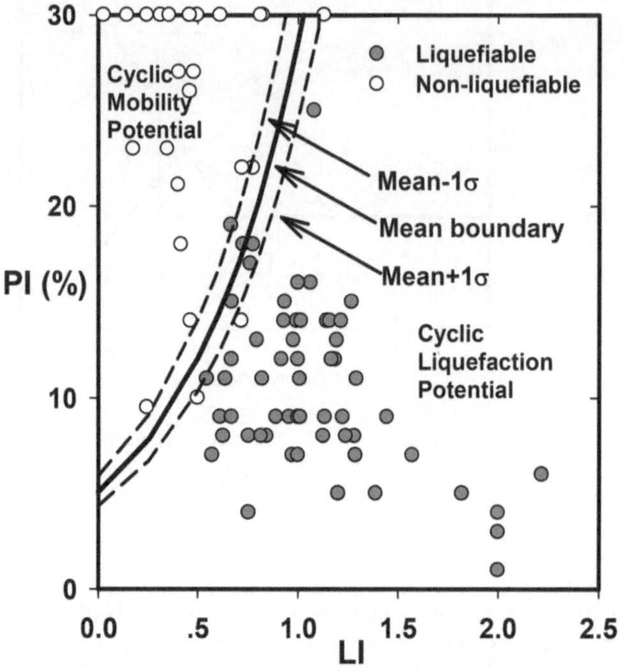

Figure 5. Liquefaction susceptibility criterion proposed by Cetin and Bilge (2014).

The authors of this manuscript believe that either fine- or coarse-grained, every soil can be liquefied, and hence potentially liquefiable, if liquefaction triggering is defined by a threshold γ_{max}, r_u, or even the existence of banana loops. The dilemma, which is yet to be solved, is the identification of cyclic stress and number of loading cycle combinations to trigger liquefaction. Hence, with increasing popularity in performance-based design practice, and available tools to assess cyclic straining and pore pressure responses of both fine- and coarse-grained soils, the elementary assessment steps of liquefaction susceptibility and triggering will be less popular and eventually eliminated. Alternatively, the assessments will directly start with the estimations of cyclically-induced strain or excess pore pressure levels. However, until this is achieved, existing liquefaction susceptibility and triggering methodologies will be used as initial screening tools.

2.2 *Assessment of liquefaction triggering*

Quantitative assessment of the likelihood of "triggering" or initiation of liquefaction is the necessary first step for most projects involving seismically-induced liquefaction problems. There exist two approaches for this purpose: the use of (1) laboratory testing of "undisturbed" samples, and (2) empirical relationships based on correlations with observed field behavior on the basis of various in-situ "index" tests.

The use of laboratory testing is complicated due to difficulties associated with sample disturbance during both sampling and reconsolidation of cohesionless soils. It is also difficult and costly to perform high-quality cyclic simple shear testing, and additionally cyclic triaxial testing poorly represents the loading conditions of principal interest for most seismic problems. Both sets of problems can be ameliorated, to some extent, by use of appropriate "frozen" sampling techniques, and subsequent testing in a high quality cyclic simple or torsional shear apparatus. The difficulty and cost of these sophisticated techniques, however, places their use beyond the budget and scope of most engineering projects. On the other hand, empirical field-based frameworks continue to be the principal approach used in engineering practice. Three in-situ index test methods; the Standard Penetration Test (SPT), the Cone Penetration Test (CPT) and the measurement of in-situ Shear Wave Velocity (V_s) have reached a level of development in the field such that their usage has been applied worldwide. These three methods and the data supporting them are discussed in detail in the recent 2016 report of the National Academies Sciences, Engineering and Medicine, Committee on Geological and Geotechnical Engineering (COGGE): "State of the Art and Practice in the Assessment of Earthquake-Induced Soil Liquefaction and its Consequences" (NAP, 2016). Advances in test procedures in a fourth method, the Becker Penetration Test (BPT) may expand the test's usage in the future. New methods are under development, yet these tests remain the cornerstone of empirical test methodologies. The oldest, and still the most widely used of these, is the SPT, and SPT-based methods will be the major focus of the following sections.

2.2.1 *SPT-based triggering assessment*

The use of SPT as a tool for the evaluation of liquefaction potential first began after the 1964 Great Alaska and Niigata Earthquakes, both of which produced significant numbers of liquefaction-induced failure case histories (e.g.: Kishida, 1966; Seed and Idriss, 1971). As discussed by the NCEER Working Group (NCEER, 1997; Youd et al., 2001), one of the most widely accepted and used SPT-based correlations is the "deterministic" relationship proposed by Seed, et al. (1984, 1985). Figure 6 shows this relationship, with minor modification at low CSR (as recommended by the NCEER Working Group; NCEER, 1997). This familiar relationship is based on comparison between SPT N-values, corrected for both effective overburden stress and energy, equipment and procedural factors affecting SPT testing (to $N_{1,60}$-values) vs. intensity of cyclic loading, expressed as magnitude-weighted equivalent uniform cyclic stress ratio (CSR_{eq}). As shown in Figure 6, the relationship between corrected $N_{1,60}$-values and the intensity of cyclic loading required to trigger liquefaction is also a function of fines content. Although widely used in practice, this relationship is dated, and does not make use of an increasing body of field case history data from seismic events that have occurred since 1984. It is particularly lacking data from cases where peak ground shaking levels were high (CSR > 0.25), an increasingly common design range in regions of high seismicity. This correlation also has no formal probabilistic basis, and so provides no insight regarding either uncertainty or probability of liquefaction. Efforts at development of similar, but formally probabilistic-based, correlations have been published by a number of researchers, including Liao et al. (1988), and more recently Youd and Noble (1997), and Toprak et al. (1999).

Cetin (2000) and Cetin et al. (2004) reassessed available case history data with improved understanding in geotechnical and earthquake engineering practice, and then recommended updated probabilistically-based liquefaction boundary curves for liquefaction triggering. Later, Cetin et al. (2018a) re-visited and updated their database, and consequently proposed new probabilistic and deterministic liquefaction triggering curves (Figure 6). Key elements in the development of Cetin et al. (2004, 2018a) were: (1) accumulation of a significantly expanded database of field performance case histories, (2) use of improved knowledge and understanding of factors affecting interpretation of SPT data, (3) incorporation of improved understanding of factors affecting site-specific ground motions (including directivity effects, site-specific response, etc.), (4) use of improved methods for assessment of in-situ cyclic shear stress ratio (CSR), (5) screening of field data case histories on a quality/uncertainty basis, and (6) use of higher-order probabilistic tools (Bayesian Updating). Bayesian updating methodology allowed for (a) simultaneous use of more descriptive variables than most prior studies, and (b) appropriate treatment of various contributing sources of aleatory and epistemic uncertainty. The resulting relationships not only provide greatly reduced uncertainty, but they also help to resolve a number of corollary issues that have long been difficult and controversial, including: (1) magnitude-correlated duration weighting factors, (2) adjustments for fines content, and (3) corrections for effective overburden stress. Moreover, non-linear mass participation factor (r_d), which is a significant component of the "simplified procedure" of Seed and Idriss (1971) (Equation 2), was re-evaluated based on the results of 2,153 seismic site response analyses. Cetin and Seed (2004) developed a relation in terms of depth (d), moment magnitude (M_w), peak horizontal ground surface acceleration (a_{max}) and stiffness of the site ($V^*_{s,12}$in m/s) (Equation 3).

$$CSR_{eqv} = 0.65 \cdot \frac{a_{\max}}{g} \cdot \frac{\sigma_v}{\sigma'_v} \cdot r_d \tag{2}$$

where CSR_{eqv}= equivalent cyclic stress ratio.

$$r_d(d, M_w, a_{\max}, V^*_{s,12}) = \frac{\left[1 + \frac{-23.013-2.949 \cdot a_{\max}+0.999 \cdot M_w+0.0525 \cdot V^*_{s,12}}{16.258+0.201 \cdot e^{0.341 \cdot (-d+0.0785 \cdot V^*_{s,12}+7.586)}}\right]}{\left[1 + \frac{-23.013-2.949 \cdot a_{\max}+0.999 \cdot M_w+0.0525 \cdot V^*_{s,12}}{16.258+0.201 \cdot e^{0.341 \cdot (0.0785 \cdot V^*_{s,12}+7.586)}}\right]} \pm \sigma_{\varepsilon r_d} \tag{3a}$$

For d ≥ 20 m,

$$r_d(d, M_w, a_{\max}, V^*_{s,12}) = \frac{\left[1 + \frac{-23.013-2.949 \cdot a_{\max}+0.999 \cdot M_w+0.0525 \cdot V^*_{s,12}}{16.258+0.201 \cdot e^{0.341 \cdot (-20+0.0785 \cdot V^*_{s,12}+7.586)}}\right]}{\left[1 + \frac{-23.013-2.949 \cdot a_{\max}+0.999 \cdot M_w+0.0525 \cdot V^*_{s,12}}{16.258+0.201 \cdot e^{0.341 \cdot (0.0785 \cdot V^*_{s,12}+7.586)}}\right]} - 0.0046 \cdot (d-20) \pm \sigma_{\varepsilon r_d} \tag{3b}$$

where $\sigma_{\varepsilon_{r_d}}$ = standard deviation of the random model correction term.

The closed-form solution of Cetin et al. (2018a) for the assessment of the cyclic resistance ratio, which involves the corrections for the influence of fines content, duration and effective stress and is defined as a function of probability of liquefaction, is given in Equation 4.

$$CRR(N_{1,60}, M_w, \sigma'_v, FC, P_L) = \exp\left[\frac{\left(\begin{array}{c} N_{1,60} \cdot (1+0.00167 \cdot FC) - 27.352 \cdot \ln(M_w) - 3.958 \cdot \ln\left(\frac{\sigma'_v}{Pa}\right) \\ +0.089 \cdot FC + 16.084 + 2.95 \cdot \Phi^{-1}(P_L) \end{array}\right)}{11.771}\right] \tag{4}$$

where P_L = probability of liquefaction in decimals (i.e. P_L = 50% is represented as 0.50); FC = percent fines content (by dry weight) expressed as an integer (e.g. 12% fines is expressed as FC = 12) with the limit of 5 ≤ FC ≤ 35); P_a = atmospheric pressure (=1 atm≅ 101.3 kPa) in the same units as the in-situ vertical effective stress (σ'_v); and $\Phi^{-1}(P_L)$ = inverse of the standard cumulative normal distribution (i.e., mean = 0, and standard deviation = 1). For spreadsheet construction purposes, the command in Microsoft Excel for this specific function is "*NORMINV*(P_L;0;1)". If a user prefers using this method for deterministic analysis (i.e., to

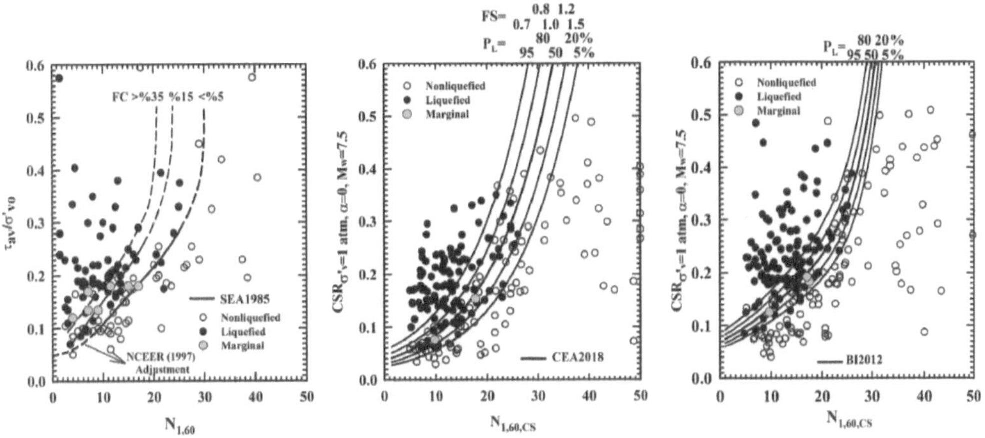

Figure 6. Liquefaction triggering relationships proposed by (a) Seed et al. (1985) (SEA1985) as modified slightly by Youd et al. (2001), (b) Cetin et al.(2018a) (CEA2018), and (c) Boulanger and Idriss (2012) (BI2012) (after Cetin et al., 2018b).

calculate factor of safety), then CRR corresponding to P_L=50% (0.5) should be used as the capacity term.

Idriss and Boulanger (2004) proposed an alternative semi-empirical framework for the evaluation of liquefaction triggering. This relationship has been modified subsequently (Idriss and Boulanger, 2006, 2008, 2010), and Boulanger and Idriss (2012) developed a probability-based framework, as well (Figure 6c).

The presence of a number of alternative liquefaction triggering methodologies is a source of confusion for practicing engineers, and is perceived as lack of consensus among researchers. For the purpose of clarifying the sources of these disagreements, Cetin et al. (2018b) conducted a detailed comparative study by revisiting the integral components of liquefaction triggering assessments. The authors highlighted followings as the major sources of the differences between existing triggering relationships: i) stress reduction factor (r_d), ii) magnitude – correlated duration correction (K_{Mw}), iii) overburden stress correction factor (K_σ), iv) fines adjustment ($\Delta N_{1,60,cs}$, and v) differences in probabilistic treatment used in the development of the triggering relationships. These issues were discussed in detail in Cetin et al. (2018b) and will not be repeated herein. However, Figures 7 through 10 were prepared to enable a visual comparison among the choices of different researchers.

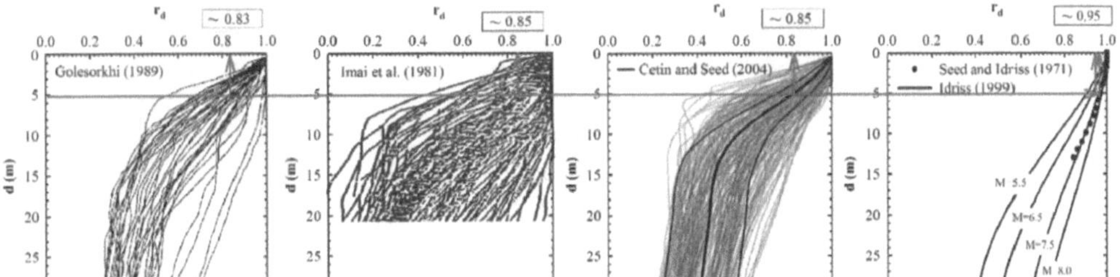

Figure 7. Plots of r_d values calculated based on seismic site analyses by Seed and Idriss (1971), Imai et al. (1981), Golesorhki (1989), Idriss (1999), Cetin and Seed (2004). (after Cetin et al., 2018b).

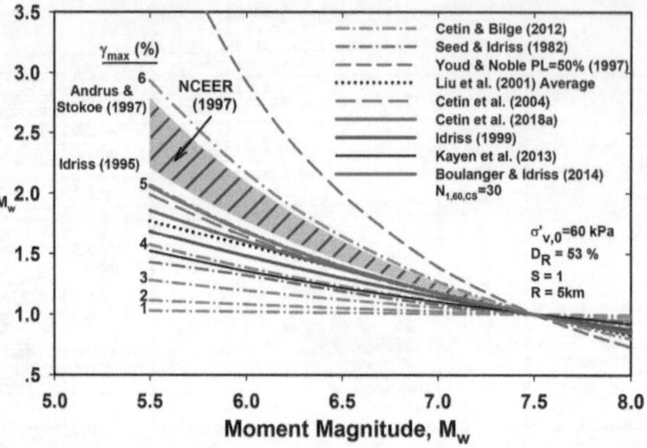

Figure 8. Existing K_{Mw} correction factors.

As revealed by these figures, these differences lead to a wide range of liquefaction triggering curves, addressing the epistemic nature of uncertainties. However, it should be noted that practicing engineers may eliminate some of the uncertainty in triggering predictions by consistently following the correction scheme of the original reference, since these corrections were consistently applied in the processing of case histories as well. Unfortunately, even consistency does not always guarantee the elimination of bias, if these models are used to predict the liquefaction performance of a site subjected to an earthquake shaking, which are different from "typical" (i.e.: median values) of the case history databases. Hence, extrapolation beyond the bounds of these relationships was discouraged.

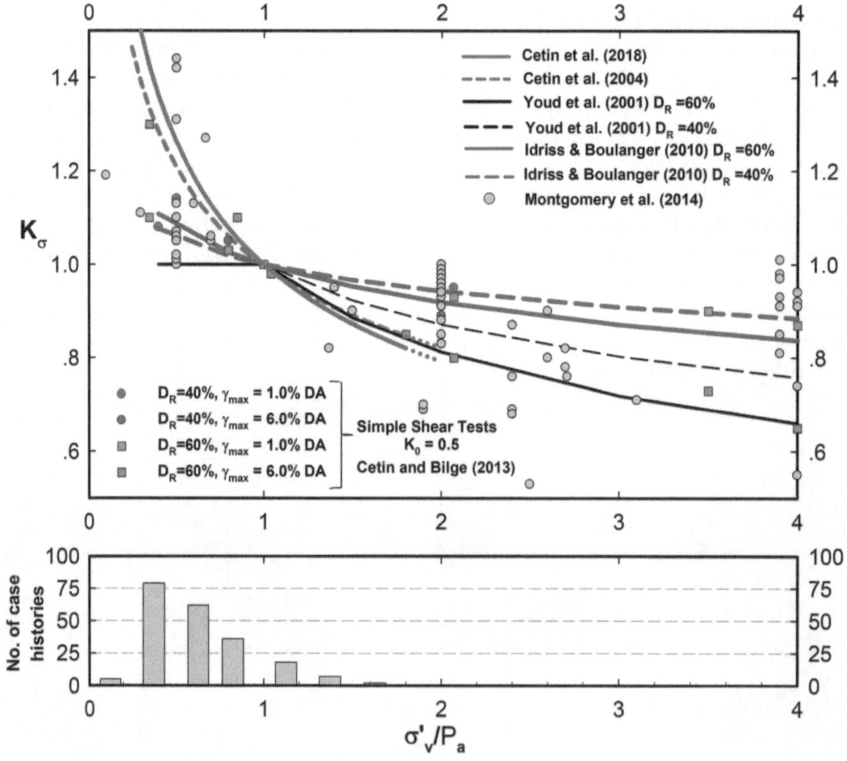

Figure 9. Recommended K_σ relationships of Youd et al. (2001), Cetin et al. (2004, 2018a) and Boulanger and Idriss (2012).

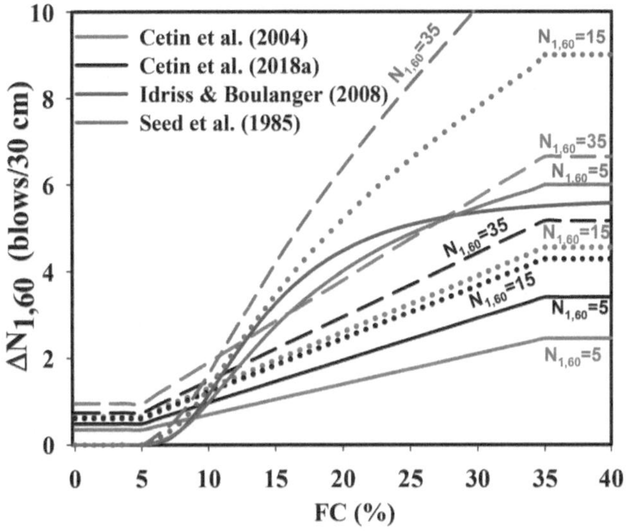

Figure 10. Comparative illustration of fines content corrections.

Within the confines of this section, due to page limitations and their wide use, only SPT-based methods were discussed. Regarding the CPT-based methods, readers are referred to the deterministic and probabilistic methods of Robertson (2009) and Moss et al. (2006), respectively. Shear wave velocity and Becker penetration test-based methods are relatively less frequently used; but readers are referred to Kayen et al. (2013) and Harder and Seed (1986), respectively, for a complete review of available literature.

It should be noted that all these methods are applicable to either clean sands or sands with limited amount of fines. As discussed earlier, silt and clay mixtures may also be susceptible to cyclic loading-induced strength loss and deformations. Unfortunately, research interest on their cyclic response picked up only recently, and hence, a comprehensive effort summarizing their cyclic performance is still missing. Yet, Boulanger and Idriss (2007) needs to be referred to as a practical tool, which is waiting to be tested via increasing number of case histories.

Following sections are devoted to the discussion of seismic strength and deformation responses of sandy soils, which allows a direct evaluation of seismic performance.

3 ASSESSMENT OF SEISMIC STRENGTH RESPONSE OF SANDY SOILS

There is a significant tendency towards the performance-based approaches in today's engineering profession. From seismic soil response point of view, this tendency puts forward the prediction of strength and deformation performances. Actually, they establish the basis of second and third level liquefaction engineering assessments, as outlined in Figure 1. For the sake of consistency, cyclic strength loss will be discussed before the discussion of cyclic straining.

Most of the previous efforts have focused on saturated clean sands and non-plastic silt – sand mixtures, as their shear strength solely rely on the effective stress state and intergranular friction. Thus, an increase in seismically – induced excess pore water pressure may cause a significant reduction in shear strength (most extreme case is "full" liquefaction) of saturated cohesionless soils. Consistent with liquefaction triggering methodologies, there exist two alternatives: i) sampling and laboratory testing, and ii) correlation of post-liquefaction strength with field case history data. The "steady-state" approach (e.g., Poulos et al., 1985), has benefitted from laboratory testing of both reconstituted and high-quality "undisturbed" samples, and a systematic basis for correction has been proposed for post-liquefaction "steady-state" strengths due to inevitable disturbance and densification effects that occurred during sampling and re-consolidation phases of undrained shearing. The method was eventually claimed to produce post-liquefaction strengths that were much higher than those back-calculated from field failure case histories (e.g. Seed et al., 1989). Hence, most research has diverted to the latter approach.

After the pioneer work of Seed (1987), many researchers (e.g., Seed and Harder, 1990; Robertson et al. 1992; Stark and Mesri, 1992; Ishihara, 1993; Wride et al., 1999) have performed extensive research to assess post-liquefaction shear strength of saturated sandy soils. Among these, Stark and Mesri (1992) had been widely accepted and used for some time. In this study, residual shear strength ($s_{u,r}$) is normalized by initial vertical effective stress, and the authors proposed a chart solution as a function of $N_{1,60,CS}$ as shown in Figure 11. After a decade, first Olson and Stark (2002), and then Idriss and Boulanger (2007) revisited the available case history database and recommended post – liquefaction shear strength relationships. SPT and CPT based correlations proposed by Olson and Stark (2002) are presented in Equations 5 and 6, respectively. On the other hand, the correlations proposed by Idriss and Boulanger (2007) are presented in Figure 12.

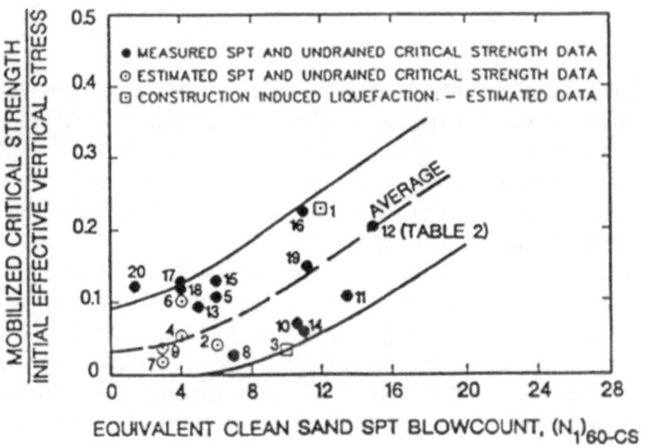

Figure 11. Post-liquefaction residual shear strength as a function of $N_{1,60,CS}$ and $\sigma'_{v,0}$ (after Stark and Mesri, 1992).

$$\frac{s_{u,r}}{\sigma'_{v0}} = 0.03 + 0.0075 \cdot N_{1,60} \pm 0.03 \quad for\ N_{1,60} \leq 12 \tag{5}$$

$$\frac{s_{u,r}}{\sigma'_{v0}} = 0.03 + 0.0143 \cdot q_{c1} \pm 0.03 \quad for\ q_{c1} \leq 6.5\,MPa \tag{6}$$

More recently, Kramer and Wang (2015) presented a hybrid model based on the existing case history data for prediction of post-liquefaction residual shear strength as a function of energy and overburden corrected SPT-N values ($N_{1,60}$) and initial vertical effective stress ($\sigma'_{v,0}$) as presented in Figure 13. An alternative probability based method is proposed by Weber et al. (2015) in terms of energy, overburden and fines corrected SPT-N value ($N_{1,60,CS}$) and $\sigma'_{v,0}$ as presented in Figure 14. The recommended probability-based boundaries are believed to be an improvement over deterministic approaches considering the scatter in case history data.

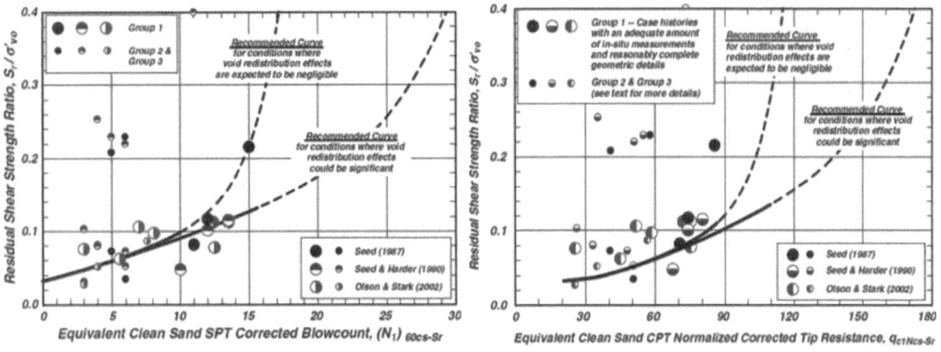

Figure 12. Residual shear strength prediction for SPT and CPT data (after Idriss and Boulanger, 2007).

Although post-cyclic strength loss is accepted to be more critical for saturated cohesionless soils, it could also be a serious threat for cohesive soils depending on intensity and duration of shaking and their undrained shear strength. Readers are referred to Cetin and Bilge (2014) for a review of available literature.

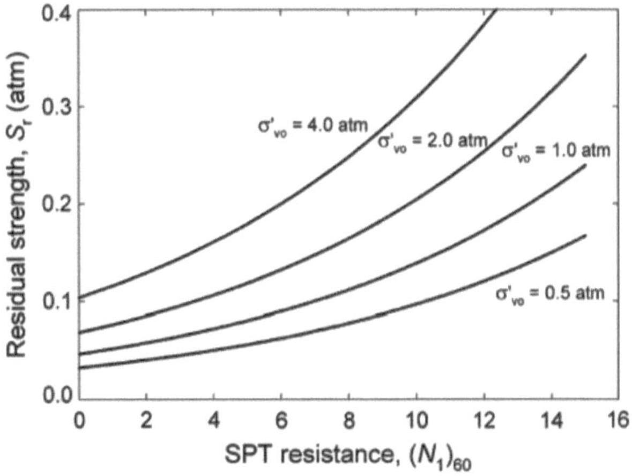

Figure 13. Residual shear strength as a function of $N_{1,60}$ and $\sigma'_{v,0}$ (after Kramer and Wang, 2015).

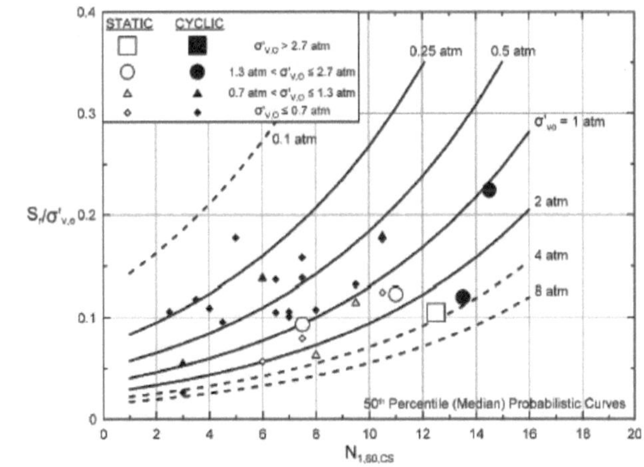

Figure 14. Residual shear strength as a function of $N_{1,60,CS}$ and $\sigma'_{v,0}$ (after Weber et al., 2015).

4 ASSESSMENT OF SEISMIC DEFORMATION RESPONSE OF SOILS

Despite major advances in soil liquefaction engineering, assessment of anticipated post-cyclic strain and deformations has remained a very "soft" area of practice. Within the confines of this chapter existing methods for assessment of cyclic-induced deformations will be discussed.

Numerous researchers have tried to quantify cyclic (or sometimes liquefaction-induced) soil straining through use of deterministic techniques based on laboratory test results and/or correlations of in-situ "index" tests with observed field performance data. Using $N_{1,60,CS}$ and CSR terms, Tokimatsu and Seed (1984) recommended a set of chart solutions for the estimation of limiting shear and post-cyclic volumetric strains based on the results of cyclic triaxial and simple shear tests performed on clean sands, further calibrated with case history performance data. Similarly, based on the results of cyclic simple shear tests, Ishihara and Yoshimine (1992) proposed cyclically-induced maximum shear and post-cyclic volumetric strain correlations, where normalized demand term was chosen as factor of safety against liquefaction, and capacity term was defined as relative density (D_R), or cone tip resistance (q_c), or SPT blow count ($N_{1,72}$). Based on the results of cyclic torsional shear tests, Shamoto et al. (1998) recommended a semi-empirical constitutive model, as well as chart solutions, for the estimation of post-cyclic residual shear and volumetric strains. Later, Wu et al. (2003) proposed cyclically-induced limiting shear and post-cyclic volumetric strain correlations based on the results of cyclic simple shear tests. Recommendations of all these four methods in the form of equi-shear or equi-volumetric strain contours are shown in Figures 15, 16 and 17, respectively. However, direct comparisons are difficult and not fair due to different definitions of demand and capacity, as well as shear strain terms adopted.

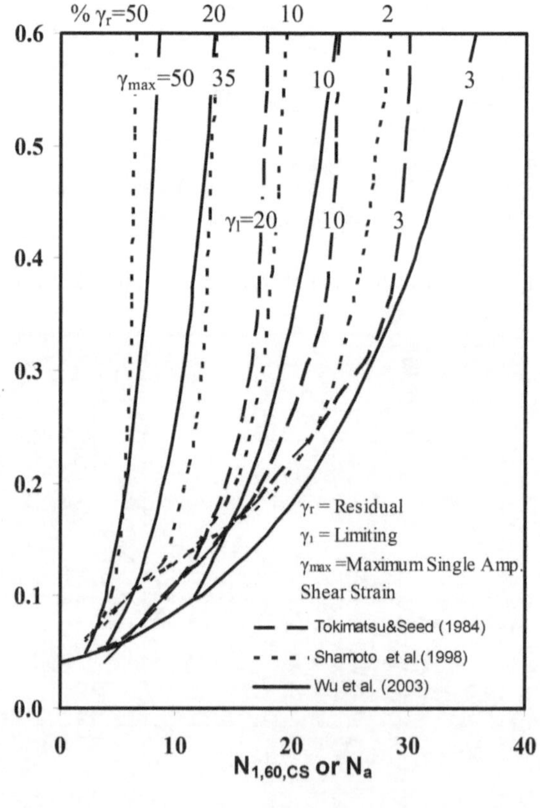

Figure 15. Existing methods for estimation of cyclic-induced deviatoric strains.

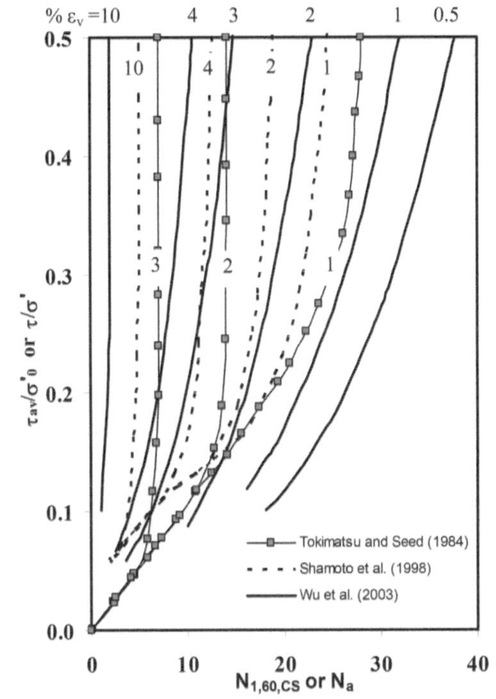

Figure 16. Existing methods for estimation of post-cyclic volumetric strains.

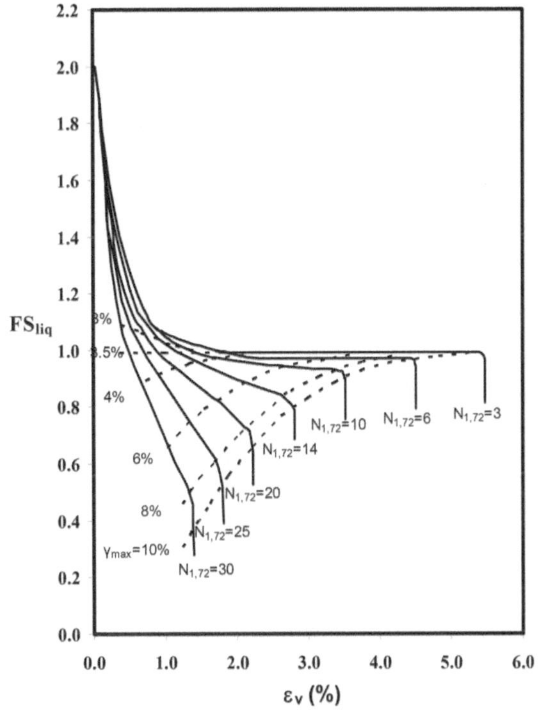

Figure 17. Ishihara And Yoshimine (1992) method for estimation of deviatoric and volumetric strains.

All these deterministic methods have been regarded as the best of their kinds and used in practical applications for many years. However, none of them considers the uncertainties associated with the nature of the problem. Cetin et al. (2009a) has introduced a new probabilistic-based framework based on the results of a comprehensive cyclic testing program. Semi-empirical models were developed for estimation of maximum cyclic shear (γ_{max}) and post-cyclic reconsolidation volumetric strain (ε_v) potentials of saturated clean sands, as presented by Equations 7 and 8, respectively.

$$\ln(\gamma_{max}) = \ln\left[\frac{-0.025 \cdot N_{1,60,CS} + \ln(CSR_{SS,20,1-D,1\,atm}) + 2.613}{0.004 \cdot N_{1,60,CS} + 0.001}\right] \pm 1.880 \tag{7}$$

where $CSR_{SS,20,1\text{-}D,1atm}$ = CSR value corresponding to 1 dimensional, 20 uniform loading cycles simple shear test under a confining pressure of 100 kPa (=1 atm), and this relation is valid for $5 \leq N_{1,60,CS} \leq 40$, $0.05 \leq CSR_{SS,20,1\text{-}D,1\,atm} \leq 0.60$ and $0\% \leq \gamma_{max} \leq 50\%$.

$$\ln(\varepsilon_v) = \ln\left[1.879 \cdot \ln\left[\frac{780.416 \cdot \ln(CSR_{SS,20,1-D,1\,atm}) - N_{1,60,CS} + 2442.465}{636.613 \cdot N_{1,60,CS} + 306.732}\right] + 5.583\right] \pm 0.689 \tag{8}$$

This relation is valid for $5 \leq N_{1,60,CS} \leq 40$, $0.05 \leq CSR_{SS,20,1\text{-}D,1atm} \leq 0.60$ and $0\% \leq \varepsilon_v \leq 5\%$.

Correction factors adopted to convert the CSR_{field} value to equivalent $CSR_{SS,20,1\text{-}D,1atm}$ are presented in Equation 9.

$$CSR_{SS,20,1-D,1atm} = \frac{CSR_{field}}{K_{md} \cdot K_{M_W} \cdot K_\sigma} \tag{9}$$

where K_{md}= correction to convert multi-directionally applied CSR_{field} value to the value of a uni-directionally applied laboratory CSR (Equation 10), K_{Mw} = correction for duration (magnitude) effects (Equation 11), K_σ = correction for varying effective overburden stress conditions (Equation 12).

$$K_{md} = 0.361 \cdot \ln(D_R) - 0.579 \tag{10}$$

$$K_{M_W} = \frac{87.1}{M_W^{2.217}} \tag{11}$$

$$K_\sigma = \left(\frac{\sigma'_{v,0}}{P_a}\right)^{f-1} \text{ and } f = 1 - 0.005 \cdot D_R \tag{12}$$

The next step following the assessment of cyclic straining potential is the prediction of soil deformations. In general, post-cyclic reconsolidation (volumetric) strains due to dissipation of excess pore water pressures are associated with settlements, whereas cyclic shear strains are associated with lateral spreads. Following discussion will be devoted to the prediction of settlements and lateral spreads.

4.1 *Assessment of post-cyclic settlements*

Currently available approaches for predicting the magnitude of post-cyclic reconsolidation settlements are categorized as: i) numerical analyses in the form of finite element and/or finite difference techniques (e.g., Martin et al., 1975; Seed et al., 1976; Booker et al., 1976; Finn et al.,1977; Liyanathirana and Poulos, 2002), and ii) semi-empirical models developed based on laboratory, field test and performance data (e.g. Lee and Albaisa, 1974;Tokimatsu and Seed,1984; Ishihara and Yoshimine,1992; Shamoto et al., 1998; Zhang et al., 2002; Wu and

Seed, 2004; Tsukamato et al., 2004). Due to difficulties in the determination of input model parameters necessary for numerical simulations, semi-empirical models continue to establish the state of practice for the assessment of cyclically-induced reconsolidation (volumetric) settlements. Even the best of their kind of these models cannot produce, at the moment, reasonably precise estimates of post-cyclic reconsolidation (volumetric) settlements.

Recently, Cetin et al. (2009b) has developed a new methodology based on their aforementioned semi-empirical post-cyclic volumetric strain estimation model. The proposed method was calibrated via 49 well-documented cyclically-induced ground settlement case histories from seven different earthquakes. Within the confines of that study, performance of the widely used methods of Tokimatsu and Seed (1984), Ishihara and Yoshimine (1992), Shamoto et al. (1998), Wu and Seed (2004) were comparatively evaluated. It was concluded that the proposed methodology, details of which will be given next, produced more accurate and precise settlement estimations compared to all other efforts.

Equation 16 constitutes the basis of the proposed method, and the estimation of $N_{1,60,CS}$ and $CSR_{SS,20,1\text{-}D,1atm}$ is the necessary first step. Next, a weighting scheme, linearly decreasing with depth, inspired after the recommendations of Iwasaki et al. (1982), is implemented. Aside from the better model fit it produced, the rationale behind the use of a depth weighting factor, is based on i) upward seepage, triggering void ratio redistribution, and resulting in unfavorably higher void ratios for the shallower sublayers of soil layers, ii) reduced induced shear stresses and number of shear stress cycles transmitted to deeper soil layers due to initial liquefaction of surficial layers, and iii) possible arching effects due to non-liquefied soil layers. All these may significantly reduce the contribution of volumetric settlement of deeper soil layers to the overall ground surface settlement. It is assumed that the contribution of layers to surface settlement diminishes as the depth of layer increases, and beyond a certain depth (z_{cr}) settlement of an individual layer cannot be traced at the ground surface. After statistical assessments, the optimum value of this threshold depth was found to be 18 meters. The proposed depth weighting factor (DF_i) is defined in Equation 13. Equivalent volumetric strain, $\varepsilon_{v,eqv}$, of the soil profile is estimated by Equation 14 and the estimated settlement, $s_{estimated}$, of the profile is simply calculated as the product of and the total thickness of the saturated cohesionless soil layers or sublayers, $\sum t_i$, as presented by Equation 15. Later, $s_{estimated}$ is further calibrated by using a factor (θ) for the estimation of field settlement values. In Equation 16, σ_ε term designates the standard deviation of the calibration model. Further discussion of the σ_ε term is presented later in the manuscript.

$$DF_i = 1 - \frac{d_i}{z_{cr} = 18m} \quad (13)$$

where d_i = mid-depth of each saturated cohesionless soil layer from ground surface.

$$\varepsilon_{v,eqv.} = \frac{\sum \varepsilon_{v,i} \cdot t_i \cdot DF_i}{\sum t_i \cdot DF_i} \quad (14)$$

$$s_{estimated} = \varepsilon_{v,eqv.} \cdot \sum t_i \quad (15)$$

$$\ln(s_{calibrated}) = \ln(\theta \cdot s_{estimated}) \pm \sigma_\varepsilon \quad (16)$$

In volumetric settlement assessment of the case histories, three situations were encountered regarding the application of DF: i) a very dense cohesionless soil layer ($N_{1,60,CS}$> 35 blows/30 cm) or bedrock or a cohesive soil layer underlying the volumetric settlement vulnerable cohesionless soil layer, ii) cohesionless soil layer continuing beyond the critical depth of 18 meters with or without available SPT profile, and iii) cohesionless soil site where the depth of boring is less than 18 m. For case (i), settlement calculations were performed till the depth to the top of the dense layer or bedrock or cohesive layer. For case (ii), potentially settlement vulnerable cohesionless layers beyond 18 meters were simply ignored due to their limited contribution to the overall ground surface settlement. For case (iii), after confirming with the geological characteristics of soil site, for the soil sub-layers without an SPT value at a specific depth, SPT values were judgmentally extended beyond the maximum borehole depth to a depth of maximum 18 m., based on available SPT blow-counts. Whenever a cohesive soil layer was encountered, it was assumed that cyclically-induced volumetric strain due to this layer was negligible. In addition, thickness of this layer was not considered in the calculation of $\varepsilon_{v,eqv}$.

For comparison purposes, each case history site (as presented in detail by Bilge and Cetin, 2007) was analyzed by using the methods of Tokimatsu and Seed (1984), Ishihara and Yoshimine (1992), Shamoto et al. (1998), Wu and Seed (2004) and finally the proposed method. The performance of the model predictions, expressed by Pearson product moment correlation coefficient, R^2, is summarized in Table 1. As a better alternative, which enabled the assessment of the model (calibration) error, predictions of each method were compared probabilistically by using the maximum likelihood analysis. Results of these analyses, a calibration coefficient (θ) which enables the model to produce unbiased predictions in the average is determined. These values are also presented in the same table along with the value of maximum likelihood and standard deviation of the random model correction term. It should be noted that higher values of maximum likelihood and lower values of standard deviation are also indicators of a better model. As the values of the calibration coefficient, θ, presented in Table 1 implies existing methods of Shamoto et al (1998), Tokimatsu and Seed (1984), and the proposed methodology under-predict the actual settlements by a factor of 1.91, 1.45 and 1.15, respectively. Similarly, Wu and Seed (2004), and Ishihara and Yoshimine (1992) over-predict settlements and need to be corrected by a factor of 0.98 and 0.90. Wu and Seed (2004) procedure produces the most unbiased settlement predictions (i.e.: the mean of the estimated settlements is about equal to the mean of the observed settlements). However, in terms of the uncertainty (or scatter) of the predictions, Wu and Seed (2004) methodology is ranked to be second to last with an R^2 value of 0.33. After scaling with the calibration coefficient, θ, the proposed model produces relatively the best predictions compared to the other four methods, also consistent with the R^2 trends presented in Table 1.

Table 1. Comparison of the performance of existing models.

Method	R^2	θ	σ_ε	$\sum$ likelihood fxn
Cetin et al. (2009b)	0.64	1.15	0.61	-19.8
Tokimatsu and Seed (1984)	0.33	1.45	1.05	-31.1
Ishihara and Yoshimine (1992)	0.42	0.90	1.12	-32.7
Shamoto et al. (1998)	0.36	1.93	1.36	-36.7
Wu and Seed (2004)	0.33	0.98	0.71	-22.9

Prediction performance of the method proposed by Cetin et al. (2009b) is also presented in Figure 18 in which predicted and observed settlements are paired and shown on figures along with the 1:2 and 1:0.5 boundary lines. Readers are referred to Cetin et al. (2009b) for the similar performance evaluation plots prepared for the other methods.

Similar to the case for the assessment of cyclic strength performance, the straining response of saturated cohesionless soils has drawn significantly more research interest. However, ground failure case histories after 1989 Loma-Prieta, 1994 Northridge, 1999 Kocaeli and 1999 Chi-Chi earthquakes have accelerated studies on assessing cyclic response of fine-grained soils. Readers are referred to Cetin and Bilge (2014) for a review of available literature and alternative strain prediction methods proposed for silt and clay mixtures.

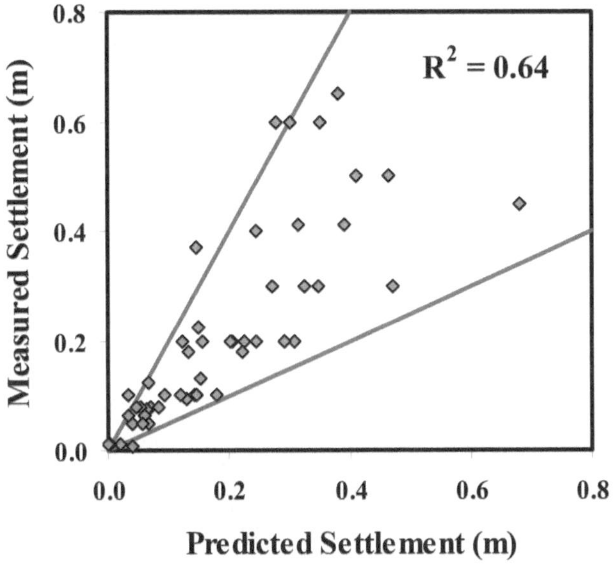

Figure 18. Comparison between the measured and predicted settlements by Cetin et al. (2009b).

4.2 *Assessment of lateral spreading*

Lateral spreading is a liquefaction-induced deformation problem identified by surficial soil layers breaking into blocks that progressively slide downslope or toward a free face during and after earthquake shaking. As opposed to the settlements, lateral ground deformations are generally more critical for the performance of overlying structures as well as of infrastructures due to their limited lateral resistance.

Currently available approaches for predicting the magnitude of lateral spreading ground deformations can be categorized as: i) numerical analyses in the form of finite element and/or finite difference techniques (e.g., Finn et al., 1994; Arulanandan et al., 2000; Liao et al., 2002), ii) soft computing techniques (e.g., Wang and Rahman, 1999), iii) simplified analytical methods (e.g., Newmark,1965; Towhata et al., 1992, Kokusho and Fujita, 2002; Elgamal et al., 2003), and iv) empirical methods developed based on the assessment of either laboratory test data or statistical analyses of lateral spreading case histories (e.g., Hamada et al., 1986; Shamoto et al.,1998; Youd et al., 2002). Due to difficulties in the determination of input model parameters of currently existing numerical and analytical models, empirical and semi-empirical models continue to establish the state of practice for the assessment of liquefaction-induced lateral ground deformations.

Hamada et al. (1986), Youd and Perkins (1987), Rauch (1997), Shamoto et al.(1998), Bardet et al. (1999), Youd et al. (2002), Kanibir (2003) and Faris et al. (2006) introduced empirically-based models for the assessment of liquefaction-induced lateral spreading. With the exception of Shamoto et al. and Faris et al., these models were developed based on regression analyses of available lateral spreading case histories. The predictive approach of Shamoto et al. (1998) and Faris et al. (2006) employ laboratory-based estimates of liquefaction-induced limiting shear strains coupled with an empirical adjustment factor in order to relate these laboratory values to the observed field behavior. Among all of these models, in addition to the pioneering study of Hamada et al. (1986), widely accepted and used Youd et al. (2002), and laboratory-based and field-calibrated model of Faris et al. (2006) will be discussed in more detail next.

Hamada et al. (1986) introduced a simple empirical equation for predicting liquefaction induced lateral ground deformations only in terms of ground slope and thickness of liquefied soil layer. This equation was based on the regression analysis of 60 earthquake case histories, mostly from Noshiro – Jana, and it was expressed in Equation (17):

$$D_h = 0.75 \cdot H^{1/2} \cdot \theta^{1/3} \tag{17}$$

where D_h = predicted horizontal ground displacement (in meters); H = thickness of liquefied zone (in meters, when more than one sub-layer liquefies, H is measured as the distance from the top-most to the bottom-most liquefied sub-layers including all intermediate sub-layers); and θ= the larger slope of either ground surface or liquefied zone lower boundary (%). Despite its simplicity and ease of use, due to limited number of case histories which established the basis of the relationship, its use should be limited to only cases with similar conditions.

Starting in the early 1990's, Bartlett and Youd (1992, 1995) introduced empirical methods for predicting lateral spread displacements at liquefiable sites. The procedure of Youd et al. (2002) is a refinement of these early efforts and the new and improved predictive models for either (i) sloping ground conditions, or (ii) relatively level ground conditions with a "free face" towards which lateral displacements may occur, were developed through multi-linear regression of a case history database. The proposed predictive models for the sloping ground and "free face" conditions are given in Equation 18 and 19, respectively.

$$\begin{aligned}\log D_h = -16.213 + 1.532 \cdot M_w - 1.406 \cdot \log R * -0.012 \cdot R + 0.338 \cdot \log S + \\ 0.54 \cdot \log T_{15} + 3.413 \cdot \log(100 - F_{15}) - 0.795 \cdot \log(D50_{15} + 0.1\, mm)\end{aligned} \tag{18}$$

$$\begin{aligned}\log D_h = -16.713 + 1.532 \cdot M_w - 1.406 \cdot \log R * -0.012 \cdot R + 0.592 \cdot \log W + \\ 0.54 \cdot \log T_{15} + 3.413 \cdot \log(100 - F_{15}) - 0.795 \cdot \log(D50_{15} + 0.1\, mm)\end{aligned} \tag{19}$$

where D_h = predicted horizontal ground displacement in meters predicted by multiple linear regression model; M_w = earthquake magnitude; S = gradient of surface topography or ground slope (%); W = the free-face ratio, defined as the height of the free-face divided by its distance to calculation point, T_{15} = thickness of saturated layers with SPT-$N_{1,60} \leq 15$; F_{15}= average fines content (particles < 0.075 mm) in T_{15} (%); $D50_{15}$= average D_{50} in T_{15}; R = horizontal distance to the nearest seismic source or to nearest fault rupture (km); and R* = a distance term calculated by Equation (20).

$$R* = R + R_0 \ and\ R_0 = 10^{0.89 \cdot M_w - 5.64} \tag{20}$$

The empirical model of Youd et al. (2002) is widely used in the engineering practice. The performance of the model was also evaluated by Youd et al., as presented in Figure 19. Reported R^2 value of 83.6% is concluded to be sufficiently high. However, it should be noted that i) an attenuation-like intensity measure in terms of magnitude and distance is adopted as opposed to an independent peak soil ground acceleration term, which further brings along the uncertainties in the predictions of these attenuation-like formulations into the lateral spreading predictions, and ii) zero lateral displacement was produced for soil sites composed of sublayers with $N_{1,60}$ to be greater than 15 blows/30 cm. Moreover, the success rate at the displacement range of 0 to 3 m, which is believed to be more critical, compared to large displacement range from performance point of view, is not satisfactorily high.

Later, Faris et al. (2006) has presented a semi-empirical model as presented in Equation 21.

$$M_{max} = \exp(1.0443 \cdot \ln(DPI_{max}) + 0.0046 \cdot \ln \alpha + 0.0029 \cdot M_w \tag{21}$$

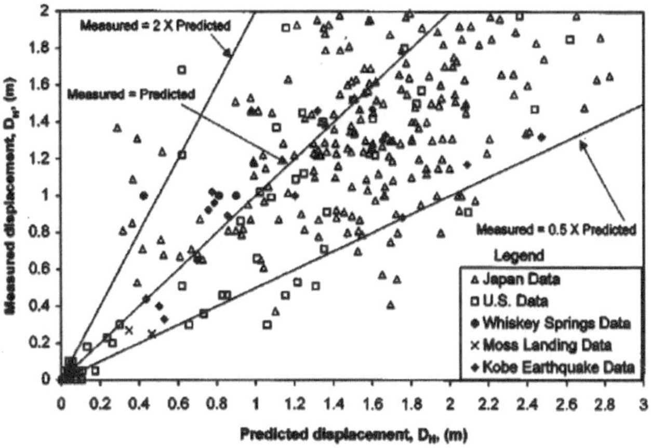

Figure 19. Prediction performance of Youd et al. (2002) method.

where H_{max} = lateral spreading (in meters); DPI_{max} = maximum cyclic shear strain potential to be determined according to Wu et al. (2003) (Figure 14); α = slope or free-face ratio; M_w = earthquake magnitude. Faris et al. (2006) has performed a performance evaluation study results of which is presented in Figure 20. Note that this framework takes into account the cyclic shear straining potential of soils, which is a physically meaningful term. However, similar to the method of Youd et al., the prediction success rate of this mode is not very high at the displacement range of 0 to 3 m.

Although these models are the best of their kind, due to large uncertainties associated with input parameters as well as model errors, more efforts are needed to achieve more precise models in the prediction of lateral spread-type soil deformations. Thus, practicing engineers are warned to be aware of the large uncertainty involved in the predictive models. A probabilistic approach addressing these sources of uncertainties could be a robust decision making approach and is strongly recommended.

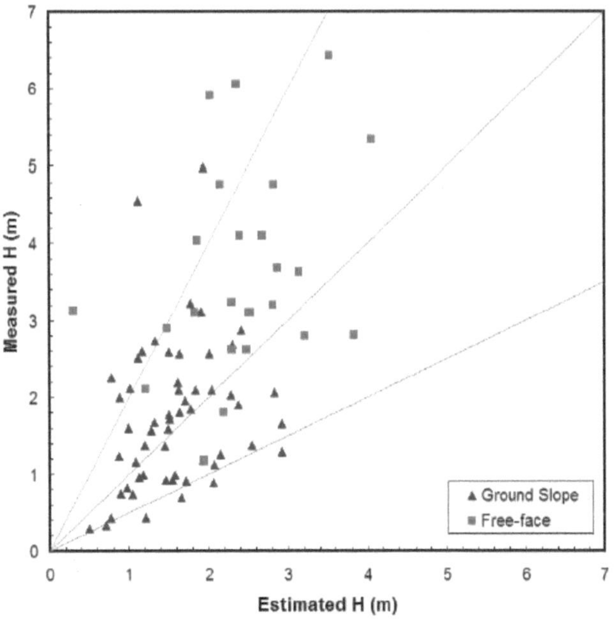

Figure 20. Prediction performance of Faris et al. (2006) method.

5 SUMMARY AND CONCLUDING REMARKS

Within the confines of this manuscript, a summary of current state of practice in seismic soil liquefaction engineering was presented. Seismic soil liquefaction engineering problems involve a five step assessment framework including the assessment of i) "triggering" or initiation of soil liquefaction, ii) post-liquefaction strength and overall post-liquefaction stability, iii) expected liquefaction-induced deformations and displacements, iv) the consequences of these deformations and displacements, v) mitigation alternatives, if necessary. The discussions presented herein are mostly focused on the first three steps of liquefaction engineering. Considering the increasing popularity of performance-based design trends, special emphasis was given on the assessment of cyclic strength and deformation performance of cohesionless soils. Available literature is summarized, along with the introduction of new frameworks. Practical recommendations were listed hoping to be helpful to engineers.

REFERENCES

Andrews, D. C. & Martin, G.R. 2000. Criteria for liquefaction of silty sands. *Proceedings of the 12th World Conference on Earthquake Engineering*, Auckland, New Zealand.

Andrus, R. D. & Stokoe, K. H., II. 1997. Liquefaction resistance based on shear wave velocity. *Proc., NCEER Workshop on Evaluation of Liquefaction Resistance of Soils*, Nat. Ctr. For Earthquake Engrg. Res., State Univ. of NY, Buffalo, 89–128.

Arulanandan, K., Li, X.S., Sivathasan, K. 2000. Numerical simulation of liquefaction-induced deformation. *J. Geotechnical Eng.*, ASCE, Vol. 126, No. 7, July, pp. 657–666.

Bardet, J. P., Mace, N., Tobita, T. 1999. Liquefaction-Induced Ground Deformation and Failure. *A report to PEER/PG&E, Task 4A - Phase 1*, Civil Eng. Dep., Univ. of Southern California, Los Angeles.

Bartlett, S. F. &Youd, T. L. 1992. Empirical Analysis of Horizontal Ground Displacement Generated by Liquefaction-Induced Lateral Spreads. *Technical Report No. NCEER-92-0021*, National Center for Earthquake Engineering Research, State University of New York, Buffalo, pp. 5-14-15.

Bartlett, S. F. & Youd, T. L. 1995. Empirical prediction of liquefaction-induced lateral spread, *J. Geotechnical Eng.*, 121(4), 316–329.

Bilge, H. T. 2010. Cyclic volumetric and shear strain responses of fine-grained soils. *Ph.D. Dissertation*, Middle East Technical University, Ankara.

Bilge, H. T., & Cetin, K. O. 2007. Field performance case histories for the assessment of cyclically-induced reconsolidation (volumetric) settlements. *METU – EERC Report No. 07 – 01*, Middle East Technical University Earthquake Engineering Research Center, Ankara.

Booker, J. R., Rahman, M. S., Seed, H. B. 1976. GADFLEA – A computer program for the analysis of pore pressure generation and dissipation during cyclic or earthquake loading. *Report No. EERC 76-24*, Univ. of California at Berkeley, Berkeley, Califonia.

Boulanger, R.W. & Idriss, I.M. 2006. Liquefaction susceptibility criteria for silts and clays. *J. Geotechnical Geoenvironmental Engineering*, ASCE, 132(11), 1413–1426.

Boulanger, R.W. & Idriss, I.M. 2007. Evaluation of cyclic softening in silts and clays. *J. Geotechnical Geoenvironmental Engineering*, ASCE, 133(6), 641–652.

Boulanger, R.W. & Idriss, I.M. 2012. Probabilistic standard penetration test-based liquefaction - triggering procedure. *J. Geotech. Geoenviron.* ASCE, 138(10), 1185–95.

Boulanger, R.W. & Idriss, I.M. 2014. CPT and SPT based liquefaction triggering procedures. *Report No. UCD/CGM-14/01*, Davis, CA: Center of Geotech. Modeling, Dept. of Civil and Envir. Eng., Univ. of California.

Bray, J. D. & Sancio, R. B. 2006. Assessment of liquefaction susceptibility of fine-grained soils. *J. Geotechnical and Geoenvironmental Eng.*, 132(9), 1165–1177.

Cetin, K.O. 2000. Reliability -based assessment of seismic soil liquefaction initiation hazard. *PhD dissertation*, Univ. of California, Berkeley, Calif.

Cetin, K.O., Bilge, H.T., Wu, J., Kammerer A., Seed, R.B. 2009a. Probabilistic models for cyclic straining of saturated clean sands. *J. Geotechnical and Geoenvironmental Eng.*, ASCE, 135(3), 371–386.

Cetin, K. O., Bilge, H. T., Wu, J., Kammerer, A., Seed, R. B. 2009b. Probabilistic model for the assessment of cyclically-induced reconsolidation (volumetric) settlements. *J. Geotechnical and Geoenvironmental Eng.*, ASCE., 135 (3), 387–398, 2009.

Cetin, K. O. & Bilge, H. T. 2012. Performance based assessment of magnitude (duration) scaling factors. *J. Geotech. Geoenv. Eng.*, ASCE, 138 (3):324–334.
Cetin, K.O. & Bilge, H. T. 2013. Stress scaling factors for seismic soil liquefaction engineeringproblems: a performance-based approach. *Proceedings of international conference on earthquake geotechnical engineering from case history to practice in honor of Prof. Kenji Ishihara*, Istanbul, Turkey
Cetin, K. O. & Bilge H. T. 2014. Recent advances in soil liquefaction engineering. *Perspectives on European Earthquake Engineering and Seismology: Volume 1*, edited by A. Ansal, Springer.
Cetin K. O. & Seed R. B. 2002. Nonlinear shear mass participation factor, rd for cyclic shear stress ratio evaluation. *J. Soil Dynamics and Earthquake Eng.* 24(2): 103–113.
Cetin, K. O., Seed, R. B., Der Kiureghian, A., Tokimatsu, K., Harder Jr., L. F., Kayen, R. E., Moss, R. E. S. 2004. SPT-based probabilistic and deterministic assessment of seismic soil liquefaction potential. *J. Geotech. Geoenviron. Eng.*, 130(12), 1314–1340.
Cetin, K. O., Seed, R. B., Kayen, R., Moss, R. E. S., Bilge, H. T., Ilgac, M., Chowdhury, K. 2018a. SPT-based probabilistic and deterministic assessment of seismic soil liquefaction triggering hazard, *Soil Dynamics and Earthquake Engineering*, 113, 75–86, 2018.
Cetin, K. O., Seed, R. B., Kayen, R., Moss, R. E. S., Bilge, H. T., Ilgac, M., Chowdhury, K. 2018b. Examination of differences between three SPT-based seismic soil liquefaction triggering relationships, *Soil Dynamics and Earthquake Engineering*, 115, 698–709, 2018.
Chen, L., Yuan, X., Cao, Z., Sun, R., Wang, W., Liu, H. 2018. *Earthquake Spectra*, Vol. 34, No. 3, 1091–1111.
Elgamal, A., Yang, Z., Parra, E., Ragheb, A. 2003. Modeling of cyclic mobility in saturated cohesionless soils. *International Journal of Plasticity*, 19(6), 883–905.
Faris, A. T., Seed, R. B., Kayen, R. E., Wu, J. 2006. A semi-empirical model for the estimation of maximum horizontal displacement due to liquefaction-induced lateral spreading. *Proc. 8th US National Conf. on Earthquake Eng.*, San Francisco, California, USA, Paper No. 1323.
Finn, L. W., Ledbetter, R. H., Guoxi, W. U. 1994. Liquefaction in silty soils: design and analysis. *Ground Failures under Seismic Conditions*, ASCE Geotechnical Special Publication, No. 44, pp. 51–79.
Finn, W. D. L., Lee, K. W., Martin, G. R. 1977. An effective stress model for liquefaction. *J. Geotech. Eng.*, 103(6), 517–533.
Golesorkhi R. 1989. Factors influencing the computational determination of earthquake- induced shear stresses in sandy soils. *Ph.D. Dissertation*, University of California at Berkeley.
Hamada, M., Yasuda, S., Isoyama, R., Emoto, K. 1986. Study on Liquefaction Induced Permanent Ground Displacement. *Report for the Association for the Development of Earthquake Prediction*, Japan.
Harder, L. F. Jr. 1988. Use of penetration tests to determine the cyclic loading resistance of gravelly soils during earthquake shaking. *Ph. D. Dissertation*, University of California, Berkeley.
Harder, L. F., Jr. & Seed, H. B. 1986. Determination of penetration resistance for coarse-grained soils using the Becker Hammer Drill. *Earthquake Engineering Research Center, Report No. UCB/EERC-86/06*, University of California, Berkeley.
Hatanaka, M., Uchida, A., Ohara, J. 1997. Liquefaction characteristics of a gravelly fill liquefied during the 1995 Hyogo-Ken Nanbu earthquake. *Soils and Foundations*, 37(3), 107–115.
Idriss, I. M. 1999. An update to the Seed-Idriss simplified procedure for evaluating liquefaction potential. *Proceedings of TRB workshop on new approaches to liquefaction, Publication No. FHWA-RD-99-165*, Federal Highway Administration.
Idriss, I. M. 1995. *Seed memorial lecture*. University of California at Berkeley.
Idriss, I. M.& Boulanger, R. W. 2004. Semi-empirical procedures for evaluating liquefaction potential during earthquakes. *Proceedings, 11th International Conference on Soil Dynamics and Earthquake Engineering, and 3rd International Conference on Earthquake Geotechnical Engineering*, D. Doolin et al., eds., Stallion Press, Vol. 1, pp. 32–56.
Idriss, I. M. & Boulanger, R. 2006. Semi-empirical procedures for evaluating liquefaction potential during earthquakes. *Soil Dyn. Earthquake Eng.*, 26(2-4), 115–130.
Idriss, I. M. & Boulanger, R. W. 2007. SPT- and CPT-based relationships for the residual shear strength of liquefied soils. *Earthquake Geotechnical Engineering, 4th International Conference on Earthquake Geotechnical Engineering – Invited Lectures*, K. D. Pitilakis, ed., Springer, The Netherlands, 1 – 22.
Idriss, I. M.& Boulanger, R. W. 2008. Soil liquefaction during earthquakes. *Monograph MNO-12*, Earthquake Engineering Research Institute, Oakland, CA, 261 pp.
Idriss, I. M. & Boulanger, R. W. 2010. SPT-based liquefaction triggering procedures. *Report No. UCD/CGM-10/02*, Dept. Civil &Envir. Eng., College of Eng., Univ. of California, Davis.

Imai, T., Tonouchi, K., Kanemori, T. 1981. The simple evaluation method of shear stressgenerated by earthquakes in soil ground. *Report no 3*. Bur.Pract.GeolInvestig.39–58.
Ishihara, K. 1985. Stability of natural deposits during earthquakes. *Proc. of the Eleventh International Conf. on Soil Mech. and Found. Eng.*, San Francisco.
Ishihara, K. 1993. Liquefaction and flow failure during earthquakes. *Geotechnique*, 43(3), 351–415.
Ishihara, K. & Yoshimine, M. 1992. Evaluation of settlements in sand deposits following liquefaction during earthquakes. *Soils Found.*, 32(1), 173–188.
Iwasaki, T., Arakawa, T., Tokida, K. 1982. Standard penetration test and liquefaction potential evaluation. *Proceedings, International Conference of Soil Synamics and Earthquake Engineering*, Southampton, Vol. 2, 925–941.
Kayen, R., Moss, R. E. S., Thompson, E. M., Seed, R. B., Cetin, K. O., Der Kiureghian, A., Tanaka, Y., Tokimatsu, K. 2013. Probabilistic and deterministic assessment of seismic soil liquefaction potential by shear wave velocity. *ASCE J. of Geotechnical and Geoenvir. Eng*. 139(3), 407–419.
Kanibir, A. 2003. Investigation of the lateral spreading at Sapanca and suggestion of empirical relationships for predicting lateral spreading. *M.Sc. Thesis*, Department of Geological Engineering, Hacettepe University, Ankara, Turkey.
Kishida, H. 1966. Damage to reinforced concrete buildings in Niigata city with special reference to foundation engineering. *Soils and Foundations*, Vol. VIII, No. 2, 38–44.
Kokusho, T. & Fujita, K. 2002. Site investigations for involvement of water films in lateral flow in liquefied ground. *J. Geotechnical and Geoenvironmental Eng.*, 128(11), 917–925.
Kramer, S. L. & Wang, C.H. 2015. Empirical Model for Estimation of the Residual Strength of Liquefied Soil. *Journal of Geotechnical and GeoenvironmentalEngineering*, 141(9).
Lasley S., Green R. A, Rodriguez-Marek A. 2016. A new stress reduction coefficient relationship for liquefaction triggering analyses. *Technical Note, Journal of Geotechnical and Geoenvironmental Engineering*, 142(11).
Lee, K. L., & Albeisa, A. 1974. Earthquake induced settlements in saturated sands. *J. Geotech. Eng.*, 100 (GT4), 387–406.
Liao, S. S. C., Veneziano, D., Whitman, R. V. 1988. Regression models for evaluating liquefaction probability. *J. Geotechnical Eng.*, 114(4), 389–411.
Liao, T., McGillivray, A., Mayne, P.W., Zavala, G., Elhakim, A. 2002. Seismic ground deformation modeling. *Final report for MAE HD-7a (year 1)*, Geosystems Engineering/School of Civil & Environmental Engineering, Geoargia Institute of Technology, Atlanta.
Liu A. H., Stewart J. P., Abrahamson N. A., Moriwaki Y. 2001. Equivalent number of uniform stress cycles for soil liquefaction analysis. *J. Geotech. Geoenviron. Eng.*, 127(12), 1017–1026.
Liyanathirana, D. S. & Poulos, H. G. 2002. A numerical model for dynamic soil liquefaction analysis. *Soil Dyn. Earthquake Eng.*, 22(9-12), 1007–1015.
Martin, G. R., Finn, W. D. L., Seed, H. B. 1975. Fundamentals of liquefaction under cyclic loading. *J. Geotech. Eng. Div.*, 101(5), 423–438.
Montgomery J., Boulanger, R.W., Harder Jr. L.F. 2014. Examination of the Kσ overburden correction factor on liquefaction resistance, *J.Geotech.Geoenviron. Eng.*, 140 (12).
Moss, R. E. S., Seed, R. B., Kayen, R. E., Stewart, J. P., Kiureghian, A. Der., Cetin, K. O. 2006. CPT-based probabilistic and deterministic assessment of in situ seismic soil liquefaction potential. *J. Geotechnical Geoenvironmental Eng.*, 132(8), 1032–1051.
NAP. 2016. State of the art and practice in the assessment of earthquake – induced soil liquefaction and its consequences. *National Academies of Sciences, Engineering and Medicine*; Washington, DC, The National Academies Press.
National Center for Earthquake Engineering Reseach (NCEER). 1997. *Proceedings of the NCEER Workshop on Evaluation of Liquefaction Resistance of Soils.* Edited by Youd, T. L., Idriss, I. M., Technical Report No. NCEER-97-0022.
Newmark, N. M. 1965. Effects of Earthquakes on Embankments and Dams. *Géotechnique*, Vol. 15, No.2, pp. 139–160.
Olson, S.M. & Stark, T.D. 2002. Liquefied strength ratio from liquefaction case histories. *Canadian Geotechnical Journal*, 39, 629–647.
Poulos S.J., Castro G., France J.W. 1985. Liquefaction evaluation procedure. *J. Geotechnical Eng.*, ASCE, 111(6), 772–792.
Rauch, A. F. 1997. An empirical method for predicting surface displacements due to liquefaction-induced lateral spreading in earthquakes. *Ph. D. Dissertation*, Virginia Polytechnic Institute and State University, Blacksburg, Virginia.
Robertson, P.K., Woeller, D.J., Finn, W.D.L. 1992. Seismic cone penetration test for evaluating liquefaction potential under cyclic loading. *Canadian Geotechnical Journal*, 29, 686–695.

Seed, H.B. 1987. Design problems in soil liquefaction. *J. Geotechnical Eng.*, ASCE, 113 (8), 827–845.
Seed, R.B. & Harder, L.F. Jr. 1990. SPT-based analysis of cyclic pore pressure generation and undrained residual strength. *Proceedings of the H. B. Seed Memorial Symposium.*
Seed, H. B. & Idriss, I. M. 1982. Ground motions and soil liquefaction during earthquakes. *Earthquake Engineering Research Institute*, Berkeley, CA, 134 pp.
Seed, H. B. & Idriss, I. M. 1971. Simplified procedure for evaluating soil liquefaction potential. *J. Soil Mechanics Foundations* Div., ASCE, Vol. 97, SM9, 1249–1273.
Seed, H. B., Martin, G. R, Lysmer, J. 1976. Pore-water pressure changes during soil liquefaction. *J. Geotech. Eng.*, 102(GT4), 323–346.
Seed, H.B., Seed, R.B., Harder, L.F., Jong, H.-L. 1989. Re-evaluation of the Lower San Fernando Dam: Report 2, examination of the post-earthquake slide of February 9, 1971. *US. Army Corps of Engineers Contract Report GL-89-2*, US. Amy Corps of Engineers Waterways Experiment Station, Vicksburg, Mississippi.
Seed, H. B., Tokimatsu, K., Harder, L. F., Chung, R. M. 1984. The influence of SPT procedures in soil liquefaction resistance evaluations. *Earthquake Engineering Research Center Report No. UCB/EERC-84/15*, University of California, Berkeley.
Seed, H. B., Tokimatsu, K., Harder, L. F., Chung, R. M. 1985. The influence of SPT procedures in soil liquefaction resistance evaluations. *J. Geotechnical Eng.*, 111(12), 1425–1445.
Seed R. B., Cetin K. O., Moss R. E. S., Kammerer A. M., Wu J., Pestana, J. M., Riemer M. F. 2001. Recent Advances in Soil Liquefaction Engineering and Seismic Site Response Evaluation. *4th International Conference on Recent Advances in Geotechnical Earthquake Engineering and Soil Dynamics*, San Diego.
Seed, R. B., Cetin, K. O., Moss, R. E. S., Kammerer, A. M., Wu, J., Pestana, J. M., Riemer, M. F., Sancio, R. B., Bray, R. B., Kayen, R. E., Faris, A. 2003. Recent advances in soil liquefaction engineering: a unified and consistent framework. *Report No. EERC 2003-06*, Earthquake Engineering Research Center, University of California, Berkeley.
Shamoto, Y., Zhang, J. M., Tokimatsu, K. 1998. Methods for evaluating residual post-liquefaction ground settlement and horizontal displacement. *Special Issue on the Geotechnical Aspects of the January 17 1995 Hyogoken-Nambu Earthquake*, No. 2, 69–83.
Stark, T.D. & Mesri G. 1992. Undrained shear strength of liquefied sands for stability analysis. *J. Geotechnical Engineering Division*, ASCE, 118(11),1727-1747.
Tokimatsu, K., & Seed, H. B. 1984. Simplified procedures of the evaluation of settlements in clean sands. *Rep. No. UCB/GT-84/16*, University of California, Berkeley, California.
Toprak, S., Holzer, T. L., Bennett, M. J., Tinsley, J. C. 1999. CPT- and SPT-based probabilistic assessment of liquefaction potential. *Proc. 7th US-Japan Workshop on Earthquake Resistant Design of Lifeline Facilities and Countermeasures against Liquefaction*, Seattle.
Towhata, I., Sasaki, Y., Tokida, K.-I., Matsumoto, H., Tamari, Y., Yamada, K. 1992. Prediction of Permanent Displacement of Liquefied Ground by means of Minimum Energy Principle. *Soils and Foundations*. JSSMFE, Vol. 32, No.3, September, pp. 97–116.
Tsukamato, Y., Ishihara, K., Sawada, S. 2004. Settlement of silty sand deposits following liquefaction during earthquakes. *Soils and Foundations*, 44(5), 135–148.
Wang, W. 1979. Some Findings in Soil Liquefaction. *Report Water Conservancy and Hydroelectric Power Scientific Research Institute* (pp. 1-17). Beijing, China.
Wang, J. G. & Rahman, M. S. 1999. A Neural Network Model for Liquefaction-Induced Horizontal Ground Displacement. *Soil Dynamics and Earthquake Engineering*, Vol. 18, pp. 555–568.
Weber, J. P., Seed, R.B., Pestana, J. M., Moss, R.E.S., Nweke, C., Deger, T.T., Chowdhury, K. 2015. Engineering evaluation of post-liquefaction strength. Dept. Civil and Env. Eng., Univ. of California at Berkeley.
Wride, C.E., Mcroberts, E.C., Robertson, P.K. 1999. Reconsideration of case histories for estimating undrained shear strength in sandy soils, *Can. Geotech. Jr.*, Vol. 36(5), 907–933.
Wu J., Seed R. B., Pestana J. M. 2003. Liquefaction triggering and post liquefaction deformations of Monterey 0/30 sand under uni-directional cyclic simple shear loading. *Geotechnical Engineering Research Report No. UCB/GE-2003/01*, University of California, Berkeley, California.
Wu, J. & Seed, R. B. 2004. Estimating of liquefaction-induced ground settlement (case studies). *Proceedings, Fifth International Conference on Case Histories in Geotechnical Engineering*, Paper 3.09, New York, NY USA.
Yegian, M. K., Ghahraman, V. G., Harutiunyan, R. N. 1994. Liquefaction and embankment failure case histories, 1988 Armenian earthquake. *J. of Geotechnical Eng.*, 120 (3), 581–596.
Youd T.L., Idriss I.M., Andrus, R.D., Arango I., Castro G., Christian J.T., Dobry R., Finn W.D.L., Harder L.F. Jr, Hynes M.E., Ishihara K., Koester J.P., Liao S.C., Marcuson W.F., Martin G.R.,

Mitchell J.K., Moriwaki Y., Power M.S., Robertson P.K., Seed R.B., Stokoe K.H. 2001. Liquefaction resistance of soils: Summary report from the 1996 NCEER and 1998 NCEER/NSF workshops on evaluation of liquefaction resistance of soils." *Journal of Geotechnical and Geoenvironmental Engineering*, ASCE, 127(10), 817–833.

Youd, T. L. Hansen, C. M., Bartlett, S. F. 2002. Revised Multilinear Regression Equations for Prediction of Lateral Spread Displacement. *Journal of Geotechnical and Geoenvironmental Engineering*, ASCE, pp. 1007–1017.

Youd, T. L. & Noble, S. K. 1997. Liquefaction criteria based on statistical and probabilistic analyses. *Proc., NCEER Workshop on Evaluation of Liquefaction Resistance of Soils*, Nat. Ctr. For Earthquake Engrg. Res., State Univ. of New York at Buffalo, 201–215.

Youd, T.L. & Perkins, J.B. 1987.Map showing liquefaction susceptibility of San Mateo County, California. *U.S. Geological Survey*, Map I-1257-G.

Zhang, G., Robertson, P. K., Brachman, R. W. I. 2002.Estimating liquefaction-induced ground settlements from CPT for level ground. *Canadian Geotech. J.*, 39, 1168–1180.

The Evolution of Geotech - 25 Years of Innovation – Hammah et al. (Eds)

Advances in data-driven subsurface mapping

Kok-Kwang Phoon
Singapore University of Technology and Design, Singapore

Jianye Ching
Department of Civil Engineering, National Taiwan University, Taipei, Taiwan

ABSTRACT: Data-driven site characterization (DDSC) is defined as any site characterization methodology that relies solely on measured data, both site-specific data collected for the current project and existing data of any type collected from past stages of the same project or past projects at the same site, neighboring sites, or beyond. One key complication is that real data is "ugly". A useful mnemonic is MUSIC-3X (Multivariate, Uncertain and Unique, Sparse, Incomplete, and potentially Corrupted with "3X" denoting three dimensional spatial variations). It is an open question whether DDSC can solve real world subsurface mapping problems based on real world MU-SIC-3X data with minimum ad-hoc assumptions. The Sparse Bayesian Learning (SBL) approach is very promising, particularly since it is nearly data-driven and it can handle a large scale 3D problem without incurring excessive cost. This 3D SBL would be made available in Rocscience's Settle3 (three-dimensional soil settlement analysis) in the near future. On the research front, the hunt is on for a "holy grail" mapping approach that is fully data-driven, MUSIC-3X compliant, and is able to exploit all available data including data from similar sites.

1 INTRODUCTION

Data-driven site characterization (DDSC) is defined as any site characterization methodology that relies solely on measured data, both site-specific data collected for the current project and existing data of any type collected from past stages of the same project or past projects at the same site, neighboring sites, or beyond. One key complication is that real data is "ugly". One example of ugly data is MUSIC-3X (Multivariate, Uncertain and Unique, Sparse, Incomplete, and potentially Corrupted with "3X" denoting three dimensional spatial variations). It is a useful mnemonic to highlight the attributes of real site data and to contrast with the highly idealized assumptions underlying classical statistics. The challenge is to draw useful inferences from ugly data (Phoon et al. 2021).

It is an open question what data-driven site characterization (DDSC) can achieve and how useful are the outcomes for practice, but this "value of data" question is of major interest given the rapid pace of digital transformation in many industries. The scientific aspects of this question are presented as three challenges by Phoon et al. (2021): (1) ugly data, (2) site recognition, and (3) stratification. These challenges are inter-related. The role of human judgment is expected to be sharpened with the advent of DDSC that can eventually progress to an artificial intelligence (cf. AlphaGeo in Phoon 2020).

The purpose of this paper is to introduce a Sparse Bayesian Learning (SBL) approach that has the potential to solve real world subsurface mapping problems based on real world MUSIC-3X data in the context of letting data speak for themselves. It is important to show

DOI: 10.1201/9781003188339-3

that the solution can be extended in reality (not in principle) to 3D at reasonable computational cost on a desktop computer. A subsurface mapping problem can include identifying geometrical features (stratification, discontinuities, anomalies, etc.), evaluating material behaviors (e.g., physical and mechanical properties) and their spatial distributions, and characterizing geoenvironmenal processes (e.g., ground water flow). The SBL approach focuses on simulating a site-specific subsurface map containing stratification and mechanical properties conditioned on the field test data and consistent with statistical (epistemic) uncertainties. A 3D version would be made available in Rocscience's Settle3 (three-dimensional soil settlement analysis) in the near future.

2 1D SPARSE BAYESIAN LEARNING

Let us denote $\mathbf{Y} = (y_1, y_2, \ldots, y_n)$ as the site investigation data observed at depths $(z_1, z_2, \ldots, z_n)$. For instance, y_i can be the corrected cone tip resistance (q_t) in the cone penetration test (CPT) at depth z_i. It can also be a transformed observation, e.g., $y_i = \ln[q_t(z_i)]$. It is common to *model* the observation y_i as a summation of a "trend" $t(Z_i)$ and a "spatial variation" $\varepsilon(Z_i)$:

$$y_i = t(Z_i) + \varepsilon(Z_i) \quad (1)$$

The trend function is parameterized by expressing it as a linear combination of basis functions (BFs):

$$t(Z) = \sum_{k=0}^{m} W_k \phi_k(Z) \quad (2)$$

where $\phi_k(Z)$ is the k-th BF, and w_k is the unknown weight. The basis functions are generally non-linear. Hence. Equation (2) is non-linear. The simplest trend function is a constant = w_0. The most widely adopted is arguably a linear trend = $W_0 + W_1 \times Z$. The spatial variation $\varepsilon(Z)$ is *assumed* to follow a zero-mean stationary Gaussian random field with standard deviation = σ and a Whittle- Matérn (WM) auto-correlation model (Ching and Phoon 2019; Ching et al. 2019):

$$\rho(\Delta) = \frac{2}{\Gamma(v)}\left[\frac{\sqrt{\pi}\cdot\Gamma(v+0.5)\cdot|\Delta|}{\Gamma(v)\cdot\delta}\right]^{v} K_{v}\left[\frac{2\sqrt{\pi}\cdot\Gamma(v+0.5)\cdot|\Delta|}{\Gamma(v)\cdot\delta}\right] \quad (3)$$

where $\rho(\Delta)$ is the auto-correlation between two points with separation distance = $\Delta = |Z_i - Z_j|$; v is the smoothness parameter; δ is the scale of fluctuation (SOF); Γ is the Gamma function; and k_v is the modified Bessel function of the second kind with order v.

It is important to emphasize here that only y_i is real. The "trend" and "spatial variation" are mathematical constructs or models. The weights of the basis functions w_k and (σ, δ, v) are parameters of the trend and random field models, respectively. The data-driven approach is to start with $\mathbf{Y} = (y_1, y_2, \ldots, y_n)$ and estimate w_k and (σ, δ, v)(including the functional form of the trend) with as few ad-hoc assumptions as possible. In this paper, we say an assumption is ad-hoc when it is not founded on physics and/or informed by data. Given that subsurface mapping is primarily based on site investigation data, an ad-hoc assumption is most likely one that is not related to real world data.

One hopeful candidate approach is a two-step Bayesian framework proposed by Ching and Phoon (2017). In Step 1, a set of suitable basis functions that parameterizes the trend function (Eq. 2) is selected using the sparse Bayesian learning (SBL). In this way, the functional form of the trend is "learned" from data rather than prescribed with no relation to data. In Step 2, an advanced Markov chain Monte Carlo method is adopted to draw posterior samples of $(\mathbf{W}, \ln\sigma, \ln\delta)$ conditioning on the $\mathbf{Y}$ data. Note that $\mathbf{W} = (w_0, w_1, \ldots, w_m)T$. In this way, the trend and random field parameters, or more precisely their posterior distributions, are "learned" from the data as well. The key ad-hoc assumption in this approach is that

($\mathbf{W}$, lnσ, lnδ) follows non-informative flat prior distributions. Nonetheless, one can argue that this SBL approach is largely or nearly data-driven.

3 3D SPARSE BAYESIAN LEARNING

The SBL approach proposed by Ching and Phoon (2017) is applicable to 1D spatial variability only. Although it laid the theoretical basis for a nearly data-driven subsurface mapping approach, its value to practice is limited. Direct extension to 3D is non-trivial. The main challenge is computational, i.e., 3D problems require numerical manipulations of very large matrices. Consider a 3D example with 20 CPT soundings, and suppose that there are 500 data points for each sounding. If the maximum likelihood method is adopted, the computation may require repeated calculations of the inverse of a $(10,000 \times 10,000)$ matrix, which is costly and the matrix determinant is errorprone. Ching et al. (2020) shows that under the "separability" assumption between the z direction and (x, y) directions in the auto-correlation structure, it only requires inversions and Cholesky decompositions for two significantly smaller (500×500) and (20×20) matrices. Therefore, the computational cost and numerical errors for 3D probabilistic site characterization are significantly reduced. The "separability" assumption is widely adopted in the literature, although the authors are not aware of studies establishing its veracity. In addition to this assumption, a second "vertically-dense-lattice" assumption is needed for conditional random field simulation. "Vertically- dense" means the sampling interval in the depth direction should be smaller than the vertical scale of fluctuation. The definition of "lattice" data is shown in Figure 1. The layout for the soundings can be arbitrary as shown in Figure 1a. This "vertically-dense-lattice" assumption can be satisfied by CPT soundings of equal lengths taken from a horizontal ground with no missing data as shown in Figure 1b. Conditional simulation is necessary, because a single most likely

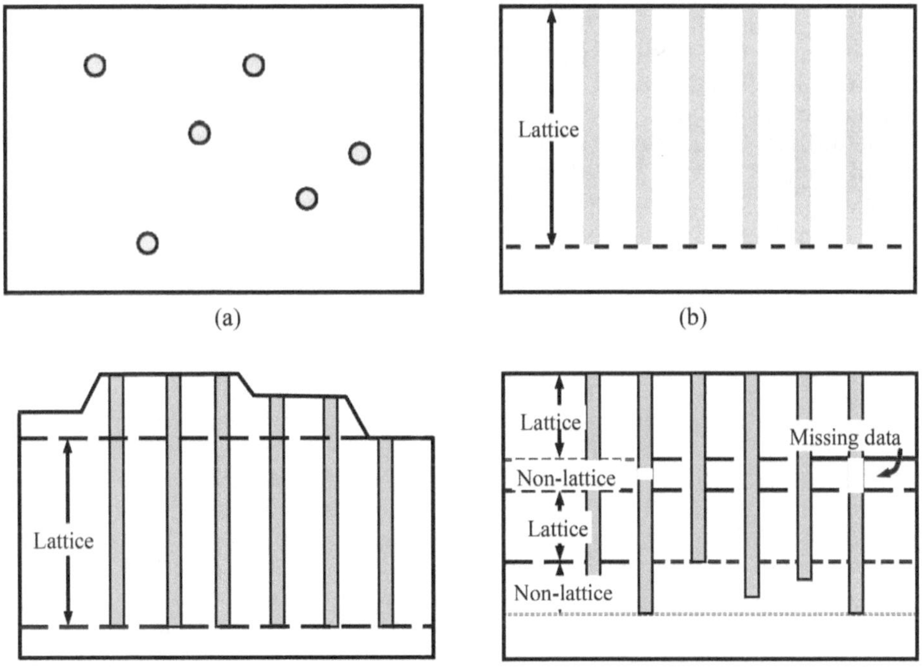

Figure 1. Illustration of (a) layout of soundings (plan view), (b) lattice data (horizontal ground), (c) lattice data (non-horizontal ground), and (d) non-lattice data with unequal soundings and missing data.

subsurface map will not alert the engineer to the presence of less likely maps that can be critical to the design. A range of simulated maps consistent with the observed soundings is a more appropriate representation of the underlying epistemic (statistical) uncertainties that can be significant under MUSIC-3X conditions.

The 3D SBL approach proposed by Ching et al. (2020) requires lattice data (all soundings are carried out to the same depth with no missing data) to take advantage of the Kronecker-product derivations. In practice, non-lattice data is more common because soundings are carried out to different depths to cater to geologic variations and/or geotechnical engineering needs. The lattice assumption implies that all CPT soundings have to betruncated to match the length of the shortest sounding. This defeats the purposes of carrying out deeper soundings, which must be important to justify the additional costs. It is also possible for the soundings to be incomplete in the sense that some sections are not recorded. These soundings do not constitute "lattice" data as well. The lattice assumption severely restricts the value of 3D SBL in practice. Fortunately, this assumption was recently relaxed by Ching et al. (2021a). In summary, research in SBL has progressed to the following stage:

1. Incorporation of the Whittle-Matérn (WM) autocorrelation model parameterized by the scale of fluctuation (δ) and smoothness parameter (ν) – this is arguably the most general monotonic autocorrelation model that includes the common single and squared exponential models as special cases (Cami et al. 2020).
2. Fully consistent characterization of statistical uncertainties for the: (a) functional form of the trend, (b) coefficients of the trend function, and (c) random field parameters: standard deviation, scale of fluctuation, and smoothness parameter.
3. Conditional simulation of the subsurface map – this is necessary to represent the effect of the statistical uncertainties on the map correctly. A less likely map is not a less important map in terms of design consequences.
4. Stratification - when **Y** denotes the soil behavior type index (I_c), the subsurface map becomes a stratigraphy. This is a special application of SBL.
5. 3D mapping – this allows all CPT soundings in a site to be used as inputs directly and equally importantly, 3D SBL is reasonably computable. There is no need to draw 2D sections. This is difficult to do in practice, because a typical CPT layout does not follow a regular rectangular grid.
6. Non-lattice data – this allows CPT soundings of unequal lengths with missing sections to be used as inputs directly. 3D SBL is less computationally efficient in the presence of non-lattice data, but it is a major step forward as it addresses the "Incompleteness" (I) attribute in "MUSIC-3X" and bring 3D SBL closer to full MUSIC-3X compliant.

4 VIRTUAL GROUND

To illustrate the performance of this 3D SBL approach, a simple 2D virtual ground is created using a 2D quadratic trend in the depth (z) direction, a 1D linear trend in the horizontal (x) direction, and a 2D zero-mean Gaussian random field with σ = 5 kPa, δ_v (vertical scale of fluctuation) = 0.5 m, and δ_h (horizontal scale of fluctuation) = 5 m. The autocorrelation model is assumed to be single exponential. Figure 2 shows the sounding data for the virtual ground. The virtual ground is "tested" in 5 locations along the x-axis as indicated in the figure, i.e., 5 soundings are assumed to be available. The SBL approach is trained by the 2 soundings at horizontal coordinates x = 0 m and x = 10 m (dark lines in the figure) and validated by the 3 sounding at x = 0.5 m, 5 m, and 9.5 m (red lines). The parameters (σ,δ_v,δ_h) as well as the trend function are treated as unknown during the SBL training.

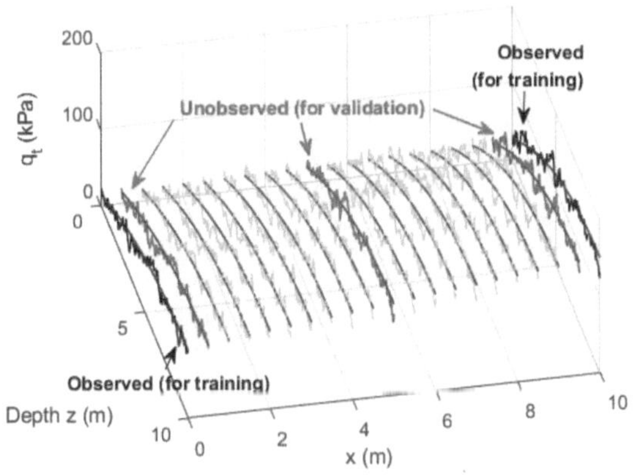

Figure 2. Sounding data for the virtual ground.

Given the training data (dark lines), the SBL approach first identifies $(\sigma,\delta_v,\delta_h)$ as well as the trend, then it simulates the conditional random fields. It is noteworthy that during this training and conditional simulation process, there is no need to prescribe the functional form of the trend (e.g., linear or quadratic) or to estimate the coefficients of the trend separately using regression. The SBL can automatically detect the optimal form and establish the coefficients consistent with Eq. (1). Figure 3 shows the samples and histogram for the identified $(\sigma,\delta_v,\delta_h)$. The red marker and line in the figure indicate the actual values of the random field parameters $(\sigma,\delta_v,\delta_h)$ used to define the virtual ground. Figure 4 shows one realization of the conditional random field. It is remarkable that the conditional random field always passes through the training data.

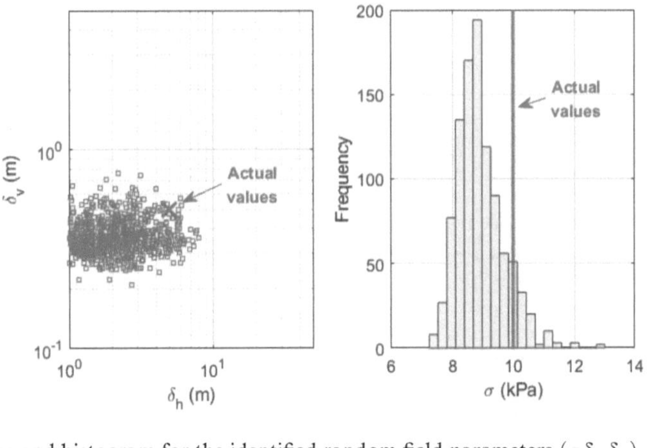

Figure 3. Samples and histogram for the identified random field parameters $(\sigma,\delta_v,\delta_h)$.

Figure 4 only shows one realization of the 2D conditional random field. One thousand such random field samples are obtained, and Figure 5 shows the resulting 95% Bayesian confidence intervals at the 3 validation soundings (x = 0.5 m, 5 m, and 9.5 m). The red lines in the figure are the actual data at the 3 validation soundings. Note that these data are treated as unknown during the SBL training. The two-step SBL approach is shown to be consistent in the well-defined sense that the resulting 95% Bayesian confidence intervals contain the actual validation sounding data with a large chance (close to 0.95, as reported in Ching and Phoon 2017).

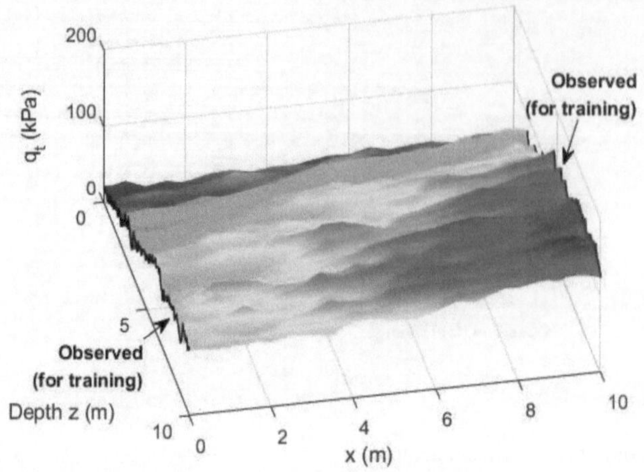

Figure 4. One 2D realization of the conditional random field.

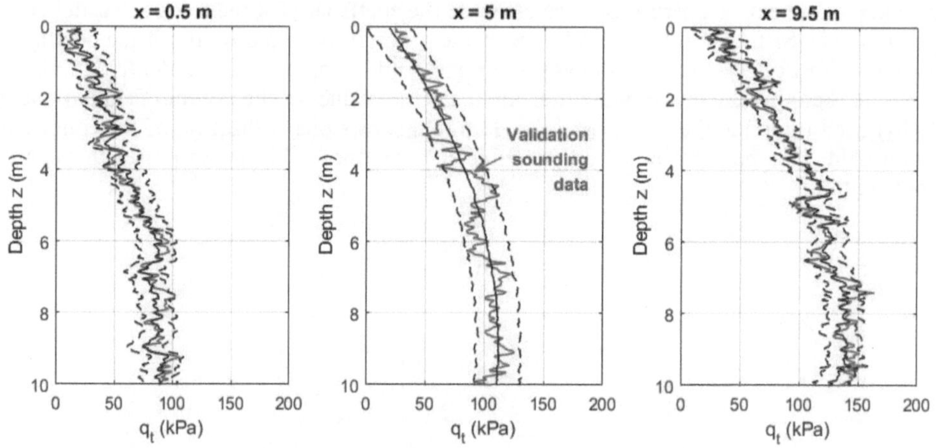

Figure 5. 95% Bayesian confidence intervals for the 3 validation soundings.

5 CONCLUSIONS

This paper introduces a 3D Sparse Bayesian Learning (3D SBL) approach that has the potential to simulate subsurface maps containing stratification and mechanical properties conditioned on field test data and consistent with statistical (epistemic) uncertainties. 3D SBL is considered to be "nearly data driven", because the following features are "learned" from data: (1) functional form of the trend, (b) coefficients of the trend function, and (c) random field parameters: standard deviation, scale of fluctuation, and smoothness parameter. In contrast, the current practice is to assume a trend function, say a linear function, compute the coefficients of the trend function using regression that contradicts spatial correlations, and characterize the random field parameters using the method of moments. Statistical uncertainties are not considered, although they are significant for MUSIC-3X data. In addition, statistical uncertainties are crucial for decision making.

The "vertically dense" assumption restricts 3D SBL to CPT data at this point, but research is in progress to remove this assumption. The objective is to consider data from the full suite of field tests in 3D SBL. More research is clearly needed, because 3D SBL is not fully

MUSIC-3X compliant, although it can handle sparse, incomplete, and spatially varying data. It is also not fully data-driven, because ad-hoc assumptions such as a separable autocorrelation model are embedded in the current version. The hunt is on for a "holy grail" mapping approach that is fully data-driven, MUSIC-3X compliant, and is able to exploit all available data including data from similar sites. A promising approach called HBM-MUSIC-3X was proposed very recently to simulate a range of subsurface maps that are consistent with site-specific MUSIC-3X data and supported by cross-correlation information from similar sites in a generic database (Ching et al. 2021b). At present, 3D SBL learnt purely from site-specific CPT data – it is unable to benefit from other test data at the same site and CPT data found in other sites.

ACKNOWLEDGMENTS

The authors thank Dr Thamer Yacoub, CEO & President of Rocscience, for his kind invitation to present this keynote paper in the Rocscience International Conference in 2021 to mark the occasion of Rocscience's 25th anniversary. In addition, the authors had frutitful exchanges with Dr Thamer Yacoub and Dr Sina Javankhoshdel that resulted in an ongoing collaboration to bring 3D SBL to practice using Settle3.

REFERENCES

Cami, B., Javankhoshdel, S., Phoon, K.K. & Ching, J. 2020. Scale of fluctuation for spatially varying soils: estimation methods and values. *ASCE-ASME Journal of Risk and Uncertainty in Engineering Systems, Part A: Civil Engineering* 6 (4): 03120002.

Ching, J. & Phoon, K.K. 2017. Characterizing uncertain site-specific trend function by Sparse Bayesian Learning. *Journal of Engineering Mechanics* 143 (7): 04017028.

Ching, J. & Phoon, K.K. 2019. Impact of auto-correlation function model on the probability of failure. *Journal of Engineering Mechanics* 145 (1): 04018123.

Ching, J., Phoon, K.K., Stuedlein, A.W., and Jaksa, M. 2019. Identification of sample path smoothness in soil spatial variability. *Structural Safety* 81: 101870.

Ching, J., Huang, W.H., and Phoon, K.K. 2020. 3D probabilistic site characterization by Sparse Bayesian Learning. *Journal of Engineering Mechanics* 146 (12): 04020134.

Ching, J., Yang, Z.Y., and Phoon, K.K. 2021a. Dealing with non-lattice data in three-dimensional probabilistic site characterization. *Journal of Engineering Mechanics* 147 (5): 06021003.

Ching, J., Phoon, K. K., Yang, Z., and Stuedlein, A. W. 2021b. Constructing a quasi-site-specific multivariate probability distribution model for soil properties using sparse, incomplete, and three-dimensional (MUSIC-3X) data. *Georisk*, under review.

Phoon, K. K. 2020. The Story of statistics in geotechnical engineering. *Georisk* 14(1), 3–25.

Phoon, K.K., Ching, J. and Shuku, T. 2021. Challenges in data-driven site characterization. *Georisk*, in press.

The Evolution of Geotech - 25 Years of Innovation – Hammah et al. (Eds)

After decades of rock engineering modelling, are we asking the right questions?

M.S. Diederichs
Queen's University, Kingston, Ontario, Canada

ABSTRACT: Rock engineering software first became available to the industry and consulting engineer a little over 35 years ago. In the two decades before that, rock mechanics analysis was evolving on mainframe computers within academic institutions. Project engineers were limited to empirical design, simple analytical tools, and the past experiences of their teams and experts. While laboratory testing of rock samples had evolved and been standardized through the 1970s, our ideas about in situ rockmass behaviour during excavation were comparatively primitive and our ability to rationally engineer in these materials was guided primarily by recently codified empirical design tools. The questions asked of these tools were simple. Will the stresses induced around (simplified) excavation shapes exceed the rock strength? What will the deformations induced be (elastic or plastic)? Will joints form simple 2D and 3D blocks that will fall out under gravity? Will my pillars yield (according to simple criteria)? How does my rockmass-excavation system compare to past successes and failures? Through the last two decades of the 20th century, and thanks to personal computer evolution and user interface development, the tools of rock and rockmass engineering evolved at a rapid pace, along the ability to use them. The toolbox continued to expand exponentially in scope, power, and complexity through the next two decades to the present day. It is important, however, to examine whether our engineering questions have evolved, whether our investigation and data collection has improved, and whether the answers obtained have improved along with the complexity of our toolset.

1 INTRODUCTION

I have chosen to write this keynote in the first person since it represents to some degree my own personal journey both as a one-time developer and as a user, in research and engineering consulting, from the beginnings of practical rock engineering modelling tools to the present day. As such the timing of my access to hardware and software may differ from the first dates of creation or commercialization. The history, of computing and geomechanical analysis tools, recounted here is from my own perspective as a student of rock mechanics through to my current roles as Professor and Consultant. I will be addressing, through this journey, the engineering questions we were asking of early code and the questions we ask of the powerful tools we use today.

The use of numerical tools to solve rock mechanics problems is both a great love of mine, when it comes to the state-of-art, but also a great frustration in terms of its use and abuse in real engineering problems. I will focus on tools for underground rock engineering in this paper and more specifically, practical rock engineering for analysis and design of mining and civil construction applications. There are not enough pages in these proceedings to cover the evolution of geotechnical modelling and simulation in terms of advances in theory and solution techniques.

DOI: 10.1201/9781003188339-4

I began my journey in rock mechanics in my 4th year of my Geological Engineering at the University of Toronto in 1986. My undergraduate initiation into rock mechanics came at the hands of Dr. John Curran (future founder or Rocscience – www.rocscience.com). In our coursework, we applied analytical solutions such as the Kirsch (1898) elastic hole in a plate solution for tunnel stresses, as well as basic ground reaction theory as summarized from numerous works by Hoek and Brown (1980). As an assignment, I wrote my first piece of geotechnical code that year, transforming the BASIC code for Boundary Element analysis of stresses, found in Hoek and Brown's "Underground Engineering in Rock" (1980), to a groundwater flow code (achieved by swapping out the Green's functions). Several years before we had been the first class to type Fortran code onto a screen and not have to use punch cards as had the classes that came before. By fourth year we had access to several IBM XT 4.77MHz 8088 computers with DOS, 640K of RAM and 360K floppy drives. Hard drives did not arrive until 1988 after I was well into my Masters degree tenure under my then supervisor and long-time colleague and friend, Dr. Evert Hoek.

While numerical methods in rock mechanics had their beginnings in the late 1960s, these tools were in the hands of a few at institutions that were fortunate to have access to the first user programable mainframe computer systems. The 1970s brought us first the handheld calculator at the beginning of the decade and the first personal computers by the end. In the middle, we saw the DEC PDP11 mini-computer make its way into universities and some lucky high schools like mine in Scarborough, Ontario. My first personal computer was a 1978 Commodore Pet 2001 with 8KB of RAM purchased second hand for $1600. I still have this icon in my office today and it is able to run the Boundary Element Code mentioned above although the output leaves me wanting.

My graduate work at the University of Toronto took me from the IBM XT (with 8088 chip and a mouse) to the Compaq Portable 286 through to the 16 MHz, 4MB RAM, 40MB HD Compaq DeskPro 386 (purchased for over $15K in the late 1980s). By 1990 the department had even acquired a Unix-driven Sun SPARC although this was not reserved for mere mortals. This gem was in the hands of none other than Brent Corkum, co-pioneer with the early Rocscience team.

Figure 1. Personal (and office) computing in the 1980s (left to right) Commodore PET 2001 (which still resides in my office), IBMXT 8088, Sun Sparc Station (pcmuseum.tripod.com).

My own early DOS coding (1987 to 1991 with the Rock Engineering Group at the University of Toronto – institutional forerunner of Rocscience) was under the joint command of Drs Evert Hoek and John Curran. The early work was founded on their shared vision that it was essential and timely, that rock engineering analysis software be made accessible to a non-academic audience, in order to elevate the practice of industry and consulting engineers. In a pre-Windows world, this development started with the creation of an in-house graphical interface, by Brent Corkum, upon which the early, but still familiar, codes were based. This coincided with a rapid development in video card configurations. My own work on DIPS as well as EXAMINE^{2D}, UNWEDGE, SLIDE, PHASE2 and CPILLAR, covered the evolution from the golden Hercules Graphics Card (my favourite), CGA (let us forget), EGA and finally VGA (640x480 x16 colours). Nevertheless, early coding was constrained by the need to fit the executable file and any support files on a 360K Floppy Disk for release. Code was stripped, routines optimized, and the value of extra features debated in order to stay under this compiled limit. The advent of 720K disks was exciting indeed!

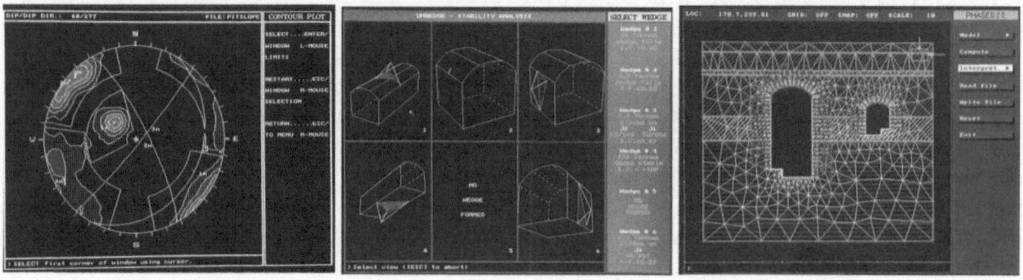

Figure 2. The DOS User Interface in 1989 (left to right) DIPS, UNWEDGE, PHASE2.

Figure 3. The 360KB Floppy (and 720KB "Stiffy") Disc Era.

During this period in the late 1980s, I was also fortunate to run early models on Itasca's FLAC2D as well as UDEC (all early versions). While earlier codes had been written, for 2 decades before, in forms accessible only to the authors and their colleagues, the late 1980s represented the birth of user-friendly geotechnical and rock engineering software and I consider myself fortunate to have been witness early in my career.

Both Itasca (www.itascacg.com) and Rocscience (rocscience.com) have gone on to produce the impressive tools we use today through Windows platforms. They have been joined by other commercial vendors like ANSYS (ansys.com), MAP3D (map3d.com), Plaxis (bentley.com), Rockfield (rockfieldglobal.com), Geomechanica (geomechanica.com), and others to create a vast array of code types, from continuum, to discontinuum, to complex hybrids (e.g. beckengineering.info) and other variants in usable platforms, for engineering practitioners as well as academics, to solve complex problems in all aspects of the discipline.

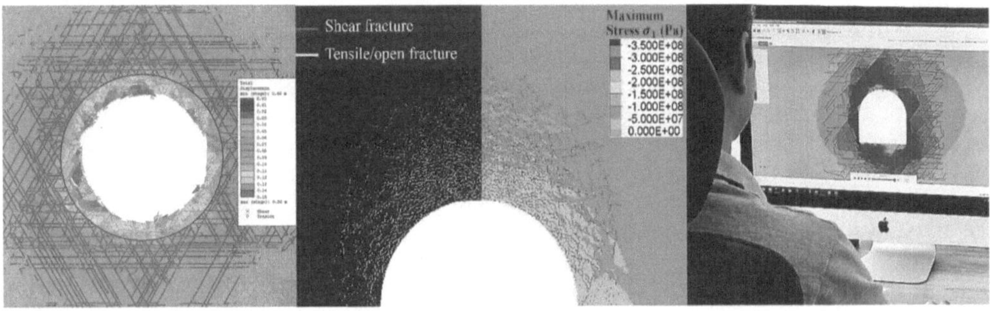

Figure 4. User-interface driven (discontinuum) tools of today (left to right) RS2 from Rocscience, UDEC from Itasca (Dadashzadeh and Diederichs 2019), Irazu (courtesy of Geomechanica).

We now perform routine engineering modelling on workstations and laptops with many dozens of GB of RAM, many TB of drive space and ever-increasing (although not as fast as we would like) CPU speeds and multiplying cores. The rise of GPU processing has brought a new wave of possibilities to practical geomechanics analysis. Software vendors and/or proprietary developers have transitioned into advanced modelling consultants capable of immense and complex 4D multi-physics modelling for rock engineering.

Throughout this personal journey through the early days of practical computing, to the endless possibilities of today, as well as the growth of powerful, user-accessible and sometimes userfriendly rock engineering software, I am compelled to ask the fundamental question:

Over the past four decades, have the questions we ask, the inputs we specify, and the answers we receive, and both the robustness and efficiencies of our designs based on our analyses grown and evolved at the same rate as the hardware and software we utilize on a daily basis.

2 CONTINUUM ANALYTICAL SOLUTIONS FOR STRESS ANALYSIS

Like any student of rock mechanics in the late 1980's, I began my journey with Brady and Brown's 1st edition (1985) and with Hoek and Brown (1980). Contained within these books were fascinating analytical solutions such as the Kirsch solution (1898) for stresses and displacements around a hole (in a plate and/or plane strain), and an adaptation of Ladanyi's (1974) ground reaction curve analysis (now normally referred to as convergence confinement analysis).

The premise of the former elastic analysis tool was simple. We asked – "how do the in situ stresses at a given depth and with an estimated or reasoned ratio between vertical and horizontal principal components (or with inclined principal stresses) get exaggerated, reduced and perturbed by the creation of a circular hole (a tunnel) perpendicular to those two stress directions?" From an engineering perspective this was a critical component of understanding. The ability to see the impact of varying the in situ stress ratio was critical to predicting the response of deep tunnels and shafts through a comparison between these elastic stresses, at the boundary and into the rock, and a suitable strength (yield) criteria such as the Mohr-Coulomb and Hoek-Brown envelopes (for intact rock or adjusted for the rockmass). It was clear that the elastic stress states predicted by the equations were ultimately invalid if yield was predicted to occur. Nevertheless, we could use this tool for design, making practical estimates of yield and therefore reinforcement densities and depths (assuming that all of the "yieldable" rock would subsequently need support). Other publications also provided solutions for other shapes such as ellipses and the curved corner rectangle (e.g. the elliptical solutions of Inglis 1913 and the rectellipse by Heller et al. 1958).

The limitation of elastic stress calculation was somewhat overcome using the ground reaction approach that calculated the limits of plasticity (and associated stresses and displacements) in an annulus around a circular hole. This gain in insight came with the limitation of isotropic stresses and circular geometry. These tools were used together to answer simple questions related to the potential for closure, extent of yield and the impact of rock strength and stiffness variations on tunnel performance and support requirements. The convergence confinement approach is still one of the most utilized forms of analysis in weak rock tunnelling today, for better or for worse.

These approaches, however, did not give us any insight into the nature of "yield" in rock, whether the rock near the outer limits of the yield zone needed reinforcement or whether spaced tendon support had any impact on behaviour in the outer extents. The tendency persisted to assume that all numerically yielded ground required throughgoing and intensive reinforcement. As a result, bolt lengths in weak rock tunnels were and still are often designed with excessive bolt lengths that likely have no impact on rockmass behaviour beyond a few metres away from the tunnel boundary. This is perhaps my earliest recollection of asking the wrong question and getting a less valuable answer from engineering analysis.

That said, convergence confinement analysis, combined with the internal relaxation technique or alternatively, longitudinal displacement profiles, allowed sufficiently robust answers to these simple questions. As a result, we could design surface liners to be installed with strategic delays that allowed for rockmass response and relaxation (an effect at the heart of the so-called NATM approach) and reduced thicknesses of concrete or smaller arches/rings. At the time, this was a disruptive shift in strategy for tunnel support, away from designing single pass support at the face to resist all of nature's demands as the tunnel advanced. This leap forward was able to gain acceptance because of these analytical tools that assisted design.

3 ELASTIC NUMERICAL SOLUTIONS AND THE QUESTIONS THEY ADDRESSED

The lingering question regarding the effects (on stresses) of different excavation shapes and multiple excavations was the subject of physical modelling prior to the 1980s. Adaptations of elegant boundary element methods, which had been previously used by researchers to great success, became available in the 1980s to the common engineer through the personal computer and source code such as that printed on no more than seven pages in an appendix of Hoek and Brown's classic text (1980). This code was used to generate parametric plots in the aforementioned text and became the basis of the program Examine2D (still popular in the Rocscience suite for education). When released, it gave the site engineer the opportunity to visualize 2D stress flow around an excavation of any shape with any in situ stress orientation. This provided some answers (albeit limited to elastic response) to questions about geometric elements and excavation aspect ratio and orientation with respect to the stress field allowing optimization of initial design.

While the subject of countless impressive but dust-collecting PhD theses prior to the late 1980s, new additions such as 2.5D displacement-discontinuities (DD), 3D elements, material boundary and fault integration, and improved solvers, made the boundary element method a critical and accessible tool for mining and cavern engineers. Codes utilizing DD elements (e.g. Nfold, Mintab, and Examtab) could be used to model thin tabular orebodies during mining extraction and predict abutment and pillar stresses. These codes made the tributary area approach, a mainstay of introductory rock mechanics, obsolete and in some cases invalid as they exposed the complexities of stress gradient.

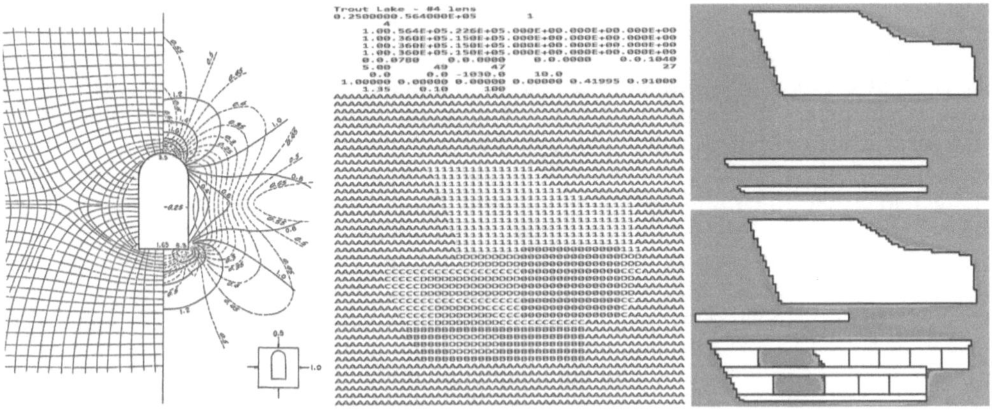

Figure 5. (left) Normalized stress plots generated using fictitious stress Boundary Elements (Hoek and Brown 1980); (middle and right) Input/output from an early Displacement Discontinuity Code for stress flow in pillars and abutments of tabular orebodies (courtesy E. Eberhardt).

Complex mathematical solutions were now available for use by the practitioner through mouse clicks and keyboard input, with results contoured in vivid colour on the computer screen (no need to wait for a pen plotter to work away). More importantly, Examine2D marked a turning point in rock engineering software as a result of something truly revolutionary – the graphical user interface. 2D programs became accessible for practicing engineers. It was possible to teach principles of engineering modelling without having to dwell on challenging and confusing control syntax. 3D programs such as Map3D and Examine3D revolutionized modelling for mining applications and evolved through the 1990s with user-friendly CAD integration for input and ground-breaking (for mining applications) output visualization capabilities. In addition, Map3D included inelastic response through innovative BEM techniques. Examine3D introduced the rock engineering world to state-of-art visualization techniques for 4D results (e.g. output-value iso-surfaces). Both programs are still in use today. With this impressive development through the 1990s, the main questions to be asked of the models, however, initially remained the same: Where would high stress concentrations occur and to what extent would strength be exceeded? For deep mining with high stress and complex sequencing decisions to be made, this question was paramount.

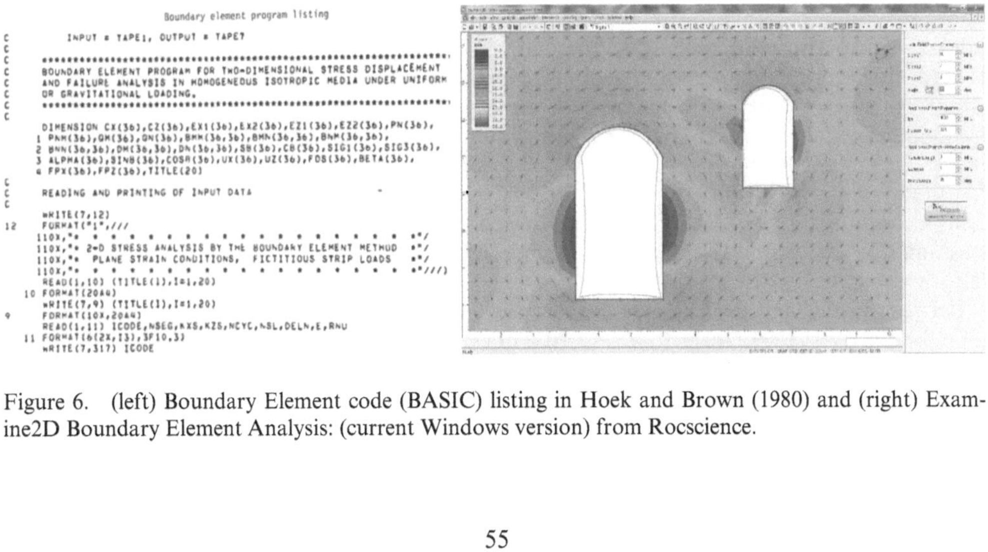

Figure 6. (left) Boundary Element code (BASIC) listing in Hoek and Brown (1980) and (right) Examine2D Boundary Element Analysis: (current Windows version) from Rocscience.

The new elastic tools did also provide opportunity to obtain precise and colourful answers to the wrong questions. For example, there were some errant attempts to use the calculated elastic deformations for support design. In mine pillar design, attempts to make larger pillars to reduce elastic stresses resulted, in reality, in more and unpredictable local overstressing in complex arrays. Modern mine analysis considers and utilizes controlled yield in these situations. With conventional emphasis on elevated stresses, another important question, "where would the stresses be relaxed to the point of allowing blocky rockmasses to unravel?" was asked less often.

Induced stress near excavations also became an input into numerous classification and empirical design tools such as Q and the Stability Graph method. These "modern" elastic stress analysis tools, therefore, became essential for providing quantitative input into simple empirical assessment and design. With the advent of microseismic monitoring in mines, there was an opportunity to correlate seismicity with various stress outputs from local and mine-wide 3D models and then use the models to assess seismic hazard in new mining areas.

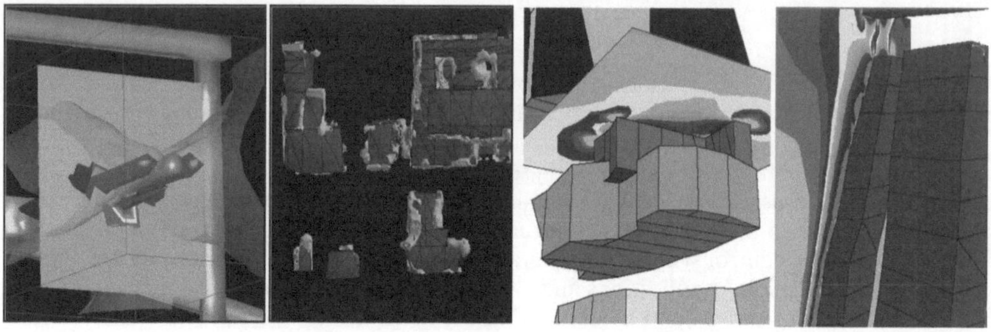

Figure 7. 3D Elastic (1990s) modelling: early versions of (left) Examine3D and (right) Map3D.

4 INELASTIC CONTINUUM, NUMERICAL SOLUTIONS

Finite element analysis was widely used by those with access to mainframe computing from the 1970s. The great challenges of meshing complex geometries were solved for other engineering analysis using simple shape primitives for "internal mesh" problems. They proved intractable, however, for rock engineering problems where the region outside the geometric void was to be meshed. Early programs required manual assignment of node locations, nodal grouping, and element numbering. The next generation of programs relied on manual reassignment of mesh coordinates from generated grids, followed by methodologies for stretching and reforming regular meshes to fit complex 2D geometries. Back at the University of Toronto, I still remember the day when Brent Corkum returned from a Siggraph Conference and brought source code for a Delauney triangulation technique into the lab and transformed the art of automated 2D finite element meshing for geotechnical modelling. Before long, PHASE2 (now RS2) had its maiden release.

Between FLAC and PHASE2, with their user interfaces and PC platforms, it was now possible for the average engineer to answer the questions previously posed in this review. Stress flow around multiple complex openings with variable stress fields, extent of yield, stress redistribution as a result of yield, meaningful displacements resulting from both elastic convergence and plastic strains, and the impact of contrasts in material properties, were all at one's fingertips.

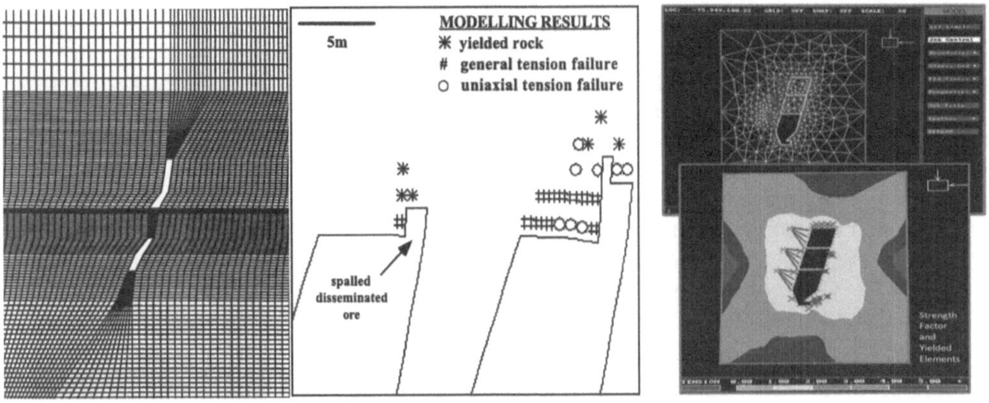

Figure 8. Early (1990s) 2D plastic continuum analysis: (left and middle) FLAC2D finite difference (after Eberhardt et al 1997) and (right) PHASE2 finite element (now RS2).

At the same time, came the recognition that laboratory rock mechanics testing, while highly advanced at this stage, did not reflect the performance of jointed rockmasses in the field. While classification systems such as RQD, Q and RMR enable a recognition of this fact and allowed for rock excavation dimensioning and support design based on past experience, a rational and repeatable tool was needed to provide rockmass strength and deformation parameters as input into these new forward analysis models. An approach for factoring rockmass strength presented in Hoek and Brown (1980) crystalized in the 1990s into the GSI system for rockmass characterization and factoring of strength parameters (Hoek et al 2002). Given the difficulty and expense in performing large-scale mechanical testing for individual underground projects, this approach offered the industry a much-needed practical tool to accompany the newly available modelling packages. With some enhancements over the next two decades, the GSI-based continuum approach to non-linear (plastic) rockmass yield and deformation became the mainstay of practical rock engineering. While there is still need for systematic verification of GSI, and in the absence of any more recently developed alternatives, it remains a key tool today.

Critical questions remain in numerical modelling, however, including the lack of reliable guidance for post peak strength, confinement-dependent rockmass stiffness, the role of dilation on stability and deformations, the limitations of GSI for sparsely jointed rock, and the challenges in obtaining primary input for very weak rockmasses. Furthermore, although continuum modelling allows for the calculation of the effect of rock deformations on support elements such as liners, it is ineffective in assessing the role of tendon reinforcing elements or adhesive surface coatings (thin shotcrete) on the rockmass. It is universally recognized that rock reinforcement holds a fractured rockmass together preserving initial and "post-yield" integrity and working to restrict local dilation and slip on the discrete joint interfaces. This critical role of support is missing in most continuum approaches, leading in many cases to overdesign of bolting or overreliance on thick lining to compensate for the apparent lack of impact of simulated bolting on rockmass behaviour.

In hindsight the question related to reinforcement design should have focused on how reinforcement changes the bulk properties of a jointed rockmass, rather than on only the physical transfer of loads and displacements within the simulated continuum. There have been attempts at providing such guidance over the years (including, more recently Moro and Stacey 2012), but these concepts have not found their way generally into mainstream modelling practice.

There were some surprise advances in understanding that arose from non-linear continuum modelling. These included the realization that excavation shapes optimized for elastic stress flow may not be optimum once yield begins. Another was an evolving understanding that the occurrence of modelled material "yield" inside a pillar does not necessarily mean "failure" of the pillar system, as post-elastic stresses and strains increase confinement at the core and increase stability for wider pillars (with greater than 1:1 width to height ratios). The question of "what degree and extent of material yield equate to engineering failure" is still an unresolved question for pre-construction pillar design. More recent advances in monitoring (displacement response, internal probing and seismic monitoring) have led to improved verification of designs and determination of damage evolution, pillar yield and true engineering failure after construction or during mine operation.

These illuminating tools answered many long-asked questions. It was tempting, faced with the wonders of non-linear continuum analysis visualized through interactive living-colour output, to forget about the elephant in the room, in the form of the following two-part query:

1. Was the plastic continuum analysis, with homogenized rockmass parameters, accurately representing the real yield mechanisms and associated behaviour in jointed rockmasses?
2. If not, what are the implications for excavation and support design?

It was understood and accepted that the first part of the question was not always the case. However, the second part question was left unanswered in the analysis for many engineering projects.

Beyond the 1990s, the evolution of inelastic continuum modelling (finite element and finite difference) was dominated by increased computing power, better visualization, enhanced front end interfaces for model construction, improved constitutive models including anisotropy, and integration with both CAD software and 3D visualization platforms. In the last decade, the ability to mesh and model, with full non-linear response, full mine-wide geometries and full complex excavation networks such as underground metro stations or power complexes has become possible and practical. RS3, FLAC3D, PLAXIS, and other public and proprietary continuum codes have fueled the 3D non-linear revolution.

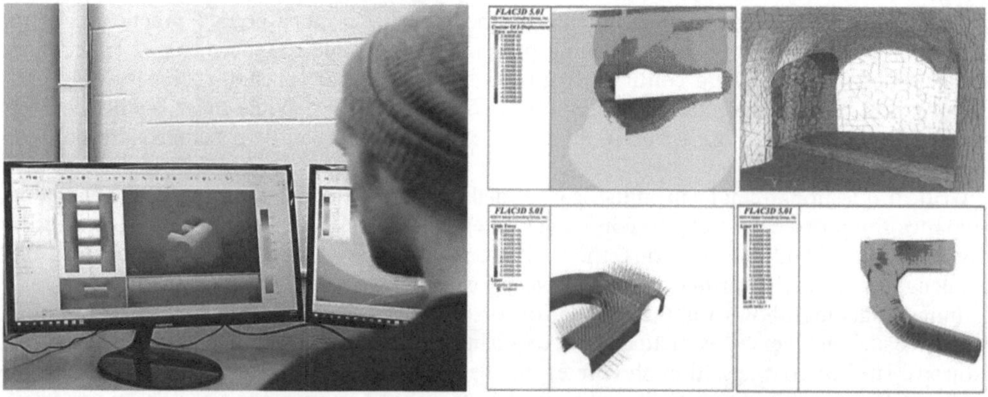

Figure 9. 3D continuum analysis of excavations and support with (left) RS3 and (right) FLAC3D.

This revolution is particularly evident in now routine mine-engineering analysis. Seismic data is now correlated with advancing yield fronts and post-yield relaxation, rather than by elastic stress concentrations (a game changing evolution in the fundamental question over two decades made possible by the tools we use today). The effect of strength loss due to fragmentation can be simulated implicitly for caving analysis at a mine scale. The complex stress paths due to mining can be combined with local yielding for infrastructure analysis. Pillar and abutment yield can be simulated along with the mine wide stress redistribution and its effects on development over the mining life, considering the post-yield strength and deformation of the rockmass.

Evolving mine-wide influences can be linked to locally detailed support simulation. Sequencing analysis includes unrecoverable deformations and yield in the affected rockmass, in contrast to the sole focus on stress magnitudes deemed so essential in the not-so -distant past. In the arena of high-stress mining applications, the critical questions are continuously evolving with the toolset. These tools have also been accompanied by enhancements and new developments in constitutive modelling, although the challenges of defining an expanding suite of rockmass parameters for non-linear continuum rockmass analysis remain, in spite of the increased sensitivities of these complex models to realistic behavioural parameters. The GSI system remains one of the only tools for systematically defining rockmass strength.

In contrast to the rockmass shear strength modelling dominated by the GSI approach and appropriate for weaker rockmasses, new constitutive tools for simulating brittle rock response in continuum models were key developments in the late 1990s and 2000s including the Damage Initiation and Spalling Limit (DISL) approach for RS2 (Diederichs 2007) and the Cohesion Weakening Friction Strengthening (CWFS) approach in FLAC (Hadjiabdolmajid et al. 2003).

Guidance for post-yield strength (and stiffness) and for dilation behaviour (flow rule) for intact rock has been provided by various authors (Walton and Diederichs 2020) while reliable and justifiable rockmass parameters remain elusive. Both represent questions we need to continuously (and perhaps more loudly) ask of our models. Without appropriate input and internal constitutive treatment, the answers remain elusive. Practical guidance for dilation, informally distilled from experience, is provided by Lorig and Varona (2013). Combining lab test data from Walton et al. (2019) with large displacement shear tests by Cai et al. (2007), it is possible to propose new practical guidance (Figure 10) for residual strength (through GSI), although more work is needed to develop robust and mechanically defensible guidance on residual rockmass strength and dilation.

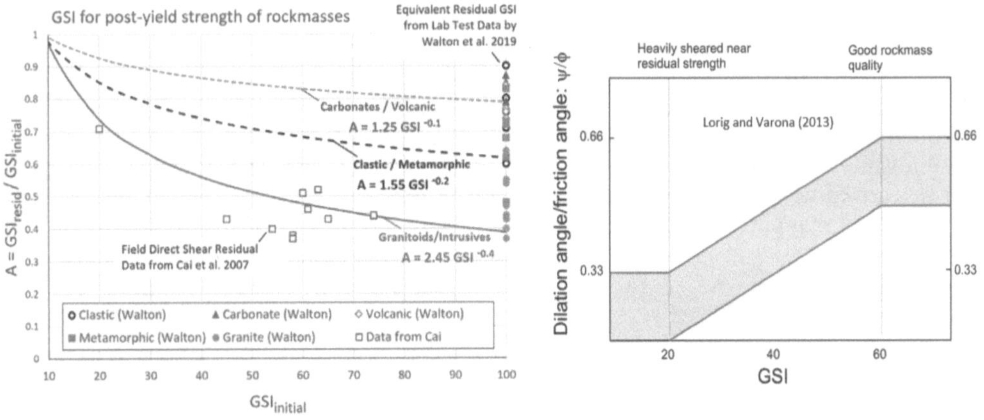

Figure 10. Guidance for (left) post-yield strength and stiffness (via GSI) and for (right) dilation.

Advanced modelling for mining applications, seismic monitoring, instrumentation and more recently lidar scanning and deformation mapping have provided a means to calibrate models on a large scale. Meanwhile, the ability to optimize the input parameters in civil applications lags. Tunnel design, which integrates rigorous observational design principles, utilizes modelling (as well as the old-school analytical deformation tools) together with closure monitoring during advance, or by deformation monitoring at the face of a tunnel. Here the engineering questions focus on closure predictions versus measurement and adjustment of model parameters so that predictions can be reassessed as tunnelling progresses. Support or sequencing can be modified based on the continuously recalibrated models. The same approach needs to be implemented for caverns although the calibration process must start early during the pilot tunnels and adit development. 4D modelling (which 3D geometry plus excavation advance), integrated with real-time monitoring, provides the ability to refine model parameters including stiffness, strength, post-yield deformability and dilation, as well as time dependent properties. In particular, advanced modelling capabilities combined with comprehensive monitoring programs, have enabled complex sequential excavation techniques to be verified and optimized.

These tools have significantly improved front and back-end interfaces and integration into other tools of the trade such as structural modelling, data block modelling, Lidar (for geometry and displacements), seismic analysis, closure monitoring surveys and tomography.

For initial support design, there is still a lack of consensus on the metrics for design acceptability. Deformation limits are set and provide for support class adjustment during construction, but at the initial design stage, the true reinforcing role of tendon support is still not typically considered. This results in unanswered questions that could impact design. In practice, the sensitivity of tunnel liners to assumptions of in situ stress ratio are often overlooked as are the impacts of structural anisotropy and heterogeneity in the rockmass at the tunnel scale. This happens even though tools for incorporating oriented strength and stiffness, as well as spatial variation of parameters, exist.

For permanent pillar design (between hydro-caverns and cross passages for example), there is no consensus on the degree of acceptable yield and deformations within a pillar. As a result of the lack of mechanistically defined modern design metrics, empirical pillar design charts from mining observations 40-50 years ago are still used in the modern construction of permanent civil works. While yield is managed in NATM tunnelling, the same philosophy is difficult to sell in large underground complexes. Modelling can help but the right questions must still be asked such as "how much post-yield pillar deformation and/or load shedding is acceptable in the context of the overall response of the cavern complex?" or "will internal core damage in the pillar increase the likelihood of time dependent deformation in the specified rockmass or is such damage benign?".

For design lives that reach or exceed 100 years, there is a need to have metrics that allow yield but constrain post-yield deformations and dilation in order to minimize the potential for long term influences on the damaged rockmass. Unlike near circular tunnels, modern caverns cannot rely on a thick permanent liner for long-term load and deformation resistance. Improved tools for assessing rock support interaction and feedback are needed. This feedback needs to include the role of tendon support in post-yield strengthening of the rockmass. On the other hand, new integrated tools for analyzing liner response and deformation resistance have advanced this aspect of design. Unfortunately, in practice, final tunnel lining design is still performed as a procedure decoupled from the rockmass modelling. This integration is a worthy goal that must be pursued.

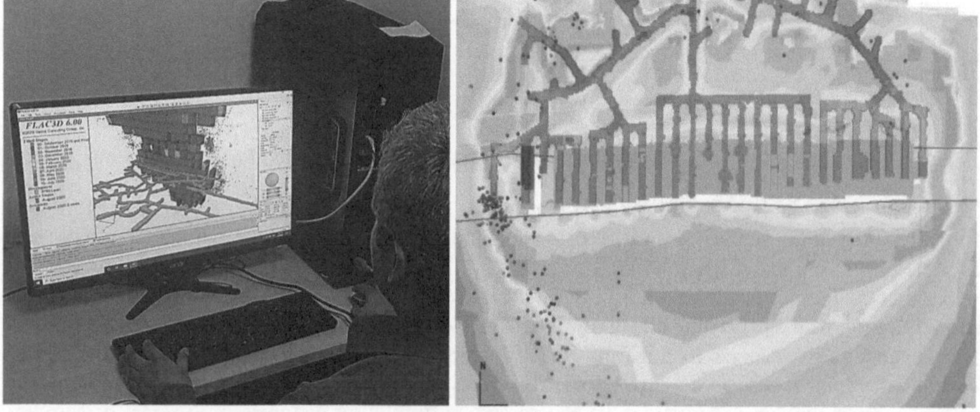

Figure 11. Modern user-friendly and user-driven 3D continuum mine-wide analysis in FLAC3D with integrated seismic data analysis (courtesy of RockEng – www.rockeng.ca).

In summary, in practical rock engineering we have at our disposal user-friendly 3D non-linear continuum modelling tools beyond our wildest dreams 30 years ago.

Nevertheless, there is often inadequate consideration of whether input data for the rock and the rockmass is adequate and appropriate. Tools exist for rigorous analysis of parametric sensitivity and variability although such analysis is underutilized in practice, in my experience. Although prescribed data collection is often aimed at the onsite selection of predetermined support classes based on analysis (which typically includes modelling), observational design is used to account for the uncertainties inherent in underground response prediction. The validity of modelling itself, however, is not always verified based on construction observations. There is a need to build into any major construction schedule intensive early monitoring of preliminary excavations to provide real feedback into the modelling process so that iterative improvements can be made.

The metrics for design acceptability, for underground works related to civil construction, based on the results of rock engineering modelling for underground works, are still ambiguous. As a result, many international designers prefer to rely on decades old generalized empirical charts for solid guidance, in some cases leaving the advanced (and expensive) modelling as a sideshow. Mining has fared better in this regard than typical civil construction such as tunnelling, due to the ongoing and mine-life nature of its engineering oversight and technical analysis.

5 DISCONTINUUM SOLUTIONS FOR ROCKMASSES

While average behaviour can be approximated with advanced continuum techniques, these approaches, on their own will never be capable of accurately reproducing the local displacement behaviour of a moderately jointed rockmass at scales where the joint-bounded blocks dominate. In the 1980s and 1990s discontinuum methods with functional user-interfaces became accessible to researchers and practicing engineers alike. While the capabilities were limited at first, the ability to investigate the workings of a jointed rockmass was exciting indeed. As author of DIPS, a stereonet program to analyze structural data, and contributor to the wedge analysis package UNWEDGE, I was particularly interested in discontinuum analysis at the start of my career.

My first vicarious discontinuum experience was in the form of discussions and demonstrations of various discrete element codes being developed for mainframe use in the Civil Engineering Department at the University of Toronto. I remember, as a Masters student in 1988, receiving the first 720K 3.5 inch "floppy" disc marked "MUDEC 1.0" from Itasca. I believe we also had a copy of FLAC 1.0 around the same time. Early tinkering with these two codes exposed the differences between continuum and discontinuum modelling. While discontinuum seemed to hold promise of more realistic modelling, unpredictable outcomes for underground problems, influenced by subtle changes in structural geometry and uncertainties in the expanded suite of input parameters rattled one's confidence in the results.

Indeed, these challenges persist today for applications in practical rock engineering where reliable answers are required and where there is no room for academic fascination with chaotic systems no matter how realistic they may be. Only recently has computing power risen to the level where multiple realizations and parametric studies can be efficiently executed on these models. There is limited guidance, from a reliability analysis perspective, on how to reduce the results of such analyses into answers to design questions. Indeed, despite the parallel development of accessible discontinuum codes and useable continuum models through the 1990s, the repeatability and predictability of continuum models in real applications proved powerful selling points for practical modelling for many years. The power of discontinuum modelling to unravel the mysteries of mechanistic rockmass response to complex stimuli cannot be understated, however. The future most certainly lies down this path. We do, however, need to up our game in terms of the engineering questions we pose to these ever more complex models.

I have always been struck by the fact that in the back pages of a practical PhD thesis entitled "The Measurement and Analysis of Accelerations in Rock Slopes" by Peter Cundall

(1971) we find source code, in BASIC, of the programs entitled "Ball Program" and "Block Program".

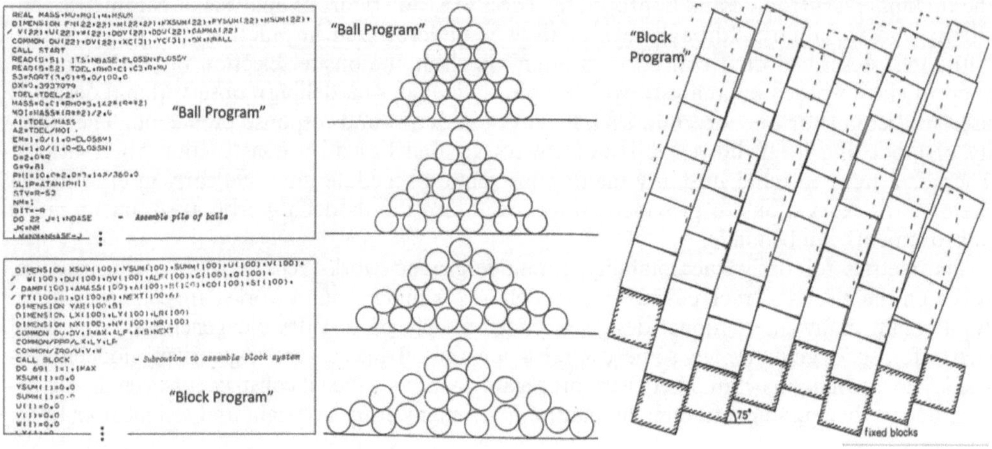

Figure 12. Source Code and Output from the PhD thesis of Peter Cundall (1971).

These utility codes were written to assist the author in understanding the effect of discontinuum slip on stresses and displacements within a slope. He started with the question that related to continuum (finite element) stress models for elastic stresses within slopes and near excavation surfaces. The stresses were rotated to resolve normal and shear stresses on unmodelled discontinuity planes and compared to the shear strength calculated for the discontinuities. He remarks:

> *The broad conclusion was that large parts of the slope exceeded the failure criterion, even though the geometry chosen was demonstrably stable. The answer to the paradox is that very small movements do occur on the joints, these movements completely altering the original stress distribution. The result of the slight shuffling of the joints is to redistribute the stresses in such a way as to make all parts of the slope stable with respect to the failure criterion . . .*
> *the obvious question is "can failure mechanism be built into the computer programs?"*

These two programs evolved into PFC2D (Potyondi and Cundall 1998) and UDEC (Cundall 1980) in their current form. I utilized both codes (including Version 1.0 of PFC2D release) extensively in my own PhD, probing these exact questions for bonded solids and the mechanics of brittle damage. A bonded particle model could answer directed questions and prove the existence of heterogenous stress flow, spatial heterogeneity of material properties and the impacts of discontinuous damage and yield on post-yield performance (Diederichs 2003). These developments in turn led to the implementation and mechanistic validation of a simplified approach to brittle damage for use in simpler and more practical continuum models (Diederichs 2007).

The same questions must be asked today before using these codes and others like them including the 3D versions in wide use today as well as the 2D and 3D FEM codes with joint elements and joint networks and the FEMDEM hybrids (Munjiza 2004) such as Elfin from Rockfield, and IRAZU from Geomechanica. How will the failure mode and rockmass response due to the discontinuum structure deviate from my comfortable continuum predictions? Will the discontinuum assembly reproduce realistic and appropriate failure modes? How will my estimation of upscaled parameters impact the outcome? Perhaps we can also add, how will a discontinuum approach change my assumptions surrounding the function and performance of support and reinforcement?

I would also argue that despite impressive developments in discontinuum modelling, including recent improvements in realistic discrete fracture networks, we are not very much closer, than we were then, to providing valid and accurate upscaled stiffness, strength, and dilation parameters for the discontinuity surfaces within these models. Creating realistically dense discretely jointed models for rock engineering design is possible on modern computers. However, our abilities to calibrate or upscale the mechanical properties become limited without the luxury of observational modelling and back analysis. Guidance in this regard, relevant even today, can be found in the landmark case study of the 62m span Gjovik arena (Barton et al. 1997) using an early version of UDEC and a joint constitutive scheme from Barton and Bandis (1990).

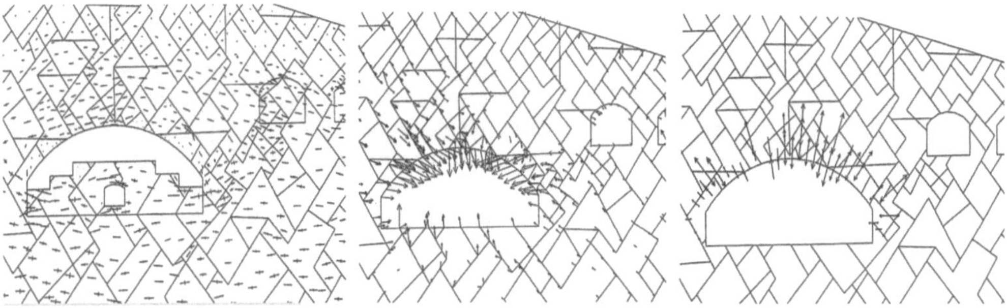

Figure 13. (from left to right): Stress flow through the blocky rockmass of the 62m span Gjovik arena modelled in UDEC; Displacements; Bolt Loadings (After Barton et al. 1997).

From the image above it is clear, that the joint pattern, as modelled, was likely far from comprehensive and realistic due to computing limitations at the time. The pproach, however, was essential to answering the following two critical questions:

1. Can rough joints, modelled with realistic confinement dependent strength properties and post-yield dilation create a stable and clamped arch, if allowed to deform gradually through sequential excavation?
2. Can reinforcement be used to knit mobile blocks together to limit localized slip and create a reliably stable arch and a viable cavern with long-term stability?

These questions could never have been answered with continuum models and this unprecedented cavern likely would not have been built. Other tools such, as empirical classification and support guidelines, were used to assess stabilizing factors and threats, while modelling aided the understanding of reinforcement and arch-building in structured ground.

From the 1990s, evolution of discontinuum models continued, from interface development and improvement, to improved geometry and structural cutting approaches and meshing algorithms, discrete fracture network generation and computer aided validation and enhanced output, including virtual reality capabilities. The impressive advances in user-interfaces from 2000 to the present has been game-changing. In 2D, the ability to create realistic joint patterns with appropriately scaled spacing and persistence within software such as RS3 is an educational boon for rock mechanics. It has also been a major bonus to practical modelling, allowing multiple realizations to be simulated in a reasonable time – a capability essential for design analysis given the highly variable response within a group of joint networks generated with the same governing parameters.

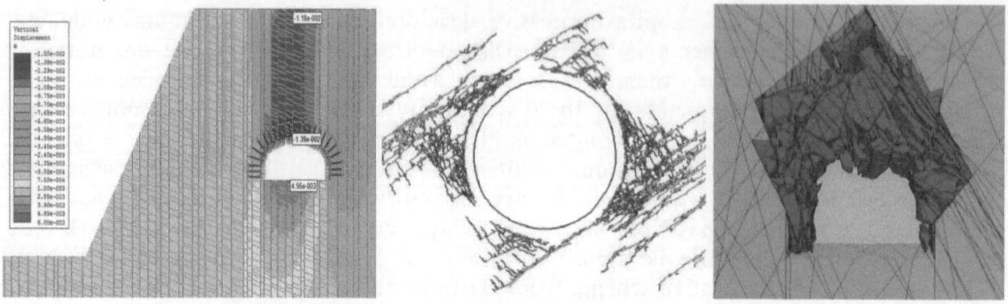

Figure 14. Practical Discontinuum Tools: RS2, IRAZU (courtesy of Geomechanica), 3DEC.

Coupled processes have been integrated with variable success and validation, along with some lingering limitations. A variety of joint constitutive models, as well as a wide variety of realistic support elements, has also evolved. Nevertheless, challenges remain in providing valid upscaled material and structural parameters to assign to model components as well as realistic joint patterns. Accurate dilation modelling is perhaps the most important aspect as this controls the evolution of stabilizing arches and governs the response of reinforcement crossing the discontinuities.

The thirst for more and less expensive parametric input has also driven some unfortunate trends in lab testing – notably the scourge of multistage shear testing. I have seen many advanced models that, for want of better shear strength data, use residual shear strength properties for design with catastrophic and unrealistic impacts on design viability. Site investigations can suffer from a disconnect with modelling – joint characteristics are often logged using RMR or Q parameters while more direct measurements such as JRC and JCS (allowing use of more sophisticated joint modelling) are not specified. There is a need to upgrade the investigation program to suit the level of modelling anticipated for projects. Continuum modelling with GSI clearly requires a more simplified lab and field data collection approach than a realistic and accurate discontinuum analysis.

6 HYBRID APPROACHES

Other than the advance of computing power since 2000 and our ability to analyze larger and larger models with greater complexity, and other than the step changes in user interfaces, the big story in modelling over the last decade is the rise of the hybrid code. There is not enough space in this keynote to delve into the development and potential of modern hybrid codes including the synthetic rockmass hybrid approach (Mas Ivars et al. 2011) with the integration of continuum and discontinuum mechanics. These models aid in answering key mining questions such as "when will two stressed and structurally controlled yield regions in a mine suddenly interact unleashing seismicity in the transitional rockmass between?"

Full hydromechanical coupling (and thermomechanical) as well as time-dependent processes, integrated with continuum-discontinuum models are rapidly advancing and will greatly aid in answering one of the greatest lingering questions of rock mechanics – "how do effective stress, drainage and pore-pressure concepts adapted from porous solid approaches translate to jointed rockmass?".

Challenges with these hyper-advanced codes include the loss of general accessibility that was achieved in the 1990s with the initial advent of graphical user interfaces and simplified program inputs and informative outputs. I see a trend back to the days of expert-driven modelling, which created a disconnect between the practical aspects of design decision-making at the front lines and the modelling supporting those decisions. I also have my doubts about the robustness, accuracy, and realism of the input parameters in these models. Probabilistic analysis or reliability approaches are inhibited in these hybrid models due to the vast array of input parameters.

Calibration and upscaling are a challenge due to the different dimensionalities of the continuum and discontinuum elements. Work by researchers such as Sainsbury and Sainsbury (2017) are examples of approaches for calibrating and upscaling complex models and using complex discontinuum codes to inform parameter selection for more accessible continuum options (such as ubiquitous anisotropy of strength and stiffness within an equivalent continuum).

The confidence in these models can only be built through back analysis and observational design during mining, or during early and mid-stages of construction (Kalenchuk 2019). While this process is now a routine mandate in complex and high-risk cave mining and large open pits, it is a challenge in civil construction often due to difficulties in integrating observational approaches, which inform design, into contractual models of construction.

In mining, effective modelling consultants, associated with the particular codes in use, develop long term relationships with their mining clients and the decision makers involved, such that a shared understanding of the inputs, assumptions, limitations, interpreted results and implications for decision making and risk management develops even if there is only a one-sided engagement in the actual modelling. An example framework for this relationship is discussed by Beck and Lilley (2011). This approach is less effective in civil tunnelling and cavern projects and, in my experience, generally fails once construction starts unless the people on the ground making both the strategic and the day-to-day decisions have some equity in the modelling program.

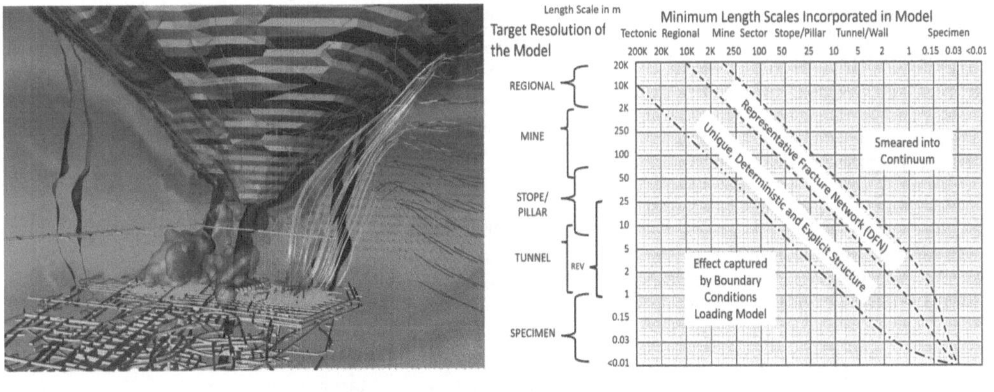

Figure 15. (Left) 4D mine-wide, fully coupled non-linear, discontinuum, transient hydromechanical simulation (courtesy of Beck Engineering) and (Right) Guidance on model selection based on component scale and output resolution (after Beck et al. 2013).

I feel that these complex hybrid codes, executed by specialists, need to be linked to simpler accessible modelling tools to be used at the front lines of a project. This can perhaps be done through updated equivalent continuum or simplified discontinuum approaches informed by the hybrid modelling. If the front-line engineers are excluded from the modelling process for design, operational decision making or risk management, then I fear that, in spite of the technical modelling advances, we would have stepped back from the goals of the barrier-free interactivity and shared technical understanding and front-line linkage we achieved in the early 1990s.

7 CONCLUSIONS

This paper has outlined a user's journey through the development and application of rock engineering modelling software from the 1980s to the present. The evolution from simple elastic models to the complex hybrid multi-physics codes of today has been impressive. I have emphasized the importance of maintaining a set of clearly defined questions to ask of the models or of the consultants who run them. While this is engineering, the scientific method must be adapted for complex modelling. The engineer must ask a series of directed and specific questions, formulate hypotheses based on other resources and test those hypotheses against the model.

Issues of appropriate input will dictate the validity of the model. Although models may provide unexpected insights to the user, these, in turn, must be investigated further with respect to model validity, completeness of the incorporated geological and geomechanical model, and the likelihood of numerical artifacts or unique outcomes specific to a single realization (mesh and discontinuum network sensitivity as well as parametric uncertainty and sensitivity). This is especially true as models become very complex, and as we return to the days of consultant-driven modelling. Simple models may still be used by project engineers to check and validate.

I will end by paraphrasing and updating sage advice given to me many years ago, while I was a young Padawan (graduate student), by the Jedi Master, Dr. Evert Hoek:

Never turn off your brain when you turn on your computer ...

... ***nor when someone else turns their computer on for you.***

ACKNOWLEDGEMENTS

Thanks are due to Evert Hoek and John Curran for introducing me to this field of Rock Engineering and Modelling. As well for this paper, Erik Eberhardt, Giovanni Grasselli, Jennifer Day, Jean Hutchinson, Reginald Hammah, Omid Mahabadi, Kathy Kalenchuk, Evan Dressel, Caitlin Fischer, David Sainsbury, Joe Carvalho, and David Beck all provided much appreciated inspiration and/or guidance on this subject.

REFERENCES

Barton, N. and Bandis, SC. 1990. Review of the predictive capabilities of JRC-JCS model in engineering practice. Proc. Int. Symp. On Rock Joints. Norway. pP603–610.

Barton, N., By, TL., Chryssanthakis, P., Tunbridge, L., Kristiansen, J., Loset, F., Bhasin, RK., Westerdahl, H., Vik, G. 1994. Predicted and measured performance of the 62m span Norwegian Olympic Ice Hockey Cavern at Gjovik. Int. J. Rock Mech. Min. Sci & Geom. Ab. V31:617–641.

Beck, D. and Lilley, C. 2011. Simulation aided engineering – Integration of monitoring, modelling and planning. 11th Underground Operators Conference, Canberra, March. 6 pgs.

Beck, DA., Lilley, CR., Reusch, F., Levkovitch, V., Putzat, G., Flattem, A. 2013. A preliminary, calibrated scheme for estimating rock mass properties for non-linear, discontinuum models. Sinorock. 6 pgs.

Cai M, Kaiser PK, Tasaka Y, Minami M. 2007. Determination of residual strength parameters of jointed rock masses using the GSI system. Int J Rock Mech Min Sci 44(2):247–265

Cundall, PA. 1971. The measurement and analysis of accelerations in rock slopes. PhD Thesis. University of London – Imperial College of Science and Technology.

Cundall, PA. 1980. UDEC – a generalized distinct element program for modelling jointed rock. DAJA37-79-C-0548, European Research Office, U.S. Army.

Dadashzadeh N. and Diederichs M. 2019. Excavation damage zone prediction for deep tunneling based on distinct element approach. Proceedings of the 14th ISRM International Congress of Rock Mechanics.

Diederichs, MS. 2003. Rock fracture and collapse under low confinement conditions. Rock Mech. and Rock Engineering Volume 36 (5): p339–381.

Diederichs, MS. 2007. CGS Geocolloquium Award Lecture: Damage and spalling prediction criteria for deep tunnelling. Canadian Geotechnical Journal. Volume 44 (9): p1082–1116

Eberhardt, E., Stead, D., Reeves, MJ., and Connors. 1997. Design of tabular excavations in foliated rock: an integrated numerical modelling approach. Geotechnical and Geol. Engineering, v15 47–85

Hajiabdolmajid, V., Kaiser, PK. And Martin CD. 2003. Mobilized strength components in brittle failure of rock. Geotechnique V53. P327–336.

Heller SR, Brock, JS and Bart, R. 1958. The stresses around a rectangular opening with rounded corners in a uniformly loaded plate. Trans. 3rd US Cong. Appl. Mech. P357.

Hoek, E. and E.T. Brown, 1980. Underground Excavations in Rock. 1st Edition, The Institution of Mining and Metallurgy, London, ISBN: 0419160302, 527p.

Hoek E, Carranza-Torres CT, Corkum B. 2002. Hoek–Brown failure criterion—2002 edition. Proc. of the Fifth North American Rock Mechanics Symposium. pp 267–273

Inglis, C.E. 1913. "Stresses in plates due to the presence of cracks and sharp corners," Transactions of the Institute of Naval Architects, Vol. 55, pp. 219–241

Kalenchuk, K. 2019. CGS Colloquium: Mitigating a fatal flaw in modern geomechanics, understanding uncertainty, applying model calibration, and defying hubris in numerical modelling. Submitted to CGJ.

Kirsch, 1898, Die Theorie der Elastizität und die Bedürfnisse der Festigkeitslehre. Zeitschrift des Vereines deutscher Ingenieure, 42, 797–807.

Ladanyi, B. 1974 Use of the long-term strength concept in the determination of ground pressure on tunnel linings. Proc. 3rd Congr. ISRM. Denver. V2b. p1150–1156.

Lorig, L. And Verona, P. 2013. Guidelines for numerical modelling of rock support for mines. Ground Support 2013 (ACG). Potvin and Brady (eds). 26 pgs.

Mas Ivars, D., Pierce, M., Darcel, C., Reyes-Montes, J., Potyondi, D., Young, P., Cundall, P. 2011. The synthetic rock mass approach for jointed rock mass modelling. IJRMMS V48:2 p219–244.

Moyo, T. and Stacey, T. 2012. Mechanism of rockbolt support in jointed rock masses. Deep Mining. 14p.

Munjiza, A. 2004. The Combined Finite-Discrete Element Method. John Wiley & Sons Ltd.

Potyondi, D., and Cundall, P.A. 1998. Modelling notch-formation mechanisms in the URL mine-by test tunnel using bonded assemblies of circular particles. IJRMMS v35(4–5), Paper No. 67. 6 pp.

Sainsbury, BL. and Sainsbury, DP. 2017. Practical use of the ubiquitous-joint constitutive model for the simulation of anisotropic rock masses. Rock Mech Rock Eng. V50. P1507–1528.

Walton, G., Labrie, D., Alejano, L. 2019. On the residual strength of rocks and rockmasses. Rock Mechanics and Rock Engineering V52:4821–4833

Walton, G and Diederichs, M. 2020. A new model for the dilation of brittle rocks based on laboratory compression test data with separate treatment of dilatancy mobilization and decay. GEGE v33(3)

Position papers for panel discussion 1

The Evolution of Geotech - 25 Years of Innovation – Hammah et al. (Eds)

From monitoring and calibration to predictive modelling

Niccolò Coli
VP MineMonitoring - Hexagon Geosystems

1 INTRODUCTION

- In modern mining, comprehensive slope monitoring program for managing potential large-scale instabilities and simultaneously detecting local scale movements is integral to geotechnical risk assessment
- Slope monitoring radars are an essential part of this process, by detecting progressive slope movements which could potentially lead to slope failure and help assess worker safety

DOI: 10.1201/9781003188339-5

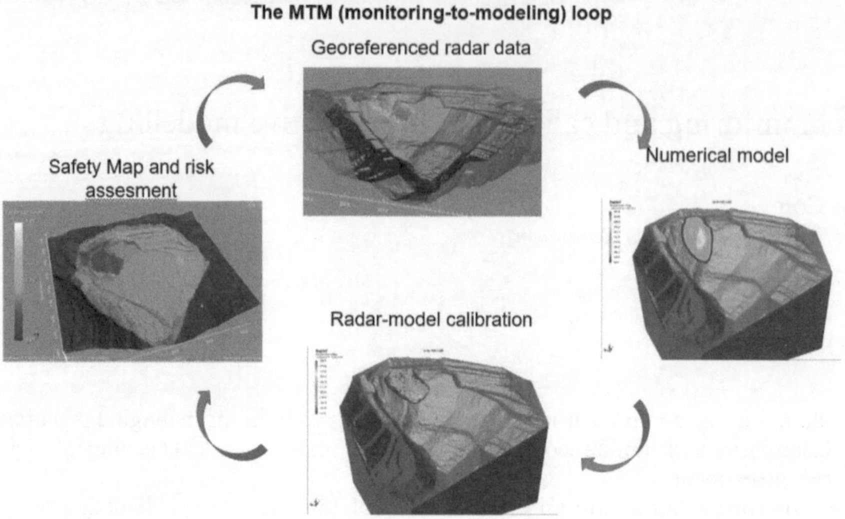

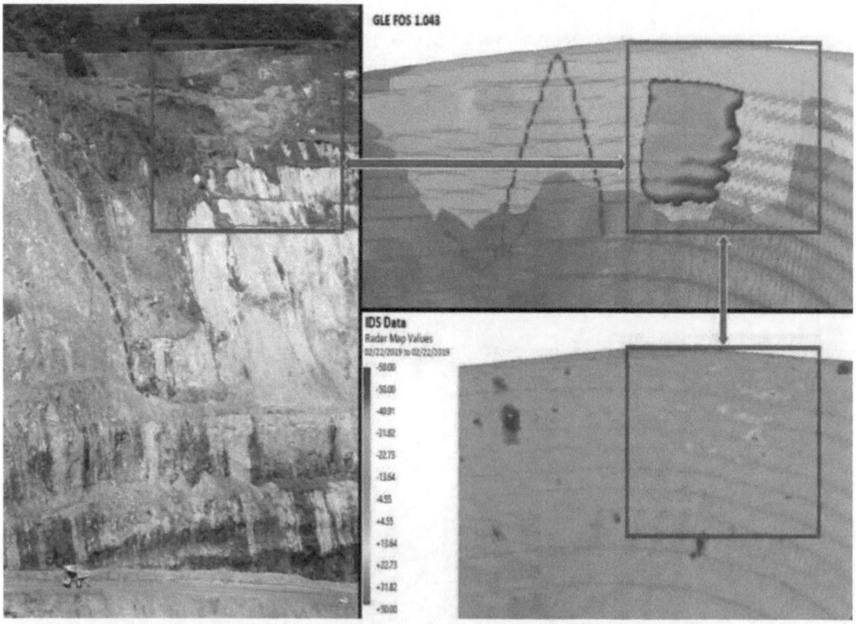

- Basic concept behind radar-numerical modelling integration
    - Match observed movements in radar data to deformations in geotechnical models
    - Better understand rock mass mechanics of slopes using accurate real-time monitoring data
- Can help achieve significant improvements to geotechnical hazard management by helping engineers better understand failure mechanisms when potential instabilities are identified by radars

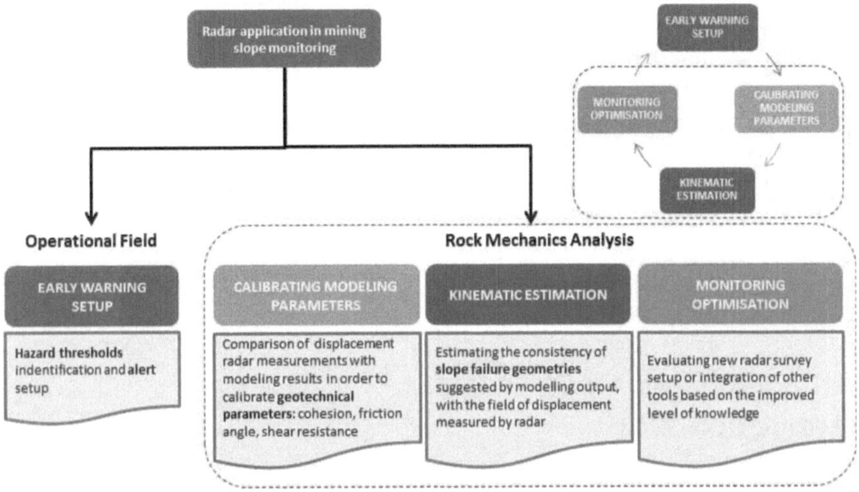

Radar application in mining slope monitoring
EARLY WARNING SETUP
MONITORING OPTIMISATION
CALIBRATING MODELING PARAMETERS
KINEMATIC ESTIMATION
Operational Field
Rock Mechanics Analysis
EARLY WARNING SETUP
Hazard thresholds indentification and alert setup
CALIBRATING MODELING PARAMETERS
Comparison of displacement radar measurements with modeling results in order to calibrate geotechnical parameters: cohesion, friction angle, shear resistance
KINEMATIC ESTIMATION
Estimating the consistency of slope failure geometries suggested by modelling output, with the field of displacement measured by radar
MONITORING OPTIMISATION
Evaluating new radar survey setup or integration of other tools based on the improved level of knowledge

The Evolution of Geotech - 25 Years of Innovation – Hammah et al. (Eds)

Thoughts on instrumentation/monitoring and numerical model calibration

"From monitoring and calibration to predictive modelling"

Will Bawden
University of Toronto and MDT Engineering

1 INTRODUCTION

- Strongly advocated for increased and expanded instrumentation in mining (since 1983)
- Progress in area was initially (and painfully) slow
- Pace of instrumentation use has increased
  - In past 10 to 15 years, explosive growth in this area, particularly with large international cave mining operations
- Also always believed in need for calibration of numerical models
- Two types of model calibration - Qualitative and Quantitative
- Both have their place
- Remember though that we still managed to successfully build and operate large, complex mines in the days when operating mines had zero computers
- Before computer processing power reached levels that allow routine use of mine-wide 3D non-linear models, we used elastic modelling
- Many today seem to have fallen into trap of thinking that only complex, nonlinear models have any value
- Many ignore the fact that the underlying inputs for such analyses are of low quality or may not be available in many cases
- While latest non-linear numerical model capabilities are truly amazing, our ability to acquire appropriate input parameters and use calibrations tools have unfortunately advanced very little

DOI: 10.1201/9781003188339-6

The Evolution of Geotech - 25 Years of Innovation – Hammah et al. (Eds)

Continuously improving models with monitoring data

"From monitoring and calibration to predictive modelling"

Neil Bar
Gecko Geotechnics

1 INTRODUCTION

- Limitations in the knowledge and understanding of ground conditions remain a major area of uncertainty and, therefore, risk in geotechnical engineering projects
- Despite major improvements in ground characterization techniques over the last decades, geotechnical engineers still face significant gaps in data availability and reliability as projects progress into execution phase
- As a result of these gaps, large extrapolations on ground behaviour are made through simplification of ground characteristics
- Ground characteristics are at basis of all geotechnical analyses, including predictive modelling
- Uncertainty is often understated and not well understood by project managers leading to inadequate or poorly defined risk mitigation measures
- Simpler models, with fewer inputs are usually easier to calibrate
- Calibration can occur in various forms, including:
  - Basic 'sense checks' including spatially correlating areas with lower factor of safety to areas of increased deformation
  - Geological structure identification and verification by spatially correlating deformation patterns and shapes of deforming areas
  - Back-analysis to confirm or update failure mechanisms
  - Detailed numerical calibration, with model matching multiple stress and strain (deformation) measurements over several excavation phases
- Model calibration
  - does not remove need to bridge gaps in ground characterization
  - does not eliminate potential for unexpected behaviour as excavations progress
  - can enable continuous improvement models such that "the models could actually be right"

DOI: 10.1201/9781003188339-7

The Evolution of Geotech - 25 Years of Innovation – Hammah et al. (Eds)

Real-time data collection from InSAR is coming soon
"From Monitoring and Calibration to Predictive Modelling"

Davide Colombo
TRE ALTAMIRA

1 INTRODUCTION

- Mining heavily relies on basic cycle of design, planning, execution and control
- Numerical modelling has played a major role in design and planning for several decades
- Monitoring has only more recently become key to execution and control
- Recent failure events, especially of tailings dams, have revealed key role of monitoring
  - Structures that should be "stable" by design have still collapsed
- Important concept now permeating discussion
  - Every model should be constantly verified through monitoring
  - Monitoring is useful only if it provides data that helps engineers update and improve their models
- Concept builds on idea that models are "living" things and depend not only on initial design parameters, but also on factual records determined with monitoring tools
- Important to identify parameters which must be monitored to ensure that relevant inputs are used to update models
- "Supply chain logic" - useful concept in monitoring which answers question
  - "What kind of monitoring is needed for a particular operation?"
- Process of answering often discusses technologies and budgets, rather than risk, root causes and optimization parameters for models
- Integration (both in terms of methodologies and software implementation) is also discussed - but generally limited to visualization of different datasets
- Numerical modelling is required to close this circle of discussions
  - It is the factor that can more effectively use monitoring data with different levels of uncertainty, and spatial (lines of sight) and time features
- InSAR is powerful monitoring that can help close the circle
- Was initially used to map very slow deformations over areas of operation
- InSAR has recently evolved towards a web platform that delivers data with every new acquisition
  - What used to be delivered in weeks just two years ago is today provided in a few hours
  - Mainly driven by needs of TSF monitoring
- Companies are about to launch new sensors that can provide daily, high resolution (less than 1 meter on the ground) measurements
- This will allow InSAR operators to provide more value to geotechnical engineers

DOI: 10.1201/9781003188339-8

Position papers for panel discussion 2

The Evolution of Geotech - 25 Years of Innovation – Hammah et al. (Eds)

Perspectives from technical support on user models

"Is numerical modelling a solution or a problem?"

Brent Corkum
Rocscience

1 INTRODUCTION

- Unique perspectives from seeing hundreds of RS2 and Slide2 customer models
- Common user statements or questions
  - "In the attached model, I'm having trouble understanding the results"
  - "Can you please review the attached model and let me know if the results are correct?"
- In many cases, model complexity makes it impossible to determine what users were trying to achieve or what questions they were attempting to answer
  - They simply add every information possible to model, making it impossible to determine whether results are meaningful
- Idea that you can build one model with all inherent complexity and meaningfully interpret results is fundamentally flawed
- Answer starts with understanding that modelling is process - sequence of steps that adds details one at a time
- Start simple
  - Simple models have great importance in understanding mechanisms
  - Especially true in models with multiple failure modes
- Even after 35 years of numerical modelling, would never jump straight into complex model with intricate loading, boundary conditions, initial stresses, material behaviour, joints, geology, etc.
- Understand level of uncertainty in input parameters
  - In geotechnical engineering, still wide gap between complexity of material models and our ability to (accurately) determine their input properties
- Simplify geometry - one of first challenges with any model
  - Now that we have block models, Lidar survey, and digital elevation maps with high accuracy, not uncommon to see external, material and excavation boundaries with thousands of vertices
  - Realize this level of detail results in needlessly huge finite element meshes with poorly shaped elements
- Initially build simple elastic single-stage models
- Set up initial conditions (before any excavation) that ensure in situ stresses, material strength, boundary conditions, pore pressure distributions and elastic material properties are correct
  - Strength factor contours from elastic analysis can indicate which model areas will fail under initial stress field
- Use simple 2D models, if they are applicable
  - Many of model setup problems are readily seen, understood, and fixed in simple 2D models
  - Simple analyses only way to look at multiple possible scenarios within reasonable time

DOI: 10.1201/9781003188339-9

- Apply new technologies that help with modelling
  - Use finite element shear strength reduction (SSR) modelling to complement limit-equilibrium (LE) analysis – helps verify results
  - Use remote sensing data (such as radar or Lidar displacement data) to calibrate and validate numerical models
- Build your story
  - Start simple
  - Progressively add complexity
  - Measuredly use available tools

The Evolution of Geotech - 25 Years of Innovation – Hammah et al. (Eds)

Numerical modelling philosophy

"Is numerical modelling a solution or a problem?"

Joe Carvalho
Golder Associates

1 INTRODUCTION

- Modelling - art of identifying mechanisms and then quantifying sensitivity to key controlling parameters
- Some rules for analyses with modelling software:
  - Do not use tool without understanding its inner workings
    - Select right tool for job
    - Understand range of application of software
    - Understand required input parameters
  - Realize that no result is absolute
  - Walk before you run
    - Start with simple model
    - Do not use complex constitutive models that require parameters you do not have
      - Corollary - if you know problem is complex and you do not have the input data, go and collect the input data
- We cannot deprive competent users of numerical tools just because a few misuse the tools
- Technological advances inevitably get used for "wrong/malevolent" purposes and numerical tools are no different
- We do have responsibility to minimize misuse
- Education (by universities, software developers, etc.) is key to ensuring user understand tools and can apply them correctly
- Often, problems are three-dimensional in nature (e.g. portals, connections, intersections of tunnels, etc)
- While 2D can be used in many instances, necessary to understand limitations, and complement with methods, such as the concept of ground reaction curves to account for 3D effects
- Not always clear whether 2D analysis will be conservative or not
- 3D model building is more complex, but tools have evolved to point where they can be used in routine engineering
- Lack of anisotropic constitutive models for continua is challenge
  - Indispensable to explicitly include joint networks
- Shear strength reduction (SSR) now state of practice when evaluating slope stability
  - Exercise caution - ensure failure mechanisms do not change during process
- In design process, calculations sometimes verified and reviewed by senior staff or review boards (both internal or external)
- Numerical models pose particular challenge to this review process because of nature of software tools

DOI: 10.1201/9781003188339-10

- Traditional reviews with calculations on paper, which can be followed and easily verified, are not readily transferred to numerical models
- Some simple steps to mitigate challenge
  - Structure models so they are easier to update and conduct QA/QC
  - Verify models have been built right
    - Ensure model is implemented correctly
    - Have a second modeller (a peer) check it
  - Ensure model appropriately represents problem
  - Understand quality and limitation of data, as well as assumptions in model

The Evolution of Geotech - 25 Years of Innovation – Hammah et al. (Eds)

Rock mechanics design process

Correspondence between Bieniawski's design process and the numerical modelling philosophy of Starfield and Cundall

Thomas R. Stacey
University of the Witwatersrand

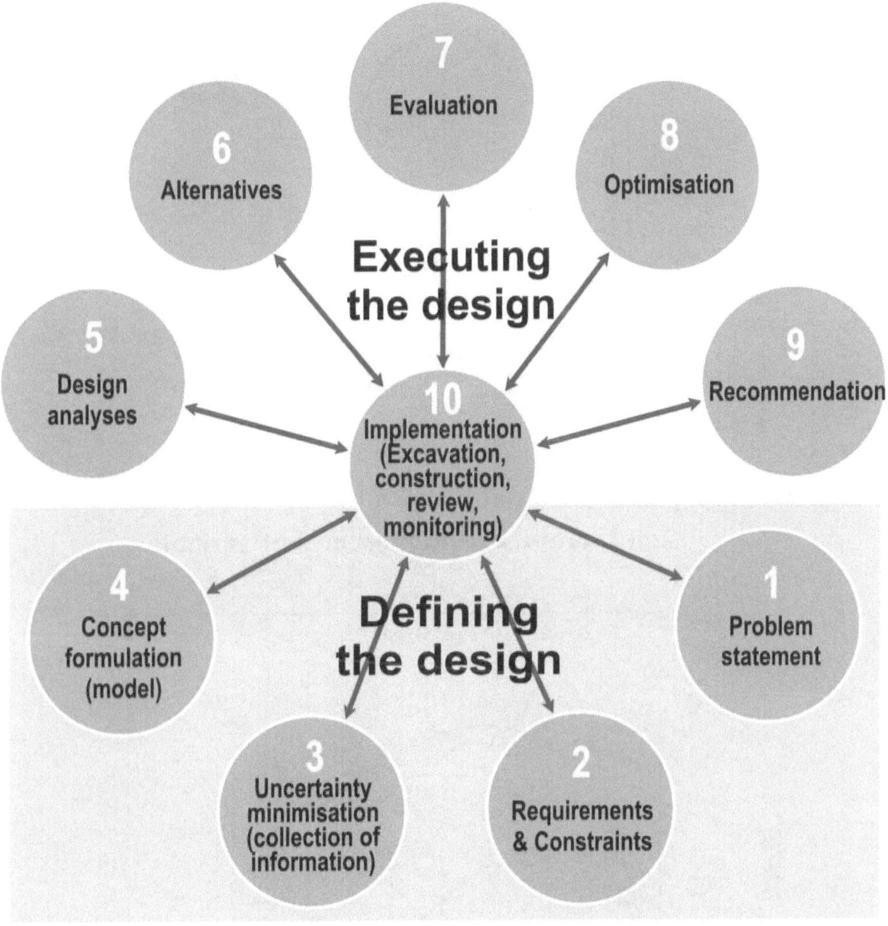

Numerical modelling philosophy (Starfield and Cundall)

Step 1 deals with the clarity of design objectives; corresponds with "Be sure, before you start, that you are quite clear about *why* you are building a model and *what* questions you are trying to answer."
Step 4 (geotechnical model) corresponds with "Look at the *mechanics* of the problem. Try to identify important mechanisms, modes of deformation and likely modes of failure."

DOI: 10.1201/9781003188339-11

Bieniawski's Simplicity Principle corresponds with "Design or borrow the simplest model that will allow the important mechanisms to occur . . .."
Steps 5 to 7 involve the running of the models.
Step 8 implies multiple models, corresponding with "It should also be noted that implicit in the above guidelines is a rule that rock engineering models should never be run only once; it is in the sensitivity of the results to changes in parameters and assumptions that the model is most informing."

The value of "simple" elastic modelling (design or borrow the simplest model . . .)

Excavation involves unloading of the rock, and behaviour is commonly linear, hence value in elastic modelling

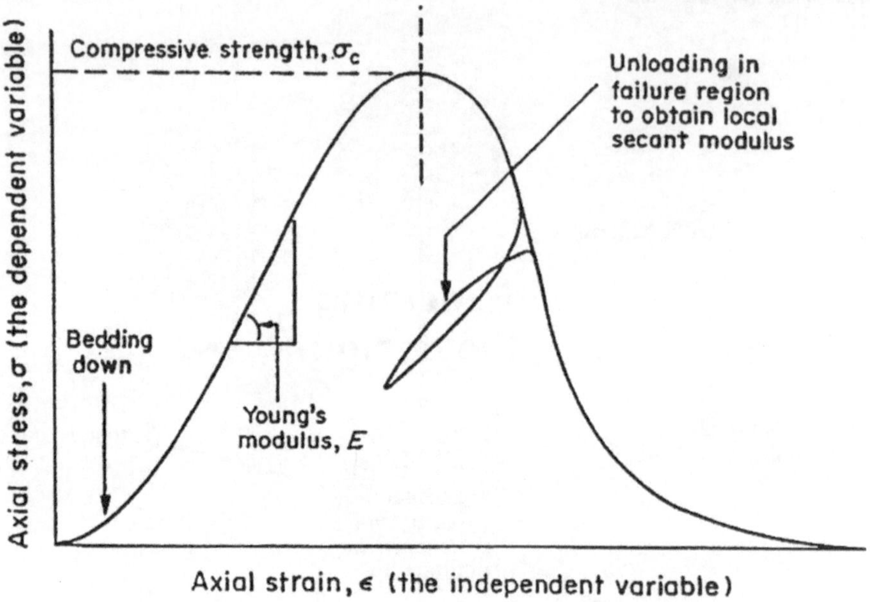

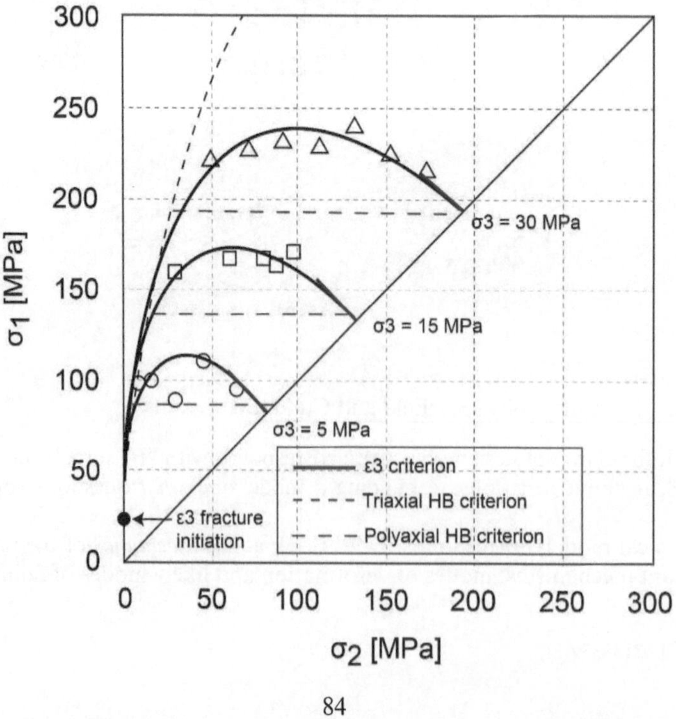

Importance of three dimensional analysis

All rock engineering problems are actually 3D – eg rock strength test results by Mogi, above. 2D is an acceptable simplification in some cases AND……

$$\begin{bmatrix} \varepsilon_x \\ \varepsilon_y \\ \varepsilon_z \\ \varepsilon_{xy} \\ \varepsilon_{yz} \\ \varepsilon_{zx} \end{bmatrix} = \begin{bmatrix} \frac{1}{E_x} & \frac{-\nu_{yx}}{E_y} & \frac{-\nu_{zx}}{E_z} & 0 & 0 & 0 \\ \frac{-\nu_{xy}}{E_x} & \frac{1}{E_y} & \frac{-\nu_{zy}}{E_z} & 0 & 0 & 0 \\ \frac{-\nu_{xz}}{E_x} & \frac{-\nu_{yz}}{E_y} & \frac{1}{E_z} & 0 & 0 & 0 \\ 0 & 0 & 0 & \frac{1}{G_{xy}} & 0 & 0 \\ 0 & 0 & 0 & 0 & \frac{1}{G_{yz}} & 0 \\ 0 & 0 & 0 & 0 & 0 & \frac{1}{G_{xz}} \end{bmatrix} \begin{bmatrix} \sigma_x \\ \sigma_y \\ \sigma_z \\ \sigma_{xy} \\ \sigma_{yz} \\ \sigma_{xz} \end{bmatrix}$$

This 3D strain-stress matrix shows 6 moduli elements and 3 Poisson's ratio elements. Therefore single values of E and ν in analyses, as commonly practised, could be misleading Synthetic rock mass analyses illustrate different strengths and moduli in different directions

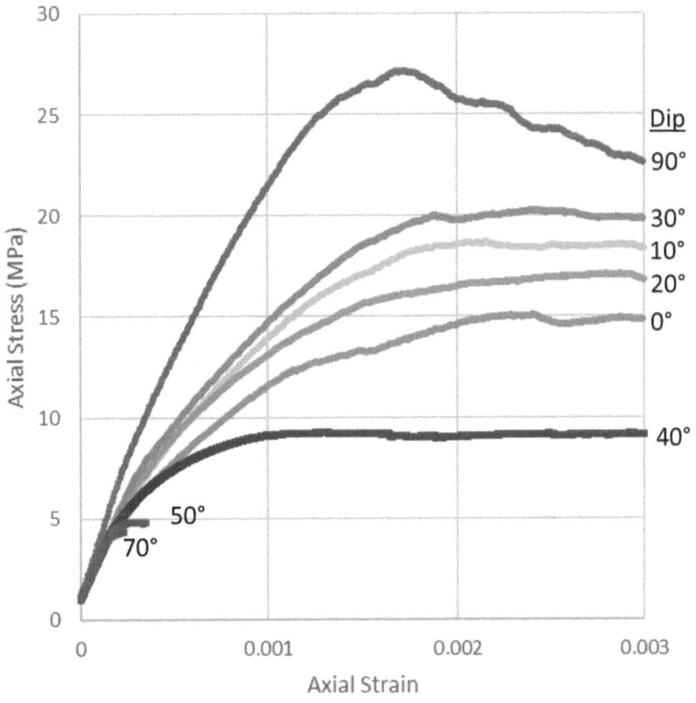

So numerical models must therefore take jointing into account in 3D. But we don't know the actual orientation, persistence or spacing in each joint set, only the statistical variations. And therefore we need to carry out numerous analyses to take into account the variability

Design and Prediction

Many publications that appear in the literature assume the mode of behaviour, and modify rock mass properties until an apparent match with observed behaviour is obtained; the assumption is then made that the modelled rock mass properties are "correct", and applicable for analysis of further situations. NOT TRUE

OK for design with a factor of safety, but not valid for prediction of behaviour in new conditions

Numerical models must take into account the mechanisms of rock fracture and failure. Fractures may initiate due to one criterion, propagate due to the same or other criteria, and failure may develop due to yet other criteria, including structural criteria such as beam bending and buckling, plate buckling etc. Prediction under such circumstances presents a significant numerical modelling challenge.

The Evolution of Geotech - 25 Years of Innovation – Hammah et al. (Eds)

Challenges in tailings dam modelling

Eng. Thiago Bretas, Msc.
BVP Geotecnia e Hidrotecnia

FUNDÃO DAM (2015)

- Breach by liquefaction;
- Loose, saturated, contractive taillings;
- Modified design (growth in the saturated condition);
- Lateral extrusion;
- Trigger mechanism;

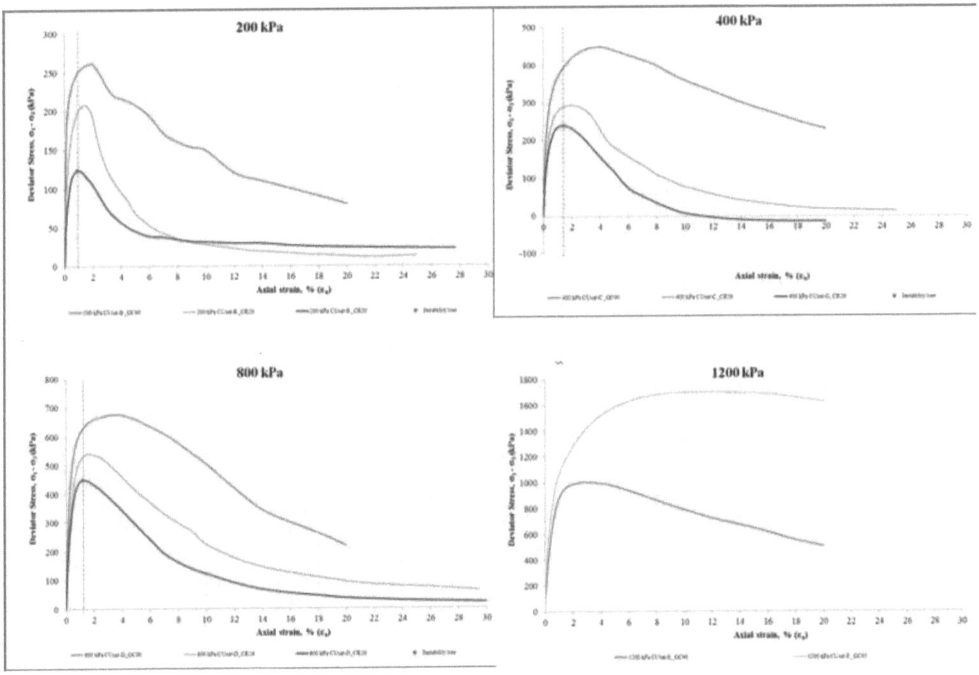

DOI: 10.1201/9781003188339-12

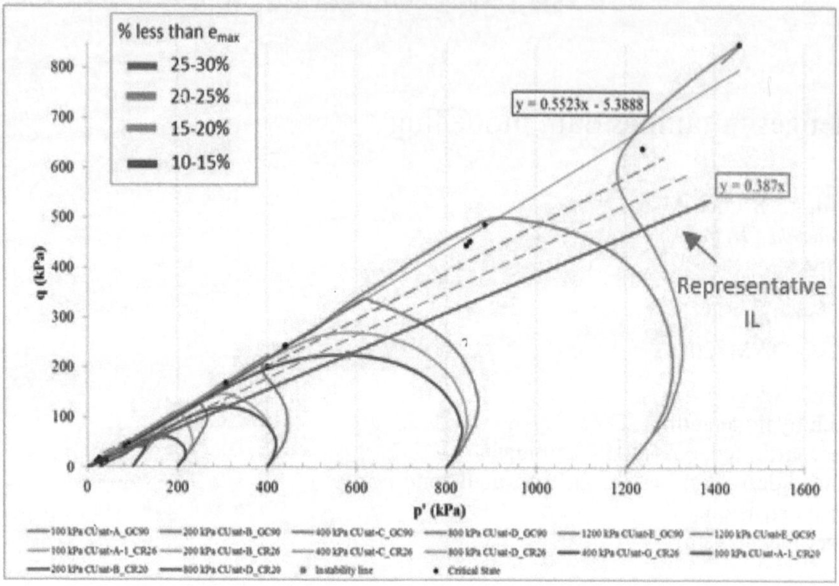

FEIJÃO DAM 1 (2019)

- Breach by liquefaction;
- Loose, saturated, contractive taillings;
- Heavy and brittle (high iron ore content);
- Amount of strain required to trigger the sudden loss of strength was very small;
- Internal strain due to creep;
- Steep uptream slopes;
- Flawed internal drainage;

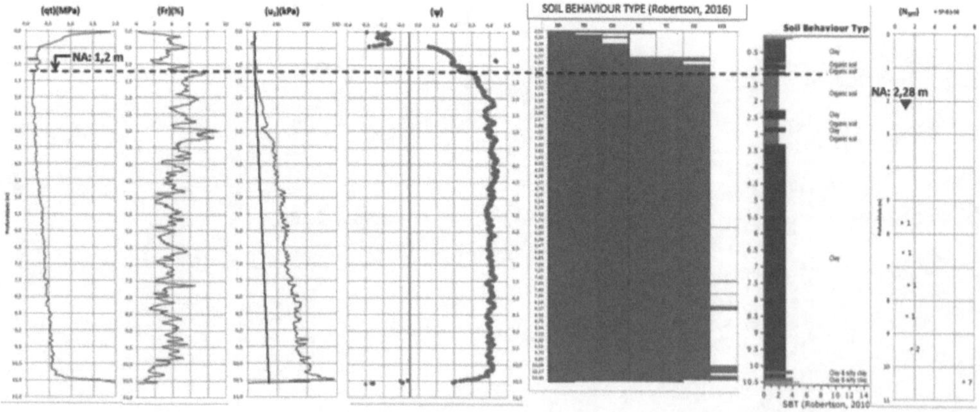

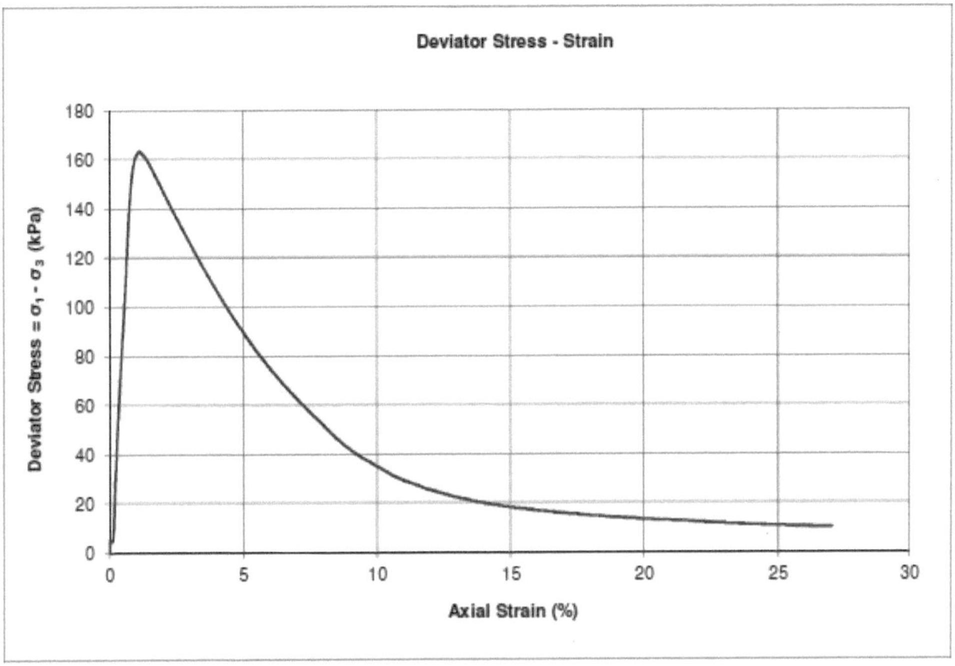

Deviator Stress - Strain
Deviator Stress = σ1 - σ3 (kPa)
180
160
140
120
100
80
60
40
20
0
0
5
10
15
20
25
30
Axial Strain (%)

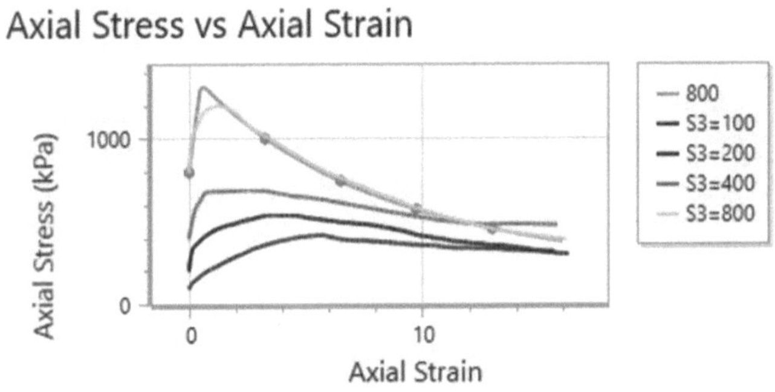

Axial Stress vs Axial Strain
Axial Stress (kPa)
1000
0
0
10
Axial Strain
800
S3=100
S3=200
S3=400
S3=800

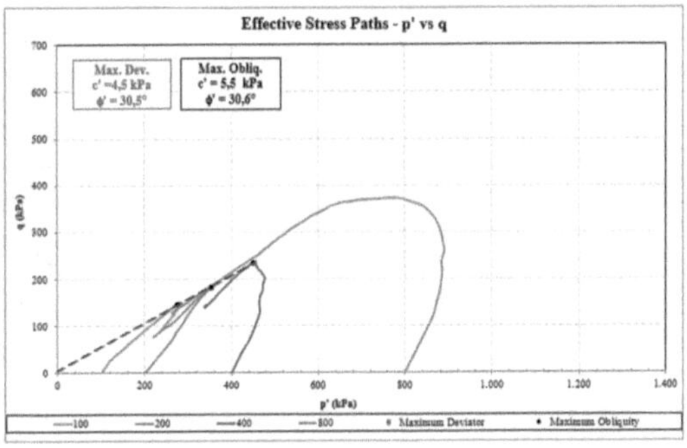

Effective Stress Paths - p' vs q
Max. Dev.
c' =4,5 kPa
φ' = 30,5°
Max. Obliq.
c' = 5,5 kPa
φ' = 30,6°
q (kPa)
700
600
500
400
300
200
100
0
0
200
400
600
800
1.000
1.200
1.400
p' (kPa)
100
200
400
800
Maximum Deviator
Maximum Obliquity

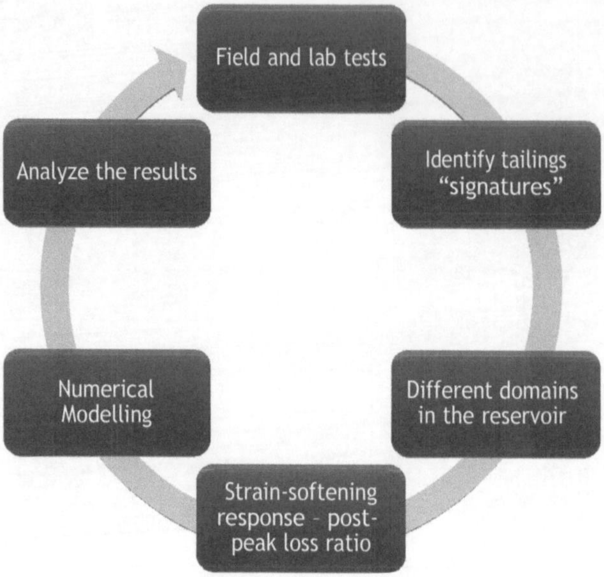
Field and lab tests
Identify tailings "signatures"
Different domains in the reservoir
Strain-softening response - post-peak loss ratio
Numerical Modelling
Analyze the results

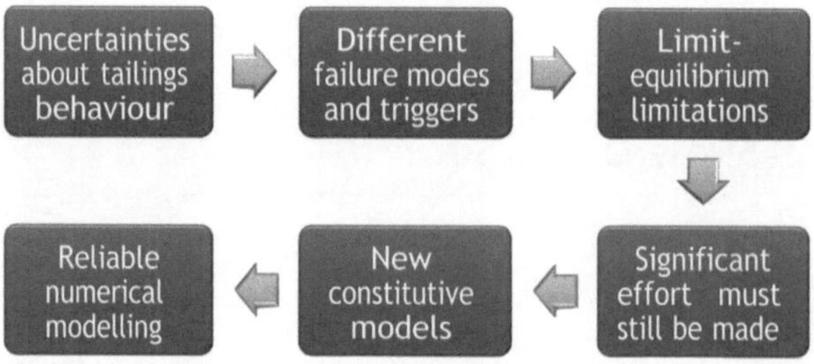
Uncertainties about tailings behaviour
Different failure modes and triggers
Limit-equilibrium limitations
Significant effort must still be made
New constitutive models
Reliable numerical modelling

Technical sessions

Session 1 - Rock mass characterization

The Evolution of Geotech - 25 Years of Innovation – Hammah et al. (Eds)

Towards improved definition of the Hoek-Brown constant m_i for numerical modelling

T.G. Carter
TGCGeoSolutions, Oakville, Ontario, Canada

ABSTRACT: The Hoek-Brown failure criterion, which is used worldwide in rock engineering design, relies on accurate assessment of two key intact rock material parameters – namely UCS_i and m_i, in addition to, for rock masses, definition of rockmass quality, through *GSI*, directly, or via a correlation relationship from *Q* or from *RMR*. Although much has been written regarding appropriate procedures for *GSI* characterization of rockmasses for Hoek-Brown evaluation, there still appears much uncertainty in reliably defining appropriate values of intact m_i. This paper thus looks at a couple of suggestions that may help practitioners improve current procedures for more accurately defining m_i than existing approaches. The two aspects specifically discussed are – (1) inferences to aid m_i estimation utilizing tensile test data, and (2) a methodology for visually assessing m_i based on simple textural characterization of fabric, grain size and interlock, conducted as part of a field core logging or lab. testing program. It is hoped that together, these suggestions will be helpful for improving estimation of m_i values, particularly for rocks with differing competence or age, but of similar character and mineralogy, hence identical names.

1 INTRODUCTION

There has been considerable discussion in the literature, since at least the mid 2000's regarding discrepancies that folk have observed between Hoek-Brown (*H-B*) m_i values determined from curve fitting of laboratory test data (per standardized methods, such as incorporated in the Rocdata® program code), compared with ranges suggested in the various tables introduced since the 1980's by Hoek & Brown (1980a), and culminating with the well-publicized Hoek, 1999 version, shown on the left-hand side of Figure 1. These discrepancies in range extent (between measured and suggested data) are clearly evident in the right-hand diagram, comparing the grey range boxes (based on the Hoek table) with the red box and whisker lines drawn on the basis of measured data, as abstracted by Richards & Read (2011 et seq.) from the Douglas, 2002 database.

2 TOWARDS BETTER DEFINITION OF *mi*

Inspection of the box and whisker range plots in Figure 1, shows that in some rocktype cases there is good agreement, but in many there is wide discrepancy. Obviously, impacts on numerical modelling of potentially predicting incorrect rock behaviour could be quite significant if wrong m_i values were selected for the modelled rocktype. But what can be done to improve this?

Part of the solution lies in just improving geological description adequacy. Utilizing solely single rocktype naming as the basis for definition of m_i values is a significant part of the problem. The other aspect is achieving representative envelope matching, particularly in the tensile region. This paper thus attempts to provide some suggestions for improving both

DOI: 10.1201/9781003188339-13

Table 4: Values for the constant m_i for intact rock,. Note that the values in parenthesis are estimates.

Rock type	Class	Group	Texture: Coarse	Texture: Medium	Texture: Fine	Texture: Very fine
SEDIMENTARY	Clastic		Conglomerate (22)	Sandstone 19	Siltstone 9	Claystone 4
				—— Greywacke (18) ——		
	Non-Clastic	Organic		—— Chalk 7 ——		
				—— Coal (8-21) ——		
		Carbonate	Breccia (20)	Sparitic Limestone (10)	Micritic Limestone 8	
		Chemical		Gypstone 16	Anhydrite 13	
METAMORPHIC	Non Foliated		Marble 9	Hornfels (19)	Quartzite 24	
	Slightly foliated		Migmatite (30)	Amphibolite 25 - 31	Mylonites (6)	
	Foliated*		Gneiss 33	Schists 4 - 8	Phyllites (10)	Slate 9
IGNEOUS	Light		Granite 33		Rhyolite (16)	Obsidian (19)
			Granodiorite (30)		Dacite (17)	
			Diorite (28)		Andesite 19	
	Dark		Gabbro 27	Dolerite (19)	Basalt (17)	
			Norite 22			
	Extrusive pyroclastic type		Agglomerate (20)	Breccia (18)	Tuff (15)	

* These values are for intact rock specimens tested normal to bedding or foliation. The value of m_i will be significantly different if failure occurs along a weakness plane.

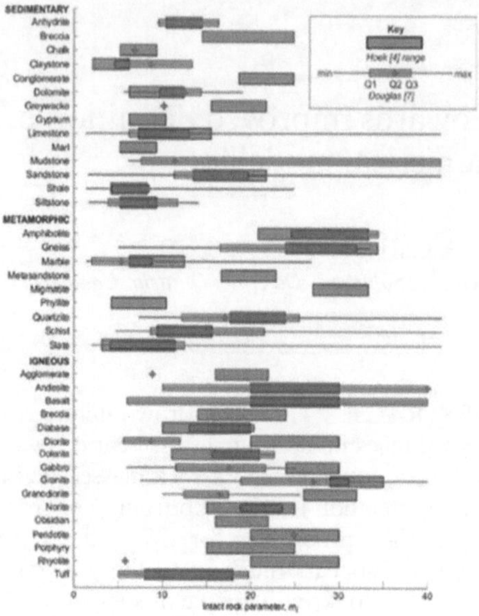

Figure 1. Hoek, 1999 tabulations of suggested m_i ranges and means for 36 different rocktypes alongside Box Plot for 39 rocktypes from Richards & Read, 2011 et seq. illustrating significant range discrepancies.

aspects, with focus later in the paper on approaches to help achieve a step change to more credibly assigning m_i values based on rocktype descriptive techniques when comprehensive laboratory data is not available.

2.1 *Problems of wrong* m_i

In an attempt to get some better clarity, it is instructive to examine the influence that m_i exerts on controlling the degree of curvature of the *H-B* envelope, and also how inclusion of "uncorrected" tensile strength data into the regression curve fits can quite easily throw off reasonable prediction of a representative value of the m_i parameter, even from good laboratory test data.

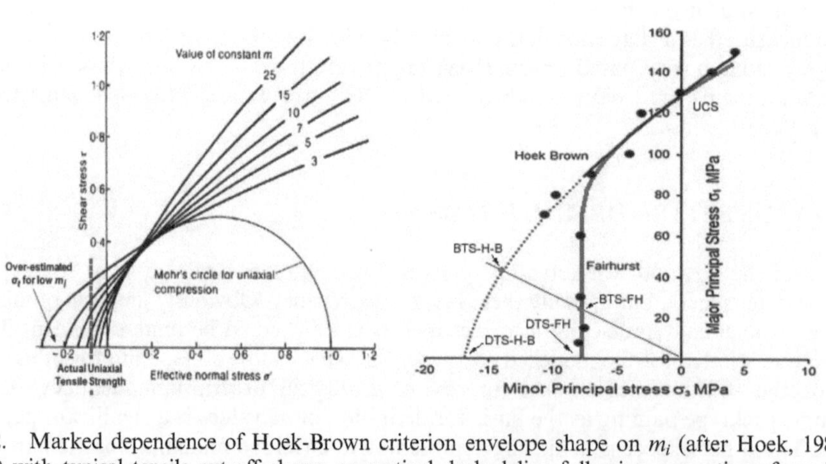

Figure 2. Marked dependence of Hoek-Brown criterion envelope shape on m_i (after Hoek, 1983), (left diagram) with typical tensile cut-off shown as vertical dashed line following suggestions from Hoek & Martin, 2014, based on considerations of wide discrepancy between x-axis intercept of Hoek-Brown and Fairhurst failure criterion envelopes, right diagram, compared with actual data points from detailed triaxial testing undertaken by Ramsey and Chester, 2004, overlaid with *DTS* and *BTS* definition stress states.

The characteristics of the parabolic *H-B* curve fit, (ref. Figure 2 left diagram) results in an increasing mismatch between the pseudo tensile strength (x-axis) intersection point of the *H-B* envelope (labelled *DTS-H-B* in Figure 2 right diagram), with decreasing *mi* values, compared with the true *DTS* value, (as labelled *DTS-FH*). While *DTS* parameter discrepancies detracting from reliable *H-B* use have been a topic of concern, (e.g., Hoek & Martin, 2014, Bewick et al, 2017), due mainly to errors in numerical modelling prediction, this mismatch has also been clouded by uncertainties arising from lack of reliable tensile strength determinations. This is not a new issue, (Fairhurst, 1964), but unfortunately, there are few alternatives available for improving Brazilian test method reliability as an expedient solution, while work continues on alleviating the many problems involved with carrying out true Direct Tension tests. To this end, there may be merit in revisiting use of pressurized cylinder testing (Detournay & Carvalho, 1989, Song et al., 2001, or bar tests, such as have been proposed in the past as alternatives to direct testing, given all the difficulties of machining dog-bone samples in anything other than good rock. Obviously, until a better solution is found, industry will continue to utilize the much cheaper and arguably, if done with correct alignment and thickness disks, the sufficiently adequate Brazilian (*BTS*) test method.

2.2 *Suggested approaches for improving definition of m_i values*

Several practical suggestions of measures to better define m_i have been published over the last decade, starting with the suite of papers by Cai, 2009, 2010, Sari, 2010, and continuing through 2011 to 2015, per Richards & Read, (2011, 2013) and Read & Richards (2012, 2015), with these latter authors essentially giving little credence to utilizing tables of geological values as a reliable approach for defining m_i. Rather, they advocate that m_i be determined only through laboratory testing, suggesting using a uniform *BTS/DTS* correction ratio of 0.9. A contrasting suite of papers, by, for example, Shen & Karakus, 2014, Sabatakakis et al, 2017 and Perras & Diederichs, 2014 have attempted to go back to geological descriptions, the latter authors making the very practical suggestion that using different *BTS/DTS* ratios for different rocktype suites may help improve estimates. Carter, 2019 and Carter & Carvalho, 2020, continue this theme, making some further suggestions for possible improvements, both theoretical and empirical. The thrust of these ideas, as illustrated in Figure 3, identify a possible new path forward towards improving definition of m_i values, by either applying enhanced laboratory data analyses (by improving regression fits utilizing "corrected" *BTS* data to derive equivalent *DTS* values), or, by making quantitative visual hand specimen observations of rock samples, (as discussed in more detail later in this paper).

The suggested adjustment for *DTS/BTS* correction per Carter & Carvalho, 2020 is:

$$DTS/BTS = \{BTS\} + \left[\frac{3}{(m_i + 1)}\right] \qquad (1)$$

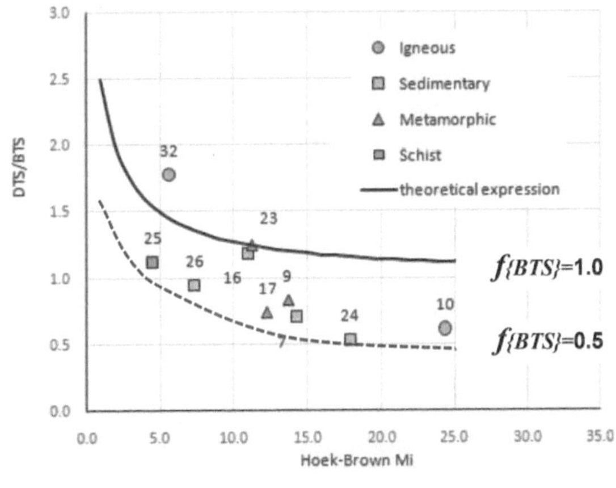

Figure 3. DTS/BTS variability by rocktype, wrt. m_i (overlaid with theoretical and empirical curve fits).

where the asymptote $f\{BTS\}$ varies from 1 = solid line plot for theoretical correction from BTS_{HB} to DTS_{HB} stress states per Figure 2, (right diagram) .. to approximately 0.5 for the dashed line allowing "best-fit" matching to the suite of laboratory test data in the tabulations in the2020 Carter & Carvalho paper.

The trends in the graph in Figure 3 raise questions regarding a uniform $BTS/DTS = 0.9$ ratio ($DTS/BTS = 1.1$) correction, particularly for low m_i values. The data suggests that a single value correction may only be valid for high, $m_i > 15$ rocktypes. At low m_i's, the evident divergence again relates to the same Figure 2 x-axis *DTS(H-B) vs DTS(FH)* mismatch issues. Unfortunately, the suggested Equation (1) cannot resolve this as it is derived directly from the quotient between *DTS-HB,* (from Hoek & Brown, 1980b, p.137, left equation), compared with the *BTS-HB* case (right equation). (Note unresolved discrepancy also exists in *BTS* positions between the criteria).

$$\sigma_{t(DTS-HB)} = \frac{\sigma_c}{2}\left[m_i - \sqrt{m_i^2 + 4}\right] \ldots and \ldots \sigma_{t(BTS-HB)} = \frac{\sigma_c}{32}\left[m_i - \sqrt{m_i^2 + 64}\right] \qquad \text{(2a \& b)}$$

While the approach using a 0.5 factor, introduced in Figure 3 (from Carter & Carvalho, 2020) provides a relatively simple way of estimating the *DTS(H-B)* pseudo tensile strength intercept, it still does not address the fundamental problem of defining this point based on actual measured tensile strength data. Some progress towards achieving this can however be made by examining the expression proposed by Brace in the 1960's between R^* defining the ratio between the crack initiation stress (σ_{CI}), the tensile strength (σ_t) and $\mu = \tan\phi$ (the coefficient of internal friction). As outlined in Hoek, 1965, Brace proposed the left expression below, based on earlier work by McClintock and Walsh (1962) following on from Griffiths' original 1921 publication, where R^* is introduced to distinguish this $\sigma_{CI}/|\sigma t|$ expression from the more common R ratio = $UCS/|\sigma t|$ (per Sheorey, 1997). The inverse expression then allows direct definition of $\phi = \tan^{-1}\mu$, via $R^* = 4/\xi$.

$$R^*(\sigma_{CI}/|\sigma_t|) = \frac{4}{\left(\sqrt{1+\mu^2} - \mu\right)} \cdots \text{giving the obverse } \mu = \frac{1-\xi}{2\xi} \qquad \text{(3a \& b)}$$

... Where $\xi = 4/R^* = \tan(|45^\circ - \phi/2|)$ defines the crack propagation geometry.

Although the exact crack controlling condition equivalent to the stress state where onset of shear occurs and friction is mobilized could range from about $0.4\sigma_C$· per Martin, 1997 to $0.48\sigma_C$

per Zhao et al, 2015, for the purpose of this attempt to develop a correction targeted solely at better defining the *DTS(H-B)* point per Figure 2, onset is assumed consistent with initiation of spalling, per the envelope defined by Diederichs et al, 2007, Carter et al., 2008, as approximated with a *H-B* envelope with exponent $a = 0.25$, such that back-calculation of ϕ, either from R^* via

ξ (as above) or directly from *mi* (as below) becomes feasible, allowing the ratio between the H-B intercept, *DTS(H-B)* and the actual tensile strength, *DTS(FH)*, to be expressed as follows:

$$DTS_{HB}/DTS_{FH} = \frac{2}{\mu} \cdots \text{ where } \mu \text{ can be derived from } \xi, \text{ or from } \mu = \frac{m_i\sqrt{1/3}}{3\sqrt{2}} \qquad \text{(4a \& b)}$$

These final semi-empirical relationships build on the Mohr circle observations from the 1960's that the equivalent intercept for the derived ϕ, is $c = 2\cdot|\sigma_t|$. Noting this, plus adopting the notion that $\sigma_{CI} = 0.4\cdot\sigma_C$ ensures consistency of these expressions with the desired tangent matching of the Griffith-Fairhurst envelope with the spalling Hoek-Brown curve (with $a \approx 0.25$), at crack initiation.

2.3 m_i = *frictional parameter in H-B equation*

It has long been argued that the *H-B* slope fit parameter, m_i is analogous to the Mohr-Coulomb friction angle ϕ. Equation *4b* above now provides some direct linkage, albeit based on 1920's and 1960's observations of crack initiation, which might not be applicable across the full range of m_i as these early observations were primarily made only on high *mi* rocktypes, Hoek, 2020. Nevertheless, it appears that the behaviour of the linkage function is at least sympathetic with other observed trends when moving to the weaker, lower strength, lower m_i rocktypes. Thus, taken together, these above relationships, perhaps will now allow raw *BTS* data (or indeed, if available, raw *DTS* test data also) to be "corrected" to yield representative pseudo *DTS(H-B)* point values of sufficient accuracy as to help properly constrain the regression curve fitting process.

It is hoped that application of the above "correction" methodology will help practitioners better constrain their regression curve fits rather than (a) ignoring use of any tensile data or (b) forcing fits with un-realistically steep regressions involving high m_i values.

This advance however only tackles the first of the two problems identified in §2. The second is less tractable – sorting out why such wide discrepancies in m_i values and ranges are being encountered for seemingly exactly the same rocktype, compared with published information.

3 IMPROVED GEOLOGICAL DEFINITION

The two compounding factors (in addition to the issues discussed above) that are most likely responsible for most of the uncertainty in the spreads of reported m_i values are (a) inaccurate test repeatability across different geological regimes, due to use only of single rocktype naming; and (b) inappropriate grouping of m_i data, due to lack of definition of material competence variability when solely retaining a single rocktype name. These two factors complicate interpretation and also cloud resolution of the discrepancies and mismatches identified in Figure 1.

The solution is remarkably simple, but requires a step change in how practitioners and how laboratories report m_i value determinations. Better, and more quantitative sample description is realistically all that is needed when reporting m_i results. Serious consideration however needs to be given to material competency, per Figure 4, in addition to just defining a rock name.

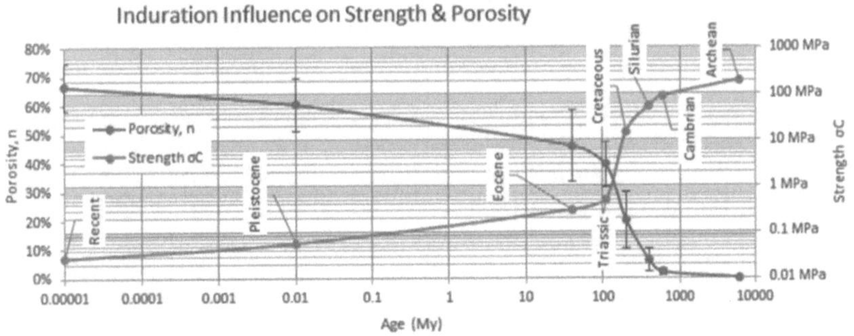

Figure 4. Characteristic Change in Material Properties – particularly in Strength and Porosity, as a result of lithification, induration and metamorphism – highlighted for many rocks of different geological age, with identical rock names, showing major differences in m_i due to induration/cementation differences.

Although the diagram in Figure 4 is conceptual, it illustrates the classic commonplace problem of mixed data in m_i summary tabulations, viz., a young (60M year old) Eocene sandstone, for example, would not be expected to exhibit the same degree of induration, competence, and strength as a 400My old Devonian sandstone – yet they would both be called "sandstone" and, using the standard Hoek, 1999 table would be given the same m_i. In this regard, of particular significance to the preceding discussion is the dramatic change that occurs at ±100M years – marking the transition when true tensile bonding develops, when engineering soils become actual rocks. Only after this point in the geological time-history can real tensile spalling manifest itself. A Mohr-Coulomb straight line strength relationship, which has exactly the same principal stress equation as the curved Hoek-Brown envelope, but with no square root term, ie., a=1, as demonstrated by Carvalho et al, 2007, constitutes the most appropriate criterion for characterizing most everything up to this onset point of significant tensile strength development. As such, it also remains applicable for most weathered rocks and highly altered rock materials. Towards the righthand end of the chart, by contrast, extreme rock brittleness generally now dominates. Here again divergence from a normal *H-B* relationship is indicated, as most extremely competent rocks are best described with the much more curved Griffiths/Fairhurst envelope or with the spalling *H-B* expression, with a=0.25 per Diederichs et al., 2007. m_i can thus be expected to increase with geological age and induration, and thus it should not be unexpected that its value can change for the same named rock; with almost the exact reverse happening with weathering and degradation. Improving quantitative sample definition and naming can help in reducing confusion due to this spread in characteristics for the same rock name, and some guidelines have already been proposed in Carter, 2019, basically as an alternative to simply updating the Hoek, 1999 table of m_i estimates and developing a more comprehensive chart than Figure 1 per Richards and Read, 2011 et seq. A basic procedure to physically estimate *mi* values from quantitative hand specimen identification has also been proposed involving visually defining five key parameters to better describe specific rock samples. The approach has already undergone some preliminary testing against known laboratory data, as discussed in Ganye et al, 2020, with promising results.

In concept the approach that has been proposed utilizes the power relationship $m_i = 1.2 \cdot R^{0.8}$ postulated by Carter & Marinos, 2020 as the central trend axis, rather than the linear expression of Hoek & Brown, 2018. Variability about the mean central fit relationship between H-B m_i and Sheorey's $R = \sigma_C/|\sigma_T|$ is then dictated for each of the five key parameters. Four of the controlling parameters are considered within the first bracket, named the Granularity term {G} and the last of the five parameters is included as a trend modifier in the exponent term, as follows:

$$m_i \cong f\left(\left(\frac{\sqrt{d}, \xi}{\psi, \varphi^g}\right) \cdot (\sigma_c/|\sigma_T|)^{-\log k_{G\phi}}\right) \quad (5)$$

The logic in formulating this expression, is that for a very wide range of rocktypes:

.. an ***increased*** numerical value of m_i would be reflected in an increase in:

d – *the rock grain (crystal) particle size* (∝ tensile strength, $|\sigma_T|$, which, per influence of the Hall-Petch effect, would scale with), (Hall, 1951, Petch, 1953, Morris, 2001);

$k_{G\phi}$ – *the degree of mixing of grain sizes* (defined by the graphic kurtosis of the grain distribution spread, per Folk & Ward, 1957); with values ≪1 reflecting an increasingly mixed distribution of grain sizes and greatest departure from uniformity in grain dimensions, and/or in;

ξ – *the degree of interlock and angularity* of the crystal grains (typically with greater influence where rocks are porphyritic or pegmatitic, with large crystals embedded in a matrix of smaller and intermediate sized grains, or where rocks have widely-graded mixtures of grain sizes);

.. and a ***decreased*** numerical value of *mi* would result from any increases in:

ψ – *any anisotropic or other similar pervasive fabric weakness*, such as foliation or layering, which characteristically is significantly weaker and more deformable across the "grain" of the rock fabric rather than parallel with the fabric; and/or. . .in

φ – *the porosity of the rock* – characterized by void ratio, but expressed as percentage "open" voids, e.g., vugginess, vesicular character, solutioning etc., and usually scaling as a $^3/_2$

to $^{5}/2$ power expression following Archie's Law, as established from resistivity profiling of porous oil reservoir rocks, typically sandstones, (Glover, 2016).

Experience to date (as presented in Ganye et al, 2020), has shown it relatively straightforward to apply the visual guidelines given in the 2019 paper for determining the Granularity term {G} based on defining the modal grain size, d, the interlock parameter, ξ, the degree of anisotropy, ψ and the approximate porosity, φ. Some problems have however been encountered with reliably being able to adequately characterize kurtosis, not just solely on a visual basis, but also at different scales. This disconnect largely arises because perception of uniformity in a rock specimen differs whether observing an outcrop, a hand specimen, or a microscope slide.

Accordingly, two modifications are suggested to the procedures outlined in Carter, 2019. The first modification involves application of a simple area-based scaling relationship to the $k_{G\emptyset}$ term, that varies depending on the dimensions of the observation base, benchmarked to a typical 150mm hand-specimen scale lump sample, as originally considered for the k_{Go} term formulation. The second improvement involves modification of the exponent log term to allow visual description using the Sorting Coefficient, S_o (the graphic standard deviation) but maintaining numerical equivalence with the original $k_{G\emptyset}$ term definitions, viz: replacing k_{Go} with ½ $(S^{-4/\pi} + 1)$. S_o in this expression can be very easily described visually based on the grain size distribution appearance of the sample, when properly benchmarked with respect to a standard reference, such as the classic suite of diagrams included in Figure 5 below, (as first published by Compton, 1962, but now widely reproduced even as clear plastic transparent field scale guides). Use of this diagrammatic benchmarking approach will hopefully help practitioners and labs better and more easily establish a visual estimate of grain sorting, equivalent to the graphic standard deviation values listed in Figure 1 of the 2019 paper, thus providing the necessary link of such S_o estimates to the kurtosis $k_{G\emptyset}$ term, based on equating the graphic grain size definitions proposed by Folk and Ward, 1957.

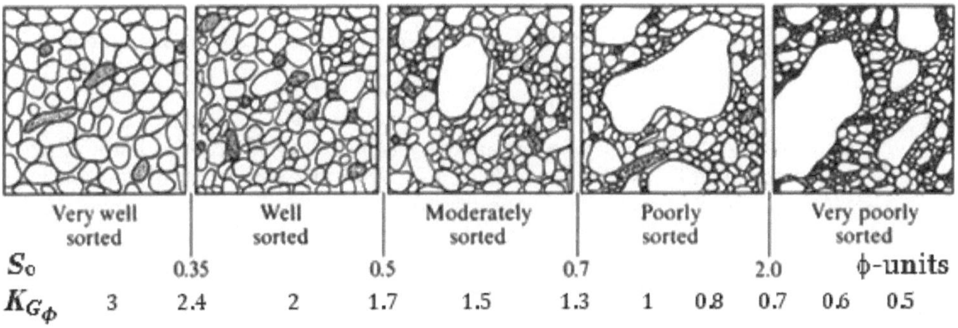

Figure 5. Typical Graphic Kurtosis & standard deviation (S_o) values for different Degrees of Sorting.

Considering the first improvement, it should be appreciated that $k_{G\phi}$ was originally defined based on assessing the appearance of a typical 4-6" (100-150mm) hand-specimen sized lump rock sample (or a similar length piece of N or H sized core), with typical rock crystal grain sizes in the range from 0.3-1mm and smaller for fine grained rocks to around 8-10mm and upwards for coarse grained rocks. Thus, each of the diagrams in Figure 5 could be considered as a 1" x 1" (25mm x 25mm) square snapshot of the side of such rock samples. The "well sorted" rock example shown in the leftmost picture (exhibiting a numerically low value of standard deviation and a high value of kurtosis) would be medium grained (with 8-10 crystals per 25mm side = 2-3mm grain size), whereas the "poorly sorted" rock at the right end of the diagrams (with wide standard deviation and low kurtosis values, consistent with mixed crystal sizes) would be almost pegmatitic, with very coarse crystals, > 20mm in length set in a fine (millimetre to sub-millimetre) groundmass.

Taking a 3-5mm circular "cut out" of each of these diagrams could now be considered as typical replication of the situation where a petrological or binocular microscope or computer equivalent were being used to examine the same rock samples, but at much greater magnification. Obviously appreciably less objectivity on the variability of grain size within the whole specimen is possible from just the nominal 3mm window, giving a quite different view of uniformity or degree of mixing than the larger scale perspective. At a very large, say metre scale, even quite mixed grainsize rocks at hand specimen size would look more uniform, thus higher kurtosis values and lower sorting coefficients would most likely be defined by any observer. By contrast, at small scale, under a hand lens or under the microscope, even the most uniform rocks will tend to look to contain "mixed" sizes of crystals, so lower $k_{G\phi}$'s and higher S_o's characteristically would tend to be defined. Figure 6 plots a suggested correction relationship to rectify this scaling mismatch.

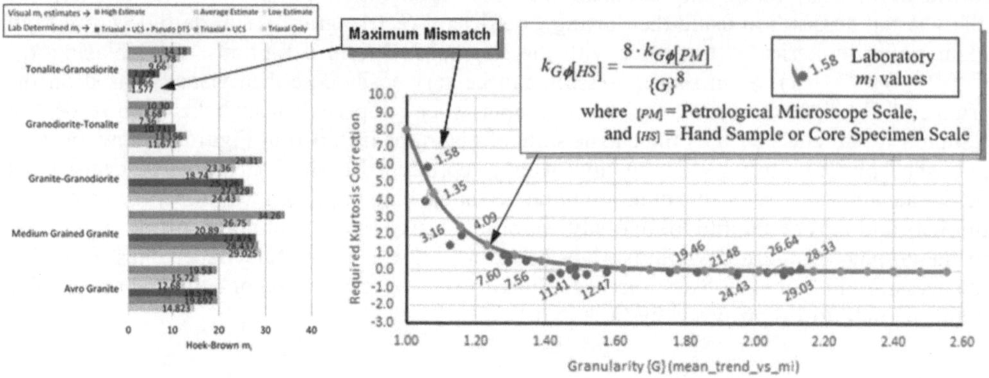

Figure 6. Proposed Kurtosis Scaling Correction, based on matching Visual m_i's to Laboratory Results.

The graph included in Figure 6 is based on an analysis of a suite of visual and laboratory m_i determinations undertaken by Ganye et al., 2020, with additional more recent data included. The y-axis plots the ratio between the reported visually-estimated $k_{G\phi\ [PM]}$ kurtosis values derived from detailed examination at laboratory/microscope scale, versus the $k_{G\phi\ [HS]}$ hand specimen (lump or core sample) kurtosis values that would be needed to adjust the basic equation input parameters to exactly match the laboratory determined m_i data. The previously reported visual m_i datapoints are plotted in the brown and orange histograms in the left diagram, with the laboratory m_i test data points plotted in the blue histograms, and also labelled on the right-hand graph.

As can be seen from the red curve, the inset equation in Figure 6 achieves >90% matching to the currently available data, thus is suggested as a reasonable correction expression so that either small (hand lens or microscope scale) or large (lump or core sample scale) observations can be utilized to determine kurtosis, either per the original $k_{G\phi}$ methodology or by defining S_o per the chart in Figure 5, then adopting the following expression to determine the m_i estimate:

$$m_i = |\{G\} \cdot R^{\left(0.8 - \log\left(1/2\left(S^{-4\pi/0} + 1\right)\right)\right)} \tag{6}$$

It should be appreciated that this revised exponent equation, with S_o incorporated, has been set up to numerically match the original ($k_{G\phi}$) term, so should yield essentially the same final m_i estimates, but hopefully more easily than previously. In the interests of brevity, however, further description of the field application of this suggested quantitative characterization methodology to improve definition of m_i values and also sample naming approaches is not presented here. Rather, the reader is referred to the original 2019 paper and to the references that follow for more detailed discussion of the suggested approaches to describe each of the key parameters, d, ξ, ψ and φ, and also the previously defined (k_{Go}) descriptor term.

REFERENCES

Bewick. R., A. Ouellet, S. Otto, and D. Gaudreau (2017). Importance of Understanding Laboratory Strength and Modulus Testing Data for Deep Mining in Hard Brittle Rocks. Proceedings of the 8th International Conference on Deep and High Stress Mining, Perth, Australia, pp. 827–842.

Australian Centre for Geomechanics: Perth, WA, Australia Brace, W.F., (1960), An extension of the Griffith theory of fracture to rocks. Journal of Geophysical Research, vol. 65, pp. 3477–3480, DOI:10.1029/JZ065i010p03477

Cai, M., (2009) A simple method to estimate tensile strength and Hoek-Brown strength parameter mi of brittle rocks. Proc. 3rd CAN-US Rock Mech. Symp., Toronto, May 2009 (Ed: M. Diederichs and G. Grasselli) 12pp.

Cai, M., (2010) Practical Estimates of Tensile Strength and Hoek-Brown Strength Parameter mi of Brittle Rocks. Rock Mechanics and Rock Engineering Vol. 43, No. 2, pp 167–184

Carter, T.G. (2019). A suggested visual approach for estimating Hoek-Brown mi for different rock types. In proc. 14th ISRM Congress on Rock Mech and Rock Eng, Paper # 14356 13pp., Foz do Iguassu, Brazil.

Carter, T.G. & Carvalho, J.L. (2020). Suggested tensile test data interpretation for estimating Hoek-Brown mi. Proc. 54th US Rock Mechanics Conf, Paper ARMA 20–1860, Golden, Colorado. 11pp

Carter T.G. & Marinos, V., (2020). Putting Geological Focus back into Rock Engineering Design. Rock Mechanics & Rock Eng. V 53, pp.4487–4508. https://doi.org/10.1007/s00603-020-02177-1

Carter, T.G., Diederichs, M.S., Carvalho, J.L., (2008). Application of modified Hoek-Brown transition relationships for assessing strength and post yield behaviour at both ends of the rock competence scale. In: Proc. 6th International Symposium on Ground Support in Mining and Civil Engineering Construction, 30 March–3April. Cape Town, South Africa. Journal of the Southern African Institute of Mining and Metallurgy, v108., pp 325–338

Carvalho, J.L., Carter, T.G., & Diederichs, M.S. (2007) An approach for prediction of strength and post yield behaviour for rock masses of low intact strength. Proc.1st Can-US Rock Symp. Vancouver. pp. 249–257

Compton, R.R., (1962), Field Sorting Chart, by Geo-Supplies Ltd, UK, as plotted for University of Leicester, derived, from original drawing on p.214 of Manual of Field Geology, Wiley, New York, 378pp,

Diederichs, M.S, Carvalho, J.L., & Carter, T.G. (2007). A modified approach for prediction of strength and post yield behaviour for high GSI rockmasses in strong, brittle ground. Proc. 1st Can-US Rock Symposium. Meeting Society's Challenges and Demands. June, Vancouver. pp.277–28

Detournay E., & Carvalho J.L., (1989) Application of the Pressurized Hollow Poroelastic Cylinder Solution to the Interpretation of Laboratory Burst Experiments. Proc. 30th U.S. Symposium - Rock Mechanics as a Guide for Efficient Utilization of Natural Resources, Khair (ed.) West Virginia, 19-22.06.1989 © Balkema, Rotterdam. ISBN 90 6191 871, pp.377–383

Douglas, K.J. (2002). The shear strength of rock masses. (Unpublished Doctoral thesis). University of New South Wales, Sydney, Australia., 516pp.

Fairhurst, C. (1964). On the validity of the "Brazilian" test for brittle materials. International Journal of Rock Mechanics and Mining Sciences & Geomechanics Abstracts, 1(4),pp.515–546.

Folk, R.L, & Ward, W.C., (1957) Brazos River Bar: A Study in the Significance of Grain Size Parameters. Journal of Sedimentary v27, pp3–26

Ganye, J., Vasileiou, A. Perras, M.A & Carter, T.G., (2020) Influence of grain size and interlock on intact rock properties. Proc. 54th US Rock Mechanics Conf, Golden, Colorado. Paper ARMA 20–2097,7pp

Glover, P.W.J. (2016) Archie's Law - a reappraisal. Solid Earth, 7, pp1157–1169

Griffith AA., (1921). The phenomena of rupture and flow in solids. The Philosophical Transactions of the Royal Society London (Series A) V221: pp163–198.

Hall, E.O., (1951). The deformation and ageing of mild steel: III Discussion of results. Phys. Soc. B v.64, pp.747–753.

Hoek, E. (1965) Modified Griffith Criterion §4.3 in Rock fracture under static stress conditions. PhD thesis, Univ. of Cape Town, S. Africa. https://www.rocscience.com/documents/hoek/references/H1965.pdf
Hoek, E. (1983) Strength of jointed rock masses. 1983 Rankine Lecture, Géotechnique. Vol.33(3): pp.187–223. doi:10.1680/geot.1983.33.3.187.
Hoek E., (1999) Putting Numbers to Geology - an Engineer's Viewpoint. QJEG v.32,1, pp.1-19(19)
Hoek, (2020) – pers comm & 35pp of unpublished notes on "Re-examination of triaxial testing of rock and the interpretation of triaxial test data", dated Dec. 2016
Hoek, E, & Brown E.T (1980a) Empirical strength criterion for rock masses. Journal of the Geotechnical Engineering Division, ASCE v. 106, n. GT9, p. 1013–1035.
Hoek, E., & Brown, E.T. (1980b). Underground Excavations in Rock, Inst. of Mining & Metallurgy, London. 527pp.
Hoek E, & Brown ET, (2018) The Hoek-Brown failure criterion and GSI – 2018 edition, Journal of Rock Mechanics and Geotechnical Engineering, 22pp doi: 10.1016/j.jrmge.2018.08.001
Hoek, E. & Martin, C.D. (2014). Fracture initiation and propagation in intact rock – A review. Journal of Rock Mechanics and Geotechnical Engineering, 6, 287–300.
Martin C.D. (1997) The effect of cohesion loss and stress path on brittle rock strength. 17th Geotechnical Colloqium, 1993, Canadian Geotechnical Journal; v34 (5): pp698–725.
McClintock F.A and Walsh, J.B. (1962) Friction on Griffith Cracks in rock under pressure. Proc. Int Cong. Appl Mech., Berkeley
Morris, J.W., Jr., (2001) The Influence of Grain Size on the Mechanical Properties of Steel, Proceedings, International Symposium on Ultrafine Grained Steels, S. Takaki and T. Maki, eds., Iron and Steel Inst. Japan, Tokyo, pp.34–41; also Lawrence Berkley Labs Report Number: LBNL-47875
Perras, MA. and Diederichs, MS. (2014). A review of the tensile strength of rock: concepts and testing. Geotechnical and Geological Engineering v32(2) pp.525–546. HTTPS://DOI.ORG/10.1007/s10706-014-9732-0
Petch, N.J., (1953). The cleavage strength of polycrystals. J. Iron Steel Institute 174, pp.25–28.
Ramsey, J.M, & Chester, F.M. (2004) Hybrid fracture and the transition from extension fracture to shear fracture. Nature V 428: pp.63–66.
Read, S. A. L., & Richards, L. (2012). A comparative study of mi, the Hoek-Brown constant for intact rock material. Proc., 12th ISRM Congr, Int. Soc. for Rock Mechanics, Beijing, China, Eds. Qian, Zhou, Paper 139, pp. 805–810.
Read, S. A. L., & Richards, L. R., (2015) Guidelines for use of Tensile Data in the Calculation of the Hoek-Brown Constant mi. Proc. 13th ISRM Int. Cong. Rock Mech., Montreal, 13pp.
Richards, L. R. & Read, S.A.L. (2011). A comparison of methods for determining mi the Hoek-Brown parameter for intact rock material. In Proceedings 45th US Rock Mechs/Geomechanics Symposium, San Francisco, USA, 26-29 June 2011, eds. A. Iannacchione et al., Paper ARMA/USRMS 11-246.
Richards, L. R. & Read, S.A.L. (2013). Estimation of Hoek-Brown parameter mi using Brazilian tensile test. Proc 47th US Rock Mechanics/Geomechanics Symposium, San Francisco, USA 23-26 June, eds, A. Iannacchione et al. (Eds.), Paper ARMA/USRMS 13-465 14pp
Sabatakakis, N., Tsiambaos, G., Ktena, S., & Bouboukas, S. (2017) The effect of microstructure on mi strength parameter variation of common rock types Bull. Eng. Geol. & the Environment, 16pp DOI 10.1007/s10064-017-1059-7.
Sari, M. (2010). A Simple Approximation to Estimate the Hoek-Brown Parameter mi for Intact Rocks. Proc. ISRM International Symposium - EUROCK 2010, 15–18 June, Lausanne, Switzerland. pp.169–172.
Shen J, & Karakus, M (2014) Simplified method for estimating the Hoek-Brown constant for intact rocks. J Geotech Geoenviron Eng 140 (6): 040140258pp. DOI:10.1061/(ASCE) GT.1943-5606.0001116.
Sheorey, P.R. (1997). Empirical rock failure criteria. 1st edition. 200 pp, Balkema, Rotterdam
Song I., Suh, M., Sik, K. & Haimson, B. (2001) A laboratory study of Hydraulic Fracturing Breakdown Pressure in Table Rock Sandstone, Geosciences Journal Vol. 5, No. 3, p. 263271, September 10pp
Zhao, X.G., Cai, M., Wang J. (2015) Objective Determination of Crack Initiation Stress of Brittle Rocks Under Compression Using AE Measurement. Rock Mechanics and Rock Engineering. V 48, pp.2473–2484

The Evolution of Geotech - 25 Years of Innovation – Hammah et al. (Eds)

Study of the effect on stability of slope variations and reserve estimation in the Kaolin mining quarry design

I.A. Chusna, N.R. Mulyaputra, V. Vergiagara & H. Suharyadi
Mining Engineering Department, UPN "Veteran" Yogyakarta, Indonesia

ABSTRACT: In the Semin District and its surroundings, Kaolin is generally formed from the hydrothermal alteration in dacitic igneous rocks and tuff breccias. The potential of kaolin natural resources in these locations is very high when exploited to support various life necessities. Reserve estimation is an important work in evaluating a mining project for guidance on Kaolin procedures' exploitation. It is an estimation of the quality and quantity of the presence of minerals that have economic value. The Reserve calculation method plays an important role in determining the efficiency in the commercial exploration of a deposit. Determining the reserve estimate requires the proposed geometry quarry mining design. Geotechnical analysis of a prospective kaolin quarry was carried out to determine the proposed slope design's safety factor and its susceptibility to failure. The geotechnical properties, such as unit weight and uniaxial compression strength parameters, were determined. The safety factor was calculated using the Limit Equilibrium Method under static and pseudo-static loading conditions with deterministic and probabilistic sensitivity analysis. This paper presents a case study in which the effects of variations of the single and overall slope angle on the safety factor (SF) and kaolin quarry reserve estimation were analyzed. A total of 14 pits were generated by each studied final pit definition methodology, and each pit had sections with a varied single slope from 20° to 80° angles with an overall height is 85 m analyzed in the stability assessment. The recommendation of Overall slope limit in kaolin is 52° with overall height of 85 meters or single slope 70° and bench height of 10 meters will produce the excavation material taken by 584,399 BCM.

1 INTRODUCTION

Minerals that are available in abundance and a variety of different uses, and the characteristics that are determined based on the mineral structure and composition are a description of clay minerals (Grim, 1950). Clay minerals are a collection of naturally occurring very small crystalline particles that belong to argillaceous materials and have earthy properties (Grim, 1962). There are various mineral clays, such as kaolin, palygorskite-sepiolite or hormites, smectite, chlorite, illite, and clay layered. (Martin-Vivaldi and Robertson, 1971; and Murray, 2000). Kaolin is a white or nearly white hydrous aluminum silicate rock which is composed of clay material that has low iron content (Clarence and Paul, 1930). There are two types of kaolin in the nature, primary kaolin and secondary kaolin. Primary kaolin is formed due to the alteration of crystalline rocks, while secondary kaolin is formed due to the sedimentary processes of primary kaolin (Prasad et al., 1990). Kaolin in the District of Semin and its surroundings is kaolin that occurs due to hydrothermal alteration in dacitic igneous rocks and tuff breccias or this process is often referred to as kaolinization. This phenomenon is formed due to a solution of heat from the remaining magma that rises to the surface through cracks or gaps in dacite rocks so that it changes the mineral feldspar becomes kaolinite (Hidayah, 2005).

DOI: 10.1201/9781003188339-14

Artisanal mining for kaolin commodity in the District of Semin has not conducted a technical study on slope design based on safety factors. Design and safety factors on slope stability are closely related to the mining activities since they can affect the amount of kaolin commodity that can be taken. In order to obtain optimal safety factor and maximum reserve, it is necessary to correlate these 2 parameters. Safety factors are used in determining slope stability because safety has become the most common method for determining slope designs, both in rock and soil. Slope stability includes activities related to costs, both direct costs in the form of stabilizing slopes, as well as indirect costs arising from handling failures. (Wyllie and Mah, 2005). This study was conducted to design a slope for kaolin which is classified as a weak category (Hawkins, 1998) in order to produce an optimal slope design and able to meet the safety criteria based on The Ministerial Decree Number 1827K/ 30/ MEM/ 2018 concerning Guidelines for Implementing Good Mining Engineering Principles, and obtaining maximum reserves. This regulation is a regulation that regulates the implementation of mining technically which includes the minimum safety factor and the probability of failure (PoF).

2 GEOLOGICAL OVERVIEW

This study was conducted in Jetak, Karangsari, Semin, Gunungkidul. The area of this study is included in the physiographic sub-zone of the Baturagung Mountains. The formation of the area of this study was the Wonosari-Punung Formation since the Wonosari Formation is directly adjacent to the Punung Formation which is located in the Southern Mountains of Eastern part (Surono et al., 1992). According to Bemmelen (1949), in the middle of the Pleistocene, the southern mountainous depositional basin experienced an uplift that formed the Java Geanticline. During the lifting of the jiwo hills, the zone of the geanticline was broken. The normal fault with the northern block of the geanticline sliding normally with a slope to the north which is located between the jiwo hills and the southern mountains. Based on the visual observations in research location, the kaolin commodity shows a grayish white color with a combination of red, wide spread, and has a fine texture. Another condition is the kaolin minerals at the top have soft properties, and become harder at the bottom. There are many folds in the rock mass (Figure 1) so that the rock mass has a disturbed/ seamy structure, while the solid surface condition is fair. The Geological Strength Index (GSI) value obtained from this description is 35.

Excavation activities carried out without a study of the stability of the planned slope dimensions can result in an ignorance of the slope dimensions that are categorized as stable and this can lead to failure potentials if digging exceeds these limits. Failures can cause losses, even death. Although the failure at the site (Figure 2) are relatively small in scale, the labors must be aware of and require good and safe mining guidance for labors. At this time, the slope geometry due to the mining activities has an average height of 10 meters with a single slope of

Figure 1. Kaolin rock mass structure.

Figure 2. Failure trace at the mining site.

Figure 3. Kaolin mining activities in Karangsari.

80o to 90o. The community carries out mining activities manually with crowbars, hammers, and hoes as rock unloading tools, while the tools used to load onto trucks are bamboo baskets (Figure 3). Each day, the average trucks in a day that can load chunks of kaolin is fourteen trucks. The kaolin in Semin is estimated at 750,000 tons (Widodo, 1995).

3 RESEARCH METHODOLOGY

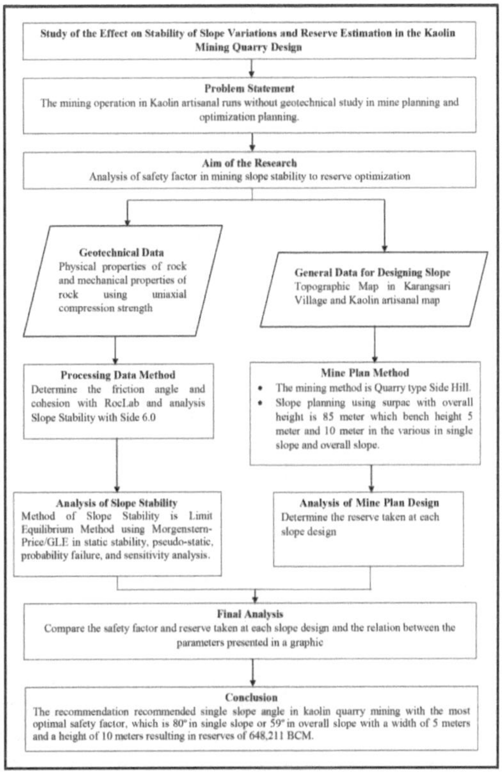

Figure 4. Research method.

Geotechnical data used as rock mass properties are listed in Table 1. The data testing from UCS needs to be approached by using parameters regarding rock mass so that the friction angle and cohesion values will be obtained through the RocLab software (Table 1). The data was used as input parameters in calculating the slope stability using Slide 6.0 software. The input seismic factor value is 0.05 based on the literature on the Indonesian seismic map. Meanwhile, the probability analysis was performed by determining the average value (μ), standard deviation (σ), and coefficient of variation. The method for slope stability was Morgenstern-Price which takes into account the amount of moment balance, force balance, and interslice force, however, this method has spatial or temporal uncertainty related to the physical and mechanical properties of the constitutive material of neglected slopes. Therefore, it is necessary to assess the reliability derived from the value of the safety factor, failure probability, and the reliability index. The level of performance of the calculation of the safety factor can be considered good if the reliability index (β) is greater than 3 (US Army Corps of Engineers, 1997). The minimum static safety factor value is 1.3, while the minimum dynamic safety factor is 1.1 (MEMR 1827K, 2018). In this study, the Monte-Carlo simulation was only considered based on the distribution of FK static and pseudo-static FK values.

Table 1. Results of uniaxial compression strength testing on intact rock.

Parameter	Result
Uniaxial Compression Strength (MPa)	3.845
Modulus Elasticity (MPa)	0.175
Nisbah Poisson	0.138
Elasticity Limit (MPa)	2.880
Unit Weigth (kN/m3)	23.86
Cohesion (MPa)	0.075
Inner Friction Angle (o)	13.920

Meanwhile, the general data was also collected through observations of surrounding conditions, such as lake locations, rivers, ex-mining areas, active mining areas, and observation of slope geometry. Determination of the ground water level on the slopes is carried out through interpolation at Semin Lake with the river at the mining site. The determination of the ground water level on the slopes was carried out through interpolation at the Semin Lake with the river at the mining site. The slope design calculated in this study uses a single slope angle of 20 degrees to 80 degrees so that the overall slope value will adjust to the single slope being modelled. On a 5- meter bench height, the benches produced is 17, while a 10-meter bench height will produce 9 benches with the bench at the pit bottom having a height of 5 meters. From the slope modelling through Surpac, it was obtained the volume of reserves that were taken based on the design. The reserve value and safety factor of each design was compared to obtain the optimal safety factor and the maximum reserve taken.

4 RESULT AND DISCUSSION

The benches on the overall slope depends on the bench height and the contours of the area. On a 5 meter bench height, 17 benches form on a single slope of 20° to 50°. Meanwhile, on a 10 meter bench height, 9 benches are formed on a single slope of 20° to 30°, while a single slope of 40° to 50° forms 8 benches. In the design of a 5 meter bench height slope with a single slope of 60° to 90° and a bench height of 10 meters with a single slope of 60° to 80°, the contour plays a dominant role in slope modelling so that the geometry formed adjusts to the contours of the mining area (Figure 7, Figure 8, Figure 11, Figure 12).

4.1 *Slope safety factor analysis*

The computational results on the 5 meter bench height model are listed in Table 2. Reserves taken and the pseudo-static safety factor have increased with the increase in single slope or the overall slope used (Figure 6). Nevertheless, the static safety factor decreased with the maximum safety factor at a single slope of 70° (Figure 5). The Kh 0.05 loading condition caused a decrease in the value of the safety factor compared to the static loading condition. On a 5 meter bench height with a single slope of 20° to 70°, the static safety factor has increased due to the effect of load shedding from excavation which is indicated by the largest possible value of the smallest safety factor for the model located in the slip plane (Figure 7). Similar conditions also occur on bench height with a single slope of 20° to 50° (Figure 11). Meanwhile, on a single slope of 80° to 90°, the static safety factor has decreased because there is an area on the slope that is steep and has a large shear load so that the lowest safety factor value in the slope

Table 2. Tabulation of slope model computation results on 5 meters bench height.

Single slope (o)	Overall Slope (o)	SF Static	SF Pseudostatic	Reserve (BCM)
20	16	1.504	1.171	111,441
30	21	1.544	1.192	253,745
40	26	1.567	1.203	345,796
50	31	1.569	1.203	386,385
60	35	1.600	1.215	461,421
70	39	1.616	1.221	504,144
80	44	1.465	1.226	542,756
90	49	1.352	1.231	552,586

model is in that area (Figure 8). Similar conditions also occur on a 10 meter bench height with a single slope of 60° to 80° (Figure 12).

The results of the computation on the 10 meter bench height model are listed in Table 3. The reserves taken have increased with the increase in single slope or the overall slope used (Table 3). Nevertheless, the static and pseudo-static safety factors decreased with the maximum static safety factor at a single slope of 60° (Figure 9) and the maximum pseudo-static safety factor at a single slope of 50° (Figure 10).

4.2 *Optimal safety factor analysis*

Based on the analysis of the safety factor in slope modeling, it was found that the greater the overall slope, the greater the reserves obtained and the slope safety factor has an optimal level. On a 5 meter bench height, the largest reserve that can be taken is 552.586 BCM, on a single slope of 90° or an overall slope of 49° (PoF: 9,2%). Meanwhile, the safety factor on the slope has met the provisions of the applicable regulations. On a 10 meter bench height, the safety factor that meets the regulations is a maximum single slope of 70° or an overall slope limit of 52° (PoF: 9,9%), with reserves taken of 584,399 BCM. The results of both deterministic and probabilistic analysis using Monte-Carlo Simulation and PDF of the FOS-1000 MCS-Morgenstern-Price LEM from those presented in Figures 13 and 14 show that a single slope of 70° at a bench height of 10 meters is the most optimal slope model since the PDF peak on the slope is higher than the slope model with a bench height of 5 meters and single slope 90°.

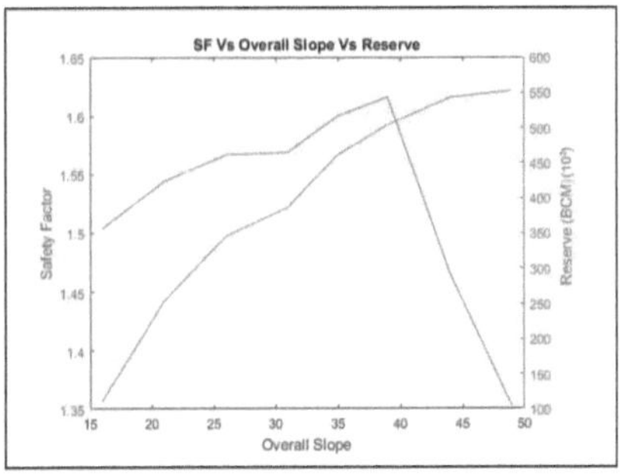

Figure 5. Graph of static analysis results.

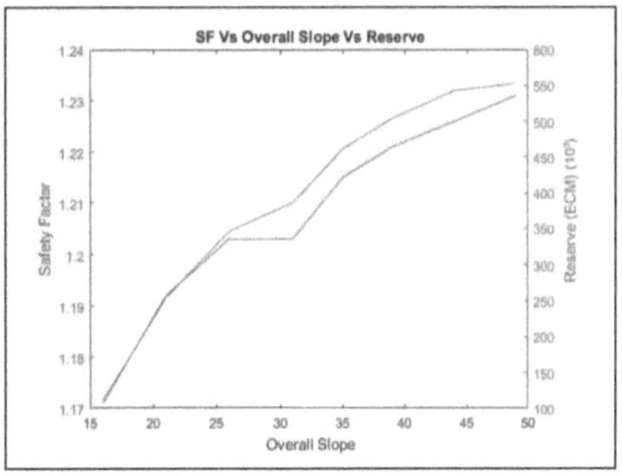

Figure 6. Graph of pseudo-static analysis results.

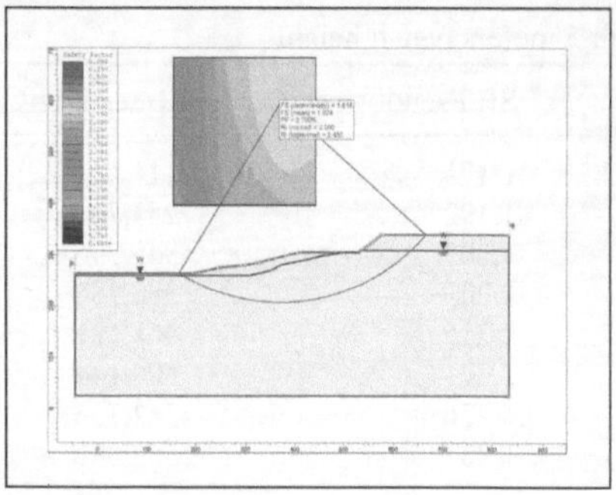

Figure 7. Model section on single slope 70°.

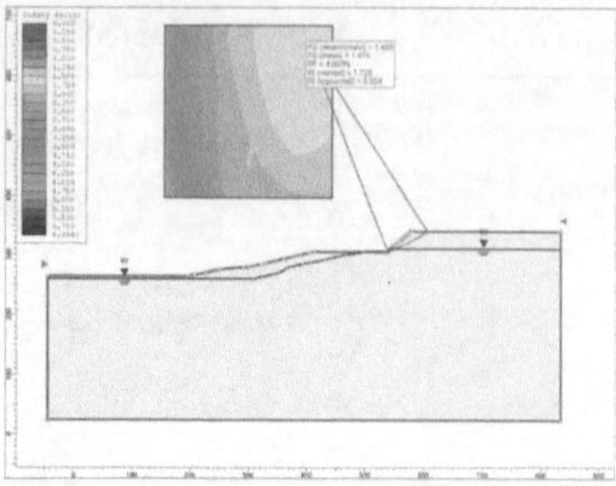

Figure 8. Model section on single slope 80°.

Table 3. Tabulation of slope model computation results on 10 meter bench height.

Single slope (o)	Overall Slope (o)	SF Static	SF Pesudostatic	Reserve (BCM)
20	17	1.509	1.174	136,886
30	23	1.510	1.167	336,907
40	32	1.594	1.213	437,137
50	38	1.614	1.221	507,172
60	45	1.506	1.229	561,393
70	52	1.312	1.188	584,399
80	59	1.135	1.041	648,211

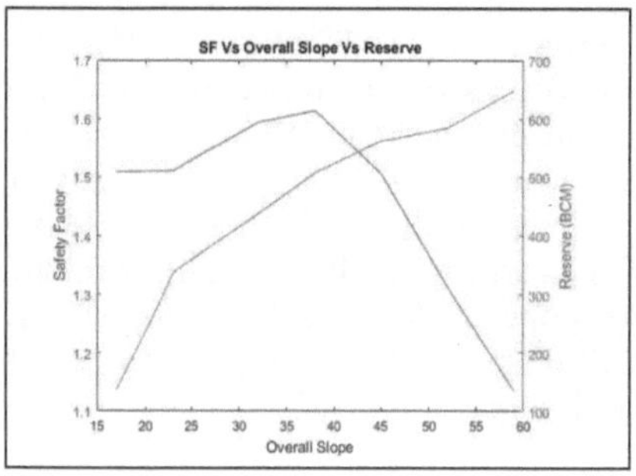

Figure 9. Graph of static analysis results.

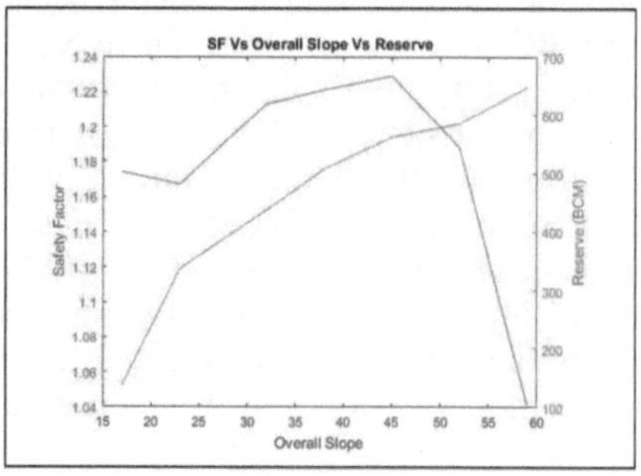

Figure 10. Graph of pseudo-static analysis results.

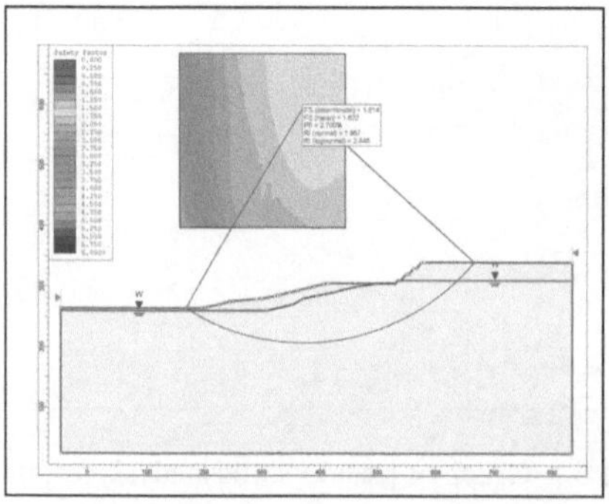

Figure 11. Model section on single slope 50°.

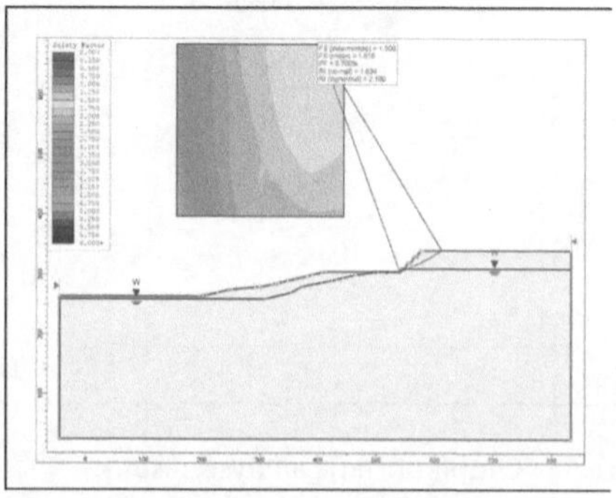

Figure 12. Model section on single slope 60°.

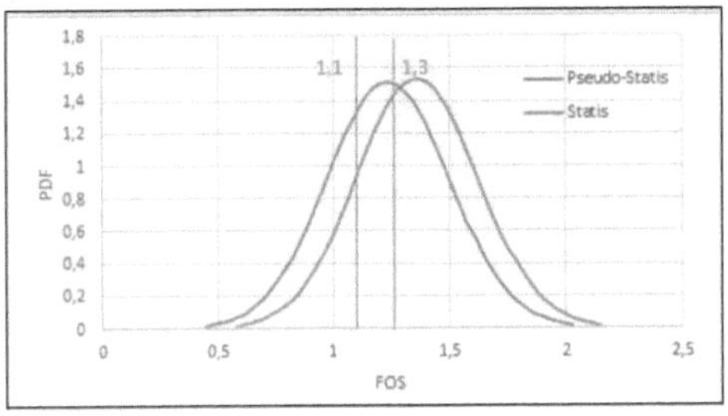

Figure 13. PDF for bench height 5 m and single slope 90°.

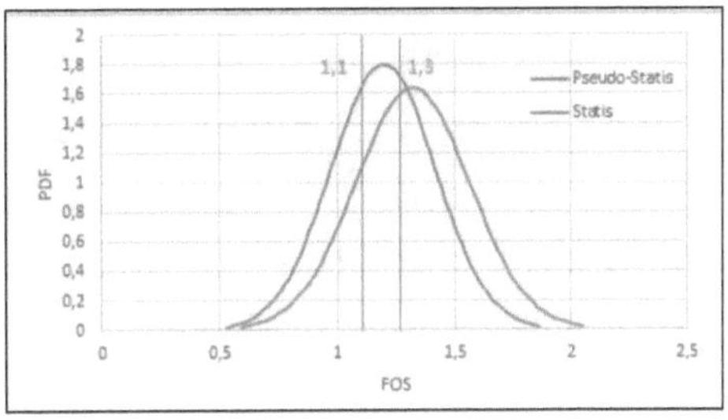

Figure 14. PDF for bench height 10 m and single slope 70°.

5 CONCLUSION

The Slope modelling in Kaolin commodity mining in Jetak, Karangsari, Semin, Gunungkidul has an overall slope limit of 52° with an overall height of 85 meters or a single slope of 70° and a bench height of 10 meters will produce excavated materials of 584,399 BCM.

REFERENCES

Bemmelen, R. W. (1949). The geology of Indonesia (Vol. 1, No. 1). US Government Printing Office.

Clarence S. Ross and Paul F. Kerr. (1930). The Kaolin Minerals. 13(3),151–160. doi10.1111j.1151-2916.1930.tb16556.x.

Grim, R.E. (1950) Modern concepts of clay materials. J. Geol., 50, 225–275.

Grim, R.E. (1962) Applied Clay Mineralogy. McGraw-Hill, New York, 422pp.

Hawkins, A.B. (1998). General Report: The nature of Hard Rocks/ Soft Soil. In Proceeding od the 2nd International Sysposium on Hard Soil-Soft Rocks, 1998, Naples, Italy. A.A.Balkema, Rotterdam 2000, 1392–1402.

Hidayah, R. A. (2005). Study of Kaolin Genesa and Its Utilization. Journal of Geological Engineering AKPRIND Yogyakarta.

Mah, C. W. and Willey, D.C. (2004). Rock Slope Engineering: Civil and Mining. 4th edition (Based on the 3rd edition by E. Hoek and J. Bray. The Institution of Mining and Metallurgy). New York: Taylor & Francis.

Martin-Vivaldi, J.I. and Robertson, R.H.S. (1971). Palygorskite and sepiolite (the hormites). Chapter in Electron Optical Investigation of Clays. Gard, J.A., ed. Mineralogical Society Monograph No. 31, London, pp. 255–275.

Ministry of Energy and Mineral Resources. (2018). MEMR Decree No. 1827K/30/MEM/2012 of 2018 concerning Guidelines for the Implementation of Good Mining Principles.

Murray, H.H. (2000) Clays. Ullmann's Encyclopedia of Industrial Chemistry, 6th Edition. Wiley-VCH Verlag GmBH, Weinheim, Germany, 30pp.

Prasad, M. S., Reid, K. J., and Murray, H. H. (1991). Kaolin: processing, properties and applications. Applied clay science, 6(2),87–119.

Surono, B. Toha, and Ign. Sudarno., 1992, Geological map sheet Surakarta-Giritontro, Java, Center for Geological Research and Development, Bandung

US Army Corps of Engineers, 1997. Engineering and Design: Introduction to Probability And Reliability Methods for Use in Geotechnical Engineering. Department of the Army, Washington, DC, Engineer Technical Letter, pp. 1110–2547.

Widodo. (1995). Quality Review of Kaolin Dsn. Jetak, Kec. Semin, Kab. Gunungkidul, DIY, AKPRIND Institute of Science & Technology, Yogyakarta, unpublished.

The Evolution of Geotech - 25 Years of Innovation – Hammah et al. (Eds)

How to incorporate variability of rockmass structures into equivalent continuum numerical models using the Composite Geological Strength Index

J.J. Day
Department of Geological Sciences and Geological Engineering, Queen's University, Kingston, Ontario, Canada

ABSTRACT: The Composite Geological Strength Index (CGSI) is an innovative solution for quantitative rockmass characterization that first considers rockmass structures individually and then calculates a single CGSI value for the whole rockmass. The application of CGSI is to replace the GSI parameter in the Generalized Hoek-Brown shear strength criterion for equivalent continuum numerical modelling. Although originally developed for rockmasses with veins and other intrablock structures, CGSI can be used for any rockmass with disparate structures. In this paper, CGSI is newly expanded to two approaches: Approach 1 is based on block volume of structure suites, while Approach 2 is based on spacing of structure sets. Examples are shown for each approach using igneous and sedimentary rockmasses, respectively.

1 INTRODUCTION

Geotechnical engineering design of underground excavations, slopes, and other infrastructure requires rockmass characterization, which includes the evaluation of both intact rock and rockmass structures through some combination of field and laboratory observations and measurements. Many modern designs implement geomechanical numerical analyses to evaluate rockmass strength and stability in the context of the planned infrastructure. In numerical codes such as the Finite Element Method (e.g. RS2 by RocScience), rockmasses are simulated as continuum materials that may or may not include explicit joint elements that simulate rockmass structures. These models require input parameters that quantify the deformability and strength of intact rocks and rockmass structures and feed into constitutive models that control the rockmass response to applied ground stress conditions. A well-established constitutive model for continuum materials is the Generalized Hoek-Brown shear strength criterion (Hoek et al., 2002), which can simulate either intact rock or rockmass strength.

In situations where explicitly (or discretely) modelling rockmass structures is not possible for reasons such as inadequate information from site investigation to determine appropriate input properties for joint elements, or excessive computational demand where too many joint elements result in software crashes, modelling rockmasses as equivalent continuum materials is a convenient solution. The Generalized Hoek-Brown criterion is well suited to this application because the input parameters that quantify rockmass structures are based on field observations through the Geological Strength Index (GSI) (e.g. Hoek and Marinos, 2000; Hoek et al., 2013). GSI is typically determined by means of a holistic assessment of the rockmass where all structures are summarized in a single GSI value. This approach becomes limiting, however, when a variety of qualities and geometries of rockmass structures is present.

DOI: 10.1201/9781003188339-15

The Composite Geological Strength Index (CGSI), originally developed for rockmasses that include hydrothermal veins (Day et al., 2019), is an innovative solution to first consider and characterize suites or sets of rockmass structures individually and then use a weighted harmonic averaging approach to calculate a single CGSI value that represents the rockmass for direct input to the Generalized Hoek-Brown strength criterion (as a replacement for the GSI parameter). The CGSI method is applicable to any rockmass where multiple sets or suites of joints, bedding, other fractures, veins, or other intrablock structures have different geomechanical properties that require individual consideration. Examples include clay infilled bedding layers versus clean vertical joints in sedimentary strata, rough joints versus quartz veins in volcanic strata, or joints versus foliations or cleavages in metamorphic strata.

The primary objective of this paper is to provide users with a straightforward guide to using CGSI in engineering practice. Two approaches to calculating CGSI are presented: (1) the original CGSI approach and calculations based on suites of rockmass structures measured in cubic volumes of rock blocks (Day et al., 2019), and (2) new CGSI calculations based on individual sets of rockmass structures measured on the basis of spacing.

2 APPROACH 1: CGSI BASED ON BLOCK VOLUME

The Composite Geological Strength Index (CGSI) was originally developed by Day et al. (2019) to address the need to consider hydrothermal veins and other intrablock structures in rockmass characterization. In this development of CGSI, the quantified GSI chart by Hoek et al. (2013) was first expanded to include a column for Strengthening Intrablock Structures at the left (high end) of surface condition of rockmass structures, along with modified text in the Very Good and Good columns to add Healed Veins and Weak Veins, respectively (Figure 1). The Massive structure category from the Hoek & Marinos (2000) version of the chart as well as cubic spacing and block volume measurements of structure by Cai et al. (2004) were reintroduced. The quantification of Infill/Surface Condition (Scale A) also required modification of the recommended Joint Condi- tion ($JCond_{89}$) input by Bieniawski (1989) to include intrablock structures ($JCond_{89}$*, Table 1).

The second part of the development of CGSI was the introduction of the equations to mathematically combine GSI values of individual suites or sets of rockmass structures into an overall CGSI value for the rockmass. This development shifted the core philosophy of GSI from relying on subjective groupings of multiple structural components to providing a quantitative tool for combining individual structures into a CGSI value for the rockmass. Although previously pub- lished research on CGSI primarily focusses on applications to rockmasses with veins and other intrablock structures (Day et al., 2019, Day et al., 2016, Day et al., 2013, Day et al., 2012), CGSI is applicable to any rockmass with disparate qualities and/or geometries of structures.

The original approach for the CGSI calculations considered rockmass structures based on block volume, where two or more very different suites of structures each create a blocky rockmass at different scales, such as a stockwork network of hydrothermal veins within jointed blocks. In this case, it is assumed the structural suites (e.g. joints and veins) each form cubic volumes. The CGSI calculations for this approach are as follows (Day et al., 2019), where A and B represent Infill/Joint Condition and Rockmass Structure (Equations 1 and 2), respectively. First, assess each suite of structure individually, *i*:

$$A_i = 1.5(JCond_{89}^*)_i \quad (1)$$

$$B_i = (20/3)Log_{10}(Vb_i) \quad (2)$$

See Table 1 for $JCond_{89}$* inputs and estimate block volume, Vb_i, directly in cubic centimetres. Second, combine A and B measurements from each suite of structure into composite values for the rockmass (A* and B* using Equations 3 and 4, respectively), for n suites (i = 1 to n), and finally calculate CGSI using Equation 5:

$$A^* = \frac{\sum_1^n (A_i/B_i)}{\sum_n^1 (1/B_i)} \tag{3}$$

$$B^* = 20log_{10}\left(1/\left(\sum_1^n 10^{-Bi/20}\right)\right) \tag{4}$$

$$CGSI = A^* + B^* \tag{5}$$

These equations are conveniently summarized in Figure 1. An example calculation of CGSI using Approach 1 (Block Volume in Suites) is presented in Section 4.

3 APPROACH 2: CGSI BASED ON SPACING OF STRUCTURES

Another application of CGSI is for any rockmass with multiple sets of structures when individual joint set information is available from drill core logs or outcrop observations. This approach is particularly useful in cases where structure conditions or spacing in a rockmass are distinct between sets. The calculations for Approach 2, CGSI based on spacing of structures, are as follows: Calculate Scale A (Infill/Joint Condition) for each set (A_i) and the rockmass (A^*) using the same equations as Approach 1 (Equations 1 and 3). Note the $JCond_{89}$* chart that includes intrablock structures (Table 1) maintains the same ratings for joints and other interblock structures as $JCond_{89}$ by Bieniawski (1989). To calculate B_i (Rockmass Structure for one set), first, calculate spacing for individual sets of structures in centimetres (S_i):

$$S_i = 100/f_i \tag{6}$$

where f_i is the fracture frequency in joints/m for that joint set i. Note that f_i should be corrected for the angle between each joint set and the drill core axis using Equation 7 (based on Palmstrom, 2005 and Terzaghi, 1965):

$$f_i^* = f_i/f_i \sin \gamma \tag{7}$$

where γ is the average angle between the joint set and the drill core axis. Next, calculate B (Rockmass Structure) for each set, i:

$$B_i = 20Log_{10}(S_i) \tag{8}$$

To determine B^* (for the rockmass), calculate volumetric joint count (J_v) for the rockmass using either of the following two approaches (Equation 9 or 10) based on data availability:

$$J_v = \sum_1^n f_i \tag{9}$$

$$J_v = (\beta/V_b)^{1/3} \tag{10}$$

where V_b is block volume and β is the block shape factor (Palmstrom, 2005). For cubic blocks, β = 27; for other blocks, consider β = 20+7R where R is the maximum aspect ratio. Also, for joints at angles between sets of other than 90°, V_b should be corrected to V_b* as follows:

$$V_b* = V_b/\sin^3\alpha \tag{11}$$

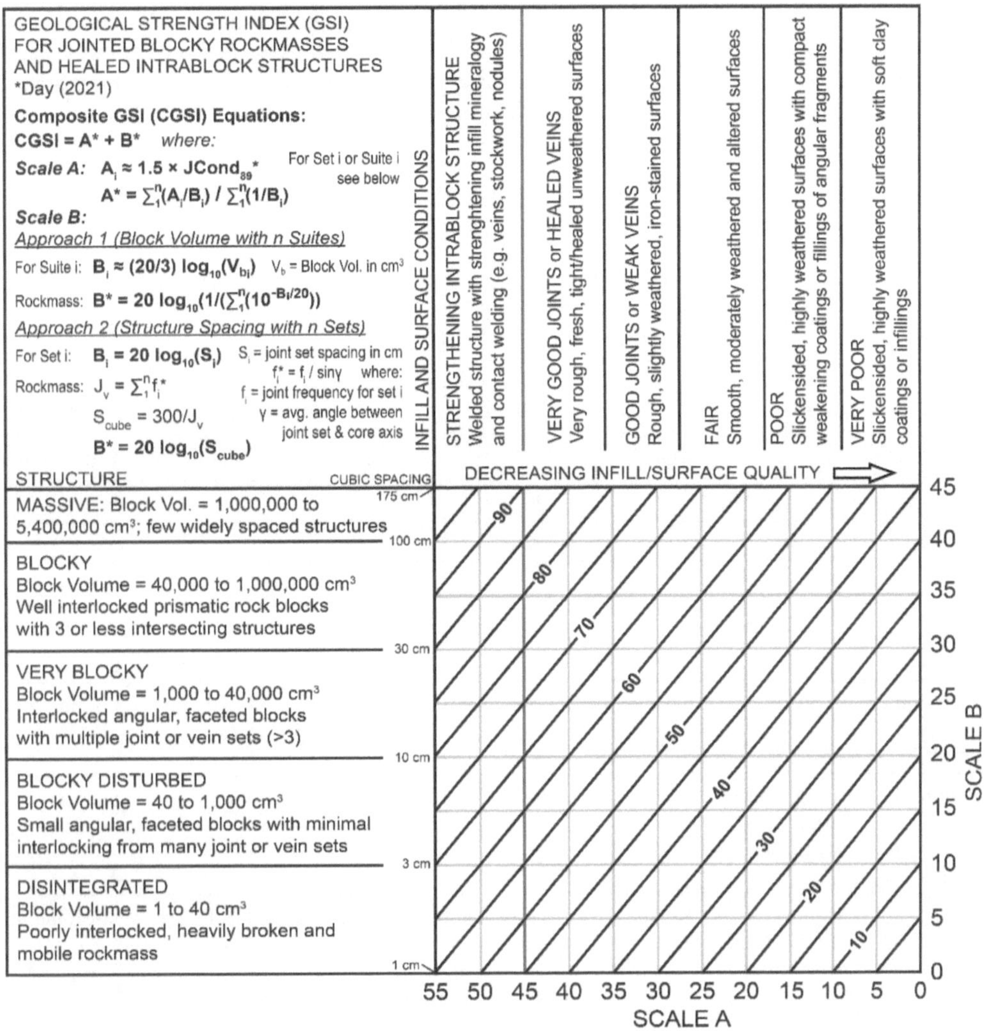

Figure 1. Geological Strength Index chart for jointed blocky rockmasses and healed intrablock structure, including equations for Composite GSI with Approach 1 based on block volume (after Day et al., 2019) and Approach 2 based on spacing of structural sets (new contribution).

where α is the average joint interset angle. Note that fracture frequency (f_i) should be adjusted in Equation 9 to account for angles between joints and core axis (Equation 7). Next, use J_v to determine overall cubic spacing (S_{cube}) with Equation 12, and use that to calculate B^* for the rockmass using Equation 13:

$$S_{cube} = \frac{300}{J_v} \tag{12}$$

$$B^* = 20Log_{10}(S_{cube}) \tag{13}$$

Note the definition of B^* in Equation 13 can be used with the normal approach of Scale B to GSI in place of RQD/2 (with A being assessed as an average of $1.5JCond_{89}$*). Quantifying

B as RQD/2 was first proposed by Hoek et al. (2013), where RQD is the Rock Quality Designation by Deere (1963). Finally, calculate CGSI for the rockmass using Equation 5 (same step as Approach 1). These equations are conveniently summarized in Figure 1. An example calculation of CGSI using Approach 2 (Structure Spacing in Sets) is presented in Section 4.

Table 1. Modified Joint Condition rating, $JCond_{89}$*, to include intrablock structures (Day et al., 2019).

Condition of discontinuities	Strengthening intrablock structure Strong infill minerals welded to wall rock	Very rough or healed surfaces Not continuous No separation Unweathered wall rock	Slightly rough surfaces or weak intra-block structure Separation <1mm Slightly weath-ered walls	Slightly rough surfaces Separation <1mm Highly weathered walls	Slickensided surfaces or Gouge < 5 mm thick or Sepa- ration 1-5 mm Continuous	Soft gouge >5 mm thick or Separation >5 mm Continuous
Overall rating	**37.5**	**30**	**25**	**20**	**10**	**0**
Guidelines for classification of discontinuity conditions						
Discontinuity length (persis-tence)	< 0.5 m	< 1 m	1 to 3 m	3 to 10 m	10 to 20 m	>20 m
Rating	**7.5**	**6**	**4**	**2**	**1**	**0**
Separation (ap-erture)	Welded	None	< 0.1 mm	0.1 to 1.0 mm	1 to 5 mm	>5mm
Rating	**7.5**	**6**	**5**	**4**	**1**	**0**
Roughness	Rough, undulating, irregular	Very rough	Rough	Slightlyrough	Smooth	Slickensided
Rating	**7.5**	**6**	**5**	**3**	**1**	**0**
Infilling (gouge)	Strong bonded vein (quartz)	None	Hard filling < 5 mm	Hard filling > 5 mm	Soft filling 5 mm	<Soft filling > 5 mm
Rating	**7**	**6**	**4**	**2**	**2**	**0**
Weathering	Strengthening by alteration	Unweathered	Slightly weath-ered	Moderate weathering	Highly weath-ered	Decomposed
Rating	**7**	**6**	**5**	**3**	**1**	**0**

4 CASE EXAMPLES

Two case examples of CGSI are presented in this section. The first case uses Approach 1 based on block volume with n suites of rockmass structure. This example igneous rockmass contains one suite of three orthogonal joint sets and one suite of a calcite stockwork vein network. The CGSI calculations and chart plots are shown in Figure 2. Individual GSI values for each structure suite are shown in addition to the resulting CGSI for the rockmass. Prior to the introduction of CGSI, a conservative approach to handle a rockmass with multiple distinct suites of structure would be to take the worst-case minimum of Scale A and Scale B. The example in Figure 2 illustrates the significant difference between worst-case GSI (43) and CGSI (53) approaches.

The second case illustrates Approach 2, based on structure spacing with n sets, for a sedimentary rockmass with clayfilled bedding, a subvertical tight, fresh joint set spaced at 0.2 m, and a subvertical moderately weathered joint set spaced at 0.4 m (Figure 3). Note this example also shows a significant difference between the worst-case GSI (36) and CGSI (54).

Approach 1 (Block Volume with n Suites) Example

Step 1: Evaluate structure suites individually

	Suite 1	Suite 2
Description:	*3 orthogonal joint sets, rough & slightly weathered*	*Calcite vein network, 15 cm block size*
Infill & surface conditions:	Good Joints (A_i = 20)	Healed veins (A_i = 40)
Structure:	Blocky (B_i = 35)	Very Blocky (B_i = 23)
GSI (individual):	55 ●	63 ●

Step 2: Calculate CGSI for entire rockmass

Scale A (Infill & Surface Condition): $A^* = \frac{(20/35 + 40/23)}{(1/35 + 1/23)} = 32$

Scale B (Structure): $B^* = 20\log_{10}((10^{-35/20} + 10^{-23/20})^{-1}) = 21$

Composite GSI (CGSI): CGSI = A* + B* = 53 ☆

Conventional worst-case GSI (minimum of A_i, B_i) shown for comparison ★

Step 3: Insert CGSI value into Hoek-Brown strength criterion in place of GSI

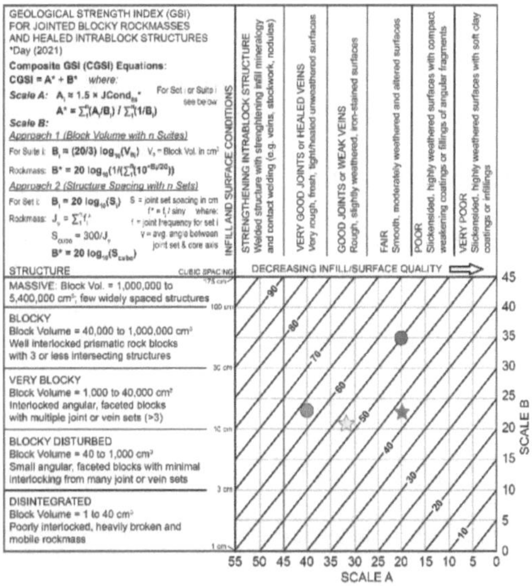

Figure 2. Example calculation and plot of CGSI Approach 1 (Block Volume with n Suites) for a rockmass with two suites of rockmass structure (joints and calcite veins).

Approach 2 (Structure Spacing with n Sets) Example

Step 1: Evaluate structure sets individually

	Set 1	Set 2	Set 3
Description:	*Horizontal bedding with clay, spaced 1.5 m*	*Subvertical Tight, fresh joint spaced 0.2 m*	*Subvertical Mod. weathered joint spaced 0.4 m*
Avg. core-joint angle (γ)	70°	30°	45°
Infill & surface conditions:	Poor-V. Poor Joints (A_i = 10)	V. Good joints (A_i = 40)	Fair joints (A_i = 24)
Structure:	Massive (B_i = 44)	Very Blocky (B_i = 26)	Blocky (B_i = 32)
GSI (individual):	54 ●	66 ●	56 ●

Step 2: Calculate CGSI for entire rockmass

Scale A (Infill & Surface Condition): $A^* = \frac{(10/44 + 40/25 + 24/34)}{(1/44 + 1/25 + 1/34)} = 27$

Scale B (Structure):	Set 1	Set 2	Set 3
Spacing, S_i (cm)	150 cm	20 cm	40 cm
Frequency, f_i	0.667	5	2.5
f_i^* (= f_i / sin(γ))	0.709	10.0	3.5

J_v = 0.709 + 10.0 + 3.5 = 14.2

S_{cube} = 300 / J_v = 300 / 14.2 = 21.1

$B^* = 20\log_{10}(S_{cube}) = 20\log_{10}(21.1) = 26$

Composite GSI (CGSI): CGSI = A* + B* = 54 ☆

Note: Conventional worst-case GSI (minimum of A_i, B_i) shown for comparison ★

Step 3: Insert CGSI value into Hoek-Brown strength criterion in place of GSI

Figure 3. Example calculation and plot of CGSI Approach 2 (Structure Spacing with n Sets) for a rockmass with three sets of discontinuities (bedding and joints).

5 DISCUSSION AND CONCLUSIONS

The Composite Geological Strength Index (CGSI) provides a quantitative approach to characterizing rockmasses that particularly have a variety of geotechnical qualities and geometries of rockmass structures. The key philosophy of CGSI is to first assess individual suites or sets of structures by their GSI components and then combine these values using a weighted harmonic average into a CGSI value that represents the whole rockmass. Although the original development of CGSI was primarily focused on rockmasses with veins and other intrablock structures, the CGSI approach and chart (Figure 1) can be used for any jointed blocky rockmasses and applied to the Generalized Hoek-Brown shear strength criterion (in place of GSI) for numerical modelling.

An important development to CGSI is presented in this paper as new equations to relate spacing of structure sets (primarily after Palmstrom, 2005) to Scale B (Rockmass Structure) on the chart, which is the basis for CGSI Approach 2 (Figure 1). This creates improved flexibility, in addition to Approach 1 (block volume and suites) by Day et al. (2019), such that CGSI can be implemented in a variety of projects that may have different measurements of rockmass structure geometry available for design. Note in both CGSI Approaches, $JCond_{89}$* (Table 1) is the basis for quantifying Scale A and calculating A^*.

ACKNOWLEDGMENTS

The author wishes to thank Dr. Mark Diederichs for his thoughtful and detailed comments.

REFERENCES

Bieniawski, Z.T. 1989. *Engineering rock mass classification.* Wiley: New York.

Cai, M., Kaiser, P.K., Uno, H., Tasaka, Y. & Minami, M. 2004. Estimation of rock mass deformation modulus and strength of jointed hard rock masses using the GSI system. *International Journal of Rock Mechanics and Mining Sciences*, 41: 3–19. https://doi.org/10.1016/S1365-1609(03)00025-X.

Day, J.J., Diederichs, M.S. & Hutchinson, D.J. 2019. Composite Geological Strength Index approach with application to hydrothermal vein networks and other intrablock structures in complex rockmasses. *Geotechnical and Geological Engineering* 35: 5285–5314. https://doi.org/10.1007/s10706-019-00980-4.

Day, J.J., Diederichs, M.S. & Hutchinson, D.J. 2016. Validation of Composite Geological Strength Index for healed rockmass structure in deep mine access and production tunnels. *In Proc.* Of Tunnelling As sociation of Canada Conference, Ottawa, ON, Canada, October 16–18.

Day, J.J, Diederichs, M.S. & Hutchinson, D.J. 2013. Incorporation of geostructural data into discrete analysis for tunnel design. *In Proc.* ITA-AITES World Tunnel Congress, May 31-Jun 7, Switzerland, 8p.

Day, J.J., Hutchinson, D.J. & Diederichs, M.S. 2012. A critical look at geotechnical classification for rock strength estimation. *In Proc.* 46th U.S. Rock Mechanics Geomechanics Symposium, ARMA, Chicago, IL, USA, 10p.

Deere, D.U. 1963. Technical description of rock cores for engineering purposes. *Felsmechanik und Ingenieurgeologie* (Rock Mechanics and Engineering Geology), 1(1): 16–22.

Hoek, E., Carranza-Torres, C.T. & Corkum, B. 2002. Hoek-Brown failure criterion—2002 edition. In Proceedings of 5th North American rock mechanics symposium, Toronto, Canada, vol 1: 267–273.

Hoek, E., Carter, T. & Diederichs, M.S. 2013. Quantification of the Geological Strength Index chart. In Proceedings of 47th US Rock Mechanics Symposium. ARMA, San Francisco, USA.

Hoek, E. & Marinos, P. 2000. Predicting tunnel squeezing problems in weak heterogeneous rock masses. *Tunnels and Tunnelling International*, Part 1—November 2000, part 2—December 2000

Palmstrom, A. 2005. Measurements of and correlations between block size and rock quality designation (RQD). *Tunnels and Underground Space Technology*, 20: 362–377.

Terzaghi, R.D. 1965. Sources of error in joint surveys. *Géotechnique*, 15(3): 287–304.

The Evolution of Geotech - 25 Years of Innovation – Hammah et al. (Eds)

Rock mass characterization of Molasse-like deposit of Dahor formation and its application to rock slope stability from Bunati area, South Kalimantan: A preliminary study

L. Zeremy
Neogen Prima Indonesia, Bandung, Indonesia

ABSTRACT: Dahor Formation in Bunati area is Mollase like deposit comprising of rockmass and sediments which deposited during/since Miocene Meratus Mountain uplift. Preliminary geotechnical investigation hasbeen conducted in Bunati Area, South Kalimantan which gave insight at depth more than 20 m its formation comprise of various rockmass such as thin intercalation of siltstone-claystone, claystone, coal and sandstone with basic GSI value range 27 from 43 with its intact rock strength range from 3 to 16 MPa. Its rock mass found have systematic isotrop oriented discontinuity orientation controlled by strike-slip deformation during or post Meratus Uplift. Furthermore, there're also discontinuities which its orientation tend to be anisotropic controlled by sedimentation, diagenetic and mechanical compaction of overlying rocks and sediments. The interaction between these two types of discontinuities with its respective orientations are very likely to trigger failure mechanisms in the rock mass which results in two types of failure in the form of step-path failure and wedge failure reponse to mechanical excavation of undercutting low wall and side wall pit slope. Finite element method and wedge block failure was utilised in order to recognized instability mechanism involving discontinuities interaction and to asses rock mass slope factor of safety and strength reduction factor utilizing rock mass strength parameters quantification.

1 INTRODUCTION

Rock mass characterization of the Dahor Formation in the Bunati area, South Kalimantan is a great example in nature which potentially induced stress due to mechanical surface excavation will influence the rock mass stability which is controlled by the presence of discontinuities in rock. Bunati area geographically is located in 3°45′30″S and 115°35′00″E.

Its discontinuity sets consist of anisotropic discontinuity sets which is controlled by the dep- ositional setting, overburden stress of overlying rocks and sediments strata, and relatively sys- tematic isotrop discontinuity sets which is controlled by strike-slip deformation during or after the sediment and rocks are deposited. The interaction between these two types of discontinuities with its respective orientations are very likely to trigger failure mechanisms in the rock mass which results in two types of failure in the form of step-path failure and wedge failure. Hammah et. al. (2008) also stated that its failures are also controlled by the scale of the discontinuity and the scale of the excavated slope.

Since rock mass classification should be fulfilling certain conditions, effectively combine the findings from observation, experience, and engineering judgment for providing a quantitative assessment of rock mass conditions (Bieniawski, 1990) therefore, this paper

DOI: 10.1201/9781003188339-16

briefly describes a comparative study based on analytic and quantitative assessments to recognize instability mechanism involving discontinuities interaction and to asses rock mass slope factor of safety in term of strength reduction factor (SRF) utilising rock mass strength parameters quantification. Finite element method (FEM) and wedge block (WB) analysis are carried out using version of the Rocscience programs which are Phase2 v8.0 and Swedge v4.0.

2 METHOD OF STUDY

Geotechnical drilling investigation was conducted using the continuous and touched core drilling with geophysical logging measurements at each boreholes. Rock mass parameters such as σ_{ci} estimation, *RQD*, discontinuity spacing (J_s) and discontinuity condition (J_{cond90}), discontinuity shear strength parameter (*JRC* & *JCS*) and Discontinuity Condition Factor (J_A, J_W & J_S) were carried out by direct measurement and observation of core samples (Bieniawski, 1989; Laubscher, 1990; Barton & Choubey, 1977 in Barton, 2017; Cai et. al., 2004). Meanwhile, the residual friction angle of bedding contact, fault, laminae and rock fracture were obtained based on the experience of back analysis results of planar and wedge failures from several open pit and geotechnical investigation reports from several open pit coal mining locations in Kalimantan, Indonesia respect to similiraty of discontinuity surface criteria. The strength parameters of intact rock were obtained from laboratory testing of rock core samples consist of physical properties test, UCS test, triaxial test, direct shear test & brazilian test. Some rock mass strength parameters quantified based on Hoek et. al. (2002), Hoek and Diederichs (2006), and rock mass tensile strength quantified using Griffith failure criterion (1924 in Gudmundsson, 2011).

Both analytic and quantitative assessments of step-path failure were carried out by FEM utilization of Mohr-Coulomb plastic failure criterion for intact rock and rock mass, while Barton- Bandis failure criterion applied to discontinuity failure mechanism. Discontinuities such as laminae and fractures are modeled using Baecher and parallel statistical joint network. Meanwhile, anisotropic discontinuity is modelled deterministically based on the results of 3D engineering geological modelling. Afterward, both analytic and quantitative assessments for the wedge failure analysis was carried out using the FEM and WB analysis.

3 ENGINEERING GEOLOGICAL CONDITION OF BUNATI AREA

3.1 *Stratigraphy & deformation style*

The Dahor Formation is tectonic mollase which was deposited in Barito Basin during the Early Miocene - Pleistocene Meratus Uplift (Satyana et. al., 1994 in Sapiie & Rifiyanto, 2017; Witts et. al., 2014). At Asem-Asem Basin its deposition took place since Pliocene (Rustandi et. al., 1995). The Dahor Formation in the Bunati area is included in the Asem - Asem Basin. In this area, the Dahor Formation is composed of rock masses and sediments. Those rock-masses comprised of sandstones, claystones, thin intercalation of siltstone-claystone and coal, whilst sediments which its mechanical behaviour tend to be considered as soil comprised of firm slightly- moderately plastic of silty to sandy clay and medium dense to dense of medium – coarse sands and will not discussed further in this paper (see Figure 1). In package of medium dense – dense sand, often found the evidence that sandstone at the bottom of its package.

Vertical bottom up succession of Mollase-like or molassic deposit in Bunati can be divided into three engineering geology unit (see Figure 1): Thin intercalation very weak claystone- very weak siltstone unit coal inserts and very weak fine sandstones whic is coarsening upward becoming medium dense -dense of coarse - medium grain sand sediments. This unit is the lower part of the Lower Marine Mollase characterized by the mm - cm scale (laminae) layering of claystonesiltstone (tempestite) with a total package thickness of up to 30 m (see Figure 2). Unconformably overlain by interbedding of medium dense to dense sand – very weak claystone Upper Marine Mollase unit with sandstone, coal and thin interlayering parallel laminated very weak claystone-siltstone inserts. This unit is a characteristic of the Upper Marine Mollase which is characterized by the thinning upward of the coarse – medium sands (Schlunegger et. al., 2007). Unconformably overlain by interbedding of medium dense-dense sand – firm slightly to moderately plastic of silty – sandy clays unit which is part of Upper Marine Mollase.

Strike-slip deformation in Bunati area inferred from its surface manifestation reflected from topograhic condition in form of NW-SE trending of dextral faults (R), NNW-SSE trending of sinistral faults with its principal development zone in form of WNW -ESE trending of dextral faults (PDZ).Those faults system interpreted as Riedel Shear (Sapiie & Cloos, 2004). Those step movements has generated strain partitions along its block movement that lead to development of N – S and NNW – SSE trending syn-sedimentation obliquenormal faults (See Figure 3 (b)) during deposition of medium dense to dense sand – very weak claystone interbedding unit that lead to relatively westward and eastward thickening of its unit and NNE-SSW and NE-SW trending synsedimentation oblique normal faults during deposition of medium dense-dense sand – firm slightly to moderately plastic of silty – sandy clay sediments interbedding unit that lead to north east ward thickening of its unit (See Figure 4). This interpretation fits with the evidence that both bedding contact and laminae discontinuity dips range from 2° to 5° which are typical initial dip during rock /sediment package during sedimentation (See Figure 3).

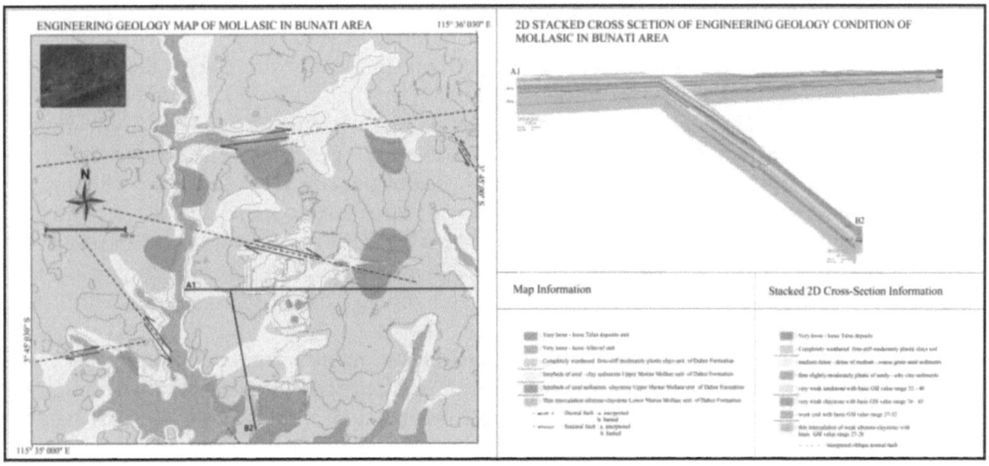

Figure 1. Engineering geology overview of mollasic deposit of Dahor Formation in Bunati Area. Map represents engineering geology unit whilst stacked 2D section represents rockmass and sediment member of each unit.

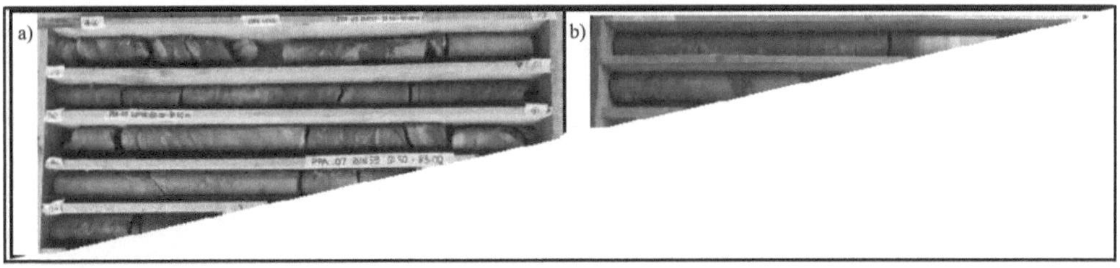

Figure 2. (a) Tempestite of Lower Marine Mollase at at depth 78 -83 m. (b) Typical upper Marine Mollase succession with thinning upward of medium dense – dense sands.

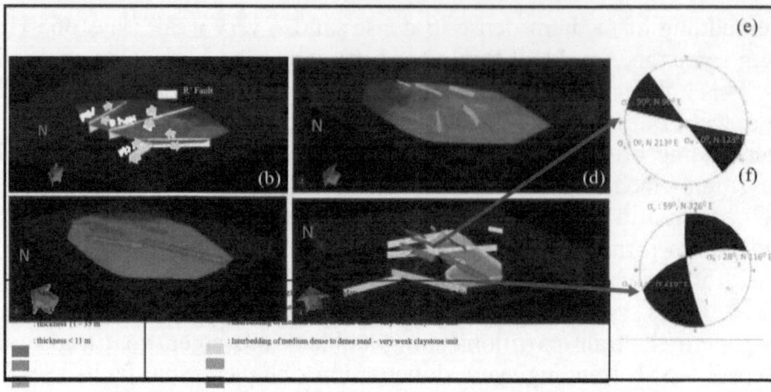

Figure 3. (a) Strike-slip faults manifestation at surface. (b) and (c) Isopach map represents thickening unit controlled by oblique normal faults at subsurface. (e) and (f) Paleostress kinematics of faults interaction.

3.2 *Discontinuity & rockmass characterization*

Based on its genesis, rock mass discontinuities of Mollasic deposits in Bunati area can be categorized into 3 main groups, such as: Firstly, discontinuities related to the sedimentation process such as bedding surface between rocks and laminae (2 – 5 cm spacing) of siltstone - claystone of tempestite-formation. Secondly, are discontinuities formed due to diagenesis and mechanical compaction due to overburden stress of overlying rocks and sediments, such as nearly horizontal joints in each rock and face cleats of coal with its dip range of 0°-7° and the last are discontinuities formed due to strike slip deformation during/after the rock mass depositional phase with dip range -25° - 75°. However, based evidence encountered in core samples, it's quite difficult to discriminate diagenetic joints /joints related overburden stress from low angle joints related strike slip deformation as σ_H related PDZ and σ_H related strain partitioning of faults move- ment (see Figure 3) had low angle plunge. In order to generate reasonable discontinuities in intact rock model, joint network modelling provided by Phase2 8.0 was utilized using parameters as shown in Table 1.

Then, GSI value for rock mass peak strength of Mohr Coulomb failure criterion can be obtained by quantified RQD and Jcond90 using equation 14 proposed by Russo & Hormazabal (2019). GSI value for rock mass residual strength seems to be fit as a function of volume block residual (V_B^r) and joint condition factor residual (J_C^R) which can be obtained using equation 13 proposed by Cai et. al. (2006) (see Table 2). Since the only way to estimate V_B^r is by determining the value of volume block peak (V_B^P), empiric correlation between V_B^P and discontinuty spacing (D_C) adapted from Figure 5 in Cai et. al (2007) can be made as shown in Equation 1.

$$V_B^P = E + 06\,(D_C)^{3.0759},\ \text{in cm}^3 \text{ with } R^2 = 0.9 \tag{1}$$

Table 1. Parameter used for discontinuity network modelling and its shear strength parameters.

Discontinuity Type	Joint Network Type	Dip Range (deg)	Shear Strength Discontinuity Parameter		
			JRC	JCS (Mpa)	ϕ_r (deg)
Bedding Contact	Deterministic	2 - 5	8	4	17
Fault	Deterministic	60 - 70	8	4	16
Joint	Baecher	-25 - 75	6	4	12
Laminae	Parallel statistical & Baecher	0 - 5	6	4	8

Both peak and residual GSI value are still needed to be adjusted since lowwall and side wall strip mining slopes design orientation are in fair conditipn relatively to the bedding orientation. Disturbance factor (D) will be set to zero as the mining sequence won't apply production blasting and and final bench slope will be constructed utilising dedicated small size excavators.

Table 2. Some Hoek – Brown parameters with adjusted GSI respect to slope fair condition, utilised to quantify mohr coulomb failure parameters of rock mass strength.

Rock Mass Type	σ_{ci} (Mpa)	mi	D	GSI_{peak}	GSI_{res}
thin intercalation of siltstone-clay-stone	5	9	0	25	15
claystone	5	9	0	30	20
Sandstone	3	9	0	30	20
Coal	16	7	0	25	15

4 ANALYSIS & RESULTS

4.1 *Low wall step path failure mechanism of low wall slope*

The preliminary mining design in the Bunati area that was submitted by the concession-holding owner sides, uses the strip mining method as main coal seam target coal has no subcrops.

Based on the preliminary strip mining design, it is known that the overall low wall and side wall slope angle is 40 degrees with its height ranging from 60 -65 m. To recognize the instability mechanism of the low wall pit slope, slope stability analysis was carried out using intactrocks with discontinuities and homogeneous rock mass slope models. The parameters of those models are shown in Table 3. The slopes for both models were adjusted to dry conditions and without seismic loads applied and weight of rockswill be set to dry. Thelow wall slopestability simulation conducted several geometries with slope fix at height of 60 m and overall slope angle set to 40°, 35° and 27° respectively. 2D Low Wall sections area marked with line A1 in Figure 4 (a).

Slope stability simulation utilising FEM results indicating that there're difference in SRF between intact rock with discontinuity and homogeneous rock mass slope models which is about only 2 - 3%. However, it is clear that the intact bridge failure can be observed in the intact rock slope model, characterized by the presence of shear dilation mechanism in intact rock-penetrated discontinuities. Meanwhile, only shearing mechanismtook placebody of therock mass. The presence of rock bridge failure in the intact rock slope model shows that shear dilation mechanism accommodated in the Barton-Bandis failure criterion. As dilation angle parameter is not set or by default its value set to zeroin rock massparameter, then its failurecriterion was not accomodating shear dilation mechanism. That's make the difference of both SRF about 2-3%. The presence of the path way of intact rock bridge failure indicates failure type low wall coud be step-path failure.

4.2 *Wedge failure mechanism of southern side wall slope design*

Wedge failure potential of southern side wall slope recognized based on the presence of a normal oblique fault cutting the southern side wall (see Figure 3 (d) & (f), Figure 4 (a) line B1). Initially wedge failure potential was analyzed using FEM. FEM simulation indicated that shearing mechanism took place in intact rock bridge with very small proportion of shear dilation mechanism (see Figure 5(a)). The presence of steep dipping fault seems will controll the shearing mechanism in response of mechanical excavation and induce shear stress along intact rock bridge and make the failure path way. Therefore GSI_R of intact rock bridge can be set to 15, so value of its residual friction angle will be 28° for its slope heigh at 23 m. More over its residual cohesion will be zero as mobilized shear degrade its discontinuities surfaces. The average dip of intact rock bridge from Figure 5 is about 22°. Using combination of fault and its shear strength parameter, then WB analysis can be done (see Figure 5 (b)).

Table 3. Intact rock and rock mass mechanical properties used in FEM simulation.

	ν	$E_i(MPa)$	$E_m(MPa)$	Peak Strength Parameters						Residual Strength Parameters			
				Intact Rock			Rock Mass			Intact Rock		Rock Mass	
				$c_p(MPa)$	$\phi_p(deg)$	$T_p(MPa)$	$c_p(MPa)$	$\phi_p(deg)$	T_p (MPa)	$c_r(MPa)$	$\phi_r(deg)$	$c_r(MPa)$	$\phi_r(deg)$
Thin inter-calation of	0.35	60	4	0.22	39	0.37	0.09	28	0.20	0.10	26	0.07	24
Claystone	0.35	60	6	0.22	39	0.27	0.12	31	0.23	0.10	26	0.08	26
Sandstone	0.40	50	4	0.18	35	0.27	0.09	25	0.12	0.09	23	0.07	22
Coal	0.35	600	36	1.77	46	1.52	0.14	33	0.33	0.18	34	0.09	29

ν : poisson ratio of rocks
E_m (MPa): rock mass modulus
E_i : modulus young of intact rock
T: tensile strength of rock

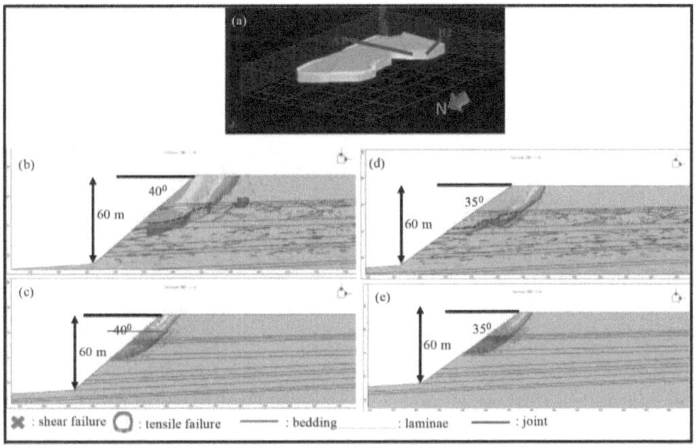

Figure 4. Comparative results of intact rock with discontinuities and homogenous rock mass slope stability simulation using FEM. Dashed blue line indicating pathway of intact rock bridge failure.

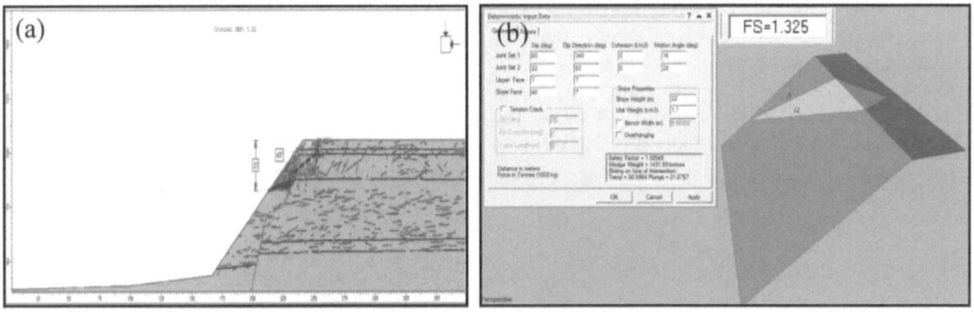

Figure 5. (a) Wedge failure potential from section B1 utilised FEM simulation, dashed blue line indicating pathway of intact rock bridge failure. (b) Safety of Factor obtained from WB analysis of southern side wall slope design utilising Swedge 4.0.

5 CONCLUSION

From both analytical and quantification assesment of mollase like deposit rockmass like encoun- tered in Bunati area, the reasonable *SRF* of step path and wedge failure type of rock mass slope could be achieved from FEM simulation. Moreover, safety factor of wedge slope failure due intact rock bridge-fault interaction could be obtained using WB analysis.

REFERENCES

Barton, N. & Choubey V. 1977. The Shear Strength of Rock in Theory and Practice. Rock Mechanics 10(1-2): 1–54.

Bieniawski, Z.T. 1989. Engineering Rockmass Classifications. Canada: John Wiley & Son Inc.

Bieniawski, Z.T. 1990. Tunnel Design By Rock Mass Classification. Washington, DC: US Army Corps Of Engineers.

Cai, M. Kaiser, P.K. Uno, H. Tasaka, Y. Minami, M. 2004. Estimation of Rock Mass Deformation Modulus and Strength of Jointed Hard Rock Masses Using The GSI System. International Journal of Rock Mechanics & Mining Sciences 41 (2004): 3–19.

Cai, M. Kaiser, P.K. H. Tasaka, Y. Minami, M. 2007. Determination of Residual Strength Parameters of Jointed Rock Masses Using the GSI System. International Journal of Rock Mechanics & Mining Sciences 44 (2007): 247–265.

Gudmundsson, A. 2011. Rock Fractures in Geological Processes. New York: Cambridge University Press.

Hammah, R.E. Yacoub, T. Corkum, B. Curran J.H. 2008. The Practical Modelling of Discontinuous Rock Masses with Finite Element Analysis. 42nd US Rock Mechanics Symposium and 2nd U.S.-Canada Rock Mechanics Symposium (08–180).

Hoek, E. Carranza-Torres, C. Corkum, B. 2002. Hoek-Brown Failure Criterion: 2002 Edition. Proc. NARMS-TAC Conference 2001(1): 267–273.

Hoek, E. Diederichs, M.S. 2006. Empirical Estimation of Rock Mass Modulus. International Journal of Rock Mechanics & Mining Sciences 2006 (43): 203–215.

Laubscher, D.H. 1990. A Geomechanics Classification System For The Rating of Rock Mass In Mine Design. Journal of The South African Institute Of Mining And Metallurgy 90(10): 257–273.

Russo, A. & Hormazabal, E. 2019. Correlations Between Various Rock Mass Classification Systems Including Laubscher (MRMR), Bieniawski (RMR), Barton (Q), and Hoek Marinos (GSI) Systems. Geotechnical Engineering in The XXI Century: Lesson Learned and Future Challenge: 2806–2815.

Rustandi, E. Nila, E.S. Sanyoto, P. Margono, U. 1995. Geological Map of The Kotabaru Sheet, Kalimantan. Bandung: Pusat Survey Geologi.

Sapiie, B. & Cloos, M. 2004. Strike-slip Faulting in The Core of The Central Range of West New Guinea: Ertsberg Mining District, Indonesia . GSA Bulletin 116(3/4): 277–293.

Sapiie, B. & Rifiyanto, A. 2017. Tectonics and Geological Factors Controlling Cleat Development in the Barito Basin, Indonesia. J. Eng. Technol. Sci. 49(3): 322–339.

Schlunegger, F. Rieke-Zap, D. Ramseyer, K. 2007. Possible Environmental Effects on The Evolution of The Alps-Molasse Basin system. Swiss j. geosci. 100 (2007): 383–405.

Witts, D. Davies, L. Morley, R. 2014. Uplift of The Meratus Complex: Sedimentology, Biostratigraphy, Provenance and Structure. Proc. IPA 38(082).

Session 2 - Tunnels & caverns I

The Evolution of Geotech - 25 Years of Innovation – Hammah et al. (Eds)

Brittle failure assessment for TBM tunneling under high overburden: A case study from Pahang-Selangor raw water transfer tunnel project, Malaysia

Heyam H. Shaalan
Iraq University College, Basrah, Iraq

Mohd Ashraf Mohamad Ismail*
School of Civil Engineering, Universiti Sains Malaysia, Pulau Pinang, Malaysia

Romziat Azit
Jabatan Kerja Raya, Malaysia

Intan Norsheira Yusoff
School of Civil Engineering, Universiti Sains Malaysia, Pulau Pinang, Malaysia

ABSTRACT: Modelling a rock mass failure around an underground excavation is essential to evaluate the form, depth, and extent of the failure zone and achieve a sustainable tunnel support system, thereby improving underground excavation safety and minimizing the impact of high construction costs. This paper highlights the application of the 2D elastoplastic FEM analysis RS2 program to analyze the tunnel's critical section under overstressing conditions at the longest water transfer tunnel construction in Malaysia. Site observations have shown that high in-situ stresses have resulted in a failure zone on the tunnel sides, which has weakened the steel fiber shotcrete liner and led to a rock spalling. The depth of the failure observed in several areas with overburden more than 1000 m is around 0.3 m. A numerical analysis is performed to check the magnitude, shape, and depth of the actual failure. The computational modeling methods include the elastic, the elastic-perfect-plastic, the elastic-brittle-plastic models, and the cohesion-softening friction-hardening (CSFH) models. Parametric analysis of the strength parameters of the CSFH model is performed to explain the strength parameters' effect on the magnitude, shape, and depth of the failure. The conclusion states that no determinisitic analysis can successfully predict real failure. However, the modeling results using back analysis of the CSFH model with C_{peak} and $C_{residual}$ at 21.14 MPa and 6.34 MPa and φ_{peak} and $\varphi_{residual}$ at 16° and 52.34° verified the observed failure's depth and form as observed by the actual phenomena.

1 INTRODUCTION

The Pahang-Selangor Raw Water transfer project's main tunnel excavation was excavated under some high rock cover sections of the Main Range at over 1,000 m for over a total length of some 4 km. Some localized rock spalling was observed in TBM-2 tunnel excavation under a high rock cover of over 1,000 m, attributed to "overstressed" rock mass under high in-situ stress conditions. The in-situ stress state of the rock mass under high overburden will induce rock mass deformation with the needs to select the suitable type and extent of necessary tunnel support and reinforcement. In-situ stress measurements through hydraulic

*Corresponding author
DOI: 10.1201/9781003188339-17

fracturing and over coring (CCBO Compact Conical-ended Borehole Over coring method) were conducted before the main tunnel drives' commencement in optional sections Adit-2 and Adit-3 with a rock cover of 227 m and 157 m, respectively. Although located at a lower rock cover, the test results had indicated the in-situ stress state at accessible locations along the tunnel alignment without interfering or delaying the TBM tunnel excavations. The in-situ stress measurements for both locations at Adit 2 and 3 indicated the average k-values (horizontal to veliical stress ratio) above 1.0 with the direction of the maximum principal stresses (horizontal stress component) parallel to sub-parallel to the tunnel axis. Both test locations indicate a k-value of over 1.0, which appears to be conclusive with numerous insitu stress measurements worldwide, showing that the ratio k tends to be high at a shallower depth and decrease at depth (Brown and Hoek, 1978). Sheorey (1994), who developed an elastostatics thermal stress model of the earth crust, provides one explanation to this complicated matter by considering the curvature of the crust and variation of elastic constants, density and thermal expansion coefficients, which provides a tool to estimate the horizontal to vertical stress ratio k. Assuming a deformation modulus of some 50GPa for the high cover tunnel section, one could assume a k-ratio of 1.0 as reasonable.

In this study, the focus is in section TBM 2 at 23048m, in which the tunnel depth is approximately 1,002. The in-situ stress conditions along the tunnel are measured, and a series of initial stress measurements are performed to determine stress magnitude and direction at different locations along the tunnel (Figure 1). The methods of hydraulic drilling and compact conical-ended overcoring of the borehole were used. Results showed that the high in-situ stress is at the centre of the mountain, particularly in the section of TBM 2 (Azit et al. 2015), where the maximum principal stress is σ_1 = 28.76 MPa, the medium principal stress is σ_2 = 10.29 MPa, and the principal minimum stress is σ_3 = 5.17 MPa. The maximum principal stress (σ_1) can be observed in the vertical direction, whereas the horizontal stress is comparatively weak. The horizontal stress ratio to the vertical stress is approximately 0.38, as recommended by Azit et al. (2015).

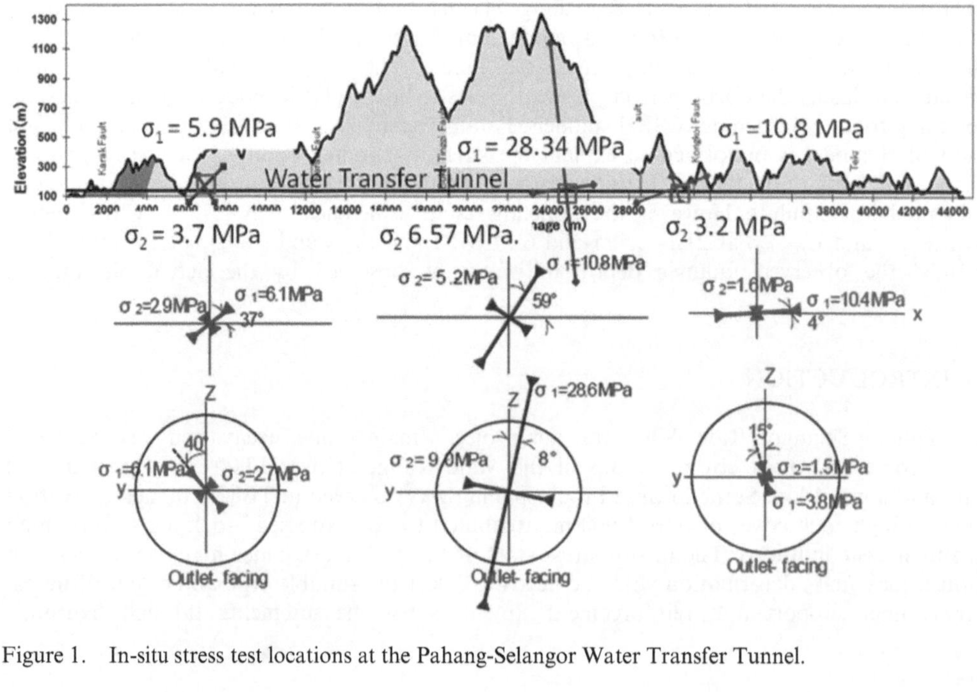

Figure 1. In-situ stress test locations at the Pahang-Selangor Water Transfer Tunnel.

2 ROCK SPALLING OBSERVATION

Rock spalling in the TBM-2 drive sidewalls (Figure 2) was observed over a length of some 20 m, mainly encountered from chainage 24+612 m to 24+592 m (TD 5550 – 5570 m) at a rock cover of 1,040 m to 1,050 m. The spalling occurred immediately after excavation after the cutter head has passed and rock fragments loosened at the finger shield area. Discontinuities and intact rock fracturing control observed Rock spalling and at given high rock cover of over 800 m and in Rock Class CI or B (Japanese Highway Rock Mass Classification) appears to be an indication for "overstressed" rock mass. The failures or "breakouts" location is mainly in the sidewalls, at and around the spring-line, indicating a maximum in-situ stress component at a vertical or near-vertical direction. The geological mapping records and rock strength tests in TBM-2 appear to be evident that spalling, the rock mass is slightly fractured, and lower rock strength was recorded. The insitu stress and induced stresses have exceeded the rock mass strength in that area.

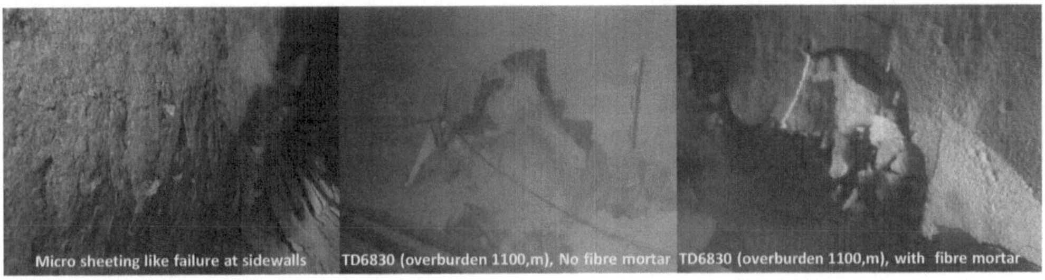

Figure 2. (a) Rock spalling at the tunnel sides (b) Rock fragments of the TBM 2 tunnel during the construction.

The rock mass surrounding a tunnel can be overstressed when either the intact Rock or the discontinuities (or a combination of both failure types) fail due to the stresses induced by the tunnel's excavation. In massive fresh Rock, tensile failure originating at flaws and defects (such as mineral grain boundaries) can occur when the maximum stress on the tunnel boundary exceeds about 40% of the uniaxial compressive strength of the Rock (Hoek and Marinos, 2010); such failures propagate along maximum principal stress trajectories forming thin plates parallel to the tunnel boundary.

Depending on the actual rock mass characteristics (strength, orientation, and intensity of discontinuities, shear/fault zones etc.) and the in-situ stress state (magnitude and directions of in-situ stress) such failures - Rock spalling and brittle rock failures are likely to be encountered over the remaining tunnel section (approximately over a length of some 3.5 km) under high rock cover (at and above 1,000 m overburden). In hard, brittle Rock, such a rock mass failure can be associated with significant energy release, known as "popping rock" or the extreme case "rock bursts".

3 NUMERICAL MODELLING OF THE BRITTLE FAILURE

RS2, the 2D FEM software used for geotechnical projects to estimate the stresses and displacements in the Rock around the underground openings, is used in this research. The geometric model is shown in the figure. The diameter of tunnel 3 is 5.2, and the model boundary extension is 10 times the tunnel diameter to confirm that it is not removed due to restraining effects (Lee et al., 2012). For the FE mesh, triangular elements are used with six nodes. Using RockLab, the rock mass strength was measured and compared to Hoek–Brown failure envelopes. The field observation showed weakness in the sidewalls up to 0.3 m depth, which weakened the steel-reinforced shotcrete lining SFRS. 4. The rock composition for the tunnel and SFRS shotcrete technological properties are shown in Table 1. Sophisticated modeling techniques are used to investigate the method of tunnel digging. In this way, the impact of stress can be spread over several parallel stages. The tunnel fill material is removed at the first stage, while the in-situ stresses are added gradually in stages. Multiple load-splitting is a technique to approximate the 3D effects of an advancing tunnel, using a 2D model. Eight stages of twelve.5% risk factor per stage are added to the model boundary.

In modeling brittle rock failure, conventional methods are adopted, namely elastic analysis, elastic-perfectly-plastic, and elastic-brittle-plastic models. In the elastic analysis model, the Hoek–Brown model with m = 0 and S = 0.11 is used. By adding the peak strength equivalent to the residual strength parameters, an elastic-perfectly-plastic model is also used. For failure simulation, the elastic-brittle-plastic model is adopted with m= 0 and S= 0.11 (as residual values). In the current research, the CWFS model proposed by Hajiabdolmajid et al. (2002) could not be used because its basic assumption demanded non-simultaneous changes in friction and cohesion.

The stress analysis program employed does not satisfy these demands. Also, it is difficult to obtain input data for such strain softening models. Therefore, with the Mohr-Coulomb model for the brittle failure simulation, an elastic-brittle-plastic model with the CSFH model presented by Edelbro (2010, 2009) is used. To demonstrate the effect of different strength parameters on the magnitude, shape, and depth of the failure, a parametric analysis for the CSFH model strength parameters is performed and investigate which parameters may capture the actual depth of failure.

Two indicators are utilized to represent the shape, magnitude, and depth of the failure. In the elastic analysis model, the strength factor in SF indicates excavation stability. The strength factor of a point is defined as the ratio of the rock mass's strength to the stress applied. A region with an SF of less than 1 may be deemed a loss. The failed shear yield elements are used as fallout indicators for the plastic models. If the shear bands connect to the excavation boundary, the region around the excavation fails (Sjöberg 1999 and Sandström 2003).

Table 1. Input parameters for the numerical modeling.

Rock Mass Properties			Shotcrete Parameters		
Term	Unit	Value	Term	Unit	Value
Value Intact Compressive Strength σ_{ci}	MPa	124	Young modulus, E	MPa	26224[1]
Material constant for intact rock m_i	–	32	Poisson ratio	–	0.2
Disturbance Factor D	–	0	Uniaxial compressive strength	MPa	30.7
Geological Strength Index GSI	–	67	Tensile strength	MPa	3.07[2]
mb	–	9.847	Steel fiber content	Kg/m^3	35
s	–	0.026	[1]According to ACI 318-08 (2008) [2]Assumption: 0.1 $f_{c,3}$ (Saurer et al. 2014)		
Friction angle ϕ_m	°	52.34			
Cohesive strength C_m	MPa	6.239			
Rock mass compressive strength σ_m	MPa	19.7			
Rock mass tensile strength σ_t	MPa	-0.322			
Young Modulus E	MPa	35516			

4 NUMERICAL RESULTS

As shown in Figure 3, the elastic-perfectly-plastic model's outcome is shown with the magnitude, including the shape and extent of the failure, which is more profound in comparison to the real results observed at the site (Figure 3a). This result agrees with Hajiabdolmajid et al (2002) results where the yield elements region failed in the shear of an elastic-perfect-plastic model to form an arch. However, this model is relatively inappropriate for modeling a brittle rock failure because the material weakening is ignored (Hajiabdolmajid et al. 2002). Figure 3b shows the result of the elastic-brittle-plastic model mr = 0 and sr = 0.11 as presented by Martin et al. (1999). This method overestimates the extent of the failure zone and cannot capture the actual depth of failure.

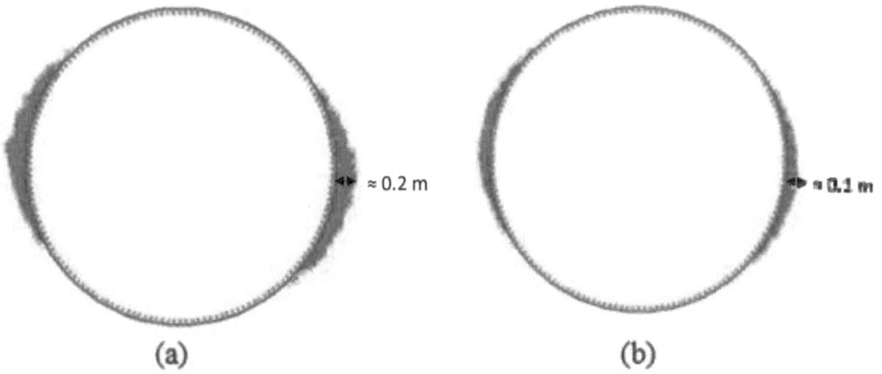

Figure 3. Yield elements failed in the shear for the (a) elastic-perfectly-plastic and (b) elastic-brittle-plastic models.

A parametric analysis is conducted to define the effect of the CSFH model's strength parameters and determine which parameters may simulate the feasible depth of failure around the tunnel. The Mohr-Coulomb model is applied to this study, and RockLab calculates the peak strength parameters. Based on the outcomes of prior research, the values are presumed.

The final simulated failure zone is shown in Figure 4, as reflected by the failed shear yield components. The input parameters for the CSFH model, as shown in Figure 5, better reflect the degree, shape, and depth of the brittle failure observed. The residual strength of cohesion is low, while the angle of residual friction is higher. This finding is considered valid by referring to Cai and Kaiser (2014), who verified that when the brittle rock fractures, the cohesive strength is lost; hence, the remaining strength part is only taken by the friction strength.

Figure 4. Simulated failure zone in the tunnel sidewalls utilizing the CSFH model.

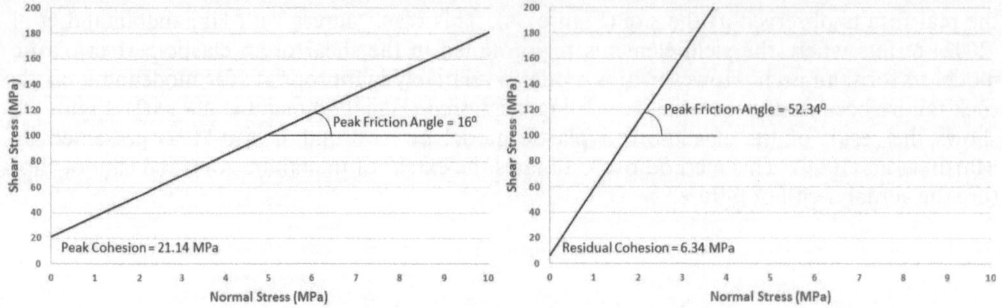

Figure 5. Peak and residual strength envelopes of the Rock utilizing the CSFH model.

5 MONITORING AND TESTING FOR ROCKBURST RISK ASSESSMENT

The extreme case of brittle failure-rockburst is less likely in this study area. However, to assess the extent and magnitude of brittle rock failure further, monitoring and investigation are suggested in the similar project such as (i) rock core testing, (ii) drilling core into the tunnel walls where failure had occurred to assess the depth of brittle failure zone (iii) strengthening of tunnel convergence measurements and (iv) in-situ stress measurement at high tunnel overburden to evaluate the magnitude and direction of the in-situ stress and to get the reliable ratio of horizontal to vertical stress. Tunnel engineers should be noted that, even with continuous monitoring, it is almost impossible to prevent the brittle failure from initiating after tunnel excavation, even using TBM. Thus, remedial measures in the zone of potential brittle rock failure should aim to retain broken Rock and control rock mass dilation, such as using shotcrete with wire mesh, shotcrete with steel fiber, and installing rock bolts.

6 CONCLUSION

The rock spalling phenomenon had occurred in the critical section of the Pahang Selangor Raw Water Transfer Tunnel under the overburden of more than 800 m with a rock class CI and B according to the JH Rock Mass Classification System. The depth of failure is approximately 0.3 m, which has damaged the steel fibre reinforced shotcrete liner installed as a countermeasure to prevent fly rock due to rockburst. Various material models are used, namely elastic analyses, elastic-perfect-plastic, and elastic-brittle-plastic models with (mr = 0 and Sr = 0.11) and the CSFH model. The strength and yield elements that failed in the shear are used as failure indicators for both the elastic analysis and the plastic models. The elastic-brittle-plastic analysis of the CSFH model with the peak friction angle16°, the peak cohesion, 21.14 MPa, residual cohesion, 6.34 MPa, and residual friction angle, 52.34° shows an appropriate agreement with the extent, shape, and depth of the failure observed at the site.

REFERENCES

Azit, R. and Ismail, M. A. M., 2016, Modeling Stress-Induced Failure for Deep Tunnel Excavation of Pahang-Selangor Raw Water Transfer Project, *9th Asian Rock Mechanics Symposium*, Bali, Indonesia.

Azit, R., Ismail, M. A. M., Syed Zainal S. F. and Mahmood, N., 2015, Rock overstressing in deep tunnel excavation of Pahang-Selangor raw water transfer project, *Applied Mechanics and Materials*, 802, 16–21.

Brown, E.T. and Hoek, E., 1978, August. Trends in relationships between measured in-situ stresses and depth. In *International Journal of Rock Mechanics and Mining Sciences & Geomechanics Abstracts* (Vol. 15, No. 4, pp. 211–215). Pergamon.

Cai, M., and Kaiser, P., 2014, In-situ Rock spalling strength near excavation boundaries, Rock Mech Rock Eng, 47, 659–675.

Edelbro, C., 2010, Different approaches for simulating failure in two hard rock mass cases a parametric study, *Rock Mechanics and Rock Engineering*, 43, 151–165.

Edelbro, C., 2008, strength, fallouts and numerical modelling of hard rock masses, PHD thesis, Lulea University of Technology Luleå, Sweden.

Edelbro, C., 2009, Numerical modelling of observed fallouts in hard rock masses using an instantaneous cohesion-softening friction-hardening model, *Tunnelling and Underground Space Technology*, 24, 398–409.

Hajiabdolmajid, V., Kaiser, P. K., and Martin, C. D., 2002, Modelling brittle failure of Rock, *Int J Rock Mech Min Sci*, 39, 731–741.

Hoek, E. and Marinos, P., 2009, Tunnelling in overstressed Rock, Rock Engineering in Difficult Ground Conditions - Soft Rocks and Karst, Vrkljan, I (ed), London: Taylor and Francis Group, 49–60.

Kawata, T., Nakano, Y., Matsumoto, T., Mito, A., Pittard, F. and Azman, A. A. S., 2014, The Relationship between TBM Data and Rockburst in Long-Distance Tunnel, Pahang-Selangor Raw Water Transfer Tunnel, Malaysia, *8th Asian Rock Mechanics Symposium*, Sapporo, 256–271.

Lee, S., Ismail, M. and Ng, S. M. 2012, The Evaluation of tunnel behaviours under high rock stress using numerical analysis method, *Electronic Journal of Geotechnical Engineering*, 17.

Matin, C., Kaiser, P. and McCreath, D., 1999, Hoek-Brown parameters for predicting the depth of the brittle failure around tunnels, *Canadian Geotechnical Journal*, 36, 136–151.

Sandström, D., 2003, analysis of the virgin state of stress at the Kiirunavaara mine, Licentiate thesis, Division of Rock Mechanics, Luleå University of Technology.

Sheorey, P.R., 1994, February. A theory for in situ stresses in isotropic and transverseley isotropic Rock. In *International journal of rock mechanics and mining sciences & geomechanics abstracts* (Vol. 31, No. 1, pp. 23–34). Pergamon.

Sjöberg, J., 1999, analysis of large scale rock slopes, Doctoral thesis, Division of Rock Mechanics, Luleå University of Technology.

The Evolution of Geotech - 25 Years of Innovation – Hammah et al. (Eds)

Use of continuum and pseudo-discontinuum FEM models in stepwise verification of the FDEM for simulating damage around tunnels in brittle rock

S.L. Markus & M.S. Diederichs
Queen's University, Kingston, ON, Canada

ABSTRACT: Numerical modelling of excavations in rock has advanced considerably in recent decades. While continuum numerical models form their basis in methods which can be verified by analytical solutions, discontinuum and hybrid numerical modelling software are challenging to verify. This necessitates the development of processes that can verify individual aspects of complex models. The hybrid finite-discrete element method (FDEM) allows for the numerical representation of progressive fracture in a simulated elastic material. The FDEM is a powerful tool for modelling instability around tunnels in brittle rock; however, significant verification of the method is required for its use in predictive modelling in critical engineering projects. To verify the FDEM for the purpose of modelling instability around tunnels in brittle rock, a multi-method and multi-scale stepwise verification approach is proposed.

1 INTRODUCTION

Utilization of underground space has grown with the demand for energy, minerals, and urban infrastructure. With the increase in underground development, and the development of new technologies, excavations are constructed in increasingly complex ground, including deep excavations subject to high in situ stress conditions. In order to produce the best possible predictions of geomechanical behaviour for critical engineering projects, the use of state-of-the-art numerical modelling tools is necessary. For highly consequential projects such as Deep Geological Repositories (DGRs) for underground nuclear waste storage, human access is restricted to only the beginning of the project design life, and thus building confidence in forward modelling is critical.

1.1 *Overview of stepwise verification*

The hybrid FDEM is a relatively new, state-of-the-art numerical modelling tool that has promising applications for DGR modelling. Due to the critical nature of DGR projects, the method must be meticulously verified. While continuum methods are easily verified through analytical solutions, the increased complexity of the FDEM necessitates a verification process that builds in complexity. The proposed verification process is shown in Figure 1.

1.2 *Lac du Bonnet granite*

For the purpose of verification of the FDEM, the Lac du Bonnet (LdB) granite was selected. Decades of laboratory research data is available for this rock type, as well as field scale experiments, including a mine-by-experiment test tunnel, at the AECL Underground Research

DOI: 10.1201/9781003188339-18

Laboratory (URL) in Manitoba. The LdB granite has a Young's modulus and Poisson's ratio of 69±5.8 GPa and 0.26±0.4, respectively, for samples taken at the URL (Martin 1993).

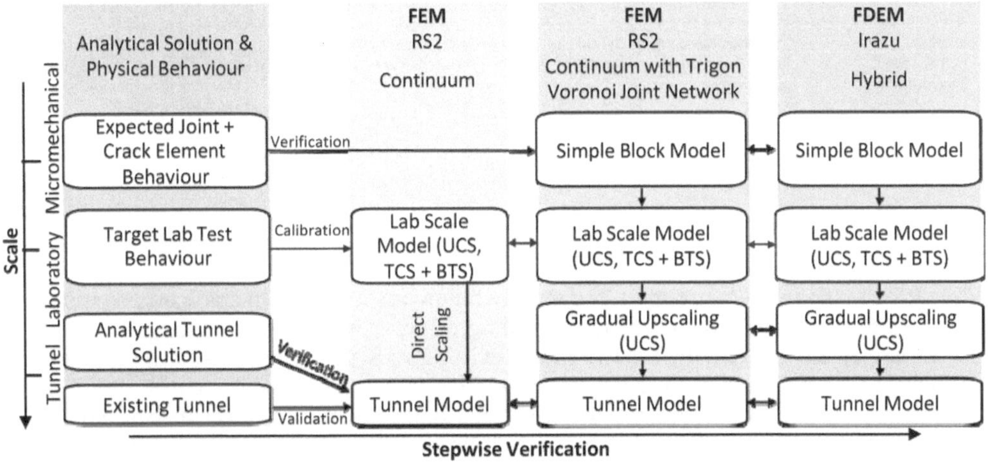

Figure 1. Stepwise verification process for modelling tunnels in brittle rock using the FDEM, including calibration, the study of scale effects, and comparison of FEM and analytical solutions.

2 LABORATORY SCALE NUMERICAL MODELLING

In FEM models, materials can be represented as a continuum, or a pseudo-discontinuum. In the latter approach, to simulate grain-based models (GBMs), discontinuities are represented by joint elements, using a voronoi or trigon representation, analogous to grains of a material at the laboratory scale. The pseudo-discontinuum approach is distinct from true discontinuum approaches because true detachment is not allowed along interfaces. In pseudo-discontinuum models, fracture propagation is represented by yield along joint elements, and detachment is represented by a prescribed decrease in joint stiffness.

2.1 *Continuum finite element approach for simulating laboratory tests*

The 2D finite element program RS2 version 9.0 (RocScience 2017) was used to simulate physical laboratory tests through a staged, pseudo-static analysis, where loads or displacements are applied incrementally. Continuum FEM models are used to compare with and verify behaviour of pseudo-discontinuum FEM models. Models of UCS, TCS and BTS tests were created with 2D specimen dimensions equivalent to laboratory testing ISRM standards of H=150 mm and D=60 mm (Bieniawski and Hawkes 1978, Bieniawski and Bernede 1979). The calibration process for laboratory scale models is shown in Figure 2.

Based on sensitivity analyses, it was found that the maximum loading rate to capture consistent material behaviour was 0.025 mm per platen, per stage. Elastic input parameters taken directly from material testing produced results, consistent with plane strain equations (emergent E=72.5 GPa & ν =0.283). In RS2, material strengths are commonly defined by the Mohr-Coulomb (MC) failure criterion. MC strength values are measured from laboratory tests and can be used directly in models as peak strength parameters (c=36.4 MPa, ϕ=52.3°, T=10 MPa). It is considered that all cohesion and tensile strength are lost at yield, but friction is maintained, as is consistent with theoretical brittle rock behaviour, as well as FDEM models. Using these MC parameters, the LdB strength envelope from laboratory testing data (Martin 1993) is matched by modelling results.

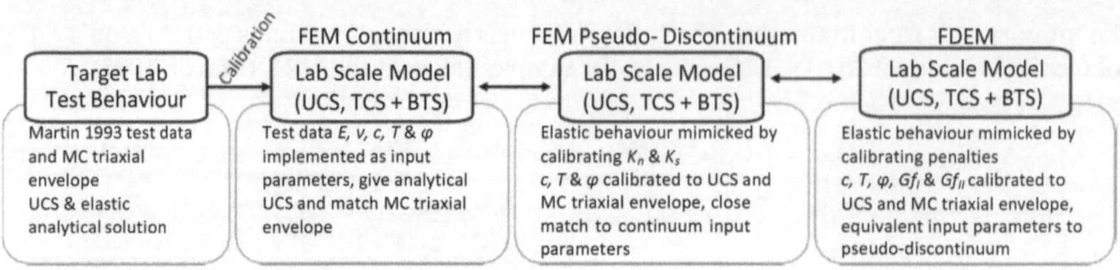

Figure 2. Calibration process for laboratory scale models.

2.2 *Pseudo-discontinuum finite element approach for simulating laboratory tests*

While RS2 is a continuum program, progressive failure associated with crack propagation can be simulated using a GBM, through the yielding of pre-existing discontinuities. Discontinuities in RS2 are conventionally used to model existing rock joints, bedding, and faults within a rockmass, so to instead implement discontinuities as potential fracture pathways, a "joint network" of voronoi or trigon-voronoi tessellated blocks is implemented in the model. The effect of discontinuities is minimized by calibrating the stiffness properties of the discontinuities (K_n and K_s) to a high enough value to not reduce the overall stiffness behaviour of the modelled material. This approach allows the jointed model to behave in an elastically equivalent manner to a continuum simulation. To determine normal and shear stiffness terms that replicate the correct elastic response for LdB granite, a series of UCS simulations with varying normal and shear stiffness terms were completed. In the case of LdB granite, an adequately correct elastic response can be simulated with joint stiffness terms 10^5 times greater than the laboratory obtained Young's Modulus (ie. K_n=K_s =69×105 GPa/m). Above this threshold, the change in overall stiffness is not significant. Below this threshold, overall stiffness is sensitive to K_n and K_s. Reducing the shear stiffness value by half, while keeping the value for normal stiffness at 10^5E, further improves stress-strain response. The selection of a K_n value of 10^5E, and K_s value of either K_n or $0.5K_n$ is sufficient for simulating LdB granite. In this work, equal values of K_n and K_s were selected to better align with FDEM tangential and normal penalty values, which are found to simulate a sufficient elastic response for LdB granite in UCS test simulations in the mm-based Irazu unit system when equal in magnitude.

To represent fracturing as a failure mechanism, the contained material within voronoi or trigon grains is modelled as elastic, while the strength parameters of the discontinuities are calibrated so that the overall model behaviour matches the desired material strength. The combination of MC joint strength parameters (c, ϕ, T) which are required to match UCS, BTS, and TCS strengths of a given material can be non-unique, and thus the selection of these parameters must be done through a combination of sensitivity analyses and comparison to laboratory observations. Consideration is given to the purpose of the study, which is to match behaviour in FDEM models.

Joint tensile strengths can be directly related to the results of a direct tension test, which can be modelled in RS2, and therefore the target tensile strength can be used as an input for *T*. Tensile strength is lost after a bond is broken, so an input of zero residual tensile strength is appropriate. The shear strength of a material is composed of cohesional and frictional components. The input values for peak cohesion, c and friction, ϕ, are calibrated to match the desired shear strength of the material, determined by the triaxial strength envelope. It is generally accepted that friction is the only parameter which retains strength post-peak, as cohesion and tensile strength are rapidly lost when material fractures (Schmertmann and Osterberg 1960, Martin 1997, Diederichs 1999, 2003, 2007). It is therefore reasonable to set both residual tension and cohesion to zero. To relate results to FDEM models, a residual friction angle equal to peak is selected. To obtain parameters which simulate the target strength envelope for the modelled material the values of peak cohesion, c_p, and friction ϕ *(*peak=residual*)*, were systematically varied.

Combinations of c_p and ϕ that simulate a specific UCS (212+/-3 MPa and 220+/-5 MPa) are related linearly as shown in Figure 3. From these input parameter values, triaxial tests are conducted to obtain the triaxial strength envelope for the target material. Peak cohesion and friction values of 70 MPa and 52°, respectively, simulate an appropriate triaxial envelope for LdB granite, and are consistent with continuum model results.

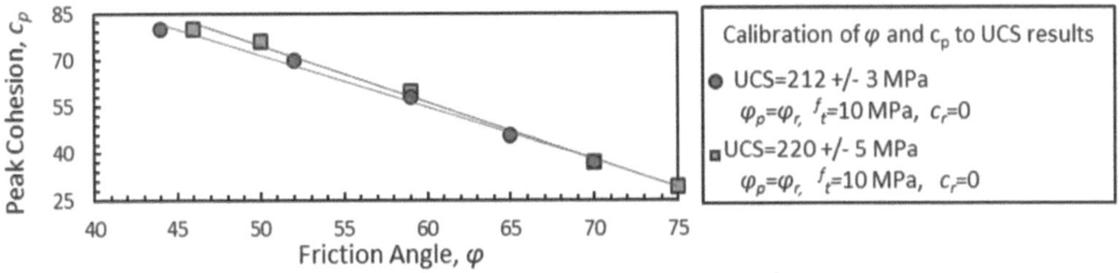

Figure 3. RS2 FEM pseudo-discontinuum calibration of ϕ and C_p to UCS results for LdB Granite.

Table 1. FDEM crack element parameter combinations to simulate the triaxial envelope for LdB granite.

UCS FDEM simulation ID	1	2*	3	4
Fracture Penalty, p_f (GPa)	690	690	6900	6900
Normal Penalty, p_n (GPa mm)	690	690	690	690
Tangential Penalty, p_t (GPa/mm)	690	690	690	690
Cohesion, c (MPa)	40	34	44	34
Tensile Strength, f_t (MPa)	10	10	10	10
Friction Coefficient, μ	1.28 (52°)	1.28 (52°)	1.54 (57°)	1.43 (55°)
Mode I Fracture Energy, G_{fI} (N/m)	10^{-3}	10	10^{-3}	10
Mode II Fracture Energy, G_{fII} (N/m)	10^{-3}	150	10^{-3}	100

Notes: *Fracture pattern for UCS-FDEM-2 is shown in Figure 4.

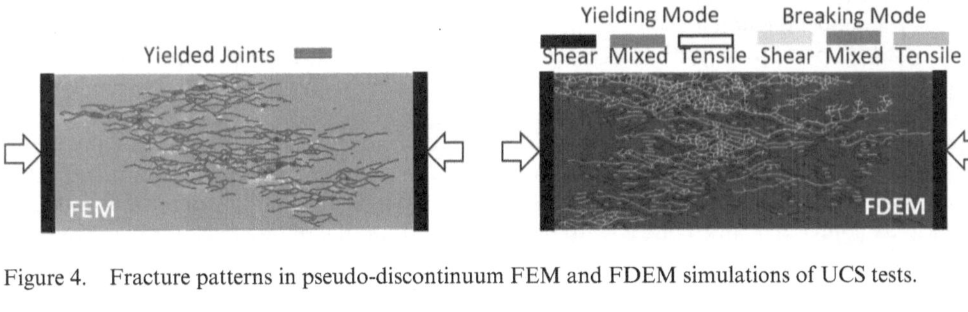

Figure 4. Fracture patterns in pseudo-discontinuum FEM and FDEM simulations of UCS tests.

2.3 *Hybrid finite-discrete element approach for simulating laboratory tests*

The FDEM software Irazu 2D (Geomechanica 2019) can be used to explicitly represent damage and fracture in material. Unlike the pseudo-discontinuum, material detachment can occur in FDEM models. The calibration process for laboratory scale models, outlined in Figure 2 incorporates the process suggested by Tatone and Grasselli (2015) for FDEM models and consists of iterative steps, including running a series of UCS, BTS and TCS models.

From sensitivity analyses, it was found that there is a trend of increasing strength with element size. An element size of 2 mm is appropriate to minimize mesh dependency. A sensitivity analysis of platen velocity showed that a velocity of 0.1 m/s per platen was satisfactory.

To determine penalty terms that replicate the correct elastic response for LdB granite, a series of UCS simulations with varying penalty terms were completed, holding the bulk elastic parameters constant at laboratory acquired values (E=69 GPa, v=0.22), following the approach described by Tatone and Grasselli (2015). Increasing the tangential penalty p_t has the effect of stiffening the lateral strain response and does not have a noticeable effect on the axial strain response. Increasing the normal penalty p_n has the effect of stiffening the axial strain response and softening the lateral strain response. Increasing the fracture penalty, p_f produces an overall stiffened response in the lateral and axial directions. For LdB granite, an adequate elastic response can be simulated with penalty terms 10 times greater than the Young's Modulus (E), and bulk elastic inputs (E and v) equal to those obtained in laboratory testing.

The strength of the crack elements is controlled by tensile strength (f_t), cohesion (c), fracture energies (Gf_I and Gf_{II}) representing the energy required to break the contact in tension (Mode I) and shear (Mode II), respectively, and a frictional coefficient (μ=tanϕ_i). Since the calibration target for modelling (triaxial envelope) is defined by just three parameters, an iterative approach to calibration was taken. It was found that multiple combinations of input parameters simulate appropriate fracture patterns for LdB granite and mimic the triaxial strength envelope in triaxial models. These combinations are non-unique and are summarized in Table 1.

3 TUNNEL SCALE MODELLING

To complete the tunnel scale portion of the stepwise verification approach for the FDEM, the same tunnel model is simulated by solutions of gradually increasing complexity, from continuum FEM, to pseudo-discontinuum FEM, to FDEM modelling. Material parameters appropriate for tunnel scale models, where element sizes are larger, are determined by employing a gradual upscaling technique, based on the UCS models of gradually increasing scale and element to diameter ratio (Markus et al. 2019). The scale adjusted parameters are shown in Table 2.

Figure 5 compares DISL FEM (Diederichs 2003, 2007) representations of the AECL URL mine-by-experiment test tunnel to pseudo-discontinuum FEM and FDEM models. There is difficulty in modelling notch geometry in pseudo-discontinuum models. As is observed in failure profiles for pseudo-discontinuum models (Ps.Disc.-FEM-1 in Figure 5), most yield occurs in tensile areas of the model, rather than areas under highly anisotropic stress (tunnel roof and floor in the URL test tunnel).

In models Ps.Disc.-FEM-1 and FDEM-1 (as shown in Table 2), near-equivalent input parameters between FEM and FDEM models do not equate to similar failure profiles. There is difficulty in including post-yield frictional strength in pseudo-discontinuum FEM models, as joint elements continue to take on load in the model, leading to fracture being reduced in the area of expected brittle spalling. While tensile fracturing occurs in the sidewalls in both models, the extent of tensile fracture in the pseudo-discontinuum FEM model is more prominent. Brittle spalling fracture occurs in the roof and floor in the FDEM model, forming a notch geometry, and does not occur in the pseudo-discontinuum FEM model.

A better agreement between pseudo-discontinuum FEM and FDEM models is shown in models Ps.Disc.-FEM-2 and FDEM-2. Limited tensile fracture occurs in the sidewalls in both models. There is close match in the depth of the modelled notch between pseudo-discontinuum and DISL models. Though the extent of fracture in the FDEM model is greater, the FEM profiles roughly correspond to notch formed by interconnected cracks in the roof in the FDEM model.

Table 2. Summary of modelling parameters used for tunnel modelling. Parameters derived from an upscaling process.

Input Parameter	Ps.Disc.-FEM-1	Ps.Disc.-FEM-2	FDEM-1	FDEM-2
Ave. Element/Voronoi size (cm)	3	3	3	3
Cohesion, c (MPa)	20	50	20	50
Friction Angle, ϕ	52°	10°	52°	*60°
Tension, T (MPa)	5	10	5	10
Mode I Fracture Energy, G_{fI} (N/m)	-	-	0.01	300
Mode II Fracture Energy, G_{fII} (N/m)	-	-	0.1	1900
Fracture Penalty, p_f (GPa)	-	-	690	690
Normal Penalty, p_n (GPa m)	-	-	690	690
Tangential Penalty, p_t (GPa/m)	-	-	690,000	690

Notes: For FEM models, all residual parameters are zero. The low friction angle in Ps.Disc.-FEM-2 is comparable to FDEM-2, as the 60° friction angle does not mobilize due to low tangential penalty.

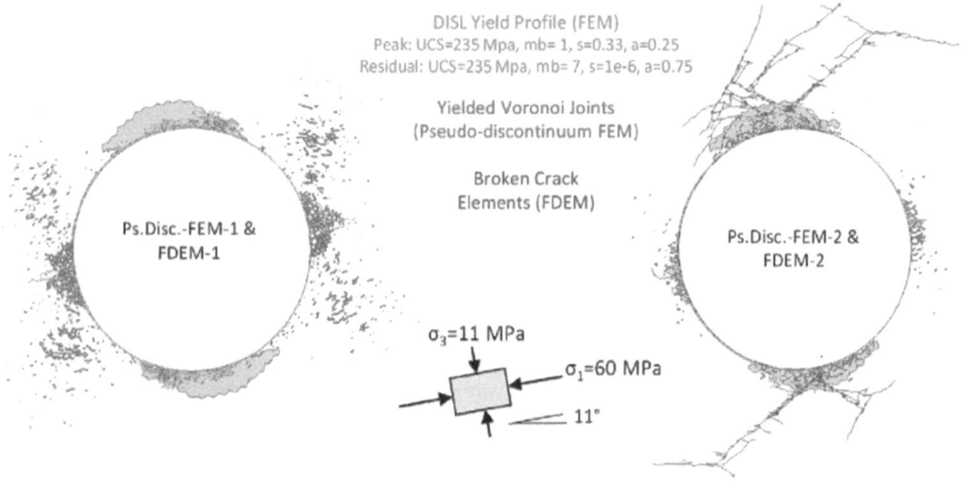

Figure 5. Comparison of fracture patterns in FEM and FDEM models around a simulated URL test tunnel.

4 CONCLUSIONS

Relating the effects of input parameters in continuum FEM, pseudo-discontinuum FEM, and FDEM on emergent behaviour is a step to verification of FDEM models of brittle rock tunnels.

On a laboratory scale, input parameters for strength were not shown to be directly equivalent between FEM and FDEM. A higher value of cohesion was required in pseudo-discontinuum FEM models of UCS tests to achieve the same result as in FDEM models. While a direct relationship is not formed, it is noted that some numerical instability is observed in pseudo-discontinuum models, which can lead the simulation to experience yield earlier than anticipated. The equivalency of the breaking strength of UCS tests modelled in Irazu can be related to a critical accumulation of cracks in equivalent pseudo-discontinuum voronoi and trigon jointed RS2 models. The correlation of crack accumulation in RS2 to UCS in Irazu is approximate; however, when equivalent input parameters are implemented in both methods, the results are comparable.

On a tunnel scale, the case study of the URL test tunnel was used as a validation tool for DISLFEM, pseudo-discontinuum FEM, and FDEM models. Tunnel models were simulated with equivalent parameters in pseudo-discontinuum FEM and FDEM models, where the effect of non-similar input parameters was minimized. Comparing multiple combinations of input parameters across both methods, the best agreement of fracture propagation was found for models with low frictional inputs. This conclusion is consistent with failure theories for rock, which suggest that friction only mobilizes after inter-grain shearing breaks cohesive bonds and the rock begins to dilate (Martin 1997, Diederichs 1999, 2003, 2007), and draws parallels with continuum numerical modelling approaches for brittle rock (Hajiabdolmajid et al. 2002, Diederichs 2003, 2007).

REFERENCES

Bieniawski, Z.T., and Bernede, M.J. 1979. Suggested methods for determining the uniaxial compressive strength and deformability of rock materials: Part 1. Suggested method for determination of the uniaxial compressive strength of rock materials. *International Journal of Rock Mechanics and Mining Sciences & Geomechanics Abstracts*, 16(2): 137.

Bieniawski, Z.T., and Hawkes, I. 1978. Suggested Methods for Determining Tensile Strength of Rock Materials. *International Journal of Rock Mechanics and Mining Sciences & Geomechanics Abstracts*, 15 (6): 99–103.

Diederichs, M.S. 1999. *Instability of hard rock masses: the role of tensile damage and relaxation*. PhD Thesis, University of Waterloo, Waterloo.

Diederichs, M.S. 2003. Manuel Rocha Medal Recipient Rock Fracture and Collapse Under Low Confinement Conditions. *Rock Mechanics and Rock Engineering*, 36(5): 339–381.

Diederichs, M.S. 2007. The 2003 Canadian Geotechnical Colloquium: Mechanistic interpretation and practical application of damage and spalling prediction criteria for deep tunnelling. *Canadian Geotechnical Journal*, 44(9): 1082–1116.

Geomechanica Inc. 2019. Irazu 2D Geomechanical Simulation Software. Toronto, Ontario, Canada.

Hajiabdolmajid, V., Kaiser, P.K., and Martin, C.D. 2002. Modelling brittle failure of rock. *International Journal of Rock Mechanics and Mining Sciences*, 39(6): 731–741.

Markus, S., Diederichs, M., & Vazaios, I. 2019. Establishing a Stepwise Verification and Upscaling Process for Modelling Brittle Failure in Rock Using the FDEM Method, *14th Annual International Congress on Rock Mechanics, Iguassu Falls, September 13-18, 2019*, London: Taylor & Francis.

Martin, C.D. 1993. *The strength of massive Lac du Bonnet granite around underground openings*. PhD Thesis, University of Manitoba, Winnipeg.

Martin, C.D. 1997. Seventeenth Canadian Geotechnical Colloquium: The effect of cohesion loss and stress path on brittle rock strength. *Canadian Geotechnical Journal*, 34(5): 698–725.

Schmertmann, J.H., and Osterberg, J.O. 1960. An experimental study of the development of cohesion and friction with axial strain in saturated cohesive soils. *Research conference on shear strength of cohesive soils. Colorado, June 1960*. New York: ASCE.

Tatone, B.S.A., and Grasselli, G. 2015. A calibration procedure for two-dimensional laboratory-scale hybrid finite–discrete element simulations. *International Journal of Rock Mechanics and Mining Sciences*, 75: 56–72.

Rocscience. 2017. RS2 9.0. Toronto, Ontario, Canada.

The Evolution of Geotech - 25 Years of Innovation – Hammah et al. (Eds)

Numerical investigation of the relationship between the inflow rate to the tunnel, block volume and block surface area

A. Shahbazi, A. Saeidi & R. Chesnaux
Université du Québec à Chicoutimi (UQAC), Chicoutimi, Québec, Canada

ABSTRACT: The relevancy of the flow rate into the tunnel to the block volume and block surface area has been investigated. A set of numerical simulations with Itasca 3DEC version 7.0 is performed in order to define each parameter's impact on the inflow rate to the tunnel. The numerical model consists of a fractured rock mass that includes three joint sets with a fixed value of the hydraulic aperture and level of water table. The tunnel has a circular cross-section and the fluid flows only through the area between the fracture planes. Different values of the block volume and block surface are achieved by variation of the orientation and spacing of the joint sets. The block volume and surface are calculated both by analytical and numerical methods. For this, a new analytical model is developed for calculation of the block surface area. Based on the results of the numerical simulations of the inflow rate, it is deduced that the inflow rate increases by decrease of the block volume and block surface. Among them, the relationship between the inflow rate and block volume is more relevant than block surface area. In addition, it is proved that the common analytical method for calculation of the block volume needs to be revised for obtaining more reliable results.

1 INTRODUCTION

Estimation and control of the water seepage during tunnel excavation (Liu, Zhao et al. 2020), failure of the tunnel support system and hindering the excavation rate (Butscher 2012) are probable risks relating to the rate of the water seepage rate into the tunnel. Various methods including analytical, numerical and empirical are developed for estimation of the water inflow rate to the tunnel. The empirical equations are mainly based on the data that gained from the field works and mostly focused on the impact of the depth on the value of the inflow rate. They showed that by increase in depth, the inflow rate increases at first and then decreases (Zhang and Franklin 1993). Analytical equations were specifically developed for the homogeneous and isotropic formations using Laplace equation (Shahbazi, Saeidi et al. 2020) that basically need to pre-determine the equivalent permeability of the rock mass. However, an analytical model was recently developed that defines the inflow rate to the tunnel without need to predefine the equivalent permeability of the rock mass (Shahbazi, Chesnaux et al. 2021). The impact of geometrical characteristics of the discontinuities and overburden load (Wang, Bi et al. 2020) as well as the water head above the tunnel and anisotropy of the formation (Shahbazi, Saeidi et al. 2020) on the seepage rate have been studied by the numerical methods.

Like other geometrical characteristics of the rock mass, block characteristics e.g. block volume and block surface could be measured by empirical, analytical and numerical methods. From the field surveying, it is possible to directly measure the value of these parameters. However, this method is limited to the surface blocks and interior blocks are not accessible for being measured (Yarahmadi, Bagherpour et al. 2015). By the analytical methods and using

DOI: 10.1201/9781003188339-19

mathematical equations, the block geometries could be specified (Palmstrom 1995). However, the simplifying assumptions that used for developing the analytical models could produce uncertainty in estimation of the block geometries of the rock mass. Other than these methods, the block geometry could be calculated using the numerical simulations (Stavropoulou and Xiroudakis 2020). The time and equipment needed for performing the numerical simulations is the limitations for using this method.

In this paper, the relevancy of the inflow rate to the tunnel to rock block characteristics such as block volume and block surface area has been investigated using Itasca 3DEC version 7.00 software. It is assumed that the fluid flows only through the fractures and rock mass include 3 joint sets with various values of orientation and spacing. A series of numerical simulations have been designed for evaluation of the relationship between inflow rate to the tunnel, block volume and block surface area. The block characteristics are calculated by analytical and numerical methods and for this purpose, a new method for calculation of the block surface area is developed by the authors.

2 BLOCK GEOMETRY CALCULATION

As it is described in section 1, block geometry could be calculated using empirical, analytical and numerical methods. In this article, the block geometry is determined by analytical and numerical methods. For this purpose, the 3DEC software is used for numerical calculations and a new analytical method is developed for determination of the block surface. Furthermore, an analytical model that is previously developed, is used for calculation of the block volume. In section 2.1 the analytical and section 2.2 the numerical methods for block geometries calculation will be explained.

2.1 *Analytical methods for calculation of the block geometries*

In this paper, rock mass includes three joint sets with fixed values of spacings, dip and dip direction and model does not contain random joints. Based on the above mentioned assumptions, the volume of the rock blocks could be defined by equation (1) (Palmstrom 1995, 2005)

$$V_b^A = \frac{S_1 \times S_2 \times S_3}{\sin \gamma_1 \times \sin \gamma_2 \times \sin \gamma_3} \tag{1}$$

Where V_b^A is the analytically calculated block volume that is constructed by the cross of 3 joint sets, S_1, S_2 and S_3 are joint sets spacings and γ_1, γ_2 and γ_3 are the angles between each pairs of joint sets.

A number of analytical equations were developed for calculation of the block volume other than equation (1) including the studies of Barton (Barton, Lien et al. 1974), Palmstrom (Palmstrom 1982) and Latham (Latham, Van Meulen et al. 2006). However, equation (1) is a commonly used equation in the investigations and field works and hence, it is used in the current study.

The surface of the block that is formed by three joint sets, could be calculated using mathematical equations. The method for developing this equation is briefly described via Figure 1.

Apart from the blocks that are cut by the model boundaries, the geometries of all blocks are identical. An intact block is a block that is not cut by the boundaries of the model and is selected for model development, as illustrated in Figure 1b. If each block edge be marked by a vector (A, B and C in Figure 1c), cross product of each pair of vectors A, B and C demonstrates the surface area of faces of the block. According to the vector multiplication rules, the surface area of the block could be calculated by equation (2).

$$S_b^A = 2 \times (|A \times B| + |A \times C| + |B \times C|) \tag{2}$$

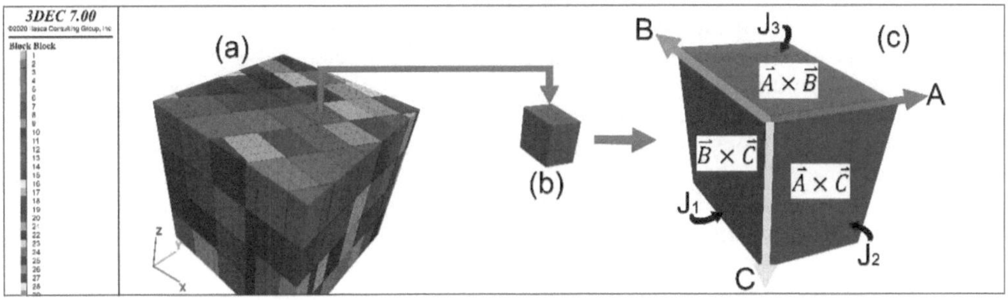

Figure 1. Analytical calculation method for block surface area. (a) a numerical model including three joint sets, (b) intact block that is not cut by the model boundaries, (c) edge vectors (A, B and C) and surface area of each side of the block.

Where S_b^A is the analytical block surface area and A, B and C are the edge vectors. All the edge vectors could be determined by knowing the dip, dip direction and spacings of three joints sets.

2.2 *Comparison of analytical and numerical block geometries*

The block volume and surface could be analytically calculated using equations (1) and (2), respectively. In addition, numerical simulation with 3DEC software is used to numerically calculate the block geometry. For this purpose, 13 cases that are listed in Table 1 are selected for comparison of numerical and analytical block geometries.

Table 1. Characteristics of discontinuities that used for comparison of analytical and numerical calculation of the block volume and block surface.

	Joint set 1			Joint set 2			Joint set 3			Block surface (m^2)		Block volume (m^3)	
Case	DIP	DD	Spacing (m)	DIP	DD	Spacing (m)	DIP	DD	Spacing (m)	Analytical	Numerical	Analytical	Numerical
1	23	30	0.34	20	10	2.05	27	320	0.14	320.2	320.2	4.31	15.14
2	90	350	0.39	20	352	0.14	90	10	0.39	1.63	1.63	0.07	0.07
3	73	25	0.35	61	352	0.35	84	8	5.91	54.15	54.16	8.55	4.6
4	22	25	0.34	90	350	0.98	60	0	0.35	49.73	49.71	0.36	3.65
5	90	350	0.39	90	10	0.39	32	50	2.05	12.08	12.07	1.03	1.08
6	90	350	5.91	90	70	0.14	90	30	5.2	N/A	N/A	10.57	N/A
7	20	350	0.34	22	25	1.37	61	354	5.2	131.54	131.53	24.9	17.02
8	54	65	0.34	90	350	0.98	90	30	0.35	4.21	4.2	0.25	0.31
9	80	0	5.91	90	70	0.14	66	292	0.34	21.35	21.35	0.45	1.04
10	60	0	3.46	67	340	5.2	90	30	0.87	278.38	278.39	89.33	85.36
11	90	30	3.46	54	295	1.37	80	0	3.94	317.93	318.04	39.91	124.94
12	90	70	0.34	43	300	1.37	73	25	5.2	54.75	54.75	3.8	7.09
13	43	60	0.14	73	335	5.2	90	10	0.39	10.15	10.14	0.53	0.51

Based on Table 1, the analytically calculated block surface is in a good agreement with the results of the numerical simulation; however, a remarkable error exists for the analytical block volume comparing with the results of the numerical method. Also, in case number 6, the analytical and numerical model for block surface as well as the numerical model for block volume could not determine any value for these parameters. However, for this case, the analytical model defines 10.57 m^3 for the block volume. As a result, the previously developed analytical model for the block volume needs to be revised and calibrated using the numerical method.

3 RELATIONSHIP BETWEEN BLOCK GEOMETRY AND INFLOW RATE TO THE TUNNEL

Average inflow rate to the tunnel excavated in a rock mass, is determined using 3DEC version 7 software for 13 cases of Table 1. For simplifying of the problem, the hydraulic aperture assumed to be 1×10^{-4} m for all joint sets. Also, a fixed level of water table is applied to all of the models.

Figure 2 illustrates the relationship between block volume and inflow rate to the tunnel. It is obvious that the inflow rate decreases by increasing the block volume. However, based on the coefficient of determination (R^2) of the trendline of diagrams, the relation of the inflow rate and numerically calculated block volume is more relevant than the analytically calculated one. The reason is the accuracy of the numerical method for block volume calculation comparing with the analytical models. Generally, it is expected that the inflow rate to the tunnel be decreased by increasing the block volume.

The relationship between block surface and inflow rate to the tunnel is illustrated in Figure 3. According to the Table 1, the numerical and analytical calculation of the block surface yield almost identical results. Therefore, both diagrams of Figure 3 are identical. As a general rule and same as Figure 2, it is expected that the inflow rate to the tunnel be decreased by increasing the surface area of the block. Although, based on the coefficient of determination (R^2) of Figure 2 and Figure 3, the relationship between block volume and inflow rate seems to be more relevant than the block surface area and inflow rate.

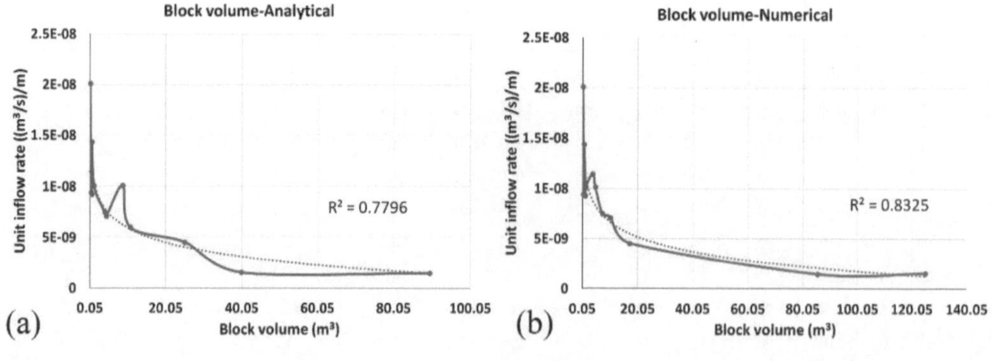

Figure 2. The relationship between inflow rate to the tunnel and block volume that is calculated using (a) analytical and (b) numerical methods.

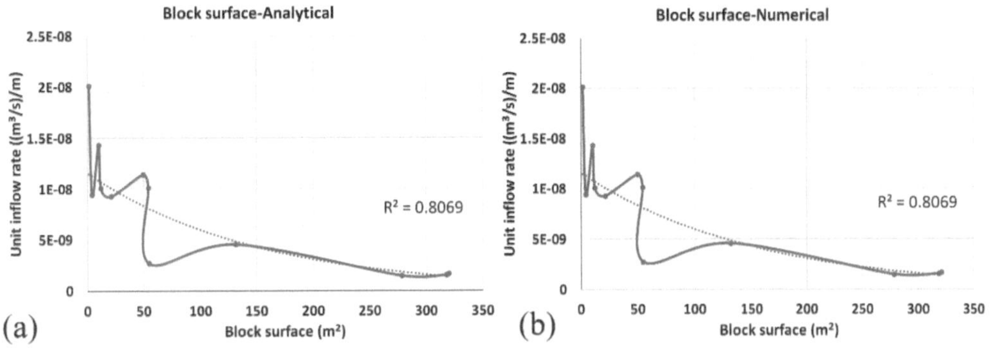

Figure 3. The relationship between the inflow rate to the tunnel and block surface area that is calculated using (a) analytical and (b) numerical methods.

4 SUMMARY AND DISCUSSION

The analytical method for calculation of the block surface is validated by 3DEC software and accurately defines unlimited surface area for the blocks that are created by 3 vertical joint sets having different values of dip directions. However, analytically calculated block volume for such cases by using equation (1) is not unlimited. The above-mentioned conflict and the discrepancies between numerical and analytical values of the block volume that are listed in Table 1, shows that the analytical model for the block volume calculation needs to be revised in order to increase its accuracy.

The intact block, i.e. the block that is not cut by the boundaries of the numerical model, should be considered for numerical calculation of the block volume and block surface. In the case that the block geometries are of interest, the numerical model should be quiet large that mostly include intact rocks. If other characteristics of the rock mass such as spacing, aperture, etc. are of interest, a new REV should be determined specifically for each parameter. In other words, the REV that is defined for each interested characteristic of the rock mass, is only applicable for that parameter only and could not be used for other characteristics of the rock mass.

Numerically determined block volume and surface have to be examined before using in the studies. The size of the numerical model could effectively affect the values of the average and maximum block size and surface area for one model with fixed values of the spacing and orientation. Furthermore, the arrangement of the discontinuities in the numerical models might have an impact on the block geometries, i.e., the start point for generation of the joint sets in numerical models. In order to prohibit this problem, the size of the numerical model should be large enough in order to mostly include the intact blocks.

According to Figure 2 and Figure 3, a relevant relationship exists between the inflow rate to the tunnel on one hand and block volume and block surface area on the other hand. According to the coefficients of determination (R^2) of each diagram, the relationship between the inflow rate to the tunnel and block volume is more relevant than block surface and hence, this parameter could be used for prediction of the inflow rate to the tunnel that is excavated in a rock mass. Since the geometries of the block could be considered as representative to spacing, dip and dip direction of joint sets, using block geometries for estimation of the inflow is more preferable than the other characteristics of the rock mass. In other word, various rock mass parameters could be summarized in one parameter, i.e. block geometry, and this is the advantage of using block geometrical parameters for estimation of the inflow rate to the tunnel.

5 CONCLUSION

An analytical model is developed for calculation of the surface area of the rock block by knowing the orientations and spacings of the 3 joint sets. The model is validated by the 3DEC software. It is demonstrated that the boundary effects might affect the values of the block geometries that is measured by the numerical models, because the blocks could be cut by the boundaries of the model. In order to avoid this error, the size of the numerical model must be large enough to include a significant number of intact blocks.

The previously developed analytical model for block volume calculation needs to be basically modified as comparison of the results of the model with output of the numerical simulation shows a significant difference. As the block size is widely used for predicting the geomechanical and hydrological characteristics of the rock mas, the accuracy of this parameter could affect the reliability of all dependent parameters.

The relationship between the inflow rate to the tunnel, block volume and block surface shows that by increasing the block volume and surface, the inflow rate to the tunnel decreases. Among them, the relationship between block volume and inflow rate is more relevant. On the other hand, block volume and surface area could be representatives of a series of geometrical characteristics of the rock mass, e.g., dip, dip direction and spacings of joint sets. In this regard, a relationship could be developed for calculation of the inflow rate to the tunnel as a function of block geometries. Practically, the rock block characteristics could be used for estimation of the inflow rate to the tunnel and as a general rule, more inflow rate into the tunnel will be gained when the tunnel is excavated in a rock mass with small blocks.

ACKNOWLEDGEMENT

The authors want to thank the Itasca Consulting Group in Minneapolis for the IEP Research Program, specially Jim Hazzard, for his valuable technical helps and advises.

REFERENCES

Barton, N., R. Lien and J. Lunde (1974). "Engineering classification of rock masses for the design of tunnel support." Rock mechanics 6(4): 189–236. DOI: https://doi.org/10.1007/BF01239496.

Butscher, C. (2012). "Steady-state groundwater inflow into a circular tunnel." Tunnelling and Underground Space Technology 32: 158–167. DOI: https://doi.org/10.1016/j.tust.2012.06.007.

Latham, J.-P., J. Van Meulen and S. Dupray (2006). "Prediction of in-situ block size distributions with reference to armourstone for breakwaters." Engineering Geology 86(1): 18–36. DOI: https://doi.org/10.1016/j.enggeo.2006.04.001.

Liu, Q., Z. Zhao, W. Nie, J. Sun and F. Xiao (2020). "An Analytical Investigation on the Estimation of Water Inflow into a Circular Tunnel Based On-site Data." Rock Mechanics and Rock Engineering 53 (8): 3835–3844. DOI: https://doi.org/10.1007/s00603-020-02114-2.

Palmstrom, A. (1982). The volumetric joint count—a useful and simple measure of the degree of rock mass jointing. International Association of Engineering Geology. International congress, India/Netherlands, A.A. Balkema.

Palmstrom, A. (1995). RMi-a rock mass characterization system for rock engineering purposes. Ph.D, University of Oslo.

Palmstrom, A. (2005). "Measurements of and correlations between block size and rock quality designation (RQD)." Tunnelling and Underground Space Technology 20(4): 362–377. DOI: https://doi.org/10.1016/j.tust.2005.01.005.

Shahbazi, A., R. Chesnaux and A. Saeidi (2021). "A new combined analytical-numerical method for evaluating the inflow rate into a tunnel excavated in a fractured rock mass." Engineering Geology 283: 106003. DOI: https://doi.org/10.1016/j.enggeo.2021.106003.

Shahbazi, A., A. Saeidi and R. Chesnaux (2020). Numerical modeling for determining the local vertical hydraulic gradient at the wall of a tunnel. GeoVirtual. Vancouver, BC, Venue West Conference Services.

Shahbazi, A., A. Saeidi and R. Chesnaux (2020). "A review of existing methods used to evaluate the hydraulic conductivity of a fractured rock mass." Engineering Geology 265: 105438. DOI: https://doi.org/10.1016/j.enggeo.2019.105438.

Stavropoulou, M. and G. Xiroudakis (2020). "Fracture Frequency and Block Volume Distribution in Rock Masses." Rock Mechanics and Rock Engineering 53(10): 4673–4689. DOI: https://doi.org/10.1007/s00603-020-02172-6.

Wang, Z., L. Bi, S. Kwon, L. Qiao and W. Li (2020). "The effects of hydro-mechanical coupling in fractured rock mass on groundwater inflow into underground openings." Tunnelling and Underground Space Technology 103: 103489. DOI: https://doi.org/10.1016/j.tust.2020.103489.

Yarahmadi, R., R. Bagherpour, L. M. O. Sousa and S.-G. Taherian (2015). "How to determine the appropriate methods to identify the geometry of in situ rock blocks in dimension stones." Environmental Earth Sciences 74(9): 6779–6790. DOI: https://doi.org/10.1007/s12665-015-4672-4.

Zhang, L. and J. Franklin (1993). "Prediction of water flow into rock tunnels: an analytical solution assuming an hydraulic conductivity gradient." International journal of rock mechanics and mining sciences & geomechanics abstracts 30(1): 37–46. DOI:https://doi.org/10.1016/0148-9062(93)90174-C.

The Evolution of Geotech - 25 Years of Innovation – Hammah et al. (Eds)
© 2022 copyright the Author(s), ISBN 978-1-032-03645-8

Severe rockburst occurrence during construction of a complex hydroelectric plant

G. Russo
Geodata Engineering (GDE), Turin, Italy

ABSTRACT: Severe rockbursts occurred during the construction of a complex hydroelectric plant in the Andean region of Chile, causing severe support failures and prohibitive work conditions. Consequently, a technical solution had to be found and fundamental safety issues achieved by completely automated support installation. A very accurate seismic monitoring was implemented, permitting the collection of basic information for understanding the seismic events associated with the rockbursts. Although extremely severe rockburst occurrences persisted, also in relation to the high natural seismicity, the close collaboration among Owner, Contractor and Designer allowed for a satisfactory damage control using effective support systems.

1 INTRODUCTION

This paper deals with the technical solution for controlling severe rockburst occurrence during the on-going construction of a complex hydroelectric plant in a seismic Andean region of Chile.

After a general setting of both the observed brittle instability phenomena, involving heavy failure of the previous installed support (Section 2) and geomechanical conditions (Section 3), the rationale and the dimensioning of new mitigation measures are described (Section 4). Brief insight into the adopted numerical model is presented in Section 5.

The effective performance of the new support system is analyzed in Section 6 in relation to successive severe rockburst occurrences.

2 GENERAL OVERVIEW

The intensity and frequency of seismic events dramatically increased while excavating one of the access tunnels to a powerhouse, just after a lithological contact between pyroclastic tuff and andesitic lava of the Abanico Western Formation (Oligocene-Miocene) with about 800 m of overburden (Figure 1).

After the first severe event reported in the Figure 1, a progressively increasing tendency to overbreaks and rock block ejection was observed in the tunnel (Figure 2).

Measures for controlling the dangerous brittle instabilities included modifications to the excavation shape (Figure 2) and adding a joint in the shotcrete in tunnel crown with the purpose of reducing stress concentrations (see Figure 3). The support system was based on two (1.5m x 1.5m) sequences of rock reinforcement (L=3m) by PM16/24 Swellex and Shell Anchored bolts (d=25mm) alternated with D-Bolts (22mm), in combination with (70+50mm) Fiber Reinforced Shotcrete (FRS) and two 6mm weld-meshes (ACMA C188).

DOI: 10.1201/9781003188339-20

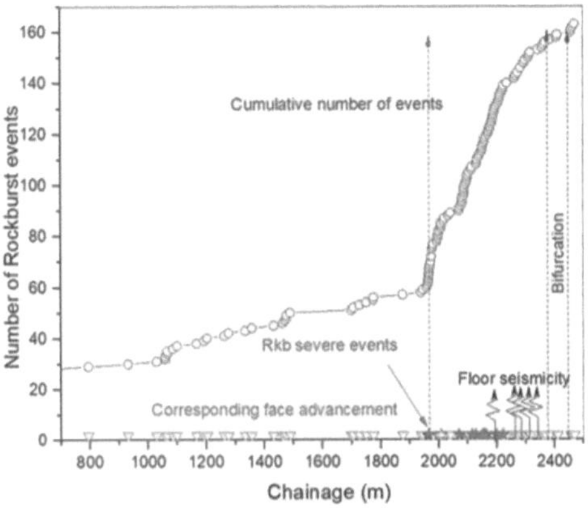

Figure 1. Cumulative number of rockburst events in access tunnel. Red dotted line coincides with lithological contact, while arrow indicate the first severe rockburst.

Figure 2. Initial modification of the excavation shape by Contractor to follow the failure mode (7.35 m of vertical dimension).

These technical solutions were however not able to control the damage from a very violent rockburst event that occurred at about Chainage (pk) 2+070 (overburden ≈ 930m), resulting in severe support failures at about 10m from the tunnel face followed by further support damage up to about 30m from the face (Figure 3).

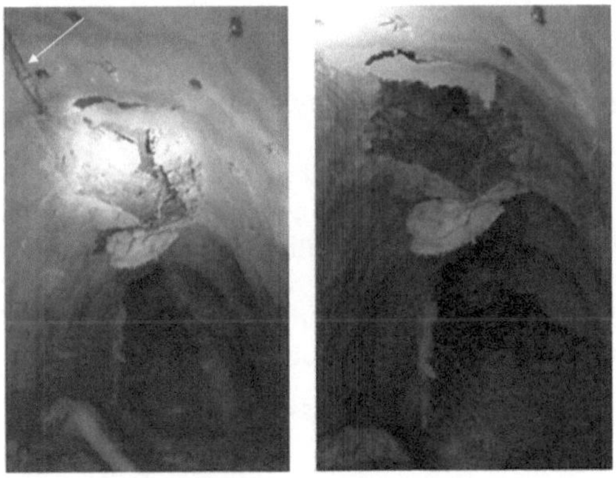

Figure 3. Support damage by severe rockburst at pk 2+070 (the arrow indicates the joint in the shotcrete).

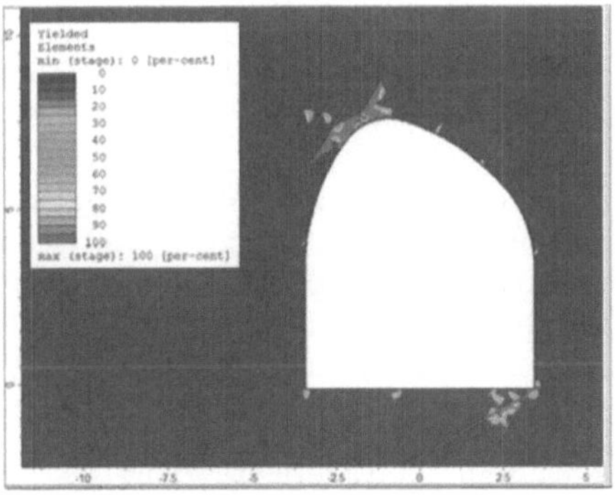

Figure 4. Comparison between results of numerical simulation based Joint Network modelling (by Rs2 software - Rocscience) and DISL approach (later described in Section 5: red lines indicate failure along joints).

In Figure 4, the result of a numerical simulation is compared with observed collapse, showing a satisfactory agreement between expected/observed failure localization and extension.

Kaiser and Cai (2013, 2018) classify this type of rockburst as "*strainbursts*", i.e. a sudden and violent failure of rock near an excavation boundary caused by excessive straining of a volume of stiff and strong rock (burst volume). If triggered by a face burst, the secondary seismic source is co-located at the damage location. If self-initiated, the primary seismic source is located at the damage location.

On the other hand, some uncertainty still exists concerning the further sub-classification proposed by the cited authors, also in relation to successive events registered during the construction of other tunnels and caverns in the same hydroelectric project. In particular, although probably the "*Self-initiated/Mining-induced*" resulted as the most frequent strainburst type, it cannot be excluded that, depending on the augmented intensity of eventual remote seismic events induced by excavation, also "*seismically triggered*" or even "*dynamically loaded*" strainburst occurred.

In dynamically loaded strainburst the energy radiated from a primary source impacts in two possible forms:

(i) it causes a dynamic stress pulse that may deepen the depth of failure, thus releasing more stored energy and, through rock mass bulking, adding strain or displacement to the rock and support; or
(ii) it transfers some of its radiated energy to kinetic energy and assists in ejecting marginally stable rock.

In the former case, according to the Canadian Rockburst Support Handbook (CRSH, 1996) the maximum dynamic stress increment that needs to be added to the tangential stress, temporarily modifying the Stress Level SL=σ_{max}/UCS (i.e. the ratio of the max tangential stress on excavation contour to the intact rock strength) and increasing the depth of failure, is:

$\Delta\sigma^d \max = \pm 4 \cdot c_s \cdot \rho \cdot PGV_s$ where: cs =propagation speed of shear waves; ρ density of rock mass

PGVs= Peak (particle) Ground Velocity of the shear waves which depends on the seismic event magnitude and distance from the source.

3 GEOMECHANICAL CONDITIONS

According to the available information, based on both CSIRO and Hydraulic fracturing in situ tests, the main regional principal stress is oriented between NE-SW and E-W, with a ratio of major horizontal to vertical stresses k=(σH/σV)~2÷2.5. Moreover, generally σV results higher than the minor horizontal stress (σh), thus indicating remarking a typical "strike-slip" condition.

In the so-called "GDE multiple graphs" of Figure 5 (Russo, 2014), the main geomechanical properties (GSI, UCS, IC=$\sigma_{cm}/\sigma_{tang(max)}$ and RMR) from about Chainage pk 1900m onwards in the access tunnel are shown in combination with the expected excavation hazards (quadrant IV) for the relative plane-strain principal stresses.

As it can be observed from this figure by proceeding clockwise from the bottom-right to the top-right quadrant (I→IV; see basic equations rationale reported outside the graph):

- I quadrant (bottom-right): in the case not implemented for GSI assessment
- II quadrant (bottom-left): GSI≈47÷66; UCS≈150÷200MPa
- III quadrant (top-left): IC≈0.05÷0.2
- IV quadrant (top-right): RMR≈50÷65. Consequently, in terms of excavation behaviour, serious overbreaks conditions, eventually associated to severe rockburst, are expected in alternative to rock wedges instability.

The assessment of rockburst hazard through the the GDE multiple graph refers to the spalling classification proposed by Diederichs et al., (2010; Figure 6), based on the empirical prediction of the depth of brittle failure as a function of the stress level SL=σ_{max}/UCS (Wiseman, 1979, reported in CRSH, 1996; Martin et al., 1999). In Figure 7, the Stress Level (SL) is detailed for an Adit stretch with reference to the increasing overburden and the localization of rockburst events.

In Figure 8, the Dynamic Rupture Potential (DRP, Diederichs, 2017) is estimated for the same UCS range reported in Figure 5, by consequently remarking an indication of possible rock ejection in case of unsupported or ineffectively supported excavations. As suggested by

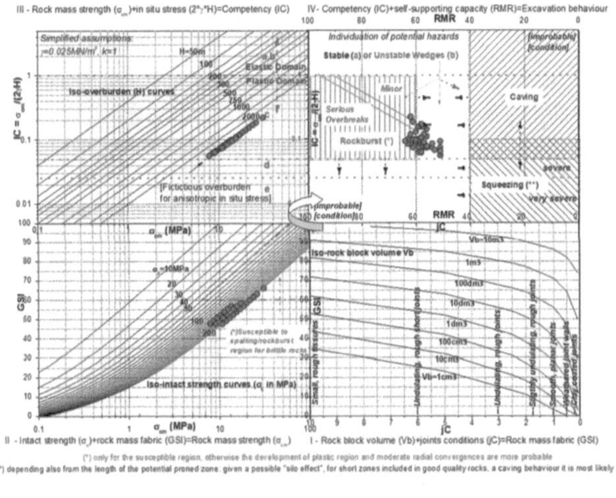

Figure 5. Application of the GDE Multiple graphs (Russo, 2014) for the 1+900÷2+070 Adit stretch. Note that“fictitious” overburden in top-left quadrant allows for considering the effective max tangential stress (3σ1-σ3) related to k≠1.

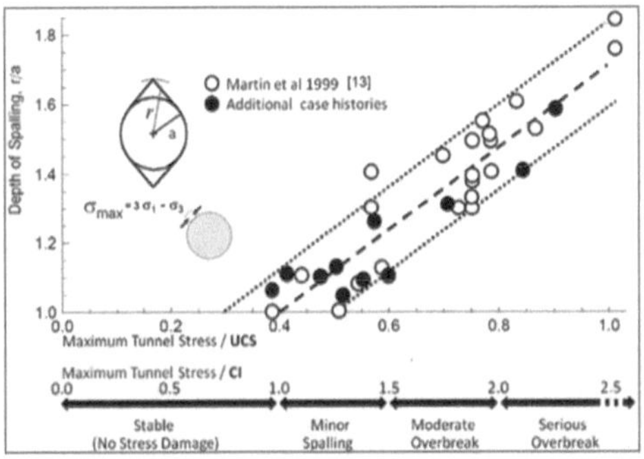

Figure 6. Empirical prediction of Depth of Spalling/ Failure=DoF) for SL<1. The figure (Diederichs et al., 2010) is based on CRSH (1996) and Martin et al., (1999). CI=Crack Initiation Threshold=0.4·-UCS in the graph. According to Nicksiar and Martin (2013), CI=0.35÷0.55 for most of rocks. The cases in figure remark the maximum registered DoF, generally for brittle failures (no rockburst).

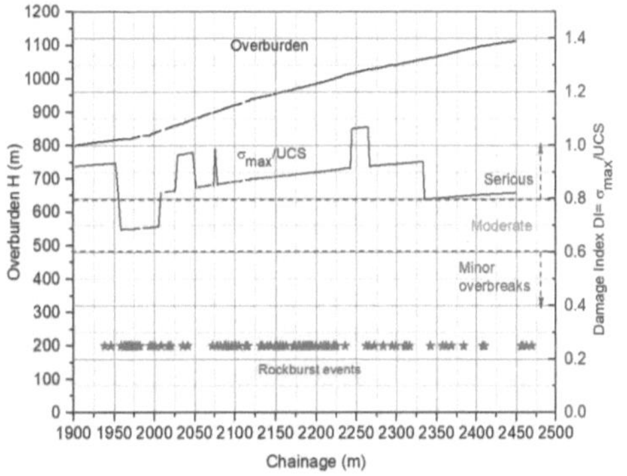

Figure 7. Relation between rockburst events and the calculated Stress Level. Note that the SL drops after Chainage pk 2+320 was hypothesized due to increases in UCS derived from laboratory tests.

Stacey (2016), ejection velocity could even be much higher than indicated in the graph and velocity in the order of 10m/s are not uncommon, thus adding uncertainty to design.

According to Nicksiar and Martin (2013), CI=0.35÷0.55 for most of rocks. The cases in figure remark the maximum registered DoF, generally for brittle failures (no rockburst).

According to Cai and Kaiser (2018), the brittleness indices in Figure 8 in the horizontal axis are indicators of StrainBurst Potential (SBP) and on the vertical axis of StrainBurst Severity (SBS). The UCS contributes to the severity of events, but other factors, particularly the “Relative Brittleness”, related to the excavation system stiffness (Tarasov and Potvin, 2013) need to be considered.

For unsupported tunnel (width and height dimension in the range of 3÷6m) Kaiser et al. (1996) defined the rockburst damage severity in terms of depth of failure as shown in the upper part of Figure 9. At the bottom of the same figure, the rockburst damage scale for support (Potvin, 2009, modified by Cai and Kaiser, 2018) is reproduced.

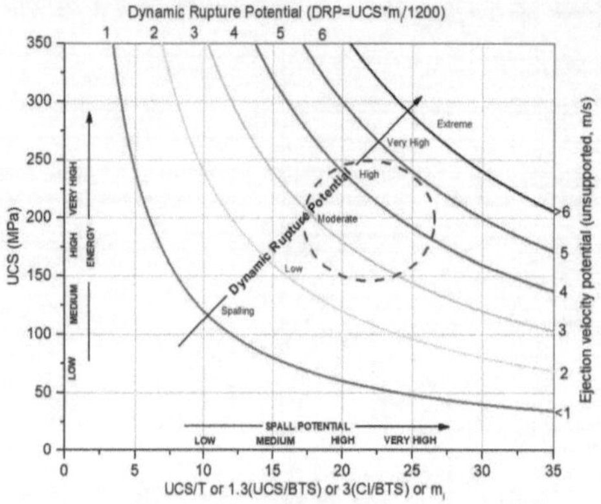

Figure 8. Dynamic Rupture Potential (DRP) for massive rock (Diederichs, 2017) with approximate indication of typical Andesitic Lavas properties.

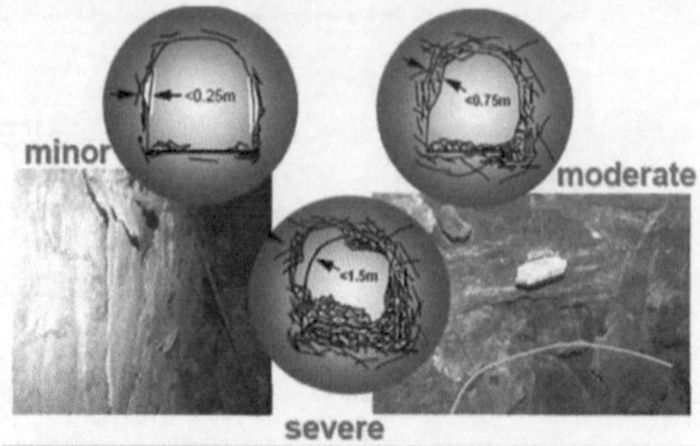

Rockburst damage scale	Rock mass damage	Damage surface area	Rock support damage
R1	No damage, minor, loose	0	No damage
R2	Minor damage, less than 1 t displaced	$< 1\ m^2$	Support system is loaded, loose in mesh, plates deformed, shotcrete cracked
R3	1–10 t displaced	$< 10\ m^2$	Some broken bolts, mesh bulged, shotcrete fractured
R4	10–100 t displaced	10 to 50 m^2	Major damage to support system; retention capacity severely compromised
R5	100+ t displaced	$> 50\ m^2$	Complete failure of support system

Figure 9. Above: Classification of rockburst damage severity for unsupported excavations (Kaiser et al., 1996). Below: Rockburst damage scale for support (Potvin, 2009; modified by Cai and Kaiser, 2018).

4 UPGRADED DESIGN SOLUTION

The extremely dangerous conditions remarked by the rockburst event at Chainage pk 2 +070 imposed a substantial revision of the applied mitigation measures and in 2016 the Contractor involved Geodata Engineering (GDE) for devising an adequate and safe technical solution.

Safety conditions for the workers, by avoiding any exposure in potentially dangerous zone, is the fundamental requisite for design in rockburst environment. In line with this basic requirement, the Contractor provided special bolting equipment for the automated installation of steel mesh and bolts.

At the same time, an accurate seismic monitoring system was deemed of paramount importance.

The design dimensioning development basically followed the approach described by Kaiser et al. (CRSH, 1996), with basic reference to "Bulking causing ejection" as the dominant damage mechanism.

Depending on the possible interaction between the different elements of the support system and relative failure occurrences, it was estimated that the described rockburst event at pk 2+070 (Figures 3-4) released energy at least for 20 kJ/m2 and up to about 30 kJ/m2. Therefore, the latter was considered the reference energy demand for rockburst design.

How to consider in rockburst design the contribution of the reinforcement and surface support in the overall system is still debated. According to Potvin et al. (2010, 2013), the surface support (shotcrete, steel mesh) could theoretically guarantee an additional safety margin but actually it often just represents the "weakest link" that will work in a "serial" function with reinforcement/holding system.

Additionally, Cala et al. (2013) remarked a variable redistribution of the energy in the different support elements as a function of the relative stiffness of the surface component. For example, according to the tests performed by Villaescusa and Player (2015) on composite support (bolt and steel mesh), the reinforcement adsorbed about 72 to 93% of the released energy.

Considering the uncertainty on the effective interaction between holding/reinforcement and retention system, the weakest link issue and the potential ejection velocity, a double-layer solution was consequently dimensioned, as shown in Figure 10.

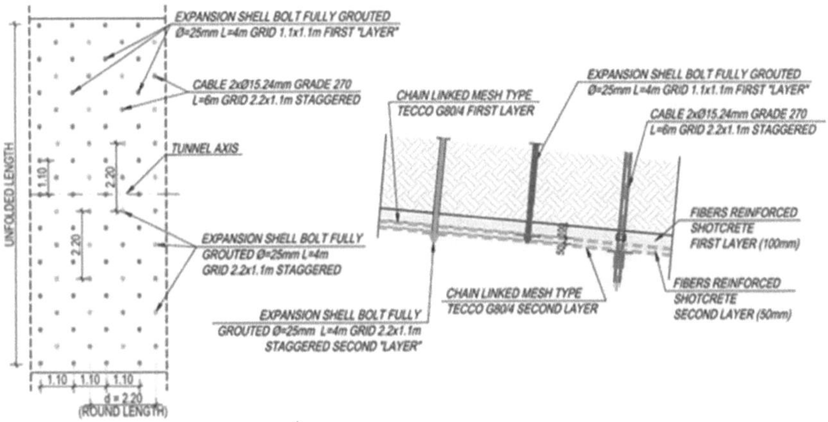

Figure 10. Double-layer solution dimensioned for severe rockburst occurrence.

The double-layer solution (crown/sidewalls) typically involves two retention system components: Fibre- Reinforced Shotcrete (FRS) and high capacity chain-linked steel mesh (Tecco G80/4). Each component was combined with a radial reinforcement by fully grouted 25mm expansion shell threadbars or twin-(15.2mm) strand cables. The latter served as partial alternative holding elements for the second retention component, depending on excavation cross section.

Figure 11. Examples of application of the Double-layer solution and tunnel face control.

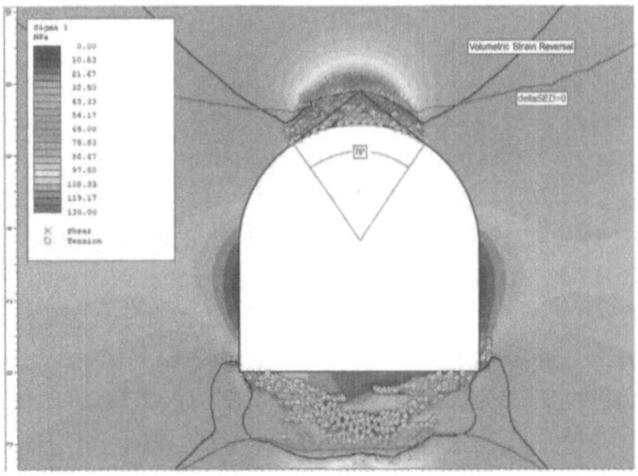

Figure 12. Example of results of numerical modelling by Rs2 in terms of major stress (Sigma1) and yielding zones. VSR and the iso-line ΔSED=0, as well as the potential brittle failure notch according to Martin et al. (1999) are additionally remarked.

Based on available results from tests on bolts and retention systems (in particular by the WASM Dynamic Test Facility, as reported by Villaescusa et al., (2015), CANMET, (2012) and *drop test* data reported by Potvin et al., (2010), including Kaiser et al. (1996) and Ortlepp and Stacey, (1997)), the reinforcement/holding system is called to adsorb in about 100-(150) mm of displacement the energy demand of reference with a Factor of Safety FS≈2, as recommended by CRSH. At the same time, the retention component guarantees a capacity itself in the order of demand, offering by the chain-link mesh component some further deformational margin.

Furthermore, as a measure for some controlled dissipation of seismic energy, a curved excavation shape was used for the invert and by temporarily leaving at the face the blasted rock, in order that invert zone results filled up to the horizontal working plane. The standard technical solution also included the retention and reinforcement of the tunnel face by FRS+welded mesh and PM24 Swellex bolts. Forepoling was applied in particular critical stretches to further control overbreak.

A representative image of the application of the Double-layer solution is presented in Figure 11.

5 INSIGHT FROM NUMERICAL MODELLING

Numerical analyses were conducted with the main purpose of understanding the potential localization of brittle failure and the extension of the stress-damage zone around the cavity.

The DISL (Damage Initiation Spalling Limit) approach (Diederichs, 2010) based on the multi-phase failure criterion was adopted.

In Figure 12, the results of an example of implementation of DISL by continuum are provided in terms of major stress (Sigma1) and yielding zone (x for shear and o for tension). Moreover, the following additional elements are highlighted:

- The Volumetric Strain Reversal (VSR): according to Perras and Diederichs (2016), VSR indicates the transition between isolated and connected damage within the rock mass surrounding the excavation and should be comparable to the depth of brittle failure investigated by Martin et al., (1999, Figure 6). Consistent with this reference, the approximate dimension of the potential brittle failure notch is as well reproduced.
- An approximate indication of the limit between zones in which Strain Energy Density (SED) reduced (close excavation) or increased in the passage from elastic peak to post-failure condition. Therefore, this limit is remarked by the union of points in which SED did not change, i.e. the iso-line $\Delta SED= SED_{post\text{-}failure} - SED_{peak}= 0$.

The Strain Energy Density is calculated by the following formula:

$SED=[(\sigma_1^2+\sigma_2^2+\sigma_3^2)-2v(\sigma_1\sigma_2+\sigma_2\sigma_3+\sigma_3\sigma_1)]/2E$ where: $\sigma_1,\sigma_2,\sigma_2$= Principal Stresses; v=Poisson Ratio; E=Young Modulus.

It can be observed that VSR and ΔSED=0 overlaps in the tunnel crown, so consistently estimating the potential thickness of unstable rock mass. Numerical model does not give the maximum (extreme) DoF as per Figure 6.

Based on the resulting VSR and the associated Depth of Failure (DoF), an estimate of the kinetic energy (E_k) can be assessed. For the example in Figure 12, considering DoF~1m and the potential ejection velocity derived from Figure 8, the previously remarked energy demand for design is approximately confirmed.

6 PERFORMANCE OF SUPPORT SYSTEM

The application of the technical solution for severe rockburst was extended to all the tunnels and caverns exposed to such a critical hazard, with some optimization depending on local geomechanical conditions, as for the dominance of tuff in place of andesitic lavas. Here, probably in relation to both lower UCS and higher CI, frequency and severity of rockburst generally show some reduction. Violent brittle failures persisted in all excavation in andesitic lithologies, as for example illustrated in Figure 13 for the zone around the powerhouse. In Figure 14, the time of occurrence of rockburst after blasting is related to the relative distance from the tunnel face. Most rockbursts occurred within 24 hours and at less than 5H distance (i.e. ~37 m) from the face (H=Height of excavation section). Only three events occurred at 40÷80 m from the face.

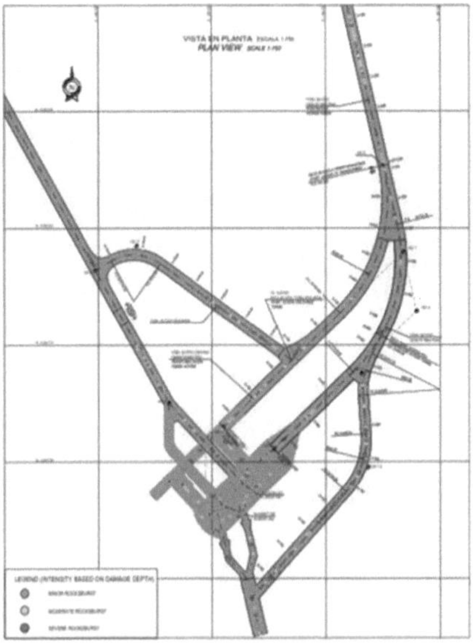

Figure 13. Rockburst locations and damage severity in the zone of the Powerhouse.

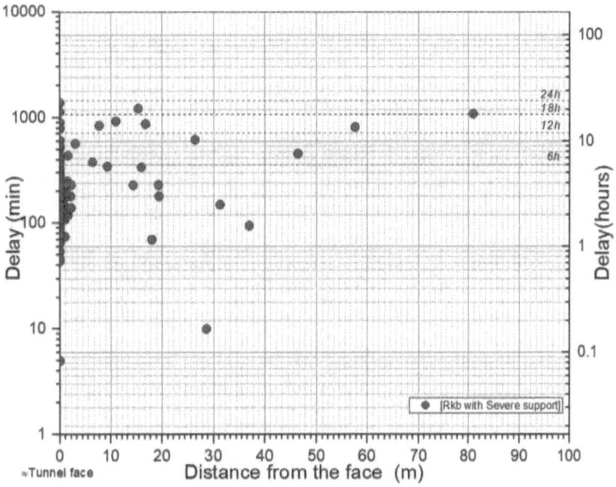

Figure 14. Time of occurrence (Delay) of rockburst after blasting vs relative Distance from the face.

The performance of the double-layer solution has been satisfactory. The support system was able to control very violent events by limiting the damages, without critical structural failure.

In occasion of the most severe events, the following type of damages have been observed (Figure 15):

- Fracturing of the shotcrete along preferred alignment, without fall-down or ejection of fragments because of the chain-link mesh protection;
- local shear cut of the threadbars. This kind of failure prevalently occurred at distance <0.5m from the bolt heads (no twin-strand cables shear failures).
- cracks in the invert zone, sometimes consequence to very impressive up-down movement of the floor, like earthquake shaking associated to seismic waves propagation (see also Figure 1).

With reference to the table in Figure 9, the R3 grade of the scale can be assessed, but observing that the associated damage surface area is much more than indicated in the classification system.

Due to the support damage, part of the support system capacity has been consumed and it is necessary to estimate the remnant capacity of the support system. Consequently, after a cautious removal of the failed elements, the support capacity can be restored by stepwise integrative measures (see Kaiser, 2013 and Cai and Kaiser, 2018). As anticipated, a key element of the upgraded design agreed with Owner and Contractor was the implementation of a seismic monitoring system. An extremely high seismicity has been observed as resulting from the D&B (Drill and Blast) tunnel advancements: in several occasions more than 10,000 seismic events per week were registered, with some hundred events with moment magnitude Mw > -1 and maximum values up to Mw = 1.4 (see details in Russo G., 2019).

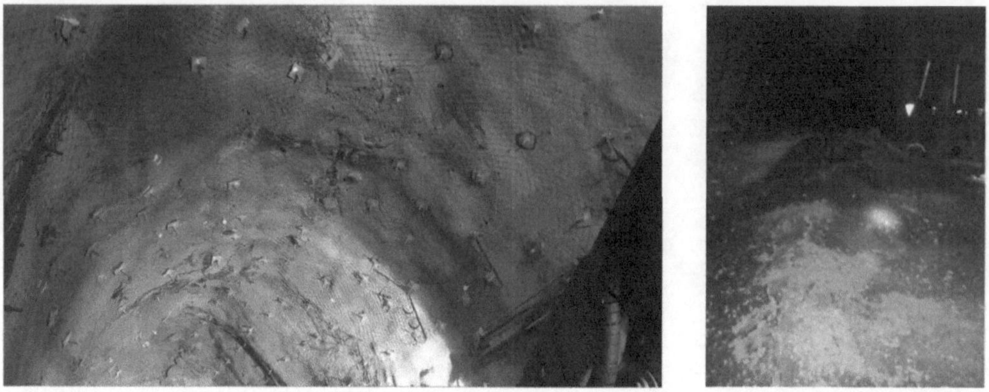

Figure 15. Examples of damage of the support systems in crown and invert for severe rockburst events.

7 CONCLUSIVE REMARKS

Rockburst probably poses the most dangerous and many times unpredictable hazard affecting underground excavation in highly stressed hard/massive rocks. This paper describes a very demanding technical challenge to control this kind of phenomenon during construction at depth of a complex hydroelectric scheme in a seismic active region.

After several attempts to implement different mitigation measures as suggested in the literature, which however could not prevent severe failures to the support system from taking place, a substantial change in construction approach in terms of equipment and technical solution had to be adopted.

In particular, given the absolute priority given to safety, automated bolters for a complete mechanized installation of support elements were introduced so as to avoid any exposure of workers.

An innovative compatible double-layer solution for reinforcement and retention system was dimensioned to achieve an adequate safety margin with respect to the rockburst energy and displacement demand.

The technical solution was then implemented at all underground excavation exposed to similar hazards. This support performed satisfactorily by limiting the support damages despite many severe rockburst occurrences in very high stress conditions.

Seismic monitoring performed a fundamental role for a comprehensive analysis and understanding of these complex and hazardous phenomena.

ACKNOWLEDGMENT

The author wishes to acknowledge the much-appreciated contributions of his colleagues Simone Addotto, Carlo Chiesa and Domenico Parisi.

REFERENCES

Cai M. and Kaiser P.K. 2018: Rockburst Support Reference Book. Available to download from: http://www.imseismology.org/imsdownloadsapp/webresources/filedownload/Cai-Kaiser(2018)-Rockburst-Support- Vol1.pdf

Cala M., Roth A. & Roduner A., 2013: Large scale field tests of rock bolts and high-tensile steel wire mesh subjected to dynamic loading. Rock Mechanics for Resources, Energy and Environment Edited by Marek Kwaśniewski and Dariusz Łydżba.

CAMIRO Mining Division, 1996. Canadian Rockburst Research Handbook (CRSH).

Diederichs, MS, Carter, T. Martin. 2010. Practical Rock Spall Prediction in Tunnel. Proc. of World Tunnelling Congress. Vancouver.

Diederichs, M.S. 2017. Early Assessment of Dynamic Rupture and Rockburst Hazard Potential in Deep Tunneling.

SAIMM 2017 in Capetown (Afrirock)

Hoek, E. and Martin CD, 2014. Fracture initiation and propagation in intact rock - A review. Journal of Rock Mechanics and Geotechnical Engineering

ISRM, 2015 Xiao et al.: Suggested Method for In Situ Microseismic Monitoring of the Fracturing Process in Rock Masses, 2015

Kaiser, P.K. and Kim B.H., 2008: Rock Mechanics Advances for Underground Construction in Civil Engineering and Mining. Keynote Rock Mechanics Symposium. Seul.

Kaiser, P.K. and Cai M., 2013: Critical review of design principles for rock support in burst-prone ground-Time to rethink! Perth - Seventh International Symposium on Ground Support in Mining and Underground Construction. Perth

Kaiser P.K. 2017: Ground control in strainburst ground –A critical review and path forward on design principles. 9th Inter. Symposium on Rockburst and Seismicity in Mines (Santiago, Chile)

Li C.C., 2010. A new energy-absorbing bolt for rock support in high stress rock masses. International Journal of Rock Mechanics & Mining Sciences 47. 396–404.

Martin C.D., Kaiser P.K, and McCreath D.R., 1999. Hoek–Brown parameters for predicting the depth of brittle failure a round tunnels. Canadian Geotechnical Journal. 36: 136–151.

Ortlepp, W.D. and Stacey, T.R., 1997. Testing of tunnel support: dynamic load testing of rock support containment systems, SIMRAC GAP Project 221.

Perras M. and Diederichs M.S. 2016. Predicting excavation damage zone depths in brittle rocks. Journal of Rock Mechanics and Geotechnical Engineering. 8, pp 60–74

Player J.R., Morton E.C., Thompson A.G. and Villaescusa E. (2008):" Static and dynamic testing of steel wire mesh for mining applications of rock surface support". SAIMM, SANIRE and ISRM: 6th International Symposium on Ground Support in Mining and Civil Engineering Construction

Potvin Y., Wasseloo J. and Heal D., 2010. An interpretation of ground support capacity submitted to dynamic loading Deep Mining 2010 — M. Van Sint Jan and Y. Potvin (eds) © 2010 Australian Centre for Geomechanics, Perth, ISBN 978-0-9806154-5–6.

Potvin Y and Wasseloo J. 2013.Towards an understanding of dynamic demand on ground support. The Journal of The Southern African Institute of Mining and Metallurgy, n.113.

Russo G. 2014. An update of the "multiple graph" approach for the preliminary assessment of the excavation behaviour in rock tunneling. Tunnelling an Underground Space Technology n.41.

Russo G. 2018: Rockburst. Lecture for Post Graduated Master Course "Tunnelling andTunnel Boring Machines" - Politecnico di Torino - Academic Year 2017–2018 (https://www.geodata.it/giordandue/)

Russo G. 2019. "Severe rockburst occurrence during construction of a complex hydroelectric plant". Lecture for the 29th Mine Seismology Seminar- IMS Institute of Mine Seismology 05–07 May 2019 – Phoenix, Arizona (USA) (https://www.geodata.it/giordandue/).

Stacey T.R. and Ortlepp W.D., 2001. Tunnel surface support capacities of various types of wire mesh and shotcrete under dynamic loading. The Journal of The South African Institute of Mining and Metallurgy.

Stacey T.R., 2016. Addressing the Consequences of Dynamic Rock Failure in Underground Excavations Rock Mech. Rock Eng. DOI 10.1007/s00603-016-0922-3

Tarasov B. and Potvin Y. 2013. Universal criteria for rock brittleness estimation under triaxial compression. International Journal of Rock Mechanics & Mining Sciences n.59.

Villaescusa E. and Player J.R. 2015. Dynamic Testing of Ground Support Systems. MRIWA Report n.312.

Session 3 - Stability analysis of mining slopes I

The Evolution of Geotech - 25 Years of Innovation – Hammah et al. (Eds)

Analytical versus numerical analysis on slope stability of surface lignite mines

A. Mikroutsikos, A.I. Theocharis & N.C. Koukouzas
Chemical Process & Energy Resources Institute, Centre for Research & Technology Hellas, Athens, Greece

I.E. Zevgolis
School of Mining and Metallurgical Engineering, National Technical University of Athens, Greece

ABSTRACT: Several failure incidents have been reported on surface lignite and coal mines, indicating that slope stability is crucial for the sustainability and reclamation of these areas. An analytical model was recently presented in the literature, concerning the evaluation of lignite mines' slope stability in the presence of a weak zone. The model is based on a bilinear failure mechanism, which is mostly controlled by a sub-horizontal sliding surface along a low strength zone. In the present work, a numerical approach for lignite mines' stability is compared with that analytical model. Assumptions of each model are discussed, and identical geometries and geotechnical parameters are implemented. The authors conclude that the analytical model is susceptible to the assumptions regarding the hydraulic conditions, indicating that an appropriate estimation of the safety factor may not be feasible under certain circumstances. Overall, the rigorousness of FEM analyses over more simplified analytical models is highlighted.

1 INTRODUCTION

Surface coal and lignite mines are significantly affected by the excavations' slope stability (Zevgolis et al. 2019). Slope failures and subsequent landslides are critical as they threaten human lives, infrastructure, and the sustainable reclamation of these areas. Initiation of these landslides is often associated with a sub-horizontal zone of low strength. This zone can be a layer or an interface between layers, named the weak zone. The immense shear strains developed on this thin zone during ground movements often lead its strength to its residual state, dominating the failure mechanism. This mechanism is frequently encountered in surface coal and lignite mines in several countries: Greece (Leonardos 2004), Turkey (Ural & Yuksel 2004), Poland (Bednarczyk 2017), the Czech Republic (Mencl 1977), and Australia (Ghadrdan et al. 2020). In addition to mining excavations, such a mechanism is also encountered in various landslide phenomena.

Various approaches have been employed to assess this type of slope stability, with analytical models being commonly used, due to their simplicity and speed (e.g. Chowdhury et al. 2010; Huang 1983; Fleurisson & Cojean 2014; Stimpson 1979). Kavvadas et al. (2020) presented a simplified analytical model to investigate the importance of critical parameters on the SF. However, the increasing computational power has set the numerical analysis as a central tool to evaluate slope stability, overcoming the limitations and disadvantages of the analytical models (e.g. Cała et al. 2020; Tutluoglu et al. 2011).

In the present work, a comparison is performed between numerical analysis and the recent analytical model of Kavvadas et al. (2020). The finite element method (FEM) is employed in RS2 software combined with the shear strength reduction technique. Identical slope geometries and geotechnical parameters, and similar groundwater conditions were considered for

DOI: 10.1201/9781003188339-21

the two methods. Overall, the present work highlights the rigorousness of numerical analysis over the analytical model. Additionally, it verifies the groundwater's dominant role in this type of slope stability and validates some conclusions of Kavvadas et al. (2020).

2 COMPARISON OF THE ANALYTICAL AND THE NUMERICAL MODELS

2.1 *Analytical model: Formulation and basic assumptions*

Kavvadas et al. (2020) presented a practical analytical tool for evaluating lignite mines' slope stability. The basic assumption relates to the bilinear sliding surface's geometry that governs the failure mechanism (Figure 1). In particular, sliding occurs along the horizontal/sub-horizontal weak zone and reaches the ground surface through a tension crack. The analytical model is derived by limit equilibrium and based on assumptions that have been validated through Greek mining experience. Two water forces, the sliding volume's weight, and the weak zone's friction contribute to the equilibrium. The SF is calculated as:

$$SF = \frac{(c_z L'/\cos\beta_z) + (W - U_1 \tan\beta_z - U_2/\cos\beta_z)\tan\varphi_z}{W\tan\beta_z + U_1} \quad (1)$$

where φ_Z is the friction angle, c_z the cohesion, and β_Z the inclination of the weak zone; W is the sliding mass' weight; U_1 is the horizontal water force along the tension crack; U_2 is the water force at the sub-horizontal slip surface (Figure 1). Note that the angle β in Figure 1 indicates the total slope angle of the excavation. The water forces U_1 and U_2 are calculated by integrating the water pressure distributions along the respective surfaces and are provided by:

$$U_1 = \lambda_1\left(\frac{1}{2}\gamma_w Y^2\right) \quad (2)$$

$$U_2 = \lambda_2\left(\frac{2}{3}\gamma_w Y L'\right) \quad (3)$$

where Y is the depth of the sliding mass at the transition point, and L' is the length of the sliding mass base, from the toe of the slope to the transition point (Figure 1). The factor λ_1 defines the height of the hydrostatic pressure in the tension crack, and the factor λ_2 the pressure at the transition point, being $\lambda_2\gamma_w Y$.

Kavvadas et al. (2020) investigated the effect of each parameter of Equation 1 on the SF. They concluded that the weak zone's inclination is the most critical parameter, while the importance of the weak zone's strength and the groundwater conditions was not explicitly examined. In a recent study, Mikroutsikos et al. (2021) presented a systematic numerical analysis on the effect of essential parameters on slope stability in the presence of a weak zone. They validated that the weak zone's strength and the groundwater conditions are particularly critical, along with the weak zone's inclination.

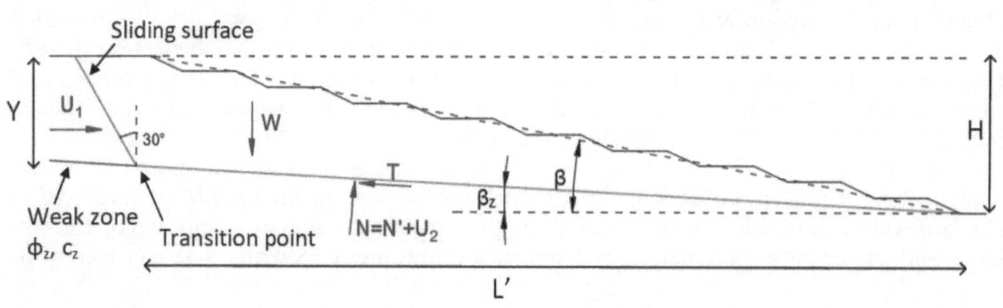

Figure 1. Geometry and forces on a typical sliding mass (modified after Kavvadas et al. (2020)).

The scope of the present work is to compare the analytical model with numerical analysis. Under this perspective, the underlying assumptions of the analytical model need to be clarified. Firstly, the overburden soil's strength is ignored as the failure surface propagates to the ground surface through a tension crack. Furthermore, the weak zone is assumed to fail, having not its residual strength, but one between its peak and residual ($\varphi_z = 22°$ and $c_z = 5kPa$). However, these two subtle assumptions are not expected to be crucial.

On the contrary, the calculation of the proposed water pressures (U_1 and U_2) is vital. Water force U_1 is the hydrostatic force inside the tension crack, with factor λ_1 theoretically ranging between 0-1; Kavvadas et al. (2020) suggested that λ_1 usually lies between 0.65-0.90. For water force U_2, the factor λ_2 was assumed to be 1, and λ_2 theoretically ranges from 0 to 1. As will be illustrated, λ_1 crucially affects SF when $\beta_z = 0°$, while factor λ_2 is important for the SF in all cases. Notice that dry conditions cannot be appropriately simulated with this analytical model as setting $U_1 = U_2 = 0$ leads to very large SF. For $\beta_z = 6°$, the SF equals 3.91, while for $\beta_z = 3°$, the SF increases to 7.78; if $\beta_z = 0°$ then SF tends to infinity.

2.2 *Numerical model: Basic assumptions*

A similar stratigraphy is assumed with a weak zone between a bedrock formation and the overburden soil (Figure 2). A two-dimensional plane strain model under drained loading conditions was used with 6-noded triangular elements. The model's boundaries were located at an adequate distance from the slope's crest and toe to minimise the boundary conditions effect (Figure 2). The slope was simulated as a staged excavation, and shear strength reduction analysis was performed at the final stage. A very fine discretisation was used, and mesh density was increased in the weak zone area (Figure 2). The Mohr-Coulomb elastic-perfectly plastic constitutive model was used for all materials. Notice that the region underneath the weak zone does not contribute to stability.

Numerical simulations were performed for slopes without benches (Figure 2). Kavvadas et al. (2020) presented a slope with benches (Figure 1); however, they derived their model for a special case (with equally distributed benches of 45° inclination) that provides identical SFs with and without benches.

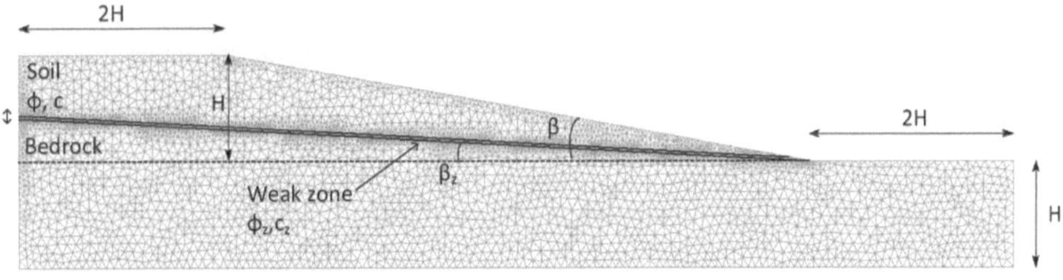

Figure 2. Geometry and main parameters of the numerical model.

The pore water pressures were calculated using the r_u coefficient approach. This simplified method is commonly used in slope stability analysis, as the groundwater regime is often challenging to identify. The pore water pressure is a fraction of the overburden vertical stress and equals:

$$u = r_u \gamma_z \tag{4}$$

where r_u is the coefficient of pore pressure ranging from 0 (zero pore pressure) to 1 (zero effective stress), γ is the soil unit weight, and z is the depth below the soil surface. If the saturated soil unit weight equals 20 kN/m^3, and the water unit weight 10 kN/m^3, r_u=0.5 is equivalent to a phreatic water table lying at the soil surface.

Identical slope geometries and geotechnical parameters were considered to compare the two approaches (Table 1). The comparison was conducted for three weak zone inclinations towards the excavation, 0°, 3° and 6° (unfavourable inclination). For the numerical model, the soil's friction angle and cohesion were φ=28° and c=185kPa (Theocharis et al. 2021), while the weak zone's thickness was d_z=5m. Note that Mikroutsikos et al. (2021) showed that the weak zone's thickness is insignificant for the overall stability.

Table 1. Geometrical and geotechnical parameters implemented for the comparison.

Parameter	Symbol	Analytical model	FEM analyses
Weak zone's friction angle	φ_z (°)	22	22
Weak zone's cohesion	c_z (kPa)	5	5
Soil's friction angle	φ (°)	Tension crack	28
Soil's cohesion	c (kPa)	Tension crack	185
Unit weight	γ (kN/m^3)	17	17
Stiffness	E (Mpa)	-	50
Poisson ratio	v	-	0.3
Dilation angle	ψ (°)	-	0
Slope height	H (m)	200	200
Slope inclination	β (°)	10	10
Weak zone's inclination	β_z (°)	0-6	0-6
Weak zone's thickness	d_z (m)	-	5

2.3 *Comparison results*

The groundwater regime is considered differently in the two approaches. The numerical analysis uses the r_u coefficient to implement a simplified approach that considers groundwater's uncertainty (Eq. 4). On the contrary, the analytical model uses U_1 and U_2, two water forces specifically defined and acting upon the two linear parts of the sliding surface (see section 2.1). Then, the uncertainty lies only on the values of λ_1 and λ_2 that quantify the details of U_1 (λ_1 defines the hydrostatic height, Eq. 2) and U_2 (λ_2 defines the hydraulic head at the transition point, Eq. 3). Although these two approaches are different, and r_u cannot be directly compared to λ_1 and λ_2, they both quantify the same problem. Hence, to compare the two approaches, various values were considered for r_u, λ_1 and λ_2.

Figure 3 presents the SF of Equation 1 for two λ_2 and various λ_1 and the present work's numerical approach. Factor λ_1 (relative water height at the tension crack) varied from 0.65 to 1, while for numerical analyses, r_u was varied to produce a similar SF range. In that vein, if λ_2=0.75, r_u ranges between 0.0 and 0.5 (Figure 3a), while if λ_2=1, r_u ranges between 0.4 and 0.7 (Figure 3b). Note that r_u=0.7 simulates pore pressures equal to 70% of the vertical effective stress.

Factor λ_2 (defines the groundwater pressure at the transition point, Eq. 3) affects the SF tremendously; increasing λ_2 from 0.75 (Figure 3a) to 1.00 (Figure 3b) decreases SF by approximately 60% regardless of β_z and λ_1. Furthermore, the two methods present a very different relation between SF and β_z. The numerical analysis demonstrate a systematic, linear decrease with β_z, regardless of the r_u. On the contrary, in the analytical model, for larger weak zone inclinations, SF is controlled mainly by β_z and is less affected by the groundwater conditions; for β_z=0° SF ranges between 2.20 and 3.38 (54%), while for β_z=6°, SF ranges between 1.38 and 1.51 (9%) (Figure 3a). Hence, λ_1 has a variant influence on the SF. If β_z=0°, the term "Wtanβ" in Equation 1 becomes zero, so U_1 governs the denominator. Thus, if β_z=0°, λ_1 affects the SF crucially, as seen in Figure 3, indicating that the analytical model might be unable to provide appropriate results for very low β_z.

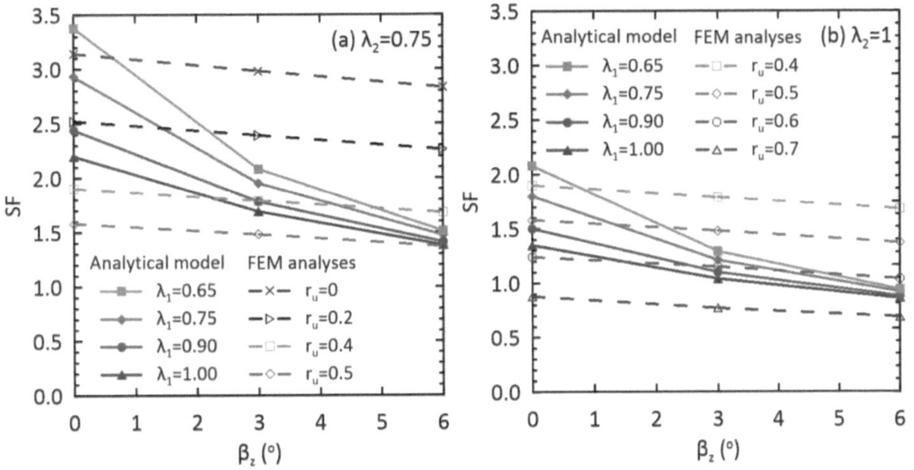

Figure 3. SF with β_z, for various r_u and λ_1, considering (a) $\lambda_2 = 0.75$ and (b) $\lambda_2 = 1$.

Therefore, the two approaches are better compared to high β_z values. For λ_2=0.75 (Figure 3a) and β_z=6°, r_u=0.5 provides the SF (1.37) closest to the analytical solutions (1.38 to 1.51). For λ_2=1 (Figure 3b), the same stands for r_u=0.65 (SF=0.87), providing the SF closest to the analytical SF (ranging from 0.86 to 0.94).

Figure 4 presents the shear strains that indicate the failure surface for FEM. Notice that no assumptions are needed for this surface's shape, but it naturally arises from the analysis. In contrast, Kavvadas et al. (2020) predefine the exact shape of the failure surface, as needed for an analytical model. In both cases, sliding occurs on the weak zone and a planar transition to the ground surface. The analytical model assumes a tension crack of 30° and a transition point that lies almost below slope's crest. Numerical results denote an almost linear transition of 60° for all r_u, and the transition point lies practically below the slope's crest.

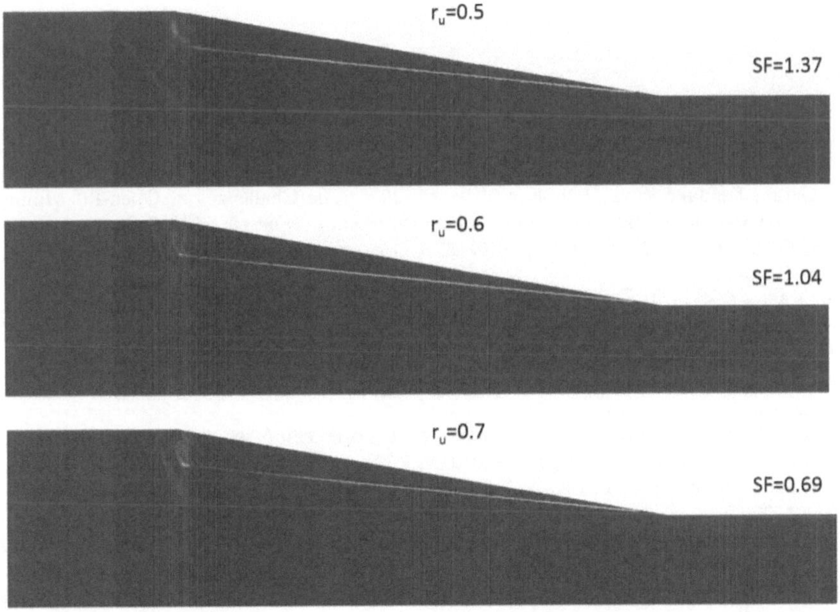

Figure 4. Failure surface and SF results for different r_u, considering β_z=6°.

3 CONCLUSIONS

Kavvadas et al. (2020) presented a practical and useful analytical model for lignite mines' slope stability in the presence of a weak zone. In the present study, a comparison was presented between a numerical approach and that analytical model. Assumptions of each model were discussed, while identical geometries and geotechnical parameters were implemented. The groundwater regime is simulated using differaent approaches in the two models; the r_u coefficient was implemented for numerical analysis, while two specific water forces were used in the analytical model. Therefore, a parametric study on the groundwater parameters was needed to compare the different approaches.

The analytical model is based on uncertain assumptions and may lead to inaccurate results, such as the magnitude of the water forces and the tension crack's geometrical characteristics.

Dry conditions cannot be appropriately simulated using this analytical model, as setting the two water forces equal to zero leads to very large SF. Simultaneously, if the weak zone is horizontal, SF tends to infinity. The analytical model's SFs indicate that for large weak zone inclinations the slope's stability is determined primarily by this inclination and is only slightly affected by the groundwater. On the contrary, if the weak zone is horizontal, groundwater influences SF dramatically.

The two types of analyses are comparable only for very high r_u values, between 0.5 and 0.65, indicating that the analytical model employs large water forces. The model's sensitivity on water forces might indicate difficulties in the SF's appropriate estimation and lacks precision compared to a more rigorous numerical analysis. Additionally, the numerical approach does not demand any assumption regarding the shape and the location of the failure surface, as it arises naturally from the analysis. Both approaches detect the strong effect of the weak zone's inclination on the SF and highlight the importance of a weak zone in lignite mines' stratigraphy.

ACKNOWLEDGEMENTS

This work has received funding from the European Union's Research Fund for Coal and Steel (RFCS) under the project "RAFF - Risk assessment of final pits during flooding slopes" grant agreement No 847299. Financial assistance by the European Commission is much appreciated.

REFERENCES

Bednarczyk, Z. 2017. Landslide monitoring and counteraction technologies in polish lignite opencast mines. Workshop on World Landslide Forum. Springer, 33–43.

Cała, M., Cyran, K., Jakóbczyk, J. & Kowalski, M. 2020. The Challenges of Open-Pit Mining in the Vicinity of the Salt Dome (Bełchatów Lignite Deposit, Poland). *Energies*, 13, 1913.

Chowdhury, R., Flentje, P. & Bhattacharya, G. 2010. *Geotechnical slope analysis*, London, UK, Taylor & Francis Group.

Fleurisson, J.A. & Cojean, R. 2014. Error reduction in slope stability assessment. *In*: Bhattacharya, J., Lieberwirth, H. & Klein, B. (eds.) *Surface Mining Methods, Technology and Systems*. 2014 ed.: Wide.

Ghadrdan, M., Shaghaghi, T. & Tolooiyan, A. 2020. Sensitivity of the stability assessment of a deep excavation to the material characterisations and analysis methods. *Geomechanics and Geophysics for Geo-Energy and Geo-Resources*, 6, 1–14.

Huang, Y.H. 1983. *Stability analysis of earth slopes*, New York, Van Nostrand Reinhold Company Inc.

Kavvadas, M., Roumpos, C. & Schilizzi, P. 2020. Stability of Deep Excavation Slopes in Continuous Surface Lignite Mining Systems. *Geotechnical and Geological Engineering*, 38, 791–812.

Leonardos, M. 2004. *Methods and procedures of monitoring, assessment and improvement of the excavated slopes stability in the deep Greek lignite exploitations*. PhD Dissertation (in Greek). Greece, Athens: School of Mining and Metallurgical Engineering.

Mencl, V. 1977. Modern methods used in the study of mass movements. *Bulletin of the International Association of Engineering Geology-Bulletin de l'Association Internationale de Géologie de l'Ingénieur*, 16, 185–197.

Mikroutsikos, A., Theocharis, A.I., Koukouzas, N.C. & Zevgolis, I.E. 2021. Slope stability of surface coal mines in the presence of a weak zone: a numerical and probabilistic approach. *Submitted for publication.*

Rocscience. 2020. RS2 Version 11.0 - Finite Element Analysis for Excavations and Slopes. www.rocscience.com. Toronto, Ontario, Canada

Stimpson, B. 1979. Simple equations for determining the Factor of Safety of a planar wedge under various groundwater conditions. *Quarterly Journal of Engineering Geology and Hydrogeology*, 12, 3–7.

Theocharis, A.I., Zevgolis, I.E. & Koukouzas, N.C. 2021. A comprehensive geotechnical characterisation of overburden material from Greek lignite mine excavations. *Submitted for publication.*

Tutluoglu, L., Öge, I.F. & Karpuz, C. 2011. Two and three dimensional analysis of a slope failure in a lignite mine. *Computers & Geosciences*, 37, 232–240.

Ural, S. & Yuksel, F. 2004. Geotechnical characterization of lignite-bearing horizons in the Afsin– Elbistan lignite basin, SE Turkey. *Engineering Geology*, 75, 129–146.

Zevgolis, I.E., Deliveris, A.V. & Koukouzas, N.C. 2019. Slope failure incidents and other stability concerns in surface lignite mines in Greece. *Journal of Sustainable Mining*, 18, 182–197.

The Evolution of Geotech - 25 Years of Innovation – Hammah et al. (Eds)
© 2022 copyright the Author(s), ISBN 978-1-032-03645-8

Practicalities when adopting a gradational damage factor within a Hoek-Brown constitutive model

P.J. Edmondson & V. Pere
Golder Associates in Perth, Australia

ABSTRACT: Mining activities and the excavation of open pits in harder rock environments inherently leads to damage of the rock mass that will form the pit walls. Both blasting and stress relaxation, due to unloading, contribute to damage mechanisms which typically include crack extension and tensile failure. This damage factor (D) is used as a strength reduction in the Hoek-Brown criterion, and how D is spatially defined has been a long running issue in current rock mechanics practice.

It is widely accepted in the literature that the degree of damage is higher at the slope face and decays logarithmically deeper into the slope. Carvalho [7] developed a method to incorporate a gradational D factor which can be applied within standard modeling packages as a shear-normal function. The rate of decay is important as the level of damage due to blasting is more significant but only impacts the near surface, typical estimates are usually ≤30m behind the face. The damage caused by stress relaxation is proportional to the depth of the total excavation and extends further into the slope at lower levels than blast damage. Commonly accepted estimates for D are 1/3 of the slope height, which is supported by modelling work carried out by Guzman and Perez [4].

The addition of the variable D parameter by depth within the Generalised Hoek-Brown criterion in Rocscience Slide2 (V9.012) uses a linear decay function to represent the reduction in damage from the slope surface. This does not account for the values of D closer to the slope face decaying at a faster rate than those at depth and as such cannot represent D for both blasting and stress relief within the same constitutive model.

Applying a gradational damage factor to a shear-normal function can be onerous and has some limitations when applied in common numerical analyses. For example, the shear strength reduction method in Rocscience's RS2 (v11.0) does not accommodate use of shear-normal functions and as such inhibits using this gradational D approach.

1 INTRODUCTION

Selecting an appropriate damage factor for input into a Hoek Brown constitutive model [1] has been a long-standing issue within standard rock mechanics practice for a number of years. Definition of geometrical zones to differentiate levels of damage is time consuming and produces step changes in material strengths that generate false boundaries. This can inadvertently influence shear plane search algorithms in standard analytical approaches. A common solution is to select a single value to represent damage from both blasting and in situ stress relief applied across the whole slope. This can result in overestimation of rock mass strength near the slope face and underestimation at higher confinements, which impacts the resulting FoS and influences pit slope design geometries. The aim of this paper is to assess recent developments in how a more representative damage factor can be applied within a Hoek-Brown

DOI: 10.1201/9781003188339-22

constitutive model, and what limitations still exist when this is applied in standard industry software packages, such as Rocscience Slide2 and RS2.

2 BLAST DAMAGE MECHANISM

Damage mechanisms due to blasting commonly include crack extension, tensile failure and spalling due to both the initial shock wave, and expansion from penetration of high-pressure gases into fractures within the rock mass. Controlled or limits blasting is often implemented as a strategy to reduce blast damage inflicted on the final pit walls. Intuitively, the level of damage due to blasting is high within the immediate vicinity of the blast hole and decreases gradationally as you move out radially.

In hard rock environments, commonly accepted distance estimates for blast induced damage are 20-30 charge diameters depending on the strength of the rock, type of explosives used and the degree of confinement [2]. Hoek (2012) provided guidance for damage behind the face for open pit applications in terms of distance (T) based on the degree of control and height of bench [3]. This work indicates that 30% of damage is due to blasting and 70% is due to stress redistribution [4].

3 DAMAGE DUE TO STRESS RELIEF

Relaxation of stresses due to excavation results in damage as remaining material expands in volume or dilates. Disturbance due to stress relief is considered the result of a complex interaction between geologic structure, hydrogeology, the rock mass and initial in situ stress conditions. It was also shown in research undertaken by Hoek (2009) that high initial horizontal stresses can also result in significantly higher levels of damage following excavation due to stress relaxation [2].

The idea of stress relief was first explored by Sakurai (1983) and Chern (1998) who found the level of strain observed during tunneling was directly relatable to the amount of ground support required for stabilisation. It was noted by Hoek during the early days of the Hoek-Brown criterion in open pit applications that a disturbance factor was required in order to fit the criterion to back analyses on failures at Ok Tedi mine in Papa New Guinea [5].

Recent work undertaken by Guzman and Perez (2015) used a synthetic slope model using software by Itasca, to derive an empirical approach to assessing damage due to stress distribution [4]. They observed a relationship between the number of cracks developing relative to excavation height.

A study by Rose [6] (2018) carried out UDEC numerical modelling to simulate the effects of rock mass disturbance and develop D factor decay curves to represent the distribution of yielded zones as a function of depth behind the slope face. Validation of results undertaken in Slide2 indicated closer alignment with UDEC yield when using a D-factor decay function (between 1-2%) compared to a zoned D factor approach (between 3-9%) using damage zoning.

4 GRADATIONAL DAMAGE WITHIN THE HOEK-BROWN CONSTITUTIVE MODEL

A method to apply a gradational D factor into a shear normal function was developed by Carvalho (2019) [7]. A decay function was incorporated into the Hoek-Brown criterion based on a half sigmoid function (equation 1) using Balmer's equations which relate principal stress space to normal stress space. It assumes logarithmic decay to account for the higher rate of reduction closer to the slope surface due to blast damage. This is in line with recommendations provided by Hoek on values for T; distance from the slope face [3]. The rate of decay decreases and merges with the D equals zero curve at a user specified distance (Dmax) to account for disturbance due to the anticipated stress relief as illustrated in Figure 1.

$$D = (2 \times Dmax)/((1 + \exp((5 \times z)/d\_zero))) \quad (1)$$

Rose (2018) recommended that the slope height be used for Dmax if no numerical analyses are available based on UDEC numerical analyses [6]. Stress relief can also be observed in RS2 finite element analyses through assessment of strain.

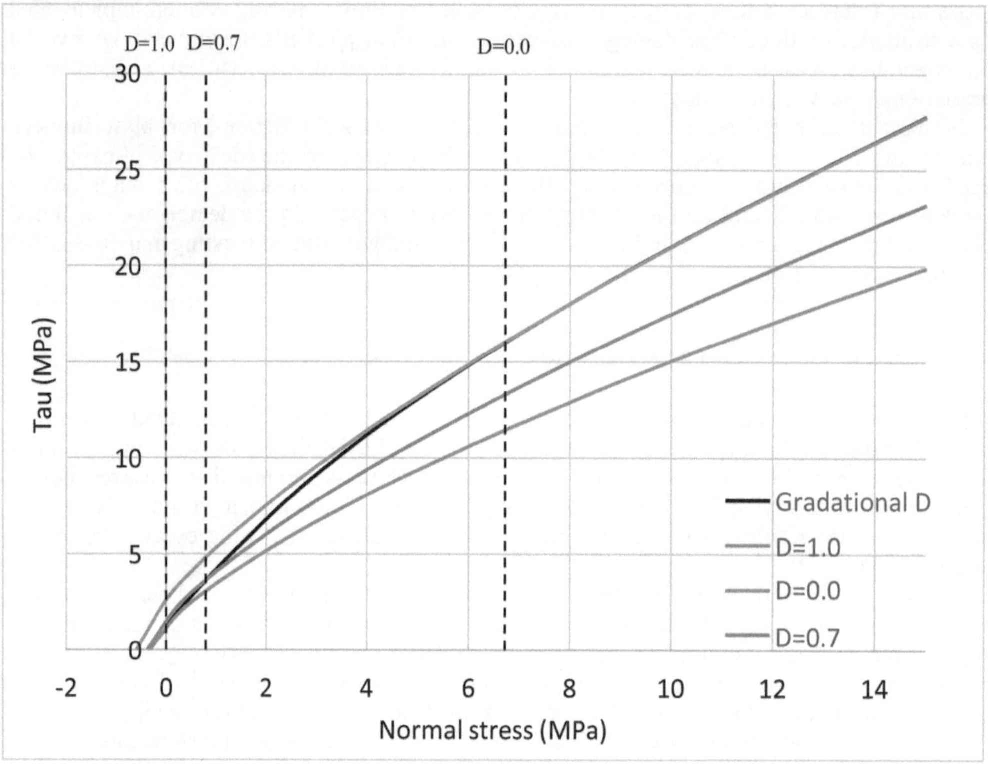

Figure 1. Gradational damage decay function applied as a shear normal function for input into Slide2.

It should be noted that the definition of the shear normal curve does not consider tensile strength that is normally calculated if using curve generating software like RocData V5. In the method of slices, tensile forces are occasionally encountered if high values of pore pressure are acting on the base of a slice or if there is a steeply inclined base angle (near vertical), for example the crest of a slope where a tensile crack is likely to form. Typically, this will not influence the calculated FoS as the default setting in Slide2 assumes zero shear strength if the base of the slice is in tension. Application of D in limit equilibrium analyses.

Incorporating a gradational decay function into a shear normal function removes the requirement to define geometrical zones. This removes bias that can be generated within search functions to inadvertently follow boundaries along material with higher damage factors applied. Research by Carvalho (2019) also indicates that failure mechanisms are more realistic when using a gradational D factor.

Rocscience introduced a feature in Slide2 (version 2018) which allows the user to vary the D parameter by depth [8] based on different reference datums, and using a linear function for reduction in D. Figure 2 shows a comparison plot of a linear approach compared to use of a logarithmic function. A linear function is unable to represent the damage associated with blasting alongside the damage due to stress redistribution further into the slope.

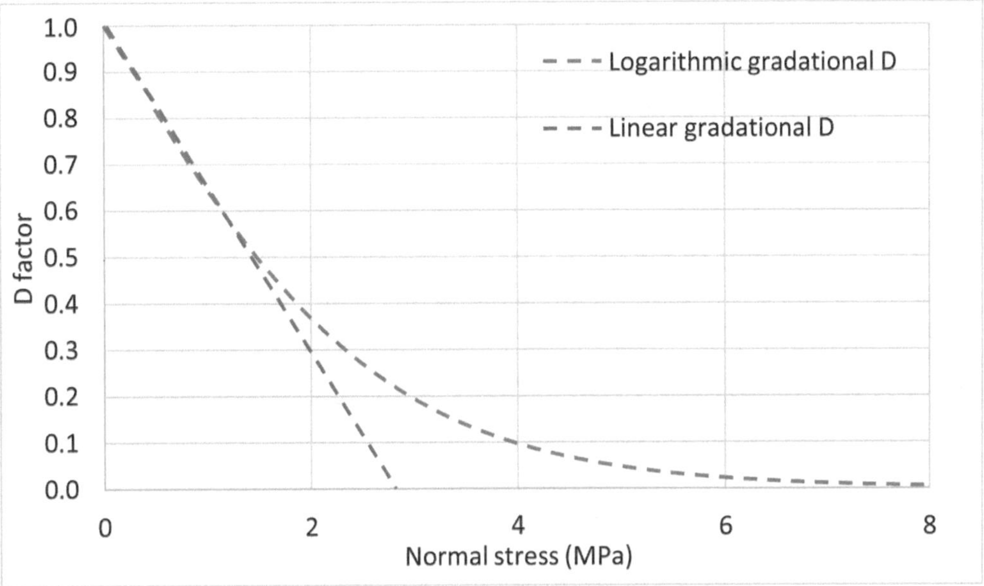

Figure 2. Example showing linear and logarithmic gradational D functions relationship to normal stress.

5 APPLICATION IN FINITE ELEMENT ANALYSIS

The use of shear strength reduction in RS2 involves reducing the strength parameters by a certain factor (SRF) until the model becomes unstable and is considered to be analogous to factor of safety (FoS) [11]. Applying a gradational damage factor within RS2 is not currently possible using a shear normal function as the SRF method is limited to this type of constitutive model.

RS2 is commonly used as a check on limit equilibrium results, produced in Slide2, to validate the failure mechanisms, and so the bulk of analyses can be run using limit equilibrium methods, which are computationally faster. The inability to use the gradational D in RS2 places limitations on being able to directly compare RS2 and Slide2 models, and requires either additional models run in Slide2 or deeper interpretation of the RS2 results.

6 DISCUSSION

The use of a logarithmic function to represent the gradation of the D factor results in a number of positive outcomes across standard slope stability analyses. Firstly, a more realistic failure mechanism is obtained due to the absence of boundary effects caused by using a zoned D factor approach. A more realistic scenario with higher levels of damage closer to the slope surface is obtained which correlates with the areas of low normal stress where most failures are likely to occur.

The onerous steps required to incorporate a gradational D by using a shear normal function within Slide2 may limit some users, especially on smaller scale projects where there are budget and time constraints. Introduction of a logarithmic decay function within the variable parameter function would solve this issue.

Similarly, the inclusion of the shear normal function in RS2 or a variable D function with logarithmic capability would add to the functionality of using the two software in unison. It would allow the user to evaluate the damage within the slope through the assessment of strain levels in different areas of the slope and validate failure mechanisms that are observed within limit equilibrium analyses.

Further work and validation of this approach is still required. The initial damage level and depth at which damage ceases to occur is largely project and material specific. It is the authors opinion that introducing these features would assist with making this type of assessment more accessible so further case studies can be evaluated.

REFERENCES

[1] E. Hoek, 'Hoek-Brown failure criterion 2002 edition' *Proceedings of NARMS-Tac*, pp. 267–273, 2002.

[2] J. Read, P. Stacey 'Guidelines for open pit slope design', CSIRO, 2009.

[3] E. Hoek 'Technical note for Roc Science. Blast damage factor 'D", Roc Science website, 2012.

[4] R. Guzman, 'Toward a mechanically based definition of the disturbance factor using the slope model lattice code', Integrating Inovations of Rock Mechanics, 2015.

[5] E. Hoek, "Rock mass properties for surface mines," *Slope stability in surface mining*, pp. 59–70, 2000.

[6] N. Rose, 'Quantifying transitional rock mass disturbance in open pit slopes related to mining excavation,' *Congresso Internacional de Energia y Recursos Minerales*, 2018.

[7] J. Carvalho, 'A method for applying a variable disturbance factor to Hoek-Brown properties in large open pits,' in *14th international congress of rock mechanics and rock engineering*, Foz do Iguassu, 019.

[8] R. Science, 'Slide2 - User manual version 9.012,' https://www.rocscience.com/software/slide2/, 2020.

[9] R. Science, 'RS2 - user manual version' 11.007, https://www.rocscience.com/software/rs2, 2020.

[10] F. Bastane, 'Predicting the extent of blast induced damage in rock masses,' *International journal of rock mechanics and mining sciences*, vol. 56, pp. 44–53, 2012.

[11] P. Persson, 'The relationship between strain energy, rock damage, fragmentation, and throw in rock blasting,' *International journal for blasting and fragmentation*, vol. 1, no. 1, pp. 99–110, 1997.

The Evolution of Geotech - 25 Years of Innovation – Hammah et al. (Eds)

Open pit 3D slope stability for anisotropic rock masses in soft iron ore deposits

L.T. Figueiredo
Anglo American Brazil, Belo Horizonte, MG, Brazil

R. Hammah
Rocscience Inc., Accra, Ghana

Y. Santos, T. Souza & C. Nogueira
Walm Engenharia, Belo Horizonte, MG, Brazil

ABSTRACT: Assessing geotechnical stability in rock masses is challenging using 2D analysis methods when the failure mechanisms are driven by 3D geometry and anisotropy. For such cases, it is necessary to use 3D analysis to correctly assess slope stability in such environments. This work aims at presenting a methodology and a case study in Serra do Sapo soft iron ore deposit using 3D Limit Equilibrium method. The paper will examine 3D implicit geological models with anisotropic rock masses. To capture the failure mechanism correctly, the work identified the main structural drivers, such as structural geological contacts and internal features or "lenses" such as filonites. Finally, the authors compared 3D limit equilibrium results to 2D limit equilibrium outcomes. The comparison showed that 3D analysis provided a reliable way to predict instabilities due to 3D interactions.

1 INTRODUCTION

The geotechnical stability assessment for Open Pit slopes for highly anisotropic rock masses is difficult using only 2D analysis methods. This challenge is due to failure mechanisms in real slopes being driven mainly by 3D geometrical interactions involving anisotropic weaknesses in the rock mass. In highly weathered rock masses, such as saprolites and saprock, this task can be even more difficult. 3D geotechnical analysis tools are required to obtain a correct understanding of slope stability in such environments. The open pit at Serra do Sapo, a soft iron ore deposit in southeast Brazil, offered a case study.

2 GEOLOGICAL SETTINGS

2.1 *Regional settings*

The Serra do Sapo iron ore deposit is hosted in a Proterozoic metasedimentary sequence in the Southern Espinhaço Ridge. The rocks were submitted to strong weathering processes that generated a friable group of lithologies. The geological domain is siliciclastic metasedimentary, belonging to the Espinhaço Supergroup, in its southern portion. It is

DOI: 10.1201/9781003188339-23

located in the Serra da Serpentina Group, where clastic and chemical metasedimentary rocks are found. Knauer & Grossi-Sad (1997), subdivide this group into three distinct lithostratigraphic units: the base, which comprises quartzites and schist quartz, followed by a unit consisting of banded iron formations and, at the top, a unit with fine phyllites and quartzites.

2.2 *Local settings*

The thickness of the mineralized body and friable ore is variable along the deposit, and is directly related to the geological structural regime in each domain. Thrust faults cause geotectonic inversions of the friable lenses and disposal of the surrounding lithologies (schist, quartzite or phyllite). A second tectonic event triggered a strike-slip fault system that thickened the weathering profile in some regions of the deposit, as shown in Figure 1.

2.3 *Geomechanical implications*

Due to the tectonic history of the deposit, the rock mass's geomechanical behaviour varies along strike. It is controlled either by the shear structures along the geological structural contacts due to the thrust fault system and the thickening of the weathering profile. The second tectonic event placed different lithologies in the footwall along the strike, causing schists, quartzites or phyllites to occur in some areas and not in others.

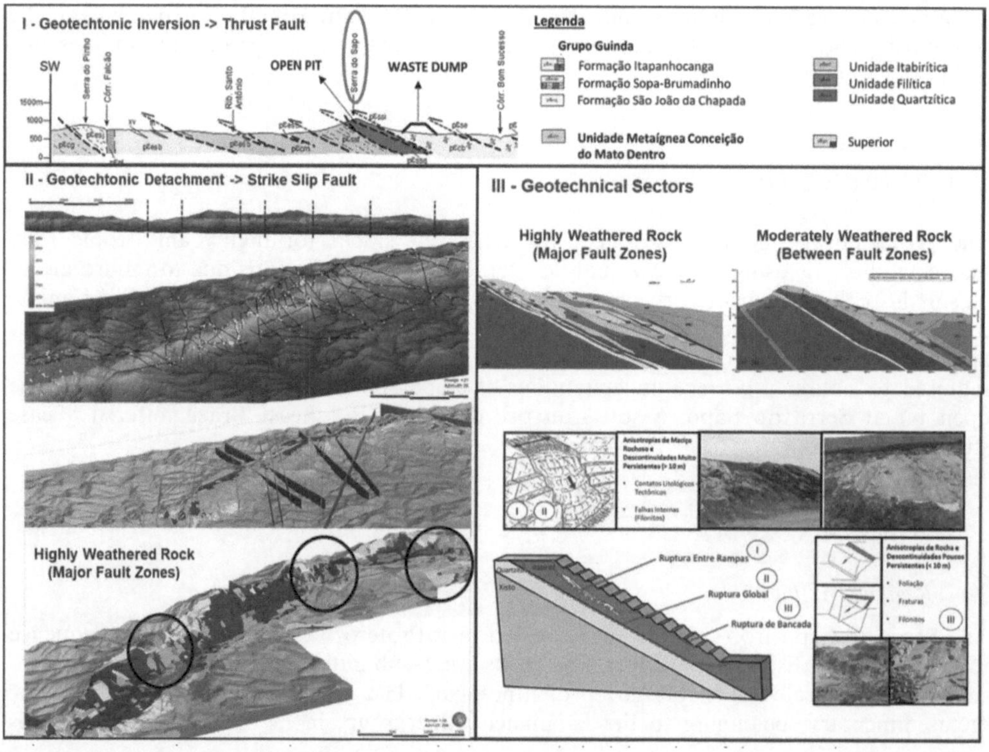

Figure 1. Tectonic stages and consequences in the geotechnical domains of the open pit.

3 ROCK MASS CHARACTERIZATION

3.1 *Rock mass classes*

The rock mass rating (RMR, Bieniawsky 1989) classification system is used to characterize core samples at Serra do Sapo. This rating is then correlated to the geological strength index (GSI, Hoek 1994) obtained from field mapping. To simplify the geomechanical model, and due to the difficulty of obtaining values for all five parameter categories used to determine RMR, the rock mass classes were grouped according to weathering profile and friability as shown in Table 1.

Table 1. Rock mass and saprolite grouping according to weathering profile and friability.

Weathering Horizon	Unit	R	RMR Class
Saprolite	Friable	R0-R1	V
Saprock	Semi Compact	R2-R3	III-IV
Compact	Unweathered	R4-R6	I -II

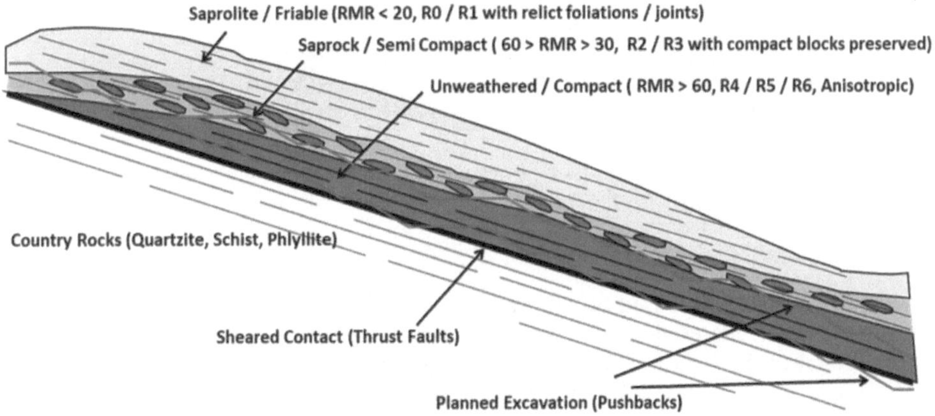

Figure 2. Schematic image over rock mass weathering profile and friability zoning in Serra do Sapo deposit.

3.2 *Discontinuities*

Generally, the geological layers trend in a North-South direction and have a low dip angle of around 30° towards the East. The foliation generated by the thrust fault regime is sub-parallel to the general trend.It expresses itself in several forms such as schistosity, gneissic foliation, and compositional banding. Filonites, which are persistent shear zones parallel to the foliation and contacts developed in the thrust faulting process, drive the west slope's primary failure mechanisms. There are also four main joint families (which are less influential in driving the major failure mechanisms). Table 2 and Figure 3 show the main discontinuity sets according to spatial orientation, spacing and persistence, Jr and Ja, roughness and alteration constants.

3.3 *Parameters*

The strength criteria for the materials were split into two groups, soil materials and rock mass materials. The saprolites and friable materials' strengths were described with the

Table 2. Main discontinuities families according to spatial orientation, spacing and persistence, as well as Jr constants in Serra do Sapo.

Family	Dip Dir (Azimuth)	Dip (°)	Spacing (m)	Persistence (m)	Jr	Ja
N-S	230 -270	70-90	1 – 2	5 – 10	1	1-2
E-W	330-30	90	1 – 2	10 – 20	1	1-2
SE-NW	150 -170	70-90	1 – 2	10 – 20	1	1-2
Filonite	70 -110	22-35	10	10 – 20	1	3-4

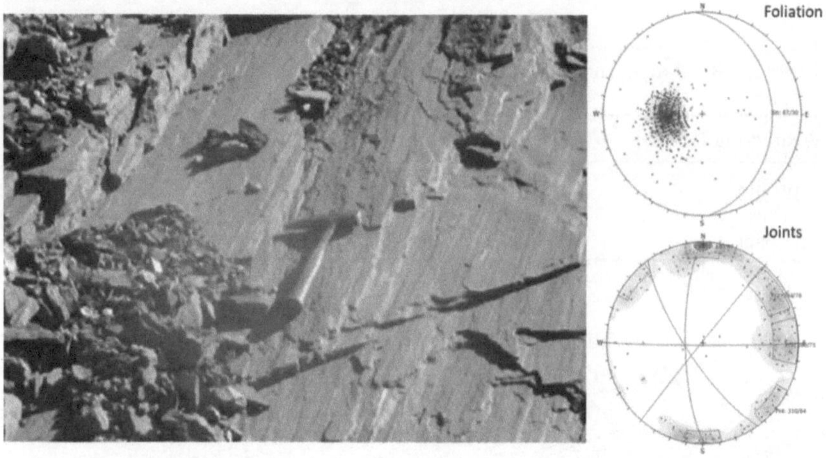

Figure 3. The picture shows an outcrop of a compact itabirite, with the stereogram of the foliation and parallel structures, filonites, geological contacts and joint sets.

Mohr-Coulomb criterion, based on laboratory tests and back-analysis results. It is important to note that all materials have anisotropic strength, with the directions being parallel and orthogonal to the foliation. The strength parameters are shown in Table 3 and Table 4. The Hoek-Brown nonlinear strength criteria were used in saprock and unweathered materials, semi-compact and compact materials. The Barton and Bandis criterion (1982) was used for the strengths discontinuities parallel to the thrust faults, foliation and geological contacts.

Table 3. Geotechnical properties for the friable materials.

Material	γ_{nat} (KN/m^3)	γ_{sat} (KN/m^3)	$C_L(KPa)$	Φ_L (°)	$C_{//}$(KPa)	$\Phi_{//}(\circ)$
Friable Itabirite (IF)	22	24	50	36	25	29
Friable Quartzite (QTF)	20	22	50	28	30	25
Friable Phyllite (FIF)	19	21	20	23	10	20
Friable Schist (XIF)	17	19	13	36	13	33

4 3D GEOMECHANICAL MODEL

Typically, geomechanical models are built from lithochemical block models. This leads to errors since block sizes are defined to address production purposes and ignore layers that are

Table 4. Geotechnical properties for the rock masses.

	Ortogonal				Parallel		
Materials	γ_{nat} (KN/m^3)	UCS (Mpa)	GSI	mi	JRC	JCS (Mpa)	$\Phi_b(\circ)$
Compact Itabirite (IC)	32	258	78	17	2	83	31
Semi Compact Itabirite (ISC)	30	40	38	17	2	12	31
Compact Quartzite (QTC)	27	172	69	14	6	48	31
Semi Compact Quartzite (QTSC)	25	30	26	14	6	10	31
Compact Schist (XIC)	27	102	57	10	1,5	35	30
Semi Compact Schist (XISC)	25	29	24	7	1,5	10	30
Compact Phyllite (FIC)	27	102	57	10	1,5	35	30
Semi Compact Phyllite (FISC)	25	29	24	7	1,5	10	30

only a few metres thick. Some of these layers can be highly persistent and have very low strength and can be the principal drivers of failure.

Another common practice is to manually generate geomechanical cross-sections and combine them into 3D solids. According to Sounders et al. (2020), "This process typically results in a biased and possibly overly linear interpretation."

It is more helpful to develop an implicit model from a geomechanical database (which does not have lithochemical wrong bias). Such efforts make the geomechanical model more reliable, practical and fast to update when new data, such as new drillholes and field mapping results, become available. The data can be interpolated in three dimensions as a function of typical spatial distribution and converted into 3D solids. Figure 4 shows the geomechanical implicit 3D model with planned excavation and the main interpolation directions and weights.

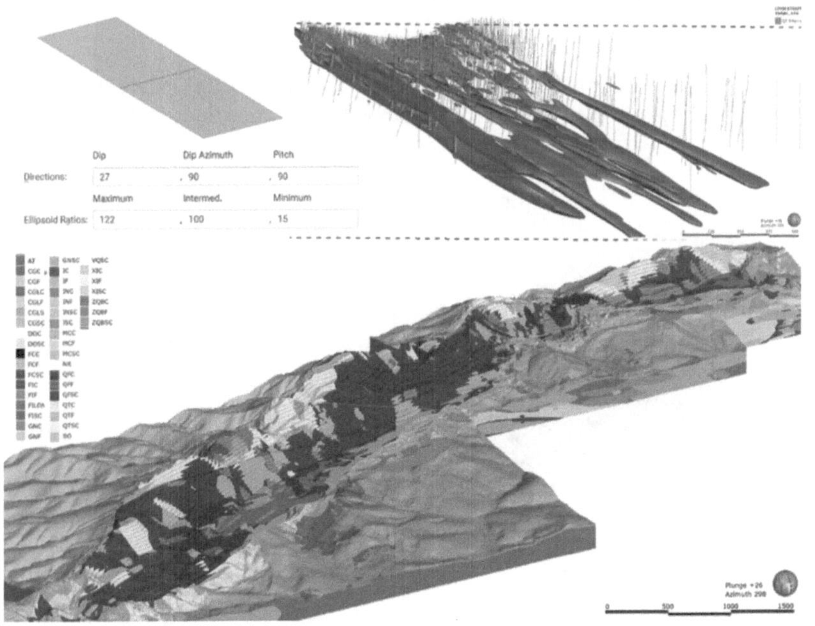

Figure 4. Below, the geomechanical implicit 3D model cut with planned excavation; and, above, the interpolant main directions and weights.

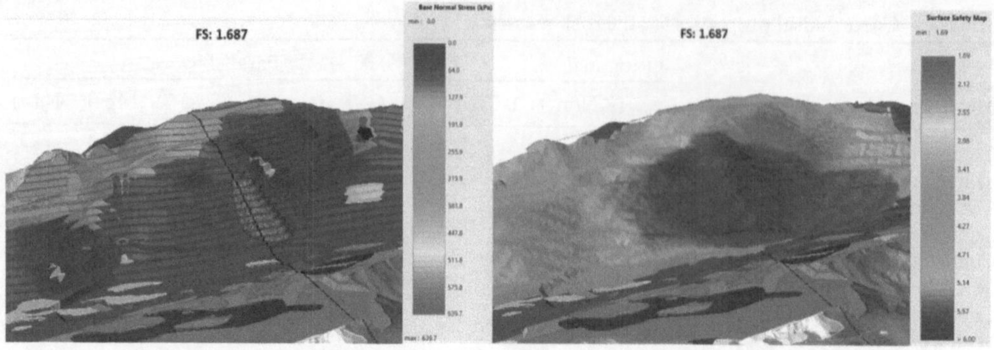

Figure 5. On left, the 3D geomechanical model with the cross-section line for 2D analysis in black and lowest factor of safety surface with the range of base normal stress; and, on right, the resulting safety map of 3D limit equilibrium analysis with Slide$^{3®}$.

5 3D SLOPE STABILITY

Once developed, the geomechanical solids were imported into Slide3® for 3D limit equilibrium analysis. Slope stability analysis using the Bishop, Morgenstern-Price and Spencer limit equilibrium methods were conducted. The model did not consider groundwater effects as the area is constantly dry. Ellipsoid surfaces, optimized iteratively with the Cuckoo search, were employed.

As shown in Figure 5, the 3D results showed a concentration of failure surface in a region where two 3D interactions were present. These areas had geological contact between itabirites and quartzites and a convex slope geometry in plan view. The 3D factors of safety were close to the results from 2D limit equilibrium analysis as shown in Figure 6; the factors of safety differed by less than 1% (with a value of about 1.6). Usually, the differences between 2D and 3D results are higher, due to the assumptions in 2D that cross-sections remain the same over long out-of-plane distances. Gitirana et al. (2008) achieved differences of 15% to 50%, when comparing the factors of safety of 2D and 3D analysis.

The similarities of the 2D and 3D factor of safety for the current case can be attributed to the 3D interactions of geological features, geomechanical properties and slope geometry.

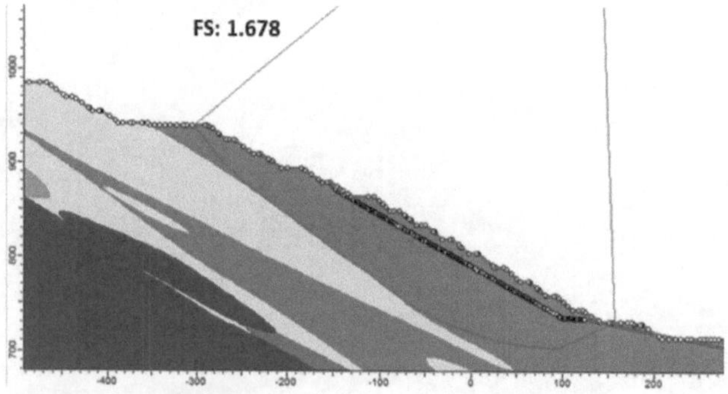

Figure 6. Slide$^{2®}$ results of the most critical cross-section.

6 CONCLUSION

This paper addressed the challenges of estimating the stability of open pit slopes in a soft iron ore highly anisotropic rock mass. Analyses of the Serra do Sapo deposit were carried out, and 2D and 3D limit equilibrium results compared to each other. The 3D stability analysis incorporated 3D interactions, including geological contacts, discontinuity anisotropy and slope geometry. The paper discussed the geological setting, essential to understanding the rock mass behaviour, and the geomechanical implications of the thrust fault system that play a crucial role in geotechnical domaining.

The geomechanical modelling was implicitly developed outside of the block model. Finally, the results for 2D and 3D limit equilibrium stability results showed that 3D interactions could lead to results that are much closer to 2D factors of safety than typically mentioned in the literature. This outcome means that 3D limit equilibrium, in addition to 2D analysis and other methods, can give deeper insights into 3D interactions of geology and geomechanical features that combine with 3D slope geometry. 3D slope stability modelling can help make open pits safer for miners.

REFERENCES

BIENIAWSKI, Z. T., 1989. Engineering rock mass classification. New York: John Wiley. 248p, 1989.

Hoek, E. (1994) "Strengths of rock and rock masses" ISRM News Journal, 2(2): 4–16.

Barton, N.R.; Lien, R.; Lunde, J. (1974). "Engineering classification of rock masses for the design of tunnel support". Rock Mechanics and Rock Engineering. Springer. 6 (4): 189–236.

Knauer, L.G. & Grossi-Sad, J.H. (1996). Geologia da Folha Serro (SE-23-Z-B-IV). Projeto Espinhaço. COMIG;

D. Martin P Stacey – 2018. Guidelines for open pit slope design in weak rocks

Gitirana, G., Santos, M., & Fredlund, M. (2008). Three-Dimensional Analysis of the Lodalen Landslide. Geocongress 2008. New Orleans

E. Saunders, A. LeRiche, T Shapka-Fels, W Barnett (2020). Characterisation of foliated rock masses using implicit modelling to guide geotechnical domaining and slope design, Slope Stability 2020 - PM Dight (ed.), Australian Centre for Geomechanics, Perth, ISBN 978-0-9876389-7-7

The Evolution of Geotech - 25 Years of Innovation – Hammah et al. (Eds)

Comparison of the probabilistic and deterministic slope stability analysis of a dolomite quarry in Hungary

G. Bögöly & F. Füzesi
Department of Engineering Geology and Geotechnics, Budapest University of Technology and Economics, Budapest, Hungary

ABSTRACT: The paper compares two approaches of the stability analyses of rock slopes using the example of an open-pit mine in Csákvár, Hungary. The aim of the study is to compare the currently used and widely accepted deterministic analysis with the probabilistic-based slope sta- bility analysis, which is presently less circumscribed and standardized in Europe. Using our pre- vious measurements in the dolomite quarry, which was mentioned above, we collected the data required for the research, such as defined characteristic and probabilistic parameters for analyzing the stability of the examined rock slope cross-sections. Our calculations were concerned with plane, wedge and circular failures, for that RocPlane, SWedge and Slide programs were used. During the comparison, the factor of safety (FoS) obtained from the deterministic calculations was examined and compared with the different results of the probabilistic approach. In addition, the optimization of the slope design was also carried out with both methods in order to see which analysis leads to more economical solution. The design of a slope (slope angle, number of benches, location of the benches) may be more flexible with the probabilistic calculations consid- ering different acceptance criteria. Therefore the optimization of the geometry was also carried out using various criteria known from the literature. The results of the different comparisons demonstrate the benefits and the difficulties of the application of the probabilistic calculations.

1 INTRODUCTION

The stability of a slope is defined by factors that show natural variation. As a consequence, deter- ministic approaches can only be valid with the completion of sensitivity analyses. The realization of the high uncertainty of the geological data used for the stability calculations created the basis of probabilistic design methods. In case of the probabilistic approaches, the stochastic nature of the input parameters is included, however the interpretation of the results might take more exten- sive consideration. Although pioneer researchers (Call 1972, McMahon 1971) have been dealing with the probabilistic slope design since the 1970s, only the recent development of computer science allowed its application for practical engineers. Since the actual acceptance criteria to be used cannot be determined from general guidelines considering the probability of failure, the con- sequence of failure should also be analyzed more thorough (Sjöberg 1996). Therefore the demand has increased lately for the determination of different acceptance criteria for specific mines and geologic conditions. The final aim of our research is to develop an acceptance criteria system for Hungarian dolomite open-pit mines by means of risk analysis. The paper presents the very first step of our planned investigations, it is concerned with the comparison of the deterministic and probabilistic slope stability analysis of one specific open-pit mine in Hungary. The optimization of the geometry was carried out with various criteria known from the literature where the factor of safety (FoS), probability of

DOI: 10.1201/9781003188339-24

failure (PoF), assessed consequence and geometric details were considered as the basis of the calculations. The results of the different comparisons demonstrate the benefits and the difficulties of the application of the probabilistic calculations. Fundamentals of the probabilistic calculations in the literature are well-established, Hoek (2007); Read & Stacey (2009); Sjöberg (1999); Wyllie & Mah (2005) served the basis of our considerations.

2 GEOLOGICAL SETTING AND QUARRYING

The study area is located in Northern Hungary near to a smaller town called Csákvár in Fejér county (Figure 1). The open-pit mine lies under the Vértes Mountain (47°24′54″N 18° 26′50″E). The approximate dimensions of the operation quarry are: 340 m × 270 m. The highest elevation of the crest is around 240 m MSL, while the elevation of the pit floor is 170 m MLS. The height of the pit wall changes between 35-70 m. The highest walls are on the southwest side of the quarry, the lowest parts are on the southeast side. Generally, pit walls are divided into benches at least in 20-25 meters for rockfall catchment. In this case, the mining company plans to excavate the walls without benches using the same steepness for all slopes. Thus, it must be stated that the geometry is not our design, we only evaluate its stability.

The Vértes Mountain is primarily made up of Upper Triassic (Carnian to Rhaetian) dolomite, that is widespread in the whole Transdanubian Range. The Carnian shallow marine platform car- bonates are represented by alternating bedded, thick-bedded lagoonal and laminated dolomites such as the Sédvölgy Dolomite and the Main Dolomite. The Main Dolomite is the thickest and most extensive formation in the Vértes Mountain. The quarry exposes this Triassic dolomite se- quence, its main raw material is the Sédvölgy Dolomite (Figure 1). Since the quarry is located on a higher spot of the region, the presence of groundwater is not expected. The closest groundwater level was observed 20 meters below the pit floor. (Budai et al., 2015)

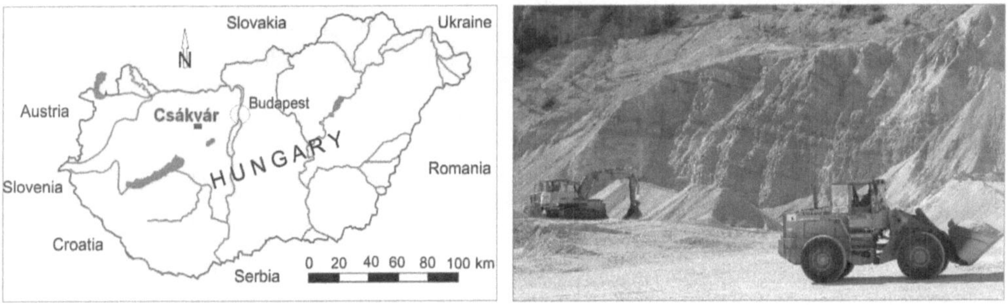

Figure 1. Location of the quarry (left), Excavated dolomite pit wall (right).

3 METHODS

3.1 *Site investigation and laboratory testing*

According to the site observations, the dolomite is mostly blocky and disturbed with moderately weathered discontinuity surfaces. Variously two or three dominant joint sets can be separated in the quarry area with other non-persistence joints. Different in situ measurements were made to provide representative data set for both calculation approaches. The orientation of sixty-three joints was measured with geological compass. The roughness profile of forty discontinuity surfaces was recorded with Profilometers (Barton comb), and the surface hardness of the joints was measured with Schmidt hammer. The typical values of the Geological Strength Index (GSI) of the dolomite rock mass were also determined (Hoek & Marinos 2000). Blocks were obtained from the quarry for further laboratory testing. Forty uniaxial compressive strength tests and thirty-three indirect tensile strength tests were carried out in accordance with the International Society for Rock Mechanics (ISRM) suggested methods (Ulusay 2015).

3.2 *Stability analysis*

The identification of the characteristic joint sets and the kinematic analysis was performed using the Dips 8.0 program based on the field measurements. The expected failure types of each pit wall were determined, and the stability analyses of the slopes were performed with the limit equilib- rium method (LEM). Each critical cross-section was examined for the different failure modes. The calculations were carried out with the SWedge 7.0, RocPlane 4.0, and Slide 9.0 programs to analyze wedge, plane and the rotational shear failure. In general, stability analyses with the SWedge and RocPlane programs are applicable mainly to benches, while the Slide analyses are more suited to determine the inter-ramp and overall slope stability. In the case of wedge and plane failure, the Barton-Bandis (Barton & Choubey 1977) shear steength model was applied, and in the case of the rotational failure, the Generalized Hoek-Brown (Hoek & Brown 2018) criterion was used. In the deterministic calculations, the rock slopes were considered stable if their FoS was 1.35 or above. This criterion is specified by Eurocode 7 (EN 1997-1, 2004) that also describes that in the calculation no practical factors should be applied neither on the resistance nor on the effect side. Characteristic values were determined statistically in relation to the 5% fractile (95% confidence interval). In the probabilistic analyses, parameters were defined as stochastic variables with their mean values, statistical distribution, standard deviation, relative minimum and maxi- mum.

3.3 *Slope design criteria for the probabilistic analyses*

In the probabilistic calculations, more acceptance criteria (Adams 2015; Kirsten 1983; Pothitos & Li 2007; Priest & Brown 1983; Schellman et al. 2006; Wesseloo & Read 2009) were used that were developed for different purposes. They are not recommended for general use, some of them are made for a specific mine or geological environment. However, as no such system of criteria has been established for Hungarian open-pit mines, the first step is to apply the international sug- gestions under the given circumstances.

A short summary is given about the main aspects of the applied criteria. Adams (2015) devel- oped a complex risk matrix that mainly takes into account the consequences of failure, the level of design confidence and lifespan of the rock slope. Based on these factors, the acceptable PoF can be determined. Wesseloo & Read (2009) likewise considers the consequences of failure, and it calculates with the size of the slope. Pothitos & Li (2007) categorizes the design elements to suggest different acceptable PoF ranges for the Ok Tedi open-pit mine (Papua New Guinea). Schellman et al (2006) acted similarly in Mantoverde mine (Chile) but took into account the mass involved in the failure, as shown in Table 1. Priest & Brown (1983) also considered the slope category and the consequences of the failure (Table 2). Contrary to the previous researches, Kris- ten (1983) did not take into account the size of the slope or the failure; his suggested PoF is related to serviceable life, public liability and monitoring.

Table 1. Acceptable PoF in Mantroverde open-pit mine (Schellman, 2006).

Mass involved in potential failure (t/m)	Factor of Safety	Probability of Failure
< 15 000	>1.20	<0.12
15 000 - 30 000	>1.25	<0.10
>30 000	>1.30	<0.08

Table 2. Acceptance criteria for rock slopes (Priest & Brown, 1983).

Category and consequences of failure	Example	Reliability index, b	Probability of failure, P
1. Not serious	Non-critical benches	1.4	0.1
2. Moderately serious	Semi-permanent slopes	2.3	0.01
3. Very serious	High/permanent slopes	3.2	0.003

4 RESULTS

4.1 *Laboratory testing and kinematic analysis*

The orientation of the joint sets indicated that wedge and plane failure might occur on the north- eastern and northwestern pit walls. Due to the height of the slopes and the fractured structure of the dolomite rock mass, analysis of rotational shear failure was also well-founded. The results of the kinematic analysis showed that toppling failure could not occur. With the evaluation of the above-mentioned laboratory tests, all input data were determined for the deterministic and prob- abilistic calculations. Table 3-4 show the results of these calculations.

Table 3. Input data for the deterministic calculations.

Deterministic parameters*		
Max. height of NE wall	[m]	44
Max. height of NW wall	[m]	70
Unit weight	[kN/m^3]	27,5
Failure plane angle	o	52
Failure plane waviness	o	0
Upper face angle	o	0
JRC	-	12
JCS	MPa	47.68
f_r	o	21.89
UCS $_{characteristic}$	Mpa	64.06
TS $_{characteristic}$	Mpa	4.61
GSI	-	35
m_i	-	5.77
D	-	0

* JRC – joint roughness coefficient; JCS – joint compressive strength; f_r – residual friction angle; UCS – uniaxial compressive strength; TS – indirect tensile strength; GSI – Geological Strength Index; m_i – material constant for the intact rock; D –disturbance factor

4.2 *Deterministic analyses*

The aim of the deterministic analysis was to establish the steepest slope angle so that it still cor- responds to the minimum safety factor of 1.35. The calculations were carried out with cautiously selected characteristic values (Table 3). The results of the calculations are summarized in Table 5 and conclude that the maximum steepness that can be applied for all walls of the quarry is 55°.

Table 4. Input data for the probabilistic calculations.

Probabilistic parameters						
	Unit	Distribution	Mean	Std. Dev.	Rel. Min	Rel. Max.
Unit weight	[kN/m^3]	Normal	27.50	0.58	2.38	0.97
Failure plane angle	°	Uniform	40.33	-	13.33	11.67
Failure plane waviness	°	Uniform	10	-	3.3	3.33
Joint 1 - Dip. Angle	°	Uniform	67	-	9	8
Joint 1 - Waviness	°	Uniform	10	-	3.33	3.33
Joint 1 - Dip. Direction	°	Uniform	91	-	9	13
Joint 2 - Dip. Angle	°	Uniform	63	-	12	10
Joint 2 - Dip. Direction	°	Uniform	191	-	9	14
Joint 2 - Waviness	°	Uniform	10	-	3.33	3.33
JRC	-	Uniform	11.9	-	7.9	7.1
JCS	MPa	Normal	53.87	23.86	38.87	43.70
f_r	°	Uniform	23.89	-	2	2
UCS	MPa	Normal	72.37	32.06	52.22	58.71
m_i	-	Uniform	8.68	-	3.00	3.00
GSI	-	Uniform	40	-	5	10

4.3 *Probabilistic analyses*

The results of the probabilistic analyses are summarized in Table 6. Based on each criterion, the maximum slope angle was determined for the different failure types. PoF of rotational failure was 0% under 77°, and it was 3.2% even for the steepest possible option (83°) of the other failure types. The calculations show that similarly to the results of the deterministic design, plane failure proved to be critical in all cases. Values of allowed PoF are significantly different, thus the cal- culated slope angles are also scattered in a wide range. Half of the applied design criteria resulted in smaller slope angles than the deterministic approach, the lowest acceptable steepness was 47°. On the other hand, some of the criteria allowed to design much higher slope angles, 21° steeper than the result of the deterministic calculation.

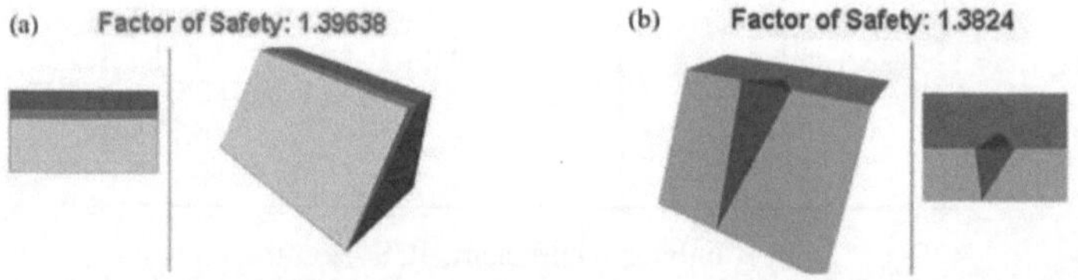

Figure 2. Results of the deterministic analyses: plane failure in RocPlane (a), wedge failure in Swedge (b).

Table 5. Result of the deterministic analyses.

Deterministic calculation			
	RocPlane	Swedge	Slide
Section	NE wall	NW wall	NW wall
Slope angle	55°	63°	63°
FoS	1.39	1.38	2.06
PoF	6.15%	0.03%	0%

Table 6. Result of the probabilistic analyses.

Results of the probabilistic calculations

Design criteria and category		Allowed PoF	RocPlane Slope angle	SWedge Slope angle	Slide Slope angle
According to Priest & Brown (1983)	Moderately serious - Semi-permanent slopes	1-2%	*47-49°*	*71-73°*	*76°* 0% PoF
Acc. Pothitos (2007)	Overall or interramp slope	1-3%	*47-50°*	*71-74°*	
Acc. Kristen (1983)	Medium-term - semi-temporary slopes	1.5-5%	*49-52°*	*72-76°*	
Acc. Wesseloo & Read (2009)	Overall slope - Medium cons.	5-10%	*52-76°*	*76-83°*	*83°* 3.2% PoF
Acc. Schellman (2006)	>30 000 t/m - FoS >1,30	<8%	*63°*	*80°*	
Acc. Adams (2015)	Permanent cut slope – Mod. cons.- Med. conf.	10%	*76°*	*83°*	

5 CONCLUSION

It was observed in our calculations that the distribution of the variables has a great influence on the outcome. Our applied distribution for the joint orientations differs from the recommendations of the literature, most researches suggest that the dip and dip dir of discontinuities are normally distributed (Coates 1981) or Fisher distribution (Priest 1993) is also accepted. However, in case of lack of information regarding the orientations, uniform distribution proved to be on the safe side. To obtain the actual distribution, more extensive sampling or additional remote sensing data collection will be necessary in our future investigations.

Probabilistic methods appear more suitable for slope stability investigations than deterministic analysis. They are capable of handle the variation of the parameters properly, formation of failures along discontinuities can be controlled more efficiently. Nevertheless, the calculations require extensive input data, and they introduce new types of uncertainties. Statistical uncertainty and sampling bias become a more important problem. These observations are in accordance with the results of other researchers (Gibson 2011; Sjöberg 1999).

Based on the results presented in the paper, it can be stated that there is clearly no better or worse design method in the comparison of the two approaches. The results show that the value of the acceptable PoF plays an essential role in the set of criteria. Design criteria that were developed for specific mines seem to be more permissive in the extent of PoF. The results showed that the values of the maximum slope angles are scattered in a wide range depending on the applied acceptance criteria. Slope angle calculated with the deterministic approach positioned in first third on the range of the probabilistic results. It can be concluded that the design approach and the applied criteria have a significant impact on the volume of the quarried raw material.

The Eurocode 7 is vague about the use of the probabilistic method. However, it states that methods that are at least as safe as the standard itself are considered appropriate. At present, there is no such criteria system for Hungarian open-pit mines therefore further investigations are needed.

REFERENCES

Adams, B.M., 2015. Slope stability acceptance criteria for opencast mine design, in: 12th Australia New Zealand Conference on Geomechanics. Wellington, New Zealand, pp. 1–8.

Barton, N., Choubey, N., 1977. The shear strenght of rock joint in theory and practice. Rock Mech. 10, 1–54. https://doi.org/10.1007/BF01261801

Budai, T., Csillag, G., Kecsmár, Z., Selmeczi, I., Sztanó, O., 2015. Surface geology of Hungary Explanatory notes to the Geological map of Hungary (1:500 000). Geological and Geophysical Institute of Hungary, Budapest.

Call, R.D., 1972. Analysis of geologic structure for open pit slope design, Doctoral Thesis. University of Arizona, Department of Mining and Geological Engineering.

Coates, D.F., 1981. Rock mechanics principles. Energy, Mines, and Resources Canada, Ottawa.

EN 1997–1, 2004. Eurocode 7: Geotechnical design - Part 1: General rules, European Standart.

Gibson, W., 2011. Probabilistic methods for slope analysis and design. Aust. Geomech. J. 46, 1–12.

Hoek, E., 2007. Practical rock engineering. Rocscience. Available Publ. Httpwww Rocscience Comhoek-Practical Rock Engineering Asp.

Hoek, E., Brown, E.T., 2018. The Hoek–Brown failure criterion and GSI – 2018 edition. J. Rock Mech. Geotech. Eng. 11, 445–463. https://doi.org/10.1016/j.jrmge.2018.08.001

Hoek, E., Marinos, P., 2000. Predicting tunnel squeezing problems in weak heterogeneous rock masses. Tunn. Tunn. Int. 132, 45–51.

Kirsten, H.A.D., 1983. Significance of the probability of failure in slope engineering. Civ. Eng. South Afr. 25, 17–27.

McMahon, B.K., 1971. A statistical method for the design of rock slopes, in: Proceedings of First Australia- New Zealand Conference on Geomechanics. Melbourne, pp. 314–321.

Pothitos, F., Li, T., 2007. Slope Design Criteria for Large Open Pits - Case Study, in: Proceedings of the 2007 International Symposium on Rock Slope Stability in Open Pit Mining and Civil Engineering. Australian Centre for Geomechanics, Perth, pp. 341–352. https://doi.org/10.36487/ACG_repo/708_21

Priest, S.D., 1993. Discontinuity Analysis for Rock Engineering. Chapman & Hall, London, UK.

Priest, S.D., Brown, E.T., 1983. Probabilistic stability analysis of variable rock slopes. Trans. Inst. Min. Metall. 92, 1–12.

Read, J., Stacey, P. (Eds.), 2009. Guidelines for open pit slope design. CSIRO Publishing, Collingwood, Australia.

Schellman, M., Sepulveda, R., Karzulovic, A., 2006. Slope design for the east wall of Mantoverde Mine, Chañaral, Chile, in: Proc. Int. Symp. on Stability of Rock Slopes in Open Pit Mining and Civil Engineering Situations. The South African Institute of Mining and Metallurgy, Cape Town, pp. 451–470.

Sjöberg, J., 1999. Analysis of large scale rock slopes, Doctoral Thesis. Lulea University of Technology, Division of Rock Mechanics, Lulea, Sweden.

Sjöberg, J., 1996. Large scale slope stability in open pit mining : a review. Lulea University of Technology, Division of Rock Mechanics, Lulea, Sweden.

Ulusay, R. (Ed.), 2015. The ISRM Suggested Methods for Rock Characterization, Testing and Monitoring: 2007-2014. Springer International Publishing, Cham. https://doi.org/10.1007/978-3-319-07713-0.

Wesseloo, J., Read, J., 2009. Acceptance Criteria, in: Guidelines for Open Pit Slope Design. CSIRO Pub-lishing, Collingwood, Australia, pp. 83–139.

Wyllie, D.C., Mah, C.W., 2005. Rock Slope Engineering: Civil and Mining, 4th Edition. Taylor & Francis Group, New York, USA.

Session 4 - Probabilistic slope stability analysis I

The Evolution of Geotech - 25 Years of Innovation – Hammah et al. (Eds)

Considering multiple failure modes: A comparison of probabilistic analysis and multi-modal optimization for a 3D slope stability case study

B. Cami, T. Ma, S. Javankhoshdel, T. Yacoub & B. Corkum
Rocscience, Toronto, ON, Canada

ABSTRACT: Limit equilibrium (LE) slope stability analysis methods are typically combined with global search methods to locate a single critical slip surface, which corresponds to the minimum factor of safety for a topographical model. However, in complicated 3D models there often exist multiple modes of failure with similar factors of safety within the topography. In such cases, it is necessary to consider multiple slip surfaces in the design process rather than just a single surface. To achieve such a goal, this paper proposes two different methods: 1) probabilistic analysis with stochastic response surfaces (SRS), and 2) the Locally Informed Particle Swarm with Radius Filter (LIPS-R) niching algorithm. SRS is a very fast and effective alternative to Monte Carlo or Latin Hypercube sampling of rock and soil material parameters for probabilistic analysis. LIPS-R is a multi-modal optimization (MMO) algorithm based on a niching method called locally informed particle swarm (LIPS). The purpose of MMO algorithms is to output multiple local minima solutions for a given optimization problem. A 3D case study is presented in this paper where both methods are demonstrated to identify multiple failure modes for an open pit mine. Despite the differences between the proposed methodologies, the failure modes identified in both cases are shown to agree.

1 INTRODUCTION

Traditional limit equilibrium (LE) slope stability analysis methods seek to locate the slip surface with the lowest factor of safety (FS), known as the critical failure surface. Various methods can be used to search the slope to determine this most critical failure surface. These range from simple brute force methods, such as a grid search, to advanced metaheuristic methods such as Particle Swarm Optimization (PSO) (Kennedy and Eberhart 1995), or Cuckoo search (Yang & Deb 2009), amongst others.

After the global search method such as PSO has found the most critical surface, an additional local optimization method is often used to modify the geometry of the surface at a local level and thus minimize the FS of that surface further, e.g. Monte Carlo random walk (Greco 1996) and Surface Altering Optimization (SAO) (Cami et al. 2018).

However, focusing the search effort on locating the single critical failure surface has its drawbacks. As pointed out by Reale et al. (2015), little research has been completed on slopes which could develop a number of critical slip surfaces with similar minimum FS. There are cases where the global minimum is of little practical importance, e.g. when the critical slip surface is too shallow to have any severe consequences, or when a slope is susceptible to multiple failure mechanisms, e.g. slopes with multiple benches and/or layers. Determination of "critical" slip surfaces is affected by the experience of the engineer or researcher, as only one failure mechanism can be identified in each trial.

DOI: 10.1201/9781003188339-25

The paper considers two very different methods of considering multiple failure surfaces and demonstrates their use through an example. The first method is probabilistic analysis, meaning an analysis that accounts for variability in the input parameters. Stochastic response surface, a method used to significantly speed up probabilistic analysis, is explored in this study. The second method is multimodal optimization (MMO). A multimodal particle swarm algorithm, LIPS-R (Li et al., 2020), is used as a means of finding multiple local minima, or critical failure surfaces. The results of both methods are examined in a three-dimensional slope model.

2 METHODS

2.1 *Probabilistic analysis with Stochastic Response Surface (SRS)*

2.1.1 *Probabilistic analysis*

The idea behind a probabilistic analysis is simple. Slope stability analysis requires the input of single parameter values, such as cohesion and friction angle for the materials in the slope; this is called a deterministic analysis. However, this means of doing the analysis does not consider any possibility of human or measurement error when obtaining these parameter values, nor does it consider the fact that strength parameters are not identical throughout the slope. Hence, it puts too much confidence in these values.

A more rational approach would be to sample values for the material parameters from a probabilistic distribution. The samples can be taken a desired number of times, and combinations of these samples are used to compute a FS for each computation. The probability of failure (PF) of the slope is defined as shown in Equation 1.

$$PF = \frac{Number\ simulations\ with\ FS < 1}{Total\ number\ of\ simulations}\ x\ 100\% \tag{1}$$

However, computing thousands of limit equilibrium analyses takes much more time than computing a single analysis, particularly in three-dimensions. Stochastic response surface is used to accelerate the probabilistic analysis in this study.

2.1.2 *SRS*

The stochastic response surface uses a small number of strategically selected computations to create a response surface of factor of safety values for various combinations of input parameters. It then predicts the factor of safety values for any combination of samples and provides an estimated probability of failure. Since a probabilistic analysis can take hours or days, this method is advantageous in significantly cutting down computation time. The SRS methodology used in this study follows that outlined by Isukapalli (1999). The steps are briefly outlined below:

<u>Step 1: Convert all variable distributions to standard normal.</u>

The initial random variables are converted from the desired distribution to standard normal random variables using transformation equations (Li et al., 2011).

<u>Step 2: Represent resulting FS in polynomial chaos expansion form.</u>

The 3rd order Hermite polynomial expansion was used in this study, shown in Equation 2.

$$F(U_i) = a_0 + \sum_{i_1=1}^{n} a_{i_1}\Gamma_1(U_{i_1}) + \sum_{i_1=1}^{n}\sum_{i_2=1}^{i_1} a_{i_1 i_2}\Gamma_2(U_{i_1}, U_{i_2}) + \sum_{i_1=1}^{n}\sum_{i_2=1}^{i_1}\sum_{i_3=1}^{i_2} a_{i_1 i_2 i_3}\Gamma_3(U_{i_1}, U_{i_2}, U_{i_3}) + \ldots$$
$$+ \sum_{i_1=1}^{n}\sum_{i_2=1}^{i_1}\cdots\sum_{i_n=1}^{i_{n-1}} a_{i_1 i_2 \ldots i_n}\Gamma_n(U_{i_1}, U_{i_2}, \ldots U_{i_n}) \tag{2}$$

In the above, F is the factor of safety and U_i is the particular combination of standard normal random variables in a simulation. The coefficient vector a must be determined.

Step 3: Use a small number of computations to determine the polynomial coefficients in Step 2.

If n random variables are defined, the number of simulations required to be computed (N) is calculated as shown in Equation 3.

$$N = 2(1 + 3n + 3n(n-1)/2 + n(n-1)(n-2)/6) \tag{3}$$

These N computations are generated using Latin-Hypercube sampling (Choi et al., 2004) to ensure that the solution space is well-covered. They are then used to determine the polynomial coefficients associated with each variable or variable combination, a.

Step 4: Generate Latin-Hypercube samples and plug them into the polynomial to estimate FS.

In this study, the number of samples required are sampled for each random variable using Latin-Hypercube sampling. Since the coefficients, a, have already been determined, the samples must simply be multiplied by a, to obtain the predicted FS for each simulation. The PF is estimated from these predicted FS values.

The SRS algorithm outlined above was used and shown to perform well in both two dimensions by Li et al. (2020) and three dimensions by Cami et al. (2021) and is outlined in more detail therein.

2.2 *Locally Informed Particle Swarm with Radius filter (LIPS-R)*

In 1995, a new evolutionary computation technique named particle swarm optimization (PSO) was proposed by Kennedy and Eberhard (1995). It is an algorithm inspired by the movement of organism in a bird flock and is widely used to solve unimodal optimization problems. The process starts with n particles, each with a random position and velocity. At each iteration, the velocity and position of each particle is updated according to equations 4 and 5:

$$V^{new} = V^{old} + c_1 * rand() * (P_b - X) + c_2 * rand() * (P_g - X) \tag{4}$$

$$X^{new} = X + V^{id} \tag{5}$$

Where X is the particle's current position, V is its velocity, P_b is the particle's personal best position, and P_g is the global best position of all the particles. Best position in this case, means the one with the optimal result (e.g. lowest FS). c_1 and c_2 are both constants and $rand()$ is a randomly generated number in the range [0, 1].

In 1998, Shi and Eberhard published a modified particle swarm optimizer. This modified PSO introduced the concept of inertia weight, w to balance local and global search ability. In this popular variant of PSO, the old velocity is multiplied by the inertia weight during the update.

While PSO has been proven to be efficient in solving unimodal problems in a wide array of disciplines, for many real-world problems it is often desirable to find several local optima. Niching methods are equipped with population-based algorithms to solve multi-modal optimization problems (Li at el. 2017). Some of the famous niching PSO methods that have been developed are ring-topology-based niching PSO (Li, 2010) and the Fitness-Euclidean distance Ratio (FER-PSO) (Li, 2007).

In 2012, Qu proposed a distance-based locally informed particle swarm (LIPS) optimizer where instead of using the personal global best positions, the particles are updated using information from their neighbour particles. This allows LIPS to eliminate the need to introduce additional niching parameters while obtaining the ability to form stable niches (Qu at el. 2012).

The LIPS-R algorithm used in this study (Li et al., 2020) inherits the key concepts from LIPS with an additional radius filter. This filter allows LIPS-R to return optima (niches) that are at least some distance away from each other, instead of many slightly different minima in the same region. If the two closest neighbours' positions for a given particle are N_1 and N_2, the particle's velocity is then updated using the formula given by equation 8 and 9:

$$V^{new} = w * \left(V^{old} + c_1 * rand() * (N_1 - X) + c_2 * rand() * (N_2 - X) \right) \quad (6)$$

$$w = \frac{2}{|2 - c - \sqrt{c^2 - 4c}|} \quad (7)$$

c_1, c_2 are both constants, typically set to 2.05 and $c = c_1 + c_2$ (Li 2006).

At the end of the iterations, LIPS-R sorts all particles by its fitness value. It then filters the results by some percentage of the model space, r. In this study r=10% was found to perform well – this suggests the optima will be at least 10% of the search space away from the others.

LIPS-R then returns all the optima that are at least r apart from each other.

3 EXAMPLE

The model considered in this study is shown in Figure 1. The slope consists of three layers of cohesive-frictional materials, each modeled using Mohr-Coulomb strength. Spencer's LE method and Particle Swarm Optimization with local Surface Altering Optimization were used in this example. The model was computed using the 3D limit equilibrium software Slide3 (Rocscience, 2021a). A finite-element seepage analysis was carried out in the 3D finite element software RS3 (Rocscience, 2021b) and the pore pressure values calculated were imported into the Slide3 model as a pressure grid.

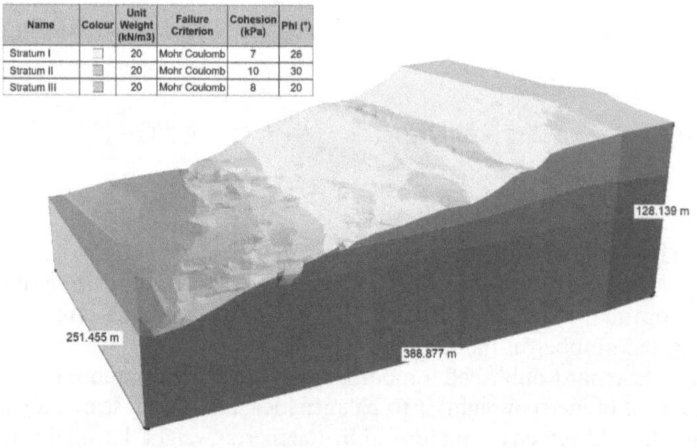

Name	Colour	Unit Weight (kN/m3)	Failure Criterion	Cohesion (kPa)	Phi (°)
Stratum I		20	Mohr Coulomb	7	26
Stratum II		20	Mohr Coulomb	10	30
Stratum III		20	Mohr Coulomb	8	20

Figure 1. The geometry and parameters of the slope model considered in this study.

3.1 *Probabilistic analysis with SRS*

In the probabilistic analysis, each material layer was considered to have normally distributed cohesion and friction angle. These random variables are listed in Table 1.

1000 samples were computed and the same methods were used for LE (Spencer) and searching (PSO with SAO) as in the deterministic case.

Table 1. Random variables in probabilistic analysis. All follow a lognormal distribution.

Material	Property	Mean	Standard Deviation
Stratum I	Cohesion (kPa)	7	2.8
	Friction Angle (°)	26	5.2
Stratum II	Cohesion (kPa)	10	4.0
	Friction Angle (°)	30	6.0
Stratum III	Cohesion (kPa)	8	3.2
	Friction Angle (°)	20	4.0

The results of the probabilistic analysis are shown in Figure 2 and Table 2. Figure 2 shows all critical surfaces found in each simulation of the probabilistic analysis. It can be seen that there appear to be two key regions where critical surfaces were found in various simulations. In Table 2, the quantitative results are summarized.

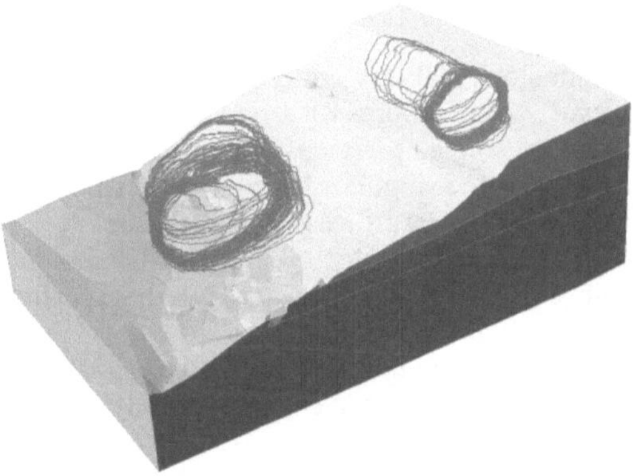

Figure 2. Results of probabilistic analysis using stochastic response surface. the red outlines indicate all the failure surfaces found through the probabilistic simulations.

Table 2. Results of stochastic response surface analysis with 1,000 simulations.

PF (%)	Mean FS
25.8	1.23

3.2 *LIPS-R*

LIPS-R with SAO was used to compute the results shown in Figure 3. By using LIPS-R, the search is not focusing its energies on finding the global minimum and has hence determined three different failure modes.

This method located two critical regions with FS ranging from 1.12 to 1.66. It is interesting to note that these modes appear to be in the same regions as the surfaces located by the probabilistic analysis, even though the means of identifying these two regions were entirely different.

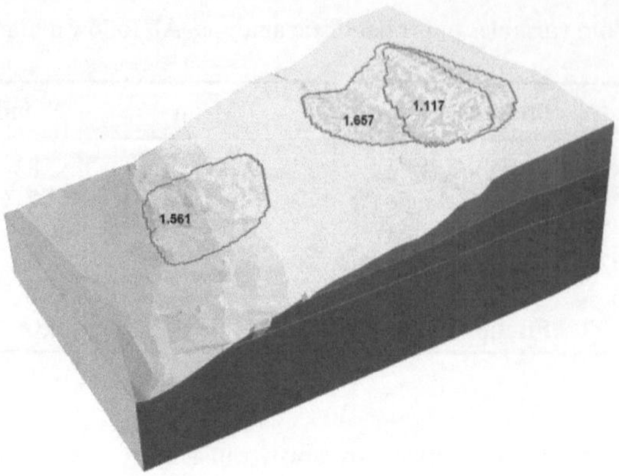

Figure 3. Results of LIPS-R multi-modal search.

4 CONCLUSION

Traditional limit equilibrium slope stability analysis methods seek to locate one single critical failure surface. However, when dealing with real world problems, it is often not sufficient to take into account the single worst case. This paper has proposed two different methods for obtaining more than the singular deterministic critical failure surface: probabilistic analysis with stochastic response surface (SRS), and locally informed particle swarm with radius (LIPS-R). SRS achieves this by taking into account the variability of input parameters. LIPS-R achieves it through an algorithm that performs local optimization instead of global optimization.

Both methods located two critical regions which a deterministic analysis would not have been able to locate. The regions were in agreement for both LIPS-R and SRS.

In conclusion, it is not sufficient to account for a single critical failure surface in LE slope stability analysis. This study has shown that using probabilistic analysis or using niching can provide the engineer with critical information about potential failure modes that would not be found with a simple deterministic analysis.

REFERENCES

Bird S. and Li X. 2006. Adaptively choosing niching parameters in a PSO, Proceedings of the 8th annual conference on Genetic and evolutionary computation (GECCO '06), New York, NY, USA, 3–10

Choi, S. K., Grandhi, R. V., Canfield, R. A., & Pettit, C. L. 2004. Polynomial chaos expansion with latin hypercube sampling for estimating response variability. AIAA journal, 42(6), 1191–1198.

Cami, B., Javankhoshdel, S., Yacoub, T. and Bathurst, R.J., 2018, November. 2D Spatial Variability Analysis of Sugar Creek Embankment: Comparative Study. In International Congress and Exhibition" Sustainable Civil Infrastructures: Innovative Infrastructure Geotechnology" (pp. 118–125). Springer, Cham.

Cami, B., Javankhoshdel, S. and Yacoub, T. 2021. Limit equilibrium probabilistic analysis of three-dimensional open pit using the stochastic response surface method. International Foundations Conference and Equipment Expo.

Greco, V.R. 1996. Efficient Monte Carlo Technique for Locating Critical Slip Surface. Journal of Geotechnical Engineering, 122(7): 517–525.

Isukapalli, S. S. 1999. Uncertainty analysis of transport-transformation models.

Kennedy, J. and Eberhart, R. 1995. Particle swarm optimization, Proceedings of ICNN'95 - International Conference on Neural Networks, IEEE, Perth, WA, Australia, 4: 1942–1948

Li, D., Chen, Y., Lu, W., & Zhou, C. 2011. Stochastic response surface method for reliability analysis of rock slopes involving correlated non-normal variables. Computers and Geotechnics, 38(1), 58–68.

Li, S., Cami, B., Javankhoshdel, S., Corkum, B., & Yacoub, T. 2020. Considering Multiple Failure Modes in Limit Equilibrium Slope Stability Analysis: Two Methods. GeoVirtual 2020. The Canadian Geotechnical Society.
Li, X. 2007a. Niching Without Niching Parameters: Particle Swarm Optimization Using a Ring Topology, IEEE Transactions on Evolutionary Computation, 14(1): 150–169
Li X. 2007b. A multimodal particle swarm optimizer based on fitness Euclidean-distance ratio. Proceedings of the 9th annual conference on Genetic and evolutionary computation (GECCO '07), New York, NY, USA, 78–85
Li, X. Epitropakis, G. M. Deb, K. and Engelbrecht, A. 2017. Seeking Multiple Solutions: An Updated Survey on Niching Methods and Their Applications, IEEE Transactions on Evolutionary Computation, 22(4): 518–538
Qu B. Y., Suganthan P. N. and Das S. 2013. Distance-Based Locally Informed Particle Swarm Model for Multimodal Optimization, IEEE Transactions on Evolutionary Computation, 17(3): 387–402
Reale, C., Xue, J., Pan, Z. and Gavin, K., 2015. Deterministic and probabilistic multi-modal analysis of slope stability. Computers and Geotechnics, 66, pp.172–179.
Rocscience, 2021a. Slide3: 3D Slope Stability Analysis. Version 3.010. https://www.rocscience.com/help/slide3/
Rocscience, 2021b. RS3: 3D Finite Element Analysis. Version 4.013. https://www.rocscience.com/help/rs3/
Shi, Y. and Eberhart, R. 1998. A modified particle swarm optimizer, 1998 IEEE International Conference on Evolutionary Computation Proceedings. IEEE World Congress on Computational Intelligence (Cat. No.98TH8360), Anchorage, AK, USA, 69–73
Yang, X.S. and Deb, S., 2009, December. Cuckoo search via Lévy flights. In 2009 World congress on nature & biologically inspired computing (NaBIC), IEEE, 210–214

The Evolution of Geotech - 25 Years of Innovation – Hammah et al. (Eds)

Assessment of asperities geometry influence on MSW landfill critical interface side-slope stability using probabilistic analysis

D. Adeleke, D. Kalumba & L. Nolutshungu
Department of Civil Engineering, University of Cape Town, Cape Town, South Africa

J. Oriokot
PSM Technologies, Johannesburg, South Africa

ABSTRACT: Geomembrane asperities are surface protrusions which distinguish smooth geomembranes from textured geomembranes. Asperities possess geometrical features such as height and concentration and are hypothesised to develop high interface shear strength, resist sliding and increases stability. To date, many textured-geomembranes with different asperity geometries have been manufactured and used in landfill linings together with geosynthetics like geotextiles. Previous studies have considered the effects of asperity geometries to geomembrane/geotextile interface shear characteristics. However, limited studies have considered the effects of asperity height and concentration on the landfill side-slope liner factor of safety (FoS) using the geomem- brane/geotextile critical interface as the point of reference. Thus, this study was aimed at investi- gating the influence of asperity geometries on liner stability. This study utilized experimental results from direct shear test (i.e. friction angle and adhesion) and performed probabilistic stability analysis using SLIDE2. Available results indicated that FoS increased as both asperity concen- tration and height increased. However, asperity-height increased beyond 1.2 mm mobilized FoS reduction. Therefore, obtaining an optimised liner stability factor is hinged on selecting the ap- propriate geomembrane asperity geometry at the critical geomembrane/geotextile interface.

1 INTRODUCTION

Presently, geosynthetics incorporation into municipal solid waste (MSW) landfill liner construc- tion is widely accepted, particularly in South Africa. The geosynthetics function of interest may include separation, filtration, drainage, barrier, protection, and reinforcement function. It should be noted that this study focuses on functions such as barrier and protection where geosynthetics surface features (asperities) have significant effects on a typical landfill liner design life. The placement of geosynthetics with other geomaterials often results in interface interaction with dis- tinct shear characteristics which is necessary to ensure the stability of the side-slope liner. In a landfill liner, possible single geosynthetics interface include geomembrane/geosynthetic clay liner (GMB/GCL), geosynthetic clay liner/compacted clay liner (GCL/CCL), geomembrane/geo- textile (GMB/GTX), and geotextile/geocomposite drain (GTX/GCD) (Bhatia & Kasturi, 1996). However, studies by Bergado et al., (2006); Xuede (2008); Bacas et al., (2015) identified GMB/GTX interface as an interface with lower frictional resistance and shear strength (i.e. "crit- ical interface"). Thus, this critical interface was the focus of this study.

The mobilization of shear strength at the critical interface (GMB/GTX) is highly dependent on surface features such as roughness and asperities. In some of the reviewed design problem

DOI: 10.1201/9781003188339-26

in literature, it has been assumed that shear mobilization is directly proportional to the measured asperity height. However, this assumption was contradicted by Blond & Elie (2006) where they concluded that, for a given range of asperity measurement, there exists an intermediate asperity property beyond which there is little or no increase to the recorded shear parameter. The assumed mobilization of shear strength dependent on asperity properties would have important implica- tions on design, particularly at the side-slope and during unfavourable environmental conditions (increased leachate). Incorrect assumptions of the shear characteristics might lead to inadequate design and eventually slope failure. Therefore, this study investigated the effect of asperities on MSW landfill critical interface side-slope stability using probabilistic analysis.

2 PROBLEM DESCRIPTION

2.1 *Overview*

In this paper, an assessment study was carried out on geosynthetics landfill liner system using finite element method (FEM) based software SLIDE2 (Rocscience) to determine the variation of peak factor of safety (FoS) as a function of altered asperity measurement. The slope geometry used in this study and illustrated in Figure 1 was adopted from Qian & Koerner (2004). The slope inclination was modified from 3H:1V to 1H:1V as the latter is more critical and typifies unfa- vourable condition. To investigate the effects of asperity geometry on the shear strength mobilized at the geomembrane/geotextile interface, asperity height and concentration were varied.

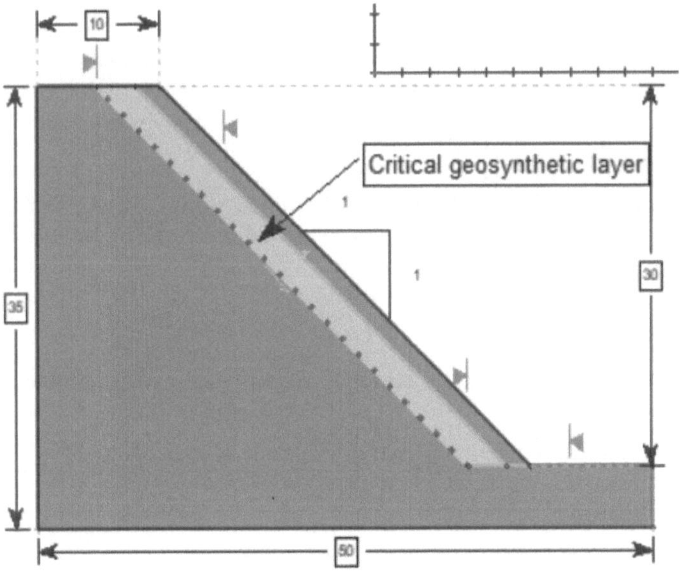

Figure 1. Slope geometry used in the study.

2.2 *Slope geometry and properties*

The selected unit weight, apparent cohesion, friction angle, Poisson's ratio, and elastic modulus of the soil used were 18 kN/m3, 30 kPa, 15°, 0.3, and 20, respectively (Qian & Koerner, 2004). A firm stiff clay layer has been assumed for the foundation soil to minimize the probability of failure surface development at the founding soil. The geosynthetic liners were designed to account for anchorage length.

2.3 *Numerical modelling of slope*

SLIDE2, a 2-D elastoplastic finite element stress analysis program was used for the modelling of this investigation. SLIDE2 is a dynamic finite element program with efficiency in slope stability analysis. Previous researchers have used SLIDE2 for stability analysis of conventional slope (with soil as a primary material) (Berisavljević et al., 2015; Pillay, 2017). An elastoplastic constitutive model available in the program was used for analyzing the slope. The GLE/Morgenstern-Price method of analysis was selected because it satisfies all conditions of equilibrium and includes reasonable assumptions (Aswathi et al., 2017; Patuti et al., 2019). Also, the circular grid search option was selected as the preferred surface option. The geomembrane/geotextile (GMB/GTX) interface was selected as the critical interface (with least resistance) and was represented in the model as the weak layer material and critical geosynthetic layer.

In the constitutive modelling, peak shear characteristics (friction angle and apparent adhesion) were taken as the weak and geosynthetic layer material strength and were described by the Mohr- Coulomb model. It should be noted that the peak shear parameters were obtained from the direct interface shear test conducted between geomembrane and geotextile in accordance to ASTM D5321 (2014) and using Shear Trac-III – a large direct shear device built by Geocomp Corpora- tion Company. Shear Trac-III top box had a cross-sectional area of 305 mm by 305 mm and a thickness of 100 mm while the bottom shear box had dimensions of 460 mm x 355 mm x 100 mm. To test the shear strength of geomembrane and geotextile interfaces, the geomembrane was affixed to the lower box while the geotextile was attached to the top box. The tests were conducted at normal stresses of 25, 50, 100, 200, and 400 kPa and together with the resulting shear stress were utilized to develop the failure envelope.

Though either polypropylene (GTX-PP) or polyester (GTX-PET) geotextile could be interfaced against the geomembrane (GMB), this study focused on examining the interaction between GMB & GTX-PET interfaces only. This is because Adeleke et al. (2019) and Adeleke (2020) reported that GTX-PET interfaces exhibit greater shear characteristics than GTX-PP interfaces. The unit weight of the GMB/GTX interface required in SLIDE2 was determined by considering the unit area of the geotextile. This is because geotextile unit area (400 g/m^2) was less than ge- omembrane unit area (9400 g/m^{2,} converted from 0.94 g/cc formulated density). GTX-PET re- ported unit area of 400 g/m^2 was converted to kN/m^3 by factoring in acceleration due to gravity (g) and per meter run length. Therefore, 4 kN/m^3 was estimated for the GMB/GTX interface. Furthermore, the weak layer friction angle and apparent cohesion were continuously altered as the geomembrane surface asperities changed, as shown in Table 1. Each geomembrane was dis- tinguished from another by adding an alphanumeric character to the label such as S, T1, T2, . . ., T7, where S and T represent smooth and textured geomembrane.

Table 1. Geosynthetic interface asperity and shear characteristics

Interface label	Asperity properties		Shear charactersitstic	
	Height (mm)	Density (knobs/area*)	Friction (°)	Adhesion (kPa)
GMB – S/GTX	0.0	0	16.8	0.8
GMB – T1/GTX	0.7	332	27.5	9.1
GMB – T2/GTX	0.7	663	30.3	17.2
GMB – T4/GTX	0.9	337	28.3	6.5
GMB – T5/GTX	1.2	306	35.2	11.7
GMB – T6/GTX	1.8	211	32.1	13.7
GMB – T7/GTX	2.0	217	31.4	13.9

* Area = 10000 mm^2.

3 RESULTS AND DISCUSSION

Although a conventional limit equilibrium method (LEM) recommended by Rouncivell (2007), ASTM D5321 (2014), and Buthelezi (2017) could have been utilized in this study, a modified approach tailored to SLIDE2 was selected instead. This was necessary to explore the added fea- tures of the Rocscience software package. Also, the conventional LEM approach is often limited to translational failure, whereas other forms of failure such as rotational, foundational, and com- posite failures are likely to occur in a landfill design life. A typical model for the slope, surround- ing soil, and critical geosynthetic interface layer is presented in Figure 2.

3.1 *Effect of asperity height on FoS*

As landfills are subjected to lower and higher confining stresses at different construction stages, and various slope positions of the landfill, a linear and/or non-linear Mohr-Coulomb failure en- velope is more appropriate to characterize the mode of failure of the liner (Sikwanda 2018). There- fore, in this section, the results obtained by computing a Mohr-Coulomb failure envelope on dif- ferent geomembrane/geotextile interfaces, depending on asperity height variation, were utilized. Additionally, to establish the degree of variation of the mobilized FoS from the smooth geomem- brane/geotextile interface, the percentage difference (PD) between the obtained FoS and the con- trol value was calculated – where the smooth geomembrane interface acted as the "control".

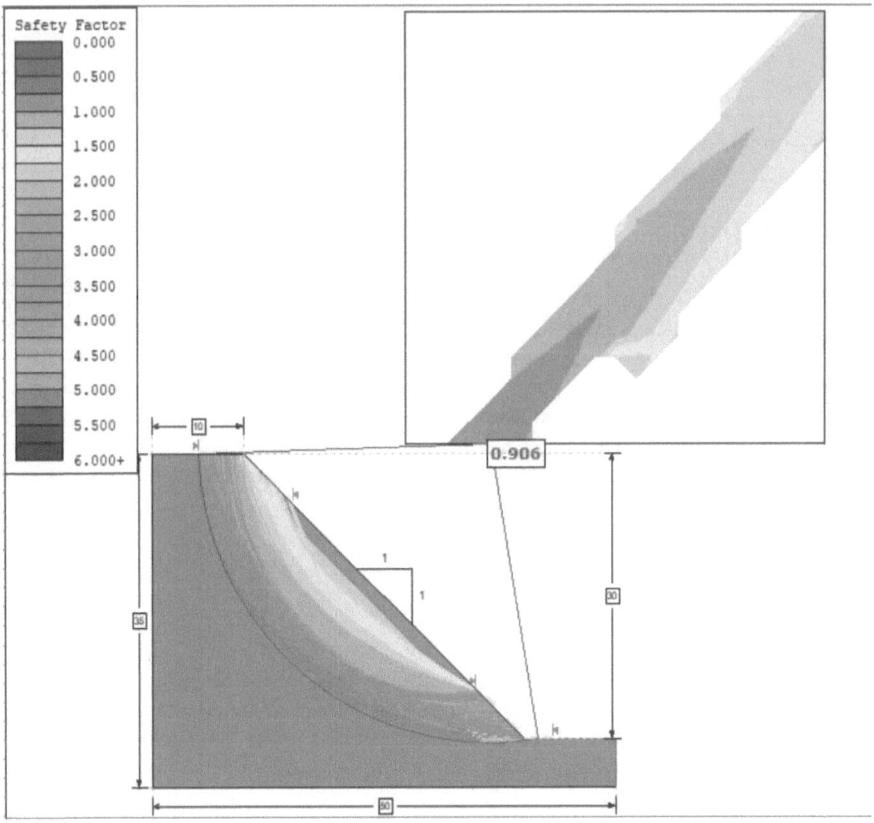

Figure 2. The numerical model of the slope prepared in SLIDE2 (Rocscience).

Considering GMB-T1/GTX relative to the smooth interface, it was evident that the inclusion of asperities produced a 69.5 % increase in the slope stability factor of safety (FoS). As regards the gradual increase in asperity height (0.70 mm, 0.85 mm, & 1.20 mm) at an average constant asperity concentration of 325 knobs/10000 mm^2 for GMB-T1/GTX, GMB-T4/GTX, & GMB- T5/GTX, it was observed that a corresponding increase in FoS was recorded, particularly at 1.20 mm asperity height (see Table 2). Though GMB-T4/GTX exhibited a slight reduction in the com- puted FoS, the reduction was attributed to other material properties such as roughness and rigidity, which were beyond the scope of this investigation.

Furthermore, it was identified that a gradual increase in asperity height for heavily textured geomembranes such as GMB-T6/GTX & GMB-T7/GTX resulted in no corresponding FoS im- provement. An apparent reason for the absence of a corresponding increase in FoS at high asperity height could be related to the mobilization of optimal hook and loop interaction between the as- perities and geotextile fibres. In summary, the presence of asperity height triggered a greater pro- pensity for slope stability but asperity height increase in heavily textured interface produced no increment to the safety factor.

Table 2. Tested geomembrane asperity and corresponding shear characteristics.

Interface label	FoS	PD*
GMB – S/GTX	0.528	0
GMB – T1/GTX	0.895	69.5
GMB – T2/GTX	0.906	71.6
GMB – T4/GTX	0.893	69.1
GMB – T5/GTX	0.906	71.6
GMB – T6/GTX	0.905	71.4
GMB – T7/GTX	0.905	71.4

* PD = Percentage difference relative to GMB-S/GTX.

3.2 *Effect of asperity density/concentration on FoS*

With the obtained data on asperity concentration variation, it was observed that the doubling of asperity concentration in (GMB-T2/GTX) produced the greatest improvement (71.6 %) on slope stability safety factor when compared with the smooth geosynthetic interface. This improvement corresponds to the optimal FoS as asperity height was varied. Besides the increase in friction angle, apparent cohesion was identified as the primary contributor to the improvement observed at GMB-T2/GTX interface. The increased cohesion was attributed to the large surface area pro- vided by asperities through which geomembranes/geotextiles interaction were mobilized.

4 CONCLUSIONS

A numerical analysis was performed to assess the shear stress mobilization within the side-slope of an MSW landfill side-slope. The following conclusion can be drawn from this study.

Peak shear strength mobilized in the "weak and geosynthetic layer" of a landfill side-slope is dependent on the individual constituent of the layer, surface features, asperity height, and asperity concentration. Also, depending on the design approach, the failure surface was slightly affected by the alteration in the asperity parameter. Furthermore, an increase in asperity height and con- centration exhibited a corresponding increase in computed FoS. While an increase in asperity height of heavily textured interface resulted in no FoS improvement.

REFERENCES

Adeleke, D. 2020. An Investigation into the Effects of Asperities on Geomembrane/Geotextile Interface Shear Characteristics. MSc Dissertation submitted to University of Cape Town, South Africa.

Adeleke, D; Kalumba, D; & Oriokot, J. 2019. Asperities effect on polypropylene & polyester geotextile-geomembrane interface shear behaviour. In E3S Web of Conferences. V. 92. 5. DOI: 10.1051/e3sconf/20199213017.

ASTM D5321. 2014. Standard Test Method for Determining the Shear Strength of Soil- Geosynthetic and Geosynthetic-Geosynthetic Interfaces by Direct Shear. ASTM Int'l. 14 (1):pp 1–11. DOI: 10.1520/D5321.

Aswathi, CK; Amalesh, J; Arindam, D; & Sreedeep, S. 2017. Stability Assessment of a Heavily Jointed Rock Slope using Limit Equilibrium and Finite Element Methods. In *Indian Geotechnical Conference*. 6.

Bacas, M; Canizal, J; & Konietzky, H. 2015. Frictional behaviour of three critical geosynthetic interfaces. Journal of Geosyn Int'l. 22(5): pp 1–11. DOI: 10.1680/gein.15.00017.

Bergado, D; Ramana, V; Sia, I; & Varun. 2006. Evaluation of an interface shear strength of composite liner system and stability analysis for a landfill lining system in Thailand. Journal of Geotextiles and Geomembranes. 24(6): pp 371–393. DOI: 10.1016/j.geotexmem.2006.04.001.

Berisavljević, Z; Berisavljević, D; Cebasek, V.; & Rakic, D. 2015. Slope stability analyses using limit equilibrium and strength reduction methods. *Journal of the Croatian Association of Civil Engineers*. 67 (10):975–983. DOI: 10.14256/jce.1030.2014.

Bhatia, S.K. & Kasturi, G. 1996. *Comparison of PVC and HDPE geomembranes (interface friction performance)*.

Blond, E. & Elie, G. 2006. Interface shear-strength properties of textured polyethylene geomembranes. In *Sea to Sky Geotechnique 2006*. Quebec. 898–904.

Buthelezi, S. 2017. Comparison of shear strength properties of textured polyethylene geomembrane interfaces in landfill liner systems. MSc Dissertation submitted to the University of Cape Town, South Africa.

Patuti, M; Rifa, A; & Suryolenono, B. 2019. Numerical Analysis of Multi-Level Gravity Walls in Tupa Village, Bulango Utara-Bone Bolango District by Limit Equilibrium Methods. In *Third International Conference on Sustainable Innovation*. V. 187. 113–117. DOI: 10.2991/icosite- 9.2019.22.

Pillay, O. 2017. Geotechnical Investigation and Slope Design Feasibility Study for the Otjikoto Gold Mine in Nambia. MSc Dissertation submitted to the University of Witwatersrand, South Africa.

Qian, X. & Koerner, R.M. 2004. Effect of Apparent Cohesion on Translational Failure Analyses of Landfills. *Journal of Geotechnical and Geoenvironmental Engineering*. 130(1):71–80. DOI: 10.1061/(asce)1090-0241(2004)130:1(71).

Rouncivell, W. 2007. Experimental Investigation Of the Shear Strength Characteristics of a Geosynthetic Clay Liner and its Application. MSc Dissertation submitted to the University of Cape Town, South Africa. Available: thesis_ebe_2007_rouncivell_wesley.pdf.

Sikwanda, C. 2018. An Investigation of the Effects of Specimen Gripping Systems on Shear Stress at the Geosynthetic-Geosynthetic Interface. MSc Dissertation submitted to the University of Cape Town, South Africa.

Xuede, Q. 2008. Critical interfaces in geosynthetic multilayer liner system of a landfill. *Journal of Water Science and Engineering*. 1(4):22–35. DOI: 10.3882/J.ISSN.1674-2370.2008.04.003.

The Evolution of Geotech - 25 Years of Innovation – Hammah et al. (Eds)

The schematization of soil properties in mathematical modeling in engineering geology and geotechnics

I.K. Fomenko, D.N. Gorobtsov, K.V. Kurguzov & M.A. Novgorodova
Russian State Geological Prospecting University n.a. S. Ordzhonikidze, Moscow, Russia

O.N. Sirotkina
Lomonosov Moscow State University, Moscow, Russia

ABSTRACT: Modeling of natural ground conditions will always require approximations due to their spatial variability. Often limited datasets describing material behavior are available resulting in the estimation of material properties. In many cases only deterministic estimates of material strength are reported or used in geotechnical analysis, despite the natural variation and uncertainty of material strength in the area under investigation. Failure to take into account the random nature and variation of material mechanical strengths can result in misleading stability calculations. This paper shows the difference in slope stability results that are calculated when soil properties are modelled as deterministic values, interpolated from known sample sites, described by a range of values (probabilistic analysis) and varied spatially and stochastically.

Keywords: engineering-geological schematization, engineering-geological element, models of property distribution, calculation of slope stability

1 INTRODUCTION

One of the main stages of building a geotechnical model encompassing the engineering-geology properties of natural ground conditions is schematization. Schematization can be categorized as generalized or special.

Generalized schematization, in this context, can be described as the process of simplifying a real natural object, with an infinite degree of complexity, to a conceptual model, limited by the framework of scientific knowledge.

Special schematization presupposes simplification of a conceptual model to a specialized scheme that maximizes simplification with minimum loss of adequacy. Special schematization can be divided into several interrelated stages: behavior schematization, structure schematization, properties schematization, and modeling object state schematization [1].

The special schematization is based on the concept of an engineering-geological element (EGE). According to Russian standards (SP 446.1325800.2019) [2] EGE is the main soil unit used to create a geotechnical model of a soil massif, including a certain volume of soil of the same type (subtype), type (subspecies) and variety when the values of soil characteristics within an element change randomly (irregularly) or with an observed pattern of changes in soil characteristics with a coefficient of variation for physical soil characteristics ≤ 0.15, for mechanical ≤ 0.30.

Despite the fact that the random nature of the distribution of properties in the EGE is present in the definition, until recently, not enough attention was paid to this fact in the

DOI: 10.1201/9781003188339-27

construction of geotechnical models. Although statistical properties describing the variation in soil properties were calculated, strength parameters applied in geotechnical models were deterministic.

However, the situation has changed in recent years. This is due to the increased computation of probabilistic analysis [3], which considers the uncertainty of material properties in geotechnical models, and can also be used in risk-based design analysis [4].

Currently, there are four main types of models that describe the distribution of soil properties geotechnical models. These are deterministic models, interpolation models, probabilistic models and models applying spatial variability.

Example applications of all four methods will be shown using a slope case from Moscow. Stability was assessed using 2D (plain strain) limit equilibrium analysis. Optimized Cuckoo search methods were applied.

2 OBJECT OF STUDY

The work site is located in Lapshinka, Moscow (Figure 1)

Figure 1. Location of the study area (red rectangle).

Geomorphologically, the study area is located within the gently undulating moraine plain, with well-developed river valleys. The relief is characterized by absolute surface elevations of the order of 180.35 - 182.05 m.

Quaternary deposits, of technogenic displaced soils, alluvial, cover, glacial and Cretaceous deposits, form the geological structure to the explored depth of 20.0 m. Hydrogeological conditions are characterized by the close occurrence of groundwater, as well as the presence of a large number of aquifers, hydraulically connected with each other.

Within the study area, the technogenic-displaced soils that compose the dump body are brown loam, slightly compacted, and refractory. The thickness of technogenic soils varies from 2.7 m to 12.0 m.

Field observations of the study area noted surficial cracking on the edge of the embankment (Figure 2).

Figure 2. Cracks on the edge of the embankment.

3 DESIGN SCHEMES AND INITIAL DATA

The physical and mechanical properties of technogenic soils occurring within the study area are very different from other soils formed in nature. The distribution of properties in technogenic soils is characterized by great heterogeneity. For this reason, the selection of layers of different physical and mechanical properties and their assignment in geotechnical models presents a certain difficulty. To assess the variation in results that can be calculated with and without consideration of the heterogeneity of the technogenic soil, slope stability was assessed using deterministic, interpolation, probabilistic and spatial variation analysis methods.

3.1 *Deterministic model*

When modelling using deterministic inputs, soil strength parameters included in the geotechnical model are the single (scalar) values for each EGE: density, cohesion, angle of internal friction The critical slip surface and factor of safety (FS) calculated using deterministic soil properties is displayed in Figure 3.

3.2 *Interpolation model*

In interpolation models, the initial parameters are numerical arrays defined for each soil property (density, specific cohesion, angle of internal friction) and coordinates of the sample sites. Based on this information, soil properties are interpolated for the remainder of the model extents using interpolation algorithms built into the analysis software Different results may be obtained using different interpolation methods.

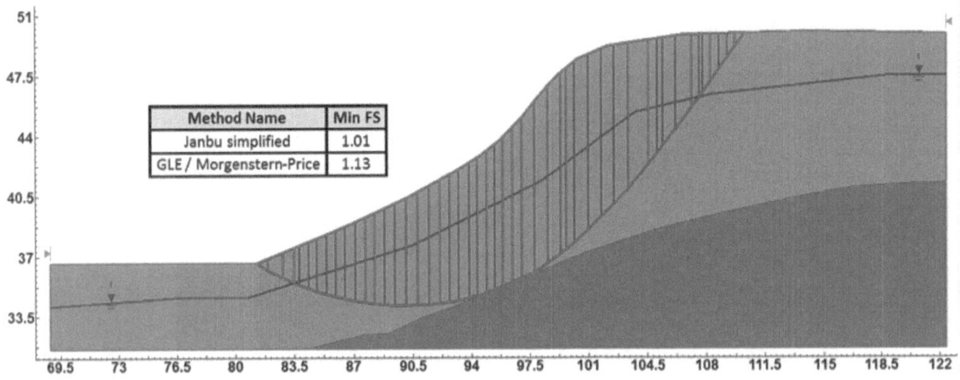

Figure 3. Model results using deterministic inputs of soil properties.

The critical slip surface and FS calculated using interpolated soil properties is displayed in Figure 4.

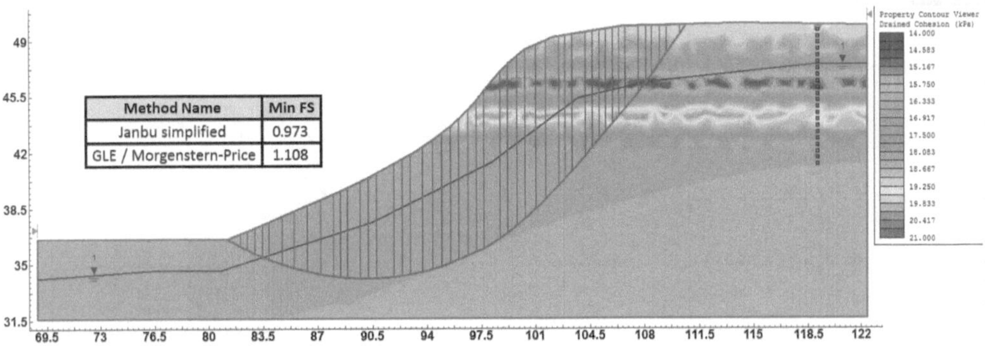

Figure 4. Model results using interpolated soil properties. Variance in soil cohesion displayed.

3.3 *Probabilistic model*

In contrast to the models described above, the initial parameters in probabilistic models are not scalar quantities, but functions of the probability distribution of soil properties [11].

The main problem in constructing a probabilistic model is to determine the distribution of material properties. In the case of a normal distribution, two parameters are required to construct a probability function: mean and standard deviation.

The output of probabilistic models is the distribution of FS, Figure 5, where several FS have been calculated using different combinations of soil parameters along the selected distribution.

The variability of the properties of soils affects the value of FS. Together with probabilistic analysis, sensitivity analysis can be completed to determine the sensitivity of FS to one unit change in material property (e.g. one unit change in cohesion over the specified range). Sensitivity analyses can assist to determine if the FS is more sensitive to changes in certain materials and/or properties in the model.

It should also be noted that the critical strength parameters are not a single pair of values (e.g. a single angle of internal friction and cohesion), but could be any combination over a range of values, Figure 6.

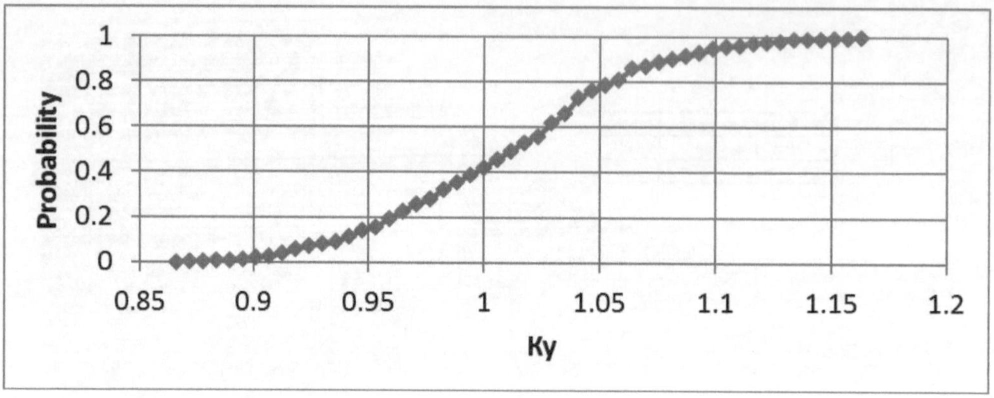

Figure 5. Cumulative distribution of FS as calculated from probabilistic model (42% of simulations had a FS less than 1).

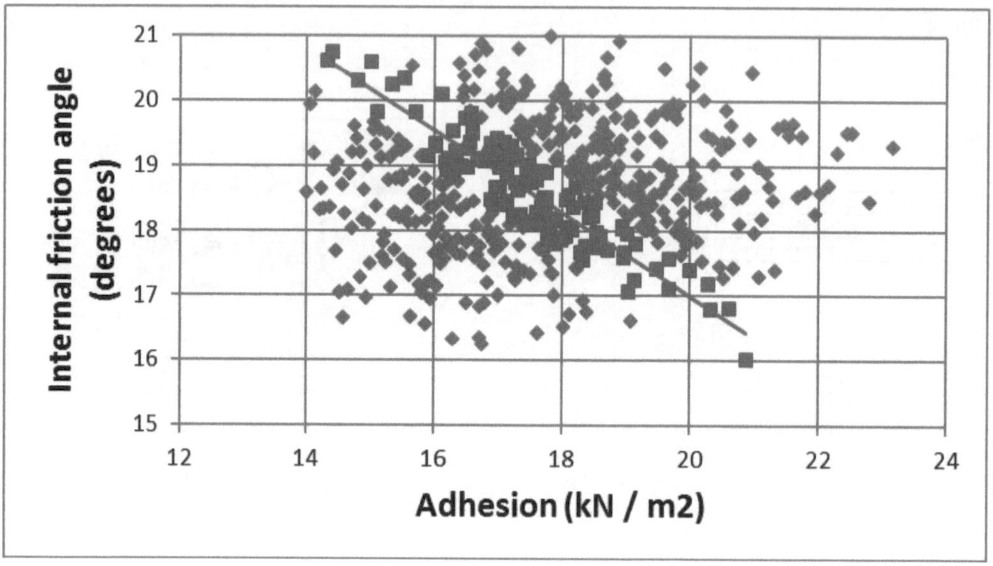

Figure 6. Two-way sensitivity analysis for the simulated slope. The red line is the dependence of the angle of internal friction on cohesion at which the FS slope is 1. Any combination of friction angle and cohesion highlighted in red will result in FS = 1.

3.4 *Spatial variation*

1. Model of variability

Not taking into account the random component in the structure of numerical estimates of the geological parameter (using purely deterministic models) often leads to erroneous results [10].

Spatial variability of soil properties can be modeled using the theory of random fields [6, 12, 13], according to which, in any area of the geotechnical model, soil properties are a random variable characterized by a probability distribution function and correlating with the values of soil properties in adjacent areas [7]. Spatial correlation of soil properties is determined based on the autocorrelation function, which can be estimated from the results of measurements of the parameter at various points according to the results of field or laboratory tests [8].

The correlation structure of a random Gaussian domain can be determined using the Markov correlation coefficient function:

$$R(\tau_x, \tau_y) = exp\left\{ -\sqrt{\left(\frac{2\tau_x}{\theta x}\right)^2 + \left(\frac{2\tau_y}{\theta y}\right)^2} \right\}$$

where $R(\tau_x, \tau_y)$ is the autocorrelation coefficient, τ_x and τ_y are the absolute distances between two points in the horizontal and vertical directions, respectively, θ_x and θ_y are the correlation distances in the horizontal and vertical directions, respectively [9].

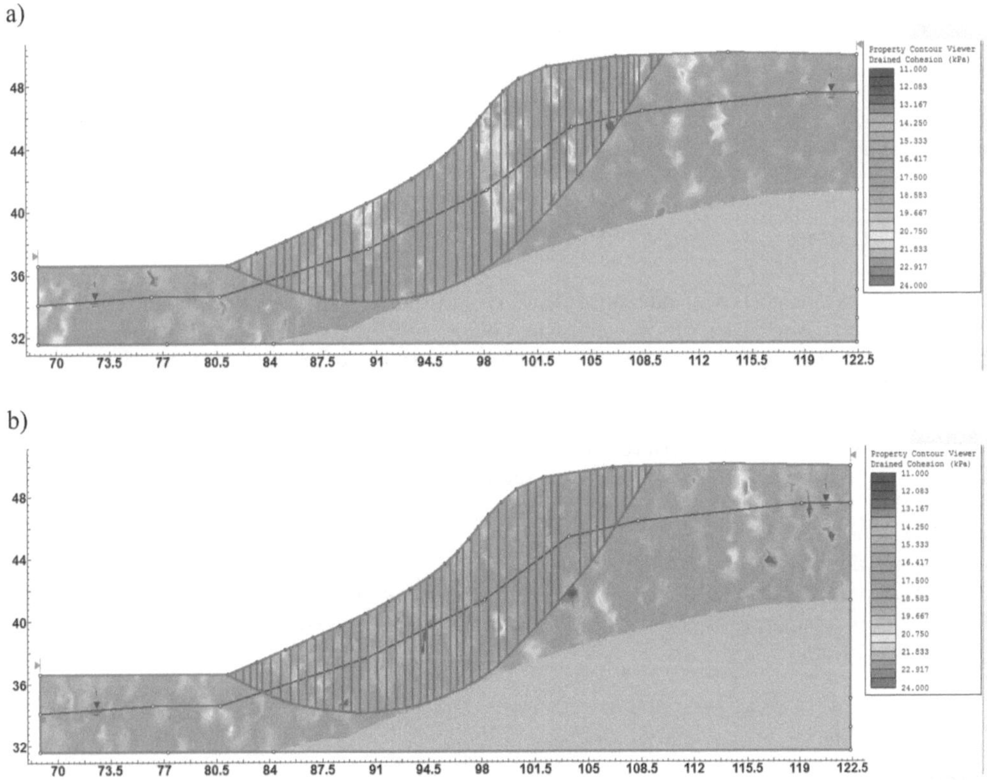

Figure 7. The model of the specific cohesion distribution for: a) the average value of Ku -1.01; b) the minimum value of Ku is 0.97. The horizontal correlation distance is 1m, the probability of a landslide process is 53.6%.

Thus, the necessary initial parameters for the variability model, in addition to the probability distribution function of soil properties, are the values of the correlation distance, θ_x and θ_y.

Stability results of the case study when spatial variation of soil properties is applied is displayed in Figure 7.

Model results show that variation in θ_x has a limited impact on FS, for all cases the average FS is 1.01). However, the probability of failure increases with decreasing θ_x, Figure 8.

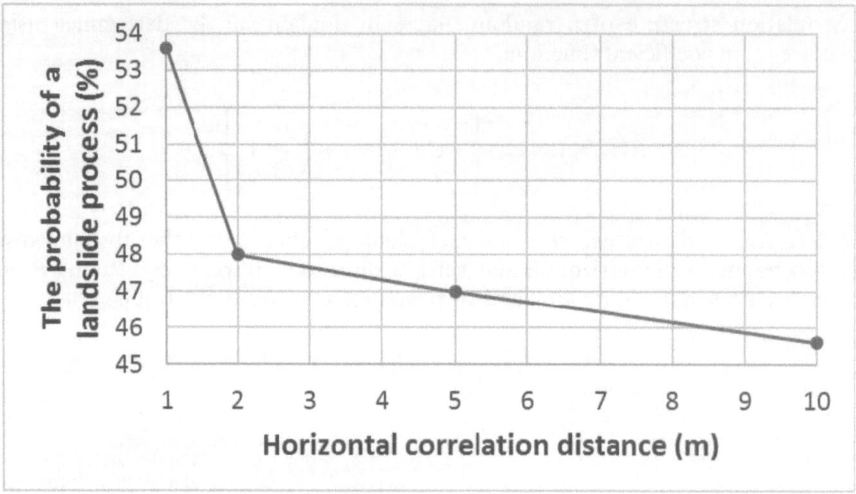

Figure 8. Variation in probability of failure with θ_x.

4 CONCLUSION

The mechanical properties of natural ground conditions are inherently variable. Often only limited is available to describe the mechanical behavior of soils, leaving much uncertainty in the true material behavior. Geotechnical models should consider this inherent variability in strength.

This paper has shown how the FS and probability of failure can vary when slope stability is calculated with and without consideration of material property heterogeneity.

A case slope from near Moscow was assessed using deterministic, interpolated, probabilistic and spatially variable analysis methods. Different results were calculated for each method for the same slope case.

Probabilistic analysis and the application of spatial variability in slope stability models can assist to account for the spatial variability and uncertainty in natural slopes. In the case study described in this paper, probability of failure was shown to increase with decreasing horizontal spatial correlation.

REFERENCES

1. Fomenko I.K. Mathematical modeling of the stress state of an engineering-geological massif composed of anisotropic rocks // abstract of a dis ... candidate of geological miner. Sciences: 04.00.07. Moscow. 2001.S. 24.
2. SP 446.1325800.2019 (Engineering and geological surveys for construction. General rules for the production of works)
3. Griffiths, D. & Fenton, Gordon. (2007). Probabilistic Methods in Geotechnical Engineering. 10.1007/978-3-211-73366-0
4. Fenton G.A., Griffith D.V. Risk assessment in geotechnical engineering - NewJersey: WILEY, 2008
5. Bufeev FK, Fomenko IK, Sirotkina on influence of methods of interpolation of strength properties of soils on the results of calculation of the stability of slopes // International scientific research journal. - 2016. - S. 127–133. DOI: 10.18454/IRJ.2227-6017
6. Vanmarcke, E., (1983), "Random Fields: Analysis and Synthesis," MIT Press, Cambridge, MA
7. El-Ramly, H. and Morgenstern, N. and Cruden, D., (2002), "Probabilistic slope stability analysis for practice," Can. Geot. J., 39 (3), pp. 665–683
8. Baecher, G.B. and Christian, J.T., (2005), "Reliability and Statistics in Geotechnical Engineering," John Wiley & Sons

9. Javankhoshdel, Sina & Cami, Brigid & Bathurst, Richard & Corkum, Brent. (2018). Probabilistic Analysis of Layered Slopes with Linearly Increasing Cohesive Strength and 2D Spatial Variability of Soil Strength Parameters Using Non-Circular RLEM Approach. 10.1061/9780784481585.014
10. Kurguzov KV, Fomenko IK, Sirotkina on probability-statistical approaches in the assessment of the uncertainty of litotechnical systems // Geoecology. Engineering geology. Hydrogeology. Geocryology. - 2020. - No. 2. - P. 80–89.
11. Zerkal OV, Fomenko IK Influence of various factors on the results of probabilistic analysis of activation of landslide processes // Engineering Geology. - 2016. - No. 1. - P. 16–21.
12. Javankhoshdel, Sina & Luo, Ning & Bathurst, Richard. (2016). Probabilistic analysis of simple slopes with cohesive soil strength using RLEM and RFEM. Georisk: Assessment and Management of Risk for Engineered Systems and Geohazards. 1–16. 10.1080/17499518.2016.1235712.
13. Izadi, Ardavan & Chenari, Reza & Cami, Brigid & Javankhoshdel, Sina. (2020). Full and Quasi-Stochastic Slope Stability Analyzes using Random Limit Equilibrium Method (RLEM).

Session 5 - New technologies in geotechnical engineering

The Evolution of Geotech - 25 Years of Innovation – Hammah et al. (Eds)

Artificial intelligence and image processing in the MIRET approach for the water detection and integrated geotechnical management of existing mechanized tunnels: Methodology, algorithm and case study

F. Foria*, M. Calicchio, A. Tarquini, G. Miceli & D. Chiaino
ETS Srl, Rome, Italy

D. Cuccato
RMT Srl, Santa Giustina in Colle, Padova, Italy

S. Rinaldo, G. Bomben, G. Rossetti & A. Allegro
ADTS Srl, Trebaseleghe, Padova, Italy

ABSTRACT: The planning and the management of existing tunnels is already a central challenge for industrialized countries. Since an increasing amount of the tunnel industry employs mechanized excavation method, the effort put into the management and maintenance of the aforementioned underground structures will significantly increase in the next years. Tunnel inspection and diagnostic are a crucial task in providing reliable predictive tunnel maintenance at cost with time consuming and error-prone process if based on human operations only. Therefore, the diagnostic process of tunnel inspection and the relevant analytical procedures are suitable to automation. ETS and its partners have carried out the diagnostic and the maintenance of existing mechanized tunnels through an innovative multi-dimensional survey system (ARCHITA), and a new approach for the Management and Identification of the Risk for Existing Tunnels (MIRET). The MIRET approach and a deep learning test are demonstrated by its application to a case study.

1 INTRODUCTION

The planning and the management of existing tunnels is already a central challenge for industrialized countries. However, careful and effective management of tunnels that includes all the activities related to inspection, planning, design, construction and maintenance is not yet usual.

Therefore, proper management would require an organized and systematic approach to evaluate and analyze all the acquired information. This approach would allow logical, effective and coordinated decision.

ETS has carried out the diagnostic and the maintenance of existing tunnels through an innovative multi-dimensional survey system (ARCHITA), and a new system for the Management and Identification of the Risk for Existing Tunnels (MIRET) (Foria, 2020; Foria, 2021). ARCHITA surveys the geometrical and structural conditions of the tunnel without influencing the traffic. MIRET is a methodology, a process and a technology made for the integration of such data and the digital design and management of the tunnels (Figure 1).

*Corresponding author
DOI: 10.1201/9781003188339-28

Nowadays, mobile mapping systems have been used for diagnostics, in order to optimize management systems and to allow fast reliable data acquisition, avoiding information loss. In fact, the current challenge in the innovation of tunnel diagnostics is to enhance the quality of acquired images, developing massive data collection and processing capabilities and analysing the data with high-level technical experience and engineering judgement. Therefore, the automatic elaboration of big datasets is mandatory. The traditional computer vision solutions, widely developed in the industry, are not suitable to process big amounts of data in semi-supervised or weakly supervised ways. On the other hand, the supervision provided by human operators is highly dependent on the operator's skill and experience. Moreover, operator's experience is so interdisciplinary and complex that it is unlikely to be well described by a traditional computer vision model. Besides, the learning process is an ongoing knowledge process, its transfer and reinforcement dynamics being hardly embeddable in traditional models. In contrast, the progression in machines computational power gives to solutions devised within the artificial intelligence (AI) framework the opportunity to capitalize on the full precious operators' experience, efficiently applying the knowledge learnt to the collected data. Such an approach has the capability to provide a baseline for the training of new operators, thus helping in the operators' generational change.

For such a reason, in order to simplify the maintenance and management of tunnel, the authors have envisaged in the MIRET framework to employ AI for the analysis of structural defects.

The process combines in a common workspace innovative mobile mapping, multidimensional survey systems, defect analysis, artificial intelligence and risk analysis (Figure 1). All the acquired and analyzed data are at the service of the technical and management table; aiming at a transparent, smart and sustainable system for planning, design and maintenance of the tunnels.

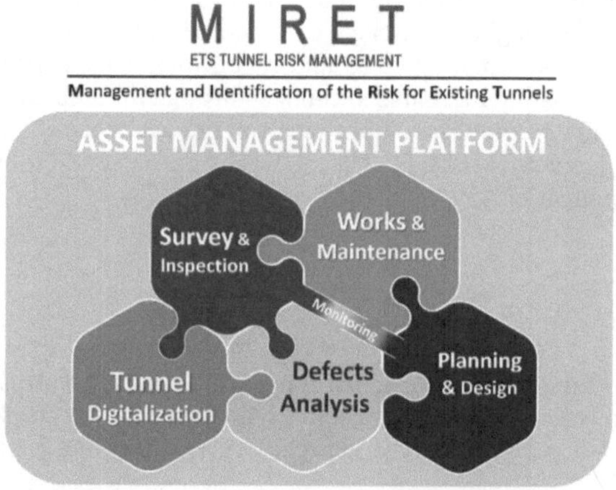

Figure 1. Puzzle chart to introduce the MIRET approach and elements.

2 INNOVATIVE INSPECTION OF EXISTING TUNNELS

2.1 *ARCHITA*

ARCHITA (Figure 2) is a multi-dimensional mobile mapping system developed by ETS consisting of linked and integrated equipment (Foria, 2019), and of survey and positioning sensors.

Key advantages of the ARCHITA approach are: the avoidance of intrusive structural surveys; the minimization of the time of traffic disruption, operating at an average speed of

15-30 km/h; the increased safety, obtained by reducing the time and number of operators working in the tunnel; the improvement in back-office capabilities, in fact the data acquired on-site is elaborated, interpreted and analyzed in the office, leaving only specific tests to be undertaken on-site; the possibility of integration with traditional methods and on-site measurements/inspections.

To complement the innovative multi-dimensional mapping of ARCHITA and its integration by non-destructive diagnosis techniques, RMT and ADTS with ETS developed a twofold technological solution: Tunnel Scan (an instrumental apparatus) and Tunnel Review (dedicate software). The mapping of the defects is carried out combining the high-resolution photos, taken by three high-definition cameras and the point cloud. The combination of the two technologies made the images measurable, with the possibility of positioning, measuring and quantifying the defects identified on the tunnel lining.

Figure 2. ARCHITA: RAIL (on the left) and ROAD (on the right) configuration.

3 DEEP LEARNING APPROACH FOR WATER DEFECTS

3.1 *Deep learning for image segmentation*

Automating the process of defects assessment in tunnel inspections is of paramount importance in the field of civil engineering. In addition, the pace of development of deep learning and convolutional neural networks techniques pushes the performances of image-based semantic segmentation. Current research trends are extending the prediction capabilities of computational approaches for automating image understanding tasks, such as defects assessment. The aforementioned learning techniques have been successfully applied to the visual data interpretation problems, with applications ranging from pixel classification of urban street scenes (Tao, 2020) to indoor scene understanding (Seichter, 2020).

3.2 *Water defect segmentation: Design*

We present a solution based on a fully convolutional deep neural network, to perform pixelwise segmentation of tunnel surfaces, detecting the presence of water. The proposed solution is adapted from the work (Ren, 2020). The authors of (Ren, 2020) devised the network architecture CrackSegNet, to achieve the efficient multiscale resolution, extraction, aggregation, and reconstruction of visual features of cracks on concrete surfaces.

In analogy with (Ren, 2020), the problem of water defect assessment is formulated as a pixelwise segmentation problem on a binary classification set. The data pipeline comprises convolutional neural network processing and postprocessing. The input of the network are the grayscale images of the internal surface of the tunnel, obtained by the Tunnel Scan operation and merged by the Tunnel Review algorithm. We label the output pixels according to the mapping: 0 – the pixel is not representative of a water defect, and 1 – the pixel is representative of a water defect on the tunnel surface. The model training is performed using pairs of image data and water defects annotations of the same shape. The defects annotations are in the form of a white background image of the same dimensions of the input image, while the pixels representing water defects are black. During the training process, the annotations are confronted with the network's output, thus providing supervision. The network output is then modified before post-processing: each element is substituted with its complement to 1. The post-processing phase is a smoothing and binarization of the network's output.

The layout of the neural network WaterSegNet has been adapted from the best performing neural network layout of (Ren, 2020), with the following modifications. Firstly, the input dimensions are set to process 512x512 grayscale images. Then, the batch-normalization layers are removed with the exception of the 4th layer, that is the final batch normalization layer. Finally, the focal loss computation has been modified to ignore contributions given by negative i.e., nondefective, samples.

3.3 *Water defect segmentation: Implementation*

The proposed approach has been implemented using Python and C++ programming languages. The Python implementation covers the network prototyping, guiding the design choices and acting as a baseline for the code structure in C++. Moreover, the functionalities of training, testing, the data pre-and post-processing, and the data analysis were implemented in Python using TensorFlow 1.15.0 (Abadi, 2016) and Keras 2.0.9 (Keras, 2015). On the other hand, the C++ implementation covers the evaluation phase, often referred to as inference phase, and the integration with the Tunnel Review software. For the C++ implementation, the frameworks OpenCV 4.4 (Journal of software tools, 2000) and Qt 5.12.10 (www.Qt.io) was used. While OpenCV was employed to embed neural network processing while adding few dependencies to the Tunnel Scan software; Qt was employed to provide an efficient and reliable graphics tool, with advanced integration and extension capabilities.

3.4 *Water defect segmentation: Training*

The proposed solution has been trained and validated on the experimental data collected by Tunnel Scan. The network is trained using backward propagation of errors, employing a stochastic gradient descent method. Random weights in network layers are initialized using He normal initializers. During training, the samples in input to the network are corrupted by varying the brightness and contrast of a fraction in the interval [0 10%], improving the generalization capabilities of the network.

The training strategy comprises two steps. Firstly, the network is trained using as loss metric the binary cross-entropy function evaluated on the pair of annotation and network's output. Then, the training is refined by using the focal loss function, which is the cross-entropy function modified to address class imbalance. Moreover, the Adam optimizer is used, with an initial learning rate of 1e-4 for the first phase of the training, that is adjusted to 1e-6 for the second phase of the training. During the training process, the network weights are saved after every epoch, provided that the associated training step improved in the loss value. The accuracy of the model was verified on a dedicated validation dataset, not seen by the network in the training process. The model with the highest validation accuracy was saved as the final model. The deep learning framework of Keras with TensorFlow as a backend has been employed to complete the training of WaterSegNet on a single Nvidia GPU.

4 CASE STUDY: GALLERIA COLLECERVO

4.1 *Introduction*

Since an increasing amount of the tunnel industry employs mechanized excavation method, the case study of Galleria Collecervo is examined. Galleria Collecervo is a mechanized tunnel extending for a total length of 3463 m, part of the works for the double-track new variant route between Andora and San Lorenzo al Mare.

The survey-inspection and the mapping of the defects along the tunnel are carried out to design an efficient maintenance plan. The activity had to comply with the following topics: the survey and the inspection of a working line with short disruptions allowed only during the night; the design and planning of the maintenance operations according to client's requirements of a working line; the need for objective data and reliable evaluations in order to speed-up the technical table among the parties (owner, tester, designer, contractor).

4.2 *Survey, defects analysis and maintenance*

On February 2019, ETS surveyed the tunnel with ARCHITA to obtain the whole state of preservation of the structures. A geometric and photographic scan of the entire of tunnels was carried out, operating on both the tracks and causing only a two-night disruption of the railway traffic.

The first step of the defects analysis process requires the choice of the defect library to be used for classification. In this case study, an ad hoc library has been elaborated for the client, classifying the defects into three macro-categories: concrete surface defects, technological defects, and previous interventions i.e. repairs carried out before the present inspection (e.g., ribs, seals etc.).

The subset of defects related to the presence of water is described as follows: category C5, presence of humidity on the concrete surface; category C6, infiltration of water through the joints; category C7, the passage of water through cracks. For the calculation of extension indices, the tunnel is divided into sectors of 210 m (Figures 3-5).

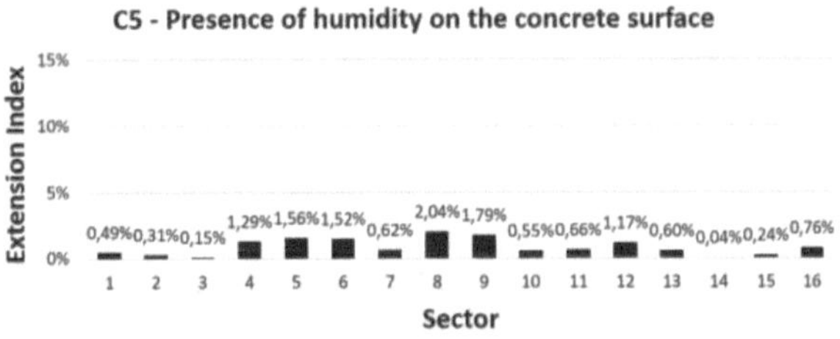

Figure 3. Example of extension index of C5 defect for each sector (Galleria Collecervo, 2019).

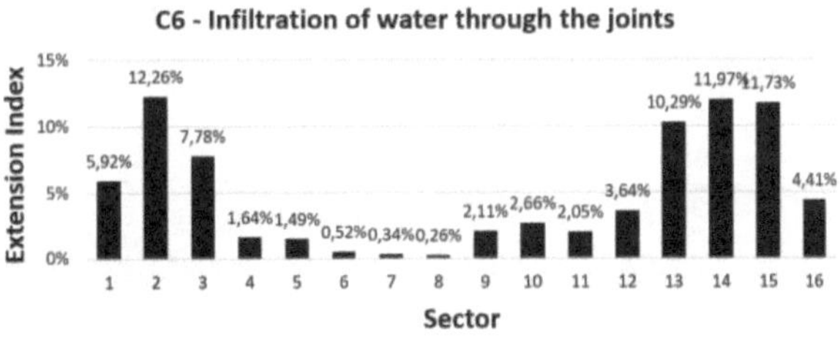

Figure 4. Example of extension index of C6 defect for each sector (Galleria Collecervo, 2019).

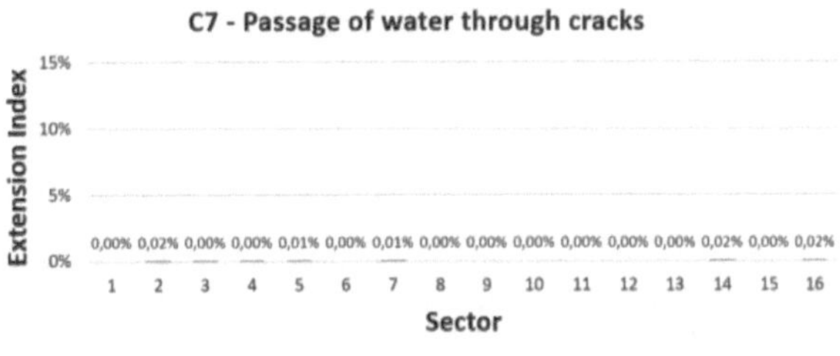

Figure 5. Example of extension index of C7 defect for each sector (Galleria Collecervo, 2019).

4.3 *Application of NN approach for water defects*

The dataset obtained during the maintenance of Galleria Collecervo in 2019 has been used to investigate the effectiveness of the proposed approach. More than 10k images were selected among the categories C5, C6, and C7. In particular: 1099 defects of type C5, 8936 defects of type C6, and 34 defects of type C7 were used. The selected data were exported by dividing the acquisition into square sections of 512 x 512 pixels. Different zoom values: x1, x1/2, x1/4 were used to obtain a multiresolution map of the tunnel. After that, another selection phase is operated, in order to exclude images corrupted by the cropping in the export process. 4783 image/label couples are kept for the resolution of 1/4x, discarding the 12% of data, 1639 couples are kept for the resolution of 1/2x, discarding the 12% of data, 427 couples are kept for the resolution of 1x, discarding the 21% of data.

Finally, the data have been augmented to reduce the dataset imbalances concerning the different types of water defects, and the amount of defective and non-defective pixels. The Python package "Albumentations" was used. The image transformation employed are: ElasticTransform, BrightnessContrast, GridDistorsion, ShiftScaleRotate, OpticalDistorsion (Buslaev, 2020). The resulting collection of approximately 15.000 images were merged in a single folder, after filtering out images corrupted by the augmentation; thus, considering all types of water defects in one general category i.e., C5+C6+C7.

The neural network has been trained on the prepared dataset, which has been split into training data and validation data in proportions of 75% and 25%, respectively. For Galleria Collecervo dataset, an average of 50 training epochs was used, with almost 7500 training rounds per epoch, and a batch-size of 2. The training was performed on a GPU NVIDIA RTX2080-Ti. The postprocessing parameters were chosen as 25 for the blur kernel strength in the smoothing phase, and a threshold of 0.5 for the binarization phase.

4.4 *Manual vs. Automatic detection*

After the training phase has been completed, the best performing model has been validated on set of manually labelled data of the same tunnel. The performances of the proposed approach have been explored, and common classification metrics have been computed and averaged over the pixel distribution. We report that the proposed approach resulted in an accuracy of 93%, and a precision and recall well above 70%. Figure 6 shows a comparison between the reference humanmade annotation and the automatic annotation obtained by the proposed approach. We note that the automatic approach is more reliable when defective visual features are simple and appear to the network in isolation, while the human approach is more suitable in situations where

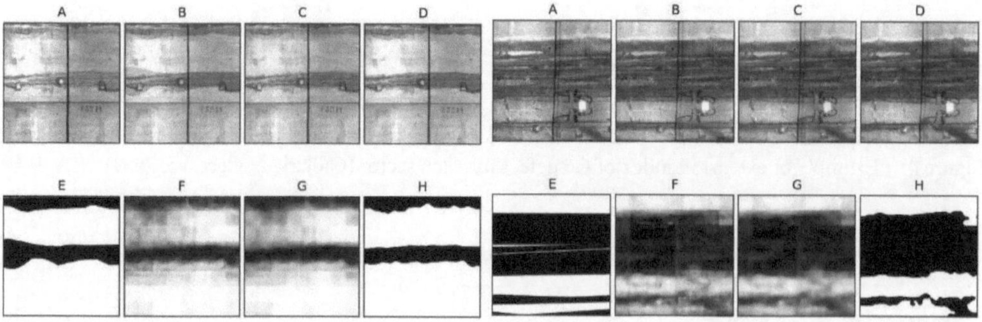

Figure 6. Example of comparison between the reference human-made annotation and the automatic annotation obtained. A – original image; B – image + human-made annotation; C – image + classification probability map; D – image + AI annotation; E – human annotation; F – classification probability map; G – blurred classification probability map; H – binarized and blurred classification probability map. The values for the accuracy, precision and recall metrics for the two examples are: 98.819%, 76.948%, 70.023%; and 90.639%, 92.544%, 91.063%; for the left and right examples, respectively.

features are of complex interpretation and/or many features overlaps. In the left picture of Figure 6, two horizontal areas with finely-grained irregular borders, representing water defects, are highlighted as human annotations. The upper water defect region possesses sound defective features, in fact the automatic annotation is more precise than the human one. In contrast, the image of the lower defect is affected by contrast and luminosity variations. Here, the shading and the presence of background features spoil the performance of the network, especially on the left part of the region, and the human annotation is more precise. In conclusion, the proposed approach showed the capability of aiding human supervision in the task of tunnel inspection.

5 CONCLUSIONS

MIRET, a new approach for the management of existing tunnels, is introduces as a generalization of the different approaches proposed by authorities, contractors and suppliers. The methodology is strongly based on large scale, quick inspection processes with multi-dimensional mobile mapping systems, such as ARCHITA. The approach allowed the implementation of a solution for the automatic assessment of tunnel defects, comprising a deep convolutional neural network and postprocessing elaboration. The results show good agreement with human annotations for water defects, demonstrating the promising capabilities of computer aided defect assessment. However, more work should be devised to strengthen the robustness of the scene understanding, and the generalization capabilities of such automatic defect assessment solutions.

REFERENCES

Abadi, M. et al. 2016. Tensorflow: A system for large-scale machine learning. 12th symposium on operating systems design and implementation.

Buslaev, A. et al. 2020. Albumentations: fast and flexible image augmentations. Information. Vol 11.

Foria, F. et al. 2019. ARCHITA: an innovative multidimensional mobile mapping system for tunnels and infrastructures. MATEC Web of Conferences. Vol. 295. EDP Sciences.

Foria, F. et al. 2020. Galleria Olmata, from survey to construction: an integrated design approach for the renewal of railway tunnels. In World Tunnel Congress 2020.

Foria, F. et al. 2021. Tunnel defects mapping and maintenance of existing tunnels with an innovative approach (MIRET): the case study of Genova-Ventimiglia railway line and Roma Metro lines. In AFTES 2021 (accepted).

Journal of Software Tools (2000) - "The OpenCV Library"

Ren, Y. et al. 2020. Image-based concrete crack detection in tunnels using deep fully convolutional networks. Construction and Building Materials. Vol. 234.

Seichter, D. et al. 2020. Efficient RGB-D Semantic Segmentation for Indoor Scene Analysis. arXiv preprint arXiv:2011.06961.

Tao, A. et al. 2020. Hierarchical Multi-Scale Attention for Semantic Segmentation. arXiv preprint arXiv:2005.10821.

The machine learning framework Keras: GitHub: "Keras", 2015. The Qt programming framework internet web site: www.qt.io

The Evolution of Geotech - 25 Years of Innovation – Hammah et al. (Eds)

An integrated geotechnical risk management approach using cloud-based risk assessments, artificial intelligence, satellite monitoring and drone technology

J. Strydom
University of the Witwatersrand, Parktown, South Africa

ABSTRACT: Open-cast slope monitoring consists of a combination of visual, drone, satellite, and radar monitoring. The size and complexity of open-cast strip mines and the amount of data generated complicates hazard identification and communication on mines. In the South African coal mining industry, geotechnical data is manually processed by the rock engineering departments. To ensure that the geotechnical data are optimally used, automated systems are required to process and display the data. An improved geotechnical risk management approach, which uses an integration of the latest technology, was proposed. This includes improvements in the geotechnical monitoring process by using artificial intelligence for crack detection, elevation differences of drone data to track slope movement, radar monitoring, and satellite monitoring. A cloud-based mobile application was developed to capture and display geotechnical risks to improve communication. This paper will focus on how these technologies have been utilized at Mine X.

1 INTRODUCTION

The arrival of the 4th Industrial Revolution (IR) introduced technology into the mining industry that can significantly improve the geotechnical hazard management process on mines. The 4th IR is recognised by the simultaneous advances in artificial intelligence (AI), robotics, the internet of things (IoT), autonomous vehicles, 3D printing, nanotechnology, biotechnology, materials science, energy storage, virtual reality, big data, and quantum computing (Skilton & Hovsepian, 2018). Due to the increased economic pressure and lower profit margins within the mining industry, innovation and the adoption of this technology is essential to reduce cost, improve safety and optimize productivity.

Mine X is situated in the Vaal Basin Coalfield, in the Free State Province of South Africa. Historic underground workings are present in all three economical coal seams. The remaining reserves are being extracted by means of an opencast dragline operation. Due to the complexity of the geology and the presence of old underground workings, slope failures and sinkholes is a major risk. Geotechnical risks are managed by a four-tier process. The four steps include the evaluation of in situ geological information, cut risk assessments, monitoring, and reconciliations. Monitoring is currently done using ground-based radars and visual inspections. The monitoring sites for radars are prioritized based on a consequence vs likelihood matrix. The advantage of ground-based radars is their broad slope coverage area, near real-time slope deformation tracking and remote monitoring ability (Dick, Eberhardt, Cebrejo-Lievano, Stead, & Rose, 2014). There is however a limitation in the ground-based radar's ability to detect displacements above the slope crest (Carla, Farina, Intrieri, Ketzimen, & Casagli, 2018). In strip mines, the radar is also

DOI: 10.1201/9781003188339-29

limited in the area of the slope that it can effectively monitor. While ground-based radars have proven to be highly successful as an early warning system for the detection of slope failures, it is not feasible to scan every slope on the mine. Further drawbacks in the current four tier risk management process, is that it is managed with Excel spreadsheets, which is time-consuming, labour intensive, and susceptible to human error. Due to the massive amount of geotechnical data that are generated daily on a mine that need to be collected, analyzed, and communicated, a degree of automation is required.

Until recently, it was considered that spaceborne monitoring systems could not yield information in the required systematic fashion to act as an early warning system for slope failures. However, Carla et al. (2019) reported that Interferometric Synthetic Aperture Radar (InSAR) data of various slope failures indicated that there was an increase in the rate of the slope movement in the time leading up to the failure, which indicates that failure is imminent. A combination of ground-based and satellite radar interferometry proved to be highly useful in monitoring slopes prone to failure (Carla, Farina, Intrieri, Ketzimen, & Casagli, 2018). Furthermore, drones have been successfully utilized in the mining industry for the mapping of geological structures and lithological contacts on high walls (Katuruza & Birch, 2019) (Kerle, Nex, Gerke, Duarte, & Vetrivel, 2019). A recent trend in the civil and mining industry is to make use of artificial intelligence and machine learning to automatically detect and highlight fractures or planes of weakness on slopes (Dorafshan, Thomas, & Maguire, 2018).

The use of mobile apps is becoming increasingly common in the mining industry. Mobile apps improve data collection and sharing, which leads to better data utilization, benefitting both production and safety (Alpha Software, 2020).

This paper will focus on how the latest technology can be utilized to improve geotechnical hazard management at Mine X. A slope failure at Mine X was used as a case study to evaluate information from satellite monitoring and drone flights to determine their efficiency as early warning systems for slope failures. To improve data management, a cloud-based risk assessment is in the process of being developed. The aim is to eliminate human error, and automate data capturing and processing to improve hazard communication by having a single, simplified dashboard to display information generated by various sources.

2 METHODS

A multi bench high wall failure occurred at Mine X on 22 October 2020. The slope height was 62 m, and consisted out of sand, siltstone and shale. A pinnacle and fault plane were visible on the high wall. It is important to note that prior to the failure and during the failure, the slope was not scanned with radars. A back analysis of the failure was conducted to evaluate the effectiveness of potential early warning systems.

2.1 *Satellite monitoring*

SkyGeo assisted with the back analysis of the failure using InSAR to measure millimeter scale displacements. The data acquired over Mine X originated from Sentinel-1 ascending orbital track T116 and descending orbital track T50. The satellite acquires new data over the same area, once every 12 days during each orbital repeat. This data is co-registered such that the data from different acquisitions (but from the same orbital track) can be compared. Co-registration is the process performed to spatially align all acquisitions. The pixels contain both amplitude and phase information of the backscattered microwaves. Temporal changes in amplitude can be caused by changes in the scattering characteristics of an object. Temporal changes in phase can be considered a combination of multiple factors, one of which being the physical change in distance between the satellite and the object. Persistent Scatterer Interferometry is the process used to correct the phase difference between acquisitions for orbital changes, flat-Earth phase, topography and atmosphere to attribute the remaining phase difference to a change in line of sight distance.

2.2 *Drone monitoring*

Routine drone surveys are done twice per week at Mine X. This drone data was used to perform a back analysis on the slope failure using aerial photogrammetry to create terrain elevation models. The data was processed in Agisoft Metashape. Elevation changes between two terrain models of the same area, was illustrated by generating contours of elevation differences. The contour threshold was set at 150 mm. This was done by comparing two tin files of the respective models. Tin files were created in MicroStation Geopack.

In addition, machine learning, and artificial intelligence was used during a proof of concept study by RocketMine to automatically detect cracks that developed on the high wall crest. After the failure occurred, the slope area was flown by RocketMine, and their crack detection algorithm was applied.

2.3 *Cloud-based risk assessments*

To digitalize and streamline geotechnical inspections and the Cut Risk Assessments (permit to mine), a mobile application (further referred to as the App) was conceptualized by the author and developed by HCL Technologies Ltd. PowerApps and PowerBi was used for capturing data and displaying reports, respectively.

3 RESULTS

3.1 *Satellite monitoring: Slope movement rates*

The slope movement detected by InSAR can be seen in Figure 1. A stable slope directly north of the failure was also monitored and plotted as a control. At 53 days before the failure, a clear increase in the rate of movement was observed. In comparison, the rate of movement on the stable slope remained consistent. Hence, the increase in the rate of movement on the failure slope could indicate the onset of failure (Carla, et al., 2019). This indicates that even with a monitoring frequency of 12 days, InSAR could have been utilized to identify potential failure areas and act as an early warning system.

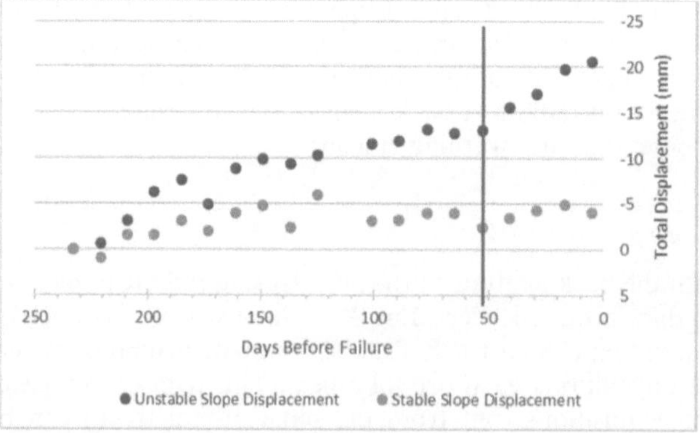

Figure 1. The total slope displacement prior to the failure was determined on stable (grey) and unstable (blue) slopes. The red line indicates the increase in the displacement rate at 52 days prior to the failure.

3.2 *Drone monitoring: Elevation differences and crack detection*

Elevation differences from aerial photogrammetry was also used to identify early sign of failure (Figure 2). In Figure 2A, 42 days before the failure, no slope movement was detected. In

Figure 2B, 21 days before the failure, downward movement on the sand crest (red contours) and outward movement on the toe of the slope (blue contours) can be seen. Figure 2C indicates the overall slope movement after the failure occurred. It was noted that the contours formed by the initial movement recorded on the sand crest (see arrows, Figure 2B), have the same outline as that of the failure (see arrows, Figure 2C). Hence, the failure could be detected with drone monitoring at least 21 days prior to the failure.

Furthermore, the cracks on the high wall were automatically highlighted in light blue using the machine learning and artificial intelligence algorithms from Rocketmine (Figure 3). While the slope had already failed in this proof-of-concept study, the automatic detection of cracks could be a powerful tool in the detection of early crack formation prior to slope failures. The process did not include the detection of fault planes on the high walls. Also, the high walls at Mine X Colliery is buffer blasted, resulting in a highly fractured high wall. It should however be noted that the processing time for the automatic crack detection amounted to 20 hours for an area of 1000 m by 120 m.

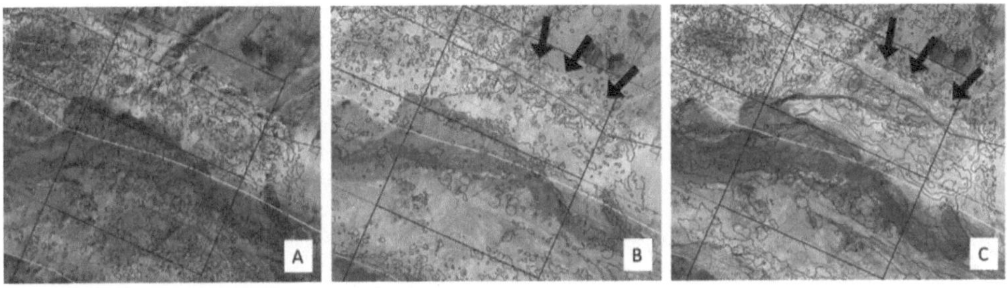

Figure 2. Elevation differences at the failure site is indicated by red and blue contours at 42 days (A), 21 days (B) and 0 days (C) prior to the failure. The black arrows point to the outline of the sand crest where the failure occurred.

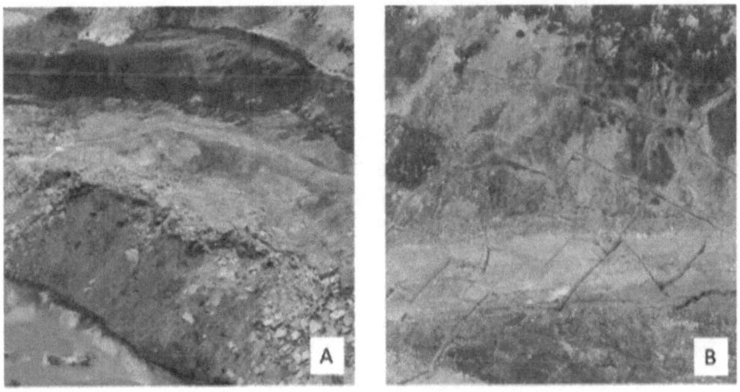

Figure 3. The automatically detected cracks are indicated in light blue. (A) 3D model of the bench, (B) zoomed in image indicating the cracks detected on the bench.

3.3 *Improved data management and hazard communication with cloud-based risk assessments*

The concept of the App is summarized in Figure 4, where the green tick boxes indicate the current completed phase of the App. Figure 5 indicates the home screen of the App and the rating forms that can be used to capture data. Detailed reports can be shared via PowerBi and a weekly or monthly summary of the residual risk of specific failure modes is also immediately available. The proposed format for the reports can be seen in Figure 5. The App has been developed in such a way to guide the user to assess specific failure modes in all the different working areas, while also verifying compliance with operational standards. In this current phase of the App, paperbased data capturing and the manual processing of this data into numerous Excel Sheets has been eliminated. The cloud-based Cut Risk Assessment is still in the development phase.

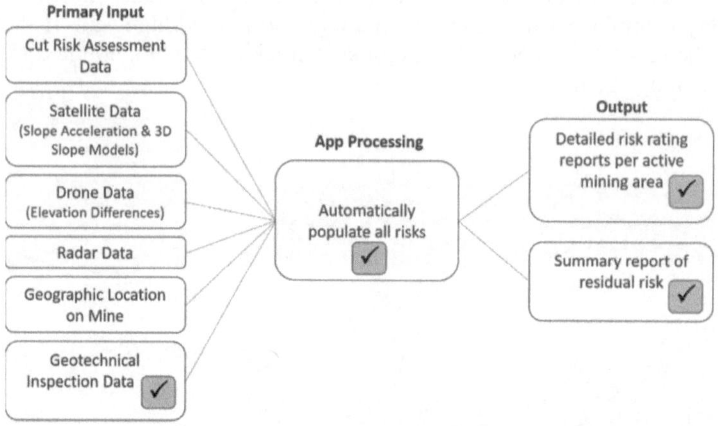

Figure 4. The flow diagram summarizes the conceptualized input and outputs of the App. The green tick boxes indicate which phases of the app development are completed.

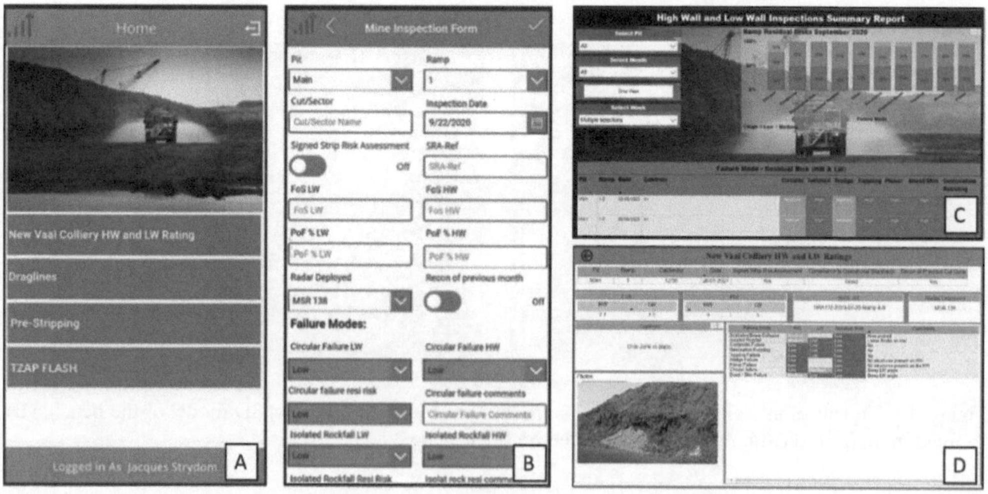

Figure 5. The App's home screen (A), the rating form used for data capturing (B), the summary report (C), and the detailed report shows the integration of the capturing and reporting of risk ratings on a single platform.

4 DISCUSSION

4.1 *An integrated geotechnical risk-management approach*

It was demonstrated that early signs of the slope failure from the case-study could be detected from the accelerated slope movement from satellite data, and the elevation differences from drone data. Due to the threshold settings used for calculating the elevation differences, the slope appeared stable at 42 days prior to the failure, while the accelerated slope movement from satellite data already detected the onset of the failure 53 days prior. The automatic crack detection algorithms were only applied to drone footage after the slope had already failed and therefore served as a proof-of-concept. The benefit in the use of InSAR is its ability to monitor large areas, acting as a mine-wide, regional monitoring system. The InSAR data can also be imported into Rocscience Slide 3 or RS3 for a stability analysis (Gervasi, 2020). The safety factor and probability of failure of the actual mined slopes can then be assessed routinely. It is realized that more case-studies need to be investigated to establish trends in terms of slope accelerations on failed and stable slopes. The author is currently working with SkyGeo on a further 10 case studies of slope failures at Mine X. This work will include using the inverse velocity technique to the InSAR data to predict time to failure. Current challenges of the crack detection algorithm and elevation differences from aerial photogrammetry are the processing and interpretation time required when analyzing large areas. The fact that the cracks and contours are highlighted in the model does assist with the identification of unstable areas, but the end-user still needs to manually work through the entire model. This is time-consuming, and the same outcome can be achieved by scrutinizing drone videos, which is available immediately.

As part of the integrated geotechnical risk management approach, a georeferenced model will be required, that can automatically display the cracks detected with AI, elevation differences detected by drones, and accelerated slope movement detected with satellites on the same model. This will indicate high risk areas to the Rock Engineer to deploy ground-based radars. The effectiveness of ground-based radars as early warning systems has proven itself over time. The challenge with ground-based radars on a strip mine is to proactively identify the high-risk areas where they must be deployed. While drones and satellites act as regional monitoring systems, their shortcomings in real time monitoring is complemented by the deployment of ground-based radars.

At the current developmental phase, the use of the mobile application to capture data during geotechnical inspections will improve hazard communication on a weekly basis. The application managed to simplify the capturing and displaying of geotechnical risks based on the currently used monitoring processes, without compromising on the quality of reporting. The ideal would be for information from the cut risk assessment, drone, radar and satellite data to be available live on the App while a risk rating is being conducted in the pit. This could be achieved by georeferencing the user in the pit and ensuring only the information relevant to the user's exact location is displayed. Ultimately, the App should provide full integration between each step of the current four tier risk management process.

5 CONCLUSION

The integration of radar, drone and satellite monitoring promised to be highly effective as an early warning slope failure system. To fully utilize the data generated by various sources, an App was conceptualized and is currently under development. The current phase of the App allows the capturing and automatic displaying of geotechnical risks on a cloud-based platform.

ACKNOWLEDGEMENTS

I would like to thank my employer for allowing me to share this information. This project would not have been successful without the help and support from SkyGeo, HCL Technologies, RocketMine and the mine Survey Department.

REFERENCES

Alpha Software. (2020). Retrieved from How Mobile Apps Benefit Mining Companies: alphasoftware.com/mining-apps

Carla, T., Farina, P., Intrieri, E., Ketzimen, H., & Casagli, N. (2018). Integration of ground-based radar and satellite InSAR data for the analysis of an unexpected slope failure in an openpit mine. *Engineering Geology*, 39–52.

Carla, T., Intrieri, E., Raspini, F., Bardi, F., Farina, P., Ferretti, A., … Casagli, N. (2019). Perspectives on the prediction of catastrophic slope failures from satellite InSAR. *Nature*, 1–9.

Dick, G., Eberhardt, E., Cebrejo-Lievano, A., Stead, D., & Rose, N. (2014). Developemnt of an early-warning time-of-failure analysis methodology for open-pit mine slopes utilizing ground-based slope stability radar monitoring data. *Canadian geotechnical Journal*, 515–529.

Dorafshan, S., Thomas, R., & Maguire, M. (2018). SDNET2018:Anannotated image data set for non-contact concrete crack detection using deep convolutionalneuralnetworks. *ELSEVIER*, 1664–1668.

Gervasi, C. (2020, February 12). *TRE ALTAMIRA*. Retrieved from TRE ALTAMIRA: https://site.tre-altamira.com/rocscience-3d-slope-stability-programs-now-integrated-with-our-satellite-insar-data/

Hermanus, M. (2017). Mining Redesigned- innovation and technology needs for the future-a South African perspective. *The Journal of the South African Institute of Mining and Metalurgy*, 811–818.

Katuruza, M., & Birch, C. (2019). The use of unmanned aircraft system technology for highwall mapping at Isibonelo Colliery, South Africa. *The Journal of the Southern African Institute of Mining and Metallurgy*, 291–295.

Kerle, N., Nex, F., Gerke, M., Duarte, D., & Vetrivel, A. (2019). Drone Geologic Mapping of an Active Sand and Gravel Quarry, Desoto County, Mississippi. *International Journal of Geo-Information*, 1–14.

Skilton, M., & Hovsepian, F. (2018). *The 4th Industrial Revolution Responding to the impact of Artificial Intelegence on Business*. Cham, Switzerland: Springer Nature.

The Evolution of Geotech - 25 Years of Innovation – Hammah et al. (Eds)

Remote sensing and mapping of geological rock-mass features employing advanced data analytics and artificial intelligence

S. Amvrazis
Mathematician, AFRY Austria GmbH, Vienna, Austria

C. Binder
Geologist, AFRY Austria GmbH, Vienna, Austria

A. Freuis
Geotechnical Engineer, AFRY Austria GmbH, Vienna, Austria

J. Patzelt
Geologist, AFRY Austria GmbH, Vienna, Austria

M. Smesnik*
Geotechnical Engineer, AFRY Austria GmbH, Vienna, Austria

ABSTRACT: The mapping and classification of geological rock mass features like discontinuities, block sizes, etc. are important information for the profound understanding and prediction of the rock mass behavior under various load conditions. These are the basis for stability assessment of rock slopes, the evaluation of geological hazards, and the design of mitigation and support measures. Manual data acquisition is often time-consuming and possesses a certain risk to human life, especially in remote areas after hazardous events. Another challenge is the mapping of large and difficult to access areas. A team of AFRY Geologists, Geotechnical Engineers and Scientists developed a novel software to map and characterize geological rock mass features based on point cloud models. The latter can be developed from various remote sensing techniques. The software uses a self-supervised neural network that provides k-d tree 3d directional search from point cloud neighbors. On the basis of the neighbor normal estimation, the best plane fit is evaluated. For each point pair, the dip angle and dip direction are calculated. The same approach can be used for the characterization of any other feature. The two characteristics/values (dip direction and dip angle) are employed to e.g. classify discontinuities in groups based on pre-specified angular ranges. This paper presents the introduced remote mapping process and results with the help of two case studies. Furthermore, insights concerning the data analysis process and evaluation approaches are provided.

1 INTRODUCTION

The mapping and classification of geological rock mass features like discontinuities, block sizes, etc. are of utmost importance besides other important information for the profound understanding and prediction of the rock mass behavior under various load conditions. The defined characteristics and information are the basis for stability assessments of rock slopes, the evaluation of geological hazards, and the design of mitigation and support measures. The manual field mapping process is often time-consuming and possesses a certain risk to human life,

*Corresponding author
DOI: 10.1201/9781003188339-30

especially in remote areas after hazardous events. Modern mapping approaches and technologies shall be employed to support the Geologist and Geotechnical Engineers in the creation of reliable geological models and a better understanding of geological conditions and rock mass behaviors. These technologies can never replace the field mapping process and should be understood as technical support. During several projects in the past the demand of such technologies raised and a Team of AFRY Geologists, Geotechnical Engineers and Scientists developed a novel software to map and characterize geological rock mass features based on point cloud models. The latter can be easily created out of surveying data collected with different methods.

2 THE SOFTWARE

The developed software can be employed to identify geological patterns in 3d point clouds. For the same, the software uses a self-supervised neural network that provides k-d tree 3d directional search from point cloud neighbors. Outliers computation is used to reduce noise in the point cloud model with a suitable given threshold in a sphere radius of 0.05 m and a maximum number of points within a sphere. After outliers are eliminated, the structure of points is improved by voxelating the model with a cube size of 0.01 m. Computation of vector Normals takes place with a consistent tangent plane method to avoid reversed directions, and k-d trees to identify the nearest neighbor of 30 points on each iteration. Therefore, all points within the model are normalized in accordance with discontinuity patterns. In the next process step, the dip directions and the dip angles are calculated based on each given vector Normals (N_x, N_y, N_z) per point pair. These two characteristics/values are employed to classify patterns and discontinuities in groups based on pre-specified angular ranges. These combinations (dip and dip direction) are used to automatically establish a dual-layered neural network. Although the angular ranges are fixed the planar surfaces will not be interrupted since k-d tree normal search will search homogeneity of each planar distribution. Therefore, all planes can be evaluated. The self-supervised neural network will then classify each plane to the corresponding dip angle and dip direction that it belongs to. Each plane class, etc. can be visualized following a defined color code.

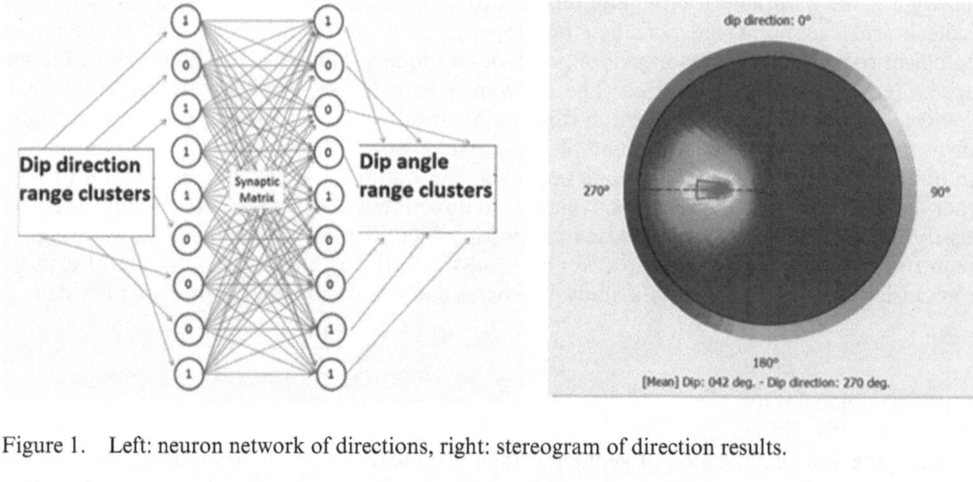

Figure 1. Left: neuron network of directions, right: stereogram of direction results.

3 CASE STUDIES

This paper discusses two case studies where geological remote sensing and mapping carried out with the developed software was successfully employed. Case Study A discusses the geological characterization of a rockface after a landslide and Case Study B introduces the employment of remote mapping technologies in the design and optimization process of a rock support systems.

3.1 *Case Study A – Geological characterization of a rockface after a landslide event*

After the occurrence of various local landslide events in the reservoir area of a storage hydropower plant, the demand for stability and risk assessments of the reservoir rims was raised. The impact of sliding masses into the reservoir triggers impulse waves. These waves and the linked fluctuation of the water levels in the reservoir can cause additional slope instabilities or can lead to overtopping of the dam. The latter can trigger, if the reservoir is formed by an embankment dam, erosion phenomena and is, therefore, a dam safety concern. For the assessment of the impact energy of possible landslide events, reliable geological models were required. A challenge in the project was that the landslide slopes were only accessible from the reservoir and partly from the top. Due to safety concerns, the geological remote mapping process was employed to enhance the available geological data. Camera equipment attached to unmanned aerial vehicles (UAV) was employed to map the rock faces. Based on captured photographs and videos a 3D point cloud model was generated. Figure 2 presents the model.

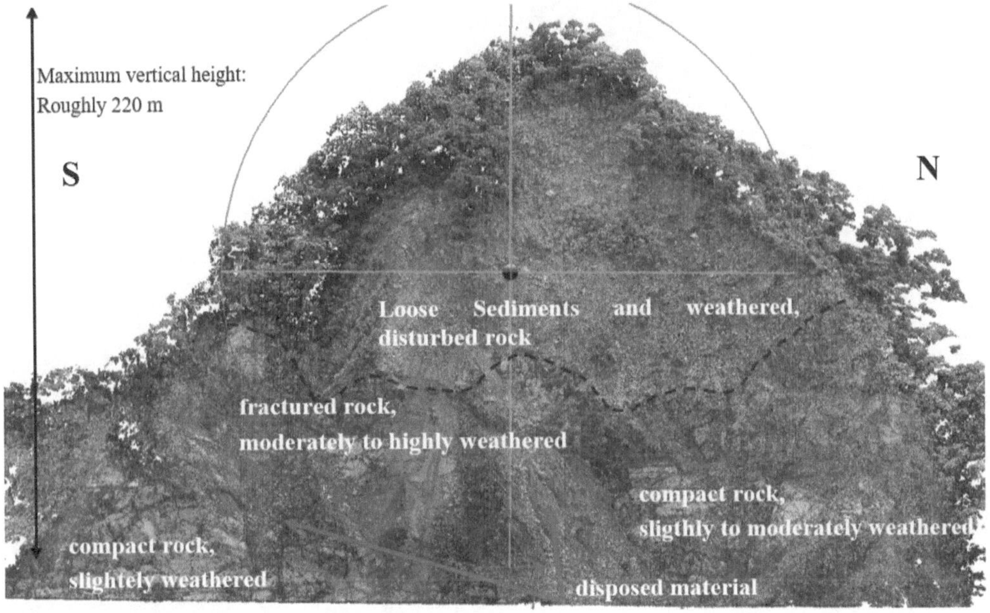

Figure 2. Point cloud model of rock face.

The discussed rock face consists of solid rock, specifically muscovite schists and gneisses, as well as highly fractured rock and loose sediments. The semi-automated analysis (automated data analysis with the software and verification/ enhancement by the geologist based on (1) of the rock face revealed four discontinuity sets which are shown in Figure 3 and described in the following paragraphs.

3.1.1 *Discontinuity sets*

The first discontinuity set (set 1) has a calculated mean orientation of 283/56 (dip direction/ dip angle). The set can be described as prominent and is clearly visible in the lower part of the rock face. The spacing is wide to very wide (2-6 m) and the persistence is classified as high (10 – 20 m) occasionally medium (3-10 m). The second distinctive set of discontinuities (set 2) has a mean orientation of 233/38. Similar to discontinuity set 1, this set of parallel planes shows very wide spacing (2-6 m) and a high to very high persistence (10-20 m and >20 m). The third discontinuity set (set 3) has a calculated mean orientation of 128/53. The spacing is wide to very wide (0.6-2.0 m and >2.0 m). The persistence is high with values between 10 and 20 m. This set is visible in the lower-left part of the rock face. The fourth set (set 4) has a calculated mean orientation of 000/79 and is mainly present/visible in the lower middle to the right part of the rock face. Figure 3 presents the four identified discontinuity sets. The white spots in the image represent a lack of data, that can occur due obstacles in the way of the desired geological feature (2). This is the consequence of insufficient camera work or flight control, as the camera's angle was not perpendicular to the general slope. The manual mapping process revealed in addition to the automated mapping process, a foliation. Unfortunately, the same could not be identified with the automated approach due to a low camera resolution. A higher resolution and better camera work would compensate for that issue. This geological feature would have been overlooked by the automated approach. This indicates the importance of the interaction between human and software algorithms. The discussed foliation is not destabilizing the rock face as it is roughly horizontal or dipping inwards the slope. Figure 4 displays the four discontinuity sets in a Stereonet.

Figure 3. Four discontinuity sets were identified.

3.1.2 *Stability analysis and result interpreatation*

The identified discontinuity sets were exported and directly analyzed with the Dips Software of RocScience. The orientation of discontinuity set 1 and 2 encourages mass movements and sliding failure. A single failure can develop progressive failure modes by undercutting the top rock face portion which consists more or less of loose material (soils) and/or highly (partly/completely) weathered and disassembled rock. Based on the observed rock/rock mass a wide distribution of grain size must be assumed for possible mass movements. In contrast to set 1 and 2, set 3 doesn't appear to be a destabilizing discontinuity set, as the dip direction faces into the slope. However, in the presence of unfavorable geometric conditions (e.g. overhanging parts) mentioned plane set can support failure. For the execution of preliminary kinematic analyses, the governing parameters like average slope dip (45°) and dip direction (280°) of the rock face have been determined. In the absence of more detailed information, a friction angle of 30° was assumed. It shall be considered, that the wetting of the discontinuity surfaces due to impounding can reduce the shear strength drastically. Furthermore, the presence of water in open joints can create significant driving forces (3).

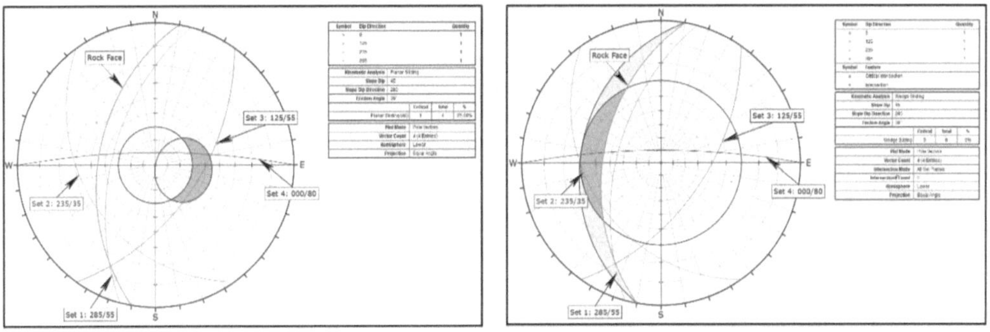

Figure 4. Left: kinematic analysis for planar sliding, right: kinematic analysis for wedge sliding.

3.2 *Case study B – Support system design for an unstable rock pillar*

Routine inspections along a rock face located in a public area revealed erosion phenomena and linked stability concerns of a conglomerate rock pillar. The exposure of the rock face and visibility raised the demand for the design and application of a safe and less visible rock support system. Besides traditional geological mapping approaches, a drone survey and geological remote mapping technics were conducted. Figure 5 presents the two main discontinuities determined during the field survey. The same are open and cutting through the rock slope forming the discussed pillar. The extend of the discontinuity is limited to approximately 20 m from the rock face crest downwards. Below no unfavorable discontinuity sets were mapped and therefore the rock mass was classified as stable.

Figure 5. Left: front view of the conglomerate rock face, right: conglomerate pillar is forming due to erosion and gravitation.

For the design of the support system, a 3d point cloud model of the unstable rock pillar and surrounding rock mass was created. The same was used for the precise determination of the rock pillar volume and resulting weight. The pillar position in comparison to the rock face and the discontinuities was used to define the design loads for the support system. Sections of the 3D model were taken and a stability analysis (toppling, sliding) were performed. A support system consisting of 4 rock bolt rows and one anchor row was determined. In section 2 introduced software was employed to define the distance between the rock pillar surface and the governing discontinuities (Figure 5, right). This information was used to select the required length of the rock bolts and anchors.

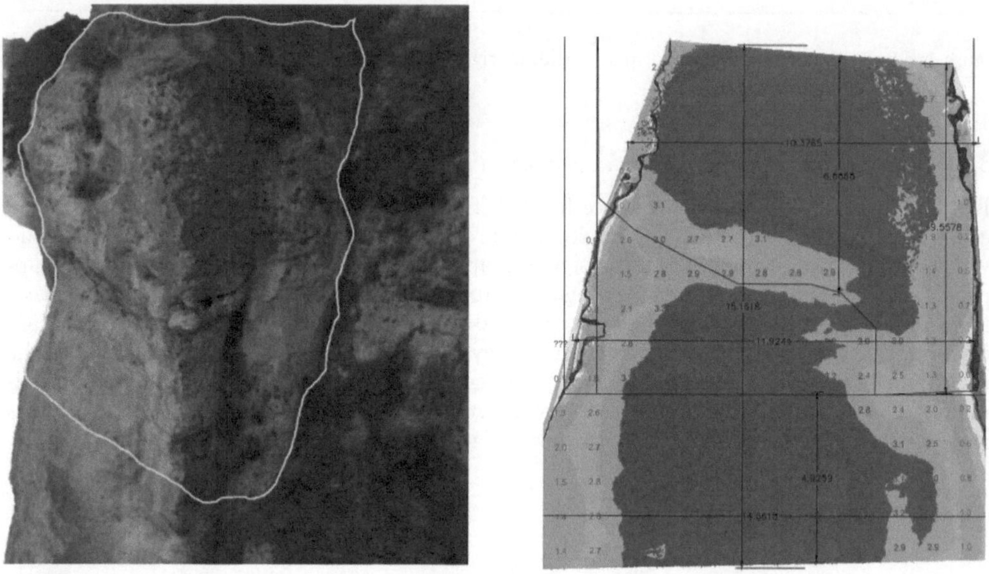

Figure 6. Left.) 3D model derived from the drone survey, right.) distance from rock pillar surface to discontinuity.

4 SUMMARY AND OUTLOOK

Remote sensing and mapping of geological rock mass features employing advanced data analytics and artificial intelligence are valuable support in the daily work of Geologists and Geotechnical Engineers. Especially in remote and dangerous areas, these approaches help to gain reliable data and increase safety at work. Another advantage is the precision and ability of pattern recognition on a small and large scale. The introduced software showed in several internal projects the capabilities and reliability of the analysis results.

REFERENCES

(1) International society for rock mechanics (1978): Comission on standardization of laboratory and field tests; suggested methods for the quantitative description of discontinuities in rock masses.

(2) Karen Kawembe Mwangangi (2019): 3D building modelling using dense point clouds from UAV, Enschede Netherlands.

(3) Prinz, H. & Strauß, R. (2006): Abriss der Ingenieurgeologie, 4. Auflage, München, Elsevier.

The Evolution of Geotech - 25 Years of Innovation – Hammah et al. (Eds)

Tunnel liner yield forecasting at Cigar Lake Mine: An input variable selection approach to understanding machine learning processes

J. Morgenroth, M.A. Perras & U.T. Khan
Department of Civil Engineering, Lassonde School of Engineering, York University, Toronto, Canada

ABSTRACT: It is often time prohibitive for geotechnical professionals to examine large datasets to investigate complex rock mass phenomena in detail. Machine learning algorithms (MLAs) are gaining momentum for data processing in rock engineering, however research into their practical applications are just emerging. These multivariate rock mass datasets are ideal for developing MLAs to forecast rock mass behaviour. An Input Variable Selection (IVS) approach is presented for a Convolutional Neural Network (CNN) that predicts tunnel liner yield due to squeezing ground conditions at the Cigar Lake Mine. A IVS method called Input Omission (IO) is modified and applied to the CNN to enhance its performance. The IO method ranks the CNN inputs in terms of usefulness for forecasting the output. The IO findings for this CNN indicate that none of the available inputs may be omitted, and that the geotechnical zones and radial tunnel displacement inputs contain the strongest signals for forecasting the severity of tunnel liner yield at Cigar Lake Mine.

1 INTRODUCTION

Rock mass data is becoming easier to collect digitally and to store in large quantities, giving rock engineers a richer dataset from which rock mass behaviour can be forecasted and underground support can be designed. Machine learning algorithms (MLAs) are becoming computational accessible and are being developed and customized for a range of rock engineering applications (Kumar et al., 2013; Song et al., 2015; Ferentinou & Fakir, 2018; Morgenroth et al., 2019). MLAs can be developed to preserve the integrity of typically collected geomechanical information, while minimizing data processing bias and capturing the nuanced rock mass behaviour. However, in engineering practice MLAs are regarded as opaque in terms of their inner workings and how they relate to the physical rock mass phenomena that they are modelling.

An approach to uncovering the underlying system of an MLA is applying an Input Variable Selection (IVS) method, which can serve to link these data-driven approaches to the physical system that is being modelled, thereby increasing user confidence that the MLA is producing results consistent with fundamental rock mechanics principles. IVS methods typically identify the most useful MLA inputs from a candidate pool of inputs, where usefulness is defined as having the maximum relevance to the output while minimizing the redundancy between the other inputs (May et al., 2011).

In this paper, a novel IVS method is developed and applied to a Convolutional Neural Network (CNN) that was developed for Cameco's Cigar Lake Mine in Saskatchewan, Canada. The Cigar Lake Mine Convolutional Neural Network (CNN) was developed to predict the yield of the tunnel liner and ground support elements in the ore extraction tunnels

DOI: 10.1201/9781003188339-31

(Morgenroth et al., submitted 2021). The CNN forecasts the liner yield using four inputs spanning from October 2015 to June 2016: geotechnical zone mapping, primary support class, ground freezing pattern, and surveyed radial tunnel displacement. The IVS method presented herein is called Input Omission (IO), which iteratively omits each input to assess its impact on the overall model performance. This approach allows for ranking of all the input variables, determination of which, if any, should be removed, and can inform where future data collection efforts should be concentrated.

2 CIGAR LAKE MINE CONVOLUTIONAL NEURAL NETWORK

The Cigar Lake Mine is located in northern Saskatchewan, Canada. It is the world's second-largest uranium mine, with an ore grade approximately one hundred times the global average. There are two main challenges facing the stability of the Cigar Lake Mine excavations: controlling groundwater inflow, and supporting areas of weak rock (Golder Associates, 2001; Paudel et al., 2012). The mine freezes the rock mass surrounding the orebody to improve rock mass properties and to restrict groundwater inflow into excavated areas. The Cigar Lake Mine was found to be an appropriate case study for a machine learning application because there are complex tunnel deformation mechanisms, namely the combination of a squeezing environment caused by poor geology and the varying imposed stress conditions caused by the ground freezing regime. The priority for the developed CNN approach was to accurately predict tunnel liner and support yield, which is critical for the safe operation of the tunnel and for forecasting rehabilitation works needed to maintain operations (Morgenroth et al., 2021).

First introduced in the late 1980s, CNNs are a type of Artificial Neural Network (ANN) that are efficient at processing spatial and temporal dependencies in image or raster datasets (LeCun et al., 1989). This algorithm type was applied to the Cigar Lake Mine dataset because CNNs are computationally efficient at processing each pixel in an image, while also accounting for the surrounding pixels and their change over time. Thus, the spatial dependencies are maintained between the static (e.g., mapped geology) and variable (e.g., displacement and ground freezing) inputs, while also capturing the temporal change in the variable inputs.

The Cigar Lake Mine CNN was developed to predict yield to the ore extraction tunnel liner and support elements (Morgenroth et al., submitted 2021). A schematic of the Cigar Lake Mine CNN architecture is shown in Figure 1. The CNN was trained to make predictions using a dataset consisting of categorical data from the digitized Ground Management Plans (GMPs) and numerical data from the radial tunnel displacement measurements. This results in four CNN inputs: geotechnical zone mapping (GEO), primary support class (SUPCL), ground freezing pattern (FREEZE), and surveyed radial tunnel displacement (DISP). The dataset was digitized from five GMPs spanning from October 2015 to June 2016, at a spatial resolution of 164 m, the length of the tunnel, by 13 elements, representing elements around the circumference of the tunnel. In other words, there is a pool of five images with dimensions 164 pixels by 13 pixels to train the Cigar Lake Mine CNN. Each image is comprised of four channels, similar to the RGB (red, green, blue) channels in a typical image, however in this case the channels are GEO, SUPCL, FREEZE, and DISP. The predicted tunnel liner yield is a categorical output, also digitized from the GMPs, and characterized by failed rock bolts, spalled shotcrete, and compressed yield packs (Morgenroth et al., 2020): Class 0 is no yield, Class 1 is yield requiring minor repair, Class 2 is yield requiring major repair, and Class 3 is yield requiring total tunnel reprofiling. The Cigar Lake Mine CNN is trained on a subset of GMP images and then tested on the subsequent GMP image, to simulate prediction of tunnel liner yield forward in time, i.e. GMP 1 was used to predict GMP 2, GMPs 1 and 2 were used to predict GMP 3, and so on. Withholding the next GMP simulates an operational environment where the next GMP represents a future point in time. Detailed information on the Cigar Lake Mine input dataset and image formatting is found in (Morgenroth et al., 2020).

The optimization of the Cigar Lake Mine CNN is discussed in detail in Morgenroth et al. (submitted 2021), and can be summarized as follows. A total of 7200 models were analyzed to determine the best combination of three CNN hyperparameters: the amount of training data,

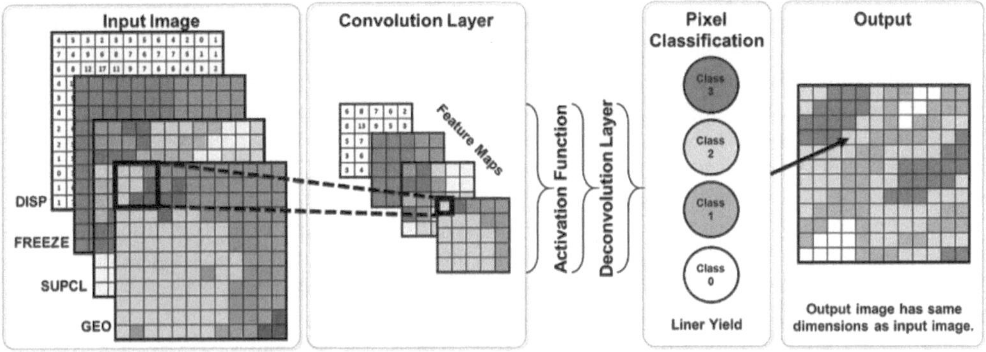

Figure 1. A schematic of the Cigar Lake Mine Convolutional Neural Network (CNN), where a 30 by 30 filter is used to convolve over the four inputs (DISP, FREEZE, SUPCL, GEO) to generate feature maps. The feature maps produced for each input are combined to make a pixel-by-pixel classification of the predicted tunnel liner yield class. The output is a tunnel map that represents a hazard classification (Class 0, 1, 2, 3) of tunnel liner yield in the format of a tunnel map.

the convolution filter size, and the error weighting scheme. It was found that more historic data, i.e. more training GMPs, generally resulted in better performance. The optimal convolution filter size was 30 pixels by 30 pixels, which is equivalent to the tunnel's zone of influence (Brady & Brown, 2006). This paper focuses on the Cigar Lake Mine CNN model that balances the overall performance for all classes of yield, called the Global Balanced model. An inverse frequency error weighting scheme, where the proportion of samples in each class is used to determine the error weights, was used to achieve reasonable prediction accuracy across all yield classes (average model accuracy >65% for all training data permutations).

3 INPUT VARIABLE SELECTION (IVS) – INPUT OMISSION (IO)

IVS is a process whereby candidate inputs are evaluated based on their usefulness in predicting the output, as measured by the chosen performance metric. IVS can be separated into two broad categories: *model-based* and *model-free*. *Model-based* methods are embedded into the training of the algorithm, while *model-free* methods are distinct from the algorithm training.

The IVS method investigated herein is Input Omission (IO), which is a model-based method. For these analyses, the chosen performance metric is the *Corrected Akaike Information Criterion* (*AIC_c*) (Hurvich & Tsai, 1989), which is based on the *Sum of Squared Errors* (*SSE*), number of samples, and number of model parameters. For this study it is important that the performance metric include the number of CNN parameters in its formulation, because the CNNs have varying numbers of parameters depending on the number of inputs used and the amount of training data. The lower the AIC_c, the better the CNN's performance.

The Input Omission (IO) method estimates the usefulness of each input by iteratively examining model performance when an input is left out from the full set on which the model has been trained (Setiono & Liu, 1997). The significance, or lack thereof, of each input can then be compared based on the error that is produced by leaving it out. Excluding redundant or irrelevant inputs from the CNN results in a model with a higher generalization capability (May et al., 2011). In order to obtain a confidence interval on the mean saliency of a network, the network is retrained with random weights at least 30 times, to create an ensemble of CNN models (Setiono & Liu, 1997). In this study, the saliency measure is the performance of the network when removing each of the inputs. This is referred to as creating an *ensemble* of models, and is considered good practice when developing a machine learning algorithm (Marsland, 2014).

The application of IO to CNNs is a novel approach that has been modified from IO approaches that have been used with other types of ANNs for time series data (Setiono & Liu, 1997; May et al., 2011; Snieder et al., 2019). It is not common to apply IO to CNNs. This is because the channels in input images are typically RGB values, which collectively form a cohesive signal that the CNN is learning in order to be able to make a classification. So, for typical image processing it is not logical nor necessary to "turn off" individual channels. However, in the case of the Cigar Lake Mine CNN, each of the channels is a discrete and independent geotechnical input that has been formatted into an image channel. In this case, the image format is being created in order for the CNN to convolve over adjacent pixels containing spatially related data. For this application, turning off individual channels using IO is a logical way to isolate the influence of each input on the predicted output.

4 RESULTS

The goal of the IO approach is to rank the inputs in terms of usefulness, and determine if any inputs can be removed entirely to improve the CNN's performance. The results of the IO approach are shown in Figure 2. Each column of subplots represents a different GMP being predicted, with an increased amount of training data in each row. Figure 2b in the second column shows GMP 3 being predicted using GMP 2, while Figure 2e shows GMP 3 being predicted using GMPs 1 and 2. Within each subplot, four boxplots are shown, where each represents an input that is being omitted – radial tunnel displacement (DISP), ground freezing (FREEZE), geotechnical zone class (GEO), and primary support class (SUPCL). For example, in Figure 2a, the first boxplot from the left shows the results for omitting DISP from the Global Balanced model when GMP 2 is being predicted using GMP 1.

The performance of the Global Balanced models is sensitive to the IO approach, i.e. omitting an input does not have a significant impact on performance. When GEO is omitted the models perform worse and have the largest variance as compared to omitting the other inputs, indicating that GEO contains information that is required for overall performance and precision. With a few exceptions (Figure 2e, h, j), removing DISP generally has the lowest impact on the performance, likely because this input is not as strongly correlated to all tunnel yield classes.

5 DISCUSSION

The goal of the IO approach is to determine whether the CNN is unaffected by the absence of one of the inputs, in which case that input could be removed from the input candidate pool. A smaller input pool results in a simpler model that achieves convergence more easily, and also translates to less input data that needs to be collected. The AIC_c for each IO scenario indicates the relative importance of each input in the training dataset for overall model performance, as identified by the change in AIC_c. This information can determine whether any of the data is redundant, thus streamlining the data acquisition process and potentially resulting in cost savings. The IO study showed that omitting GEO from the Global Balanced models had the highest impact on the model performance, indicating the CNN is most sensitive to this input. This indicates that GEO has crucial information the CNN must learn in order to make a prediction of tunnel liner yield. It is intuitive in the context of rock engineering that the geotechnical domain has the highest sensitivity in the model for a squeezing ground environment, as it is the poor geology that is the main driver for the failure mechanism. After GEO, the next most sensitive input was DISP, which is the finest resolution input temporally. DISP is the only input that measures the tunnel liner deformation directly and has a relatively low level of subjectivity associated with it.

The IO approach reveals that of the four categories of inputs used to develop the Cigar Lake Mine CNN, none of them can be omitted entirely in order to achieve better performance. The results also indicate that the special emphasis should be placed on collecting the

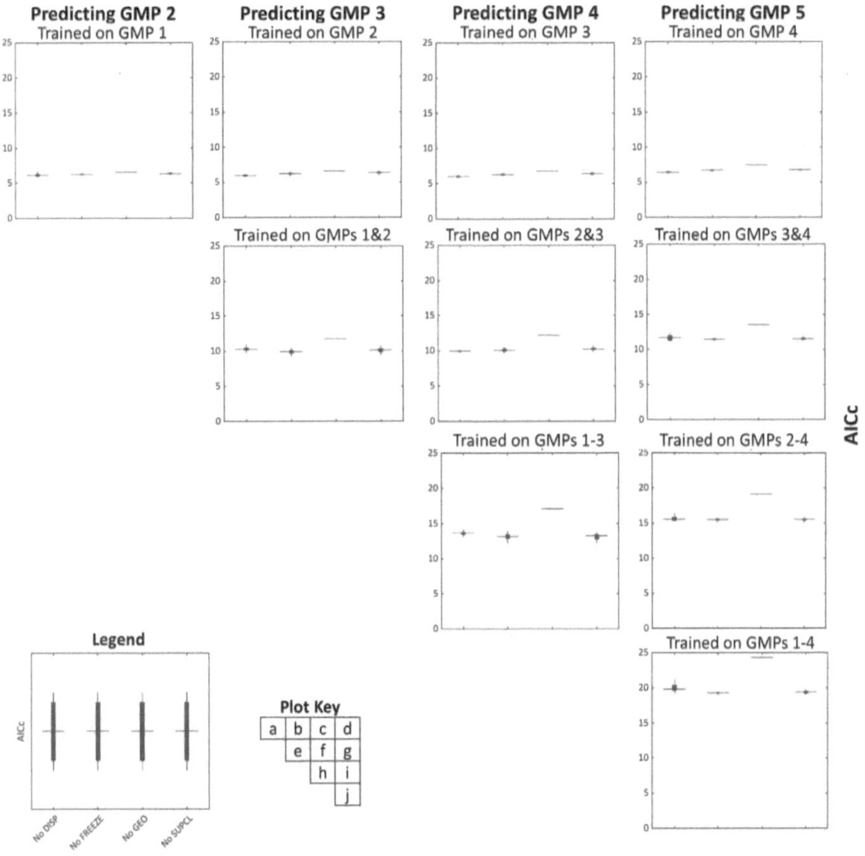

Figure 2. Results of the Input Omission (IO) Input Variable Selection (IVS) approach for Cigar Lake Mine. The model performance, AIC_c, is shown for the Global Balanced models for each permutation of training and testing Ground Management Plans (GMPs). AIC_c is a metric that allows comparison between models with different numbers of hyperparameters, where a lower AIC_c indicates better model performance.

geotechnical zones in detail and the radial tunnel displacements with high temporal regularity. For example, the geotechnical zone could be collected in a finer spatial resolution and the radial tunnel deformation could be collected in a finer temporal resolution. Alternatively, the displacement data could be interpolated temporally in order to provide more data in the spatial realm. The drawback of this is that there is a risk of injecting bias by using interpolated instead of raw data.

This study focused on one model-based IVS method, but the application of model-free methods may offer further insights into how the correlation between each input and the output may be used to inform the CNN architecture and hyperparameters. Findings from further IVS studies may result in more efficient CNN development for similar datasets. Conducting these analyses in advance of developing the algorithm could reduce the amount of trial and error needed to develop an appropriate architecture.

6 CONCLUSIONS

It is important to continue to research the internal mechanics of machine learning algorithms applied to rock engineering, including their architecture, hyperparameters, and the underlying assumptions. One way of doing this is by implementing an IVS approach that is suitable for the dataset available and the chosen MLA type. IVS allows for critical examination of the data that is being used for forecasting, and its relevance to the desired prediction. In this study, IO was chosen as an appropriate method of measuring the effects of removing each of the four inputs, and thereby determining whether there was redundancy in the CNN.

The IO results for the Global Balanced models showed that removing GEO entirely caused the model performance to worsen as compared to the other inputs, indicating that GEO has important information that the other inputs are not able to convey to the CNN. The IO study found none of the inputs could be removed completely – i.e. all available data in the temporal realm has to be included to produce the best possible model performance, for both the Targeted Class 2/3 and the Global Balance model architectures.

REFERENCES

Brady, B. H. G., & Brown, E. T. (2006). *Rock Mechanics for underground mining* (3rd Editio). https://doi.org/10.1007/978-1-4020-2116-9

Ferentinou, M., & Fakir, M. (2018). Integrating Rock Engineering Systems device and Artificial Neural Networks to predict stability conditions in an open pit. *Eng Geo*, *246*, 293–309. https://doi.org/10.1016/j.enggeo.2018.10.010

Golder Associates. (2001). *Numerical Analysis to Estimate Stresses in Crosscut Linings at Cigar Lake Mine*. Calgary, Canada.

Hurvich, C., & Tsai, C. L. (1989). Regression and time series model selection in small samples. *Biometrika*, *76*(2), 297–307. https://doi.org/10.1093/biomet/76.2.297

Kumar, M., Samui, P., & Naithani, A. K. (2013). Determination of uniaxial compressive strength and modulus of elasticity of travertine using machine learning techniques. *Int J Adv Soft Comput Appl*, *5*(3).

LeCun, Y., Boser, B., Denker, J. S., Henderson, D., Howard, R. E., Hubbard, W., & Jackel, L. D. (1989). Backpropagation applied to handwritten zip code recognition. *Neural Computation*, Vol. 1, pp. 541–551.

Marsland, S. (2014). Machine learning: An algorithmic perspective. In *Machine Learning: An Algorithmic Perspective, Second Edition*. https://doi.org/10.1201/b17476

May, R., Dandy, G., & Maier, H. (2011). Review of Input Variable Selection Methods for Artificial Neural Networks. In K. Suzuki (Ed.), *Artificial Neural Networks - Methodological Advances and Biomedical Applications*. https://doi.org/10.5772/16004

Morgenroth, J., Khan, U. T., & Perras, M. A. (2019). An overview of opportunities for machine learning methods in rock mechanics. *Geosci J*, *9*(12), 504–524.

Morgenroth, J., Perras, M. A., & Khan, U. T. (2020). Convolutional Neural Networks for predicting tunnel support and liner performance: Cigar Lake Mine case study. *Proceedings of the 54th US Rock Mechanics/Geomechanics Symposium*. Golden, CO.

Morgenroth, J., Perras, M. A., & Khan, U. T. (2021). A Convolutional Neural Network approach for predicting tunnel liner yield at Cigar Lake Mine. *Rock Mech Rock Eng*, (ARMA 2020 Special Edition), RMRE-D–20–01226.

Paudel, B., Jafarpour, M., & Brummer, R. (2012). *Cigar Lake Mine MDS Tunnel Liner Loading Analysis at Cameco*. Sudbury, Canada: ITASCA International Inc.

Setiono, R., & Liu, H. (1997). Neural-network feature selector. *IEEE Transactions on Neural Networks*, *8* (3), 654–662. https://doi.org/10.1109/72.572104

Snieder, E., Shakir, R., & Khan, U. T. (2019). A Comprehensive Comparison of Four Input Variable Selection Methods for Artificial Neural Network Flow Forecasting Models. *J Hydrol*, 124299. https://doi.org/10.1016/j.jhydrol.2019.124299

Song, Z., Jiang, A., & Jiang, Z. (2015). Back Analysis of Geomechanical Parameters Using Hybrid Algorithm Based on Difference Evolution and Extreme Learning Machine. *Math Probl Eng*, *2015*. https://doi.org/10.1155/2015/821534

Session 6 - Surface excavations

The Evolution of Geotech - 25 Years of Innovation – Hammah et al. (Eds)

Evaluation of theoretical liquefaction of compacted tailings piles based on the state parameter approach: A case study

E. Muguet
Geotechnical Civil Engineer, Statum Geotecnia, Brazil

R. Silva
Water Resources Civil Engineer, Statum Geotecnia, Brazil

ABSTRACT: The goal of this article is to present a Stack Break study using Rocscience Slide 3 software to identify the rupture mechanisms of a Waste Mining and Filtered Tailings combined stack. A methodology will be presented for the evaluation of the undrained strength of tailings based on the uniqueness of the *critical state line* concept developed by Jefferies and Been (2016), where the parameters of the constituent tailings model was estimated through the analysis of laboratory tests results performed in samples of trial embankments executed prior to the construction of the pile. The state parameter is estimated using as a reference the compaction control of the trial embankment and correlations between theoretical stresses, determined from stress-strain models and laboratory tests. The study results in the determination of the flood induced by the hypothetical stack break of the resulting surface wedge using RiverFlow 2D finite element software. As a result, a simplified approach is proposed to estimate the undrained shear strength of waste mining and tailings piles during the design phase, when liquefaction susceptibility information based on geotechnical field tests is not available.

1 INTRODUCTION

Liquefaction is defined as the phenomenon in which a soil mass loses most of its shear resistance when subjected to monotonic loading, cyclical or dynamic, and flows in a manner similar to a fluid (Morris, 1983).

The typical methods for quantifying liquefied soil strength are based on geotechnical field investigations (Olson, 2001; Shuttle and Cunning, 2007; Robertson, 2016). For the application of these methods, CPTu tests or percussive boreholes with SPT measurement must be conducted on materials already deposited in the field, which creates a problem when developing engineering project designs that have yet to be realized, and therefore lack the existing field data necessary for study development.

Because of this, a methodology based on theoretical computational models is recommended for the evaluation of undrained strength of liquefied tailings. The methodology is applied as part of a theoretical failure by liquefaction of a tailings pile using the *Rocscience Slide 3* software which will provide a global rupture surface. This surface will serve as an input for the Stack break model and the impact assessment, using the RiverFlow2D software.

2 THEORETICAL LIQUEFACTION EVALUATION

The shear behavior of non-plastic granular soils is controlled by the *state parameter* (ψ).

DOI: 10.1201/9781003188339-32

This parameter is defined as the difference between the in-situ void ratio *(e)* of the soil and the critical void ratio *(e_c)* under the same average stress applied. Negative ψ values indicate dilative soil behavior, or negative pore pressure generation under undrained shear conditions, while positive values of ψ suggest contractive behavior, or positive pore pressure generation under undrained shear conditions.

Jefferies and Been (2016) and Shuttle & Cunning (2007) suggest that when the soil has ψ greater than *-0.05*, softening and loss of resistance during shear is expected and, therefore, the occurrence of liquefaction is likely. However, for the formation of liquefaction processes, the soil must also be saturated and have post-peak strain softening behavior with a high brittleness index (Robertson, 2017).

Among the available methods for estimating the liquefied shear strength, Jefferies and Been (2016) report that this can be determined through the direct application of *critical state line* (CSL) uniqueness concept and its value depends solely on the state parameter and CSL constants. Based on these premises, the authors develop the following analytical equation:

$$\frac{S_{u,liq}}{\sigma'_{v0}} = \frac{1+2K_0}{3}\frac{M}{2}\exp\left(\frac{-\varphi}{\lambda_e}\right) \tag{1}$$

Where K_0 is the lateral earth pressure coefficient at rest, M is the critical friction ratio (q/p_0 ratio in large deformations) and λ_e is the CSL slope at the Napierian base.

The challenge in applying the analytical method is in determining the field void ratio. Next, a simplified procedure will be described, with analysis based on data from a trial embankment constructed during the design phase of a filtered tailings pile in Brazil. The proposed calculation procedure is an adaptation of the liquefaction evaluation procedure suggested by Poulos et. al (1985).

Step 1: *Estimate the initial void ratio of the filtered tailings based on compaction data from trial embankments.*

According to classic Soil Mechanics weight-volume relationships, the *void ratio* (e_i) is determined from the dry unit weight of the soil (γ_d), according to equation 2.

$$\gamma_d = \frac{G_s \gamma_w}{1+e_i} \tag{2}$$

Where G_s is the specific gravity of soil solids.

Using the *minimum relative compaction* (R_c, min), guaranteed by quality control during construction, the void ratio before material consolidation can be determined by rewriting equation 2.

$$e_i = \left(\frac{G_s \gamma_w}{R_{c,\min} \gamma_{d,\max}} - 1\right) \tag{3}$$

Where γd, max is the maximum proctor-normal specific weight of the compacted layer.

Step 2: *Calculate the field voids ratio from step 1 and by theoretical volumetric strains of tailings determined in FEM analyses.*

The tailings undergo shear stresses and volume variation due to the stresses imposed as the tailings pile is deposited. If crushing of soil grains does not occur, the volumetric strains of the tailings mass are related to variations in the in-situ void ratio as follows:

$$d\in_v = -\frac{de}{1+e_i} \tag{4}$$

Step 3: *Determine the critical voids ratio in the computational model, based on the CSL of the tailings obtained from test results of trial embankment samples and from the field stresses calculated by stress-strain analysis.*

From the stress-strain models, the volumetric strain ε_v of the tailings is determined, and the *in-situ* void ratio (e_f) is estimated using equations 3 and 4. The critical void ratio e_c is calculated by the CSL equation:

$$e_c = \Gamma + \lambda_e \cdot \ln(p'_c) \tag{5}$$

Where p'_c is the mean critical state stress.

Step 4: *Assess the liquefaction potential of the pile by state parameter approach, calculated from the results of steps 2 and 3.*

Step 5: *Perform limit equilibrium stability analyses applying the liquefied strength ratio calculated using equation 1.*

3 STRESS-STRAIN ANALYSIS

In the authors' experience, for realistic estimates of stresses and volumetric strains in the field, computational models must employ:

- Non-linear soil constitutive models calibrated using laboratory data;
- Stage modeling compatible with order of construction, in which initial stresses and strains of each stage are determined from previous stage.

3.1 *FEM model and filtered tailings constitutive model*

As a case study, the result of a stress-strain analysis performed using the nonlinear hyperbolic model (Duncan and Chang, 1970) will be demonstrated. Although there are more sophisticated and representative models of silty soils (i.e. NorSand, PM4Silt), the hyperbolic model is popular in the technical field in Brazil due to its simplicity and the solid database built from the country's dams and tailings projects.

One of the limitations of the hyperbolic model is that negative volumetric (dilative) deformations are not captured, which implies incorrect results of displacements, stresses, and strains in shallow layers or in very compact soil layers.

The adjustment of the theoretical model was conducted based on CID and CIU triaxial tests, performed on undisturbed samples collected from a trial embankment. Table 1 shows a summary of the model parameters and Figure 1 shows comparisons between the laboratory results and the model response at 2 levels of confining stresses. Figure 2 shows the FEM model developed for the filtered tailings pile under study.

Table 1. Duncan-Chang models parameters adopted in stress-strain analysis.

Parameter	Value	Parameter	Value
Modulus Number	472	Bulk Modulus number	200
Modulus exp (n)	0.49	Bulk Modulus exp (m)	0.50
Unloading Modulus number	600	Friction angle (°)	31.5
Failure Ratio (Rf)	0.80		

The Duncan-Chang model well-represents samples of loose tailings and contractile behavior. This limitation results in exaggerated volumetric strains in those elements with dilating

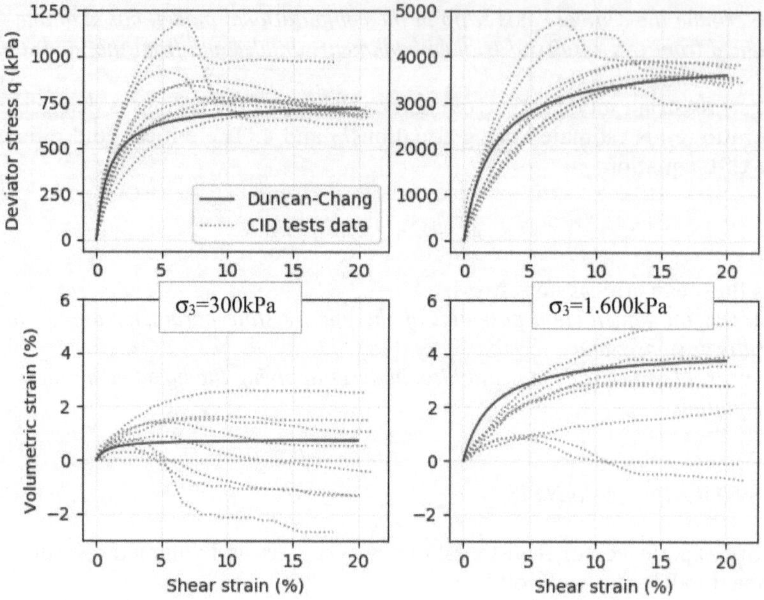

Figure 1. Comparison between data from CID triaxial tests and calibrated Duncan-Chang model.

Figure 2. Finite element model - Mesh and geometry.

behavior or those that are deformed with a constant volume, leading to estimated field void ratios that are smaller than the actual ones.

3.2 *Trial fill compaction data*

For the pile under study, different construction methods were tested to achieve compaction control within a test area. As an example, the initial field void ratios estimated using 546 proctornormal test samples of material from the trial embankment will be presented.

Figure 3 shows the statistical distribution of void ratios calculated with the application of equation 3 for each value of maximum dry unit weight of the sample.

There is a broad range of data, resulting from the heterogeneity of the tailings. For the application of the method, a deterministic void ratio value was adopted corresponding to a 5% percentile, which means that 95% of the calculated void ratios are less than the determined value.

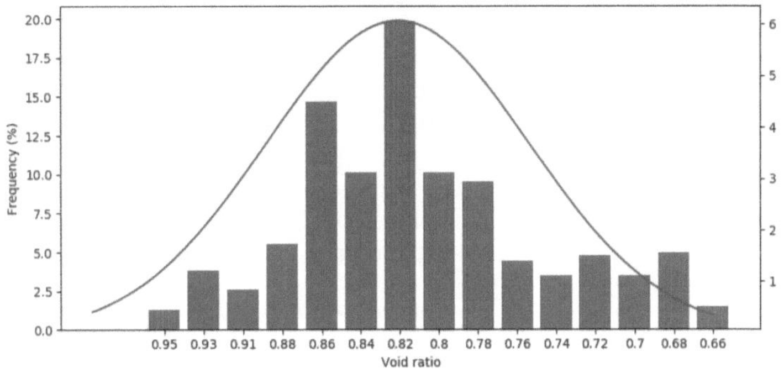

Figure 3. Initial void ratio determined from trial embankment laboratory tests.

3.3 *Results*

Figure 4 shows the state parameter values calculated in the numerical analysis considering the relative compaction R_c= 95% (with corresponding e_i = 0.76).

As can be seen, the values of ψ vary between 0.00 (at the bottom of the pile) and -0.025 (at the top) which indicates that the tailings have low liquefaction potential. However, the values are slightly higher than the limiting value of ψ_{lim} = -0.05 suggested by Jefferies & Been (2016) and Shuttle & Cunning (2007).

Table 2 shows the results of the analyses for a range of relative compaction between 85% and 98%.

Table 2. Analysis results to a range of R_c (%).

Rc (%)	ei	ψ	Liquefaction susceptibility	$S_{u,liq}/\sigma'v0$ (from eq. 1)
85	0.91	+0.125	Very high	0.004
90	0.86	+0.075	High	0.029
92	0.82	+0.050	High	0.073
95	0.76	+0.000	Moderate	0.456
98	0.71	-0.050	Low	-

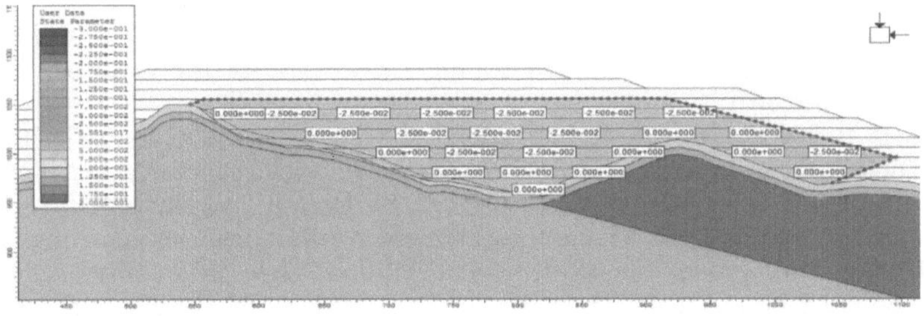

Figure 4. State parameters for Rc = 95%.

4 3D STABILITY ANALYSIS

Numerical modeling was conducted using Rocscience Slide 3 software, which performs tridimensional limit equilibrium stability analysis with the method of columns. This approach is an improvement on the traditional 2D limit equilibrium analysis, which considers the balance of forces and moments of soil masses dividing the soil masses into slices.

The critical surface presents an estimated mobilized volume of 4.717 Mm^3 and a safety factor equal to 0.67 and is shown in Figure 5. The low safety factor presented on the failure surface is associated with the liquefaction of the entire portion of tailings disposed in this region of the pile. This consideration, while conservative, is that which provides the greatest potential risk and for this reason it was adopted for the purposes of Stack Break studies. In practice, rock drains are constructed to prevent development of a phreatic surface within stockpile tailings.

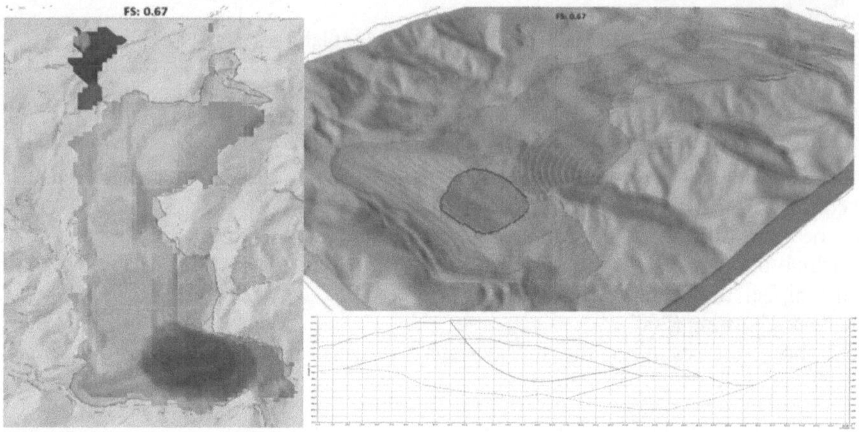

Figure 5. Safety map derived from analysis considering tailings liquefaction and potential rupture surfaces.

5 STACK BREAK ANALYSIS

For structures composed of tailings, it is important to consider the influence of solids on hypothetical rupture events, since their concentration is inversely proportional to propensity to flow. According to ARMANINI (2013, 11p), the hyper-concentrated flow caused by the rupture of a tailings disposal structure or a natural slide is formed by 2 different fluids, the interstitial fluid, which follows the principles of fluid mechanics, and the granular flow, represented by the solid phase of the material, which follows specific rheological relationships.

The Stack Break study was developed using the RiverFlow2D model, a Finite-Volume Numerical Solution and flexible mesh model, with the simulation inputs being the pile surface, the calculated rupture surface, the Digital Terrain Model with spatial resolution of 5 x 5 m and the Manning coefficient of the terrain as the roughness parameter. In order to cause the geotechnical failure, high shear displacements and deformations must be triggered, therefore the resistance to movement and propagation of the mass in the ground is controlled by the residual friction angle of the material and its susceptibility to the generation of excess pore pressures. For the analysis of the tailings flow, the following parameters were adopted:

- Constant volume friction angle = 27°;
- Saturated unit weight = 20kN /m^3.

The mass flow model used was "*Turbulent, Coulomb and Yield*", which considers the turbulent stress (τ_t), coulomb-type frictional stress (τ_f) and yield stress (τ_f) in the definition of fluid propagation, according to equation 6. τ_b is the bed stress calculated within the finite-volume mesh.

$$r_b = r_t + \min(r_y, r_f) \tag{6}$$

The results are presented in the Figure 6. The maximum extent of the resulting flood zone was equal to 2.50 km, and the maximum depth was equal to 29.00 m.

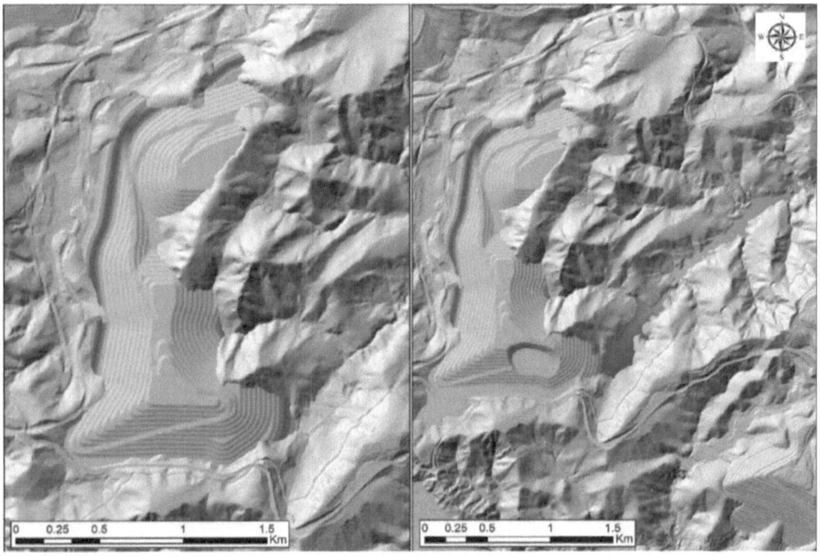

Figure 6. Stack break flood zone.

6 CONCLUSIONS

This article has presented a framework for liquefaction evaluation as applied to dry-stack tailings stockpiles. The presented methodology is particularly useful in feasibility and preliminary design, when data from geotechnical field investigations is not available, and allows evaluation of the influence of the following aspects on the overall safety of structures analysis:

- Geomechanical behavior of tailings (resistance and deformability);
- Foundation geometry, internal stack arrangement, and order of construction.
- Field void ratio and material compaction control parameters.

Furthermore, analysis of compaction data in this case study highlights the heterogeneous nature of mining tailings. Because of this, it is the authors' opinion that the probabilistic approach in selection of admissible relative compaction seems to be the most appropriate. In addition, the prediction of the field state parameter combined with three-dimensional stability analyses allows the engineer to identify more realistic failure mechanisms in stack-break studies.

REFERENCES

Armanini, Aronne. "Granular Flows Driven by Gravity". Journal of Hydraulic Research, 51: 2, 111–120. April 20th 2013, 11p.

Hydronia, 2020. RiverFlow2D, Two-Dimensional River Dynamics Model. Reference Manual. Pembroke Pines, Hydronia LLC, 2020, 296p.

Jefferies, M.; Been, K. (2016), Soil Liquefaction - A critical State Approach. MORRIS, DV (1983), A note on earthquake-induced liquefaction. Geotechnique.

Olson, S. M. (2001), Liquefaction analysis of level and sloping ground using field case histories and penetration resistance. PhD. Thesis, University of Illinois at Urbana- Champaign, Urbana, IL Poulos, S. J.; Castro, G.; France, J. W. (1980) Liquefaction Evaluation Procedure. Journal of Geotechnical Engineering, v. 9410, n. July.

Robertson, PK (2017) Evaluation of flow liquefaction: influence of high stresses. Proceedings of the 3rd International Conference on Performance Based Design (PBD-III).

Shuttle, DA; Cunning (2007), J. Liquefaction potential of silts from CPTu. Canadian Geotechnical Journal.

The Evolution of Geotech - 25 Years of Innovation – Hammah et al. (Eds)

Lessons learnt from 3D soil-structure modeling of a peanut-shaped cofferdam for cut & cover tunnel

A.K.L. Kwong
Adjunct Professor, The University of Hong Kong

ABSTRACT: The design and construction of infrastructures, whether it is in a recently reclaimed land or in a congested urban environment, are becoming more and more complex due to excavation of the deep grounds in soft and heterogeneous soils. To arriving a safe and economic design where the ground responses can be fully comprehended in a 3-dimensional configuration, 3D geotechnical modeling of soil-structures interaction has become popular and a tool that is indispensable.

The use of a peanut-shaped cofferdam for deep excavation of the cut-and-cover tunnel is gaining popularity. The cofferdam consisted of 3 numbers of truncated circular cells of 21 m long and 22 m wide for each of the cell connecting together. Each cell is formed by connecting perimeter D-wall panels in an arc shape which resists the lateral earth pressure by hoop forces. The hoop forces induced on the perimeter D-wall panels are then transferred at the junction of the cells to some heavy-duty D-walls designated as Y-Panels, which are transversely supported by some reinforced concrete struts and cross-walls.

A 3D finite element computer program, RS3, is used in this paper to model the construction sequences with generic soil properties adopted to illustrate the key lessons learned from excavation involving 3-dimensional geometry. The bending moment and deformation of the D-wall panels, Y-Panels, and struts are presented in this paper to illustrate the mechanism developed in the composite structure.

In this study, it is found that the soil spring constant deduced from the simulation is variable at different unloading stages and will approach a relatively constant value which is higher than the prescribed value usually adopted in structural analysis.

1 INTRODUCTION

The design and construction of infrastructures, whether it is in a recently reclaimed land or in a congested urban environment, are becoming more and more complex due to excavation of the deep grounds in soft and heterogeneous soils. To arriving a safe and economic design where the ground responses can be fully comprehended in a 3-dimensional configuration, 3D geotechnical modeling of soil-structures interaction has become popular and a tool that is indispensable.

The use of a peanut-shaped cofferdam for deep excavation of the cut-and-cover tunnel is gaining popularity. The principle of the peanut-shaped cofferdam structure is that it makes use of the development of arching effect in the connecting perimeter D-walls which are more efficient in resisting compression than in bending. In this numerical study, the cofferdam consisted of 3 numbers of truncated circular cells of 21 m long and 22 m wide for each of the cell connecting together. Each cell is formed by connecting perimeter D-wall panels (1.5 m thick) in an arc shape which resists the lateral earth pressure by hoop forces. The hoop forces induced on the perimeter D-wall panels are then transferred at the junction of the cells to

DOI: 10.1201/9781003188339-33

some heavy-duty D-walls designated as Y-Panels, which are transversely supported by some reinforced concrete struts (2 m wide by 2 m deep) and cross-walls (1.5 m thick). The cross-walls top level is 5 m to 8 m above the final excavation level (FEL) and is cut off to the FEL to make room for the permanent tunnel box construction. The perimeter D-walls are embedded into the stiff ground to ensure stability against base heave, piping, and uplift failures. Figure 1 shows the 3D view and the load path of the Peanut-shaped Cofferdam.

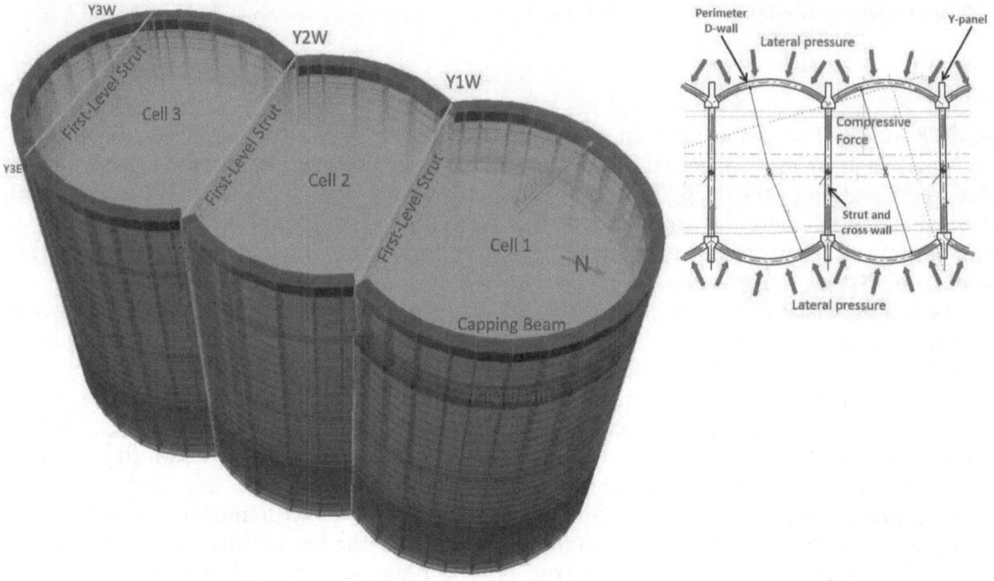

Figure 1. 3D view and load path of a Peanut-shaped Cofferdam.

During the excavation stage, the absence of heavy steel struts, king posts, and waler beams offer a drastic reduction in the construction program. It allows greater ease of machine plant movement and vertical lifting, leading to a much faster overall excavation cycle. The permanent structure can be cast in open space instead of confined by multiple strut layers that require the conventional re-propping sequence. It also reduces the amount of ground treatment below the excavation level due to toe stability.

A 3D finite element computer program, RS3, is used in this paper to model some typical construction sequences (Table 1) with generic soil properties (Table 2) adopted to illustrate the key lessons learned from excavation involving 3-dimensional geometry.

2 LOADING PATH

The modeling sequence closely follows the method and sequence to be adopted during the construction stages, and they are briefly summarized as follows:

1. The initial stage of the model consists of the forming of the capping beam, the 1st layer of struts at ground level, the perimeter D-walls, and the cross-walls.
2. Then followed by 1st stage – excavation to the bottom of the 2nd layer of struts.
3. 2nd stage – installation of the 2nd struts and also the ring beam at the 2nd strut level. Excavation to the bottom level of the 3rd layer of struts.

Table 1. Modeling sequences match with construction phases.

Stage	Modeling Activities	Stage	Modeling Activities	Stage	Modeling Activities
1	Assign elastic element to generate in-situ stresses	12	Install Strut in Fill and Ring Beam at -7mPD (Second-Level Strut)	23	Excavate to -27mPD
2	Change from elastic to plastic element	13	Excavate to -9mPD	24	Excavate to -29mPD
3	Install Y-Panel, D-wall, Capping Beam and Cross Wall	14	Excavate to -11mPD	25	Excavate to -31mPD
4	Install Top Strut at 5.5mPD (First-Level Strut)	15	Excavate to -13mPD	26	Excavate to -33mPD
5	Excavate to +3.5mPD	16	Excavate to -15mPD	27	Excavate to -35mPD
6	Excavate to +1.5mPD	17	Excavate to -17mPD	28	Excavate to -37mPD
7	Excavate to -0.5mPD	18	Install Strut in MD at _17mPD (Third-Level Strut)	29	Excavate to Final Excavation Level (-37.5mPD)
8	Excavate to -2.5mPD	19	Excavate to -19mPD	30	Install Strut in Alluvial Clay at -17mPD (Fourth-Level Strut)
9	Excavate to -4.5mPD	20	Excavate to -21mPD	31	Demolish Cross Wall
10	Excavate to -6.5mPD	21	Excavate to -23mPD		
11	Excavate to -7mPD	22	Excavate to -25mPD		

4. 3rd stage – installation of the 3rd layer of struts and excavation to the final excavation level (FEL), but without the demolition of the cross-walls (keeping the cross-walls at 8 m above the FEL at the Y-Panels location.
5. 4th stage – installation of the bottom struts at FEL, adjoining the cross-walls supporting the Y-Panels. Demolition of all the cross-walls down to the FEL.

Results of the analysis are presented in Figures 2 to 6.

Figure 2 shows the sectional view of the structural arrangement of the Y-Panel with the struts and the cross-walls. Very high bending moment and the high compressive force of about 117,000 kN-m and 160,000 kN are experienced by the Y-Panel and the struts respectively.

Figure 3 shows the deformation pattern of the Y-Panel and the D-wall. It is noted that with the very high bending moment in the Y-Panel, the lateral deformation of the Y-Panel and the Dwall is moderate of about 80 mm which is considered to be within the serviceability limit.

3 SOIL STRESS PATH

From the assessment of the deformation pattern and magnitude of the D-wall panel, it would appear that the structural system is very rigid. The element stresses at the excavation side are extracted from the model below the third strut level (-20 mPD) and the FEL (-37.5 mPD) for examination (Figure 4). As the excavation of the soil increases with depth, the reduction of the vertical stresses is not constant during the unloading because of the stress redistribution, and interaction with the deformation of the D-wall panel.

A detailed examination of the deformation pattern along the depth profile of the D-wall at the major axis of the 3-cells is presented in Figure 5, which shows that the upper part of the D-wall panel deflects away from the excavation side whereas the lower part deflects in the opposite direction in the form of a bi-axial mode.

Table 2. Soil and structure material properties.

Soil	Cohesion	Friction Angle	Youngs Modulus	Poisson`s Ratio	Unit Weight	
	kPa	Degrees	kPa		kN/m^3	
Fill	0	33	22500	0.25	20	
Marine Deposit	40	0	50400	0.25	16	
Alluvial Clay	60	0	20000	0.25	19	
Alluvial Sand	0	35	30000	0.25	19	
Completely Decomposed Granite	5	32	60000	0.25	19	
Structural Member	Thickness	Area	Youngs Modulus	Poisson`s Ratio	Moment of Inertia Max axis	Moment of inertia Min axis
	m	m^3	kPa			
Y-Panel (Cells 1 and 2)			3E+07	0.2	49.98	8.36
Y-Panel (Cell 3)			3E+07	0.2	47.31	7.377
Wall	1.5		2.9E+07	0.2		
Capping Beam	2		3E+07	0.2		
Ring Beam	2		2.9E+07	0.2		
First Level Strut	Width 2m Depth 2m	4	2.9E+07	0.2	1.333	1.333
Second Level Strut	Width 2m Depth 2m	4	2.9E+07	0.2	1.333	1.333
Third Level Strut	Width 2.5m Depth 2.5m	6.25	2.9E+07	0.2	3.255	3.255
Fourth Level Strut	Width 1m Depth 1.5m	3	2.9E+07	0.2	0.5625	0.25
Cross Wall	Width 1m Depth 1.5m	1.5	3E+07	0.2	0.28125	0.125

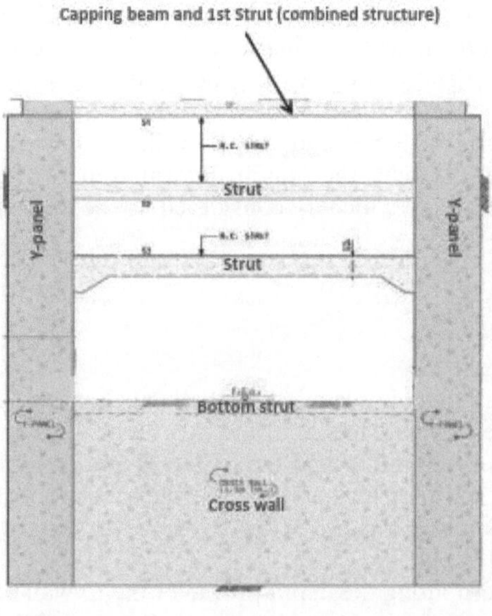

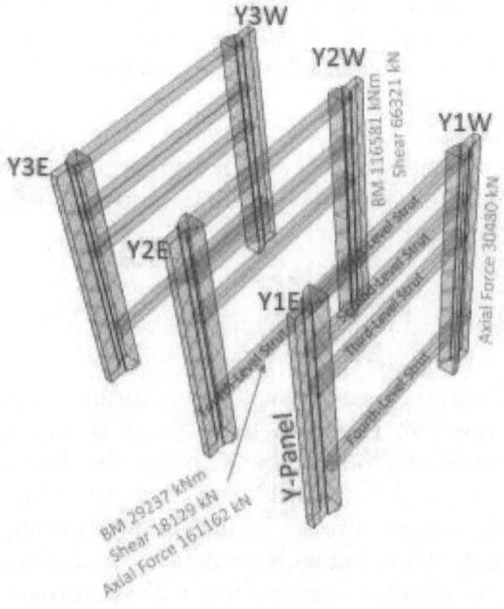

Figure 2. Sectional View of a Y-Panel.

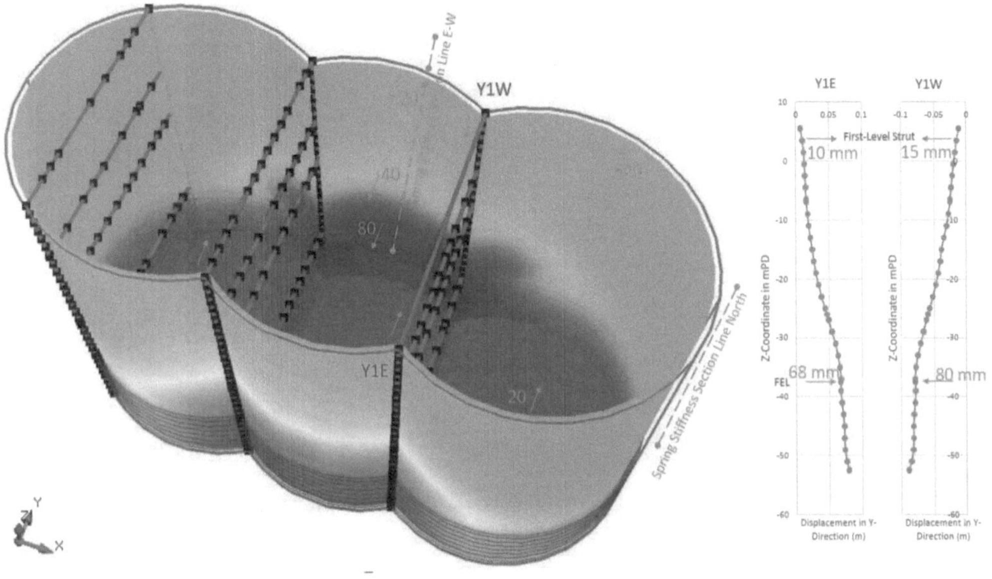

Figure 3. Deformation of the Y-Panel and the D-wall.

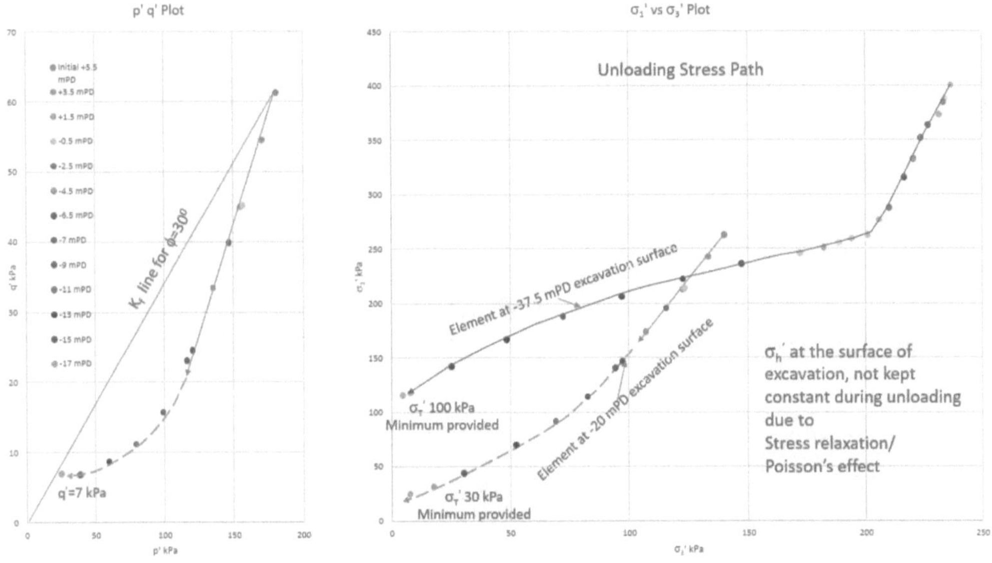

Figure 4. Unloading stress path.

In a typical structural analysis, a ground spring stiffness of the geotechnical materials was usually assumed from a handbook (e.g., Geoguide 1, 2020) for carrying out the detailed reinforcement design of the major structural system consisting of the perimeter D-wall panels, Y-Panels, struts, cross-walls, ring beams, capping beam and corbels, with each modeling sequence follows closely to the stages of construction. The perimeter D-walls, cross-walls, struts, and Y-panels were represented by a series of thin-shell elements. The ground medium surrounding the perimeter D-walls was represented by a series of area springs perpendicular to the shells. Figure 6 extracts the spring stiffness (horizontal force divided by horizontal deflection) along the major and minor axes of the 3-cells along the D-wall panels.

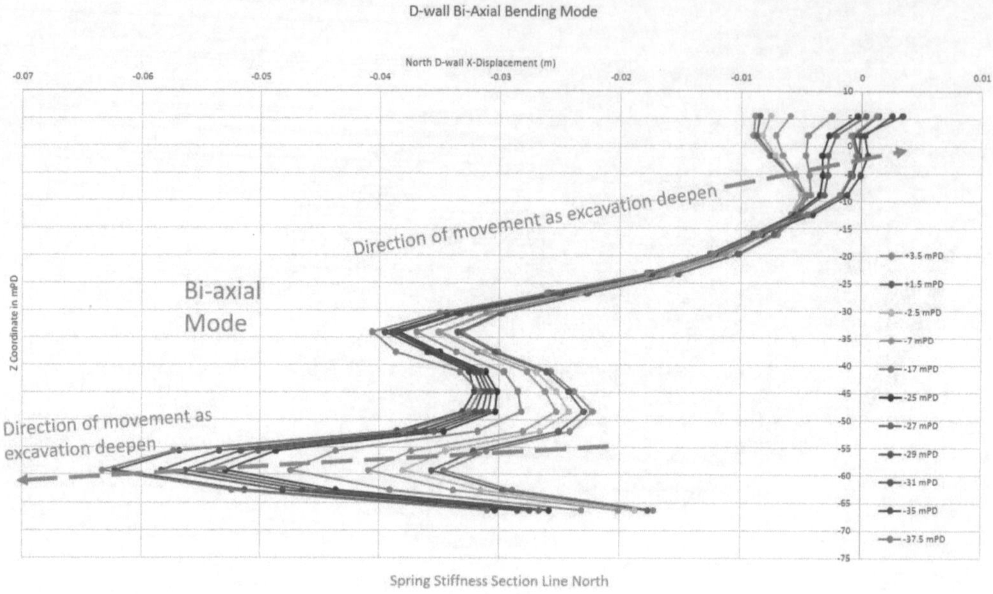

Figure 5. Bi-axial mode of deformation.

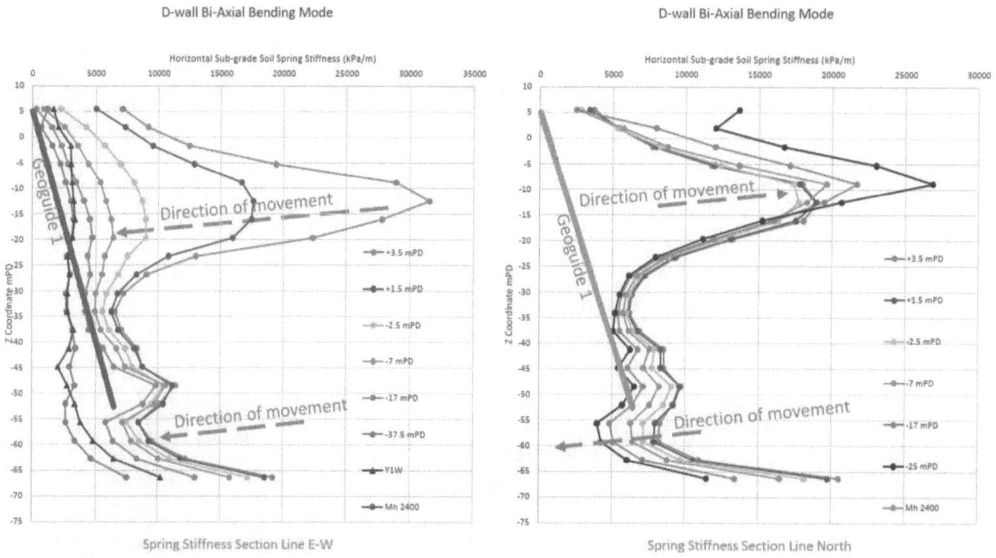

Figure 6. Distribution of the horizontal subgrade soil spring stiffness.

It can be seen that the soil spring stiffness is variable at different unloading stages and not linearly increases with depth.

4 CONCLUSIONS

A 3D finite element computer program, RS3, is used in this paper to model the construction sequences with generic soil properties adopted to illustrate the key lessons learned from excavation involving 3-dimensional geometry. It is found that the soil spring stiffness is variable at different unloading stages and not linearly increases with depth and is higher than the prescribed value usually adopted in structural analysis.

REFERENCES

Geoguide 1 (2020). Guide to retaining wall design. Geotechnical Engineering Office, CEDD, The Government of the Hong Kong SAR. 245 pages.

RS3. https://www.rocscience.com/software/rs3

The Evolution of Geotech - 25 Years of Innovation – Hammah et al. (Eds)

Field instrumentation-based performance verification of an improved soil site

A.A. Yunatcı & H.T. Bilge
GeoDestek Ltd. Şti., Ankara, Turkey

E. Çakır
Middle East Technical University Department of Civil Engineering, Ankara, Turkey

Y.U. Doğan
GeoDestek Ltd. Şti., Ankara, Turkey

B. Söylemez & K.Ö. Çetin
Middle East Technical University Department of Civil Engineering, Ankara, Turkey

ABSTRACT: This study presents a real case where a combination of site investigation and monitoring techniques were used for assessing the performance of a soil site. The site was reserved for construction of an industrial facility where a large paper production plant containing a range of settlement sensitive equipment was to be constructed. Weak, saturated alluvial deposits were improved using a carefully optimized pattern of rigid columns. Implementation of site characterization efforts was augmented with a limited scale dead load test, reaching the design loads for the most settlement sensitive unit within the facility. The field-testing program included remote settlement monitoring under the test loads, until a stage where the initial compression and consolidation settlements were evaluated as complete. The test results were used for calibrating the engineering parameters for modelling the behaviour under the design load, applied over the actual raft dimensions used in the project.

1 INTRODUCTION

1.1 *Description of study area*

The study site was reserved for constructing a privately owned industrial facility with a 2500 tons per day capacity paper production plant. The plant has a range of settlement sensitive equipment. The facility consists of numerous production units, a power plant, warehouse units, silos and auxiliary structures with static load magnitudes varying between 150 and 300 kPa. The anticipated raft dimensions were also highly variable, with foundation widths ranging from 18 m to 165 m. The length-to-width ratios for these structural elements were reported to rangefrom 1.25 to slightly over 10. Located in western Turkey at a near-fault site prone to high seismic activity levels, the design load magnitudes were assessed to be demanding in terms of static, live, machine and seismic load combinations. Results of an independent site-specific seismic hazard study for the project location proposed a T_r=475-year return period and peak ground acceleration values of 0.470 g and 0.416 g for representative V_{s30} values of 187 m/s and 760 m/s, respectively. A carefully optimized pattern of drilled concrete rigid columns, impact-type rammed aggregate piers and GeoConcrete columns were used for

DOI: 10.1201/9781003188339-34

improving the soil site to mitigate the effects of seismic soil liquefaction triggering, liquefaction-induced settlements, as well as differential settlements.

1.2 *Overview of geotechnical data and performance assessment*

A total of 21 boreholes were drilled, ensuring characterization of up to 60 m below surface. The uncorrected SPT blow counts, assuming a 45% hammer energy efficiency, were predominantly below 10 blows/30 cm over the upper 10 m. They were in a relatively more confined interval of 10 – 20 blows/30 cm along the 10-20 m interval, and barely reached an overall average of 20 blows/30 cm after that. The borehole investigations were supplemented with pressuremeter testing at a sparse distribution; to reveal the Menard pressuremeter modulus values of 5 – 7 MPa for the first 10 m from the ground surface. The preliminary stage of geotechnical investigations revealed that 20 cm of topsoil with very high organic content was followed by unengineered artificial fill of 2.5 m thickness, logged at 5 borehole locations in the south of the site. Deep alluvial soil deposits followed these layers.

Borehole investigation and laboratory results indicated that medium-stiff – stiff clays with low plasticity (CL) and occasionally sandy, low plasticity (mostly non-plastic) silts (ML) were present at the site. A representative average plasticity index of 20 was assigned for the idealized profile, due to near-uniform distributions over the PI=17-23 range. Additionally, gravelly, clayey, silty sand layers (SM, SW) were recorded at shallow depths. The site investigation campaign was supplemented by 8 CPT-u based field tests to correlate borehole study findings with continuous characterization along the alluvial basin. The results were concluded to be in conformity. A characteristic tip resistance of around 1 MPa and 1-1.5% friction ratios was recorded in weak cohesive layers.

A detailed geotechnical assessment quantified the safe net bearing capacity values of the shallow foundation dimensions assigned for each structure. Since the stress distribution underneath a single foundation was expected to be affected by other foundations in proximity, the analyses were performed by defining the foundations and loads within a single Rocscience Settle3 model. Elastic and consolidation settlements revealed that the anticipated settlements were expected to be in on the order of 67 – 105 cm, and the differential settlements in the range of 6 – 67 cm. These analyses were complemented with liquefaction triggering analyses for the sand – silt mixtures 20 m from surface. It revealed that the average factor of safety against liquefaction triggering for the idealized profile at the project site was far below 1.0; with values ranging from 0.2 to 0.8 (Çetin, 2018).

2 OVERVIEW ON GROUND IMPROVEMENT DESIGN AND IMPLEMENTATION

The project required an iterative approach. A suite of ground improvement methods was chosen to meet the target foundation performance while maintaining constructability and reaching a cost-effective solution. Impact-type rammed aggregate piers of approximately 50 cm diameter and 18 m length, GeoConcrete columns of 60 cm diameter and 18 m length and drilled concrete rigid columns of 80-100 cm diameter having 40 m length were used. The center-to-center spacing of the abovementioned improvement elements was carefully optimized for each subgroup of design loads. The ultimate design approach included improvement of the layers to meet the acceptable foundation performance in terms of reducing elastic and consolidation settlements within the tolerable limits. The calculated sublayer consolidation settlements of magnitudes 35 to 40 cm below the 40 m zone were considered to occur uniformly and create minimal interaction with the raft systems.

Following the completion of the ground improvement system, implementation of site characterization tools including seismic cone penetration testing (sCPT-u) and surface geophysical surveys, completed during pre- and post-improvement phases of the construction, allowed indirect assessment of improvement performance (Çetin et. al., 2019).

3 AREA LOADING TESTS FOR VALIDATING IMPROVEMENT PERFORMANCE

Based on the fact that the structural design phase required the validated performance and design basis subgrade reaction moduli, representing the improvement pattern before foundation casting, additional efforts were carried out to test a limited span of the improved site under full scale dead loading. The site has been reported to have been re-elevated via engineering fill material of up to 1 m, prior to foundation casting. Site observations revealed that the engineering fill was mostly granular in character. Before placement of the test load, the area was re-compacted using a vibrating roller and levelled using survey tools. The selected test area was a relatively quiet area of the construction site to avoid any interference with ongoing activities.

The test apparatus consisted of a 3000 mm x 3000 mm x 25 mm base steel plate to enable uniform distribution of test loads, and 18 concrete blocks for dead weights. It also comprised four steel assemblies to generate the fixed reference points for surface deformation settlements, four linear potentiometers for measuring displacements, and a data acquisition system with associated software.

A parametric study was carried out to determine the minimum distance required for correct placement of the surface deformation reference points. The results of the Settle3 model revealed that the reference points should be fixed at least 3 m away from the edges of the loading geometry to avoid interactions and misinterpretation of measured surface deformations.

A set of rigid steel platforms were designed with cross-sections and connection details to enable a cantilever frame resistant to wind effects and self weight. The setup was also designed to allow for zero contact conditions on the displacement sensors and the steel cantilever platform during or after the loading operation. These would randomly alter the references and cause erroneous recordings. The platforms were balanced by dead counterweights placed on the fixed ground points.

The sensors and the data acquisition system were set up immediately after the preparation of the reference conditions. The system was run overnight (3.12.2018) and the collected data was evaluated on 4.12.2018.

The test instrumentation consisted of four linear potentiometers (5k ohms) having variable measurement capacities of 100 mm and 150 mm. The sensors met the accuracy and precision requirements of the project by allowing detection of movements on the order of 0.01 mm. The sensors were carefully aligned and fixed at the steel frames extruded from the base steel plate. Environmental protection was achieved using a protective cover around the sensors and corrugated flexible tubing installation over the signal cables.

The acquisition system was composed of a conventional data logger, a PC located inside a housing unit at the site, AC power connection in conjunction with a DC battery backup, and a series of software. These software programs were setup to collect data and store it within a built-in memory module of the data logger, synchronize data to the PC located on-site, automatically push the collected data to a previously setup FTP account and manage it through a custom cloud software via internet connection established using both on site WiFi and a 3G GSM device. Thus, the collected data was accessible in processed form remotely; both by the project owner and the instrumentation specialists.

The loading program was executed on 04.12.2018, after fixing the reference readings for all data channels. A total of 18 concrete blocks were placed without imposing any eccentricity within a half working day period. The total weight of the dry concrete blocks was calculated to be 1718 kN (175,180 kgf) based on the individual weights reported by the project owner. The plan dimensions of the blocks were slightly less than 3000 mm x 3000 mm to avoid any clashes with the reference frame during placement. The height of each block was approximately 500 mm. Total maximum vertical stress imposed by the dead weight blocks (excluding the weight of the steel plate) was calculated as 190.9 kPa.

Loading stages were completed within 5 hours with irregular loading intervals. Sampling rate from all channels were selected as 3 minutes. The raw form of collected data during the 4.12.2018 – 25.12.2018 period included traces of exposure to days with high winds and intense rainfall. Close examination of data revealed that minor scatter in data diminished after the wind effects (possibly acting both on the frame and the 9 m high concrete block tower) reached a calm period. Further reduction or correction was not deemed necessary. Other factors influencing the collected data can be listed as effects of precipitation, which caused concrete blocks to be partly or fully saturated temporarily. Any adverse effect on the fixed frame footings (counterweight area) was not observed during periodic visual checks made after the rainy periods and prior to demobilization.

Figure 1 presents the settlement build-up as recorded by four sensors installed on the mid-span of the edges of the loaded area and a general view from the test site. Positive values indicate settlement whereas the negative values indicate heave, which indicates minor angular distortion of the plate under the design loads. The degree of rotation of the base was calculated as 0.08° along the north-south axis, and 0.04° along the east-west axis, respectively. These distortions were negligible. It was also observed that 90% of the settlements recorded until 25.12.2018, occurred within 41 hours from the time of first loading, and 36 hours from full application of the design basis load.

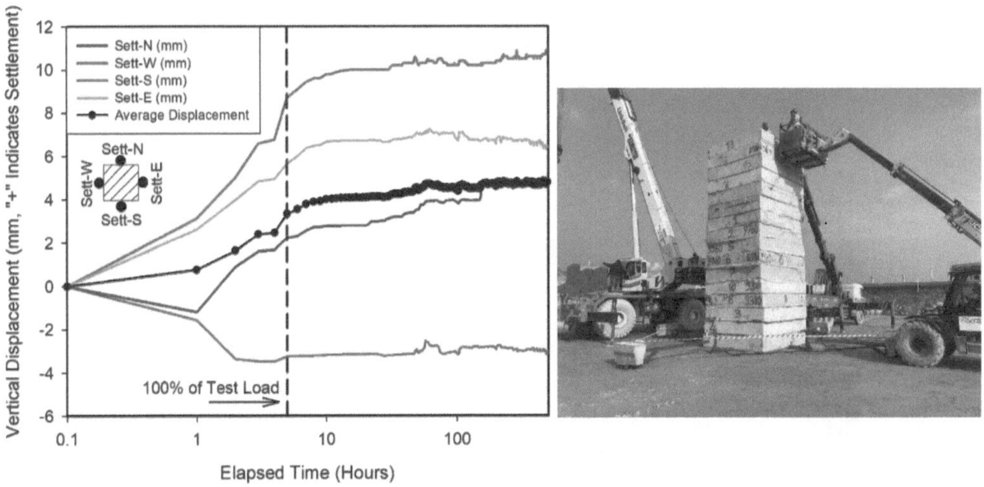

Figure 1. Cumulative settlement buildup and general view from the test setup of the downscaled area.

4 RESULTS OF MODEL VALIDATION USING FIELD DATA

To validate the effectiveness of ground improvement application, results retrieved through field test loading were evaluated. The assessment included quantification of subgrade reaction moduli representing improved conditions under static and live loads, as well as machine vibration induced effects. The small strain magnitudes expected to be generated by operating machinery for two different loading scenarios and seven sub sections of the facility were reported to be in the range of 2.30×10^{-7} to 3.79×10^{-9}.

Due to proportionality of the size of a generated stress increment zone and foundation size under a uniformly loaded area, a critical re-evaluation of scaled loading tests to avoid misinterpretation of results was deemed necessary. Composite deformation moduli of the improved soil site were back analyzed to ensure compatibility with field settlement results for the 3 m x 3 m platform subjected to gross 190 kPa vertical loading. The resultant calibrated parameter was used to assess the performance, settlement analysis and determination of subgrade reaction modulus for the following combination of i) concentric loads acting on 3 m x 3 m sized footings or smaller, ii) uniformly distributed loading over 4 m x 12 m sized foundation, iii) 64 m x 19 m sized main production line foundation and iv) 500 m x 500 m slab approximately representing the main facility.

Settle3 was used to simulate the field load test. The ground improvement pattern preferred at the region where the test load was applied, was deliberately chosen to include a 80 cm diameter, 40 m long vibrex pile centered on the loaded area, surrounded by 50 cm diameter, 18 m long rammed aggregate piers spaced at 2.5 m grid.

The back analyses confirmed that a composite deformation modulus of E=90 MPa was appropriate for maintaining compatibility, with elastic deformations measured to be in the order of 5.1 mm (Figure 2). The model was cropped at 30 m depth from the surface. Consolidation settlements below the improved zone were considered to be uniformly distributed.

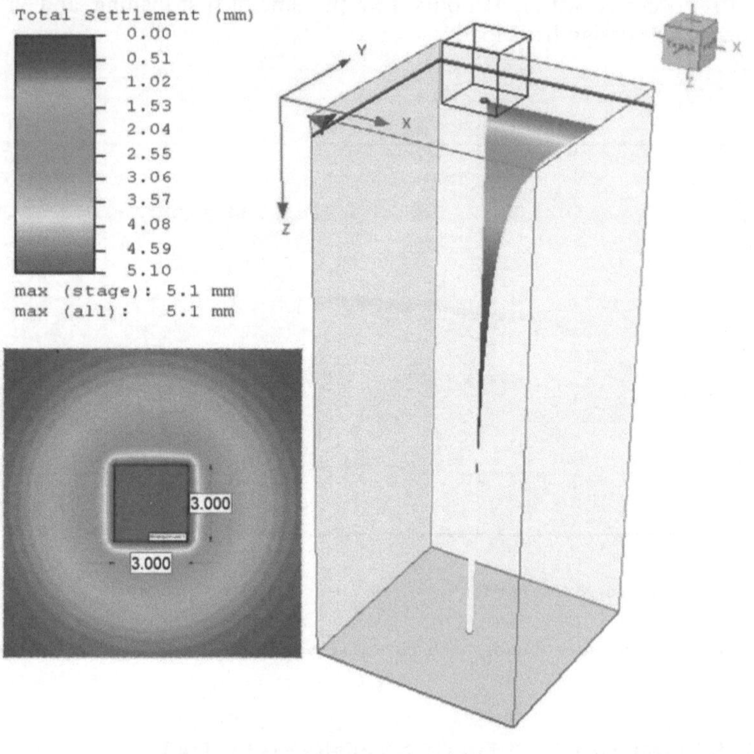

Figure 2. Settle3 model used for calibrating the stiffness properties of the improvement zone.

Table 1 presents the results of numerical analysis for the field calibrated model. Elastic settlements for 4 m x 12 m sized foundation, 64 m x 19 m sized main production line foundation and 500 m x 500 m slab are summarized in Table 1. Elastic and consolidation settlements below the improved zone were neglected. Modulus of subgrade reaction is simply defined as the ratio of applied foundation loading to calculated (or measured) settlements as expressed in Equation 1.

$$k_v = \frac{\text{Applied Foundation Loading(kPa)}}{\text{Amount of Settlement(mm)}} \tag{1}$$

The modulus of subgrade reaction values for the respective loading scenarios are presented in Table 1.

Being aware that modulus of subgrade reaction is strain dependent as well, an engineering approach was developed for the evaluation of subgrade reaction modulus of the same footings under machine vibration induced strains, reaching 10^{-5} to 10^{-6} % or less. The first set of subgrade reaction moduli were on the order of 10^{-1} to 10^{-2} %, which is typically valid for design considerations under static and live load combinations. A typical modulus degradation curve developed for predominantly cohesionless soils (Seed & Idriss, 1970) clearly states that at 10^{-1}% cyclic shear strain level, the modulus degradation reaches approximately 30% of its value characterized over the small strain range. This statement is also valid for a modulus degradation curve developed for cohesive soils (Vucetic & Dobry, 1991), for plasticity index defined as PI=20, parallel to laboratory findings from site investigation phase. Thus, the proposed moduli were suggested to be scaled with a factor of 3 for utilization in the structural design of rafts under cyclic machine loading effects (Figure 3).

Table 1. Summary of geotechnical assessment for design basis static foundation loading scenarios.

Raft Size	Settlement mm	k_v MN/m^3
3 m x 3 m	5	37
4 m x 12 m	11	17
19 m x 64 m	27	7
500 m x 500 m	45	4

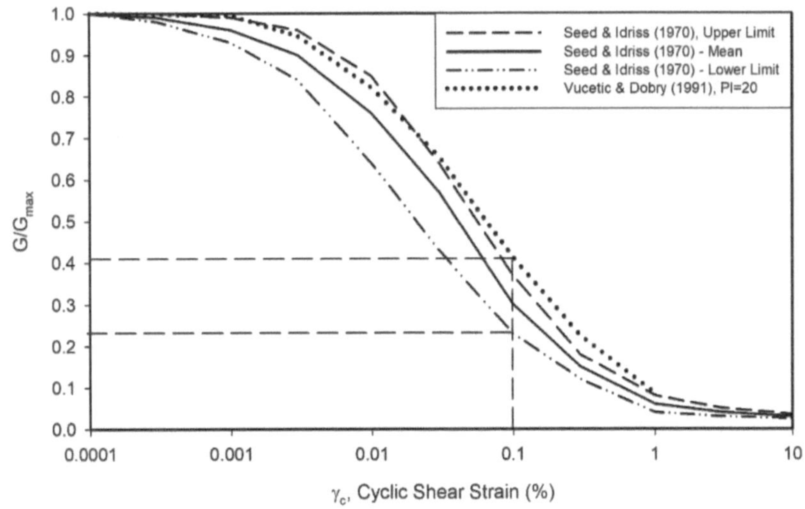

Figure 3. Typical shear modulus degradation of granular soils and clays with PI=20.

5 CONCLUSION

This case presented an overview of site-detailed characterization, interpretation, parameter estimation and instrumentation efforts to build an efficient numerical model for validating geotechnical design. This case highlights the seamless integration of reasoning during preliminary and final design phases, a judicious tailoring of quality control program, and calibration of field data for validating, combined with models to managing overall risks during the service life of the facility.

REFERENCES

Çetin, K.Ö. 2018. Ground Improvement Design Report for Aydın Province, Söke County, Söke Organized Industrial Zone Papermill Factory.

Çetin, K.Ö., Bal, E.K., Öner, L. & Arda, S. 2019. CPT-Based Assessment of Densification After Ground Improvement with Rigid Inclusions and Rammed Aggregate Piers® (RAP). *8th International Geotechnical Symposium*, 13-15 November 2019, İTÜ Süleyman Demirel Culture and Congress Center, İstanbul.

Seed, H.B. & Idriss, I.M. 1970. Soil moduli and damping factors for dynamic response analyses. *Report No.EERC 70-10, Earthquake Engineering Research Center*, University of California, Berkeley, California, 40p.

Vucetic, M. & Dobry, R. 1991. Effect of Soil Plasticity on Cyclic Response. *J. Geotech. Engrg.*, *117(1)*, 89–107.

Yunatcı, A.A., Doğan, Y.U. 2018. Field Load Testing Factual Report for Aydın Province, Söke County, Söke Organized Industrial Zone Papermill Factory.

Session 7 - 3D slope stability analysis

The Evolution of Geotech - 25 Years of Innovation – Hammah et al. (Eds)

Finite element analysis of a deep excavation: A case study ground response due to deep excavations in Sydney sandstone

P. Hewitt
Technical Executive, Geotechnics, WSP Australia Pty Ltd, Sydney NSW

M. Kitson
Principal Tunnel and Geotechnical Engineer, WSP Australia Pty Ltd, Sydney NSW

ABSTRACT: A deep basement was recently constructed in Hawkesbury Sandstone for a property development on Sydney's North Shore, in Australia, that is located adjacent to critical transport infrastructure. With a depth of over 43m, the excavation is among the deepest basement excavations in Australia. This paper presents the role of Finite Element Method (FEM) as a key design tool in the Observational Method (OM), to make accurate predictions and back-analysis, to overcome geotechnical design challenges, improve construction outcomes, and allow important decisions that may lead to cost savings through an innovative design and construction approach. Applications of smart technologies were use in the instrumentation and monitoring that provided accurate data that was incorporated into the finite element back-analysis.

When unusual displacement was observed adjacent to the transport corridor due to structural uncertainties and construction activities, the OM and verification process provided a flexible framework within which to reassess ongoing movements and effects on adjacent infrastructure to ensure construction could proceed safely. The case history demonstrates the benefit of adopting the OM for excavations to achieve savings in time and cost, and to react to unexpected movements during construction.

1 INTRODUCTION

Ground movements during deep basement excavations may potentially seriously impact adjacent infrastructure and utilities. We need to consider the displacements that deep excavations induce to assess the impact on those assets and the necessary mitigation measures. This paper describes the excavation for a property development on Sydney's North Shore. The project involved excavating over 43 m deep, and is among the deepest building basement excavations in the world. The excavation took place adjacent to critical road and rail transport infrastructure. The influence of excavation on existing transport infrastructure was an important design consideration. An added complexity was the potential impact that building excavation and imposed building loads could have on nearby utilities and underground structures.

2 PROJECT BACKGROUND

The project development (known as the "Eighty Eight") is currently under construction at 88 Christie St, St Leonards, NSW. Figure 1 shows the proposed development and an aerial perspective of the development.

DOI: 10.1201/9781003188339-35

The new development includes two residential towers (maximum 47 storeys high) and a commercial tower (15 storeys) over a large retail precinct with 10 levels of below-ground basement up to 43 m deep. The structure for the building is founded on pad footings. The 8,000 m2 basement excavation extends to the Sydney Trains boundary along the west of Lithgow Street with Civic Plaza on the ground level, and is located adjacent to major rail and road infrastructure as shown in Figure 3.

2.1 *Adjacent infrastructure*

The site boundary is surrounded by sensitive buried utilities, road and rail infrastructure, and nearby buildings for which ground movement was a key consideration. An initial scheme of the development was about 14 m from the earlier-planned CBD Rail Link. The earlier-planned rail transport project comprised twin tunnels connecting south of the city in the Redfern/Airport region to the north at St Leonards beneath the western edge of site and track quadruplication that added two more tracks, one on each side of the existing lines and within the existing protection corridor, as shown in the schematic diagram in Figure 2.

Figure 1. Aerial view perspective of the excavation.

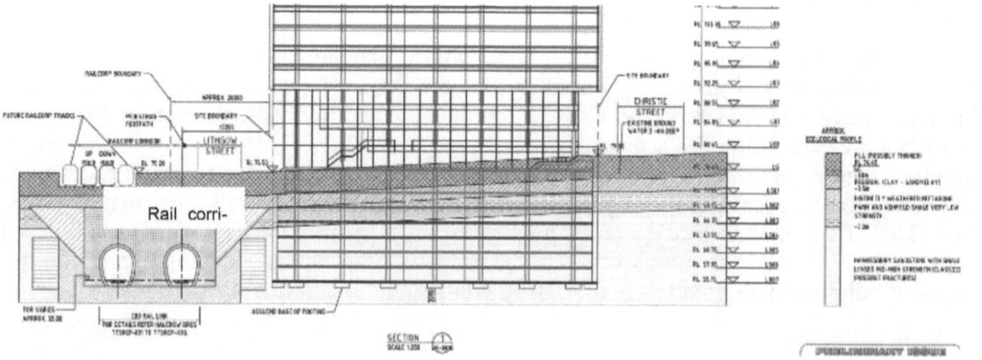

Figure 2. Schematic diagram with initial development scheme and earlier-planned CBD rail link.

The earlier-planned CBD rail tunnel project was abandoned and a different route selected for CBD rail link which is now the Sydney Metro City & Southwest project (under construction).

Figure 3. Location plan showing adjacent transport infrastructure and aerial view of excavation progress.

3 EXCAVATION DESIGN

3.1 *Geology*

The site geology comprises the upper sedimentary formations of the Sydney Basin stratigraphic sequence, which consists of sub-horizontal beds of Triassic-aged rock comprising (youngest to oldest) Mittagong Formation and Hawkesbury Sandstone.

3.2 *Subsurface conditions*

We assessed ground conditions from geotechnical information, including 52 borehole locations, from investigations carried out by WSP and others within the proposed development and surrounding areas. Several rock strength index tests (unconfined compressive strength, UCS and point load) on recovered rock core samples were tested. In addition to the geotechnical borehole investigation, we used downhole geophysical surveys/borehole imaging to learn details of orientation, spacing, aperture, and infill characteristics of various rock mass defects including joints and bedding partings.

The development site is underlain by uncontrolled fill and residual soil to maximum 1.0m deep, followed by Mittagong Formation and Hawkesbury sandstone of varying weathering and strength. The Mittagong Formation is characterised by interlaminated sandstone and siltstone comprising fine-to-medium grained, light grey sandstone with dark grey siltstone bands that are (generally) extremely weathered and extremely low strength. The Hawkesbury Sandstone is characterised as medium-to-coarse grained, grey, with cross bedding and medium-to-high strength. The subsurface conditions encountered are summarised in Table 1.

The rock classification adopted the Sydney classification system (Pells et al, 2019). The Sydney classification system, which was developed for foundations, is based on UCS of saturated substance (i.e. intact sandstone or shale), defect spacing and percentage of seams within a defined vertical interval of the near-horizontal bedded rock. Both strength testing and borehole imaging identified a weaker shale/ laminite band at about 40m depth close to final excavation depth. The summary of rock strength data is shown in Figure 4.

Table 1. Subsurface condition.

Material/origin	Material description	Thickness (m)	Top of the unit, RL (m AHD)
Fill and residual soil	Clay and Sand	1.5 - 2.0	80.5 - 75.0
Mittagong Formation	Class IV Sandstone and Shale	4.5 - 7.5	78.5 - 73.0
Hawkesbury Sandstone	Class IV Sandstone	2	74.0 - 68.5
	Class II Sandstone	3.5	72.0 - 66.5
	Class I Sandstone	19.6 - 24.0	68.5 - 63.0
	Class II Interbedded Sand and Siltstone	1.5	44.5 - 42.8
	Class I Sandstone	Not penetrated	43.0 - 41.3

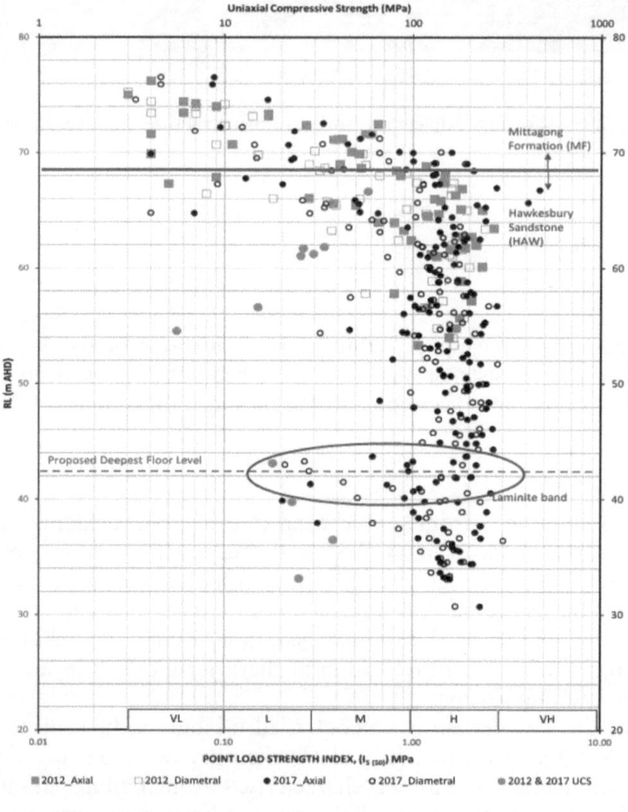

Figure 4. Rock strength data (WSP, 2018).

3.3 *Excavation restrictions*

Restrictions by the adjacent asset owners, TfNSW (ASA, 2015) and RMS (RMS, 2012) for the development included:

- Anchor systems cannot be used in the rail easement.
- Construction and operation of external developments shall not affect the stability and integrity of railway infrastructure through loading from the development and ground deformation.

- Maximum displacement of 30mm on the Pacific Highway.
- Monitoring of ground movements due to bulk excavation and monitoring of and track structures required.

The selection of retention systems for the excavation to satisfy these criteria and avoid both temporary and permanent anchors within the rail corridor and road along the project boundaries had the major influence on basement geometry and the choice of retention systems.

3.4 *Excavation support*

Most of the basement excavation face retains a sequence of weathered shale and sandstone. The temporary shoring system supporting the ten levels of basement along the Sydney Trains boundary (Figure 5) consisted of:

- A contiguous concrete pile wall in combination with an anchored shotcrete wall located within the middle of the site.
- Ground anchors within a 20m wide square buttress of rock on the south corner that was not to be excavated.
- Ground anchors for 20m length in the north corner adjacent to the Pacific Highway railway bridge. The north corner was not fully excavated which allowed angled ground anchors to be installed within the remaining triangle of rock.

Figure 5. Excavation support along Sydney Trains (west) boundary.

The middle section of wall along the railway side was about 40m along, that could not be anchored. We used two piled walls to avoid installing long anchors under the railway line, where the upper soldier piled wall was a permanent wall installed on the railway boundary, comprising 750mm diameter piles spaced at 2.5m founded in min. Class II/III Sandstone up to 4.5m height. The lower contiguous piled wall was temporary, comprising 600mm diameter piles spaced at 1.5m also founded in min. Class II/III Sandstone anchored with short rock dowels 4.5m long in every pile. This lower wall provided the passive toe support to the upper wall.

We adopted the following proposed excavation sequence along the Sydney Trains boundary:

- Install upper permanent piled wall on railway boundary to 4.5m depth.
- Install lower temporary piled wall 3m from Sydney Trains boundary.
- Excavation to commence. Install rock dowels in every pile and then shotcrete within piled wall.
- Geotechnical engineer/ geologist to inspect exposed rock face every 2.0m intervals of excavation depth and advise requirements for localised rock bolting and shotcrete.
- Excavate to a maximum of 500mm below the next level of rock dowels and install, then install strip drains and shotcrete. Repeat for each level of anchors until reaching high strength rock.
- Continue excavation in this manner until basement excavation level.
- Construct basement structure including all pad footings, column, walls and slabs back up to the loading dock slab LB01.
- Once loading dock slab has reached full strength continue construction.

The shoring system to the three remaining sides of the excavation comprised shotcrete concrete walls spanning to 600mm diameter soldier piles. The piles are spaced 2.5m apart with ground anchors providing temporary support. The ground control solution with a pattern of rock bolts and ground anchors was selected to support the excavation below 10m to 12m below the surface level.

4 IMPACT ASSESSMENT

The impact assessment involved geotechnical analysis that considered the effects of the proposed development including basement excavation and building loads, and the effects of key values and the correction of the natural stress field based on rock mass quality.

The numerical assessment to assess the rock mass responses, and installed support subject to the design variables. Initially, a numerical two-dimensional finite element model using 2D finite element numerical modelling program Phase 2 for the Development Application (DA) submission that provided initial estimates of the potential impacts on the existing infrastructure.

The assumptions and limitations of a two-dimensional model were too restrictive for the model to provide detailed estimates of ground movements and their impacts on the existing infrastructure. The numerical assessment using 2D and 3D finite element numerical modelling programs for the detailed design and to support ground movement estimates that consider the proposed excavation's 3D geometry and provide more realistic estimates of the impact of construction on the existing infrastructure. These numerical analyses included continuum (using RS2 and FLAC 3D) and discontinuum analyses (using RS2).

4.1 *Design parameters*

The geotechnical design parameters adopted for this impact assessment were selected based on the Sydney classification system, results of the geotechnical investigation, the intact parameters, the estimate of GSI for each rock mass class, published data on sandstone and shale strength and modulus, and the excavation depth of proposed development.

The adopted geotechnical design parameters are summarised in Table 2. To address sensitivity, we reduced the values of cohesion and tensile strength for Sandstone IV and Mittagong Formation to less than 50% of design values (see values in brackets).

Table 2. Adopted design parameters.

Material type	Unit weight (kN/m^3)	UCS (MPa)	Mass modulus (MPa)	GSI	Mohr-coulomb criterion Tensile Strength (kPa)	Friction Angle (deg)	Cohesion (kPa)
Sandstone I	24	30	3000	75	300	55	1000
Sandstone II	24	25	2000	65	100	50	500
Sandstone IV	24	10	500 (200)	45	25 (0)	45 (35)	250 (30)
Shale II	24	15	1000	50	60	40	250

4.2 *In-situ stress*

The field in-situ stresses have a significant impact on both deep excavation conditions and induced ground movements in the immediate area of the excavation works, due to high in-situ lateral stresses, which can be 'locked in' within the bedrock stratum.

We incorporated adjacent deep excavations, including "The Forum" building, located north of the development (Figure 10) within the modelling as part of the impact assessment to provide a holistic approach to the major and minor stress distribution within the subsurface geological units adjacent to the excavation.

We adopted the following in-situ stress relationship based on WSP's work on the reference design for the nearby Sydney Metro City & Southwest project:

- Upper bound:

$$\sigma_{H(NS)} = 1.0\text{MPa} + 3.5\sigma_v; \sigma_{H(NS)}/\sigma_{h(WE)} = 1.5$$

- Lower bound:

$$\sigma_v = \sigma_H = \sigma_h = 1.0$$

(stress field assumed to be lithostatic)

We used the upper-bound stresses applied to fresh, good quality sandstone and shale (Class I and II). In poorer quality rock masses, the horizontal stresses are expected to be less, and the lower bound stresses were applied. The minor horizontal stress was applied perpendicular to the excavation's eastern and western walls; the major horizontal stress is applied perpendicular to the excavation's northern and southern walls.

4.3 *Predicted ground performance*

The proposed retention system wall design addressed the following displacement mechanisms which have been observed to cause ground surface deformation adjacent to the excavation and could affect the railway.

- Lateral earth (soil) pressure acting on the shoring system causing it to deflect.
- Relaxation of the rock mass resulting from a reduction in lateral stress (stress relief).
- Anchor hole drilling and installation.

The proposed shoring system with soldier piles and anchors was designed to control the ground surface deformation due to lateral soil pressure in the upper parts of the proposed excavation (mechanism 1). We chose the layout and stiffness of the shoring system to minimise the ground movements and the impact on the railway tracks, and rail overbridge.

The relaxation of the rock mass due to stress relief (mechanism 2) from the deeper parts of the basement excavation will happen irrespectively of the shoring system type. The numerical assessment was calibrated against monitoring results from various deep excavations around Sydney, including monitoring results of the Embassy Residences (see Figure 3) basement excavation located on the western side of the rail corridor. Moving the boundary of the proposed development footprint and installing anchors just outside of the rail corridor will not significantly decrease ground movements of the rail corridor caused by rock stress relief, which are a result of an adjacent deep excavation.

Figure 6 shows the east-west cross sections of the development with the predicted ground movements from 2D numerical assessment due to the proposed excavation. As shown in Figure 6, the basement excavation below LB01 at RL69.5m AHD was extended to 3.1m inside the boundary to temporarily support the cantilever pile wall along the western boundary. This revised basement plan was adopted to address Sydney Trains' preference, such that no (temporary or permanent) anchors are constructed inside the rail property. The wall configuration was based on a similar cantilever post-tensioned pile wall which was successfully adopted for the Gore Hill Freeway widening at Artarmon at a location where project boundary constraints dictated that no ground anchors could be installed (Hewitt et al, 2008).

The adopted loadings for the 2D numerical assessment are more conservative than actual loading environments and the realistic loading reduces the impact of 3D effects. The predicted deformation from 3D numerical assessment indicates that the predicted deformations and stress concentrations obtained from the 2D assessment are likely to be conservative and reduces the displacement by up to 40% due to the 3D "buttressing effects". Figure 7 shows the site's north east quadrant with the predicted 3D ground movements within the rail corridor, Pacific Highway and rail overbridge.

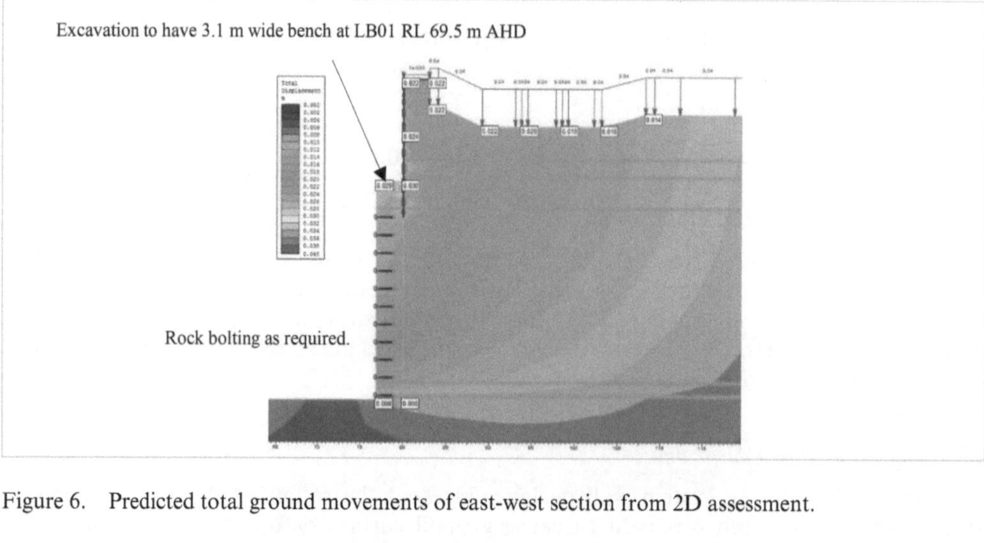

Figure 6. Predicted total ground movements of east-west section from 2D assessment.

4.3.1 *Deformation within rail corridor*

The maximum predicted total vertical and horizontal deformations below the existing railway tracks after excavation completion are approximately 2mm and 6mm respectively (Figure 6). The maximum differential vertical and horizontal settlements below the existing rail track within the rail corridor due to the excavation were calculated to be less than 1 mm and 2mm. Trigger levels for "Line Alarm Level 1" were 10mm for 60km/h track and 14mm for 40km/h track. Lateral movement affects the line value, with the line value determined by three track locations over 8m.

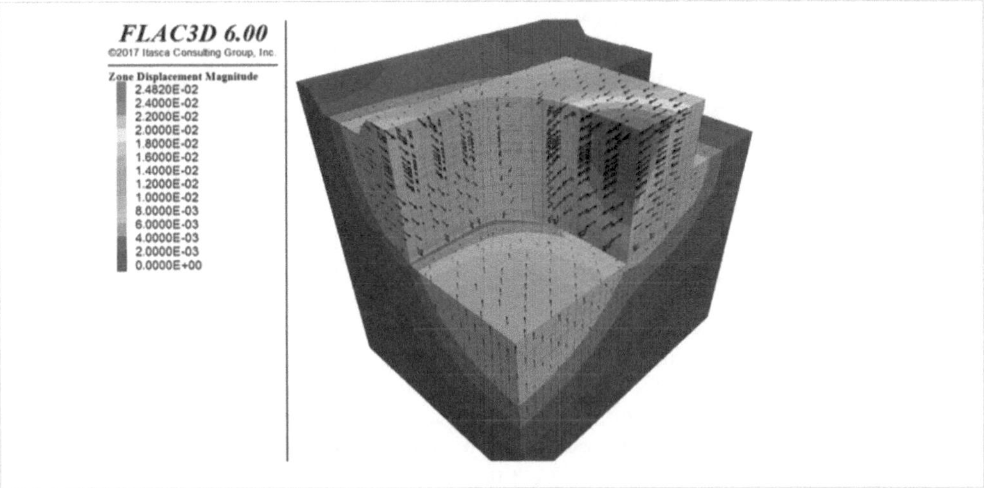

Figure 7. Predicted ground movements from 3D assessment.

The estimated ground performance from the numerical assessment, a database of movements for walls using published case history data, and monitoring data on of other nearby projects (Hewitt et al, 2008) indicates that lateral wall movements are generally in the 0.5mm to 2mm range per metre depth of excavation in rock. Figure 8 indicates the field performance of the Embassy development adjacent to the "Eighty Eight" development and the typical rates of movement observed in northern Sydney, in similar ground conditions. The lateral wall movement at the adjacent Embassy development was approximately 0.5mm or less per metre depth of excavation in rock. Trigger levels addressing total serviceability deflection (lateral displacement) of the wall in any one direction were 30mm adjacent to Pacific Highway.

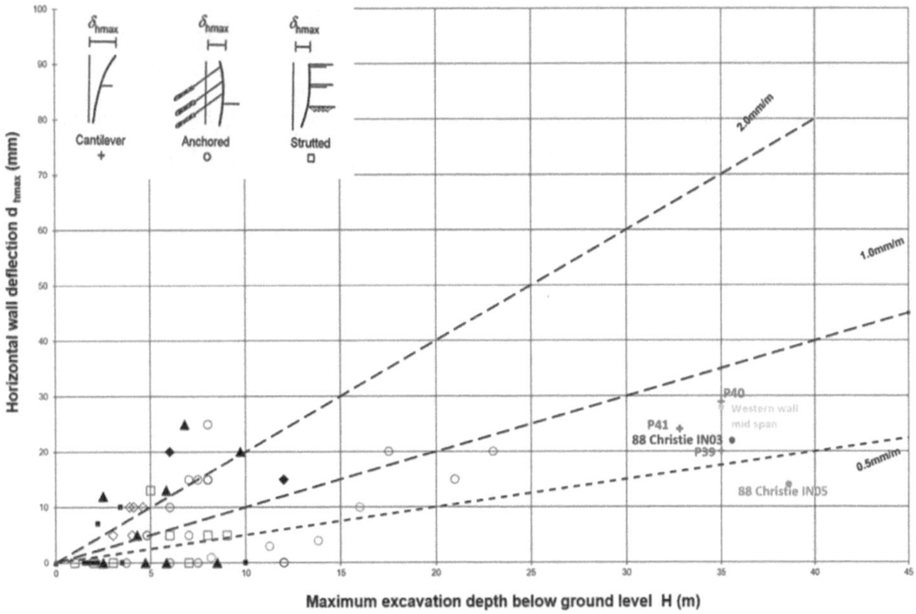

Figure 8. Field performance with different retention systems.

5 CONSTRUCTION PERFORMANCE

We adopted the observational method with some contingency measures to prevent a SLS or ULS from occurring as described in CIRIA C760 (Gaba et al, 2017), including the following:

- The installation of a temporary high stiffness anchor at a high level early in the excavation sequence to control ground movements due to wall deflection, with waler and allowance for additional pre-stressing.
- Excavation start in the south-eastern corner of the site to obtain evaluation of actual wall performance, recalibration of ground and analytical models and identification of recalibrated parameters.
- Along northern boundary, excavation to proceed in 6m to 9m wide vertical panels for horizontal distance of 6m to south of Pacific Highway retaining wall to provide "berm" effect.
- Allowance for additional anchoring/cable bolts along potential sub-horizontal shear/ laminite planes identified from borehole imaging and instrumentation installation, ideally prior exposure/ displacement (adopt endoscope methods if necessary).
- Limiting temporary excavation depths along northern boundary if SLS trigger level approached.
- Trigger limits were identified at key construction stages to enable appropriate and timely decisions and interventions to made by the project team in relation to how the site retention scheme is performing and how movements are developing compared to the recalibrated and SLS characteristic predictions.

5.1 *Basement excavation*

The bulk excavation progress, rock condition and shoring adopted along the south and northern (Pacific Highway) boundaries are shown in Figure 9, as at July 2020.

Figure 9. Bulk excavation works - view to south and north.

5.1.1 *West (railway) wall*

The investigation indicated there could be weaker shale/laminite bands at about 40m depth. (see Figure 4). We therefore planned the installation of a temporary high stiffness anchor to control ground movements due to wall deflection, depending on the observed displacement. During the bulk excavation, the automated inclinometer measured more pronounced horizontal sliding movement on these two shale bands (Figure 10). The sliding movement on the horizontal shale bands was caused by the release of in-situ stress within the sandstone that allowed the block of sandstone above the shale band to move more than had been predicted. We used finite element analysis using RS3 by Rocscience. Initially we used a slice model to calibrate the model to match the movement in the inclinometer (Figure 10 to Figure 12). The shale band was modelled as a ubiquitous joint model in conjunction with the Mohr-Coulomb parameters. The dip direction was set at 10 degrees into the excavation and the friction angle was reduced to 24 degrees to calibrate with the inclinometer results. Once the model was calibrated we assessed various option to reduce the movement and provide additional support to the wall, so that movements would not impact the railway.

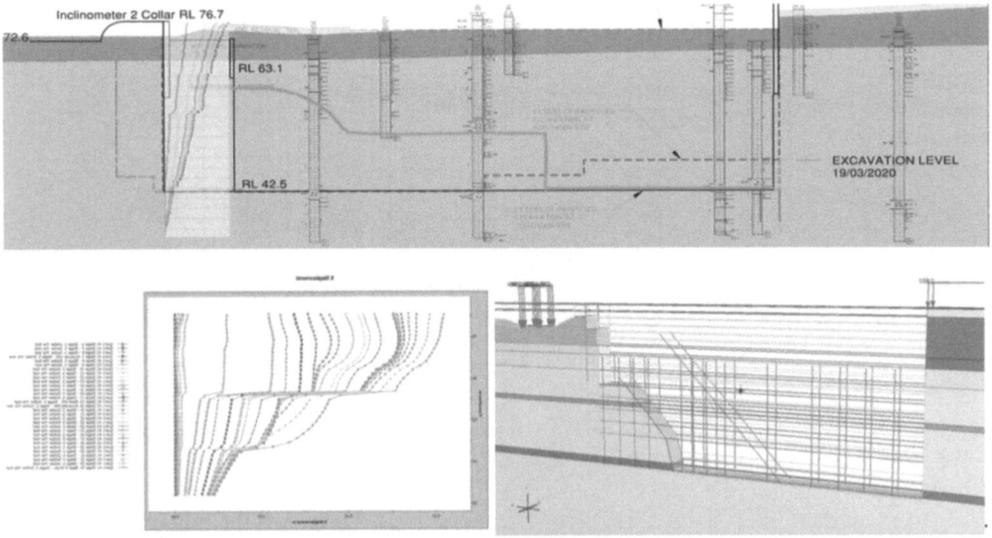

Figure 10. Bulk excavation progress from west to east and RS3 slice model including shale bands calibrated with inclinometer measurements – 14 November 2019 to 19 March 2020.

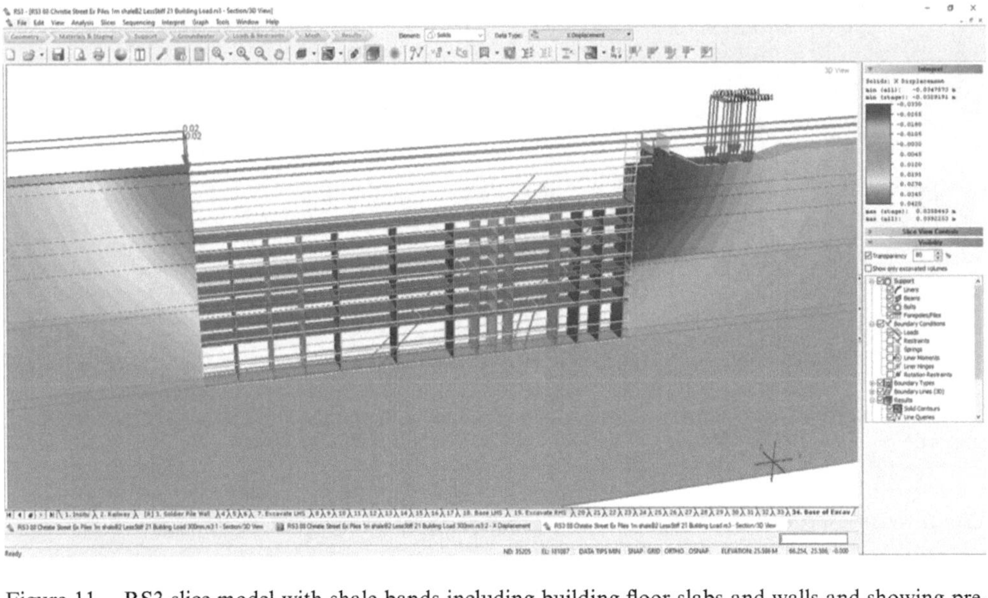

Figure 11. RS3 slice model with shale bands including building floor slabs and walls and showing predicted movement of the railway.

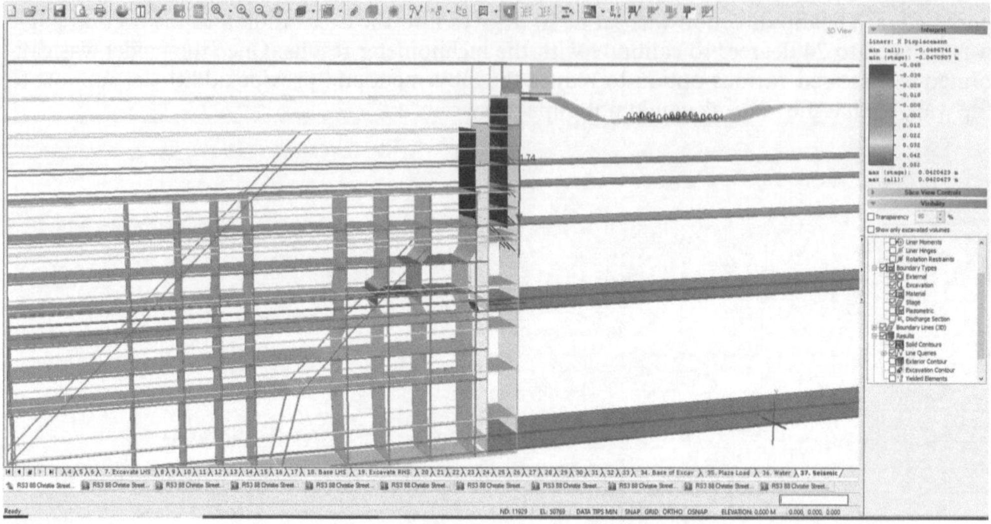

Figure 12. RS3 slice model with shale bands calibrated showing effect on movement of building floor slabs and walls.

We used a 3D model to assess the bulk excavation staging and it was decided to leave a rock buttress against the wall to be excavated last, with corner propping to be installed against the southern rock buttress (Figure 13 to Figure 15). To reduce the potential for movement on the railway we modified the anchor length of the southern 20m wide rock buttress anchors to have an adequate bond length below the shale band, and installed hydraulic corner propping with 2.5MN force applied to the wall (Figure 5). Additional anchors were installed on the opposite corner to counteract this force.

Figure 13. RS3 full excavation model with pile and anchor support and additional corner propping support.

5.2 *Instrumentation and monitoring*

A project team objective was to streamline data collection to maximise system and project integration, and shorten the review and decision-making process to improve construction safety. As a general trend, advances in construction monitoring are moving away from physical measurements at a limited number of points, towards widely distributed, wirelessly connected sensor networks, and to digital scanning techniques. This data allowed the excavation

Figure 14. RS3 full staged excavation model leaving a rock buttress below the railway wall to be excavated last.

contractor to optimise construction processes and increase project safety performance. Monitoring for enhanced safety was also critical due to the project scale and proximity to existing infrastructure (e.g., rail bridge, underground utilities and road pavement).

Methods included surveys of deflections and rotation of the walls, laser wall scanning (Figure 16), ground settlement/heave and rail track. The analyses helped set trigger values based on the 'traffic light' principle to ensure we could anticipate and control excessive ground movements.

As part of controlling the excavation process, instrumentation and monitoring points were adopted as part of the excavation protection strategy as shown in Figure 17. The frequency of instrumentation monitoring was based on the excavation pace and was conducted in conjunction with regular visual observation. Table 3 shows the instrumentation schedule. Walls were monitored to check the actual design performance during construction and to provide data

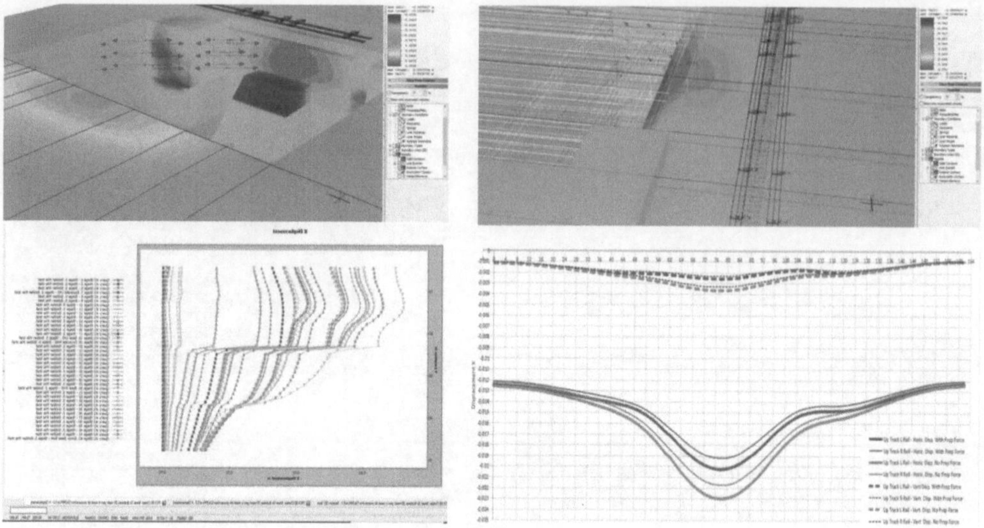

Figure 15. RS3 full staged excavation model predicted effects on the railway line leaving a rock buttress below the railway wall to be excavated last and introduction of corner propping and additional anchoring.

for reviewing design and performance, and for risk management. Monitoring satisfied the designer that the geotechnical models employed in the design were representative, that predictions of the ground and rock support behaviour were accurate, and verified compliance with the design requirements. The maximum measured horizontal wall movement was 28mm at the mid-point of the west (railway) wall, which addressed restrictions by adjacent asset owners and demonstrated excellent agreement with design predictions.

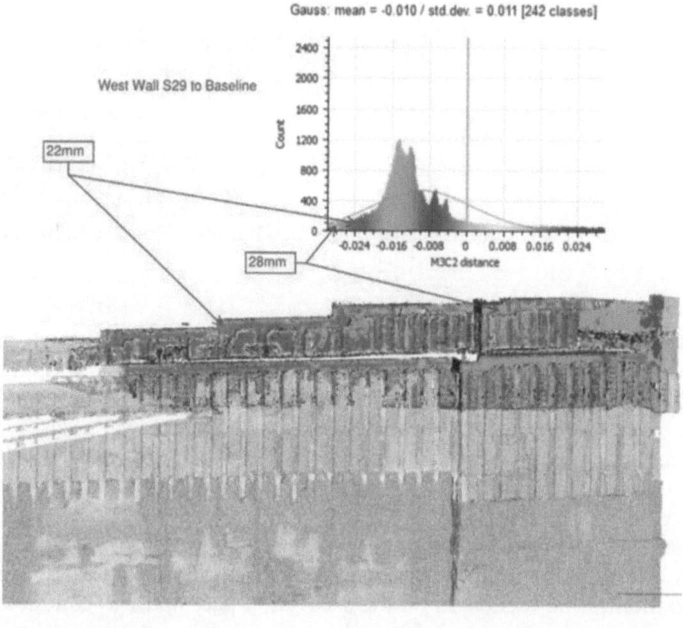

Figure 16. Laser wall scanning on west wall indicated mean displacement of 10 mm and maximum 28mm.

Table 3. Instrumentation schedule.

Instrument	Number	Remarks
Inclinometers	5	3 manual, 2 real-time, remote GeoFlex type. See Figure 15
Displacement Points	112	Around site perimeter and buildings
Bridge displacement	3	Tiltmeters
Laser scan		Individual point accuracy of 2mm on site perimeter
Track monitoring	30	Automated remote monitoring and precise manual monitoring of rail corridor to west of site
Vibration Monitoring	5	Site perimeter
Crack gauge	~30	Visual check

Figure 17. Instrumentation plan and measured displacement.

Automated Remote Monitoring and Precise manual surveying was undertaken of the railway track and associated infrastructure on the North Shore Line as shown in Figure 18. The monitoring network extends from approximately 8.20km to 8.31km. Monitored assets include the Up Track North Shore Line, Pacific Highway bridge and piers, OHW structures and upside crib wall. The Track Geometry was monitored in accordance with ASA Standard ESC

Figure 18. Automated Remote Monitoring provided Real Time Continuous Track Monitoring.

210 Track Geometry and Stability. A Track Certifier was engaged by JQZ in the early stages of the bulk excavation to inspect the track as a baseline, and then subsequently inspected the track during various stages of the bulk excavation.

6 SUMMARY AND CONCLUSIONS

The construction of the project "Eighty Eight" required excavation to over 43 m below ground level and adjacent to existing railway and highway assets. The project demonstrates how appropriate numerical analysis can be a valuable method in assessment of the influence between underground infrastructure and high-rise building foundations/deep basement excavations. Design, excavation and construction of the site retention system incorporated several critical issues, including addressing stringent settlement and angular distortion criteria, construction safety, constructability, and the constraints of defined road and rail reserves. The design process was successful and effective in addressing the concerns of all parties involved in the project.

Limited construction data available to-date indicates the displacement has been significantly reduced to achieve the performance criteria and validate the design implemented for the project. Monitoring data to date indicates the pre-construction geotechnical models and design parameters were appropriate, and that an observation-based approach allows selection of adequate retention support design to manage the risks associated with elevated stress conditions.

- Finite difference analyses were carried out to assess the behaviour of the rock, site retention system, and adjacent infrastructure.
- The accuracy of this type of interaction assessment was significantly influenced by ground model parameters and rock mass properties. This emphasises the value of detailed ground investigations prior to modelling.
- The laser is a state-of-the-art development for monitoring wall movement. It offered precision and broad area coverage of wall movements through rain, dust and smoke (which occurred due to Sydney bushfires in January 2020).
- The real-time display of the movement of basement walls allowed continuous management of deformation during excavation, with the remote monitoring inclinometer system providing early detection of horizontal sliding along shale bands.
- The real-time, automated remote monitoring and precise manual surveying of the railway track geometry and engaging a Track Certifier early in the bulk excavation works assisted in addressing impacts on track geometry.
- The maximum measured horizontal wall movement was 28mm at the mid-point of the west (railway) wall.

ACKNOWLEDGEMENTS

The work reported in this paper was carried out during studies by WSP (formerly Parsons Brinckerhoff) from 2010 to 2020, for various schemes. ABC Consultants Pty Ltd were structural engineers for the project. Moit and Sons (NSW) Pty Ltd carried out the main excavation. JQZ, which owns the site, generously allowed us to publish this work. The authors gratefully acknowledge the contributions of fellow staff.

REFERENCES

ASA Standard, External Developments, Ref: T HR CI 12080 ST, Version 1.0, February 2015, and Civil Revision 1.0 10/ 11/2017

Gaba A, Hardy S, Doughty L, Powrie W & Selemetas D. (2017) Guidance on embedded retaining wall design. CIRIA Report C760, London, UK

Hewitt P, Burkitt S & Baskaran B. (2008) Design and construction of retaining structures for Lane Cove Tunnel, 10th ANZ conference on geomechanics, Brisbane. 626-631; and Australian Geomechanics (2008), 43(1). 55–60

Parsons Brinckerhoff (2012) Report on Geotechnical Investigation St Leonards Commerce Centre, 88 Christie Street, St Leonards. Report 2108360A_PR_290 Rev A.

Pells P.J.N, et al. (2019) "Classification of sandstones and shales in the Sydney region; A forty-year review". Australian Geomechanics, 54(2)

Roads & Maritime Services (2012) Technical Direction GTD 2012/001, Excavation adjacent to RMS infrastructure https://www.rms.nsw.gov.au/business-industry/partners-suppliers/documents/technical-directions/gtd_2012-01.pdf

Transport for NSW (TfNSW). (2019) Engineering Specification Track SPC 207 - Track Monitoring Requirements for Undertrack Excavation, Version 1.5, Issued April 2013, Reconfirmed 03 July 2019

WSP (2018), 88 Christie Street, St Leonards. Geotechnical report for JQZ Pty Ltd. Report PS106114-GEO-REP-608A.

The Evolution of Geotech - 25 Years of Innovation – Hammah et al. (Eds)

North wall stability analysis of Vasilkovskoye open pit Kazakhstan

D. Dossymbek & A. Mortazavi
Nazarbayev University, Nur-Sultan, Kazakhstan

ABSTRACT: This work focused on the stability analysis of Vasilkovskoye open pit north wall slope. The current height of the north wall is about 270 m and within the north wall area there are several critical areas that have the potential for instability at the scale of two to three benches. These areas are characterized by significant jointing and form a blocky rock mass. Because the main haulage ramp goes through this location, any wall failure would jeopardize the operational safety and productivity of the pit. Stability analysis of the wall was conducted to assess the wall stability and determine the wall failure mode as well as shape and depth of probable failure plane. Knowledge of failure plane geometry and depth is a key parameter in selecting the stabilization method. A complete rock mass characterization of north wall aiming at determining design parameters was carried out and a comprehensive numerical analysis of the wall critical areas was conducted. Moreover, with regard to the calculated shape and depth of probable failure plane at wall critical areas, recommendations for appropriate support system were suggested. Thus, slope stability analysis of this area will significantly contribute to maintaining safety and productivity of the mine. The results of the study will provide the Vasilkovskoye with valuable information on stability for safe and sustainable operations on the north wall of the mine.

1 INTRODUCTION

In the developing stages of deposits in an open-pit mine, it is very important to ensure the stability of both individual benches and the whole pit walls. To prevent their deformation and collapse during the entire period of operation of the pit, stability analysis of all mine walls is essential. The stability of the pit walls depends on different factors, for instance: structural and geological features of the deposit (including fracturing), physical and mechanical properties of the rocks, stressstrain state, pit wall design, hydrogeology, blasting technology, weathering, seismicity, etc. The stability of the pit walls is associated with issues of safety and technology of work in mines, as well as determining the ultimate pit angle, the maximum depth of the pit and the possibility of drainage of the deposit. The aims of this project are to evaluate the stability of the critical pit walls of Vasilkovskoye open-pit gold mine, based on numerical simulations in RS3 environment.

2 GEOLOGY OF VASILKOVSKOYE OPEN PIT

The Vasilkovskoye gold deposit is located in the Akmola region of Kazakhstan, nearly 20 km northwest of Kokshetau town. The ore deposit was discovered in 1963, and mining activities began in 1980 and were carried out by open pit mining.

DOI: 10.1201/9781003188339-36

The main observed rock types are as follow:

- Megacrystic granodiorite with abundant K-feldspar megacrysts. This is the main host rock for the mineralization.
- Quartz diorite and gabbro-diorite. These rock types contribute to a lesser extent to the mineralization.

Mineralization is hosted mainly by megacrystic granodiorite and it is associated with hydrothermal quartz and quartz-arsenopyrite veins.

3 ROCK MASS CHARACTERIZATION

The rock mass characterization is a major step to determine rock mass design parameters to assess the stability of a north wall slope at its current state. With regard to the previous investigations of Vasilkovskoye mine, the laboratory data provided by Golder Associates Consulting Company, SRK Consulting and AMC Consultants Companies were analyzed. Furthermore, rock mass characterization was performed and rock mass design data, as input for 3D modelling, were obtained. Accordingly, rock mass data were collected for the main geotechnical domains and critical areas of the pit. In general, the uniaxial compression strength of the main lithological units in different pit sections exceeds 150 MPa. However, the top portion of the pit is consists of weak weathered rocks that can be visually distinguished from the base rocks. The thickness of these rocks varies and encompasses several benches at higher elevations. The compressive strength of the weak rocks is less than 50 MPa. The following types of laboratory tests were performed: Uniaxial Compression Test, Triaxial Compression Test, Tensile Strength Test, Shear Box Tests, Residual Joint Strength tests. A much less weathered rock mass underlies the above described weathered rock mass. Based on field observations, a GSI index greater than 70 was attributed to the deeper rock mass of the open pit (approximately below 150m). Borehole camera analysis confirmed these observations and indicated a general GSI index improvement with depth down to -395 m. This level is equivalent to the geotechnical domains named 2-7 in northeast and 2-12 in south-west sectors (AMC Consulting, 2008).

The following three main rock types are identified at Vasilkovskoye:

- Granite and granodiorite;
- Quartz diorite/mix-diorite;
- Gabbrodiorite;

The rock type distribution map observed in the open pit is based on data provided by Vasilkovskoye: gabbrodiorite and quartz-diorite outcrop in the northern side of the open pit and they represent approximately 30% of the total rock mass. Granites and granodiorites outcrop in the central and southern parts of the open pit, and represent approximately the 70% of the north wall rock mass.

For the rock mass characterization, samples were collected from several boreholes that were located in the north-east side of the pit. Boreholes were oriented toward N171° with an inclination to the horizontal of 60° (at starting point). In the south-west side of the pit the boreholes were oriented toward N26° with an inclination to the horizontal of 60° (at starting point) to conduct a laboratory tests. Consequently, to determine the rock mass rating (RMR) and geological strength index (GSI), various elevations within the north wall were taken as representative locations to assess rock mass design parameters of the main geotechnical domains. The recorded rock mass parameters were calculated for each geotechnical units using the laboratory test results. Table 1 illustrates a prepared designed data of rock mass for 3D modelling.

Table 1. Summarized Intact Rock mass parameters.

Rock Type	Unit Weight (t/m3)	Young's Modulus (GPa)	Pois. Ratio	UCS (MPa)	Cohesion	GSI	Friction°	RMR	mi	mb	s
Granodiorite	2.6	58.9	0.24	92.9	8.18	70	40.6	62/72	8.39	2.875	0.035
Quartz Diorite	2.68	86.7	0.28	114.9	13.57	70	38.9	76	8.75	2.99	0.0357
Gabbro Diorite	2.9	86.7	0.28	100	13.57	70	36.9	76	8.5	2.9	0.035

4 ROCK MASS DESIGN PARAMETERS

After completing the data analysis, it is necessary to transfer the laboratory data to the in-situ values for the modeling. In order to determine the rock mass design data the RocLab software was used.

RocLab results for granodiorate, quartz diorite and gabbro-diorite are presented in Figure 1, 2 and 3.

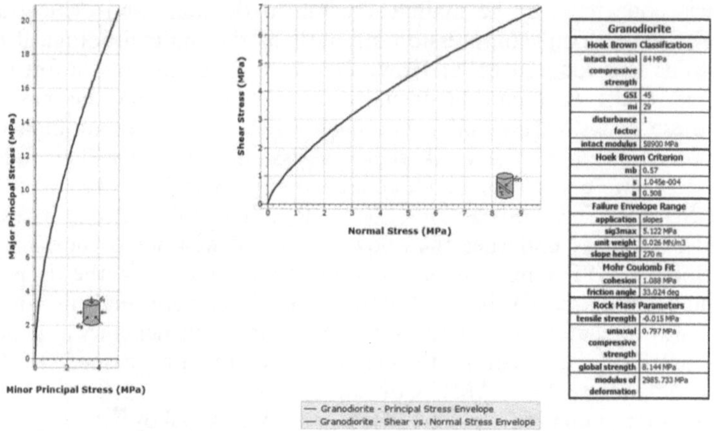

Figure 1. Rock mass designed data determined for Granodiorite geotechnical unit.

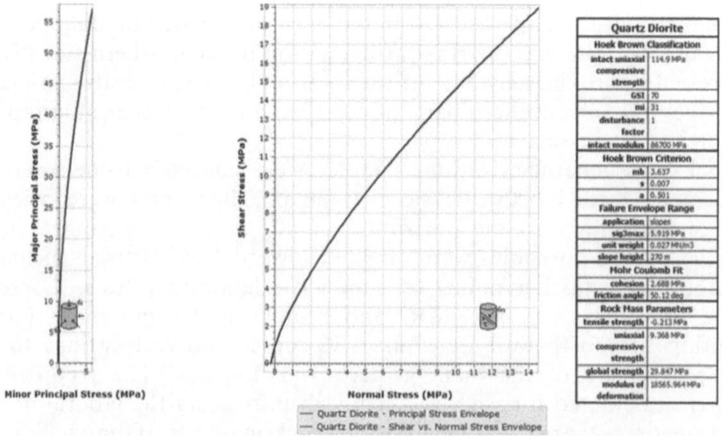

Figure 2. Rock mass designed data determined for Quartz Diorite geotechnical unit.

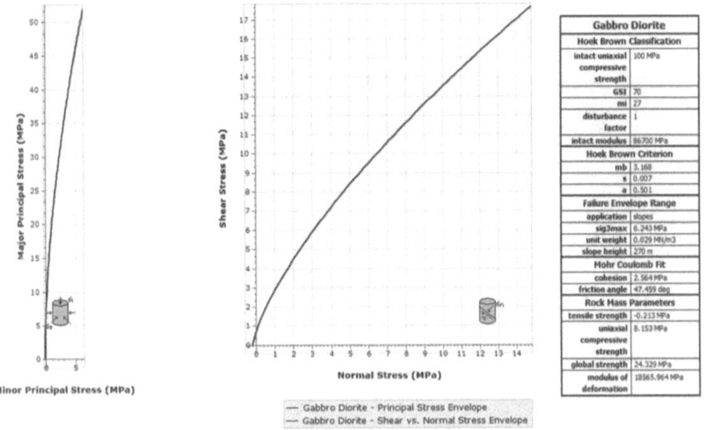

Figure 3. Rock mass designed data determined for Gabbro Diorite geotechnical unit.

The parameters which have the strongest influence on the slope reliability were summarized and presented in Table 2.

Table 2. Determined Designed Data of the rock mass as input for numerical analysis.

Rock Type	Unit Weight (t/m3)	Tensile strength (MPa)	UCS (MPa)	Global strength (MPa)	Modulus of deform (GPa)	Cohesion (MPa)	Friction angle°	mb	s	a
Granodiorite	2.6	0.015	0.797	8.144	2.986	1.088	33	0.57	0.0011	0.508
Quartz Diorite	2.68	0.213	9.368	29.847	18.566	2.688	50	3.64	0.007	0.501
Gabbro Diorite	2.9	0.213	8.153	24.329	18.566	2.564	47	3.17	0.007	0.501

5 3D NUMERICAL SIMULATION OF VASILKOVSKOE NORTH WALL

With regard to limitations in computational capabilities, the model dimensions were selected as illustrated in Figure 6. Since this work in progress, for future publications, the model boundaries will be extended to minimize the probable errors associated with boundaries. Having prepared the data (Table 2) for the modelling, 3D model of the north wall of Vasilkovskoe open-pit was con- structed within the RS3 environment. The north wall region was employing an average mesh size of 1m. Key geological materials were assigned as appropriate, RS3 software has a different approach for restraining the model. This model was restrained with auto surface restrain tool. Figure 4 and Figure 5 illustrates the determined maximum and minimum principal stress distribution within the simulated pit region.

Thus, the maximum principal stress (sigma 1) and the minimum principal stress (sigma 3) were obtained from the simulation results using RS3. The calculated maximum principal stress was 7.56 MPa (Figure 4) and the minimum principal stress was found to be 7.14 MPa (Figure 5). Both stresses pose the greatest compressive stress on the basement of the north wall indicating lithostatic load, because of the mass of the overlying layers in rocks, and gravity that increases pressure on each layer of rock that increases with increasing depth.

Distribution of total displacements within the North wall is shown in Figure 6. The contour of total displacement presented for a representative section of the wall in Figure 7. The results of the overall rock mass stability analyses indicated that the proposed design for pit walls are

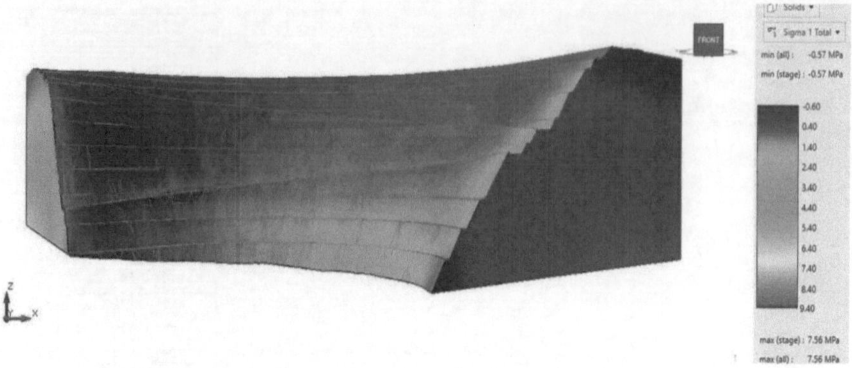

Figure 4. The calculated maximum principle stress distribution within north wall (front view).

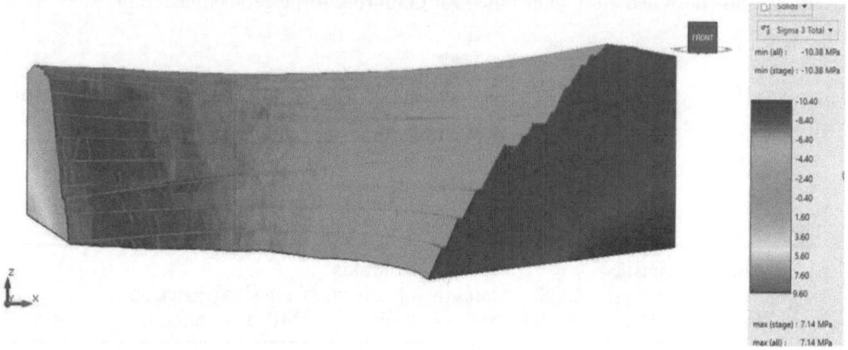

Figure 5. The calculated minimum principle stress distribution within north wall (front view).

expected to have instability with respect to potential failure mechanisms that could involve multiple benches.

As illustrated in Figure 6, the central section of the north wall possesses the highest instability potential within the whole structure. The overall size of the most unstable area corresponds to about 120 meters in length, 50 meters in height and 10 meters in depth and the probability of the maximum displacement is about 0.11 meters.

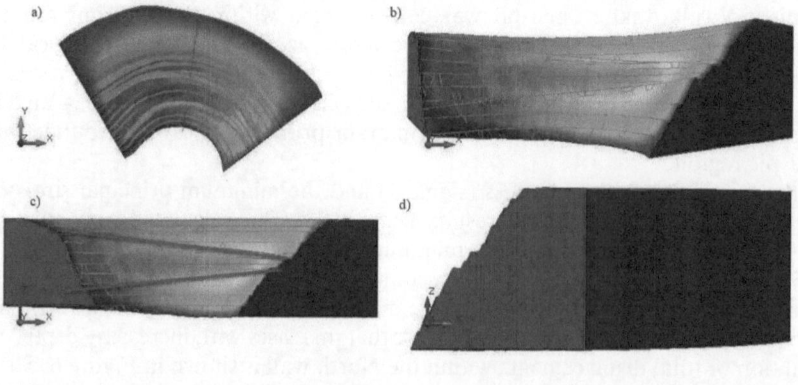

Figure 6. The calculated total displacement of the north wall (from four different perspectives: a) top view, b) isometric view, c) front view and d) side view).

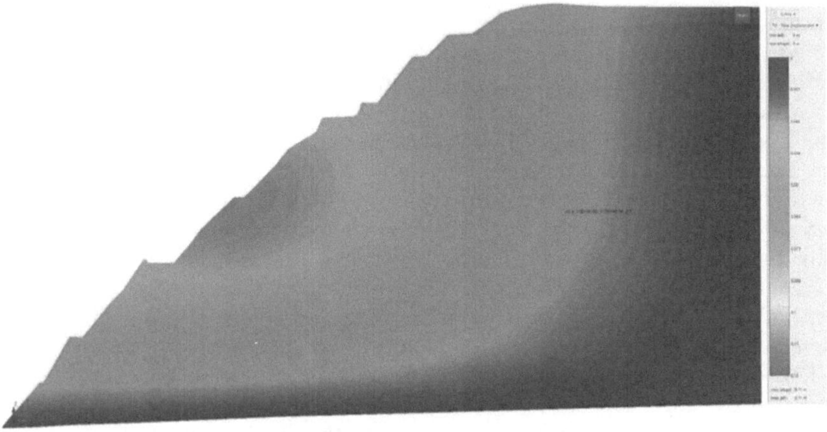

Figure 7. Total displacement contours on the north-south cross section.

With the help of RS3 feature of displaying displacement vectors Figure 8 was obtained.

This model simulation allows us to predict possible direction of the rock movement. As vectors point to forward direction towards the center of the pit, it can be stated that there is a risk of several benches' failure. We have shown within 3D numerical modelling results that the north wall has the potential to fail, due to the wall is destressing potential at the middle

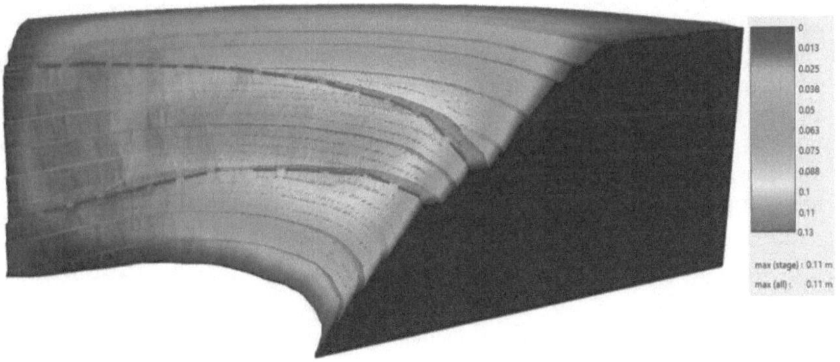

Figure 8. Displacement vectors direction.

portion. With regard to the jointed nature of the wall as the potential to fail at those areas, we recommend stabilization to the pit.

6 DISCUSSION

Stability analyses were conducted for the overall pit walls considering the potential failure mechanisms. The finite element program RS3, was used to confirm that the proposed pit walls are comparatively stable with instability potential in central part. Numerical modelling approach allowed a better method to evaluate the stability of the slope than the conventional methods, it also takes into account complex geometry and non-linear behavior.

The conducted 3D modelling of the north wall, that was constructed using RS3, allowed us to determine the induced wall stress field and its influence on slope stability (in dry wall condition). These values indicate that the effect of these stress directly contribute to potential slope instability in the central region of the north wall. As it is stated in the 3D numerical simulation of Vasilkovskoe north wall section both stresses pose the greatest strain on the basement of the north wall indicating the maximum principle stress as 7.56 MPa and the minimum principal stress as 7.14 MPa lithostatic load. Because of the mass of the overlying layers in rocks, and gravity, as being an increasing pressure on each layer of rock section, Figures 4 and 5 illustrate the base of the pit as the most stressed portion of the wall.

As illustrated in Figure 6, the central section of the north wall possesses the highest instability among the whole structure. The main source of wall center instability is the distressing of wall and weak rock mass properties. The size of the predicted unstable area was found to be about 120 meters × 50 meters × 10 meters with an average displacement of 0.11 meters. Other simulation model predicts possible direction of the rock movement towards the center of the pit which poses a risk of several benches' failure. This indicates the risk for further operations on those benches. Bench, inter-ramp and overall slope geometries were studied by means of numerical simulation. It is suggested to follow the specific design criteria that will support the stability of the slope such as rock inter-ramp should comprise no more than about three benches and should have a vertical height of no more than 90 meters. These bench stacks should be separated by a 25-meter ramp or geotechnical berms.

7 CONCLUSION

Pit wall stability analysis is of great interest to the Vasilkovskoye open pit due to the fact that the safety and profitability of the mine directly depends on the stability of the pit walls. The results of this work provided advantages for Vasilkovskoye, for example: determination of critical wall (the north wall) stability, rock characterization and 3D modeling. The results of the work are of a great significance as the safety and stability of the north wall is vital. The scope of this paper is to evaluate the stability of critical areas in the north wall. Through numerical modeling, it was possible to examine the north wall in detail and determine possible scenarios of deformations and failures, and propose remedial support.

REFERENCES

AMC Consulting, 2008. Open pit mining sustainability JSC Vasilkovskoe zoloto. West Perth. Golder Associates, 2016. Steeper Slope Angle Evaluation at Vasilkovsky Open Pit. Atyrau: Golder Associates Kazakhstan LLP

Karaganda State Technical University, 2009. Substantiation of stable parameters of open-pit bench slopes and walls at Vasilkovsky GOK. Karaganda.

SRK Consulting, 2019. Justification of the stability of benches and berms of the Vasilkovskoye open pit. Moscow.

The Evolution of Geotech - 25 Years of Innovation – Hammah et al. (Eds)

3-D stability analyses in soft clays with strain-softening

L.S.M. Silva & M.P. Pacheco
State University of Rio de Janeiro, Rio de Janeiro, Brazil

ABSTRACT: Stability analyses by the finite element method for slopes in soft clays with strainsoftening behavior are an important tool for a better understanding of the behavior of soft clays when compared to limit equilibrium methods. In finite element analyses, the post-peak residual stresses are activated when the deformations corresponding to the maximum peak stresses are exceeded, what is not possible to account for in limit equilibrium analyses. Vane tests performed in slopes of soft soils in the Amazon River indicated strain-softening behavior in the investigated clays. Three-dimensional analyses by the finite element method were performed with the aid of RS3 software for a slope in soft clay considering the loss of post-peak strength. The Mohr-Cou-lomb model with post-peak residual strength and the influence of stepwise loading stages were considered in the analyses. The results obtained by RS3 accounting for strain softening show an important decrease in the safety factor comparably to SLIDE 3.

1 INTRODUCTION

Saturated clays mobilize the undrained shear strength when an increment of loading is applied at a rate greater than the capacity of soil to allow the dissipation of excess pore pressure. This dissipation capacity is associated to the clay low permeability.

Stability analyses are usually processed by two categories: i-limit equilibrium analyses and ii-stress-strain analyses (usually by the finite element method).

Stability analyses by the limit equilibrium method (LEM) consider a soil wedge delimited by a failure surface, which is assumed as a rigid body, under static equilibrium conditions. The soil is considered a perfect elastic-plastic material and failure occurs when the mobilized stresses equals the shear strength of soil.

According to Duncan (1996), there is an implicit assumption in limit equilibrium analyses, that the soil has a ductile stress-strain behavior, that is, the shear strength does not reach a peak value to decrease thereafter. According to Chen and Liu (1990), for the application of the limit equilibrium method, the material should exhibit perfect elastic-plastic behavior (Figure 1). In most cases the effects of soil softening or hardening are disregarded.

The finite element method (FEM) considers a physical continuum discretized into small-volume elements. Each element has a certain number of nodes, for which a global system of equations is solved. The system unknowns are the nodal soil displacements and nodal pore pressures.

In addition to soil displacements and pore-presures, an additional advantage of performing stability analyses of slopes with the aid of FEM is the calculation of the safety factor (SRF, strength reduction factor). In this type of analysis, SRF is obtained by gradual degradation the soil strength parameters until the slope reaches numerical failure.

The strain-softening behavior of clays was presented by Skempton (1964), who pointed out that the soil reaches a peak shear strength value and, as the displacements continue to increase, the strength gradually drops to a residual value.

DOI: 10.1201/9781003188339-37

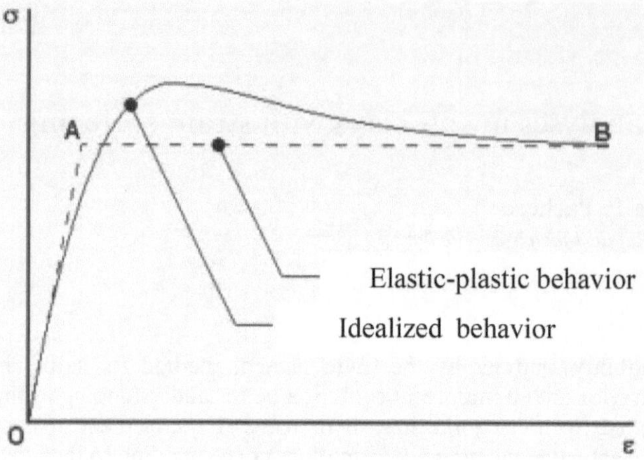

Figure 1. Stress-strain for perfect elastic-plastic behavior (A) and real behavior idealized (B).

This behavior cannot be simulated in limit equilibrium analyses. To compensate this effect, a correction factor is applied to the undrained shear strength, S_u, to reduce it as a function of the plasticity index. In this paper, it is shown that a reduction factor equivalent to Bjerrum`s correction factor is given by the ratio between the safety factors without and with strain-softening.

2 CASE STUDY

The case study is a slope failure in the Amazon River, State of Amapá, Brazil. The failure of the submerged slope resulted in human and material losses, in addition to a large volume of displaced mass into the Amazon River. After the failure, a complementary soil investigation was carried out to obtain additional data from subsoil conditions.

Soil investigations carried-out before and after the failure indicated a thick layer of soft clay, responsible for the strain-softening behavior identified by Vane and CPTu tests. This field instrumentation indicated a low to medium sensitivity ranging between 1.5 and 4.

The failure site is a port for ore exportation. Few months before the accident, due to equipment breakdown, ore stacks were inadvertently placed beyond the safe limit of resistant soil and advanced through the soft soil, triggering the failure. The failure scenario is shown in Figure 2.

Figure 2. Failure scenario.

The surcharge placed by the ore stack was assumed as a circular loading, with four loading stages for analyses by RS3. The maximum 150 kPa surcharge represents a 6 m-high ore stack with 25 kN/m³ unit weight.

The three-dimensional model for RS3 and SLIDE3 has twenty-four cross sections and two boreholes to define the soil stratigraphy (Figure 3).

Figure 3. Three-dimensional model of the slope (RS3).

3 STABILITY ANALYSES BY THE FINITE ELEMENT METHOD (RS3)

The stability analyses performed by RS3 took into account the strain-softening behavior of the clay, using the Mohr-Coulomb constitutive model with residual stresses. However, this model assumes an abrupt strength loss from peak to residual stress that differs from the continuous strength decay observed in all vane tests.

For this reason, Pereira Pinto (2017) introduced the concept of equivalent sensitivity (S_t^*), where:

$$S_t^* = \frac{S_u}{S_{u,r}^*} \tag{1}$$

Where S_u = peak undrained shear strength; and $S_{u,r}^*$= equivalent residual shear strength. Figure 4 shows the equivalent sensitivity concept.

For appropriate S_u peak values and small to medium clay sensitivity S_t, the equivalent sensitivity S_t^* is determined by back analysis, to reach SRF = 1, according to Table 1. Pereira Pinto (2017), based on RS2 analyses, found the equivalent sensitivity S_t^* = 1.4, which is roughly one half of the average sensitivity S_t determined from vane tests. In this work, based on RS3, the Authors found S_t^* = 1.6, only slightly higher than 1.4 (RS2), as expected.

The ponded water load was applied for the total head related to the river lowest tide. A graduated mesh with 4-noded elements was used.

The analyses considered four stages: the initial stage, comprising the intact slope with no loading, and three subsequent stages of 50 kPa, totaling the full stack loading (150 kPa). Only 30% of the total extent of the stack was supported by resistant soil, while the 70% remaining was inadvertently placed on the weak soil.

Table 1 presents the back analysis of the equivalent sensitivity. Figure 5 shows the contours of the maximum shear strains corresponding to a cross section passing thorough the stack pile, which compares reasonably well to the failure surface observed in the field. Figure 5 confirms that the portion of the stack pile that advanced through the weak soil triggered the failure.

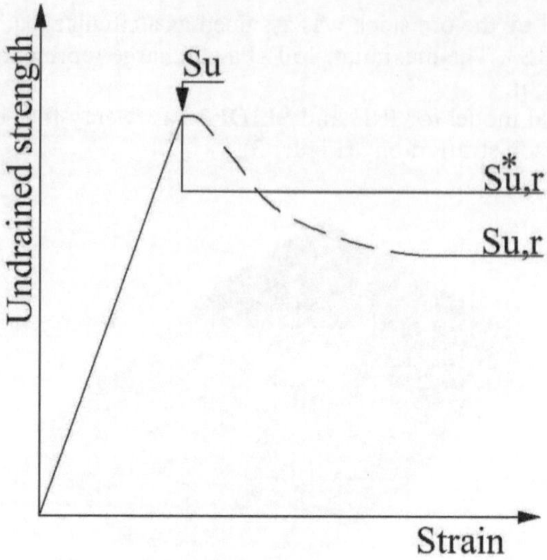

Figure 4. Stress-strain curve for equivalent sensitivity concept.

Table 1. Back analysis of the equivalent sensitivity (RS3).

S_t^*	*SRF*
1.0	1.54
1.4	1.13
1.5	1.08
1.6	1.00

Figure 5. Maximum shear strain contours at failure (S_t^* =1.6), RS3.

4 STABILITY ANALYSES BY THE LIMIT EQUILIBRIUM METHOD (SLIDE3)

Stability analyses by LEM were performed with the aid of SLIDE3. The three-dimensional model was the same used in the finite element analyses. The constitutive model was Mohr-Coulomb, and the analyses performed in two categories: one considering the reduction of the undrained shear strength by a correction factor μ and the other considering the undrained shear strength without correction (average strength obtained by field tests). The Morgenstern-Price's method was selected to obtain the safety factors and the corresponding failure surfaces. The analyses were performed without loading and with the stack loading that triggered the failure (150 kPa, radius equal to 25.0 m).

According to Schnaid and Odebrecht (2012), for Brazilian clays, the average value for the Bjerrum's correction factor μ is around 0.65. Alternatively, the correction factor can be estimated from the previous RS3 analyses according to Equation 2:

$$\mu_r = \frac{SRF_{ss}}{SRF_{nss}} \quad (2)$$

Where SRF_{ss} = strength reduction factor at failure with strain-softening; and SRF_{nss} = strength reduction factor with no strain-softening.

Considering the safety factors shown in Table 1, the equivalent reduction factor obtained by backanalysis is μ_r = 1.0/1.54 = 0.65, which is in accordance with the recommended value for the Brazilian small to medium sensitivity soft clays.

Table 2 presents parametric variations of the safety factors by SLIDE3, with the reduction factor ranging from 0.60 to 0.75.

Table 2. Results by SLIDE3.

μ	Scenario with loading	Scenario without loading
1.0	1.52	1.69
0.75	1.13	1.26
0.70	1.06	1.17
0.65	1.04	1.13
0.60	0.98	1.03

Figure 6 presents the failure surface provided by SLIDE3, considering the reduction factor μ_r = 0.60, and the application of the stack loading. According to the results, this scenario triggered the failure. The corrected results in Table 2 indicate that the slope was already under critical stability before application of the stack loading. However, without correction ($\mu = 1$), the calculated safety factors are deceiving (FOS=1.69), leaving the false perception of a safe slope.

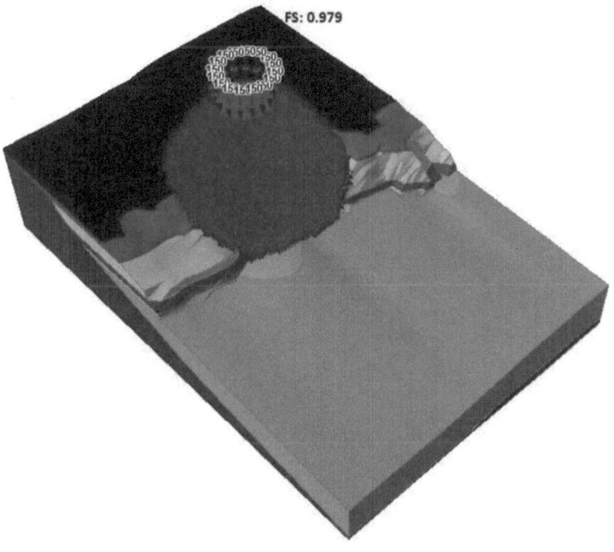

Figure 6. Failure scenario for μ_r= 0.60 (SLIDE3).

5 CONCLUSIONS

The effect of strain-softening represents an important issue that cannot be disregarded in engineering design. The Mohr-Coulomb constitutive model with post-peak residual stresses provides a simple way to account for stress-softening by the program RS3.

Both RS3 (with strain-softening) and SLIDE3 (with reduction factor μ) provided very close results and represented satisfactorily the soft clay failure of the case history presented. Neglecting strain-softening may lead to misleading, non-conservative results in stability calculations in soft clays.

The concept of equivalent sensitivity proposed by Pereira Pinto (2017) is essential to determine the equivalent residual stresses to load the model. This procedure is deemed necessary because of the simple assumption of abrupt post-peak loss of strength in the model. Very conservative results will be obtained if the fully disturbed strength determined from vane tests is used for the residual strength.

For small to medium sensitive soft clays, the equivalent sensitivity is generally in the range of 40% to 60% of the average sensitivity determined from vane tests, where 50% is a reasonable approximation for practical purposes.

REFERENCES

Bjerrum, L. 1973. Problems of Soil Mechanics and Construction on Soft Clays. *Proceedings of the 8th International Conference on Soil Mechanics and Foundation Engineering* 3: 111–159.

Chen, W.F. & Liu, X.L. 1990. *Limit analysis in soil mechanics*. Amsterdam: Elsevier Science Publishers B.V.

Duncan, J. M. 1996. State of the art: limit equilibrium and finite-element analysis of slopes. *Journal Geotechincal Engineering* 122: 577–597.

Marcos da silva, L. S. 2021. *Análise tridimensional de taludes em argilas moles* (Thesis). Rio de Janeiro: State University of Rio de Janeiro.

Pacheco, M. P. et al. 2014. *Technical Report – Parecer sobre a ruptura do talude do Porto de Santana*, State Univesity of Rio de Janeiro.

Pereira pinto, G. 2017. *A influência da sensibilidade na estabilidade de solos moles brasileiros* (Thesis). Rio de Janeiro: State University of Rio de Janeiro.

Schnaid, F & Odebrecht, E. 2012. *Ensaios de campo e suas aplicações à engenharia de fundações*.São Paulo:Oficina de Textos.

Skempton, A.W. 1964. Long-term stability of clay slopes. *Géotechnique* 14(2): 77–102.

The Evolution of Geotech - 25 Years of Innovation – Hammah et al. (Eds)

On the comparison of 2D and 3D stability analyses of an anisotropic slope

A. McQuillan
Rocscience, Inc.

N. Bar
Gecko Geotechnics

T. Yacoub
Rocscience, Inc.

ABSTRACT: This paper summarizes the results of a parametric study that demonstrates the variability in Factor of Safety that is calculated for a highly bedded open pit mine when 2D and 3D limit equilibrium stability methods are applied. Factors of Safety are calculated for linear (Mohr-Coulomb) and non-linear (Generalized Hoek-Brown, Shear-Normal, Barton-Bandis) material models as well as different search settings. Results of this parametric study show that the Factor of Safety is consistently higher for 3D analysis, compared to 2D analysis. This increase in Factor of Safety, in this example, is attributed to the correct inclusion of end effects and varying directional strength, which is not possible to include in 2D analysis.

1 INTRODUCTION

Three-dimensional (3D) limit equilibrium (LE) modelling is increasingly being applied to assess the stability of open pit excavations. Previously, kinematic and two-dimensional (2D) LE modelling techniques were the most routinely applied to assess slope stability (Stacey 2007, McQuillan et al. 2019). However, as 3D LE software becomes more widely commercially available and more user friendly, geotechnical engineers are realizing the benefits of assessing stability in real world, three-dimensions (Bar et al. 2019a, Bar et al. 2019b, Bar et al. 2020. McQuillan et al. 2020).

3D stability analysis will always provide a more accurate representation of slope behavior and slope stability, especially where discontinuities strike more than 20 to 30° from the excavated face (Lorig and Varona 2007, Stacey 2007, Wines 2015, Cala et al. 2020). This is because the failure mechanism being modelled in 2D is always, at least to a degree, 3D in nature, so 2D analysis does not model the true mechanics of rock slope conditions and material strengths. For this reason, in many instances, material strengths derived from 2D back-analyses are higher than 3D-derived material strengths.

Variance in Factor of Safety (FOS) calculated between 2D and 3D methods will be more pronounced when the rock mass under investigation is anisotropic and/or the slope design includes confining geometries (Bar and Weekes 2017, Bar and McQuillan 2018). This is where 3D analysis can calculate a more realistic failure surface (e.g. model a slip surface that includes both sliding along bedding or foliation at the base of the slip surface and failure through rock mass bridges at all extremities of the slip surface – i.e. usually the toe, sides and crest).

DOI: 10.1201/9781003188339-38

Historically, 2D models were the most commonly constructed, due to their relative ease of model construction and rapid computation time (McQuillan et al. 2019). Sjoberg (1999) and Wines (2015) state that for relatively long open pits with basic geological conditions, 2D models can be justified. However, 2D LE analysis neglects the normal and horizontal side resisting forces along the sides of the sliding mass (i.e. end effects), which can lead to a conservative estimate of FOS (Shamsoddin Saeed et al. 2015). Neglect of this effect, or simply by applying a rule of thumb adjustment to predict the 3D FOS equivalent, is a serious shortcoming of the current state of the art (Stacey 2007).

2D analyses also produce conservative indications of slope stability where worst-case scenarios are most often selected for 2D analysis. These worst-case scenarios will typically not be representative of all slope conditions (Sjoberg 1999, Duncan 2015, Wines 2015, Dana et al. 2018).

McQuillan et al. (2019) state that to reliably predict the performance (e.g. propensity for failure) and critical failure mechanism (including spatial location) of slope failure, geotechnical engineers must select appropriate tools to complete slope stability assessments. 3D LE is such a tool that can adequately account for the failure mechanisms typically observed in highly anisotropic geological settings.

This paper presents a parametric study that demonstrates the variability in FOS that is calculated for an anisotropic rock mass when 2D and 3D LE methods are applied. Sensitivity analysis are completed for linear and non-linear material models (e.g. Mohr-Coulomb, Generalized Hoek-Brown, Shear-Normal, Barton-Bandis) and slip surface search methods.

2 CASE STUDY

2.1 *Geological setting*

An approximately 1280 m section of slope from a design in an iron ore mine has been used in this case study, Figure 1. The iron ore mine is located in the Pilbara region of Western Australia. This case study slope intersects the Newman formation, which displays anisotropic strength, with weaker strengths observed parallel and sub-parallel to bedding. Planar sliding along adversely orientated bedding planes and/or interbedded, weaker shale bands is the most common failure mechanism in the Pilbara region (Bar 2012, Seery 2015, Bar and Weekes 2017). Failures of this type occur at single bench to overall slope scale.

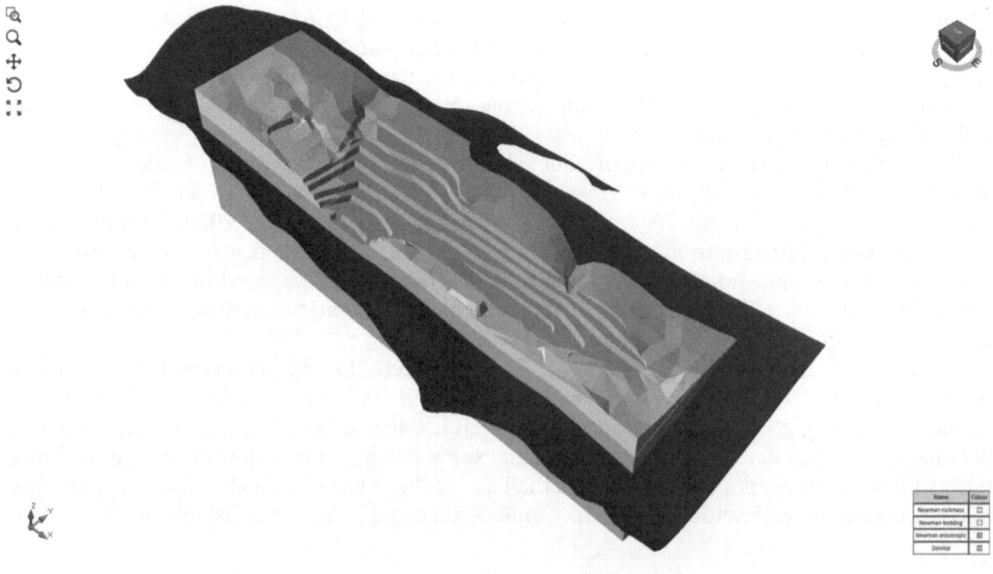

Figure 1. Parametric view of case study slope.

2.2 *Parametric analysis*

To quantify the difference in FOS calculated using 2D and 3D LE methods, a series of models were computed with the following variables:

1. 3D models using the mine design, natural surface topography and lithological surfaces exported from the geological model (Figure 1). The deposit was above the water table, so pore pressures were not considered. Anisotropy and true dip are included in 3D LE models;
2. 2D models using a 2D section cut through the middle of the critical FOS calculated from 3D modelling (Figure 2). Apparent dip is inherently included in 2D LE models;
3. 3D models derived by extruding the 2D section at varying lateral slope lengths (50 m, 100 m, 150 m, 200 m, 400 m). True anisotropy is not included in 2D extruded models. Instead, three dimensions are added by uniformly laterally extruding a 2D section;
4. Cuckoo vs Particle Swarm slip surface search methods, at varying search options (i.e. number of search surfaces and search depth limits); and
5. Linear (i.e. Mohr-Coulomb) and Non-Linear (Shear-Normal, Generalised Hoek-Brown and Barton Bandis) material strengths.

Rocscience, Inc.'s (2020) Slide2 (2D LE) and Slide3 (3D LE) software have been used to calculate the FOS of slope stability in this parametric study. Slide2 uses the method of slices, and Slide3 the method of columns, to sum the forces acting on the failure surface (i.e. mobilized stress) and compare these to the available shear strength.

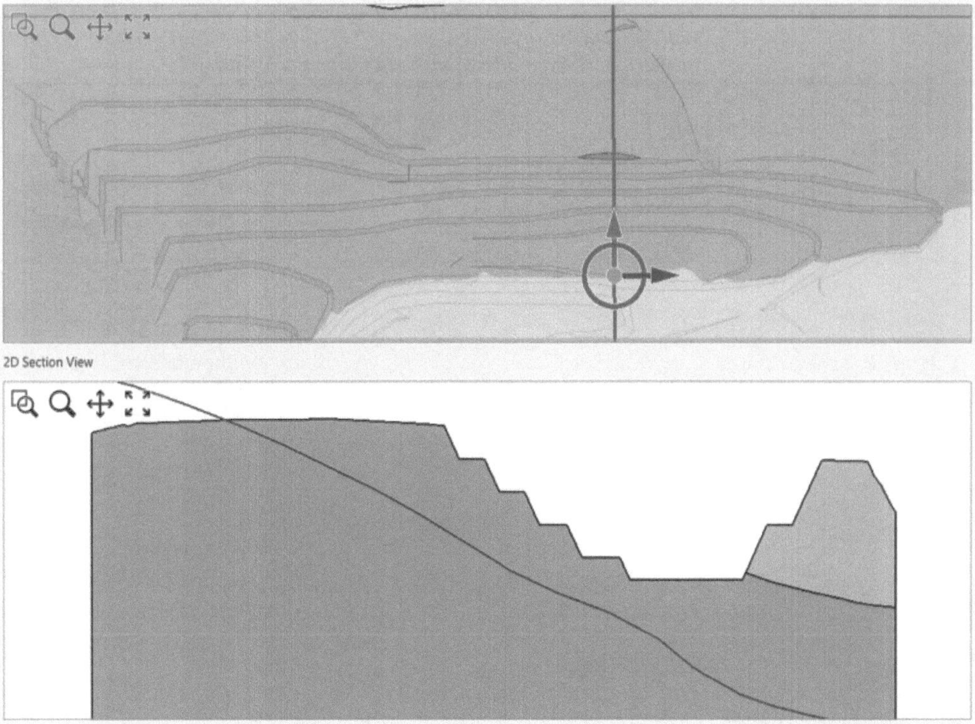

Figure 2. Plan view map showing location and cross-section cut for Slide2 analysis from the Slide3 model.

Cuckoo and Particle Swarm are two metaheuristic search method algorithms built into Slide2 and Slide3 to search for the slip surface with the lowest FOS within the model search extents. Slide2 and Slide3 allows the user to vary the number of slip surfaces generated for each search method. The user can also define the minimum depth that slip surfaces are generated below the external boundary to test multi-bench stability. These options were enabled in the sensitivity analysis to determine the change in FOS calculated with: (i) varying search method; (ii) varying number of slip surfaces; and (iii) varying slip surface location to be able to simulate multi-bench slope failures.

Material strengths applied to LE models are summarized in Table 1 and Figure 3. Non-linear (Shear-Normal equivalents of Generalized Hoek-Brown and Barton-Bandis) material models were applied to LE models. Linear (Mohr-Coulomb) approximations of non-linear shear-strength envelopes were also applied to LE models to test the sensitivity of FOS to varying strength criteria. Linear material models have historically been applied to 2D LE models analyzing inter-ramp or overall slope failure so material models that cannot use non-linear shear strength inputs can be applied. However, non-linear material models are preferred, as their linear-equivalent approximations: (i) overestimate apparent cohesion and total shear strength at low normal stresses (e.g. ≤ 0.3 MPa); (ii) overestimate apparent friction and total shear strength at high normal stresses (e.g. ≥ 1.1 MPa); and (iii) underestimates shear strength at a normal stress of approximately 0.7 MPa (Bar and Weekes 2017), Figure 3.

Table 1. Modelled material strengths.

Unit	Material Model	Model Parameters
Newman Shale – Bedding	Non-linear: Barton-Bandis (BB) and/or Shear-Normal	JRC = 2, JCS = 9, ϕ = 24 °
	Linear Approximation: Mohr-Coulomb (MC)	c = 9 kPa, ϕ = 26°
Weathered Newman Member – Rock mass	Non-linear: Generalised Hoek-Brown (GHB) and/or Shear-Normal	σ_c = 35 MPa, GSI = 40, mi = 10, D = 0.5
	Linear Approximation: Mohr-Coulomb	c = 210 kPa, ϕ = 44°

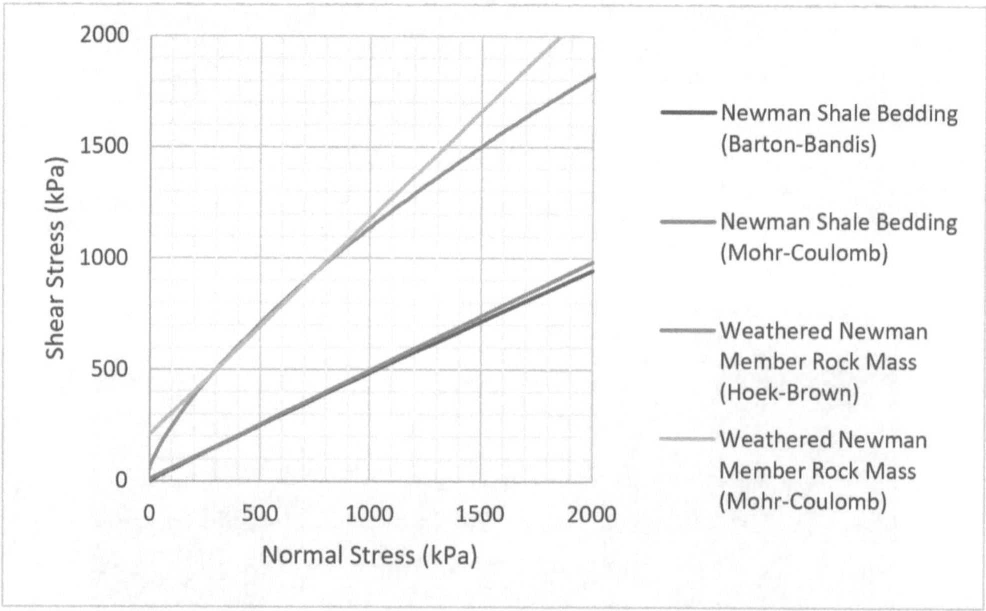

Figure 3. Graphical display of shear strengths applied to Newman Shale bedding and rock mass units.

Anisotropy was included in 2D and 3D models using the Snowden Modified Anisotropic Linear (Mercer 2012, Mercer 2013) material model in Slide2 and Generalized Anisotropic material model in Slide3. These material models allow differential strengths for bedding and rock mass to be applied on the slip surface. A-values of 5° and B-values of 30° were applied in both 2D and 3D models, based on Bar and Weekes (2017).

2.3 *Results*

Parametric study results are summarized in Table 2 to Table 4, Figure 10 and Figure 12. FOS are reported for the GLE calculation method.

Table 2. Parametric study results – 3D LE analyses.

Dimension	Material Model	Search Method	Number of Nests/ Particles	Slope Depth Limit	Critical FOS_{GLE}	Reference Figure
3D	Non-linear (GHB + BB)	Cuckoo	20	None	0.90	Figure 4 A
		Cuckoo	80		0.89	Figure 4 B
		Particle Swarm	20		0.90	Figure 4 C
		Particle Swarm	80		0.81	Figure 4 D
3D	Non-linear (GHB + BB)	Cuckoo	20	15 m	1.13	Figure 5 A
		Cuckoo	80		1.00	Figure 5 B
		Particle Swarm	20		1.08	Figure 5 C
		Particle Swarm	80		1.06	Figure 5 D
3D	Linear (MC)	Cuckoo	20	None	1.10	Figure 6 A
		Cuckoo	80		1.11	Figure 6 B
		Particle Swarm	20		1.06	Figure 6 C
		Particle Swarm	80		1.11	Figure 6 D
3D	Linear (MC)	Cuckoo	20	15 m	1.12	Figure 7 A
		Cuckoo	80		1.10	Figure 7 B
		Particle Swarm	20		1.14	Figure 7 C
		Particle Swarm	80		1.09	Figure 7 D

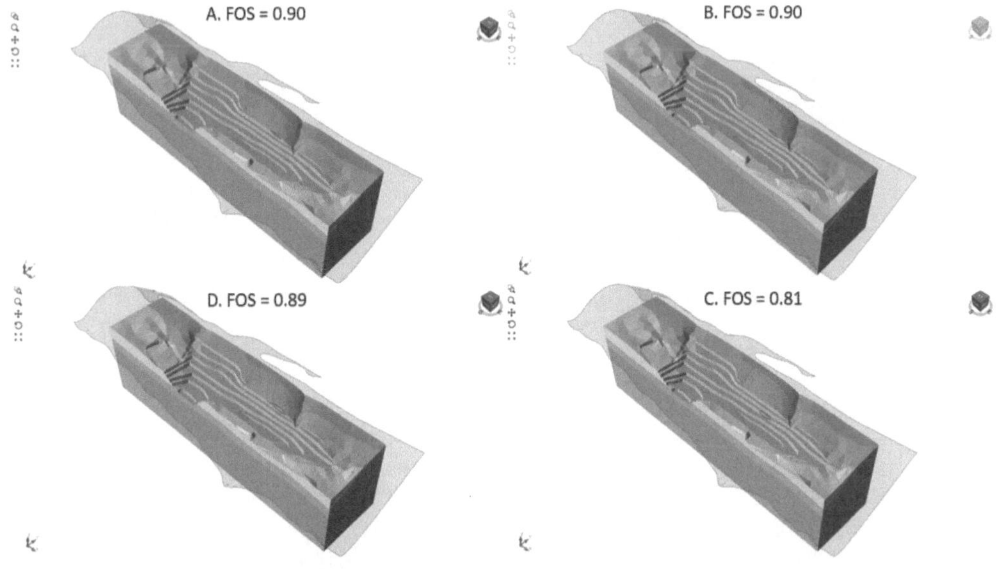

Figure 4. 3D LE non-linear GLE FOS, unrestrained search limit.

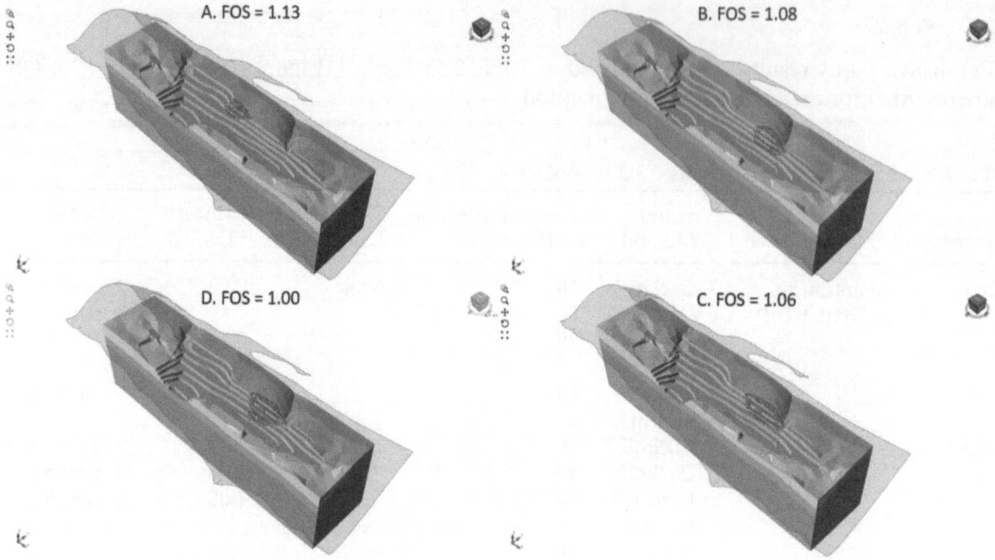

Figure 5. 3D LE non-linear GLE FOS, 15 m depth search limit.

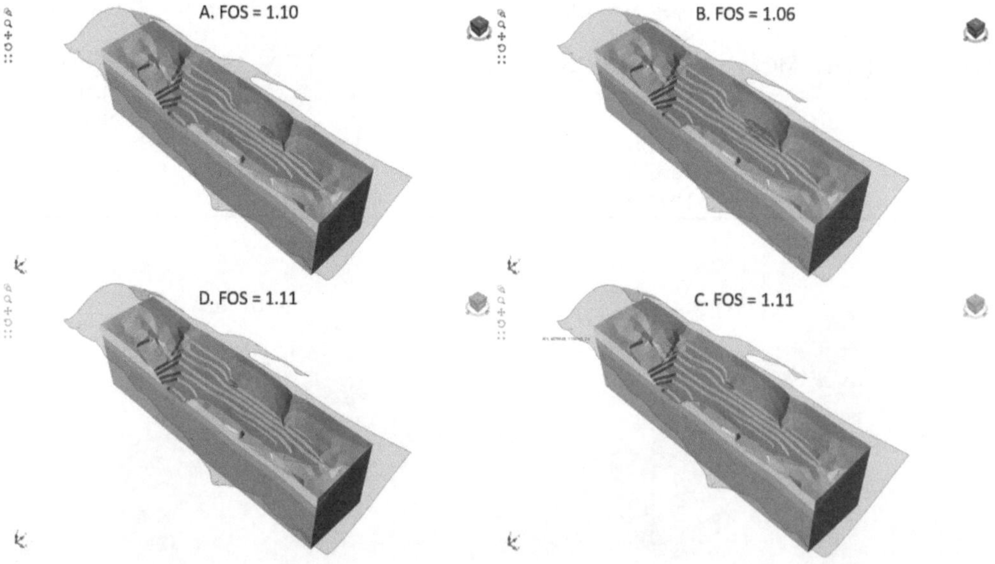

Figure 6. 3D LE linear GLE FOS, unrestrained search limit.

3 DISCUSSION AND RECOMMENDATIONS

Sensitivity analyses of 2D and 3D search methods, non-linear and linear material models, search method, number of slip surfaces, and search limit depth of an anisotropic rock mass indicate the following:

Sensitivity analysis with different search methods and different numbers of slip surfaces should be completed to determine the critical FOS. The number of slip surfaces calculated

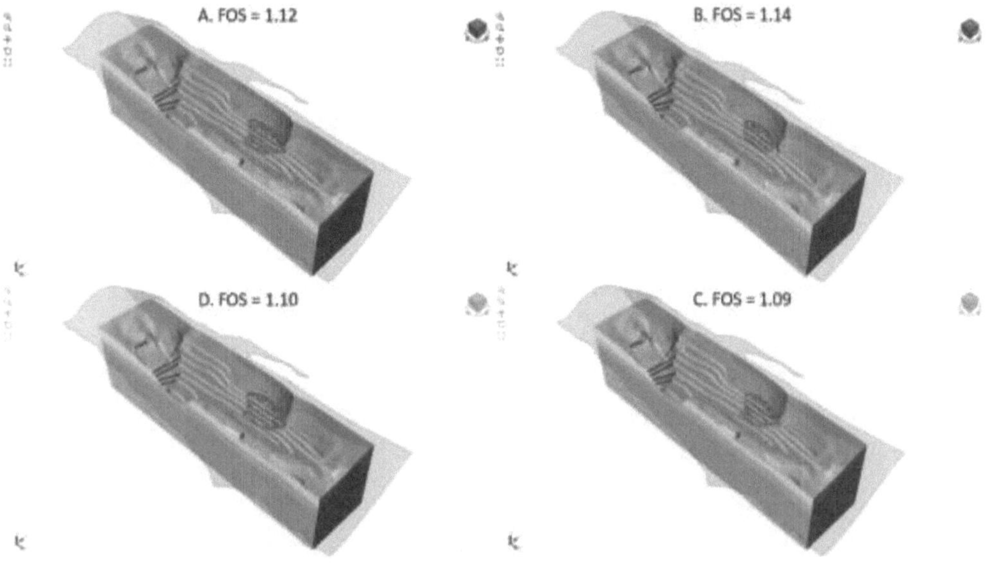

Figure 7. 3D LE linear GLE FOS, 15 m depth search limit.

Table 3. Parametric study results – 2D LE analyses.

Dimension	Material Model	Search Method	Slope Depth Limit	Critical FOS_{GLE}	Reference Figure
2D	Non-linear	Cuckoo	None	0.73	Figure 8 A
	(GHB + BB)	Particle Swarm		0.84	Figure 8 B
	Linear (MC)	Cuckoo		0.84	Figure 8 C
		Particle Swarm		0.88	Figure 8 D
2D	Non-linear	Cuckoo	15 m	0.80	Figure 9 A
	(GHB + BB)	Particle Swarm		0.80	Figure 9 B
	Linear (MC)	Cuckoo		0.85	Figure 9 C
		Particle Swarm		0.85	Figure 9 D

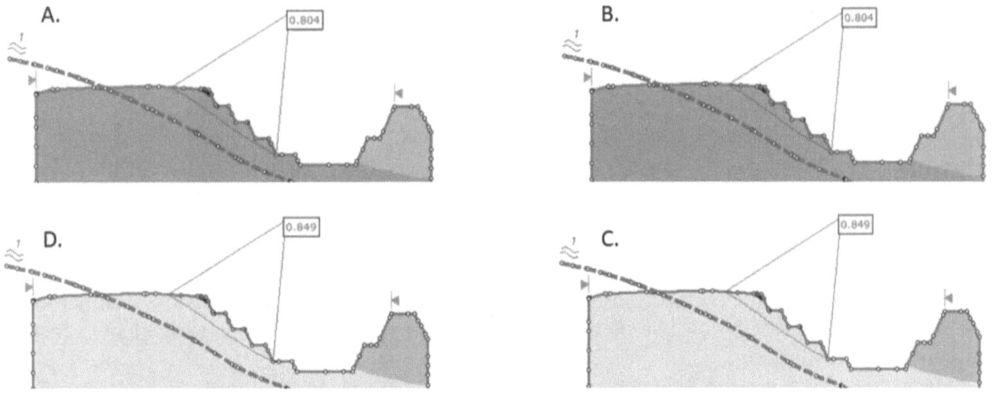

Figure 8. 2D LE non-linear GLE FOS, unrestrained search limit.

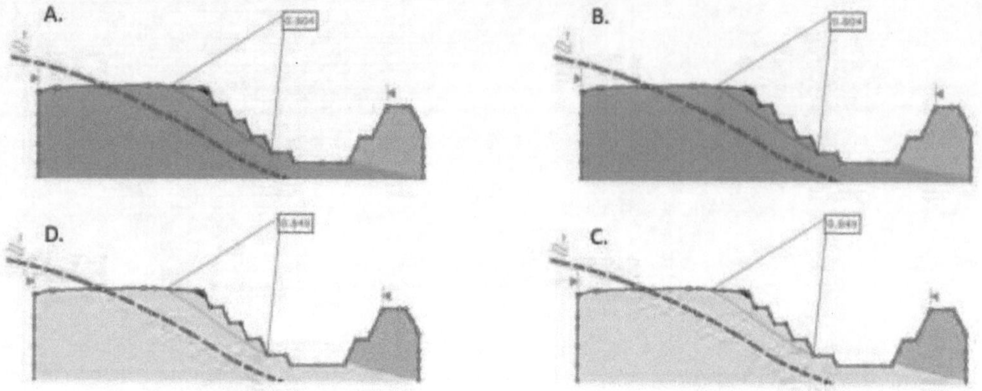

Figure 9. 2D LE non-linear GLE FOS, 15 m search limit.

Table 4. Parametric study results – 3D (2D extruded) LE analyses.

Dimension	Material Model	Search Method	Slope Depth Limit	Number of Slip Surfaces	Critical FOS_{GLE}	Reference Figure
3D (extrusion length = 50 m)	Non-linear	Cuckoo	None	826	0.85	Figure 10
3D (extrusion length = 100 m)				1665	0.86	
3D (extrusion length = 200 m)				3288	0.85	
3D (extrusion length = 400 m)				6569	0.80	
3D (extrusion length = 50 m)	Non-linear	Cuckoo	15 m	845	1.15	Figure 10, Figure 11A
3D (extrusion length = 100 m)				1665	0.97	Figure 10, Figure 11B
3D (extrusion length = 200 m)				3306	0.97	Figure 10, Figure 11C
3D (extrusion length = 400 m)				6568	0.95	Figure 10, Figure 11D

should adequately cover the section of slope being analyzed. Default software settings may not produce an adequate number of slip surfaces in the search to find the critical FOS in the slope under investigation.

Non-linear material models generally result in a lower FOS being calculated, compared to linear material models. Previous publications by Bar and Weekes (2017) support these findings that non-linear material models are better suited to model anisotropic rock masses where shallow failures are expected along pitwards dipping structures.

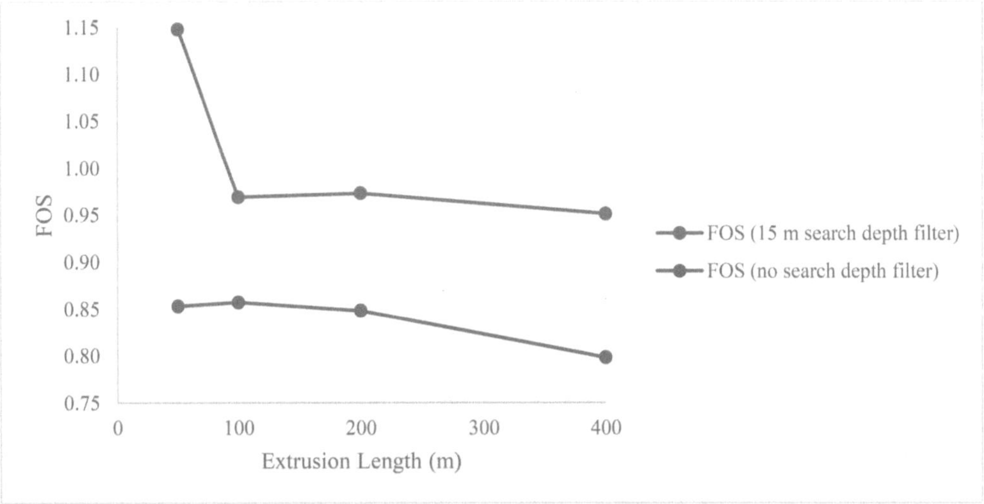

Figure 10. 2D extruded non-linear GLE FOS.

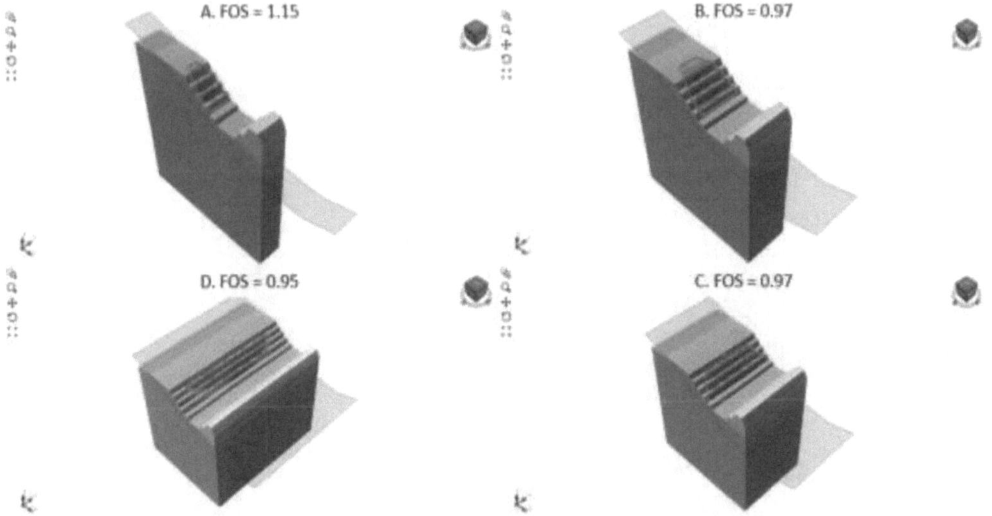

Figure 11. 2D extruded FOS, 15 m depth search limit. Slip surfaces are constrained at edge of Slide3 model at 50 m, 100 m and 200 m extruded depths.

Applying search depth limits provides an estimate of multi-bench stability, which generally have higher FOS compared to scenarios where no search depth limits are applied.

The length of 3D models, created from the extrusion of 2D sections, will impact the FOS calculated. The shorter the width (i.e. the more confined) the extruded model, the higher the FOS will be. Slip surfaces should be reviewed to ensure their extents are not constrained to the boundaries of the 3D model. If this is condition is observed, the critical FOS may not have been found.

The FOS calculated from 2D analyses are generally lower (average 25%) than the FOS calculated from 3D analysis. The difference in FOS is more pronounced in models with linear material models applied (average FOS 30% lower). In this parametric study the increase in 3D

FOS is attributed to the inclusion of rock mass strength at the sides of the failure surface (i.e. end effects) in the anisotropic rock mass, Figure 12, Figure 13 and Figure 14. Such forces will not be included in 2D analyses.

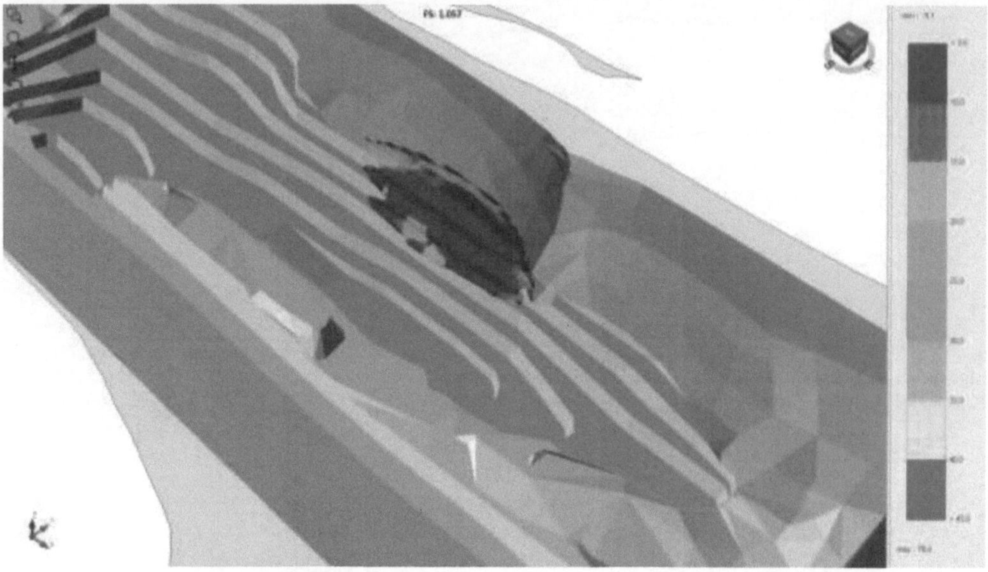

Figure 12. Perspective view of critical slip surface showing variation in cohesion applied to column bases relative to applied Generalised Anisotropic strength function.

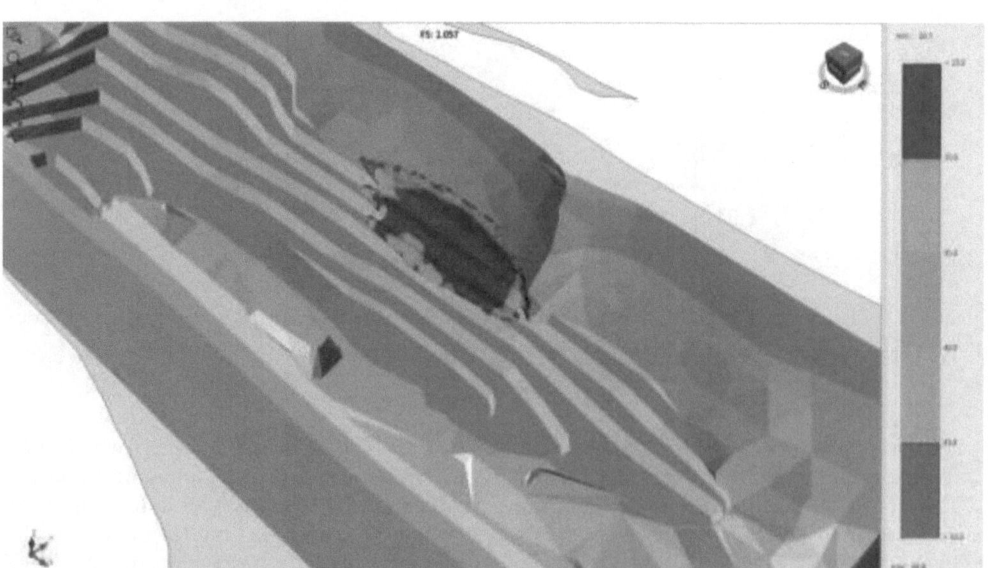

Figure 13. Perspective view of critical clip surface showing variation in friction angle applied to column bases relative to applied Generalised Anisotropic strength function.

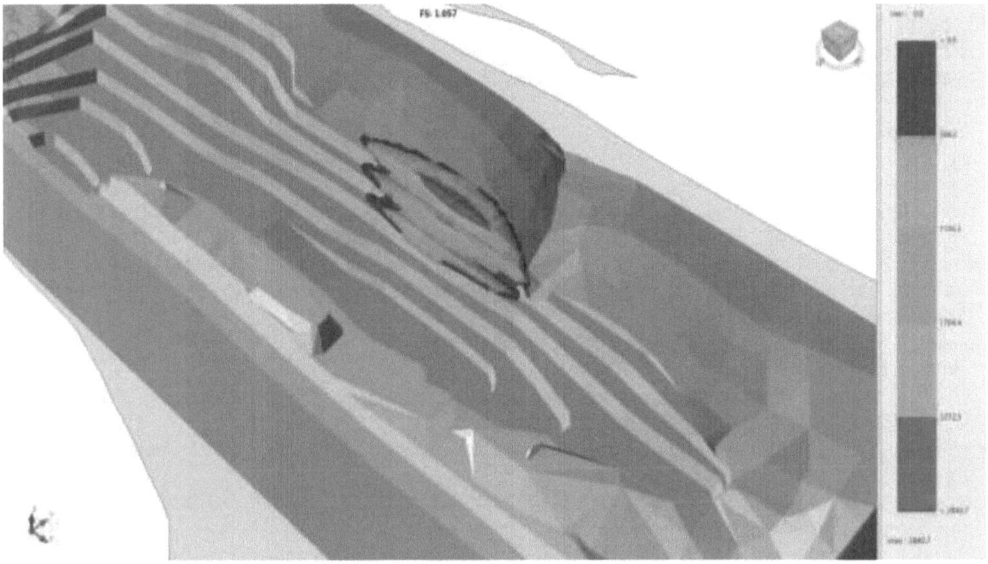

Figure 14. Perspective view of critical clip surface showing variation in base shear force along column bases.

4 CONCLUSIONS

3D models allow geotechnical engineers to model the true ground conditions which are generally always anisotropic in nature. This paper has demonstrated the variance in FOS when calculated using 2D and 3D LE methods. The case study utilized in the parametric study has shown that for anisotropic rock masses, 3D LE consistently produces a higher FOS. This is expected where 3D analysis allows a more accurate representation of failure mechanism, as the slip surface that is created considers the strength of the sides of the slip surface, which are not considered in 2D analysis.

3D LE software is now widely commercially available. Such tools provide geotechnical engineers a means of rapidly predicting slope performance to assist in the risk management and slope optimisation open pit slope designs.

REFERENCES

Bar, N. 2012. Performance driven slope management, design and optimization at Brockman Operation, Rio Tinto Iron Ore. In Proceedings: 9th Young Geotechnical Professionals Conference (9YPGC), St Kilda, Australian Geomechanics Society, pp. 19–24.

Bar N, Weekes G. Directional shear strength models in 2D and 3D limit equilibrium analyses to assess the stability of anisotropic rock slopes in the Pilbara region of Western Australia. Australian Geomechanics Journal 2017; 52(4); 91–104.

Bar N, McQuillan A. 3D Limit Equilibrium Slope Stability Analysis for Anisotropic and Faulted Rock Masses in Australian Coal and Iron Ore Mines, In: Proc. ISRM International Symposium - ARMS10, Singapore, 2018; 12p.

Bar, N., Kostadinovski, M., Tucker, M., Byng, G., Rachmatullah, R., Maldonado, A., Pőtsch, M., Gaich, A., McQuillan, A., Yacoub, T. 2020. Rapid and robust slope failure appraisal using aerial photogrammetry and 3D slope stability models. International Journal of Mining Science and Technology. https://doi.org/10.1016/j.ijmst.2020.05.013

Bar N, Ryan C, Yacoub TE, McQuillan A, Coli N, Leoni L, Harries N, Rea SA, Bu J. Integration of 3D limit equilibrium models with live deformation monitoring from interferometric radar to identify and manage slope hazards. Proc. ISRM 14th International Congress of Rock Mechanics, Iguassu Falls, Brazil, 2019a.

Bar N, Yacoub TE, McQuillan A. Analysis of a large open pit mine in Western Australia using finite element and limit equilibrium methods. Proc. ARMA 2019: 53rd US Rock Mechanics/Geomechanics Symposium, New York, United States of America, 2019b.

Bromhead E. Landslide slip surfaces: their origins, behavior and geometry, Landslides: Evaluation and Stabilization 2004, Taylor & Francis Group, London.

Cala, M., Cyran, K., Kakóbczyk, J., Kowalski, M. 2020. The challenges of open pit mining in the vicinity of the salt dome (Bełchatów Lignit deposit, Poland). Energies. Vol. 13. doi:10.3390/en13081913

Cheng Y, Yip C. Three-dimensional asymmetrical slope stability analysis extension of Bishop's, Janbu's, and Morgenstern-Price's techniques. J. Geotech. Geoenviron. Eng 2007; 133(12); 1544–1555.

Dana, H., Kakaie, R., Rafiee, R., Bafghi, Y. 2018. Effects of geometrical and geomechanical properties on slope stability of open-pit mines using 2D and 3D finite difference methods. Journal of Mining & Environment. Voly. 9(4), pp. 941–957.

Duncan J. State of the art: Limit equilibrium and finite-element analysis of slopes. J. Geotech. Engng 1996; 122(7); 577–596.

Lorig L, Varona P. Numerical Analysis, Rock Slope Engineering, Civil and Mining, 4th Edition. C. Wyllie and C.W. Mah (eds). Spon Press; 2007.

Mercer, K. 2012. The history and development of the anisotropic linear model: part 1. Australian Centre for Geomechanics, Perth. Western Australia.

Mercer, K. 2013. The history and development of the anisotropic linear model: part 2. Australian Centre for Geomechanics, Perth. Western Australia.

McQuillan A, Canbulat I, Oh J. 2020. Methods applied in Australian industry to evaluate coal mine slope stability. International Journal of Mining Sciences and Technology. Vol 30(2), pp. 151–155. https://doi.org/10.1016/j.ijmst.2019.11.001

Seery, J. 2015. Limit equilibrium analysis of a planar sliding example in the Pilbara Region of Western Australia – comparison of modelling discrete structure to three anisotropic share strength models. In Proceedings: SAIMM Slope Stability 2015, Cape Town, pp. 681–696.

Shamsoddin Saeed, M., Maarefvand, P., Yaaghubi, E. 2015. Two and three-dimensional slope stability analyses of final wall for Miduk mine. International Journal of Geo-Engineering. DOI 10.1186/s40703-015-0009-0.

Sjoberg J. Analysis of Large Scale Rock Slopes, PhD Thesis, Lulea University of Technology Department of Civil and Mining Engineering Division of Rock Mechanics, Sweden; 1999.

Stacey, T. 2007. Slope stability in high stress and hard rock conditions. In Proceedings: International Symposium on Rock Slope Stability in Open Pit Mining and Civil Engineering (Slope Stability 2007). Perth, pp. 187–200.

Wines D. A comparison of slope stability analyses in two and three dimensions. In: Proc. SAIMM Slope Stability 2015, Cape Town, South Africa, 2015; 305–316.

Wines, D. A comparison of slope stability analyses in two and three dimensions. The Journal of the Southern African Institute of Mining and Metallurgy. Vol. 116, pp. 399–406.

Session 8 - Numerical slope stability analysis I

The Evolution of Geotech - 25 Years of Innovation – Hammah et al. (Eds)

Numerical profile modeling for transient groundwater flow at pit slope

E.V. Leonteva
Associate Professor at the Department of applied geology and mining, PhD in Geography, Belgorod State University, Belgorod, Russia

R.Yu. Sapachev
Chief specialist (Hydrogeologist, Rocs Mechanics Engineer, Engineering geologist), EuroChem-Project LLC, Saint-Petersburg, Russia

ABSTRACT: Geomechanical and Hydrogeological tasks often require knowledge of groundwater level at different time-points. In usual hydrogeological practice, groundwater flow modeling is based on steady-state flow. As such, modeling results can provide analysis of groundwater final state only. More detailed results can be obtained if hydrogeological modeling is made for non-steady state (i.e., transient) groundwater flow. This paper describes approach of applying numerical profile modeling for transient flow to determine groundwater level, height of seepage and hydrodynamic flow parameters at various time-points for pit slope stability analysis using Slide2 software tools (Rocscience Inc.).

1 INTRODUCTION

Transient simulations are rather cumbersome, and 3D models are thought to suit better for transient models, whereas profile models - for steady-state flow simulations (Anderson et al. 2015). However, application of 3D numerical flow simulation software for this purpose requires much time, and is sometimes impossible due to lack of input data. In this case Slide SW complex shows promising results when applied to numerical profile modeling for non-steady state groundwater flow and uses minimum initial data. Methodology of non-steady state flow profile modeling at pit slope, as proposed, contains four main steps. Time-periods for modeling should be adjusted following the scenario applied.

2 METHODOLOGY OF NON-STEADY STATE FLOW PROFILE MODELING AT PIT SLOPE

Step One. This step addresses a number of key issues, i.e. setup of modeling area external boundaries, boundaries for areas of hydraulic properties variations; assigning permeability model and hydraulic parameters.

External Boundary. 2D non-steady state flow profile modeling starts from setting external boundary conditions. Model upper external boundary is usually in align with topography. Lowest external boundary is set by preference - considered impermeable, it is usually 500-1000 m more from the deepest point of mining (Beale & Read (2013)). Left boundary at open-pit slope is the most relevant model boundary. Model right boundary is set as boundary of

DOI: 10.1201/9781003188339-39

constant head in order that hydraulic effects could have no impact on it. This means, model right boundary should be placed at distance of open-pit radius of influence.

To calculate open-pit radius of influence, formulas below can be used:

1) Kusakin-Kerkis equation is applied in terms of unconfined water flows and minings with very large cross-section (open-cast minings, open pits, etc.), and significant drawdowns (Abramov et al. 1979).

$$R = \sqrt{r_0 + 30kmS_0(1 + 0.00015r_0^2)} \tag{1}$$

where r_0 = open-pit radius, m; k = hydraulic conductivity m/day; m = saturated thickness of unconfined aquifer, m; S_0 = drawdown in the pit, m.

2) Equation implements recharge for calculation (Mironenko et al. 1965):

$$R\sqrt{\lg R - \lg r_0 - 0.217} = 0.66\sqrt{\frac{k}{\varepsilon}(2m - S_0)S_0 - 0.5r_0^2} \tag{2}$$

where k = hydraulic conductivity m/day; ε = recharge, m/day; m = saturated thickness of unconfined aquifer, m; S_0 = drawdown in the pit, m; r_0 = open-pit radius, m. If not available, recharge can be replaced with the value of 10% from subsurface water flow module.
Based on the method described, open-pit radius of influence can be calculated automatically in ANSDIMAT SW.

3) Equation applies recharge and groundwater (GW) inflow to the open pit (Kresic 2006).

$$R = \sqrt{\frac{Q}{\pi\varepsilon}} \tag{3}$$

where Q = total inflow to the open pit, m^3/day; ε= recharge, m/day.

When calculated, model right boundary is placed at distance of open pit radius of influence.

Material Boundary. Material Boundary in Slide2 is used to identify areas of interest inside external model boundaries and assign them various properties on the basis of hydraulic properties, which are different for different soils (materials). For both External and Material Boundaries, Slope Stability tab is used.

Define Hydraulic Properties. Permeability model Simple should be selected. As far as rocks water saturation is 1, saturated permeability (Ks) is considered the same as hydraulic conductivity (k). The volumetric water content is the volume of water as a proportion of the total volume. You can think of water content as the porosity times the degree of saturation where the degree of saturation ranges from 0 (dry) to 1 (saturated).

Step Two. This step contains discretization and assigns boundary conditions.

Mesh and Discretization Setting. In Mesh Setup dialogue – number of elements and their type are assigned. The function “Use Discretization Region” can be applied for more detailed segmentation, if necessary.

Boundary Conditions (BC). In Steady State Groundwater tab (Slide2) boundary conditions are assigned. When applied to numerical profile modeling for non-steady state groundwater flow, the following BCs are used: impermeable boundary None/Remove BC; BC Total Head; BC Unknown. BC Total Head is applied to model right boundary, pit slope and pit bottom. As such, these BCs allow modeling initial GW level. Upper boundary should have either Vertical Infiltration or Unknown condition. Lowest boundary to be set as impermeable.

Step Three. This step includes transient boundary conditions setup and assign of Discharge function to different model sections in order to calculate the flow rate at various time-points.

Transient Boundary Conditions. Transient BCs are set in Transient Groundwater tab. BC Unknown is assigned for pit slope. Left boundary and pit bottom should have BC Total Head representing the position of the groundwater level, other conditions are constant and have no difference with initial ones.

Discharge function. Used as indicator of the flow rate stabilization, Discharge function is applied to identify the time-period, as required for GW level to achieve its steady state. Flow rate calculations are performed for different model sections and time-points.

Step Four. It is final interpretation of modeling results. Numerical profile modeling of transient groundwater flow is applied to determine final GW level, height of seepage at pit slope and hydrodynamic parameters of the flow - at different points of time. If required, time-period for the flow to achieve its steady state can also be calculated.

3 COMPARATIVE ANALYSIS. TEST CASE

Comparative analysis is based on Test case for the tasks as follows: to determine final GW level, seepage height for left model boundary and to calculate time-period for GW flow to achieve its steady state using Discharge function. As applied to one and the same test case to decide on saturated thickness in different model sections at one and the same moments of time, comparison of Slide2 modeling results and results of analytical calculations is provided.

Following Step One of methodology described, modeling area was identified and external boundaries for upper, lowest, right and left model boundaries were set. For the purpose of comparative analysis accuracy, rectangular model was created, for which an analytical solution can be provided. Left boundary was placed at distance of 1357 m from the right one after pit radius of influence was calculated, using equation of Kusakin-Kerkis (1) based on pit radius value (r_0) (that is 258 m), and had constant head (90 m). In terms of the whole model, hydraulic conductivity parameter of 0.69 m/day was assigned, water content saturated was 0.04 and K2/K1 = 1. Permeability model - Simple.

Step Two: discretization and type of element (6 Noded Triangles) were selected. Initial BCs assigned as follows: upper boundary – BC Unknown, for left and right boundary – as responsible for initial GW level - BC Total Head (90 m) was set (Figure 1).

Step Three should provide transient boundary conditions. As such, transient BC Unknown was assigned only to the left boundary. Other conditions were constant and had no difference with initial ones. Modeling was performed for 10 time-periods (stages). To calculate the time-period, as required for GW flow to achieve its steady state, Discharge function was applied for three model sections (as indicator of flow rate stabilization).

Interpretation of modeling results (Step Four) for test case is represented in Figure 1 and Table 1.

Should Test Case modeling results be followed, stabilization of GW level starts in 3000 days and then reaches final steady state. Height of seepage at left boundary is 2.82 mss and has no changes after 3000 days period.

Verification calculations for saturated thickness at distance x are based on application of analytical calculations for confined flow with constant head at remote boundary if water level is instantaneously changed (Verigin 1975):

$$U_x = U_e + (U_2 - U_1)\overline{Z_1}(\overline{X}, \tau) \tag{4}$$

where Ue = head function for GW natural level (model right boundary); U_1 = head function for GW initial level (model left boundary); U_2 = head function for GW level after instantaneous decrease of GW level (model left boundary); U_x = head function at x distance to model left boundary; $Z_1(\bar{x}, \tau)$ = dependence function; $\bar{x}$ = non-dimensional distance; τ - non-dimensional time.

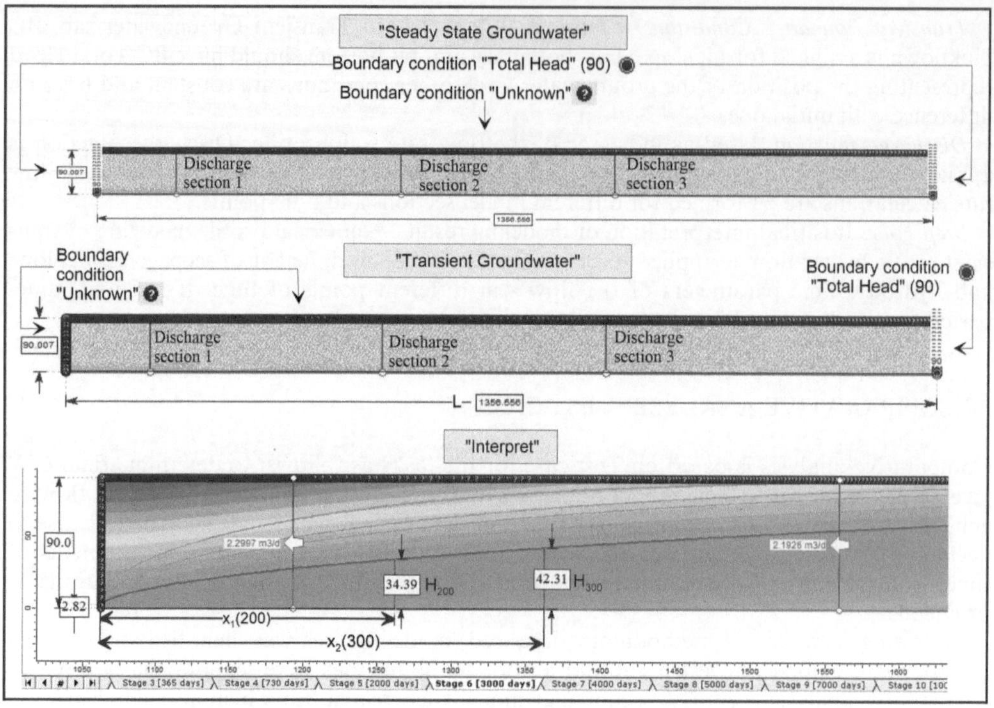

Figure 1. Test case. Modeling results (Slide2).

Table 1. Test case. Modeling results (Slide 2).

Name	Time (days)	Discharge section, m^3/d			Height of seepage meters subsea, mss
		№ 1	№ 2	№ 3	
Stage 1	100	8.08	1.11	0.09	13.19
Stage 2	250	5.96	2.26	0.56	8.49
Stage 3	365	5.09	2.53	0.90	6.92
Stage 4	730	3.79	2.55	1.48	5.05
Stage 5	2000	2.59	2.27	1.99	3.17
Stage 6	3000	2.30	2.19	2.10	2.82
Stage 7	4000	2.21	2.17	2.13	2.82
Stage 8	5000	2.18	2.16	2.15	2.82
Stage 9	7000	2.16	2.16	2.16	2.82
Stage 10	10000	2.16	2.16	2.16	2.82

Head function at any point of layer is calculated as follows:

$$U_i = 0.5H_i^2 \tag{5}$$

where Hi = saturated thickness in any cross-section, m. To calculate $\overline{x}$ and τ the following equations are used:

$$\overline{X} = \frac{X}{L}; \tau = \frac{at}{L^2} \tag{6, 7}$$

where x = is the distance from model left boundary to dedicated cross-section, m; L = model length, m; a = hydraulic diffusivity of unconfined aquifer, m^2/day; t = time-period, days.

For $\tau \geq 0.2$ with 1% accuracy it can be assumed that:

$$\overline{Z} = 1 - \overline{X} - \frac{2}{\pi} e^{-\pi^2 \tau} \sin \pi \overline{X} \tag{8}$$

Input parameters for calculations are provided in Figure 2.

Formula (5) was applied to calculate saturated thickness at distances of 200 m and 300 m from model left boundary, other conditions - the same as in Slide2. H2 – is height of seepage, as obtained after modeling in Slide2. Time period for calculations – 3000 days. Calculation results are represented in Table 2.

Test case comparative analysis shows high convergence of modeling results, as performed in Slide2, and results of analytical calculations (Table 3).

Table 2. Input parameters and analytical calculation results for saturated thickness at x distance from model left bundary.

Parameter	Symbol	X=200	X=300
Head function for H_1	U_1	4050	4050
Head function for H_2	U_2	3.976	3.976
Head function for H_e	U_3	4050	4050
Head function for H_x	U_x	600.29	898.45
Saturated thickness at x distance	H_x,m	34.65	42.38

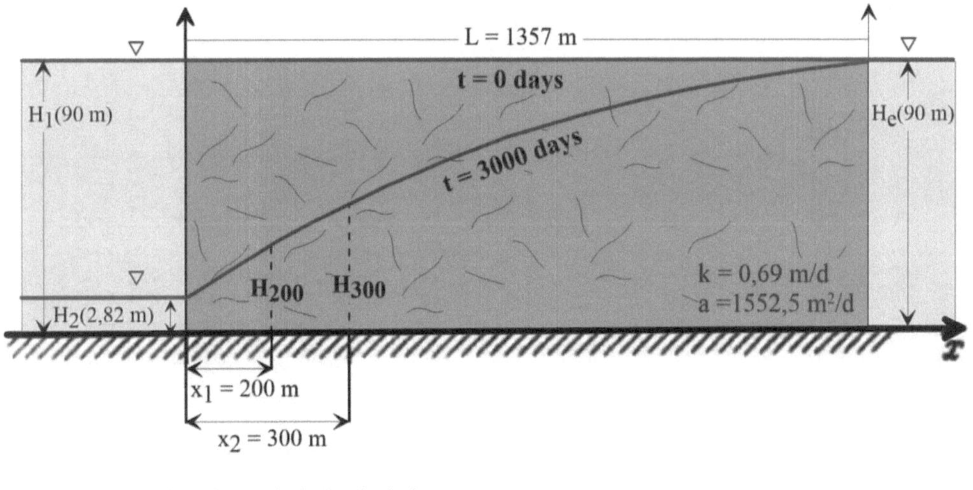

Figure 2. Input data for analytical calculations.

Table 3. Results of comparative analysis for head of flow at x distance from model left boundary in 3000 days.

Parameter	Slide2	Analytical calculations	Discrepancy
Saturated thickness at distance of 200 m (H_{x1}), m	34.39	34.65	0.26
Saturated thickness at distance of 300 m (H_{x2}), m	42.31	42.38	0.07

4 NUMERICAL PROFILE MODELING FOR TRANSIENT FLOW AT PIT SLOPE. SITE

4.1 *Methodology Assessment*

Numerical profile modeling for transient flow at pit slope was performed for 14 time-periods (stages), the last stage corresponds to mine lifetime and is equal to 55 years. Modeling area was identified on the basis of pit geometry, topography and radius of influence. Open-pit radius of influence was calculated via formula (1). Thus model right boundary was placed at distance of 9468 m from the pit. Based on hydraulic properties variations (as shown by hydraulic conductivity damping with depth increase), 7 (seven) aquifers were identified. Their hydraulic properties are represented in Figure 3.

Under Step Two, model discretization into 8000 elements was performed (type of element - 6 Noded Triangles). To increase accuracy of calculation in terms of seepage height more detailed segmentation for pit slope area was provided. Natural GW level was modeled for initial BCs (BC Total Head), as applied to model left and right boundaries. For upper boundary – BC Unknown was set, as infiltration at this site had no significant impact on GW level. Lowest model boundary was set as impermeable.

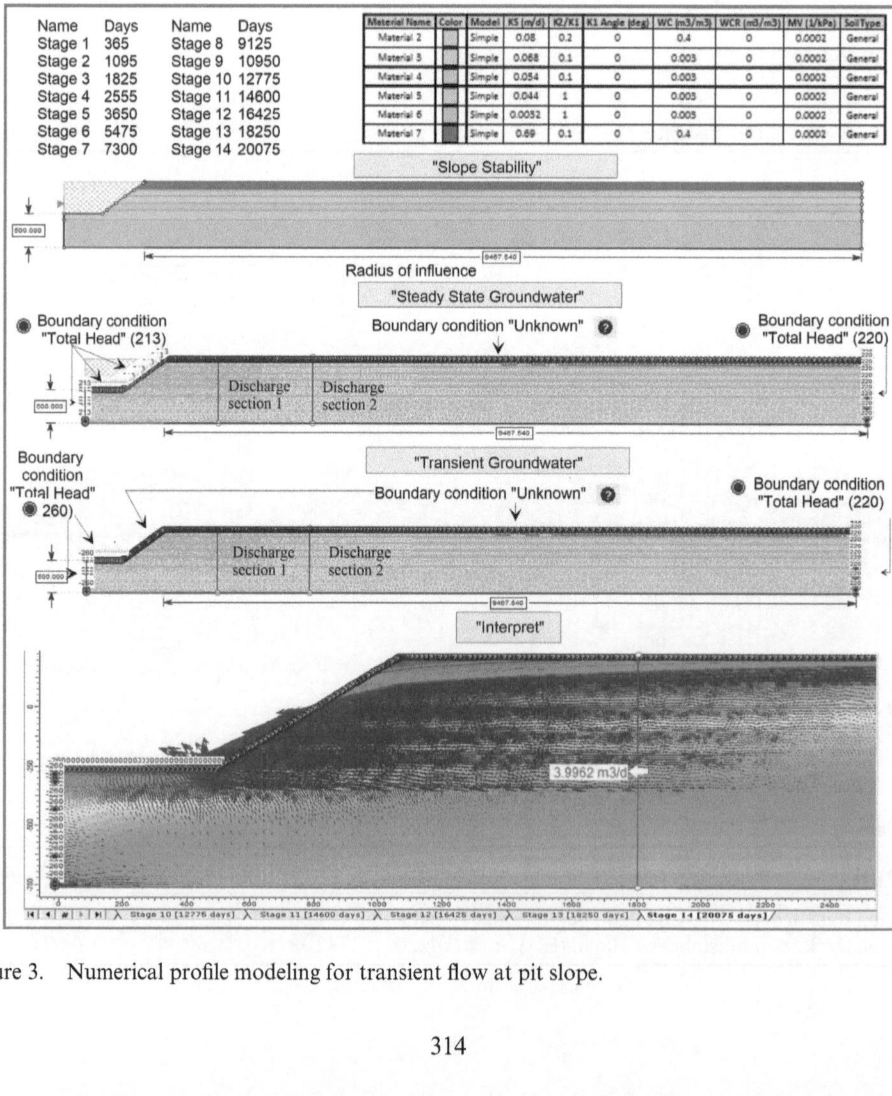

Figure 3. Numerical profile modeling for transient flow at pit slope.

Step Three required only adjustment of model left boundary BC, other conditions had no difference with initial ones: model left boundary (pit slope) - BC Unknown assigned. For model right and pit bottom (mss) – BC Total Head was applied. Two Discharge sections were created for flow rate control.

Interpretation of modeling results (Step Four) for site is represented in Figure 3. Numerical profile modeling of transient groundwater flow at pit slope has provided final GW level, height of seepage at pit slope, rate and hydrodynamic parameters of the flow for 14 time-periods (stages). Comparison of modeling results with field measurements, as performed via observation well (230 m away) shows discrepancy in 2.5% from the observed data. Modeling results are represented in Figure 3.

5 SUMMARY

Test case comparative analysis shows high convergence of modeling results, as performed in Slide2, and results of analytical calculations. Thus, numerical profile modeling of transient groundwater flow in Slide2 allows to quickly obtain the height of seepage at pit slope, GW level at various time-points and other hydrodynamic parameters of flow for the purpose of slope stability analysis, using minimum initial data. Application of 3D numerical flow simulation software for the same purpose requires lots of time, and is sometimes impossible due to lack of input data. In this case Slide2 SW complex shows promising results when applied to numerical profile modeling for transient flow in complex geological and structural situations and in conditions of hydraulic properties variations.

REFERENCES

Abramov S.K. et al. 1979. Spravochnoye rukovodstvo gidrogeologa [Guidelines for Hydrogeologist] Vol. 1, 2 In V.M. Maksimov (ed.). Leningrad: Nedra.

Anderson M. P., Woessner W.W., Hunt R.J. 2015. *Applied Groundwater Modeling: Simulation of Flow and Advective Transport*. Academic Press.

Beale G., Read J. 2013. *Guidelines for Evaluating Water in Pit Slope Stability*. Collingwood: CSIRO Publishing.

Kresic Neven. 2006. *Hydrogeology and Groundwater Modeling, Second Edition*. CRC Press.

Mironenko V.A., Norvatov Yu.A., Bokiy L.L. 1965. *Filtracionnye raschety osusheniya karernyh polej [Filtration calculations for drainage of quarry fields*Materials for Methodical Guide on Drainage of Mineral Deposits, subject to open-cut mining/State Committee on Fuel Prospecting at Gosplan of USSR. All-Union Scientific and Research Institute of Mining Geomechanics and Mine Surveying "VNIMI"; Part 2. Leningrad. (in Russian)*]* .

Verigin N. 1975. *Filtratsiya iz vodokhranilishch i prudov [Flows from Water Storage Facilities and Ponds]*. Moscow: Kolos. (in Russian).

The Evolution of Geotech - 25 Years of Innovation – Hammah et al. (Eds)

Soil-structure interaction of Mechanically Stabilized Earth (MSE) retaining walls subjected to construction sequencing and seismic loading

Lance Leighton T. Clemente*, Giancarlo P. Ventura & Pher Errol B. Quinay
Institute of Civil Engineering, College of Engineering, University of the Philippines, Diliman, Quezon City, Philippines

ABSTRACT: Mechanically stabilized earth (MSE) wall systems are a retaining wall system consisting of facing elements, soil reinforcement, and backfill material. To accurately model this soil-structure, soil structure elements or interface layers are critical to modelling since the load transfer from geosynthetics to backfill material is affected by backfill properties, geogrid length, geogrid spacing, and material properties of reinforcement. Utilizing 3D finite element software, simulation of general construction sequencing and seismic loads is conducted to determine the complex nature of 5 m and 10 m MSE walls. A parametric study was completed, varying backfill and geogrid spacing. This study indicated that limiting fine content of backfill was 24.60%, 47.61%, 53.83% for 5 m wall height while 12.78%, 14.50%, and 34.33% for 10 m wall height, when relating fine content to maximum lateral displacement. The dynamic characteristics, i.e. approximate natural frequency range through frequency sweep, was approximated to be about 4.00 Hz to 4.60 Hz for 5 m MSE wall (for peak response in velocity and acceleration, respectively). For 10 the m wall a peak response for both velocity and acceleration was obtained at 3.0 Hz. Because of the clayey soil characteristic of the unreinforced area, it is recommended that ground improvement techniques be applied before construction of MSE walls.

Keywords: Mechanically Stabilized Earth (MSE) Walls, Backfill, 3D Finite Element Dynamic Analysis, Construction Sequencing

1 INTRODUCTION

1.1 *Background of the study*

Widespread seismic and volcanic activity is experienced in the Philippine archipelago as the country is located in the Pacific Ring of Fire where the Philippines' oceanic plate and other small tectonic plates are converging, creating a source of frequent earthquake activity. A 6.9 magnitude earthquake struck Davao del Sur on December 15, 2019, causing 210 people to be injured, 13 deaths, 131,000 displaced citizens in Region XI (Davao), and Region XII (SOCKSARGEN). In addition to this, the earthquake damaged nearly 26,000 homes, 400 schools, and over 60 health facilities, the majority in the province of Davao del Sur (Bidder and Maramag 2019). According to research on the probability of high magnitude earthquakes (which was explored and evaluated using a frequency-magnitude distribution model), the highest possible maximum magnitude, return period, and the probability of earthquake occurrence in the Philippines indicate that the greatest possible magnitude is around 8.0 Mw over 50 years

*Corresponding author
DOI: 10.1201/9781003188339-40

(Pailoplee and Boonchaluay 2016). This insight leads to a great need in designing structures against seismic activities in the Philippines. This study assesses the impact of seismic events on retaining wall, particularly MSE wall systems, that provide slope or erosion control, and perform well under seismic loading.

2 REVIEW OF RELATED LITERATURE

2.1 *Introduction to design and construction of MSE walls*

Since the introduction of MSE walls in the Philippines is relatively new, and MSE wall systems vary significantly in criteria depending on its components, the structure's overall and internal stability may vary. Although there are no direct guidelines in the design and seismic analysis of MSE walls in the country, other institutions specialize in regulating the structural integrity of these structures. For this study, the authors have referenced the National Structural Code of the Philippines (2015) for the local specifications in precast walls and Design of Mechanically Stabilized Earth Walls and Reinforced Soil Slopes – Volume I (Berg, R., 2007) of the Federal Highway Association that is per American Association of State Highway Transportation Officials (AASHTO) LRFD specifications that provide recommendations for the design and construction of these structures. Case studies in this research also assume that the structure is internally stable in the problem. To counter check procedures are stated in FHWA-NHI-10-024 Sec 4.4.7 by assuming a Rankine failure surface (FHWA, 2017).

2.2 *Construction sequence of MSE wall*

MSE walls are generally classified into two based on facing: (i) MSE walls with rigid facing (e.g. pre-cast panel facing); and (ii) MSE walls with flexible facing (e.g. geosynthetic reinforced, welded wire mesh reinforced or gabions). For this research project, rigid facing construction was implemented. For rigid facing construction, a concrete panel must initially be installed, after which the granular backfill is placed up to the specified spacing height. Reinforcements are then placed at the specified spacing before the backfill is compacted up to 95% maximum dry density. This sequence continues to the top of the desired height of the wall.

2.3 *Soil-structure interactions in MSE wall*

Another important parameter in MSE walls is the interface elements which define the behaviour between two surfaces. The case study presented in this paper will incorporate the following interfaces or interaction elements: Concrete – Backfill, Reinforcement – Backfill, Concrete – Concrete. Two of these involve interaction between soil-structures: Facing – Backfill, Reinforcement – Backfill. These parameters are important as they affect the transfer of load between two surfaces. No laboratory experiments are used to identify design parameters of the interface parameters. Instead, parameters for interface elements will be taken from the manufacturers' design guidelines and from FHWA 2009 design manual. A material-dependent failure criterion was used since this was the equivalent failure criterion closest to the Plaxis Rinter value that was used in related literature. Rinter is a strength reduction factor, a counterpart to the interface coefficient. In a study conducted by Moradi (2014), a Rinter strength reduction factor of 0.6 was applied (for backfill-reinforcement). The model generated was able to predict the lateral deformations of the wall within a reasonable tolerance. Figure 1 illustrates the soil-structure interaction elements considered in this case study.

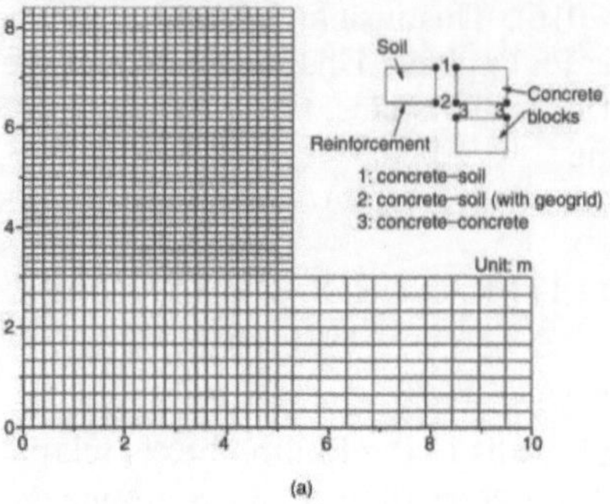

Figure 1. Model for SSI Analysis in MSE Walls (Ling and Leshchinsky 2003).

3 METHODOLOGY

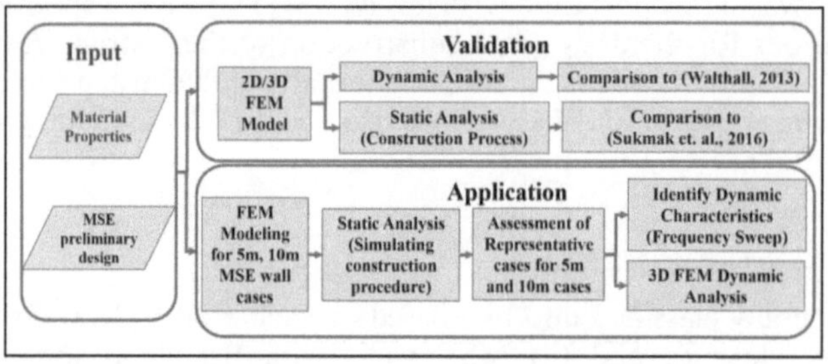

Figure 2. Methodological framework.

The methodology mainly consists of two tracks: the validation track and application track. The validation track focuses on reproducing the results of the chosen studies related to the research. For the application track, the gathered data for Marikina, the area of interest, will be used for the static and seismic analysis. Several cases were analyzed to find the most suitable MSE wall model configuration. The cases are discussed in the application section. Based on FHWA 2009 standards, selection criteria were set to determine the best performing wall in terms of lateral deflection, effective vertical stress, and base settlement of base. After which a representative model configuration, for both a 5 m and 10 m case was selected for dynamic analysis. The representative models then underwent a frequency sweep of a series of Ricker wavelets to identify their natural frequency. After determining the dynamic characteristics of the MSE wall, the model underwent static and dynamic analysis. For the dynamic analysis, the earthquake motion applied was the one that had a dominant frequency closest to the natural frequency of the wall.

4 RESULTS AND DISCUSSION

4.1 *Determination of representative case*

To determine the representative case, the performance criteria were tabulated for each criterion (Table 1). The researcher then selected the corresponding case to represent the 5 m wall for dynamic analysis. The cell highlighted in yellow means that it has failed the specific criteria. For the 5 m wall height, Case 4 with geogrid spacing of 0.5 m and backfill material of poorly graded sand was chosen to be the representative case. This was because it passed all the criteria having the least settlement and reasonably small lateral displacement.

Table 1. 5 m wall case assessment.

Case Number	Max Lateral Displacement (mm)	Max Effective stress at Base (kPa)	Max Settlement (mm)	Max Geogrid Tensile Strength (kN/m)	Max Geogrid Shear Strength (kN/m)
1	25.62	520.00	-12.82	0.696	0.790
2	47.59	600.36	-12.76	0.909	1.142
3	64.66[1]	635.92	-12.86	1.287	1.306
4	17.97	422.95	-11.28	2.423	0.945
5	32.98	471.11	-12.46	3.725	1.454
6	43.95	500.46	-12.31	4.234	1.637
7	12.39	366.21	-11.48	0.535	0.446
8	26.79	410.11	-12.68	0.844	0.901
9	39.62	424.01	-12.59	1.384	1.307

[1] Exceeded the criteria.

4.2 *Significant parameters for MSE walls*

To identify significant parameters in the model, a simple linear regression analysis was conducted to investigate the sensitivity of the parameters to the model. The data points are taken at the mid-section of the structure, where critical lateral displacements occur. Parameters that had correlation coefficients (R2) greater than 0.8 were determined to have a linear relationship.

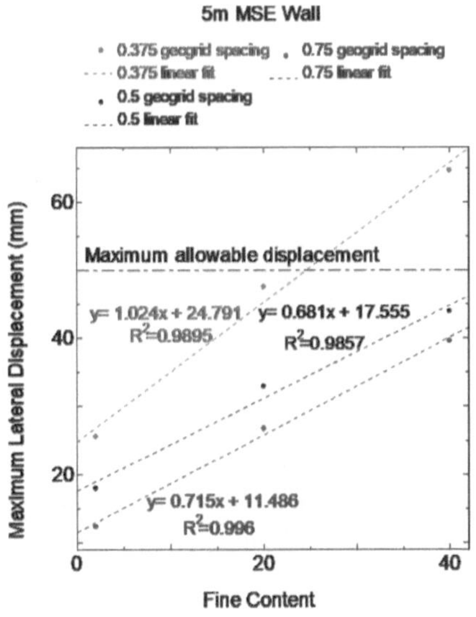

Figure 3. Maximum Lateral Displacement versus Fine Content (5 m).

Although there are not enough variations in soil fineness content to reach an acceptable amount of observation samples, a simple linear regression analysis was made to highlight the linear fit of maximum lateral displacement and fineness content of soil. The horizontal displacement increased by 85.75% to 219.77% as fine content increased by 18% to 38% for the 5 m MSE wall. As for the 10 m MSE wall the percent increase of lateral displacement was 92.32% to 192.66%.

4.3 *Dynamic characteristics determination by frequency sweep*

For the 5 m representative case, a frequency sweep was done on a range of (0-10 Hz) and it was determined that the peak amplitudes occur between 2 to 4 Hz. The frequency which gained the highest amplitude at a certain frequency are an approximate value for the model's natural frequency. This value can then be analyzed using an earthquake motion nearest to that frequency. The output was taken from the response of the topmost node of the wall. The 10 m MSE wall also underwent the same process with the results summarized in Table 2 and Table 3 below.

Table 2. Frequency Sweep Summary.

Wall Height (m)	Parameter	Maximum Fourier Amplitude	Corresponding Frequency (Hz)
5	Velocity (Y)	0.123	4.0 Hz
	Acceleration (Y)	3.16	4.6 Hz
10	Velocity (Y)	0.220	3 Hz
	Acceleration (Y)	3.87	3 Hz

Table 3. Summary of Time-History Displacements.

Wall Height (m)	Location	Highest Lateral Displacement in Time History (mm)	Time occurred (s)	Residual Displacement (mm)
	Top	2.321	40.740	0.173
5	Middle	1.271	41.370	0.077
	Bottom	0.433	41.210	0.003
	Top	7.174	41.240	-1.382
10	Middle	3.808	41.190	0.441
	Bottom	0.927	40.660	0.042

5 CONCLUSION

The research presented in this paper has been able to conduct a parametric study, frequency sweep, and dynamic analysis on MSE walls. Results of this study are summarized below.

1. When selecting limit criteria at post construction conditions:
   a) Lateral displacement influenced backfill content of the soil. Using the current data, limiting fine content of backfill was found to be 24.60%, 47.61%, 53.83% for 5 m wall height while 12.78%, 14.50%, and 34.33% for 10 m wall height by linear regression relating fine content to maximum lateral displacement. The behaviour observed was a zigzag configuration. This is because no plaster was modelled in between wall panels enabling movement at the junctions of the wall panel. There should be relevant detail in modelling these wall panel joints to prevent error. Due to this, limiting the fines content of the backfill may be overestimated or underestimated.
   b) The maximum effective vertical stress was observed at the base of the wall and then decreased suddenly until it reached the reinforced fill region, wherein the stresses were very small. This change in stress can be attributed to the following: (i) 4 to 7 kN/m^3 difference in the unit weight between the concrete and backfill material; and (ii) the rotation experienced at the bottom of the panel. This means that if the concrete panels fail in bearing capacity failure, there is a narrow region in which it will fail due to the abrupt change in pressure.

c) Points of highest settlement are at the unreinforced soil behind the mid-section of the wall. A 2D contour plane at the midsection of the wall showed that the points of maximum settlement are at the base of the reinforced soil fill and the middle of the soil fill in between reinforced layers. At the post-construction stage, it appears that settlement at the mid-span of the reinforced fill should be monitored to prevent excessive settlement.
d) Geogrid tensile and shear stresses deviate only slightly in the post-construction state. Besides this, there are observable localized tensile stresses near the wings of the wall. In 10 m walls, the geogrid tensile stress exceeds allowable tensile stress at about fine content of 20% for a geogrid spacing of 0.375 m. This is due to the simplification of the compaction effort to 8 kPa load, which also may lead to discrepancies to the true nature of the MSE wall. The critical areas in which significant displacements occur are at the unreinforced fill area just at the end of the geogrid length and the MSE wall's corners. Therefore, it can be inferred that using the assumptions provided by the study, the limiting fine content (15%) by the NSCP 2015 and geogrid spacing range of 0.375 m to 0.75 m will yield a safe MSE wall. To obtain a more accurate model, values should be counter-checked with experimental tests.

2. Regarding the determination of dynamic characteristics, i.e., approximate natural frequency range through frequency sweep, was approximated to be about 4.0 Hz to 4.60 Hz for the 5 m MSE wall (for peak response in velocity and acceleration, respectively). For 10 the m wall a peak response for both velocity and acceleration was obtained at 3.0 Hz.
3. The dynamic analysis confirmed that the representative cases from assessment performed well under the seismic load applied, with PGA of approximately 0.5 g. The residual top wall deflections relative to post-construction were found to be 0.173 mm for the 5 m wall and 1.382 mm for the 10 m wall which is apparently minimal. The motion of both MSE walls exhibited tilting inwards into the reinforced fill. The highly stressed and displaced sections are located at the unreinforced backfill area and right wing at wall panel connection for the specified ground motion. Because of the clayey soil characteristic of the unreinforced area, it is recommended that ground improvement techniques be applied before the construction of MSE walls.

REFERENCES

Newmark, N. M. 1965. Effects of earthquakes on dams and embankments. Geotechnique, 15 (2) 139–160.

Kramer, S., 1996. Geotechnical Earthquake Engineering. New Jersey: Prentice Hall.

Terzaghi, K., Peck, R. & Mesri, G., 1996. Soil Mechanics in Engineering Practice. 3rd ed. New York: John Wiley & Sons, Inc..

Ling, H. I. & Leshchinsky, D., 2003. Finite element parametric study of the behaviour of segmental block reinforced-soil retaining walls. Geosynthetics International, 10(No 3), pp. 77–93.

Wartman, J., Rondinel-Oviedo, E. & Rodriguez-Marek, A., 2006. Performance and Analyses of Mechanically Stabilized Earth Walls in the Tecomán, Mexico Earthquake. Journal of Performance of Constructed Facilities, pp. 287–298.

Berg, R. R., Christopher, B. R. & Samtani, N. C., 2009. Design and Construction of Mechanically Stabilized Earth Walls Volume I, s.l.: Federal Highway Administration.

Das, B. M., 2011. Geotechnical Engineering Handbook - References, Florida: J. Ross Publishing, Inc.

Vaziri, M. & Hartlib, T., 2012. ICE Manual of Structural Design: Buildings. London: Institution of Civil Engineers.

Walthall, R. et al., 2013. Finite-Element Analyses of Mechanically Stabilized Earth Walls Subjected to Midlevel Seismic Loads. Journal of Performance of Constructed Facilities, pp. 171–179.

Abbasi, O., Ghanbari, A., & Hosseini, S. 2014. An analytical method for calculating the natural frequency of reinforced retaining walls with soil-structure interaction effect. Geosynthetics International, 53–61.

Bathe, K., 2014. Finite Element Procedures. 2nd ed. Watertown: Prentice Hall, Pearson Education, Inc.

Kibria, G. &. H. & Khan, M., 2014. Influence of Soil Reinforcement on Horizontal Displacement of MSE Wall. International Journal of Geomechanics., Volume 14, pp. 130–141.

Moradi, G., 2014. Seismic Response Analysis of Geosynthetic Reinforced Soil Retaining Wall. Electronic Journal of Geotechnical Engineering, Volume 19, pp. 3819–3835.

Pailoplee, S. & Boonchaluay, N., 2016. Earthquake activities in the Philippines Islands and the adja-cent areas. Geosciences Journal, 20(6), pp. 877–889.

Sukmak, K., 2016. Numerical parametric study on behavior of bearin reinforcement earth walls with differen material properties. Geosynthetics International No. 6.

Carangan, F. M. R., 2017. Development of an Octreee-Based Tool for Generation of Three-Dimensional Structure Models Using PSM Approach.

FHWA - Office of Research, D., 2017. https://www.fhwa.dot.gov/. [Online] Available at: https://www.fhwa.dot.gov/publications/research/infrastructure/ structures/bridge/15034/002.cfm [Accessed 13 June 2020].

Ereño, M. N. 2018. Developing 3D FEM Tool for Modeling Structure-Soil-Structure Interaction (3SI) Observed in the NUPEC Field Experiment. Quezon: University of the Philippines Diliman.

Bidder, M. & Maramag, G., 2019. FLASH UPDATE NO. 3 - Philippines: Davao del Sur 6.9-magnitude earthquake. Available at:https://reliefweb.int/sites/reliefweb.int/files/resources/20191224%20Flash%20Update%20no.%203%206.9magnitude%20earthquake%20Da-vao%20del%20Sur%20final.pdf [Accessed June 4 2020].

GarBia, S. a. G., 2019. Sub-Surface Soil Exploration for Construction of Retarding Tank and Pumping Station (GSGS-SI-0118-19), Manila: Metro Manila 1st Engineering District Office.

Rocscience, 2020. RS3 Theory. [Online] Available at: https://www.rocscience.com/help/rs3/Theory/Theory_Overview.htm [Accessed 10 June 2020].

Rocscience, n.d. https://www.rocscience.com/. [Online] Available at: https://www.rocscience.com/help/rs3/Support/Define_Joint_Properties.htm [Accessed 13 May 2020].

The Evolution of Geotech - 25 Years of Innovation – Hammah et al. (Eds)

2D and 3D numerical study of geosynthetic mechanically stabilized earth GMSE walls

E.A. Ardila Montilla & E.R. Esquivel
Department of Geotechnical Engineering, São Carlos School of Engineering EESC, University of São Paulo USP, São Carlos, SP, Brazil

F.H. Martins Portelinha
Department of Civil Engineering, Federal University of Sao Carlos UFSCar, São Carlos, SP, Brazil

S. Javankhoshdel
Rocscience Inc., Toronto, Ontario, Canada

ABSTRACT: Numerical modeling in geotechnical engineering has generally proved feasible when analyzing the behavior and performance of reinforced soil structures, due to their low cost, versatility, and practicality with respect to physical models. One of the most significant challenges in this process is the choice of appropriate models and parameters that are accurate in describing reality. This study presents numerical modeling analyzes of a Geosynthetic Mechanically Stabilized (GMSE) wall built, instrumented and monitored in laboratory by Hatami & Bathurst (2005). The numerical modeling was carried out in 2D and 3D using the finite element method (FEM). The behavior of the soil was represented by the Linear Elastic Perfectly Plastic (Mohr Coulomb) and Elastic Hyperbolic (Duncan Chang) constitutive models. The geosynthetics and facing elements behaviors were represented by the Linear Elastic constitutive model. In these analyzes, the construction stages and the subsequent wall loads were taken into account. The objective of this study was to validate the model parameters in such way the assessed face displacements, foundation stresses, reinforcement strains were close to those corresponding the instrumented wall. This validation was performed through the comparison of model results with measured values. With the found It was obtained a good agreement between numerical modeling and physical test results in soil structures reinforced with geosynthetics.

1 INTRODUCTION

The reinforced soil technique consists of introducing reinforcement elements into the soil mass in order to improve its strength. These elements are usually metallic strips, steel bars or geosynthetics. Thus, this technique allows to design safe and reliable earth structures, such as, landfills, slopes, bridge abutments and other engineering applications. Geosynthetic reinforced soil structures are advantageous in many aspects when compared to conventional retaining structures. They are more economical, generate less environmental impact, their executive process is simple and their maintenance is not complex (Bathurst & Hatami, 2001; Wu, 2019).

Different researchers have investigated the behavior of soil structures reinforced with geosynthetics through numerical simulations. Hatami & Bathurst (2005) performed numerical analyses to simulate the behavior of an instrumented full-scale prototype of a GMSE, under working stress conditions using finite difference method during the construction stages until

DOI: 10.1201/9781003188339-41

the end of the construction. The analyzes results showed that numerical simulations accurately predicted the behavior of the wall. Adopting a linear elastic–plastic model for the soil, wall displacements and toe boundary stresses showed a good agreement. However, the analyzes showed that this model is not appropriate to represent the strain distribution in the reinforcements. On the other hand, the adoption of a nonlinear elastic–plastic model for the backfill soil showed a good agreement with the prototype wall behavior. Subsequently, Hatami & Bathurst (2006) performed numerical analyzes simulating the construction and overloading stages of a GMSE. The results showed that more detailed considerations of the wall components and executive processes are required, such as, the soil-reinforcement interaction and the effect of soil compaction.

Using the FEM approach, Guler *et al.* (2007) carried out parametric analyzes of geosynthetic reinforced soil walls, considering cohesive and granular soils, vertical separation distance between reinforcement layers and reinforcement length. It was shown that the potential failure surfaces assumed in analysis methods using the limit equilibrium approach (LEM) to assess the safety factors of GMSE structures are correct and that both cohesive and granular soils performances are as adequate as backfill materials for the modelled cases.

Huang *et al.* (2009) considered three constitutive models for the soil in their numerical simulations: linear elastic-plastic Mohr-Coulomb, modified Duncan-Chang hyperbolic model, and Lade's single hardening model. Their stress predictions were within measurement accuracy corresponding to the end of construction and surcharge load levels. They found that the three models were typically within the range of measured values of the physical model with an outstanding performance with the Duncan Chang model. This model has shown to be adequate to predict the behavior of reinforced soil walls under typical operational conditions, provided that the soil reinforcement, interfaces, boundaries, construction sequence, and soil compaction are properly modeled. Damians *et al.* (2014) conducted a numerical study on the influence of foundation and reinforcement stiffness on the behavior of reinforced soil walls. which showed that the reinforcement stiffness has a higher influence on the performance of the structure compared to the type of foundation.

Although several numerical studies have been done, demonstrating the good performance of reinforced soil walls, there are no studies in literature comparing 2D and 3D analysis of these structures, nor there are any models using the software RS2 and RS3 (Rocscience, 2020). The present study aims to approach the numerical simulation of the behavior of GMSE walls and their components, through 2D and 3D analysis. In the numerical modeling, two constitutive soil models were used to simulate the construction and surcharge loading response of a constructed, instrumented, and monitored reinforced soil wall. This will allow engineers and researchers to understand the behavior of this modeling approach and to verify the relevance of numerical modelling of geosynthetic reinforced soil walls using the FEM approach.

2 NUMERICAL MODELING

The physical model adopted is part of the Royal Military College of Canada study of real scale walls Hatami & Bathurst (2005). The named Wall 1 was adopted in this study, with a height of 3.6 m, base width of 3.3 m and base length of 5.95 meters. The wall has rigid face with a slope of 82° and was built with concrete blocks of 0.15 meter-high each. The reinforcement consists of a polypropylene biaxial geogrid with a length of 2.52 meters. The RS2 and RS3 software were used to develop the numerical model of reinforced soil and simulate its stress-strain response during, at the end of construction and at overload, as showed in Figure 1. The analysis of the model was done in a plane strain with Gaussian elimination solution. A graded mesh was generated, discretely by quadrilaterals elements of eight nodes, in a total of 3114 nodes for 2D model, and by tetrahedral elements of four nodes, in a total of 23810 for 3D model.

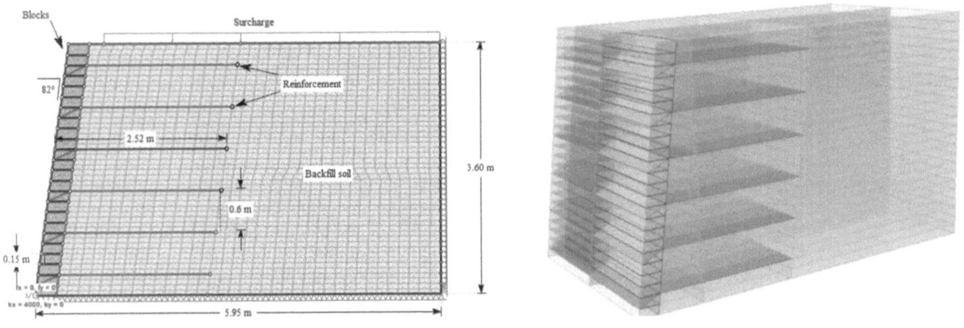

Figure 1. 2D and 3D numerical models.

The foundation of the wall in the physical model is rigid, therefore, in the numerical model it was simulated through supports with horizontal and vertical restriction. The base of the face was simulated with supports with vertical restriction. The foot of the face was contained horizontally by a very rigid spring element with a stiffness of 4 MN/m, simulating the fixed reaction beam. To simulate the construction and subsequent overload, the wall was simulated in 32 stages. The first 25 stages correspond to the construction of the wall, in which, for each stage, the equivalent of a face block, i.e., 150 mm, was modeled and a compaction load of 8 kN/m was applied, successively until the construction of the wall was completed. Subsequently, 7 stages of increasing surcharge load of 10 kPa were simulated until reaching the surcharge of 70 kPa, according with Hatami & Bathurst (2006).

The backfill of the wall and face blocks were modeled with continuous materials, using two constitutive soil models: linear elastic-plastic Mohr-Coulomb and Duncan-Chang hyperbolic model, the former, is the most common in geotechnical modelling and the latter is based on the stress-strain curve in triaxial compression tests that can be approximated by a hyperbolic function with variable Young's Modulus. The parameters used, based on the data presented by Huang *et al.* (2009), are presented in Table 1.

Table 1. Soil parameters.

Soil parameters	MohrCoulomb	Duncan Chang
Young's Modulus (kPa)	40.000	-
Cohesion (kPa)	2	2
Friction angle (°)	44	44
Poisson's ratio	0.25	0.25
Dilation angle (°)	11	-
Elastic modulus number	-	800
Unloading-reloading modulus number	-	960
Elastic modulus exponent	-	0.5
Failure ratio	-	0.86

To simulate the interaction between the face blocks, the joints were modeled with *the Material Dependent* sliding criteria. In addition, the interaction between the face blocks and the backfill soil was modeled adopting *Mohr Coulomb's sliding criterion*, with the parameters published by Damians *et al.* (2014). The reinforcement geosynthetic layers elements with Structural Interfaces to simulate the interaction between the geosynthetic and the soil. The geogrid was considered as elastic material, with the Young's modulus based on the stiffness data from Hatami & Bathurst (2006). The tensile strength of the joint between the soil and the reinforcement was considered as the ultimate geosynthetic strength equivalent divided by the average geogrid thickness (5 mm). The joint friction angle was made equal to friction angle of the soil, and considering that the joint was perfectly fully bonded. The joint characteristics were adopted from Damians *et al.* (2014).The parameters of reinforcement, face and joint are showed in the Table 2.

Table 2. Parameter of reinforcement, face and joints.

	Parameter	Value	Parameter	Value
Reinforcement	Geogrid tensile modulus (kN/m)	115	Ultimate strength (kN/m)	14
Face blocks	Unit weight (kN/m3)	22	Young's Modulus (GPa)	23
	Poisson's ratio	0.15	-	-
Soil- reinforcement joint	Joint Tensile Strength (kPa)	2800	Normal stiffness (kPa/m)	1000
	Cohesion (kPa)	1000	Shear stiffness (kPa/m)	1000
	Friction angle (°)	44	-	-
Block- block joint	Interface coefficient	1	-	-
Soil- block joint	Cohesion (kPa)	1	Normal stiffness (kPa/m)	100000
	Friction angle (°)	44	Shear stiffness (kPa/m)	1000

3 RESULTS

Numerical results for the wall were obtained from 2D and 3D modeling using the linear elastic-plastic Mohr-Coulomb and hyperbolic Duncan-Chang constitutive models to simulate soil behavior and match the values with the physical test of reference. Figure 2 presents the maximum lateral displacements of the facing wall with the increase of the surcharge load, for 2D and 3D models, considering the two different constitutive models. In this figure, it can be seen that the wall displacements obtained with the numerical model show good agreement with the displacements measured in the physical model.

In all considered cases, the wall displacements obtained with numerical analyzes increase with the surcharge level, consistently with the pattern of the measured values. It is observed that just at the end of construction of wall, when surcharge is 0 kPa, the maximum wall displacement is about 0,15 % of the wall height, while for the highest surcharge level, when surcharge is 70 kPa, the displacement corresponds to about 1% of the total wall height, demonstrating the good performance of the GMSE, even under high overload conditions. In general, it can be observed that in the surcharge stages, the estimate displacements are very close or slightly above to the expected values, which is considered a good prediction and in favor of safety.

In addition, it is worth to note that the Duncan Chang model, both in the 3D simulation and in the 2D simulation provided displacements that are closer to the real wall displacements. Probably, this is due to the hyperbolic nature of the model, which is better suited to the stress-strain behavior of the soil. Also, it should be observed the difference of the displacements corresponding to each constitutive model in respect to the 2D or 3D model. It is evident that in the lower range of surcharge loads the displacement values are similar, while in the higher range, higher displacements correspond to the 3D model.

Regarding the vertical pressure on the foundation was verified at end of construction. Figure 3 shows the vertical pressures on the foundation (σ_v), normalized with the soil unit weight (γ_s) and the wall height (H). In this figure, the pressure values also show satisfactory agreement with measured values, capturing the high pressure generated by the weight of the block face, as well as the reduction trend of the pressures in back of facing, due to the mobilized forces from the reinforcement connection to the face. It is also observed that the vertical pressure distribution on the foundation is satisfactory, regardless the constitutive model or the 2D 3D modeling adopted. Finally, the reinforcement strains, as a general observation, the model predicted strain values are in good agreement with the measured values, in both cases

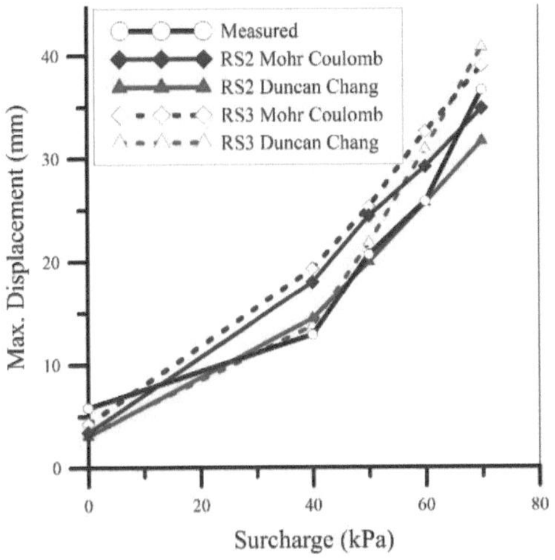

Figure 2. Maximum face displacement in surcharge load stages.

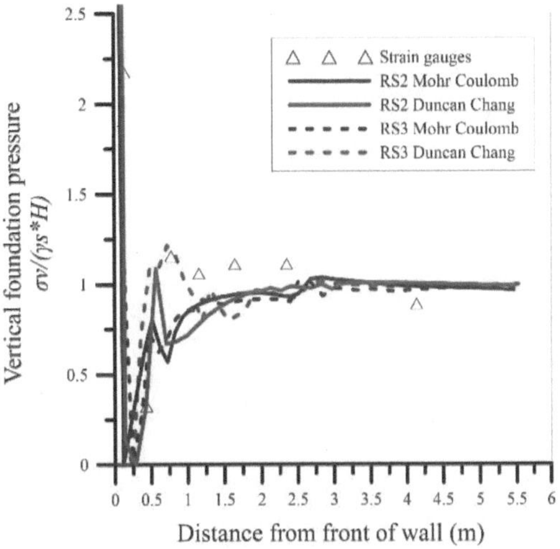

Figure 3. Vertical foundation pressure at end of construction.

considering and not considering surcharge. Figure 4a shows the reinforcement strains corresponding to the end of the wall construction, and Figure 4b shows the reinforcement strains corresponding to 50 kPa surcharge load stage. As shown in Figure 4a, the model reinforcement strain values show a good agreement with the instrumentation measured values, especially for layers 4, 5 and 6, for maximum strain magnitudes and positions at the end of construction in all simulations. For layers 1, 2 and 3, although the strains are within the expected range, the predicted values are not as close as in layers 4, 5 and 6, at the end of wall construction. It is also noted that the models using the Duncan Chang Model tended to overestimate the reinforcement strains.

According with the Figure 4b, for the 50-kPa surcharge stage, the reinforcement strain values, predicted by the numerical models, are in a good agreement with those corresponding to the reference physical model. However, in layer 5, the numerical model predicted values are underestimated, below the measured values in the physical model. It should also be pointed out that, in any case, no reinforcement strain, higher than 2%, was observed.

4 CONCLUSIONS

This paper describes the FEM analyses of a GMSE wall, during its construction process and after the application of a surcharge load, The analyzes were conducted using the software RS2 and RS3. The following conclusions were drawn based on the analysis results:

a) The models were able to predict the physical measurements of facing displacements, reinforcement strain distribution and vertical pressure on the foundation within reasonable ranges, during construction and surcharge load simulation.
b) The constitutive model that best predicted the face displacements was the Duncan Chang model. however it was shown that a simple constitutive model, such as the Mohr Coulomb elastoplastic model, is suitable to simulate stiffened face GMSE walls during construction and operation.
c) Reinforcement strains were satisfactorily predicted at the end of construction and in the surcharge stages. However, it should be noted that the Mohr Coulomb elastoplastic model

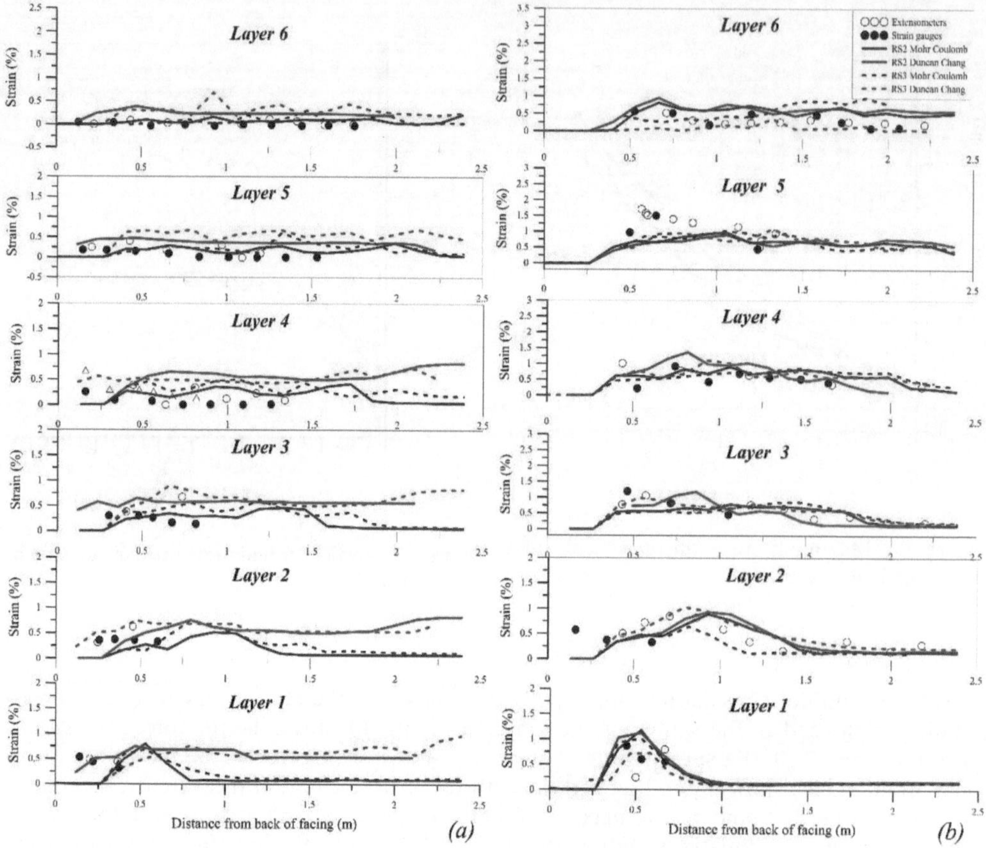

Figure 4. Reinforcement strain a) Surcharge 0 kPa b) Surcharge 50 kPa.

showed a better performance, since it was able to characterize the soil dilation angle, generating a volumetric variation of the soil.

d) A satisfactory prediction of the vertical pressure at the foundation was obtained, regardless the constitutive model adopted.

e) It has been shown that both 2D and 3D models provide adequate simulations of the behavior of GMSE walls. However, the elaboration and analysis of 3D models is much more complex, requires much more time, require higher data processing resources. Therefore, it is more convenient to perform stress-strain analysis of geosynthetic reinforced soil walls by means of 2D models.

ACKNOWLEDGEMENTS

The authors express their gratitude to the Brazilian National Council for Scientific and Technological Development CNPq (Grant No. 830855/1989-0) for the scholarship granted to the first author.

REFERENCES

Bathurst, R. J., & Hatami, K. (2001). Review of numerical modeling of geosynthetic reinforced soil walls. In Proc., 10th Int. Conf. on Computer Methods and Advances in Geomechanics (pp. 1223–1232).

Damians, I. P., Bathurst, R. J., Josa, A., & Lloret, A. (2014). Numerical study of the influence of foundation compressibility and reinforcement stiffness on the behavior of reinforced soil walls. *International Journal of Geotechnical Engineering*, *8*(3), 247–259.

Guler, E., Hamderi, M., & Demirkan, M. M. (2007). Numerical analysis of reinforced soil-retaining wall structures with cohesive and granular backfills. Geosynthetics International, 14(6), 330–345.

Hatami, Kianoosh, & Bathurst, R. J. (2005). Development and verification of a numerical model for the analysis of geosynthethic-reinforced soil segmental walls under working stress conditions. Canadian Geotechnical Journal, 42(4), 1066–1085.

Hatami, K, & Bathurst, R. J. (2006). Numerical model for reinforced soil segmental walls under surcharge loading. Journal of Geotechnical and Geoenvironmental Engineering, 132(6), 673–684.

Huang, B., Bathurst, R. J., & Hatami, K. (2009). Numerical study of reinforced soil segmental walls using three different constitutive soil models. *Journal of Geotechnical and Geoenvironmental Engineering*, *135*(10), 1486–1498.

Rocscience. (2020). RS2, 2D Finite Element Analysis Software, Toronto, Ontario, Canada.

Rocscience. (2020). RS3, 3D Finite Element Analysis Software, Toronto, Ontario, Canada.

Wu, J. T. H. (2019). Characteristics of Geosynthetic Reinforced Soil (GRS) Walls: an Overview of Field-Scale Experiments and Analytical Studies. *Transportation Infrastructure Geotechnology*, *6*(2), 138–163.

The Evolution of Geotech - 25 Years of Innovation – Hammah et al. (Eds)

2D and 3D FEM modeling of the initiation of progressive landslides

Rajib Dey
Stantec Consulting Ltd, Toronto, ON, Canada

Sina Javankhoshdel
Rocscience Inc., Toronto, ON, Canada

ABSTRACT: Large-scale landslides in sensitive clays are common in eastern Canada. Failure occurs quite rapidly, essentially in undrained conditions. Based on initiation and progression of the sliding surfaces or shear bands, the failure of slopes can be divided into two categories, namely upward and downward progressive failures. Upward progressive failure might be initiated by excavation, erosion or small slides in the riverbank slopes. Field observations show that the progressive failure propagating on the upslope direction is initiated by the formation of a quasi-horizontal shear band from the toe of the slope. Several studies in the literature have shown the initiation of the quasi-horizontal shear band in 2D models. In this study, the initiation of the quasi-horizontal shear band is compared in 2D and 3D models. It is shown that 3D models can capture the initiation as well, and the results of 2D and 3D models are in good agreement.

1 INTRODUCTION

Large-scale landslides in sensitive clays could be in the form of spreads which involves dislocation of the soil mass in blocks having horst and graben shapes (Figure 1) (Dey et al. 2015). In such landslides, the failure is generally triggered near the toe of the slope and the failure surface or shear zone progresses upslope quasi-horizontally into the deposit (Demers et al. 2013). Owing to significant displacement, the soil above the failure surface then starts to move almost laterally, forming several horst and graben shaped blocks and results in spread (Figure 1). The shear strains are mainly concentrated around the surfaces of the horsts and grabens with minimal strains inside these blocks. The movement of the soil occurs rapidly which indicates that the water pressure generated during the landslide does not have enough time to dissipate during the failure (undrained condition). Limit Equilibrium method (LEM), used for analysing the stability of these landslides, initially shows a factor of safety above one (Demers et al. 2013).

The progressive failure cannot be predicted using LEM approach due to the elastic-perfectly plastic nature of the material models in the LEM analysis. In this study, elastic-perfectly plastic materials are assumed in the first step, and LEM (Slide2) and Finite Element Method (FEM) (RS2) are carried out to show the stability of the initial condition of the slope (before failure).

To explain upward progressive failure and spreads, the presence of a horizontal weak layer surface was recognised by researchers several decades ago (Odenstad, 1951; Carson, 1977). Examining field evidence, in the recent studies (e.g., Locat et al., 2008, 2011a; Quinn et al., 2011, 2012) it is shown that the initial horizontal shear band might form due to erosion,

DOI: 10.1201/9781003188339-42

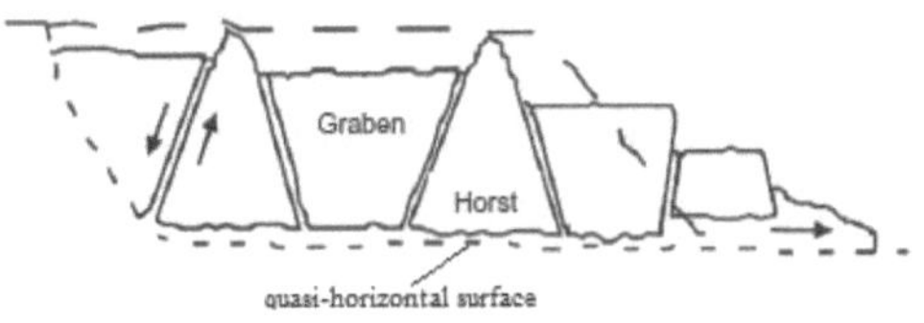

Figure 1. Sketch of spread type landslide in sensitive clay (After Dey et al. 2015).

excavation or small slides near the bank, and these landslides were explained with a progressive failure mechanism. In such a mechanism, the soil exhibits a stress–strain behaviour, including post-peak strain-softening, to propagate shear stresses and deformations along a shear zone (Bjerrum 1967; Bernander 2000). Locat et al. (2011) and Quinn et al. (2011, 2012) showed that the failure generally initiates near the toe of the slope and propagates essentially horizontally into the intact deposit, reducing horizontal stresses in the deposit. The failure of the slope will occur when the length of the shear band becomes sufficiently large, such that the downward unbalanced force causes active shear failure. Very large strains and displacements are developed around the shear bands and the soil above the initially developed quasi-horizontal shear band displaces almost laterally forming horsts and grabens. Spreads are therefore the results of progressive failure and active failure of an extensive part of the soil mass above the failure surface (Dey et al. 2015).

In this study, modelling of initiation of such landslides was attempted in the second step where the stiffness and strength of the toe of the slope was reduced in several stages using the 2D FEM software, RS2 (Rocscience 2020). It should be noted that due to the limitation of handling very large deformation problem such as landslides, efforts were limited to capture the triggering of slope failure, development of the quasi-horizontal shear zone and the first failure region.

There is limited to no study in literature (to the authors best knowledge) that investigates the behavior of such landslides in 3D models. Therefore, the last step of this study was to capture the behavior explained above, using the 3D FEM software RS3, considering the plain strain behavior (to be able to compare it with the 2D results).

2 PROBLEM DEFINITION

The geometry of the model considered in this study is shown in Figure 2 (After Dey et al. 2015). In this model, the slope of the riverbank $\beta = 30°$ was assumed. The thickness of the crust below the level ground surface was $H_c = 3$m. Under the crust, there was a thick layer of sensitive clay (thickness $H_{st} = 16$), followed by a base layer of stiff clay. It is assumed that the base layer was very thick; however, in this study only 3m of this layer was modelled, as it did not have a significant effect on the results. For simplicity, the groundwater table was assumed to be located at the ground surface and the river was full. It is assumed that the upward progressive failure was initiated by erosion and/or excavation of the toe of the riverbank slope. In the FEM analyses, the erosion or excavation was modelled by weakening a block of soil near the toe of the slope, as shown by the shaded zone in Figure 2 (eroded block).

For the initial step of the model (2D LEM and FEM analysis before failure), material properties of the eroded block were assumed, like the crust material. For the next steps, 6 stages were assumed to simulate the toe erosion, reducing the strength (unit weight) and stiffness (Young Modulus) of the eroded block by 90%, 80%, 60%, 40%, 20%, and 10%.

It is also assumed that the erosion occurred relatively fast, such that the deformation failure of the remaining soil happened in the undrained condition. The length of the soil domain in the present FEM model was 140 m.

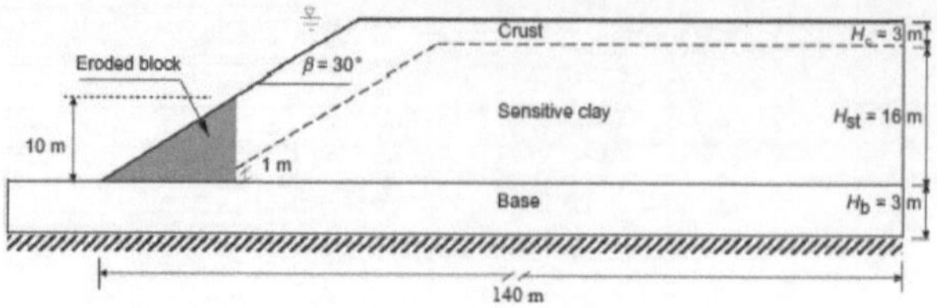

Figure 2. The base model considered in this study.

3 LEM AND FEM MODELLING

The elasto-plastic FEM analysis and LEM are two powerful approaches used for slope stability analysis.

LEM is the conventional method in the slope stability analysis. However, in recent years, more advanced LEM analysis have been developed by employing the advanced metaheuristic approaches such as Cuckoo search (Yang and Deb 2009) or Particle Swarm search methods (Kennedy and Eberhard 1995). These methods are able to get the same Factor of Safety (FS) and critical slip surface as the more advanced FEM analysis for the case of elastic-perfectly plastic material models in the FEM analysis, e.g. Mohr Coulomb with the same value for peak and residual strength.

Most of the FEM models available in the literature for slope stability analysis are developed in a Lagrangian framework (e.g. Griffiths & Lane, 1999). The non-convergence of the solution is considered as one of the conditions of failure in some studies (e.g. Griffiths & Lane, 1999). The advantage of FEM over LEM is the ability to consider more complicated constitutive models, such as softening-hardening model in the slope stability analysis. Also, FEM makes it possible to report a stress-strain relationship in addition to FS for the slope stability analysis.

In the first step of this study, it is assumed that the materials follow the elastic-perfectly plastic constitutive model. Thus, FEM and LEM analyses were carried out to calculate the FS before the erosion occurred at the toe of the slope.

Laboratory tests (e.g. Bjerrum & Landva, 1966; Bernander, 2000) show that the undrained shear strength of sensitive clay decreases with plastic shear strain. Following Anastasopoulos et al. (2007), it is assumed that equal shear strains develop throughout the whole depth of the soil specimen until it reaches the peak strength. After that, a softening behavior is expected where the shear strength decreases with shear strain. However, for simplicity, it is assumed in this study that the sensitive clay follows the elastic-brittle plastic behavior as it is shown in Figure 3. The peak shear stress (Figure 3) drops to a mobilized (residual) shear stress when the strain becomes larger than the shear strain at peak.

Table 1 shows the geotechnical parameters used in this study. These parameters were estimated from laboratory tests, interpretation of test data, constitutive model development and numerical studies on landslides in sensitive clays that were available in the literature (Dey et al. 2015, 2016).

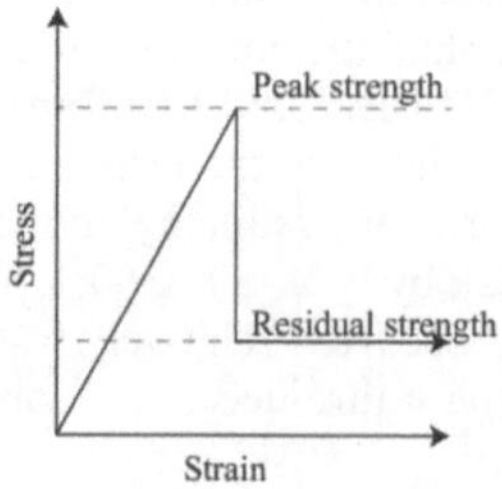

Figure 3. Elastic-brittle plastic material model (used for step 2 and step 3 in this study).

Table 1. Soil parameters for different steps.

Parameters	Values		
	Sensitive Clay	Crust	Base
Peak undrained shear strength, Su, kPa	37.5	60	-
Mobilized undrained shear strength (Step1), Sur, kPa	37.5	60	
Mobilized undrained shear strength (Step 2 and 3), Sur, kPa	7.5	60	
Young's Modulus, E, MPa	7.5	10	100
Poisson ratio, ν	0.495	0.495	0.495
Unit Weight, γ, kPa	18	19	21

4 MODELLING STEPS AND RESULTS

4.1 *Step 1. LEM and FEM analysis for the initial condition*

As mentioned above, the elastic-perfectly plastic material models (peak strength = residual strength) were considered for both sensitive clay and crust materials. Therefore, it is possible to compare the LEM and FEM analysis while calculating the initial FS before failure. The Particle Swarm Optimization (PSO) method was used to search for the critical slip surface together with the Spencer LEM method in Slide2 to calculate FS.

Figure 4a and 4b show the results of LEM and FEM analysis. The FS calculated using both methods was FS = 1.86 and it can be seen in Figures 4a and 4b that the shape of the failure surface is the same using both LEM and FEM methods, respectively. Considering elastic-perfectly plastic material models in a slope program confirms that it is possible to get the same failure surface and FS in FEM analysis using LEM method with more advanced metaheuristic approaches.

a)

b)

Figure 4. a) LEM analysis in the Slide2 software, FS = 1.86 b) FEM analysis in the RS2 software, SRF = 1.86.

4.2 *Step 2. 2D FEM analysis after erosion of the block*

In the second step of the analysis, the strength and stiffness of the eroded block were gradually reduced in 6 stages as mentioned earlier. Figure 5a shows the results of the 90% of the initial strength of the eroded block (the block was assumed to be an elastic material and the strength depended only on the Unit Weight and Young Modulus of the material). It can be seen in Figure 5a that the quasi-horizontal shear band has been developed. This confirms the behavior reported by Dey et al. (2015). Figure 5b and 5c show the two next stages where the maximum shear strain appears in the model connecting to the quasi-horizontal shear band after the strength of the eroded block was reduced to 80% and 60%. After the 80% point and generation of the band with maximum plastic shear strain, this band gets stronger in other stages.

Also, the total displacement of the model changed from 0.1 m in stage 1 to 0.28 m in stage 2 (creation of the quasi-horizontal shear band) and then 0.8 m, 1.57 m, 2.76 m, 5.25 and 10 m for the stages 3, 4, 5, 6 and 7, respectively, when the failure slip surface (shear band) was formed.

4.3 *Step 3. 3D plain strain FEM analysis*

The 3D model was created in the 3D FEM software RS3 (Figure 6). In this model, the extrusion depth was assumed to be 1m with the restraints along the Y-direction to satisfy the plain strain condition and to be comparable to the results of the 2D FEM analysis. The same material properties, and the same number of stages with the staged material properties were assumed in the 3D model.

Figures 6b and 6c show the maximum shear strain in stages 2 and 3. Figure 6b indicates that the quasi-horizontal shear band was generated in the 3D model like 2D model.

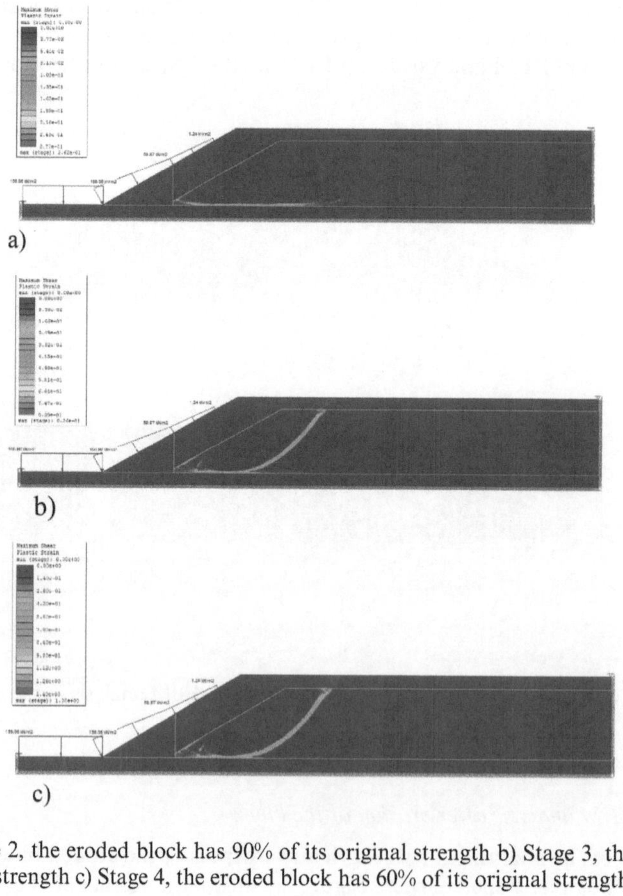

Figure 5. a) Stage 2, the eroded block has 90% of its original strength b) Stage 3, the eroded block has 80% of its original strength c) Stage 4, the eroded block has 60% of its original strength.

Also, with the similar reduction in the material strength for the eroded block, similar maximum shear strain band was generated connecting to the quasi-horizontal shear band in Figure 6b and 6c.

The same amount of displacement was computed in the 3D model as well, which confirms the results of the 2D FEM analysis.

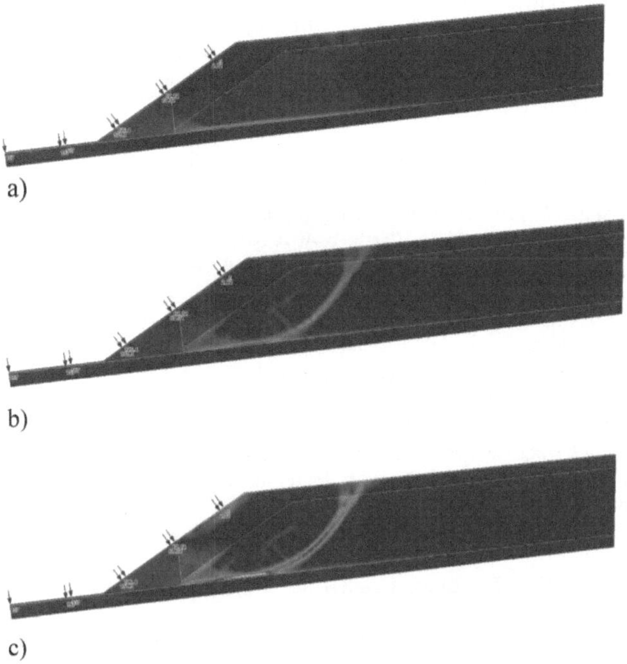

Figure 6. a) Stage 2, the eroded block has 90% of its original strength b) Stage 3, the eroded block has 80% of its original strength c) Stage 4, the eroded block has 60% of its original strength.

5 CONCLUSION

In this study an upward-type progressive landslide is modelled in three steps. In the first step (before failure), the 2D LEM and FEM analysis were compared, and the results showed a good agreement. Then, in step two, using 2D FEM analysis, the material at the toe of the landslide became weaker in several stages and the progressive failure showed the initial quasi-horizontal shear failure and then the shear strain failure band in the model. And finally, the plain strain 3D model was used to model the same stages as step 2 and the results of the 3D analysis showed a good agreement between 2D FEM and 3D FEM.

REFERENCES

Anastasopoulos, I., Gazetas, G., Bransby, M. F., Davies, M. C. R. & El Nahas, A. 2007. Fault rupture propagation through sand: finite-element analysis and validation through centrifuge experiments. J. Geotech. Geoenviron. Engng, ASCE 133, No. 8, 943–958.

Bernander, S. 2000. Progressive failure in long natural slopes: formation, potential extension and configuration of finished slides in strain-softening soils. Licentiate thesis, Luleå University of Technology, Luleå, Sweden.

Bjerrum, L. & Landva, A. 1966. Direct simple-shear tests on a Norwegian quick clay. Géotechnique 16, No. 1, 1–20.

Carson, M. A. 1977. On the retrogression of landslides in sensitive muddy sediments. Can. Geotech. J. 14, No. 4, 582–602.

Carson, M. A. 1979. On the retrogression of landslides in sensitive muddy sediments: reply. Can. Geotech. J. 16, No. 2, 431–444.

Demers, D., Robitaille, D., Locat, P. & Potvin, J. 2014. Inventory of large landslides in sensitive clay in the province of Quebec, Canada: preliminary analysis. In Landslides in sensitive clays – from geosciences to risk management (eds J.-S. L'Heureux, A. Locat, S. Leroueil, D. Demers and J. Locat), pp. 77–89. Dordrecht, the Netherlands: Springer.

Dey, R., Hawlader, B., Phillips, R. and Soga, K., 2015. Large deformation finite-element modelling of progressive failure leading to spread in sensitive clay slopes. *Géotechnique*, *65*(8), pp.657–668.

Dey, R., Hawlader, B., Phillips, R. and Soga, K., 2016. Modeling of large deformation behaviour of marine sensitive clays and its application to submarine slope stability analysis. Canadian Geotechnical Journal, 53 (7).

Griffiths, D. V. & Lane, P. A. (1999). Slope stability analysis by finite elements. Géotechnique 49, No. 3, 387–403.

Kennedy, J. and Eberhart, R. 1995. Particle swarm optimization, Proceedings of ICNN'95 - International Conference on Neural Networks, IEEE, Perth, WA, Australia, 4: 1942–1948

Locat, A., Leroueil, S., Bernander, S., Demers, D., Locat, J. & Ouehb, L. 2008. Study of a lateral spread failure in an eastern Canada clay deposit in relation with progressive failure: the Saint-Barnabé-Nord slide. Proceedings of the 4th Canadian conference on geohazards: from causes to management, Québec, Canada, pp. 89–96.

Locat, A., Leroueil, S., Bernander, S., Demers, D., Jostad, H.P. & Ouehb, L. 2011. Progressive failures in eastern Canadian and Scandinavian sensitive clays. Can. Geotech. J. 48, No. 11, 1696–1712.

Quinn, P. E., Diederichs, M. S., Rowe, R. K. & Hutchinson, D. J. 2011. A new model for large landslides in sensitive clay using a fracture mechanics approach. Can. Geotech. J., 48, No. 8, 1151–1162.

Quinn, P. E., Diederichs, M. S., Rowe, R. K. & Hutchinson, D. J. 2012. Development of progressive failure in sensitive clay slopes. Can. Geotech. J. 49, No. 7, 782–795.

Rocscience Inc. 2020. RS2 – 2D Finite Element Analysis. www.rocscience.com, Toronto, Ontario, Canada.

Rocscience Inc. 2020. Slide2 – 2D Limit Equilibrium Slope Stability Analysis. www.rocscience.com, Toronto, Ontario, Canada.

Rocscience Inc. 2020. RS3 – 3D Finite Element Analysis. www.rocscience.com, Toronto, Ontario, Canada.

Yang, X.S. and Deb, S., 2009, December. Cuckoo search via Lévy flights. In 2009 World congress on nature & biologically inspired computing (NaBIC), IEEE, 210–214

Session 9 - Liquefaction & foundation analysis

The Evolution of Geotech - 25 Years of Innovation – Hammah et al. (Eds)

Bearing capacity and settlement analysis of closely spaced shallow foundations with various footing geometry on multi-layered soils

M.E.C. Amorin & G.P. Ventura*
Institute of Civil Engineering, University of the Philippines Diliman, Quezon City, Metro Manila, Philippines

ABSTRACT: Rapid urbanization coupled with scarcity of construction sites forces the foundations of structures to be placed close to each other to meet functional requirements. This leads to a potential overlapping of stressed zones and alter the load-settlement behavior of the system of footings. The footing geometry, spacing configuration, and the type of soil are varied to study the interference effects in terms of the variations of the Bearing Capacity Ratio and settlement ratio with the spacing ratio. Numerical analysis results show that the bearing capacity, and total settlement of an isolated footing are significantly altered by the interference phenomenon. For footings on cohesionless soil and cohesive-friction (c-ϕ) soil media, the magnitude of the ultimate bearing capacity increases as the clear spacing between the foundations decreases. For footings on clay medium, the variation of the ultimate bearing capacity is not significant; however, the permissible bearing pressure corresponding to 50 mm settlement is found to decrease with the decrease in clear spacing. Moreover, the present analyses indicate that the settlement of closely spaced footings is found to be higher than that of a single isolated footing, and the settlement further reduces with increase in the clear spacing ratio. In general, for $S/B \geq 3.5$, the load-settlement behavior of each of the adjacent footings is almost the same as a single footing. This study will ultimately be relevant to future endeavors associated with the design of closely spaced shallow foundations in enhancing the bearing capacity, and controlling the deformation of soils.

1 INTRODUCTION

Foundation is an integral part of a building whose stability determines the stability of the entire structure, as it acts as a medium through which loads are transmitted to the soil or rock below. In addition to the foundation material, the stability of a structure also depends on the properties of the soil or rock on which the structure stands. Thus, a foundation engineer deals with the coupled problem of evaluating the ability of the soil to support the superstructure and designing the proper foundation to effectively transmit the loads to the ground.

In designing a foundation, it is necessary to know the type of soil, the bearing capacity, and settlement behavior of the soil. The foundations must be designed such that the stresses induced in the soil are within its capacity. If the soil is overstressed, a shear failure may occur causing the soil to slide from underneath the structure and cause its failure. Thus, the estimation of the load carrying capacity of a footing is essential in the design process.

This study focuses on the evaluation of the two important parameters required in shallow foundation design, these are the bearing capacity, and the total settlements of soil.

*Corresponding author
DOI: 10.1201/9781003188339-43

Several theories have already been established to study the behavior of shallow foundations (i.e. bearing capacity, settlement, failure surface, etc.). Mostly, conventional theories deal with ideal situations. However, inadequacies of those theories have been experienced by engineers when faced with situations which deviate from those assumed in the theories.

In particular, foundations encountered in practice are often closely spaced and are rarely isolated. When two or more foundations are placed very close to each other, their respective stress isobars overlap, leading to a phenomenon called interference, which affects the settlement, bearing capacity, and failure mechanism of the footings. The behavior of neighboring foundations is greatly affected by the spacing between them and differs from that of an identical isolated footing. Such an influence of the closeness of the footings in the bearing capacity and settlement is defined as the effect of interference of footings.

The interference of footings may lead to excessive settlement and severe damage to structures if not properly controlled. Thus, neglecting the interference effect in the design of adjacent footings may result in unconservative footing design.

Considering the possible interference effect of closely spaced foundations along with other factors such as multi-layered soil strata and non-linear behavior of earth's materials, the load-settlement response of these foundations may be difficult to predict, thus, making foundation design a challenge in projects.

2 FINITE ELEMENT METHOD FOR FOUNDATION ANALYSIS

The key in using FEM to solve the ultimate bearing capacity is how to determine whether the foundation reaches the ultimate destructive state based on the results of finite element calculations. There are several types of soil instability criteria as discussed by Du et al. (2017). However, this study uses the load-displacement curve from the FEM software to determine the ultimate bearing capacity. The load corresponding to the inflection point on the load-displacement curve is used as the criterion of critical damage state. At present, many scholars tend to choose this criterion because the load-displacement curve gives the most intuitive reaction of foundation instability (Du et al., 2017).

The problem is formulated in *RS3* where the footings are placed on top and of the center of the soil domain with dimensions 40m (length) by 40m (width) by 20m (depth). The thicknesses of the upper and lower layers are the same with those of the 2D model in Figure 1. The two footings of width *B*, placed at a clear spacing *S*, rest on top of the two-layered soil deposit

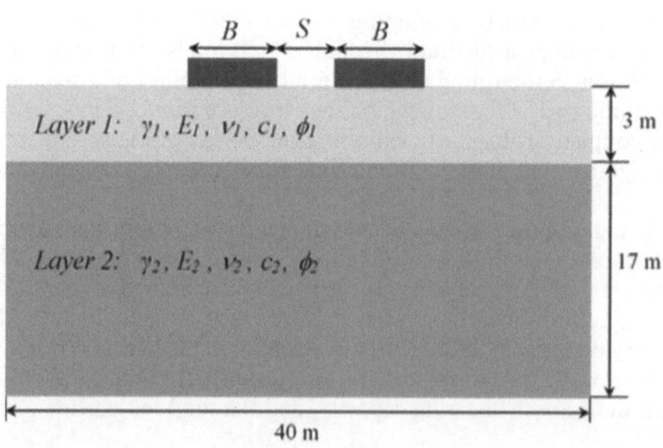

Figure 1. The geometry of interfering footings in 2D.

The ground surface or the top boundary in the model is free in all the directions. Full fixity is placed at the base of the geometry, that is, horizontal and vertical displacements are restricted in the bottom boundary. Meanwhile, the vertical boundaries are restricted only in the horizontal direction. The 10-noded tetrahedron elements are used in *RS3*. The mesh refinements are done automatically using a graded mesh as illustrated in Figure 2. Here, the element density is focused around the footings. A load-controlled approach is utilized in *RS3* where an incremental load is applied step by step until failure. The *Automatic* solver type is chosen to perform the required calculations in *RS3*.

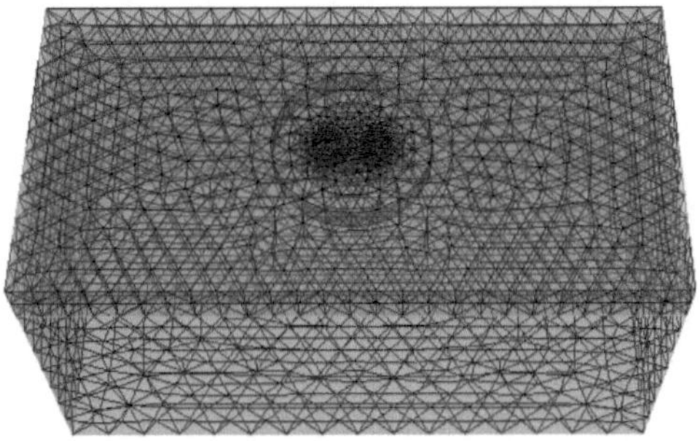

Figure 2. 3D mesh generation in RS3.

Furthermore, the soil input parameters for the three soil test cases are shown in Table 1.

Table 1. Mohr-Coulomb parameters of the two-layered soil profiles.

	Case 1: Sand over Clay		Case 2: Clay over Sand		Case 3: Layered c- ϕ soil	
	Layer 1	Layer 2	Layer 1	Layer 2	Layer 1	Layer 2
Mohr-Coulomb soil parameters	Loose Sand	Medium Clay	Soft Clay	Dense Sand	Sandy clay	Dense Sand
γ (kN/m^3)	18	18	16	18	17	18
E (kN/m^2)	13,000	4,000	2,000	40,000	15,000	40,000
ν	0.3	0.3	0.3	0.3	0.3	0.3
c (kN/m^2)	1	35	40	1	5	1
ϕ (°)	30	0	0	40	20	40

3 RESULTS AND DISCUSSION

In order to understand the variation of bearing capacity and total settlements of interfering footings, two dimensionless terms are used. These are the Bearing Capacity Ratio, BCR, and the settlement ratio, ξ_δ, defined in Equations 1 and 2, respectively.

$$BCR = \frac{q_{u(int)}}{q_{u(single)}} \tag{1}$$

where $q_{u(int)}$ is the ultimate bearing capacity of the interfering footings, and $q_{u(single)}$ is the ultimate bearing capacity of an identical isolated footing.

$$\xi_\delta = \frac{\delta_{v(int)}}{\delta_{v(single)}} \tag{2}$$

where $\delta_{v(int)}$ is the total displacement of the interfering footings corresponding to $q_{u(int)}$, and $\delta_{v(single)}$ is the total displacement of an identical isolated footing corresponding to $q_{u(single)}$.

3.1 *Variation of the bearing capacity with the spacing ratio S/B*

All the load-displacement curves are generated based on the results of the finite element calculation. In particular, Figure 3 shows the l-d curves for the cases of circular, square, and strip footings on sand overlying clay soil.

The Tangent Intersection Method, proposed by Trautmann & Kulhawy (1988), is used to find the ultimate bearing capacities. The peak load, q_u, is obtained as the intersection point of the initial and final tangent slopes of a load-displacement curve. The BCR values are obtained for all the footing cases. These values are summarized in Table 2.

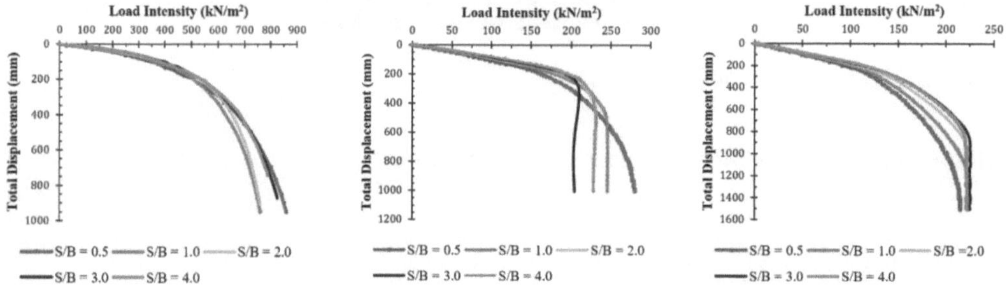

Figure 3. Load-settlement curves for interfering circular, square and strip footings on sand over clay soils.

Table 2. Comparison of BCR for closely spaced footings on different soil profiles.

		S/B									
Soil Case	Shape	0.5	1.0	1.5	2.0	2.5	3.0	3.5	4.0	4.5	5.0
Case 1: Sand over Clay	Circle	1.185	1.037	1.000	1.037	1.037	1.000	1.000	1.000	1.000	1.000
	Square	1.232	1.102	0.986	1.040	0.990	0.937	1.026	1.026	1.000	1.006
	Strip	0.973	1.014	0.995	1.000	1.009	1.018	0.995	0.991	1.009	1.002
Case 2: Clay over Sand	Square	0.746	0.822	0.886	0.900	0.925	0.948	0.973	0.994	0.998	1.001
	Strip	0.917	0.942	0.956	0.968	0.973	0.979	0.986	0.990	0.991	0.994
Case 3: Layered c-ϕ soil	Square	1.033	1.024	1.022	1.024	1.015	1.016	1.008	1.000	1.001	0.999
	Strip	1.142	1.058	1.026	1.025	1.021	1.018	1.017	1.014	1.013	1.011

It can be observed that when footings are placed close to each other (i.e. S/B = 0.5, 1.0), BCR is greater than one. This indicates that at close spacing, there is an increase in the ultimate bearing capacity (UBC) of interfering footings compared to the isolated footing. By increasing S/B, the UBC reduces, and for $S/B \geq 3.5$, the UBC of the footings approaches that of the single isolated case.

Both the interference of footings on sand over clay and layered sandy clay soil media are in consensus with previous works dealing with closely spaced footings on cohesionless soil medium. Hence, at close contact, footing interference is significant in terms of UBC.

On the other hand, the change in UBC is not significant for footings on clay over sand medium. Saran & Varma (1988) and Nainegali & Ekbote (2019) showed that the UBC of footings on clay is barely affected by the interference; in fact, for undrained condition, this can be ignored.

However, the effect of interference on the bearing pressure corresponding to an allowable settlement of 50 mm (as per Indian standard code, IS 1904-1986) is analyzed. The bearing ratio is less than one at close spacing. Hence, the permissible bearing pressure of interfering footings is less than that of the isolated footing case. The bearing ratio increases with increase in *S/B*, attaining the value of one at $S/B \geq 3.5$.

3.2 *Variation of the total settlement with the spacing ratio S/B*

The variation of the total settlements is also investigated to observe the settlement behavior of interfering footings. The plots of the settlement ratio against *S/B* are presented for all the footing cases as shown in Figure 4.

It can be seen that at close spacing, the settlement ratio is greater than one, signifying that the total settlement of interfering footings is greater than that of the isolated footing. The settlement ratio decreases with the increase in *S/B* and approaches the value of one at $S/B \geq 3.5$, wherein footings can be specified to be acting individually. This is true for all studied cases.

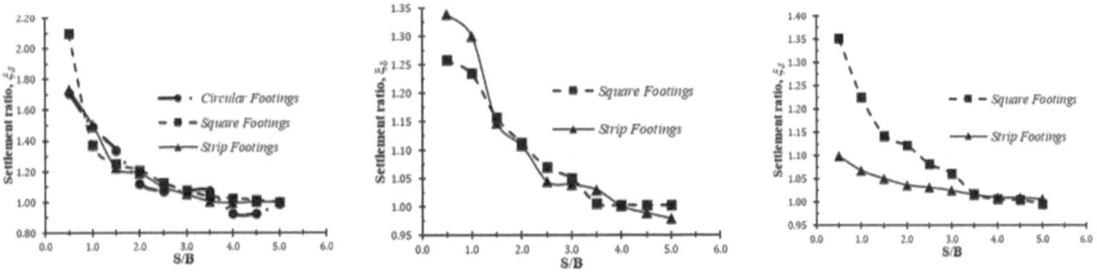

Figure 4. (A) ξ_δ vs. S/B for footings on sand over clay (B) ξ_δ vs. *S/B* for footings on layered sandy clay (C) ξ_δ vs. *S/B* for footings on clay over sand.

3.3 *Effect of soil media on the behavior of closely spaced footings*

The type of soil media where the footings are founded affects the interference behavior of closely spaced foundations. In particular, the interference effect in a cohesionless soil is different than in a cohesive soil and in cohesive-frictional soil.

For a two-layered soil profile, where the thickness of the upper layer is greater than the width of the foundation ($H_1 > B$), the load-displacement behavior of the footings greatly depends on the parameters of the upper soil layer. The effect of the lower layer on the footing behavior is negligible.

For these cases, the footings resting on the two-layered profiles can be treated as footings on a single homogeneous layer of soil considering the parameters of the upper layer. Hence, the three cases of two-layered soil profile can be simplified as follows: (1) the sand overlying clay soils can be treated as homogeneous sand; (2) the clay overlying sand soils can be considered as purely cohesive soil; and (3) the layered sandy clay soils can be considered as homogeneous sandy clay soil.

4 CONCLUSIONS

In the present study, numerical finite element analyses are carried out to determine the interference effect of closely spaced shallow foundations resting on the surface of various two-layered soil profiles. The footing geometry, spacing configuration, and the type of soil medium in the finite element models are varied to study the interference effects in terms of the variations of the bearing pressure and total settlements with *S/B*.

In a two-layered soil profile, where the thickness of the upper layer is larger than the foundation width (*H1* > *B*), the interaction behavior of the footings greatly depends on the parameters of the upper soil layer and not much on the lower layer.

For circular and square footings on cohesionless soil medium, the magnitude of the ultimate bearing capacity increases as the clear spacing between the foundations decreases. For strip footings on sand, the ultimate bearing capacity is barely affected by interference.

On sandy clay soil medium, the ultimate bearing capacities of interfering square and strip footings also increase as the footings come closer to each other. At the critical spacing *S/B* = 0.5, the ultimate bearing capacities are found to be greater than the ultimate bearing capacities of the isolated cases by 3.3% and 14.2% for the square footings and strip footings respectively.

The effect of interference of footings on clay medium is very negligible with respect to the change in magnitude of the ultimate bearing capacities. However, the effect is noticeable on the permissible bearing pressure measured in the allowable settlement of 50 mm. The bearing pressure reduces as the spacing decreases so that the allowable settlement is maintained at a tolerable value. The interference is particularly significant for square footings wherein the decrease in bearing pressure at *S/B* = 0.5 is 25.4%.

It is found that for *S/B* ≥ 3.5, the bearing capacity of the interfering footings is almost the same as that for single isolated footing. This means that for a clear spacing more than 3.5 times the footing width/diameter, no significant interference effect is observed, and the footings can act independently.

The peak value of total settlements occurs at *S/B* = 0.5 for all test cases. Among the three cases of soil media, the most significant increase in total settlements can be found in the case of footings on cohesionless soil medium. For footings on sand medium, the increase in total settlements at *S/B* = 0.5 are 70.7%, 110% and 73.6% for circular, square, and strip footings respectively.

REFERENCES

Alwalan, M. F. Interaction of Closely Spaced Shallow Foundations on Sands and Clays: A review. *International Journal of Advanced Engineering Research and Science*, 5(9).

Brinkgreve, R. B. J., Broere, W., & Waterman, D. (2002). *PLAXIS 2D-version 8 reference manual.* Delft University of Technology&Plaxis bv, The Netherlands.

Das, B.M., Puri, V. K., & Neo, B. K. (1993). Interference effects between two surface footings on layered soil. *Transportation research record*, (1406).

Du, P., Liu, X., & Zhang, Y. (2017). Discussion of the Method to Determine the Ultimate Bearing Capacity of Soil Foundation. *IOP Conference Series: Earth and Environmental Science* (Vol. 100, No. 1, p. 012007). IOP Publishing.

Ghosh, P., & Sharma, A. (2010). Interference effect of two nearby strip footings on layered soil: theory of elasticity approach. *Acta geotechnica*, 5(3), 189–198.

Hanna, A. M. (1982). Bearing capacity of foundations on a weak sand layer overlying a strong deposit. *Canadian Geotechnical Journal*, 19(3), 392–396.

Kumar, J., & Ghosh, P. (2007). Ultimate bearing capacity of two interfering rough strip footings. *International Journal of Geomechanics*, 7(1), 53–62.

Nainegali, L., & Ekbote, A. G. (2019). Interference of Two Nearby Footings Resting on Clay Medium. In *Geotechnical Applications* (pp. 59–67). Springer, Singapore.

Rocscience, 2013. *3D Meshing Customization Developers Tips*. Rocscience Inc., Toronto Canada, 8.

Stuart, J. G. (1962). Interference between foundations, with special reference to surface footings in sand. *Geotechnique*, 12(1), 15–22.

Ti, K. S., Huat, B. B., Noorzaei, J., Jaafar, M. S., & Sew, G. S. (2009). A review of basic soil constitutive models for geotechnical application. *Electronic Journal of Geotechnical Engineering*, 14, 1–18.
Trautmann, C. H., & Kulhawy, F. H. (1988). Uplift load-displacement behavior of spread foundations. *Journal of Geotechnical Engineering*, 114(2), 168–184.
Verma, G. S., & Saran, S. (1988). Interference effect on the behavior of footings. *Indian geotechnical journal*, 18(2), 171–185.
West, J. M., & Stuart, J. G. (1965). Oblique loading resulting from interference between surface footings on sand. In *Soil Mech & Fdn Eng Conf Proc/Canada/*.

The Evolution of Geotech - 25 Years of Innovation – Hammah et al. (Eds)

Behaviour of rigid block on uniform sand under horizontal base acceleration

N. Trbović, N. Čeh & V. Jagodnik
Faculty of Civil Engineering, University of Rijeka, Croatia

ABSTRACT: The behavior of rigid structures on rigid base or deformable media such as soil has been addressed by many researchers in the past. The behavior of a rigid body on the soil under static or cyclic actions, such as self-weight, wind or earthquake, is of great concern for both structural and geotechnical engineers. The engineers have to take into consideration the effects of such forces on the structure itself and on the soil as well. Here, a very significant role is in the soil-structure interaction. The paper presents the behavior of a rigid block on a uniform, medium dense, sand under horizontal acceleration of the base. A 6 cm wide rigid aluminum block with different height to width ratio (H/B) has been used. Hight to width ratio used in this research was 2.25, 3.75 and 4.5. The block was either placed on the sand (D = 0 cm) or was embedded in the sand prior to subjecting it to horizontal excitation using small shake table. The embedment depth D varied from 0 to 6 cm. The horizontal excitation was a sinusoidal displacement function with different amplitude and frequency. Frequency varied from 0.5 to 3 Hz while the amplitude varied from 0.5 to 5 cm. Combination of various frequencies and amplitudes were used to determine the critical combination of amplitude and frequency for a given height to weight ratio which was later used for more detailed analysis and numerical verification. Displacements of a rigid block and sand below the rigid block were measured using a contactless measurement system, ARAMIS. The results obtained through ARAMIS system were later used as a reference displacement for numerical analysis. Numerical simulations were performed using Rocscience RS2 software. Mohr-Coulomb model was used in numerical simulation.

1 INTRODUCTION

One of the most common problems that arises in geotechnical engineering is the calculation of soil bearing capacity. In order for soil bearing capacity to be determined, it is necessary to determine the soil bearing capacity factors. Chowdhury and Dasgupta (Chowdhury and Dasgupta, 2017), among others, defined soil bearing capacity in the case of dynamic load on the foundation. Following an earthquake, the foundation will begin to vibrate along with the surrounding soil, resulting in the appearance of inertial forces in the horizontal and vertical directions. Due to the increase of the dynamic load on the foundation soil, there is a reduction of the soil bearing capacity factor during seismic action. Very few experiments have been performed to determine bearing capacity factors under dynamic load.

The aim of this paper is to present a specifically designed experimental model of a rigid foundation on a deformable media such as soil and to investigate the possibility of determining the dynamic bearing capacity factors. The main focus of this research is on the behavior of rigid block on deformable media with special emphasis on the use of optical measuring system and simple numerical models to validate the block behavior.

DOI: 10.1201/9781003188339-44

2 METHODOLOGY

Laboratory experiments were performed in the Structural laboratory at the Faculty of Civil Engineering, University of Rijeka, Croatia. Experiments were performed on small shaking table which was used to simulate dynamic loading. text. Dimensions of the testing model are presented with Figure 1. For the purpose of conducting the experiment, a special box 100 cm long and 50 cm high was made, in which the uniform sand was placed, as shown in with. The width of the box can be changed depending on the width of the foundation and in the range of 6 - 18 cm. In the conducted laboratory experiments, a block with floor plan dimensions of 6 x 6 cm and a height of 27 cm was used, which gave the width to height ratio of the foundation H/B = 4.5. Due to the limitation of the load-bearing capacity of seismic platforms, the box is made of aluminum "L" profile of 25 mm, wall thickness 2 mm, which results in a sufficiently rigid, and at the same time light box.

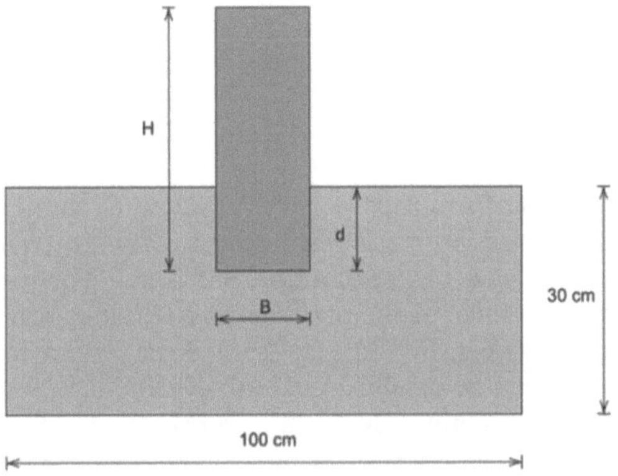

Figure 1. Dimensions of the testing model.

Figure 2. Testing box for shake table test.

The front of the box is made of transparent acrylic plastic (plexiglass) to allow tracking of displacements by the non-contact optical method. The soil used in the experiments is classified as uniform sand, (Jagodnik et al., 2020) named DrOS018. The relative density of sand Dr of 50% is defined by the pore coefficient which in this case is 0.789. The sand was built-in inside the box in layers to achieve a uniform density of the material using the undercompaction method developed by Ladd (Ladd, 1978). Table 1 shows the parameters of the dynamic load, amplitude (A), frequency of the harmonic function (f) and embedment of the block (d) for each experiment.

Table 1. Test ID and model parameters.

Test ID.	$f[HZ]$	$A[cm]$	$d[cm]$
EXP14	3	0.5	2
EXP15	2.5	0.5	0
EXP16	2.5	0.25	0
EXP17	3	0.25	0
EXP18	4	0.25	0

shows a prepared box with built-in sand, a set of block and cameras, and lighting in front of the box, ready to start measuring. The markers needed to track the displacement are placed on the block and soil media.

3 EXCITATION SYSTEM AND OPTICAL MEASURING SYSTEM

The Quanser STI-III seismic platform used, which simulated dynamic loads, was driven by an electromagnetic motor controlled by a LabView-based software. Within the computer program interface, a sinusoidal dynamic load function is defined. The top plan dimension of the platform is 625 x 625 mm, the travel of the platform in each direction is 15 cm, and the range of operating frequencies is between 0 - 20 Hz. With a maximum mass on the platform of 130 kg, an acceleration of 1 g can be produced in both directions in the horizontal plane.

The behavior of the block was monitored with GOM Aramis 5M optical system able to measure displacements and strains on the surface of the model. The system consists of a set of high-speed cameras with a resolution of 2400x1728 pixels. Each experiment was filmed with 50 frames per second and post-processed to calculate the horizontal displacement corresponding to the midpoint of the base of the block.

4 NUMERICAL ANALYSIS

To analyze the behavior of the block under dynamic loading, the computer program Rocscience 2 (hereinafter RS2) was used. RS2 is based on the finite element method. Numerical simulations were performed for five different models, based on laboratory experiments, which differ in frequency F and amplitude A of the dynamic load and the embedment depth of the block (foundation depth d), as shown in Table 2.

Table 2. Soil material characteristics.

Material characteristic	Value
Friction angle ϕ [°]	30
Cohesion c $[kPa]$	0
Minimum void ratio e_{min}	0,627
Maximum void ratio e_{max}	0,951
Unit weight γ $[kN/m]$	21

The Mohr Coulomb criterion was chosen for the constitutive model of sand. The associated flow rule was used in the analysis to simplify the calculations (e.g. Pietruszczak, 2010; Potts and Zdravković, 1999).

Numerical analysis was performed in phases in such a way that the phases are divided into time intervals, a total of 16. Displacement data were recorded at the end of each phase. The first phase, denoted by the time t = 0 s, represents the static conditions when the dynamic load has not yet been applied.

Figure 3 shows a numerical model created in the computer program RS2 used for numerical simulations. Isoparametric four-nodded quadrilateral finite elements were used in the simulation (eg. Bathe, 2008; Carter et al., 2002; Desai and Kundu, 2001; Lees, 2016; Puzrin, 2012). The minimum length of the element defined by wave velocity was taken into account based on the suggestions made by Kuhlemeyer and Lysmer (Kuhlemeyer and Lysmer, 1973). In the first phase of the simulation, standard boundary conditions were used, horizontal shifts on the sides were prevented, as well as vertical and horizontal shifts on the lower side of the model. In the dynamic phases of the simulation, dynamic boundary conditions were used on the sides of the model to allow equal movement of the points on both sides of the model (tied bound condition), and on the underside of the model conditions that absorb dynamic load waves (absorb boundary condition). The dynamic excitation is given at the bottom of the model, and the load is given in the form of a time shift. The stated dynamic excitation is taken as the input from the measured data of the optical measurement system. Figure 2 also shows the point P3 in the middle of the contact surface between the base of the block and the surrounding soil for which the data were analyzed and the "joint" element (orange) (Bathe, 2008), which determined the conditions at the contact between the block and the soil beneath. A normal stiffness of 1000 kPa per meter and a tangential stiffness of 5 kPa per meter are defined on the "joint" element. Also, Figure 2 shows an aluminum block, whose mechanical characteristics used in the models are shown in Table 3.

Table 3. Mechanical characteristics of aluminum block.

Mechanical characteristic	Value
Unit wighty $[kN/m^3]$	26.29
Poisson coefficient v	0.34
Young modulus E $[GPa]$	68.3

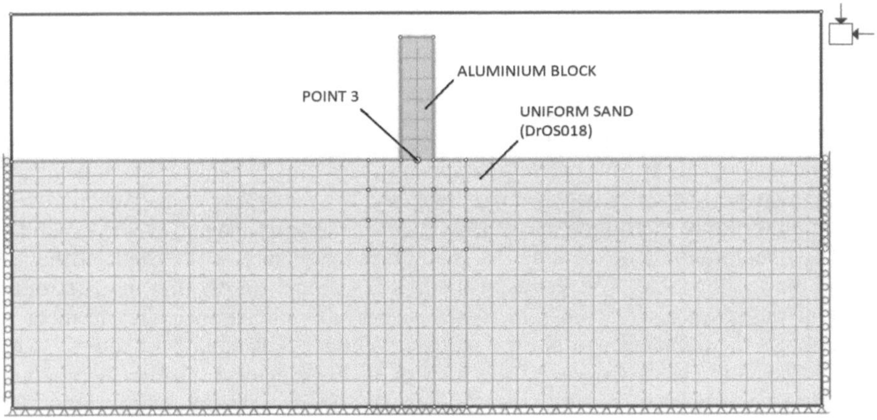

Figure 3. Numerical model in Rocscience RS2 according to dimensions on Figure 1.

5 RESULTS

The results of two characteristic tests are presented in this section due to the limit in paper length:(i) EXP14 together with MRS14 and (ii) EXP17 together with MRS17. The results are presented in Figure 4. The results for the horizontal displacements are shown for the point P3 located in the middle of the contact surface between the block and the surrounding soil. In the EXP14 experiment, the block was buried in the substrate 2 cm and subjected to a dynamic load with frequency 3 Hz, and amplitude 0.5 cm, which caused overturning behavior. In the EXP17 experiment, the block placed on the sand surface subjected to the dynamic load with frequency of 3 Hz and amplitude of 0.25 cm remained stable throughout the experiment.

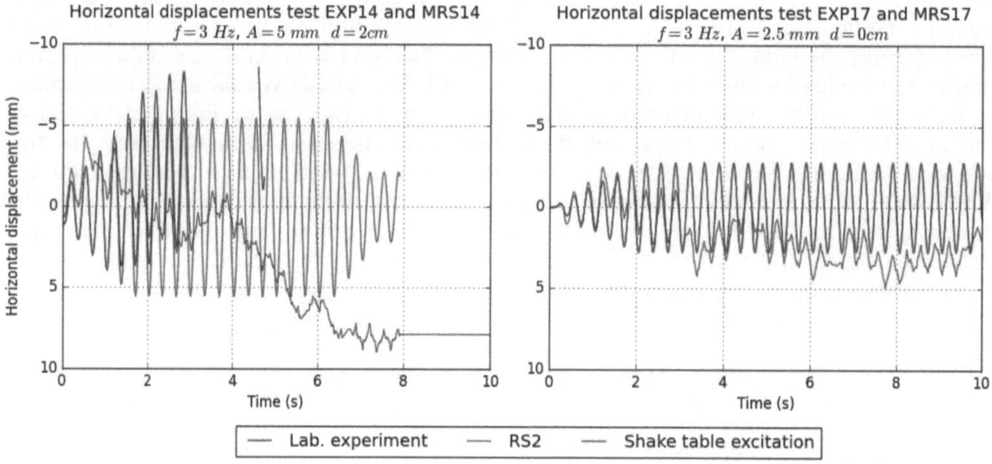

Figure 4. Example of test results for experiments EXP14 and EXP17 compared with Rocsciene RS2 simulations.

6 DISCUSSION

In the EXP14 experiment, the block overturned at time of t = 4.6 s. The overturning of the block is manifested by a sudden increase in the horizontal displacement of the point P3, presented on Figure 4. The block begins to lose stability when the displacement curve of point T3 begins to deviate from the horizontal displacement curve of dynamic load. Loss of displacement data records occurs because during larger displacements, when the block begins to roll over, sand penetrates the space between the block and the plexiglass, thus covering the markers points monitored by the optical measuring system. Towards the end of the experiment, it is visible that the amplitude of the dynamic load decreases, which is the result of the gradual stopping of the seismic table. The results of the MRS14 model, shows that the horizontal displacement of the block after t = 4 s increases intensively. The sudden displacement of the block determines the occurrence of block instability. The block behavior in the EXP17 experiment and MRS17 model is shown by the horizontal displacement diagrams, presented on Figure 4. The coincidences of the curves of horizontal displacement of the point P3 and dynamic load are shown, which proves that the block was stable during the experiment. The horizontal displacement of the P3 point on the block in the MRS17 model, after the time of t = 3 s begins to deviate from the dynamic load curve but does not increase significantly (compared to MRS14 model). Over time, the block slightly translates in the direction of positive horizontal displacements.

7 CONCLUSIONS

The presented paper analyzes the experiments of the rigid foundation on a deformable media such as soil using non-contact measurement methods. A series of dynamic experiments of a rigid block on a deformable media was carried out using a shake table platform. The critical behavior of the block was determined by a series of experiments, and two critical experiments were selected for detailed analysis. Two selected experiments were also numerically modeled using Rocsceince RS2 software. Data collected by non-contact measurement were used as input data for numerical models as well as to compare the results of performed numerical analyzes.

The two characteristic tests presented in detail are: EXP14 and EXP17 along with numerical simulations, MRS14 and MRS17.

Performed research proved that the results obtained by the calculation in the computer program Rocscience RS2 largely depend on the correct choice of the constitutive soil model, which can be clearly seen through deviations of the numerically obtained results from the laboratory measurements. The results of laboratory experiments are relevant, but also sensitive to the change in excitations and block embedment. Furthermore, RS2 cannot capture some behavior of the block such as separation from the soil media that occurred during the overturning of the block in the conducted laboratory experiments. In the numerical analysis, the Mohr-Coulomb strength criterion was used, which is very widespread in geotechnical practice. The paper shows how limited it is in depicting the behavior of vertical deformation. The authors suggest using better constitutive models that have better defined stress – strain law along with volumetric behavior.

In the results of laboratory experiments, the overturning of the block can be observed by the deviation of the horizontal displacements of the block from the displacement of the shake table and the sudden increase of the vertical displacement. The results of the numerical RS2 models do not lead to the conclusion that the block overturned, but irregularities can be detected in the diagrams indicating the existence of instability (eg sudden displacement changes or displacement function that does not follow the dynamic load displacements in the horizontal direction).

By carefully planning the experiments and selecting the appropriate measuring equipment, it is possible to determine the stresses below the block, and with the advanced use of non-contact measurement, it is possible to determine the deformation field below the block and thus the bearing capacity factors under dynamic loading.

The application of the measurement methods presented in this paper will greatly help in the methodology of preparation of scaled laboratory experiments to further investigate dynamic bearing capacity.

ACKNOWLEDGMENTS

This research was partially supported by the following research projects:

- Bilateral project between Croatia and Germany "Rigid body rocking on a flexible structure non-smooth contact-dynamics approach and experimental validation"
- UKF project "Collisions in rocking multi-body systems – experimental and numerical investigation"
- Ministry of Science, Education and Sports of the Republic of Croatia under the project Research Infrastructure for Campus-based Laboratories at the University of Rijeka, number RC.2.2.06-0001. Project has been co-funded from the European Fund for Regional Development (ERDF),
- Student Science Fund "SIZIF" (Project code: N-SZF 1/2020; Project financier: Student assembly of the University of Rijeka and University of Rijeka Foundation, Republic of Croatia)

REFERENCES

Bathe, K.-J., 2008. Finite Element Method. In: Wiley Encyclopedia of Computer Science and Engineering. John Wiley & Sons, Inc., Hoboken, NJ, USA.

Carter, J.P., Desai, C.S., Potts, D.M., Schweiger, H.F., S.W.Sloan, 2002. Computing and Computer Modelling in Geotechnical Engineering. In: Unknown (Ed.), Unknown.

Chowdhury, I., Dasgupta, S.P., 2017. Dynamic Bearing Capacity of Shallow Foundation Under Earthquake Force. Indian Geotech. J. 47, 35–46.

Desai, C.S., Kundu, T., 2001. Introductory Finite Element Method, Mechanical and Aerospace Engineering Series. Taylor & Francis.

Jagodnik, V., Kraus, I., Ivanda, S., Arbanas, Ž., 2020. Behaviour of uniform Drava river sand in drained condition-A critical state approach. Appl. Sci.

Kuhlemeyer, R.L., Lysmer, J., 1973. Finite element method accuracy for wave propagation problems. J. Soil Mech. Found. Div 99.

Ladd, R., 1978. Preparing Test Specimens Using Undercompaction. Geotech. Test. J. 1, 16–23. Lees, A., 2016. Geotechnical Finite Element Analysis: A Practical Guide. I C E Publishing.

Pietruszczak, S., 2010. Fundamentals of Plasticity in Geomechanics, A Balkema Book. Taylor & Francis Group.

Potts, D.M., Zdravković, L., 1999. Finite Element Analysis in Geotechnical Engineering: Theory, Finite Element Analysis in Geotechnical Engineering. Thomas Telford.

Puzrin, A., 2012. Constitutive Modelling in Geomechanics: Introduction, Constitutive Modelling in Geomechanics: Introduction. Springer.

The Evolution of Geotech - 25 Years of Innovation – Hammah et al. (Eds)

Skin friction displacement relations for load settlement behavior of bored piles in Dubai

Ahmed Al-Mufty
Rocscience, Toronto, Ontario, Canada

ABSTRACT: From a piling project in Dubai, static load tests were performed on piles of different diameters and lengths using traditional Kentledge procedure. The results of these tests are used to back calculate and estimate skin friction – soil displacement curves which can be applied with low risk to assess the load settlement behavior of long bored piles socketed in Sandstone and Siltstone evaporites of Dubai. RSPile software is used in testing of the estimated side shear-displacement relations. The idea may be generalized to other areas to establish a data base for other rock types to fill the lack of this type of data. The results may be used to optimize the pile lengths saving lots of cost and time.

1 INTRODUCTION

1.1 *Predicting the settlement of a bored pile with a linear soil model*

The traditional method to predict the settlement of a pile under static load is to sum the shortening of the pile body w_e, the displacement due settlement of the ground layers surrounding the pile w_s, and the displacement at the pile base due to settlement of the layers beneath the pile tip w_b. All components are usually calculated using simplified elastic models depending on the pile's elastic modulus and the modulus of deformation estimated for each ground layer. The elastic shortening of the pile may be divided into two parts, the first one, w_f, is for the free length above the ground (or within a soil of insignificant resistance such as very loose sand) and the second part for the pile length that is having a skin friction with the surrounding soil which may be a portion of the pile length or the full length of the pile and is denoted as w_p.

$$w_{tot} = w_f + w_p + w_s + w_b \tag{1}$$

For a load Q applied at the pile head, as shown in Figure 1, the equations for the pile elastic shortening are:

$$\mathrm{w_f} = \frac{\mathrm{QL_f}}{\mathrm{AE_p}} \tag{2}$$

$$w_p = \frac{\lambda \mathrm{Q}_s \mathrm{L}_p}{\mathrm{AE_p}} + \frac{\mathrm{Q}_b L}{\mathrm{AE_p}} \tag{3}$$

Where Q is the applied load and divided into skin friction Q_s and end bearing resistance Q_b. L_f and L_p are the lengths of the pile portion above ground and the pile portion subject to skin friction respectively, while L is the total length of the pile. The modulus of elasticity of the pile E_p is the Young's modulus of the concrete adjusted for the composite effect of the

DOI: 10.1201/9781003188339-45

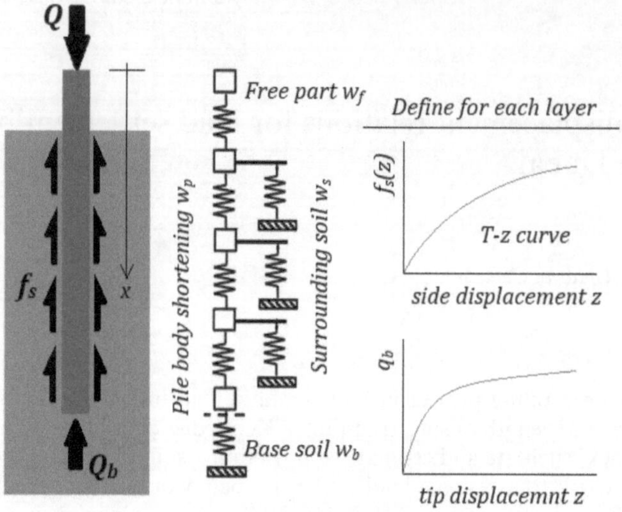

Figure 1. Elastic settlement components and soil nonlinear modeling explained.

reinforcement. The parameter λ is dependent on the assumed distribution of the skin friction along the pile and can be derived from the axial stress-strain theory. The most common values for λ are shown in Figure 2, see Vesic (1977). Other values may be derived from different distributions.

The next component w_s is found from, Vesic (1977):

$$w_s = \frac{Q_s B}{p L_p E_s}(1 - \mu^2) I_s \tag{4}$$

The Poisson's ratio μ and the modulus of deformation E_s are for the soil surrounding the pile and the author suggests using the harmonic average for the modulus through the depth of the pile. Whereas B and p are the breath and the perimeter of the pile and the influence factor I_s, is defined as $2+0.35\sqrt{(L_p/B)}$.

The fourth component of the settlement w_p takes place when some or all load is transferred to the base and calculated as:

$$w_b = \frac{q_b B}{E_b}(1 - \mu^2) I_b \tag{5}$$

The modulus E_b is for the effective soil beneath the base and q_b is the mobilized unit end bearing Q_b/A. The influence factor I_b is a function of rigidity of the pile, its cross-sectional shape, and the depth to breadth ratio of the pile. For long rigid piles, values in Table 1 may be used for I_b, see Perloff (1975). Perloff's values are already corrected for depth in the table for different Poisson's ratios.

Table 1. Influence factor I_b for rigid long piles ($L_p \geq 10B$).

			Rectangular with width to breadth ratio of			
Poisson's ratio μ	Circular	Square	1.5	2	3	5
0	0.58	0.73	0.85	0.97	1.13	1.35
0.25	0.62	0.79	0.92	1.04	1.22	1.46
0.49	0.68	0.86	1.00	1.13	1.32	1.59

1.2 *Prediction of pile settlement using nonlinear soil models*

The so-called T-z and Q-z models are widely accepted for estimating the load settlement behavior of piles. The idea of these models, T-z being the side friction load transfer relation to soil displacement defined through testing and Q-z, the curve defining the relation between the load transferred to the pile base and settlement of the base, is depicted in the right side of Figure 1. The modeled ground springs and examples to their nonlinear T-z and Q-z curves are shown.

Although the method has a semi-empirical nature, but with increasing size of data base, the results are getting significantly more acceptable. This method consists of solving the differential equation of axially loaded member with side traction and point load at the base which can be written as:

$$E_p A \frac{d^2 z}{dx^2} = f_s(x, z)p \tag{5}$$

where $f_s(x)$ is the upward unit skin friction at depth x measured from the head of the pile downwards and z is the displacement at the depth x positive downwards. At the pile tip, a boundary load exists as Q_b which is equal to unit end bearing pressure multiplied by the area of the base, $q_b A$. The values of f_s and q_b are functions of the displacement z taking place at the level in consideration and defined by the aforementioned, T-z and Q-z curves. The latter curves are obtained and defined for each soil type (and possibly location also) and p is the perimeter which can vary with x as well.

Derivation of Equation 5 can be found in standard textbooks of mechanics of solids such as Popov (1968). The solution of the equation may be by finite differences or finite element methods. In any of the two, there will be an iterative procedure at each load step to reach balanced displacement values on both sides of the equation. Hence, the defined T-z and Q-z curves are used in every iteration to find the corresponding resistance to the displacement value obtained until the error becomes tolerable. A commercial software such as RSPile is used for solving the model. In RSPile, even the pile body is modeled with nonlinear concrete stress strain relation.

1.3 *Aim of the study*

The traditional static load testing of bored piles is the best way to predict a soil model and to put forward some representing T-z and Q-z curves. For cohesionless soils and cohesive soils several models may be found in literature. The one that will be adopted here as T-z curve for cohesionless soils is presented by Reese and O'Neill (1988). The situation is not the same for weak rocks. Literature lacks T-z curves for weak rocks while for end bearing Q-z curves, the author found only one graph for Sandstones in CIRIA Report 181, Gannon et al. (1999).

The aim of this study is to determine a suitable T-z curve for the weak rocks of Dubai to enable engineers to predict settlement better. As the piles are long enough, no load transfer to the base is expected in the range of applied loads in the study.

2 THE PROJECT AND THE GROUND LAYERS

2.1 *The project description*

The project is located at Sheikh Zayed Road near interchange 1 in Dubai, UAE. A skyscraper tower of more than 60 floors with 260m height at its highest, with an adjacent but lower twin tower, were constructed on a foundation made of 242 Piles, out of which 118 piles were of 1.8m in diameter and the rest were having a diameter of 1.2m.

A trial first design for pile lengths required 30m and 31m socket lengths into rock for the 1.2m and 1.8m diameter piles, respectively. Unfortunately, the rock layers found lower than the cut off levels COL, and that all sand and decomposed Sandstone sandy gravel layers penetrated, denoted here as SG1 and SG2 respectively, were not added to the length in the first estimate. Also, an SG2 layer that was included in the pile length was treated as Sandstone while they were cohesionless decomposed rock sandy gravel. The Sandstone will be denoted hereafter as SS. The geotechnical investigation report did not show that SPT was carried out at deeper SG2 layers. Instead, some small core recoveries were recorded which suggested that deeper SG2 layer was SS continuation, so the first trial design was misled.

The first design length was examined by a test pile and the resulted settlement was high and unacceptable. Later, the pile lengths were designed again to penetrate a deeper layer of Calcisiltite, which denoted here as CS and two additional pile tests were conducted and the results were this time acceptable.

2.2 *Ground layers and their properties*

Six boreholes were drilled, and the ground layers with their thicknesses and sequences are shown in Figure 2. The borehole locations are shown in Figure 3 with the test piles as well. The figure also shows a three-dimensional view of the ground layers. These figures have been produced by the same commercial software RSPile. All elevations are based on the local Dubai Municipality Datum, DMD.

Table 2 lists the relevant soil and rock average properties that can be used in the analysis. The SPT average blow count N is used to obtain a design value for the ultimate skin friction f_{sult} for the SG1 layer as $2N$ MPa based on CIRIA 143, Clayton (1995), that suggests $1N$ to $3.3N$ for bored piles in cohesionless soils. For SG2, as N is higher than 50, AASHTO-FHWA method cited in Abu-Hejleh (2003) is followed. This method suggests a design ultimate skin resistance as:

$$f_{sult}(\text{in kPa}) = 150 + 2(N - 53) \qquad N > 53 \tag{6}$$

To estimate the ultimate skin resistance for the bored piles in the weak rock layers, Williams and Pells method is applied. Williams and Pells (1982), suggested that design ultimate skin resistance may be calculated as:

$$f_{sult} = \alpha\beta q_{uc} \tag{7}$$

The factor α is related to the average unconfined compressive strength of the rock q_{uc} through a graph presented in their article. The values of α tare 0.31 and 0.2 for the corresponding unconfined strengths of 2MPa and 4MPa, respectively. The reduction parameter β is a function of the rock quality designation RQD of the rock and is equal to 0.65 if RQD is less than or equal to 50. The resulting ultimate skin resistance corresponding to each layer is shown in the table. There are many methods to estimate the skin resistance of bored piles socketed in weak rocks but a recent study on the rocks of Dubai, Latapie et al. (2019), suggested that f_{sult} may be taken equal to $0.42\sqrt{q_{uc}}$ (in MPa). Still, the author feels that relating the skin resistance to RQD is more reliable although the values obtained from Williams and Pells method are more conservative than the latter suggestion.

The mass modulus of deformation E_m of the SG layers is determined following minimum suggestions of CIRIA 143. An estimated E_m of $2N$ (MPa) is used for SG1 and SG2. In fact, the higher the factor multiplied, the lesser the settlement calculated by elastic methods. The elastic method is significantly sensitive to the value of the modulus used. The decision for the modulus to be used for the rock layers is even more difficult without experience and back calculation of field test results. The modulus of deformation may be estimated from an equation recommended by BS8004 (1985),

$$E_m = j\ M_r q_{uc} \tag{8}$$

It is basically a reduction of the intact modulus of the rock $M_r\ q_{uc}$ with the mass factor j which is in turn a function of RQD (basically varies according to joints frequency in the rock). Anyhow, it is suggested in many references that j is taken as 0.2 for RQD values less than 50%. The modular ratio M_r is the ratio of the intact rock modulus to the unconfined strength, and it is suggested by BS8004 (1985) to be in the range 75 to 600 based on weathering conditions. It is found by the author's personal experience that the value of M_r is between 200 to 400 in UAE Sandstones according to the range of the unconfined strength. For low q_{uc} in the vicinity of 1 MPa, 200 will work fine increased to 400 or more when q_{uc} increase to higher than 2MPa. In this project, the intact modulus is assumed as $400q_{uc}$ which is still lower than what can be obtained from pressure meter tests or suggested by others, cf. Latapie et al. (2019), but for the author it gives reliable results for highly loaded piles. In the above suggestions, the modulus is assumed as the secant modulus of deformation at a factor of safety of around 2.0, i.e., at a stress level of about half of the unconfined strength.

The piles were long enough to transfer no or insignificant load to the base so the end bearing resistance is not discussed or presented. The pile lengths calculated to consume all the load in skin friction based on the ultimate skin resistance values in Table 2, were much lower than the applied ones.

Table 2. Ground layers properties needed for design.

		RQD	q_{uc}	f_{sult}	E_m
Layer type	SPT N (av.)	%	MPa	kPa	MPa
SG1	18	–	–	36	36
SG2	75	–	–	195	150
SS	–	0 to 50	2.0	403	160
CS	–	0 to 50	4.0	532	320

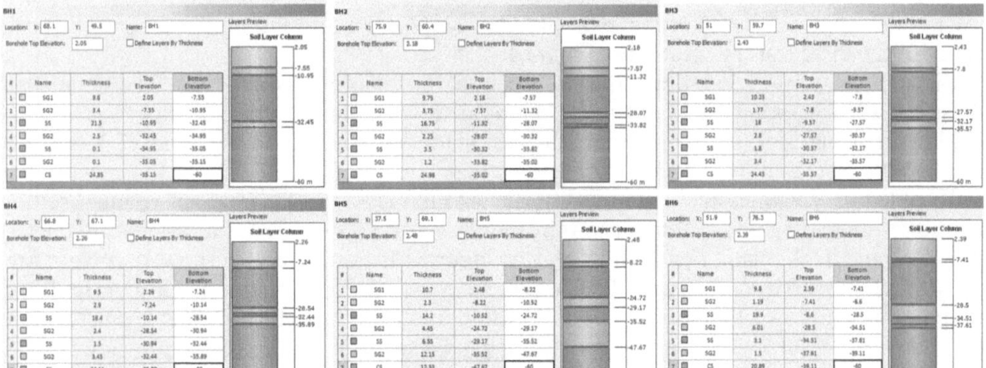

Figure 2. The ground layers in the six boreholes of the project.

3 RESULTS OF PILE LOAD TESTS

As mentioned earlier, three piles were tested, TP1, TP2, and TP3. The first two piles were 1.2m in diameter. The length of pile TP1 was 30m which was not enough to penetrate into the CS layer. The next two piles TP2 and TP3 were lengthened to 42m to ensure a penetration length in the CS layer. The characteristics of the tested piles are listed in Table 3.

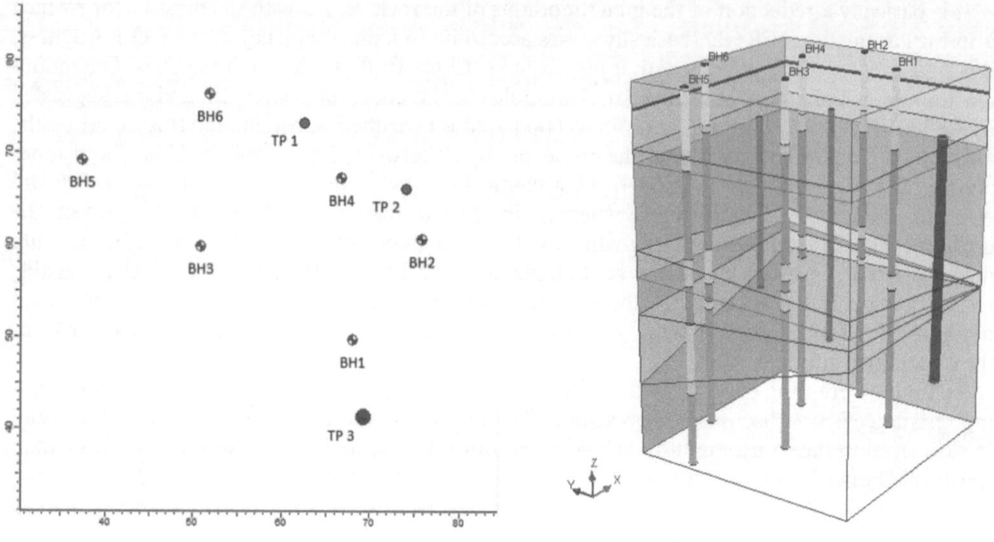

Figure 3. The borehole layout and test pile locations plan and 3D.

The piles were lightly reinforced. TP1 and TP2 had a top cage of 1% reinforcement ratio followed by a second cage of 0.5% reinforcement ratio and the rest of the pile length was left unreinforced. While TP3 was reinforced with 0.8% of the cross-sectional area at the top followed by half of that in the second cage and no reinforcement afterwards. The reinforcement was 25mm deformed bars with Young's Modulus of E=200GPa and a yield strength f_y =420MPa. The concrete cube strength was 60MPa for all piles or a cylinder strength f_c'= 50MPa for analysis.

The piles were tested following ICE SPERW (1996). The results of the load settlement curves of the three tests are plotted in Figure 4.

It is obvious from Figure 4 how the settlement of TP1 was excessively high. The settlement at the specified working load SWL was 23mm, at the maximum test load MTL was 40mm, and left a residual settlement of 31mm after test load was removed. In TP2 and TP3, the pile lengths were increased from 30m to 42m ending their tips at -50m DMD. With the new length, TP2 yielded only 9.4mm and 16.5mm settlement at SWL and MTL. The net residual settlement after unloading was 7.2mm. The third graph is for TP3 which has a different diameter than TP1 and TP2, 1.8m. The bottom plot of Figure 4 is normalized to show how close are the normalized curves of TP2 and TP3, which have the same pile length but of course differ a little in ground layer thicknesses due to location as shown in Figure 3. This normalization will show close results when the pile load is applying a similar level of average pressure on pile cross section. The average stress at TP3 was less than the average stress on TP2. That is why TP3 yielded less normalized settlement although close to the normalized settlement for TP2. The settlements of TP3 were 12.5mm, 20.0mm, and 11.1mm for SWL, MTL and fully unloaded.

4 PREDICTION OF SETTLEMENT USING T-Z CURVES

The settlement was calculated with the elastic method explained in section 1.1. The settlements were calculated using the elastic moduli given in Table 2. Results were as follows:

It can be traced from Table 3 that the pile length that exceeds the minimum length required to consume all the applied load within the skin friction has insignificant effect on settlement

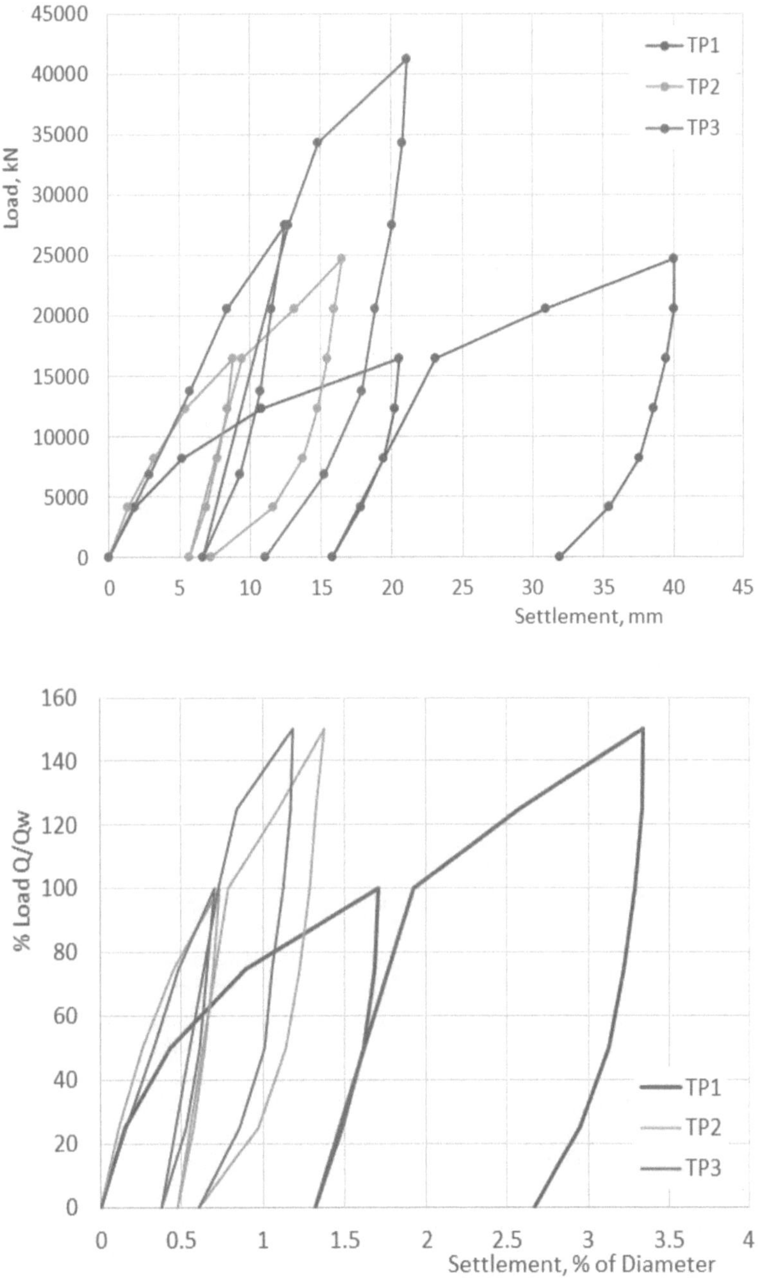

Figure 4. Load settlement plots of the test piles. The bottom plot presents the load and the settlement normalized to working load and diameter of pile, respectively.

calculated by linear elastic method, although little increase may be detected with increased length. Theoretically, at these lengths, no load is transferred to the base. The small increase with length is the effect of length in pile shortening equation. Another issue is realized as well. The settlement at MTL of 1.5SWL will always cause a settlement of 1.5 times the one at SWL

Table 3. Characteristics of the test piles.

Pile designation		TP1	TP2	TP3
Coordinates	X, m	62.6	74.1	69.3
(adjusted from E and N)	Y, m	73.1	65.9	41.2
Diameter D	m	1.2	1.2	1.8
Test pile COL	DMD (RL)	-8.0	-8.0	-8.0
Pile tip level PTL	DMD (RL)	-38.0	-50.0	-50.0
Test pile length	m	30	42	42
Specified working load SWL	kN	16500	16500	27500
Maximum test load MTL	kN	24750	24750	41250
Average stress at SWL	MPa	14.6	14.6	10.8

Table 3. Characteristics of the test piles.

Pile dia. B (m)	Pile length (m)	Load (kN)	Settlement (mm)	Estimate error
1.2	21	16500	8.2	
		24750	12.3	
	30	16500	8.3	-63.0%
		24750	12.4	-69.0%
	42	16500	8.6	-8.5%
		24750	12.9	-21.8%
1.8	36	27500	13.9	
		41250	20.8	
	42	27500	14.4	15.2%
		41250	21.5	7.5%

if all are calculated based on linear elastic method. The error in calculated settlement is added to Table 3 as the right column.

The linear method will result in a straight-line relation between load and settlement and the curved nature of the relation cannot be simulated. Here comes the role of presenting a T-z curve for SS and CS layers to construct a load settlement curve similar in shape to the real behavior. As there is no previous estimate, several curves have been tried and the two chosen curves are presented in Figure 5 and were found acceptable.

These curves were used as a user defined T-z relation for the SS and CS layers in the RSPile software and the results obtained for the load settlement curves with several runs at different load levels are plotted in dashed lines in Figure 6 along with the original test results.

The shapes of the load settlement curves were accurately simulated but of course there is still some deviation from the real settlements. Nevertheless, the error was around 10% to 25% with the maximum at TP1 but in general, the results are promising and adoptable and similar in shape to the real curves.

5 CONCLUDING REMARKS

If the moduli of deformation are chosen with suitable accuracy, the linear elastic method will be adequate for long piles having at least skin resistance equal to twice the applied load. While linear methods do not construct a load settlement curve similar in shape as T-z Q-z methods do.

A T-z model for Dubai Sandstone and Siltstone is presented. The results were highly acceptable, but the curves were tested on a single project. These curves must be widely checked and applied to many field load tests and be adjusted or subcategorized if needed before being adopted for Dubai rocks in general.

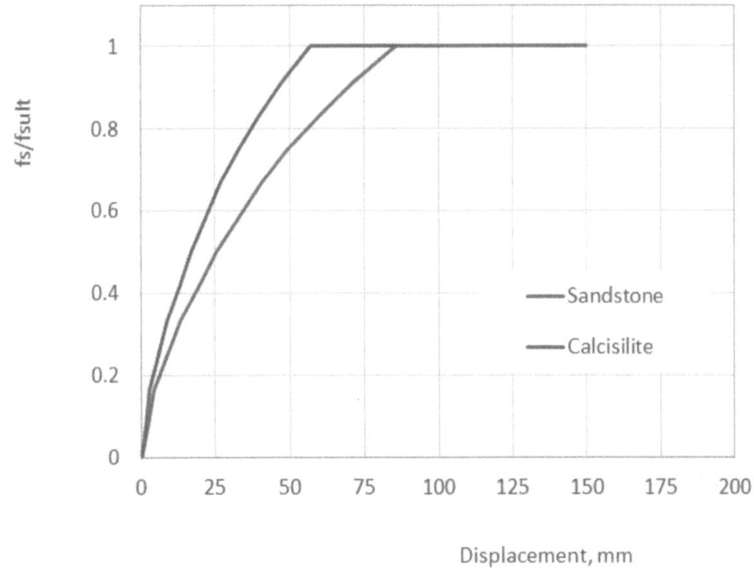

Figure 5. T-z curves suggested for SS and CS layers.

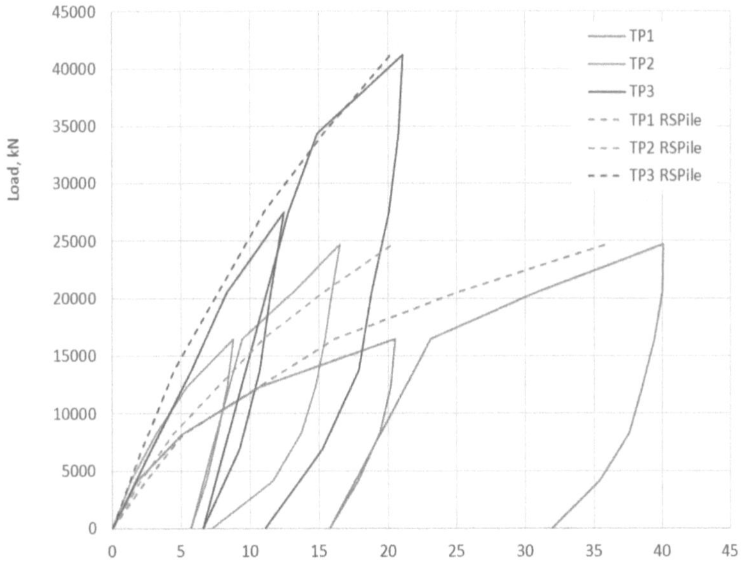

Figure 6. RSPile results using suggested T-z curves for SS and CS layers along with real load test results.

Additional research should be done for Q-z curve estimates. This needs special tests to get the tip load and displacement relation separately. Plate load test results may be useful.

The effect of normalized displacement is not checked here, and absolute displacement was used to establish the T-z curve. The idea is correct to a great extent but needs to be checked with more testing.

REFERENCES

Abu-Hejleh, N., O'Neill, M.W., Hanneman D. and Atwooll, W.J. 2003. *Improvement of the Geotechnical Axial Design Methodology for Colorado's Drilled Shafts Socketed in Weak Rocks.* Report No. CDOT-DTD-R-2003-6. Colorado Department of Transportation. Colorado, USA.

BS 8004. 2015. *Code of Practice for Foundations.* British Standards Institute. London, UK.

Clayton, C.R.I. 1995. *The Standard Penetration Test (SPT): Methods and Use.* CIRIA Report 143. CIRIA, London, UK.

Gannon, J.A., Masterton, G.G.T., Wallace, W.A. and Wood., D Muir. 1999. *Piled Foundations in Weak Rock.* CIRIA Report 181. CIRIA, London, UK.

ICE 1996. *Specifications for Piling and Embedded Retaining Walls.* Institution of Civil Engineers. Thomas Telford, London, UK.

Latapie B., Albelda, R.A., Abou Samra M., Alzaylaie M. and Sumputh J. 2019. A Review of Piling Industry Practices in Dubai, UAE: proposed UCS-based correlations. *Geotechnical Research*, Vol. 6, Issue 2, 103–129. ICE Publishing. https://doi.org/10.1680/jgere.18.00021.

Perloff, W.H. 1975. Pressure Distribution and Settlement. Chapter 4. In H.F. Winterkorn and H.Y. fang (eds) *Foundation Engineering Handbook.* Van Nostrand Reinhold, New York, USA.

Popov, E.P. 1968. *Introduction to Mechanics of Solids.* Prentice-Hall, New Jersey, USA.

Reese, L.C. and O'Neill, M.W. 1988. *Drilled Shafts, Construction Procedures and Design Methods.* Report FHWA-HI-88-042. US DOT, Federal Highway Administration, McLean, Virginia, USA.

Reese, L.C. and Van Impe, W. 2011. *Single Piles and Pile Groups Under Lateral Loading.* CRC Press, Taylor & Francis Group, London, UK.

Vesic, A.S. 1977. *Design of Pile Foundations.* NCHRP Synthesis 42, Transportation Research Board TRB, Washington D.C., USA.

Williams, A.F. and Pells, P.J.N. 1981. Side Resistance Rock Sockets in Sandstone, Mudstone and Shale. *Canadian Geotechnical Journal*, 18, No.4, 502–512.

The Evolution of Geotech - 25 Years of Innovation – Hammah et al. (Eds)

Implementation of constrained differential evolution in reliability-based evaluation of soil-structure interaction-influenced liquefaction potential

C.R. Guanzon & G.P. Ventura
Institute of Civil Engineering, University of the Philippines Diliman, Quezon City, Metro Manila, Philippines

ABSTRACT: Liquefaction potential is commonly evaluated by utilizing deterministic methods and by discounting the influence of soil-structure interaction (SSI). Though these simplify liquefaction assessment, lapses may occur in situations where they emerge as significant. In this paper, Constrained Differential Evolution (CDE) algorithm was applied in reliability-based evaluation of SSI-induced liquefaction potential. The developed CDE algorithm was validated using the dataset of a First Order Second Moment-based study on liquefaction potential. Validation results showed that the difference between results of CDE algorithm and FOSM-based approach was not statistically significant. The reliability-based CDE method was utilized in a case study in Baseco, Manila. Results showed that foundation increases the surrounding vertical stresses, thereby decreasing liquefaction probability. However, in deeper layers, SSI analysis was found to be more conservative than free-field analysis. Results also indicate that soil layers deemed supposedly safe by deterministic methods, i.e., FS>1, may still have considerable probabilities of liquefaction.

1 INTRODUCTION

Liquefaction is a phenomenon typically observed during cyclic undrained loading (e.g. seismic loading) where saturated soil loses shear strength and temporarily behaves as a viscous liquid. In practice, liquefaction potential is assessed without considering the effect of uncertainties arising from soil variability (i.e. deterministic approach) and of the structures overlying the soil (i.e. free- field condition). Deterministic methods which make use of factor of safety (FS) rather than probabilities are commonly used to quantify liquefaction potential. According to National Center for Earthquake Engineering Research (NCEER), FS against liquefaction is obtained from dividing the cyclic stress ratio (CSR) due to the seismic loading by the cyclic resistance ratio (CRR) which determines the soil capacity to impede liquefaction. FS less than 1 means liquefaction while FS greater than 1 means non-liquefaction. CSR for free-field sites is given by Equation 1:

$$CSR = 0.65\left(\frac{a_{max}}{g}\right)\left(\frac{\sigma_v}{\sigma_v^F}\right)(r_d) \tag{1}$$

where a_{max} is the maximum horizontal ground acceleration caused by earthquake, g is the acceleration due to gravity, σ_v is the total vertical stress at the bottom of the soil column, $\sigma_v^{'}$ is the effective vertical stress, and r_d is the stress reduction factor to account for soil deformability. CRR of a soil layer is determined from the corrected blow counts $(N_1)_{60}$ of standard penetration test using the equation from Rauch (1997):

DOI: 10.1201/9781003188339-46

$$\mathrm{CRR} = \frac{1}{34 - (N_1)_{60}} + \frac{(N_1)_{60}}{135} + \frac{50}{\left[10(N_1)_{60} + 45\right]^2} - \frac{1}{200} \tag{2}$$

Additional parameters may be incorporated to depict soil liquefaction potential in certain conditions. K_σ correction factor from Hynes and Olsen (1999) accounts for the effect of confining pressure and is given by Equation 3:

$$K_\sigma = \left(\sigma_v^F\right)_{p_a}^{f-1} \tag{3}$$

where p_a is the atmospheric pressure and f is the exponent that is a function of site conditions. Magnitude scaling factor (MSF) corrects the CRR curves for earthquake magnitudes other than Magnitude 7.5 and is given by Equation 4:

$$\mathrm{MSF} = \frac{10^{2.24}}{M_w^{2.56}} \tag{4}$$

Current practice of liquefaction potential estimation is dominated by free-field analyses, and the focus barely zeroes in on the effects of soil-structure interaction (SSI). Wang et al. (2015) pointed out that applying methods that assume free-field conditions is "impossible" unless structure effects are considered in such methods. Moreover, deterministic approaches like factor-of-safety method compartmentalizes liquefaction potential into safe and unsafe regions by demarcating them with a definite FS whose value that can confidently assess soil safety is even debated by engineers (Bagheripour et al., 2012).

With the prospect of exploring a probabilistic approach and augmenting SSI perspectives on liquefaction, this paper utilized a reliability-based method associated with Constrained Differential Evolution (CDE) algorithm in evaluating liquefaction potential while considering the influence of structure. In finding a solution to the optimization problem presented by the reliability-based method, CDE algorithm was used.

2 CONSTRAINED DIFFERENTIAL EVOLUTION

CDE is a variant of Differential Evolution (DE) algorithm which makes use of crossover, mutation, and selection operations (Storn & Price, 1997). The premise of DE is simple: a generation of solutions is "evolved" with strengths obtained from previous generation. What makes it different is its mutation operation which generates random modifications (perturbations) in the population coupled with its recombination (crossover) operation which crosses the characteristics from two population vectors with another vector to produce a child vector with improved traits. Figure 1 shows the general computation scheme of DE.

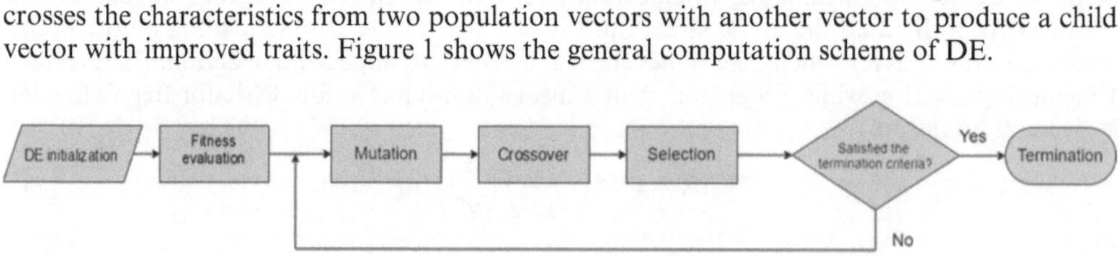

Figure 1. General flow of DE.

At the start ($g = 0$), a population composed of N_p D-dimensional $\mathbf{X}_{i,g}$ vectors of real-valued parameters is initialized through random number generation subject to specified lower (L) and upper (U) bounds, whose boundary values are collected in D-dimensional initialization vectors, b_L and b_U:

$$\mathbf{X}_{i,g=0} = X_{i,j,g=0} = \text{rand}_j(0,1) \cdot (b_{j,\text{u}} - b_{j,\text{L}}) + b_{j,\text{L}} \quad (5)$$

A mutant vector $\mathbf{V}_{i,g}$ is obtained according to:

$$\mathbf{V}_{i,g} = \mathbf{X}_{r1,g} + F \cdot (\mathbf{X}_{r2,g} - \mathbf{X}_{r3,g}) \quad (6)$$

where $r_1, r_2, r_3 \in \{1, 2, \ldots, N_p\}$ are randomly generated integers and F is the scaling factor. Recombination (crossover) makes mutant vectors cross with current population members to diversify the population and let resulting trial vectors acquire characteristics different from the parent population. DE operates recombination according to

$$\mathbf{u}_{i,g} = u_{j,i,g} = \begin{cases} V_{j,i,g} \text{ if } \text{rand}_j(0,1) \leq CR \text{ or } j = j_{\text{rand}} \\ X_{j,i,g} \text{ otherwise} \end{cases} \quad (7)$$

where $\mathbf{u}$ is called the trial vector corresponding to individual i in generation g, j is the index of every parameter in a vector, j_{rand} is a randomly selected integer from 1 to N_p, and $CR \in [0,1]$ is the crossover probability. If the trial vector $\mathbf{u}_{i,g}$ yields an equal or lower fitness function value than that of the target vector $\mathbf{X}_{i,g}$, $\mathbf{u}_{i,g}$ replaces the target vector in the next generation. Otherwise, the target vector survives and retains its position in the population. This process called selection is expressed by:

$$\mathbf{X}_{i,g+1} = \begin{cases} \mathbf{u}_{i,g} \text{ if } f(\mathbf{u}_{i,g}) \leq f(\mathbf{X}_{i,g}) \\ \mathbf{X}_{i,g} \text{ otherwise} \end{cases} \quad (8)$$

where $f(\mathbf{u}_{i,g})$ represents the fitness function value of the trial vector and $f(\mathbf{X}_{i,g})$ is the fitness function value of the target vector.

DE provides a reliable technique in managing multi-dimensional spaces abundant in real-world applications (Hegerty et al., 2009). However, DE is essentially designed for unconstrained optimization and needs modifications to handle constraints in constrained optimization problems (Wazir et al., 2016). Hence, constraint handling strategies are incorporated to modify the basic DE into a CDE.

3 CDE IN THE CONTEXT OF RELIABILITY-BASED LIQUEFACTION ANALYSIS

3.1 *CDE implementation*

Reliability can be incorporated in liquefaction analysis by studying the likelihood of the system, which is the soil layer, to succumb to failure, i.e., liquefaction. The performance of a soil layer in resisting liquefaction is given by the function Z = CRR − CSR. The performance function is considered safe if the value is greater than 0, impending failure if it yields 0, or unsafe if it is less than 0. The probability of failure P_F is defined as the probability that the random variables CRR and CSR constitute a Z value in the failure region, i.e, $Z < 0$. In reliability-based method, P_F is given by Equation 9:

$$P_F = \phi(-\beta) = 1 - \phi(\beta) \quad (9)$$

where $\phi[.]$ is the cumulative distribution function of the standard normal distribution and β is the reliability index. The reliability index β is defined as the minimum distance between the origin to the limit state surface at the most probable point (MPP). It represents the closest distance of the current system combination of variables to the MPP which has the "worst possible combination of input parameters" (Manoj, 2016). In this study, reliability index was calculated using the Hasofer-Lind approach. Hasofer and Lind (1974) provided the reliability index in matrix terms as

$$\beta = \min_{X \in F} \sqrt{\left[\frac{X_i - \mu_i^N}{\sigma_i^N}\right]^T [\boldsymbol{R}]^{-1} \left[\frac{X_i - \mu_{Ni}^N}{\sigma_i}\right]} \tag{10}$$

where F represents the failure surface, $\boldsymbol{R}$ is the correlation matrix, and σ_i^N and μ_i^N are the equivalent normal standard deviation and mean of each original random variable X.

A code for CDE that uses random mutation strategy with one difference vector and binomial recombination was developed using MATLAB. Since minimizing the reliability index is bounded by Z = CRR − CSR = 0 constraint, a static penalty function was employed to handle the constraint.

3.2 *Validation*

To see if the developed CDE algorithm is reliable in solving the optimization problem, the algorithm was tested using the dataset and results from Janalizalde et al. (2015). First Order and Second Moment (FOSM) reliability-based approach was conducted in Janalizalde et al. (2015) to evaluate soil liquefaction potential in Chalos, Iran. Using a dataset of 190 records, the authors were able to execute FOSM in determining the reliability index and consequently, the probability of liquefaction. Parameters, data, and coefficients used in Janalizalde et al. (2015) were adopted in this validation to maintain consistency.

Figure 2 plots the results from Janalizalde et al. (2015) and from the CDE algorithm along with trendlines and R-squared values. Inspection of R-squared values shows that the two datasets show a good association. Nevertheless, to further support this inference, Mann-Whitney U Test was conducted to compare the two data groups. Upon performing the test on the two data groups' reliability indices, p-value of 0.323 was calculated which means that the difference between the two results is not statistically significant. The same was conducted on the liquefaction probabilities, and p-value of 0.298 was computed which means also that the difference is statistically insignificant.

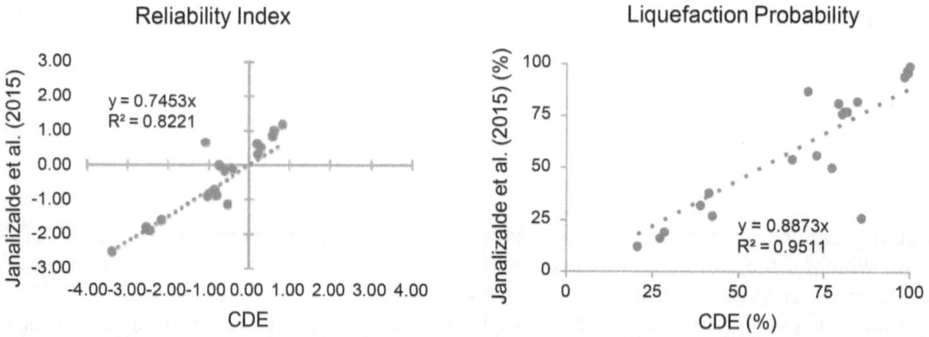

Figure 2. Results comparison between Janalizalde et al. (2015) and CDE algorithm.

4 CASE STUDY OF BASECO, METRO MANILA, PHILIPPINES

3 boreholes were examined to characterize the subsurface soil layers of Baseco study area. 4 different cases, each with varying dimensions and depths of isolated footings, were analyzed for each borehole through a finite element analysis program to obtain the induced soil stresses. Table 1 lists the parameters used for each case number. Distributed load of 4000 kPa was applied on the 0.5 m x 0.5 column base and was kept the same for all cases.

Table 1. Footing Cases per Borehole.

Case Number*	Depth of footing
BHx-1	1m
BHx-2	1m
BHx-3	2m
BHx-4	2m

* x in BHx pertains to numbers 1, 2, and 3 depending on the borehole being analyzed

Figure 3 plots the total and effective stresses along the depth for each case of BH1. As shown, inclusion of structure induces an increase in total and effective vertical stresses. However, the stresses seem to stabilize and approach the stresses under free-field condition along the depth.

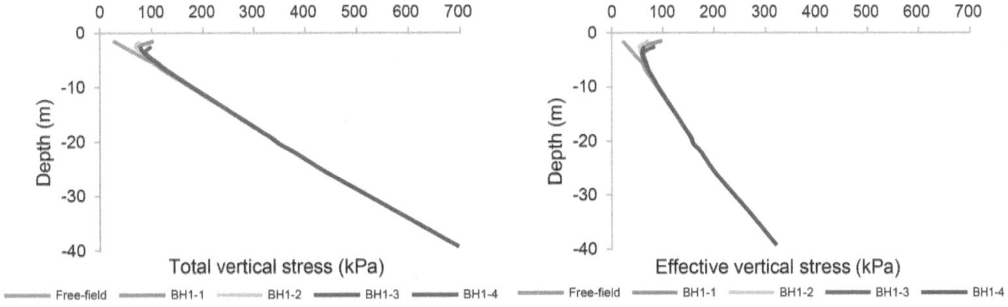

Figure 3. Variation of (a) total and (b) effective vertical stresses along the depth of BH1 cases.

Prior to reliability-based evaluation, deterministic factor-of-safety approach based on NCEER procedure was performed so that a baseline estimate of safety level of each soil layer was determined. Since SSI is incorporated in the analysis, FS was calculated while including correction factors that describe SSI and is given by Equation 11:

$$\mathrm{FS} = \frac{\mathrm{CRR}}{\mathrm{CSR}} K_\sigma \mathrm{MSF} = \frac{\mathrm{CRR}}{\frac{\mathrm{CSR}}{K_\sigma \mathrm{MSF}}} = \frac{\mathrm{CRR}}{\mathrm{CSR_{SSI}}} \tag{11}$$

CSR, K_σ, and MSF are consolidated into one variable called $\mathrm{CSR_{SSI}}$ which captures the corrected seismic load due to the presence of SSI. Input motion applied to the model was selected to be the 2010 Sierra El Mayor earthquake in California, US which generated a moment magnitude of 7.2 and maximum intensity of IX. Ground motion recorded in Calexico Station, California revealed that the maximum acceleration was 0.27g.

Figure 4 shows the variation of FS along the depth. According to NCEER (Youd & Idriss, 2001), soils having $(N_1)_{60} \geq 30$ and clays are not potentially liquefiable soils and hence, are considered as non-liquefiable.

SSI Analysis showed that soils near the foundation recorded higher FS values due to inclusion of K_σ as it is highly dependent on effective vertical stress. Thus, free-field estimates which registered lower FS values were more conservative for soils near the foundation. Yet, in deeper layers, SSI estimates which incorporated K_σ correction factor were observed to be more conservative than free-field estimates. As effective vertical stress

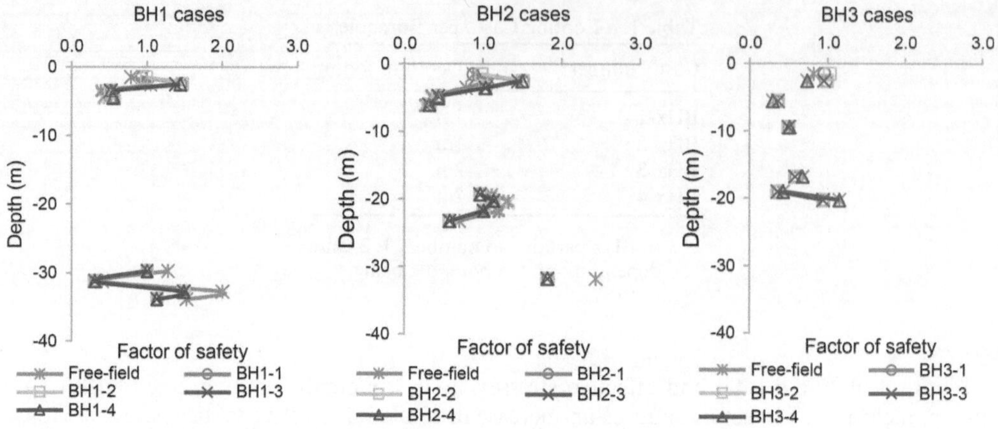

Figure 4. Variation of *FS* along the depth.

increases along the depth, K_σ decreases. In SSI Analysis, dividing the free-field CSR by a lower value of K_σ in deeper layers produces a higher CSR_{SSI}, thereby yielding a lower factor of safety.

The probability of liquefaction occurring in each soil layer was determined through the reliability-based CDE method. The limit state function considered was $Z = CRR - CSR_{SSI}$. Variables α_{max}, M_w, $(N_1)_{60CS}$, σ'_v and σ_v were selected to be the random variables in the analysis as these are the fundamental variables in which dependent variables CSR_{SSI}, CRR, MSF, and K_σ are functions of. Table 2 presents the coefficients of variation while Table 3 shows the correlation coefficients used.

Table 2. Coefficients of Variation for Baseco Case Study.

Parameter	Coefficients of Variation*
$(N_1)_{60cs}$	0.30
σ'_v	0.15
σ_v	0.10
α_{max}	0.15
M_w	0.05

* Juang et al. (1999), Uzielli et al. (2006), Bagheripour et al. (2012)

Table 3. Correlation Coefficients for Baseco Case Study.

Parameter	$(N_1)_{60CS}$	σ'_v	σ_v	α_{max}	M_w
$(N_1)_{60CS}$	1	0.3	0.3	0	0
σ'_v	0.3	1	0.9	0	0
σ_v	0.3	0.9	1	0	0
α_{max}	0	0	0	1	0.9
M_w	0	0	0	0.9	1

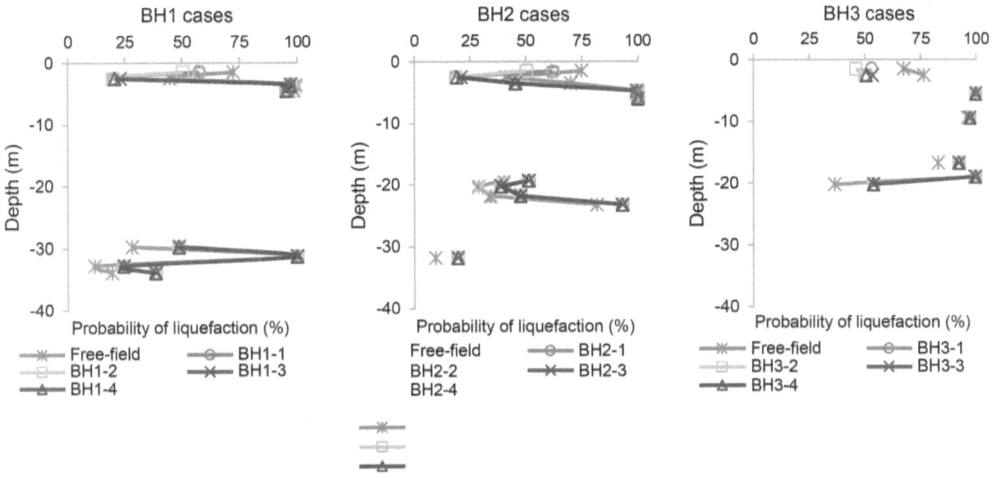

Figure 5. Liquefaction probability along the depth for Baseco Case Study.

Figure 5 shows that at depths 1 m to 5 m, liquefaction probabilities considering free-field condition were more conservative rather than those from SSI Analysis. The opposite was observed in soil layers deeper than 30 m; SSI Analysis produced more cautious probabilities of liquefaction because of the higher CSR_{SSI} values induced by the decrease in K_σ correction factor.

5 CONCLUSIONS

The study was able to apply CDE in reliability-based assessment of liquefaction potential. In executing the reliability-based analysis, liquefaction was viewed as system failure wherein the system being assessed was the soil layer. The reliability index needed to calculate the liquefaction probability was obtained by solving the constrained optimization problem by means of CDE algorithm. Accuracy of CDE was validated and compared with actual results from Janalizalde et al. (2015). Validation outcome showed that the difference between the results given by CDE and those from the FOSM-based approach in Janalizalde et al. (2015) was not statistically significant.

Findings from Baseco study showed that employing a free-field analysis produces more conservative results for upper soil layers near the foundation. However, SSI analysis was observed to generate more conservative estimates for deeper soil layers. Reliability-based analysis showed that soil layers, despite having FS greater than 1 from deterministic methods, still incur liquefaction probabilities.

REFERENCES

Bagheripour, M. H., Shooshpasha, I., & Afzalirad, M. 2012. A genetic algorithm approach for assessing soil liquefaction potential based on reliability method. *Journal of earth system science*, *121*(1), 45–62.

Hasofer, A. M., & Lind, N. C. 1974. Exact and invariant second-moment code format. *Journal of the Engineering Mechanics division*, *100*(1), 111–121.

Hegerty, B., Hung, C. C., & Kasprak, K. 2009. A comparative study on differential evolution and genetic algorithms for some combinatorial problems. In *Proceedings of 8th Mexican international conference on artificial intelligence*, 9–13.

Hynes, M. E., & Olsen, R. S. 1999. 'Influence of confining stress on liquefaction resistance.' *Proc., Int. Workshop on Phys. and Mech. of Soil Liquefaction*, Balkema, Rotterdam, The Netherlands, 145–152.

Janalizade, A., Naghizadehrokni, M., & Naghizaderokni, M. 2015. Reliability-based method for assessing liquefaction potential of soils. In *5th ECCOMAS Thematic Conference on Computational Methods in Structural* Dynamics *and* Earthquake Engineering.

Juang, C. H., Rosowsky, D. V., & Tang, W. H. 1999. Reliability-based method for assessing liquefaction potential of soils. *Journal of Geotechnical and Geoenvironmental Engineering, 125*(8), 684–689.

Manoj, N. R. 2016. 'First-order Reliability Method: Concepts and Application', Thesis, Delft University of Technology, <http://resolver.tudelft.nl/uuid:c4c941fa-a9c1-4bd4-a418-afc54bb6d475>.

Rauch, A. F. 1997. *EPOLLS: an empirical method for predicting surface displacements due to liquefaction-induced lateral spreading in earthquakes*, Doctoral dissertation, Virginia Polytechnic Institute and State University.

Storn, R., & Price, K. 1997. Differential evolution–a simple and efficient heuristic for global optimization over continuous spaces. *Journal of global optimization, 11*(4), 341–359.

Uzielli, M., Lacasse, S., Nadim, F., & Phoon, K. K. 2006. Soil variability analysis for geotechnical practice. *Characterization and engineering properties of natural soils, 3*, 1653–1752.

Wang, G., Wei, X., & Liu, H. 2015. Liquefaction evaluation of dam foundation soils considering overlying structure. *Journal of Rock Mechanics and Geotechnical Engineering, 7*(2), 226–232.

Wazir, H., Jan, M. A., Mashwani, W. K., & Shah, T. T. 2016. A penalty function based differential evolution algorithm for constrained optimization. *Nucleus, 53*(1), 155–166.

Youd, T. L., & Idriss, I. M. 2001. Liquefaction resistance of soils: summary report from the 1996 NCEER and 1998 NCEER/NSF workshops on evaluation of liquefaction resistance of soils. *Journal of geotechnical and geoenvironmental engineering, 127*(4), 297–313.

Session 10 - Stability analysis of mining slopes II

The Evolution of Geotech - 25 Years of Innovation – Hammah et al. (Eds)

The effect of water filling on slope stability of open pits: A numerical investigation

I.E. Zevgolis
School of Mining and Metallurgical Engineering, National Technical University of Athens, Greece

A. Mikroutsikos, A.I. Theocharis & N.C. Koukouzas
Chemical Process & Energy Resources Institute, Centre for Research & Technology Hellas, Athens, Greece

ABSTRACT: One of the most common practices for valorisation of abandoned open-pit mines is flooding them to form pit lakes. The present work numerically examines stability issues for a horizontal water table being elevated from the bottom to the top of an open-pit, initially with the limit equilibrium method. During the water table's elevation, the safety factor gradually decreases until the water's height reaches 40% of the total height. It then increases and becomes larger than the initial one when the water table reaches full height. The maximum safety factor decrease is approximately 10% and thus, flooding does not affect significantly the slope stability Furthermore, limit equilibrium analysis results are compared and validated with the finite element method. Finally, analyses of slopes with and without benches, but of the same overall angle, are performed and compared. The omission of benches is always more conservative, leading to smaller SFs by 6%-14%.

1 INTRODUCTION

For many decades, coal and lignite (brown coal) mining have contributed significantly to global energy needs. However, many surface coal and lignite areas will be abandoned soon, as several countries pass in the post-lignite era. One of the most common reclamation practices is the formation of pit lakes. During this process, old excavations are flooded with water to create a lake offered to the local societies, mainly for recreational purposes. Failure incidents and landslide phenomena have frequently been linked with coal and lignite mines, leading to casualties and social and economic disruption (Zevgolis et al. 2019). Thus, the reclamation of these areas poses engineering challenges related to slope stability. The water table's elevation during flooding affects the slope safety by increasing the pore water pressures of the submerged soil layers, while erosion and liquefaction phenomena may occur. However, the increase of the water body inside the open-pit also acts as a supporting force, and the pit lake's creation has been reported to improve slope stability (e.g. Desjardins et al. 2020; Faur et al. 2020).

The present work examines possible slope stability issues during the pit lake's creation process through numerical analysis. A horizontal water table is elevated from the bottom of the excavation to the crest, as the open-pit is flooded. In this framework, the influence of the water table's elevation on the safety factor (SF) is examined for various soil strength parameters. The limit equilibrium method (LEM) is employed and then compared with the finite element method (FEM), as implemented in Slide2 and RS2 of Rocscience, respectively. Finally, slopes with and without benches are compared, and the error from the omission of benches is quantified. Identical slope geometries, geotechnical parameters, and groundwater conditions were considered for the comparisons.

DOI: 10.1201/9781003188339-47

2 NUMERICAL ANALYSIS

2.1 *The effect of water filling on slope stability by the LEM*

In surface coal/lignite mines, slopes commonly consist of layers of sterile materials and coal/lignite seams that are frequently (but not always) characterised by similar shear strength. As a result, under certain circumstances, considering the various soil formations as a single material for slope stability analysis is a reasonable assumption (Mikroutsikos et al. 2021). Moreover, a bedrock formation with high strength is commonly encountered close to the excavation bottom. Figure 1 presents a typical soil stratigraphy for coal and lignite mines, with a homogeneous slope and a bedrock formation at the bottom of the excavation. This stratigraphy is governed by a complex failure mechanism and a non-circular failure surface, due to the bedrock's presence. Note that in practice, mining slopes consist of several benches, which are omitted in the present numerical analysis models (Figure 1). This geometry simplification does not critically affect the results, as will be quantified and discussed on the sequel (section 2.3).

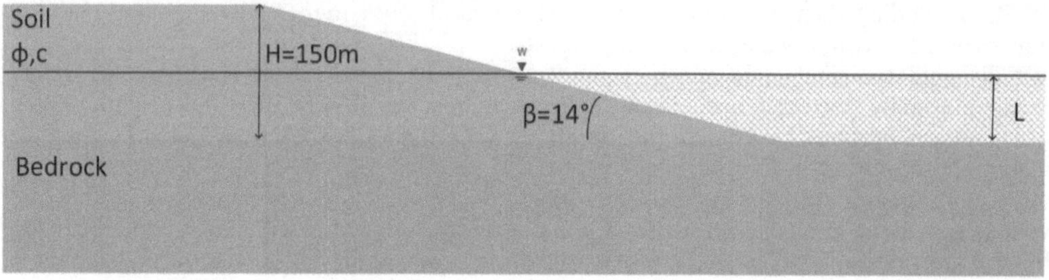

Figure 1. Geometry and stratigraphy of the numerical model.

The slope stability of this stratigraphy was analysed based on the LEM (using Slide2). Providing simplicity and speed, the LEM has been for many decades the most common slope stability method. However, most LEM software require the knowledge (or estimation) of the shape and the location of the failure surface to provide reliable results. This is mainly the case for noncircular failures, as in the present study. The advanced "cuckoo" search method was used in this work, which does not demand the failure surface's location or shape, combined with Spencer's method. The Mohr-Coulomb failure criterion was employed for modelling soils' shear strength. The water level's rise in the pit (leading to the lake's formation) was simulated with a horizontal phreatic water table elevating from the bottom of the excavation (and the interface between bedrock and overburden soil) to the pit's crest. In Figure 1, L refers to the water table's elevation measured from the bottom of the excavation; initially L=0. The final elevation was achieved in 11 stages, with L being risen by 15m in each stage. The slope's height and inclination are H=150m and β=14°, respectively, and the unit weights are γ_{moist}=17kN/m^3 and γ_{sat}=20kN/m^3 for all materials. The bedrock's strength is significantly higher than the soil's strength and does not affect the results, since the failure mechanism is not encountered in this formation.

The friction angle and cohesion of the homogeneous soil layer may vary in a wide range. A recent study shows that the average characteristic peak friction angle and cohesion for Greek lignite mines are 25° and 111kPa, respectively (Theocharis et al. 2021). However, for individual mines, these values can be as low as 22° and 20kPa. Similar values of friction angle and cohesion are common for coal and lignite mines globally (e.g. Bednarczyk 2017; Ulusay et al. 2014), while cohesion is frequently assigned values even close to zero, due to its high uncertainty. Hence, six different φ-c combinations were considered in the present study, with φ and c varying between 20° and 25°, and between 5kPa and 110kPa, respectively.

For all φ-c combinations, increasing L from 0 to 60m leads to the gradual decrease of SF by approximately up to 10% (Figure 2). However, a further increase of L (from 60 to

150m) favours the stability and the SF increases by 14%-28% with respect to the L = 60m case, and by 2%-16% with respect to the initial, dry conditions. In all cases, the lowest differences are encountered for the lowest cohesions. In other words, for this specific stratigraphy and hydraulic conditions, the lowest SF (corresponding to the most critical condition) is computed when the water table rises at 60m above the bottom of the excavation, i.e. at 40% of the slope's height. The most stable conditions are when the water table reaches the slope's crest. At this final flooding stage, the free water surface results in an immense supporting force that exceeds the pore water pressure's influence on the SF, and sequentially in higher SFs than for dry conditions. For example, if $\varphi = 20°$ and $c = 40kPa$, SF equals 1.81 for dry conditions ($L = 0$), decreases to 1.61 for $L = 60m$, and then increases to 1.98 when the lake reaches its final level.

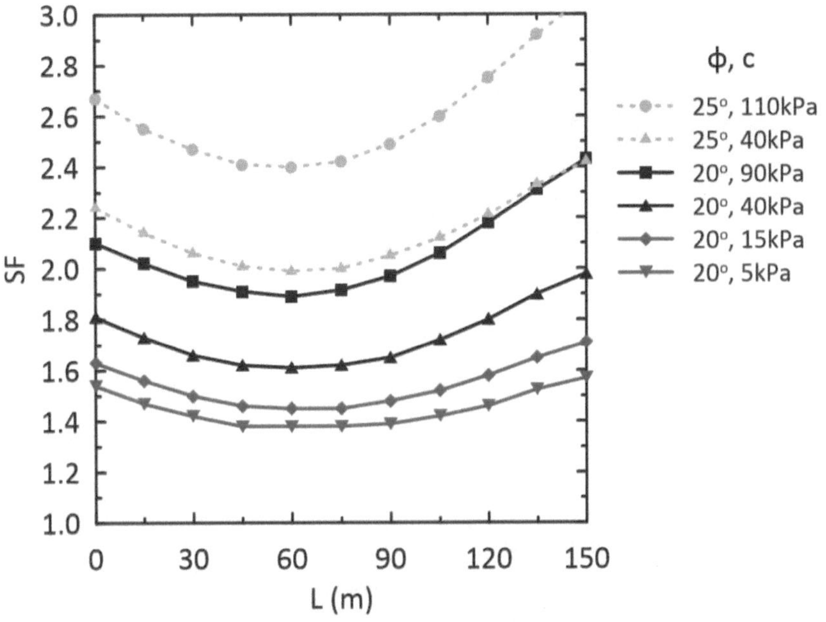

Figure 2. SF with water elevation L, for various φ-c combinations.

2.2 *The effect of water filling on slope stability by the FEM*

In this section, FEM simulations are conducted for comparison purposes considering the model of Figure 1 and following the same flooding process. Two-dimensional plane strain and drained conditions were used, while simulations were performed for 6-noded triangular elements with a very fine discretisation in RS2. The vertical boundaries were located at a 2H distance from the slope's crest and toe, respectively; the horizontal bottom boundary was at a distance H from the slope's toe, where H stands for the overall height (Figure 1). The slope's creation was simulated as a staged excavation, and shear strength reduction analysis was performed at the final stage to calculate the SF. The Mohr-Coulomb elastic-perfectly plastic constitutive model was used for all materials.

Slope geometries, geotechnical parameters, and groundwater conditions were identical with the previous section. The increase of water table elevation was performed in 11 stages, and a single pair of φ-c values was examined ($\varphi = 20°$ and $c = 40kPa$). Stiffness for the overburden soil is 50MPa, while Poisson's ration and dilation angle are $\nu = 0.3$ and $\psi = 0°$ respectively, for all materials. Additionally, the strength and stiffness of the bedrock do not affect the results.

In principle, the FEM provides a robust model of a slope's behaviour but demands several parameters to be defined and increased computational time. Nevertheless, herein FEM results are practically in total agreement with LEM results for all L values. The SFs from FEM are 0-1% lower than the respective results from LEM, and the failure mechanisms are almost identical. Figure 3 illustrates two indicative simulations for the case of the lowest SF for both methods (L = 60m); the calculated failure surface from LEM and the shear strains that indicate the failure surface position for FEM are presented. It is concluded that the LEM is adequate for stability problems as the one in the present study - considering a simplified stratigraphy and a horizontal water table - and more advanced FEM simulations are not necessary. Notice that for more com-plicated conditions, e.g. groundwater flow, FEM might be more advantageous, and a similar comparison would be needed.

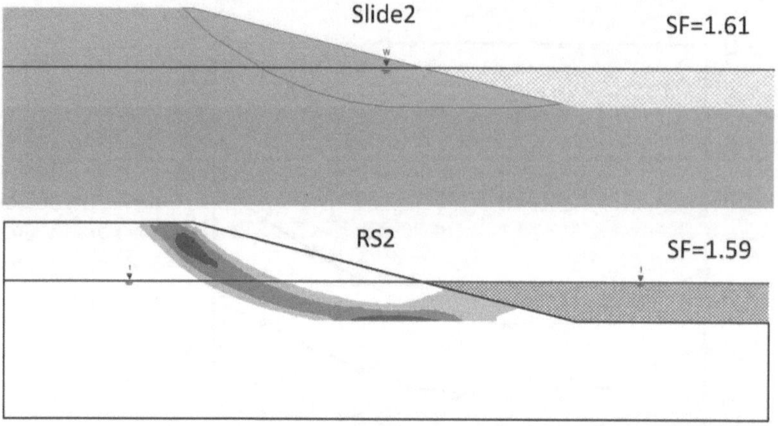

Figure 3. The calculated failure surfaces and SFs from the two methods, considering L = 60m. In FEM, the illustrated shear strains indicate the position of the failure surface.

2.3 *The role of benches on the stability*

Slopes of open-pit coal and lignite mines typically consist of several benches. The number and the geometry of benches may vary widely, depending on the excavation method and the mine exploitation scheme. Therefore, benches are often neglected from slope stability analysis since the systematic study of slope stability with benches is complicated. However, it is essential to quantify the error of such an assumption. Hence, in this section, slopes with and without benches are compared, in the flooding process.

Employing FEM analysis for an internal dump in a coalfield, Verma et al. (2013) concluded that omitting benches was always more conservative (i.e. led to smaller SF) with an SF error within 3-6%. Additionally, in a recent study referring to lignite mines, Mikroutsikos et al. (2021) conducted similar FEM analyses considering the presence of a weak zone. In that case, SF's differences did not exceed 11% (models with benches provided higher SF).

FEM simulations demand an increased mesh density and the assignment of a linear elastic material to the benches to prevent local failures. To avoid this simulation complexity and since FEM and LEM were found to provide practically identical results, the comparison is performed using the LEM. The area of the model where the failure surfaces can appear was appropriately constricted to avoid minor bench failures insignificant to the overall slope stability (by modifying the tools "slope limits" in Slide2).

Four different slope geometries with different number of benches (n) were created; their characteristics are summarised in Table 1. Identical total slope inclination and height, geotechnical parameters, and groundwater conditions were considered for the comparison. The water table elevation was performed in 11 stages as in the previous sections, while φ and c are 20° and 40kPa, respectively.

Table 1. Benches geometry for the four configurations.

	Bench height	Bench face inclination	Bench width
Number of benches, n	m	degrees	m
1	75	20	188
2	50	30	170
4	30	40	105
5	25	45	90

Simulations with benches provide higher SFs, and the models without benches are always more conservative by 6%-14% for any number of benches and water table levels; the failure surface is similar for all models. The relation between SF and water table level (L) for each configuration is similar to the simulations without benches; the most critical state during the flooding process is when L equals 60m (40% of H), while the most stable when L=150m (L=H). Figure 4 presents indicative slope stability results with and without benches for L=30m, where the highest SF differences between the different configurations were encountered (10%-14%). The highest number of benches provide results closer to the simulations without benches, for all cases examined; in Figure 4, the slopes with n=4 and n=5 provide critical SFs and failure surfaces closer to the reference results (n=0). This conclusion is reasonable because as the number of benches increases, the slope's overall geometry resembles the no-bench case.

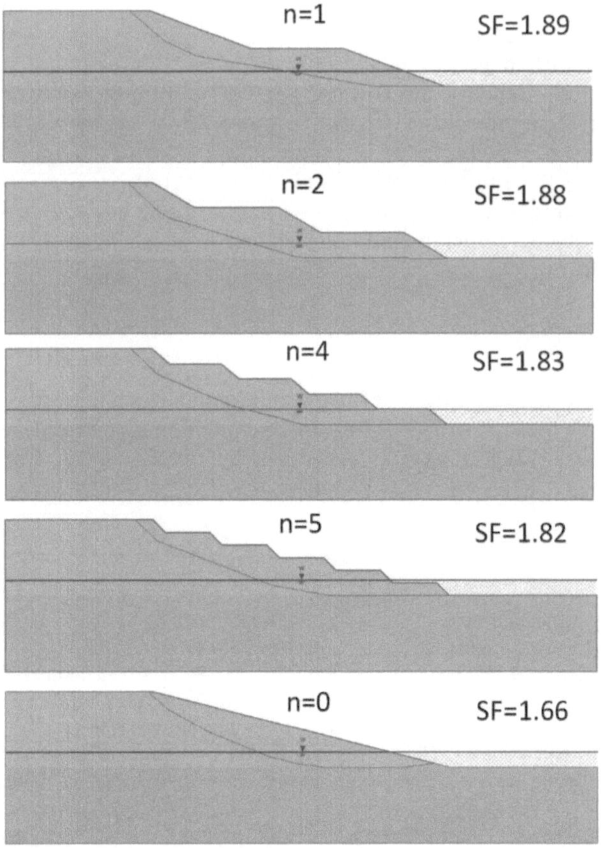

Figure 4. Failure surfaces and SFs for five slopes with different number of benches (n), considering the same water level and overall slope angle.

3 CONCLUSIONS

In this work, open pits' slope stability during flooding for lake creation was investigated through numerical analysis. A homogeneous soil slope was considered with bedrock at the bottom of the excavation. The pit lake's formation was simulated by rising a horizontal phreatic water table from the bottom to the pit's crest (with a simultaneous water table rise within the slope's soil mass). The effect of flooding on the SF was quantified and evaluated through the LEM, implementing an advanced search method (the so-called "cuckoo" of Slide2) to capture the non-circular failure surface.

The slope height and inclination are 150m and 14° respectively, while the water table was elevated stepwise in 11 stages, increased by 15m in each stage. Various shear strengths were ex-amined for the homogeneous soil. In all cases, the lowest SF (indicating the most critical conditions) is when the water table lies at 60 m above the bottom of the excavation (i.e. at 40% of the slope's height). In this case, SF is reduced by approximately 10% with respect to initial, dry conditions. The highest SF is when the water table reaches the crest of the pit, because in this case, the water mass results in an immense supporting force. Notice that SF is higher at this final flooding stage than for initial, dry conditions by 2%-16%. Overall, it is concluded that the flooding process does not affect significantly the slopes' stability.

Results from the limit equilibrium analyses were compared with equivalent finite element analyses in RS2. Identical slope geometries, geotechnical parameters, and groundwater conditions were considered. The two methods present practically identical results; calculated SFs from FEM are 0-1% lower than from LEM, and the failure mechanisms are almost identical. It is concluded that the LEM is adequate for the present study (that uses a horizontal water table rising and remaining horizontal) and could be preferred for similar analysis, being simpler and quicker than the FEM.

Moreover, slopes with benches and without benches, but with the same overall slope angle, were compared. Benches are always part of a mining excavation, so numerical models considering them are closer to the real problem's geometry. However, omitting benches from the model saves time, is simpler, and provides a more generic analysis' framework. Based on the present work, simulations without benches provide similar failure surfaces and lower safety factors by 6%-14%. Thus, benches' influence can be ignored for this stability problem, noticing that this type of analysis is always more conservative and on the safe side.

ACKNOWLEDGMENTS

This work has received funding from the European Union's Research Fund for Coal and Steel (RFCS) under the project "RAFF - Risk assessment of final pits during flooding slopes" grant agreement No 847299. Financial assistance by the European Commission is much appreciated.

REFERENCES

Bednarczyk, Z. 2017. Landslide monitoring and counteraction technologies in polish lignite opencast mines. Workshop on World Landslide Forum. Springer, 33–43.

Desjardins, M., de Graaf, P.J.H., Beale, G. & Rougier, M. 2020. Geotechnical risk management for Victor Mine closure. Proceedings of the 2020 International Symposium on Slope Stability in Open Pit Mining and Civil Engineering. Australian Centre for Geomechanics, 399–414.

Faur, F., Apostu, I.M. & Lazăr, M. 2020. Flooding Speed of Former Lignite Open Pits Regarded as a Determining Factor on the Stability of Final Slopes. *Emerging Trends in Engineering Research and Technology*, 11, 166–176.

Mikroutsikos, A., Theocharis, A.I., Koukouzas, N.C. & Zevgolis, I.E. 2021. Slope stability of surface coal mines in the presence of a weak zone: a numerical and probabilistic approach. *Submitted for publication.*

Rocscience. 2019. Slide2 Version 7.0-2D Limit Equilibrium Slope Stability Analysis. www.rocscience.com. Toronto, Ontario, Canada.
Rocscience. 2020. RS2 Version 11.0 - Finite Element Analysis for Excavations and Slopes. www.rocscience.com. Toronto, Ontario, Canada.
Theocharis, A.I., Zevgolis, I.E. & Koukouzas, N.C. 2021. A comprehensive geotechnical characterisation of overburden material from Greek lignite mine excavations. *Submitted for publication.*
Ulusay, R., Ekmekci, M., Tuncay, E. & Hasancebi, N. 2014. Improvement of slope stability based on integrated geotechnical evaluations and hydrogeological conceptualisation at a lignite open pit. *Engineering Geology*, 181, 261–280.
Verma, D., Kainthola, A., Gupte, S.S. & Singh, T.N. 2013. A finite element approach of stability analysis of internal dump slope in Wardha valley coal field, India, Maharashtra. *American Journal of Mining and Metallurgy*, 1, 1–6.
Zevgolis, I.E., Deliveris, A.V. & Koukouzas, N.C. 2019. Slope failure incidents and other stability concerns in surface lignite mines in Greece. *Journal of Sustainable Mining*, 18, 182–197.

The Evolution of Geotech - 25 Years of Innovation – Hammah et al. (Eds)

Geotechnical analysis of cut and fill mining sequence on dumping materials re-handling X pit by using slide3

Handanawarih Pancamanto, Zidni I. Munthaha & Chandra H. Saputra
Geotechnical Section, Geology Department PT Kaltim Prima Coal (KPC)

ABSTRACT: In an effort to optimize remaining coal reserves below dumping material. The general action that could be taken is the re handling of dumping materials in whole or in part. In X Pit Case, re-handling would be taken in part to reduce the volume of dumping material which needed to be removed. Partial removing will be conducted by undercut of dumping material where there is going to have the consequences of reducing the resisting force in slope stability. For minimizing the risk due to reduced restrain force, therefore a mining sequence will do by cut and fill method. This method is going to divide workspace become some smaller block workspace. Each smaller workspace would be conducted cut by excavating until mine out and then back filled before excavating next block to manage reduction of resisting force. Using slide3, we could do a two-dimensional (2D) and three-dimensional (3D) stability analysis with limit equilibrium method in the same model. The 2D force calculation is performed per slice on base and 2 side strength of slice. At the same time, 3D force calculation is performed per column on base and 4 side strength of column. The analysis result show when each block have cut until mined out have Factor of safety (FoS) value on 2D calculation less than 1.2. It also have FoS value less than 1.2 on 3D when re-handle sequence without cut and fill. Whereas 3D analysis result which considering 4 side strength of column calculation and limitation of reduction resisting force by cut and fill mining sequence show FoS value greater than 1.2 where have fitted based on the company standard which refer to government regulation and Stacey criterion.

Keywords: Slide3, Dump Stability, 3D Analysis

1 INTRODUCTION

Optimizing remaining coal reserves below dumping material sometimes became good choice of the dynamic coal prices so long as economics. Geotechnical analysis is significant parameter to calculate coal and overburden which would be mined as stability indicator. In X Pit case, excavating to rehandle of dumping material only be undertaken on toe of dumping material for minimizing dumping material removal to achieve optimal stripping ratio.

Excavation on toe of existing dumping materials are going to have impact steeper geometry compared before excavation and reducing resisting force. This activity has higher risk where the activity lied on face of slip failure potential with steeper geometry. Geotechnical analysis would be performed by using Slide3 with limit equilibrium method to figure FoS value in 2D and 3D calculation models. For these reason, the model could be deemed have fitted based on company standard which refer to government regulation or not in slope stability views.

DOI: 10.1201/9781003188339-48

2 METHOD

Slope stability analysis is taken with 2 type of mining model by using Slide3 where 2D and 3D calculation could be done at same time with limit equilibrium method and spencer calculation. The 2D force calculation is performed per slice on base and 2 side strength of slice whereas 3D force calculation is performed per column on base and 4 side strength of column. Force acting in detail for both individual slice and column are displayed in Figure 1.

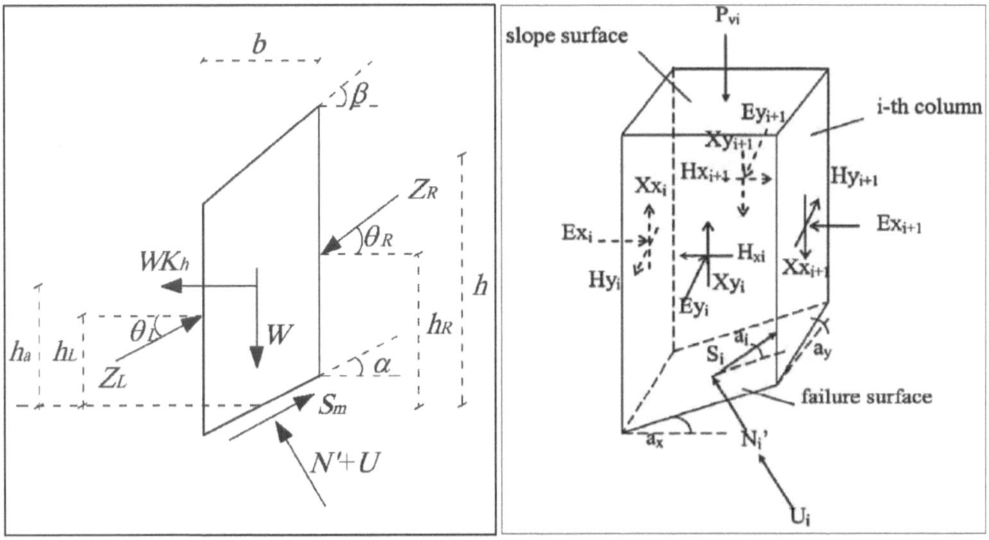

Figure 1. – Force acting comparison between 2D (Spencer, 1967) and 3D (Chang and Yip, 2007) calculation for Limit Equilibrium Method analysis.

The first mining model would be done without mining sequence cut and fill method. It means, excavation would be conducted along of toe dumping material directly. The other mining model would be conducted with mining sequence cut and fill by staging. On this model would be dividing 5 block excavation and filling. Each block would be represented a section to analyse 2D calculation and excavation is conducted until mined out and then doing filling with geometry as plan before continue excavation the next block to limit reducing resisting force (Figure 2). Determination of line section is based on most critical geometry for each stage

FoS value for re-handling dumping material in X Pit is deemed safe and fitted is greater than 1.20 based on company standard which refer to Read & Stacey (2008) criterion combine with government regulation after consider slope scale and consequence of failure (Figure 3).

3 ANALYSIS AND RESULT

Analysis result by using slide3 on design without mining sequence cut and fill shown both factor of safety value on 2D and 3D less than 1.2. The other side, design with mining sequence cut and fill which divide design become 6 stages shown FoS value on 2D value less than 1.2

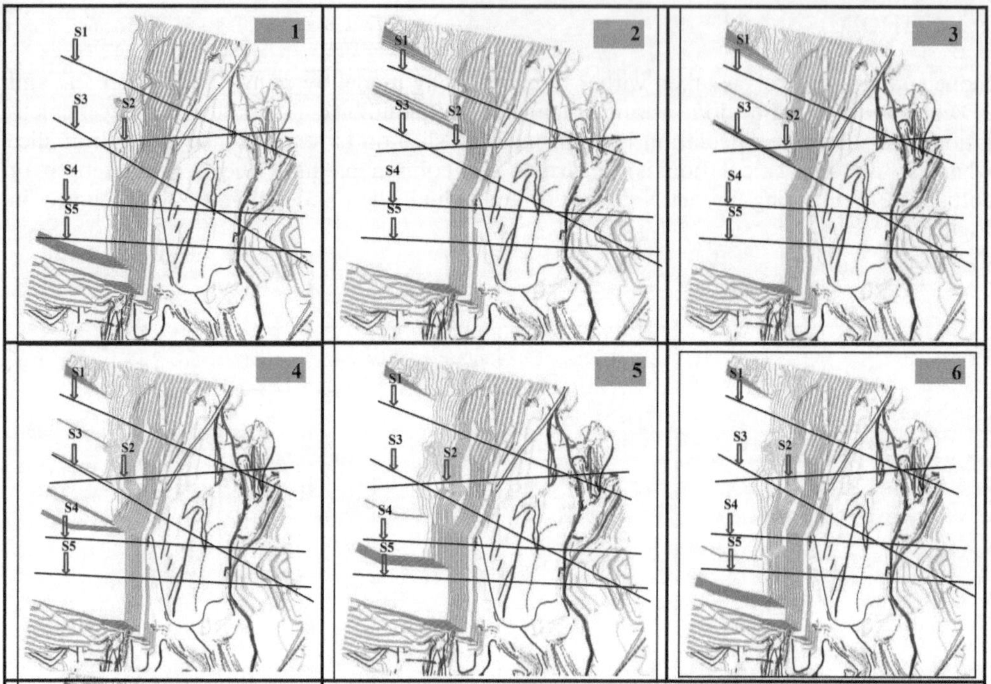

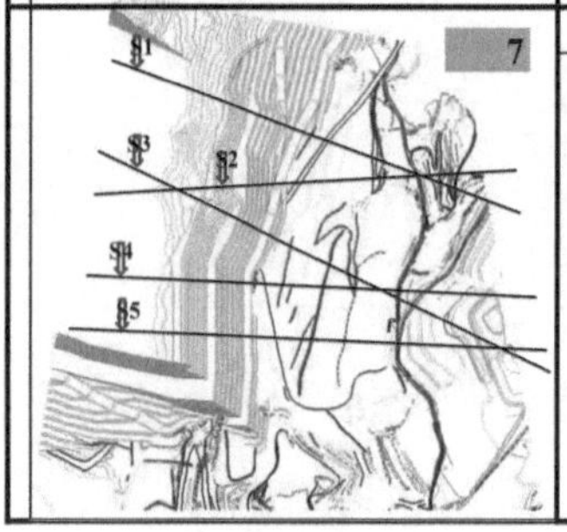

NO	STAGE	ACTION
1	Design without cut and fill	Excavation along toe dumping material directly
2	Design with cut and fill stage 1	Excavation on block 1 until mined out
3	Design with cut and fill stage 2	Filling block 1 and then excavation on block 2 until mined out
4	Design with cut and fill stage 3	Filling block 2 and then excavation on block 3 until mined out
5	Design with cut and fill stage 4	Filling block 3 and then excavation on block 4 until mined out
6	Design with cut and fill stage 5	Filling block 4 and then excavation on block 5 until mined out
7	Design with cut and fill stage 6	Filling block 5

Figure 2. Design for excavation re-handling dumping material.

		Acceptance criteria[a]		
Slope scale	Consequences of failure	FoS (min) (static)	FoS (min) (dynamic)	PoF (max) P[FoS ≤ 1]
Bench	Low–high[b]	1.1	NA	25–50%
Inter-ramp	Low	1.15–1.2	1.0	25%
	Moderate	1.2	1.0	20%
	High	1.2–1.3	1.1	10%
Overall	Low	1.2–1.3	1.0	15–20%
	Moderate	1.3	1.05	10%
	High	1.3–1.5	1.1	5%

Figure 3. Typical FoS and PoF acceptance criteria values (Read and Stacey, 2008).

each block when the area is excavated until mined out. On contrary result on 3D still have FoS value greater than 1.2 on each stage. Detailed analysis result using Slide3 for all mining sequence stage is displayed in Figure 4.

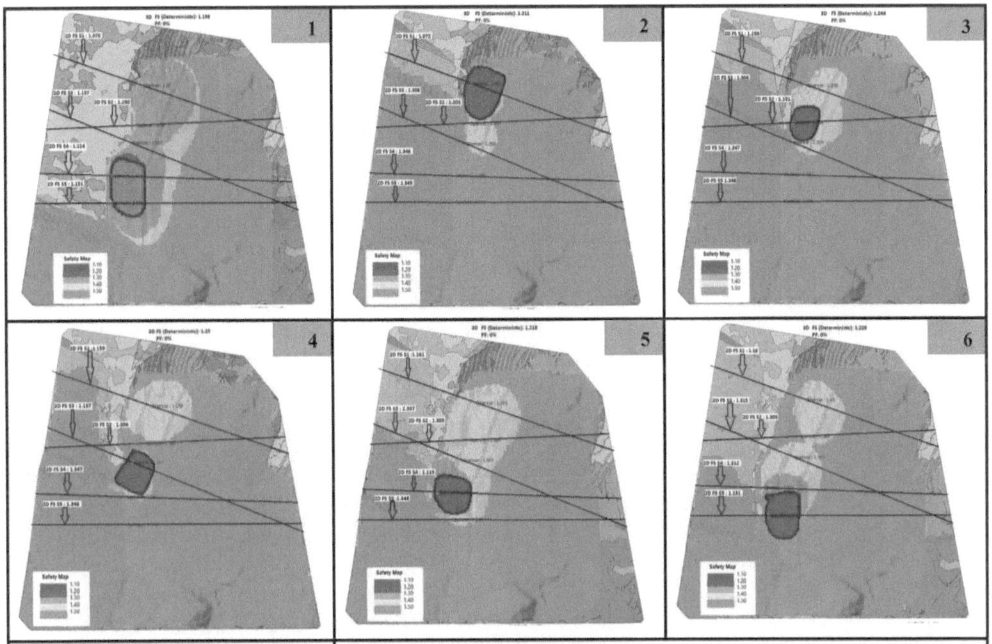

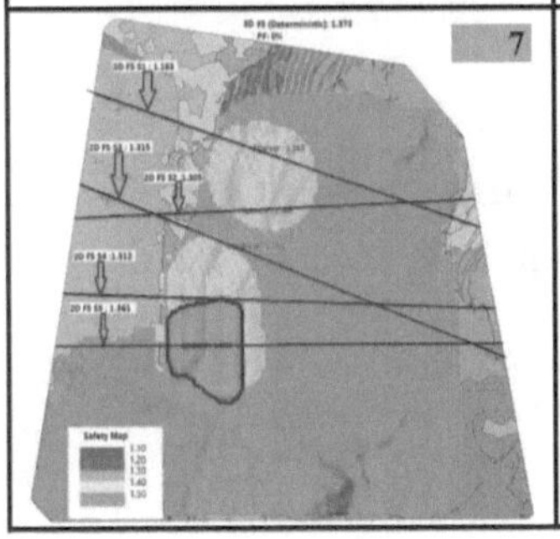

NO	ANALYSIS	3D		2D FS				
		FS	PF	S1	S2	S3	S4	S5
1	Analysis without cut and fill	1.198	0 %	1.070	1.150	1.157	1.114	1.151
2	Analysis with cut and fill stage 1	1.211	0 %	1.072	1.203	1.306	1.346	1.349
3	Analysis with cut and fill stage 2	1.244	0 %	1.158	1.151	1.304	1.347	1.348
4	Analysis with cut and fill stage 3	1.250	0 %	1.159	1.306	1.157	1.347	1.346
5	Analysis with cut and fill stage 4	1.218	0 %	1.161	1.305	1.307	1.115	1.348
6	Analysis with cut and fill stage 5	1.228	0 %	1.160	1.305	1.315	1.312	1.151
7	Analysis with cut and fill stage 6	1.373	0 %	1.163	1.305	1.315	1.312	1.361

Figure 4. – Comparison of Slide3 calculation for each mining sequence stage.

4 CONCLUSION

Utilizing Slide 3 which calculates all forces in 3 dimensions, gives us the opportunity to plan the mining sequence by dividing it into several stages with cut and fill method where each stage until the final design will have a factor of safety comply with company standards where it could not be done by using 2D analysis and 3D analysis without cut and fill sequences.

REFERENCES

Cheng, Y. and Yip, C. (2007). Three-dimensional asymmetrical slope stability analysis extesion of Bishop's, Janbu's, and Morgenstern-Price's techniques. J. Geotech. Geoenviron. Eng, 133(12), 1544–1555.

J. Read and P.F. Stacey, Guidelines for Open Pit Design, CSIRO Publishing, Melbourne, 2009.

Spencer E (1967). A method of the analysis of the stability of embankments assuming parallel inter-slice forces. Geotechnique 17, 11–26.

The Evolution of Geotech - 25 Years of Innovation – Hammah et al. (Eds)

3D slope stability analysis of a filtered iron ore tailings dry stacking

A.S. Freire, T.C. Bretas, T.B.G. Moreira, M.C. Medeiros & F.F.M Vilela
BVP Geotecnia e Hidrotecnia, Belo Horizonte, Minas Gerais, Brazil

ABSTRACT: This paper presents the stability assessment of a filtered, dry-stacked, iron ore tailings storage facility using 3D limit equilibrium methods. Slope stability assessments were performed using Rocscience's Slide3 software. For comparison, 2D stability analysis of a cross-section derived from the Slide3 model was also completed using Rocscience's Slide2 2D limit equilibrium software. Both analyses demonstrate the ease and speed at which Slide3 and Slide2 can model and analyze even the most complex 3D geologic structures. The results show the importance of 3D and 2D stability analysis where undulating topography forms the foundation of tailings storage facilities.

1 INTRODUCTION

Following recent catastrophic tailings storage facility (TSF) failure events that occurred in Minas Gerais State, Brazil, the ANM ("Agência Nacional de Mineração") published "Resolution N° 13, of August 8 (2019)." It prohibits using 'upstream' TSF dams (i.e., dams constructed from the raising of dumped slopes) throughout the national territory.

In response, the mining industry has developed alternative filtered TSF's to allow the continuation of mining and processing iron ore in the country. Filtered TSF's require dry stacking of tailings material. 3D limit equilibrium analysis has become an essential tool for verifying the stability of constructed slopes.

The filtered iron ore tailings dry stacking (FTDS) methodology was designed to improve the future operation of waste disposal, especially in rainy periods. FTDS construction is divided into three stages: dry period, preparation for the wet period and wet period, with these definitions being made from a history of rainfall events.

2 OBJECTIVE

This paper presents the stability assessment of a filtered, dry-stacked, iron ore TSF using 3D limit equilibrium methods. The 3D limit equilibrium model was constructed using Slide3, geological data exported from Leapfrog, and the FTDS design generated in Civil3D.

3 METHODOLOGY

3.1 *Geological framework*

The Iron Quadrangle Region (Dorr, 1969), located in the southeast part of the São Francisco Cráton (Almeida, 1977), is an important Brazilian metallogenetic province composed of archaean to paleoproterozoic rocks. These rocks evolved from multiple deformational events, arranged over different periods (Chemale Jr et al.,1994).

DOI: 10.1201/9781003188339-49

The geological-geotechnical characterization of the foundation and surroundings of the tailings stack was performed from the integration between mapping data and geotechnical drilling information and converged to the individualization of the following materials: (i) Colluvium - clayey detrital cover that generally covers the entire area with relatively constant thickness of one meter; (ii) Residual Soil - mostly claysoil, with a final n'SPT of less than 8 strokes and devoid of reliquiar structures; (iii) Saprolitic soil - mostly claysoil, with final n'SPT between 8 and 30 strokes and presence of reliquiar structures; and (iv) Saprolite - clayey material, typically grayish and derived from shales and Phyllite. It has a final n'SPT greater than 30 or impenetrable and well-marked foliation.

A 3D geological model was subsequently constructed for these materials using Leapfrog Geo software (Figures 1A and 1C). The tailings pile foundation was also defined in the 3D geological model (Figure 1C), where a surface representing the transition between materials of n'SPT greater than or less than 8 strokes was interpolated.

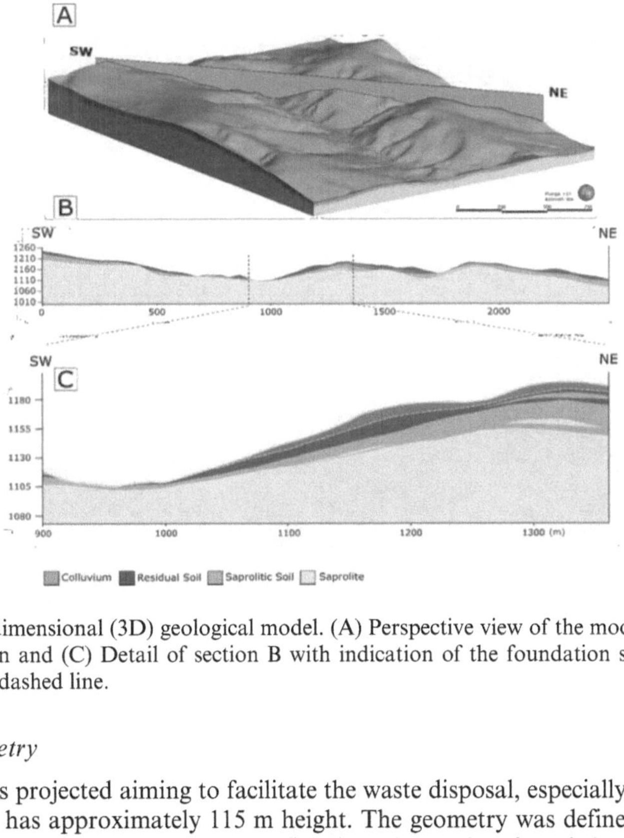

Figure 1. Three-dimensional (3D) geological model. (A) Perspective view of the model, (B) Representative NE-SW section and (C) Detail of section B with indication of the foundation surface (n'SPT = 8) represented by the dashed line.

3.2 *Stack geometry*

The structure was projected aiming to facilitate the waste disposal, especially on the rainy seasons. The FTDS has approximately 115 m height. The geometry was defined considering the waste generation that comes from the beneficiation plant. Therefore, it has a capacity 62,65 Mm^3 and it can operate for 17 years.

The final geometry has a bench height of 10.0 m and bench width of 8,0 m. Furthermore, the bench face angle established was 18°.

An overview of the geometry is shown in Table 1. Also, a general view of the structure is shown on Figure 2.

3.3 *Model parameters*

Material properties applied to 3D stability modelling of FTDS are shown in Table 2. The undrained resistance of the tailings material was defined according to the methodology proposed by Olson (2008) based on the results of the Consolidated Isotropic Undrained (CIU) saturated triaxial tests, at an 85% degree of test compaction (worst case scenario).

Table 1. FTDS Geometry Overview.

Base Elevation (m)	1.085,00
Top Elevation (m)	1.200,00
Total Height (m)	115
Overall Slope Angle	15°
Bench Face Angle	1V:3H (18°)
Maximum Bench height (m)	10
Minimum Bench Width (m)	8
Total Volumetric Capacity (m^3)	62.645.026,26
Operational Life	17 anos
Ocupation Area (m^2)	1.444.117,22
Vegetal Supression Area (m^2)	1.232.000,00

Figure 2. General view of the FTDS.

3.4 *3D stability analysis*

3D stability analyses of filtered iron ore tailings dry stacking were performed using Slide3 (v2018). Different scenarios were analyzed considering peak and residual parameters of filtered tailings in the "*off spec*" zone and a minimum search surface depth of 5 m.

The Slide3 model was constructing by importing the FTDS' design generated in Autodesk's Civil3D. Material properties were then assigned to each volume representing a different lithological unit. The searches for critical slip surfaces were performed using the "Cuckoo Search" algorithm, with ellipsoidal asymmetric surfaces, and the Bishop analysis method. According to Rocscience Inc (2019), ellipsoidal surfaces cover a larger span of slip surface shape.

The 3D stability analyses were performed considering the post-liquefied (residual) strength of the tailings deposited in the regions outside the shell zone, which are tailings discharged with high water content (rainy season), for all years of stack construction. To simulate the piezo metric levels "perched" in the FTDS, an Ru coefficient of 0.545 was applied for the "off spec" tailings.

Stability analyses were conducted based on Standard 13 (ANM, 2019), which recommends a minimum Factor of Safety (FOS) equal to or greater than 1.3 for undrained liquefaction at peak resistance tailings strength.

Table 2. Material properties applied in the 3D stability analysis.

Material	γ(kN/m3)	c′(kN/m2)	ϕ′(o)	Vertical Strength Ratio
Colluvium	18,0	0	26	0,33
Residual Soil	18,0	0	25	0,46
Saprolitic soil	18,0	10	27	-
Saprolite	20,0	20	34	-
Compacted tailings	23,0	0,0	31	
Tailing (Peak Resistance)	23,0	-	-	0,25
Tailings (Liquefied Resistance)	18,0	-	-	0,06
Rockfill	21,0	LEPS, 1970		

Also, as the Brazilian standard does not establish guidance for the FOS of the liquefaction analysis for stack projects, the international guidelines recommended by the Canadian Dam Association (CDA, 2014) were adopted. They state that the post-liquefaction FOS should be greater than 1.2.

4 RESULTS

In this section, the results obtained from the 3D stability analysis of FDTS will be presented.

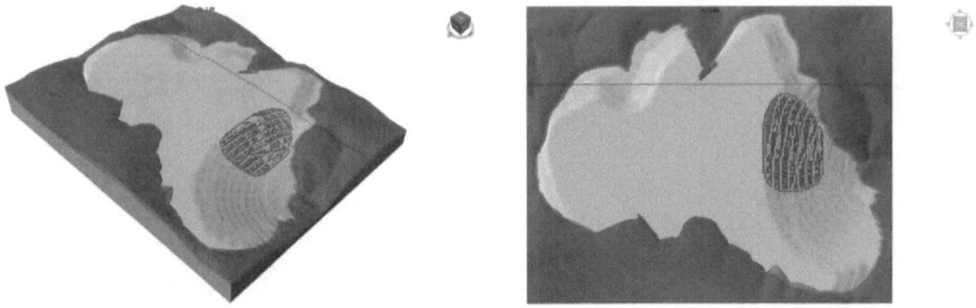

Figure 3. Perspective and plan view of the results obtained from 3D stability analysis at Residual material strengths. Critical FOS = 1.85.

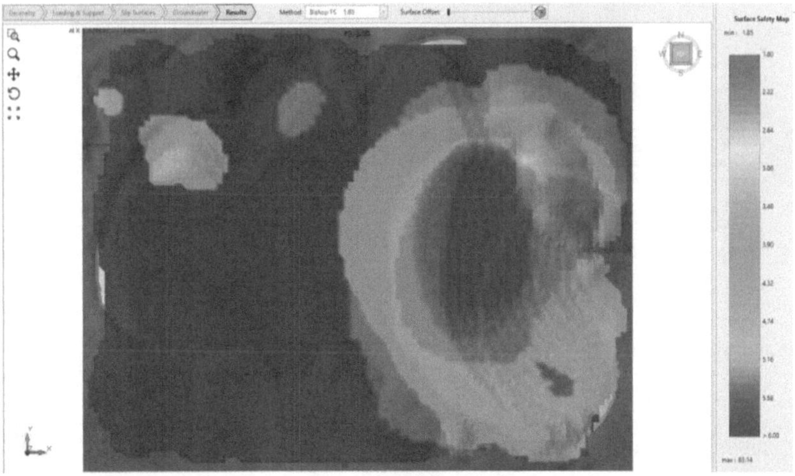

Figure 4. Safety Map view of the results obtained from 3D stability analysis at Residual material strengths.

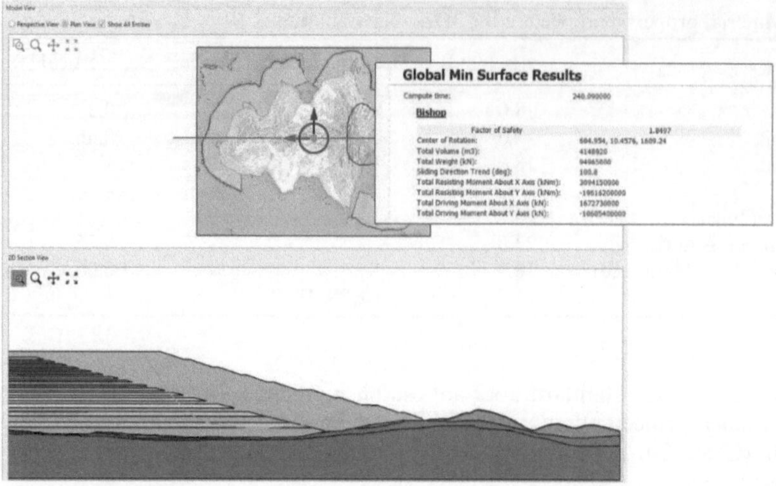

Figure 5. 2D section view of the results obtained from 3D stability analysis at Residual material strengths.

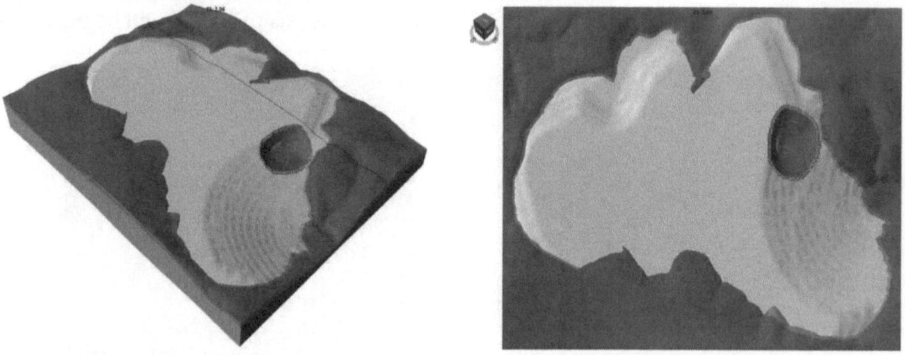

Figure 6. Perspective and plan view of the results obtained from 3D stability analysis at Peak material strengths. Critical FOS = 2.04.

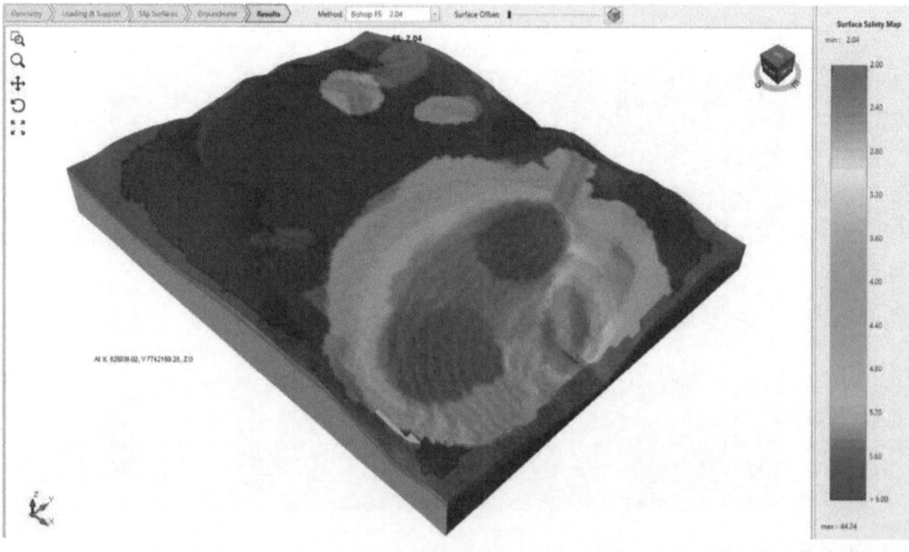

Figure 7. Safety Map view of the results obtained from 3D stability analysis at Peak material strengths.

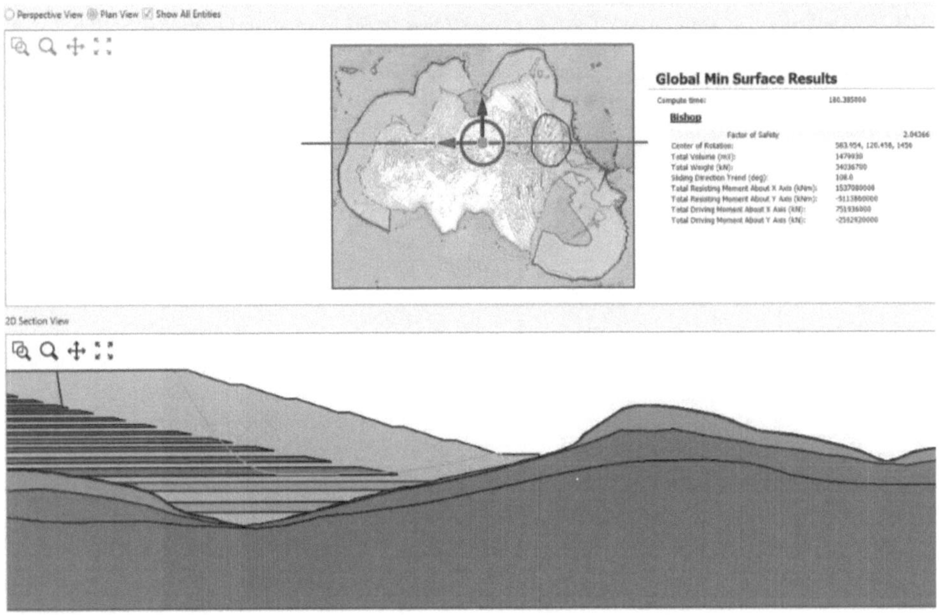

Figure 8. 2D section view of the results obtained from 3D stability analysis at Peak material strengths.

5 DISCUSSION

Slope stability analysis of the proposed FTDS demonstrates that design acceptance criteria, set by the ANM (2019) and CDA (2014) of a minimum FOS of 1.3 is met at the applied water levels and at both the peak and residual material strength levels.

The results demonstrate the importance of modelling in 3D for risk analysis, especially when the variation in local topography forming the FTDS foundation has a significant influence on the stability of the structure. In addition, 2D analysis of sections cut from the Slide3 3D model helped evaluate the depth and characteristics of the critical slip surface.

6 CONCLUSION

This paper summarized the stability analysis of a proposed FTDS using 3D and 2D limit equilibrium analysis methods. Calculated FOS were compared to regulatory requirements to ensure the design met minimum industry FOS requirements.

Slide3's interoperability with other industry software allowed the fast model build from geological surfaces exported from Leapfrog and design generated in Civil 3D. Internal to the software, Slide3's ability to create 2D sections and assess 2D stability within the software allowed the fast validation of 3D results with 2D results.

Further, from Slide3 analysis, it was possible to obtain the critical slip surface volume, plus range in FOS across the FTDS design, as a surface safety map, for detailed risk analysis of the proposed design.

REFERENCES

Almeida, F.F.M. 1977. *O Cráton Do São Francisco*. Rev. Bras. Geoc., 7: 285–295.

BRASIL. Resolução N° 13, de 08 de Agosto de 2019. Brasil: ANM

CDA.2014. *Application of Dam Safety Guidelines to Mining Dams.* Canadian Dam Association, Canadá.
Chemale, F. Jr., Rosière, C. A. & Endo, I. 1994. *The Tectonic Evolution Of The Iron Quadrangle*, Minas Gerais, Brazil. Precambrian Research, 65: 25–54
Dorr II JVN 1969. *Physiographic, Stratigraphic And Structural Development Of The Iron Quadrangle*, Minas Gerais, Brazil. US Geological. Survey Professional. Paper, 614(a):1–110.
Leps, TM 1970. *Review of Shearing Strength of Rockfill.* Journal of the Soil Mechanics and Foundations Division, ASCE, Vol. 96: 1159–1170.
Olson, S.M & Mattson, B.B. 2008. *Mode of Shear Effects on Yield and Liquefied Strength Ratios.* Can. Geotech. J. Vol. 45.
Rocscience INC. 2019. *Surfaces Search in Slide3 Principles and Practices.* Rocscience Guidness.

The Evolution of Geotech - 25 Years of Innovation – Hammah et al. (Eds)

Slope angle optimization applied for geometry design of an open pit iron ore mine

A.S. Freire, A.P. Guimarães, B.A. Alemão M., F.F.M. Vilela, M.C. Medeiros, T.B.G. Moreira & T.C. Bretas
BVP Geotecnia e Hidrotecnia, Belo Horizonte, Minas Gerais, Brazil

ABSTRACT: During the excavation of a mining open pit, it is common to encounter several types of materials, which have resistance characteristics specific to each lithotype and, usually, have different dipping angles. These materials might have great resistance or low resistance and, due to difference between the materials behavior, a sectorization is necessary with different angles. Thus, during the elaboration of the open pit design, many stability analyses are necessary to optimize the slope angle to get better usage of iron ore, with less generation of waste. This way, the Slide 2D "Change Slope Angle" tool assists in the stability analysis to obtain the minimum safety factor considering the maximum slope angle. This paper shows the case of an open pit mine, which was sectorized according to the geomechanical behavior of the foundation materials. Then, different slope angles were tested in different scenarios to attend the minimum Factor of Safety required.

1 INTRODUCTION

Slopes design is one of the major challenges when planning an open pit mine. It is important to define a good geometry for the open pit, considering variables such as safety, ore recovery, sterile excavation and financial return.

The design must guarantee a certain level of safety as well as accomplish the economic expectations of the owners. Therefore, the ore recovery needs to be maximized and the waste stripping should be as low as possible. To do this, the slopes angles need to be optimized, so that they reach an admissible Factor of Safety and improve the volume of material that needs to be removed.

2 OBJECTIVE

The aim of the study presented in this paper to determine an optimal excavation configuration for the open pit mine. The best solution is the one with less sterile stripping e more ore excavated. In this sense, it was necessary to establish a geometry with slopes as steep as possible attending the minimum to Factor of Safety (FoS) required by the Brazilian legislation.

3 FACTOR OF SAFETY ADMISSIBLE

There is not a specific legislation for designing open pit mines in Brazil. Therefore, to establish an acceptable FoS the book "Guidelines for Open Pit Design" (Read & Stacey, 2009) was used. It is presented in the book a table by Swan & Sepulveda that shows the acceptance

DOI: 10.1201/9781003188339-50

criteria for the FoS together with the probability of failure (PoF) associated with the consequences of the failure.

Regarding the conditions of the open pit mine in this study, the minimum FoS required by Swan & Sepulveda is 1.20 for inter-ramp slopes and 1.30 for overall slopes.

4 METHODOLOGY

Different scenarios were analyzed based on the open pit proposed geometry. Firstly, studies of each lithotype were carried out to determine the greater angle achievable to reach a Factor of Safety of 1,3. For this preliminary study, a simple slope geometry was used (Figure 1). Then the angle maximum bench face angle and the maximum overall angle were obtained for each lithotype. To optimize the slope angle the “Change Slope Angle” Slide2 editing tool was used.

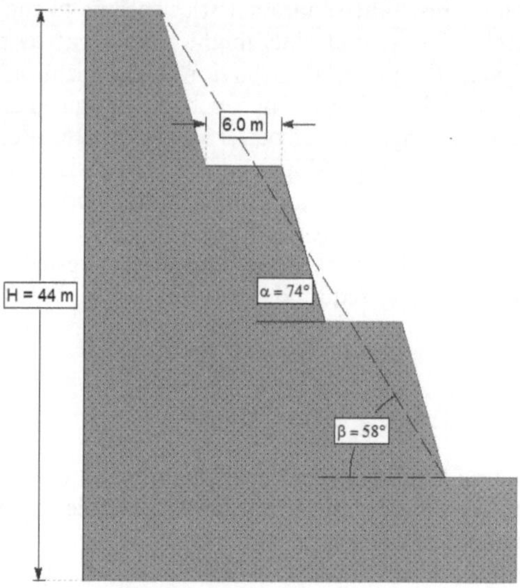

Figure 1. Schematic slope structure used in preliminary studies.

After the preliminary study, the angles were applied in the open pit’s design, and this way the stability analyses considered the field conditions that were mapped. When the different lithologies were analyzed together, there were also a few adjustments in the slope angles that needed to be made in order to reach a satisfactory FoS.

In the following items, it will be presented the main steps used in the stability analyses.

4.1 *Lithology mapping*

To give subsidy to the stability analysis, a field campaign was conducted, mapping the main lithologies, structural orientation, and geomechanical parameters on a 1:2000 scale. The field-collected attributes together with the laboratory analysis were used to understand the rock mass. The main rock masses are composed of Silica Itabirite (Figure 2a) in the interior of the pit, with resistance ranging between R1 and R2, Phyllite (Figure 2b) on the outer limits of the pit with resistance from R1 up to R3, Aluminous Itabirite and Dolomitic Itabirite from R0 to R1. For the rock mass classification, the RMR system was adopted, using the formulas proposed in 2011 by Bienawski and the “Weak Rocks” modification proposed in 2019 by Cella et al. Most rock mass were defined as class V on the surface and class IV and III in the final depth of the open pit.

The main observed geological structures that may influence on the geomechanical behavior of the rock masses are the compositional bedding in Itabirites and the foliation in Phyllites both dipping around 35 and 40 degrees between NE and SE.

This work stage allowed a correlation between the rocks seen in the field and their correspondence on the borehole data, granting a coherent geomechanical section building.

Figure 2. a) Silica Itabirate; b) Phyllite.

4.2 *Sections*

To define the sections of analyses, the open pit was sectorized according to the geological occurrence and structural disposal of the materials and their respective geomechanical attributes. Then, for each sector a critical section was chosen considering the slope's total height. These sections were built using both interpolations between the available boreholes and the knowledge acquired during the mapping, regarding the structural orientations of the rock bedding and foliation and the surface disposal of the lithologies.

4.3 *Materials and parameters*

After mapping all the materials present in the area, laboratory tests were specified. For the more resistant materials and the ones with anisotropic behavior, direct shear tests were made.

Table 1. Resistance parameters.

Lithotype	Resistance	Class	Specific Weigth (kN/m³)	Effective Cohesion (kPa)	Effective Friction Angle (°)	Parallel Effective Cohesion (kPa)	Parallel Effective Friction Angle (°)
Phyllite	R1	V	17	87.7	28	65	25
	R2	IV	18	87.7	30	65	28
Dolomite	R1	V	15	33	25	-	-
Dolomitic Itabirate	R1	V	16	32.9	27.1	-	-
Manganese Itabirate	R0-1	V	18	32	34	-	-
Silica	R1	V	20	37.6	33.4	31.6	33.4
Itabirate	R2	IV	27	117	40	60	36

For the ones that showed lower resistance in the field and isotropic behavior, undrained triaxial tests were carried out. This way, the resistance parameters obtained are shown in Table 1.

4.4 *Criteria and premisses*

All the analyses were made without considering the water level, due to the drawdown that will be made before the excavations.

4.5 *Stability analysis*

The analyses were made using Slide2 software by Rocscience. The software is used to do two-dimensional stability analyses and determines the structure's FoS using the limit equilibrium analysis procedures. The method chosen was GLE/Morgenstern-Price that uses the equilibrium of forces and moments to calculate the FoS. Also, it was considered only searching for noncircular slip surfaces, which often follows weak layers or weak interfaces between different materials (Duncan, 2014). Consequently, the slip surface tends to have an irregular shape. The search algorithm used was "Cuckoo Search" developed exclusively for noncircular wedges.

5 RESULTS AND DISCUSSION

The preliminary study resulted in the maximum angles showed in Table 2. Important to say that the number of benches presented is the highest that occurs in the field for each lithotype.

Table 2. Overview optimization per lithotype.

Lithotype	Number of benches	Bench Face Angle α (*)	Total Height H (m)	Overall Angle β (*)	FoS
Tertiary Deposit	6	59	72	45	1,32
Aluminous Itabirite R1 V	5	54	60	41	1,30
Silica Itabirate R2 IV	8	54	96	41	1,32
Phyllite R2 III	17	53	204	39	1,32
Phyllite R1 V	6	64	72	48	1,31
Phyllite R1 V [(1)]	8	43	96	34	1,36
Silica Itabirate R1 V	11	44	132	34	1,31

(1) Dip 40° unfavorable to slope stability

After these studies, several scenarios were analyzed according to the needs of the mining management. Each scenario with a different geometry due to changes in the mining development plan.

Since there were many analyses of different sections, it will be showed next an example of what was done in all the sections. Figure 3a shows the stability analysis while Figure 3b shows the same section optimized. The Slide "change slope angle" tool allowed to easily change the bench slope angle as well as the overall angle without changing the benches width. Figure 4 shows the detail of the changes in geometry. Important to highlight that for the analyses with a lower FoS than 1.30, the tool was used to reduce the slope angle. However, for the cases where the FoS was way higher than 1.30, the mentioned tool was used to increase the slope angle.

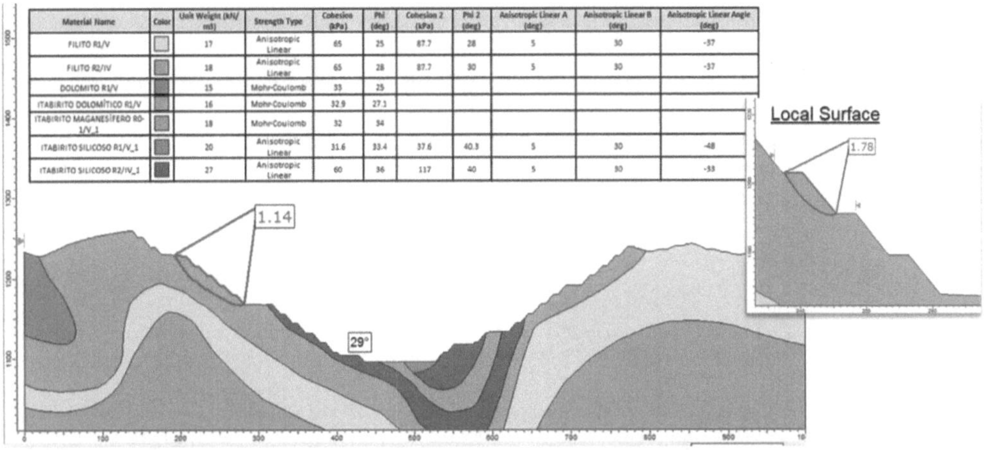

Material Name	Color	Unit Weight (kN/m3)	Strength Type	Cohesion (kPa)	Phi (deg)	Cohesion 2 (kPa)	Phi 2 (deg)	Anisotropic Linear A (deg)	Anisotropic Linear B (deg)	Anisotropic Linear Angle (deg)
FILITO R1/V		17	Anisotropic Linear	65	25	87.7	28	5	30	-37
FILITO R2/IV		18	Anisotropic Linear	65	28	87.7	30	5	30	-37
DOLOMITO R1/V		15	Mohr-Coulomb	33	25					
ITABIRITO DOLOMÍTICO R1/V		16	Mohr-Coulomb	32.9	27.1					
ITABIRITO MAGANESÍFERO R0-1/V_1		18	Mohr-Coulomb	32	34					
ITABIRITO SILICOSO R1/V_1		20	Anisotropic Linear	31.6	33.4	37.6	40.3	5	30	-48
ITABIRITO SILICOSO R2/IV_1		27	Anisotropic Linear	60	36	117	40	5	30	-33

Figure 3a. Stability Analysis – FoS = 1.14.

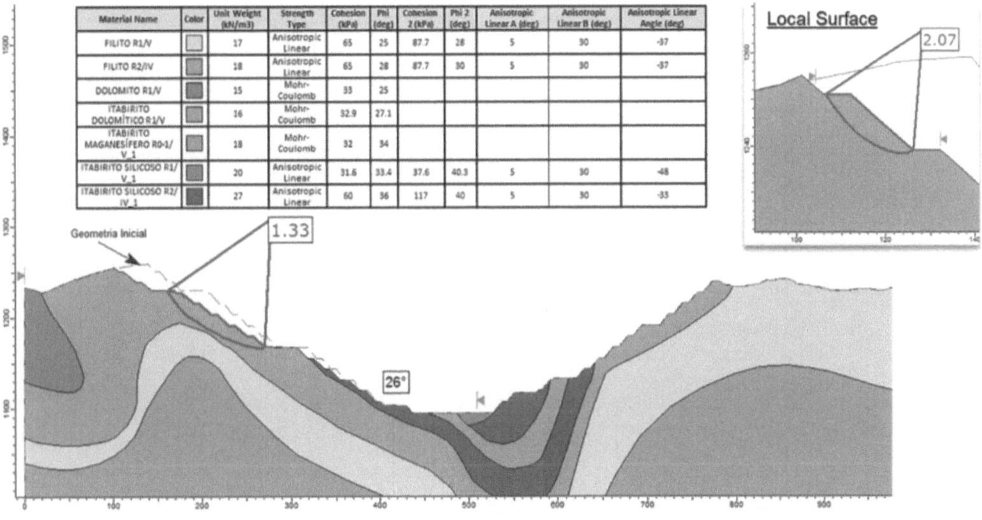

Material Name	Color	Unit Weight (kN/m3)	Strength Type	Cohesion (kPa)	Phi (deg)	Cohesion 2 (kPa)	Phi 2 (deg)	Anisotropic Linear A (deg)	Anisotropic Linear B (deg)	Anisotropic Linear Angle (deg)
FILITO R1/V		17	Anisotropic Linear	65	25	87.7	28	5	30	-37
FILITO R2/IV		18	Anisotropic Linear	65	28	87.7	30	5	30	-37
DOLOMITO R1/V		15	Mohr-Coulomb	33	25					
ITABIRITO DOLOMÍTICO R1/V		16	Mohr-Coulomb	32.9	27.1					
ITABIRITO MAGANESÍFERO R0-1/V_1		18	Mohr-Coulomb	32	34					
ITABIRITO SILICOSO R1/V_1		20	Anisotropic Linear	31.6	33.4	37.6	40.3	5	30	-48
ITABIRITO SILICOSO R2/IV_1		27	Anisotropic Linear	60	36	117	40	5	30	-33

Figure 3b. Stability Analysis – Optimized – FoS = 1.33.

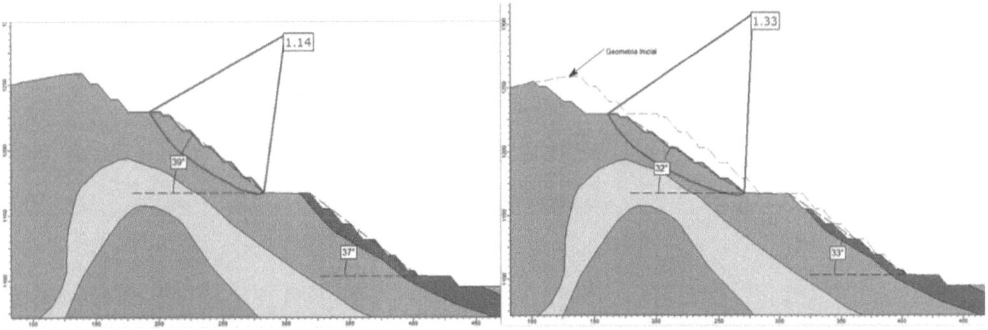

Figure 4. Detail of geometry changes.

6 CONCLUSION

In the present work, the stability and optimization of a pit was evaluated using the slope abatement tool in Slide2.

The results of the analyses showed a great improvement in the design of the open pit mine. As an iterative process, defining the geometry using the Slide2 tool spent less time than if using other software to import geometry each time.

REFERENCES

Bieniawski, Richard Z. T. 2011. *Misconceptions in the applications of rock mass classifications and their corrections.* ADIF Seminar on Advanced Geotechnical Characterization for Tunnel Design. Madrid, Spain, 29 June, 2011.

Cella, P., Castro, L. & Carter, T. 2019. *Mining Slopes in Weathered and Weak Rocks.* Chapter 24 of the Soft Rock Mechanics and Engineering, ISRM, Ed. Kanji, He and Souza.

Duncan, J. M., Wright, S.G. & Brandon, T.L. 2014. *Soil Strength and Slope Stability.* Hoboken, New Jersey: John Wiley & Sons.

Read, J. & Stacey, P. 2009. *Guidelines for Open Pit Slope Design.* Australia: CSIRO.

Session 11 - Probabilistic slope stability analysis II

The Evolution of Geotech - 25 Years of Innovation – Hammah et al. (Eds)

Back analysis for landslide in mine waste dump slope using probabilistic analysis

Gilang Firmanda* & Arthur Gemas Pradhana Nayoan
PT Lereng Nusantara Konsultan, West Jakarta, DKI Jakarta, Indonesia

ABSTRACT: The purpose of this research is to use back analysis for mine waste dump slope stability and to determine the causes of the landslides that occured. The research method that we use is the Limit Equilibrium Method and Probabilistic Analysis with the help of the Rocscience SLIDE2 application. In this case study, it can be seen that the slope has a 1.299 static FoS (Factor of Safety) value. This value is categorized into the safety condition FoS category on slope stability if we referring to FoS classification i.e from Read and Stacey (2009). However, in fact the land- slide occur on the slope itself even has a safety FoS category. Therefore probabilistic analysis was carried out to to be follow up with back analysis for the mine waste dump slope in this case study. In the back analysis of the waste dump slope it is assumed the landslide will occur if the FoS value is 1. From the probabilistic analysis, it is found that the closest possibility for the slope landslide failure is when the cohesion and unit weight of the materials decrease to 88.97 kPa and 24 kN/m^3. This shows that the amount of water pressure from the recharge area that supply the slope body can reduce the strength and the slope factor of safety. The material cohesion (c) is reduced by 9.54 % and increase the material unit weight value by 3.52 %.

Keywords: Mine Waste Dump Slope, Factor of Safety (FoS), Back Analysis and , Probabilistic Analysis

1 INTRODUCTION

1.1 *Background*

Open pit mining is a mining method by excavating overburden and mining the desired ore (Gold, Silver, Coal, etc.). Open pit mining method in coal mining in Indonesia is still very common, considering that almost all coal mining in Indonesia (i.e. Sumatra and Kalimantan) uses this method. In addition, the geological conditions in the Indonesia on coal mining are very supportive for open pit mining and generally related to syncline-anticline fold structural geology. The aim of any open pit mine design is to provide an optimal excavation configuration in the context of safety, ore recovery, and financial return (Read & Stacey, 2009).

Slope stability analysis is a must in an open pit mining (Figure 1) to avoid landslide in mining and require field data like material strength parameter and groundwater condition. The term 'land- slide' includes a wide variety of slope movements, such as soil slips, deep-seated slides, mud flows, debris flows, rockfalls, etc. (Varnes, 1978). A landslide or slope failure implies that the factor of safety of the slope at the moment of failure is unity (Zhang

*Corresponding author
DOI: 10.1201/9781003188339-51

Figure 1. One of the open slope mining on coal mine in Indonesia.

et al, 2010). Many factors can affect landslides or slope failure, but in general the factors are slope geometry, materials (Soil or Rock), orientation of geologic feature (Strike & Dip), geological structure (Fault and Fold), groundwater, seismic activity or static & dynamic load, and weathering.

The limit equilibrium method is used to analyze a slope in mining. However, keep in mind that the weakness of this method is that there are still many assumptions used in its calculations, so that it requires an appropriate correlation with field. In this case study, the authors conducted a back analysis on slopes that had failure, but with a good factor of safety value based on the Read and Stacey (2009) to know the cause of the slope failure.

1.2 *Methods*

In analyzing slope stability requires a factor of safety (FoS) value. Basically FoS is a deterministic measure of the ratio between resisting forces (capacity) and driving forces (demand) of the system in its considered environment (Read & Stacey, 2009) (Equation 1).

$$FoS = \frac{C}{D} \tag{1}$$

Where C = resisting forces (capacity); D = driving forces (demand).

There have been quite a lot of FoS calculation formulas made and develop by several researcher, considering that the FoS formula is quite old and was developed since the early 20th century. Fellenius (1936), Bishop (1955), Janbu (1954), and Spencer (1967) are some of the researchers who have developed the FoS formula. In this case, the authors use the Spencer (1967) formula in calculating FoS in the research area. The Spencer's method or formula is basically a modified and extended version of the Bishop's simplified method (Agam et al, 2016). In summary Spencer formula considers both shear and normal interslice forces, satisfies both resisting & driving forces, and assumes a constant interslice force function resulting a more realistic and acceptable FoS.

If we make a simple comparison than other formula in Fellenius (1936) there are many assumption in its formula that make the FoS result is unrealistic. In Bishop's simplified formula consider normal interslice forces, but ignores interslice shear forces and satisfies over all resisting forces, but not overall horizontal driving forces. Then in Janbu's simplified formula considers normal interslice forces, but ignores interslice shear force and satisfies over all horizontal driving forces, but not over all resisting forces. Because of this the FoS result when using Bishop's formula is always greater than when using Janbu's formula.

Photogrammetry analysis was used to identify and analyze the types of landslide, so as to determine the cause of the landslide descriptively with the help of DroneDeploy software. To perform back analysis, it is need probabilistic analysis to recognize that there might be numerous combinations of such parameters, but their relative likelihoods are different, which can be quantified by probability distributions. According to Zhang et al (2010) major advantages of probabilistic back analysis methods include: (1) it provides a logical way to incorporate information from other sources in the back-analysis and (2) it is capable of back-analyzing multiple sets of slope stability parameters simultaneously.

Main research problem in this study case is the author use back analysis for the mine waste dump slope to know about the cause of the slope failure. The authors have a hypothesis that the slope failure (FoS = 1) occurs due to a decrease or increase in material properties on the slope. Slope stability analysis and the back analysis processed using the help of Rocscience SLIDE2 software.

2 DATA & DISCUSSION

2.1 *Research area condition*

The research area in this case study is in the mine waste dump of highwall coal mine in East Kalimantan. In the regional geology of the East Kalimantan area around Samarinda, according to Supriatna et al (1995), there are several rock formations such as the Balikpapan and Pulau Balang formations which consists coal lithology with varying thicknesses, sandstone, and claystone with a terrestrial depositional environment, transition (Deltaic) to shallow marine with miocene age with geological structures in the form of anticlinorium folds generally trending northeast-south- west and accompanied by several faults. Mine waste dump slope in the authors research area located in the limb of anticline.

2.2 *Photogrammetry analysis*

From the photogrammetry analysis the authors can separating the landslide in mine waste dump into 3 sections that are crest, body, and toe (Depositional area) of the slope based on elevation difference from the digital elevation model map as shown in Figure 2-3 below. Each landslide section on the slope has a different description, in the crest section of the slope there is an indication of main scrap which reflects the symptoms of circular type landslides on the crest section (Figure 4).

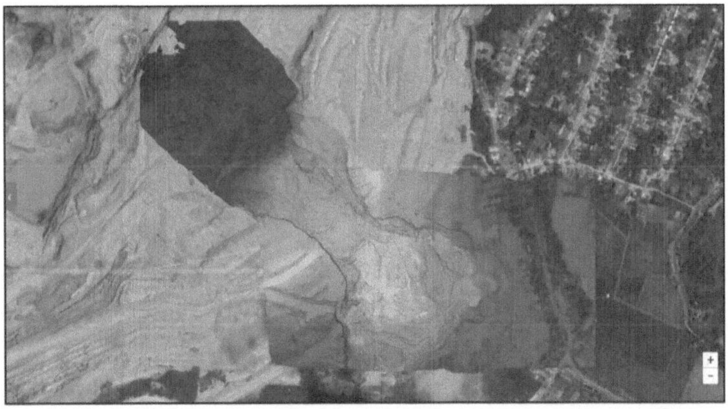

Figure 2. Digital elevation model (DEM) map which show a difference elevation to divide landslide section.

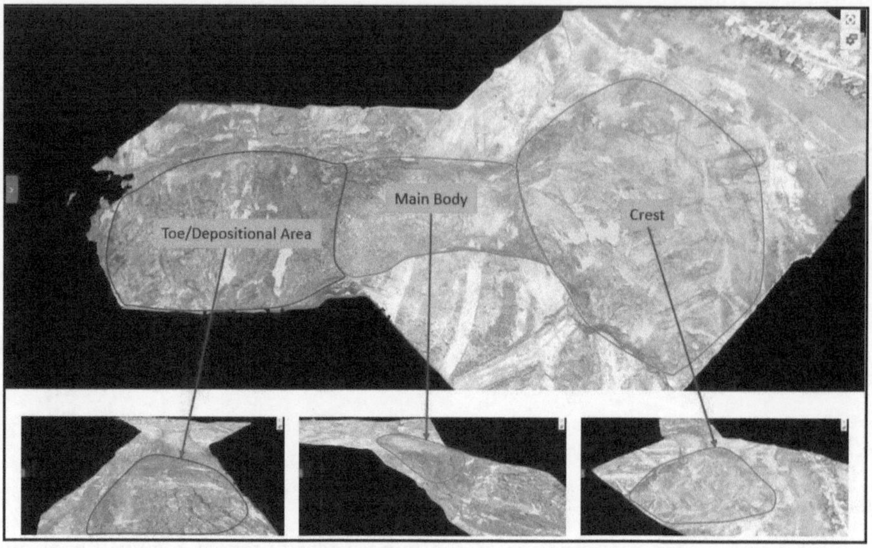

Figure 3. Landslide section which consists of crest, body, and toe of the slope in the field.

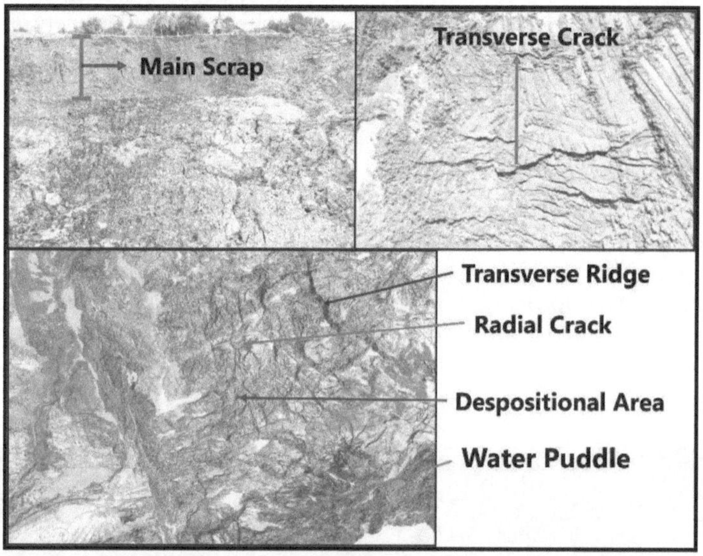

Figure 4. Landslide features in the authors research area in each section.

This section is a narrowing of the landslide area and is known as the body section of the landslide. From the photogrammetry analysis the authors find landslide features like minor scrap, transverse cracks, dan transverse ridge (Figure 4) in this section and indicated flow type landslide characterized by the presence of water puddles. In the toe section of the landslide or referred to as depositional area which the landslide energy begins to decrease and make them spread of the landslide material extends like a fan (radial crack). There are many water puddles in the toe section of the landslide. From the landslide anatomy with the help of photogrammetry the landslide type that occurred is Circular-Earth Flow type based on Varnes (1978) classification.

2.3 *Back analysis*

In normal circumstance research area mine waste dump slope has a good static Factor of Safety (FoS) value = 1.299 (Figure 5) using material properties as shown in Table 1 below. But in reality slope failure still occurred in this mine waste dump slope despite having a good FoS value. The foundation material and mine waste dump on the slopes are soil. This can be known from the UCS (uniaxial compressive strength) surface sample as well as sub-surface sample (wellsite) that have < 1 MPa from the laboratory test result.

Table 1. Slope material for calculate static FoS in authors research area.

Material Name	Unit Weight (kN/m^3)	Strength Type	Cohesion (kPa)	Phi (deg)
Waste Dump (Material Timbunan)	20.48	Undrained	99.48	-
Foundation Material (Material Insitu)	21.93	Mohr-Coulomb	779	28.65

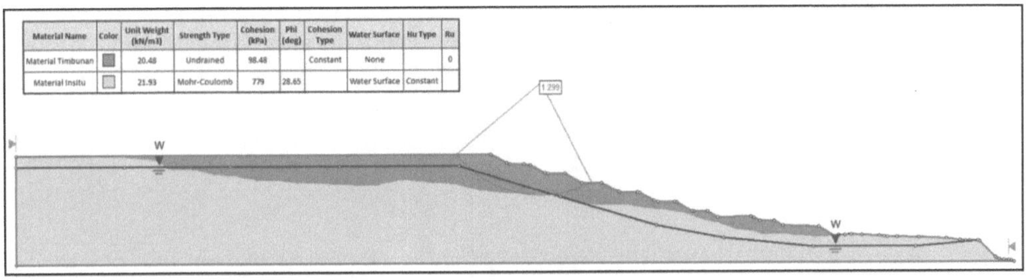

Figure 5. Static FoS analysis in authors research area with a FoS = 1.299.

Waste dump material is an undrained material which is material condition when the pore water is unable to drain out or the rate of loading is much quicker than the rate at which the pore water is able to drain out. When performing back analysis in mine waste dump slope of the research area, it is assumed that a landslide occurs when FoS = 1. With using probabilistic analysis in the waste dump material, the result is the closest possibility for the slope landslide failure is when the cohesion and unit weight of the materials decrease to 88.97 kPa and increase to 24 kN/m^3 (Figure 6-7).

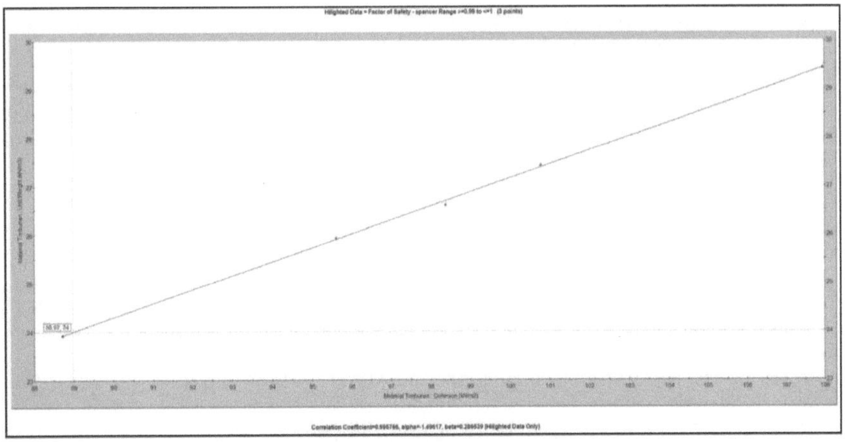

Figure 6. Probabilistic analysis to know the changes in research area material parameter values.

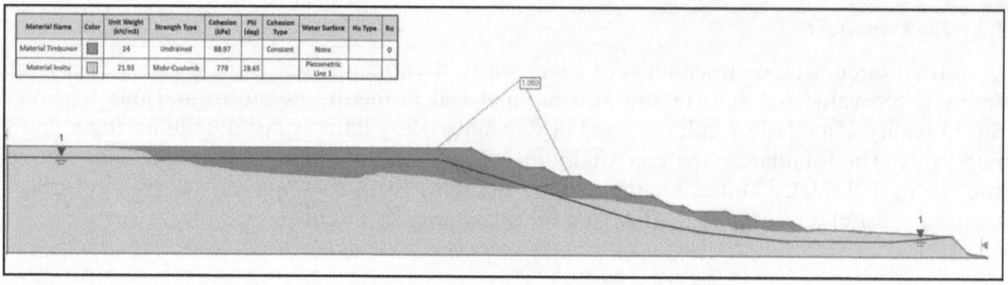

Figure 7. Back analysis in authors research area with assumption FoS = 1.

3 CONCLUSIONS

Mine waste dump slope failure back analysis gives a possibility decreasing or increasing material properties that makes slope failure. It also answers the hypothesis that the authors initially determined earlier in this research. In this case study the closest possibility for the slope landslide failure is when the cohesion and unit weight of the materials decrease to 88.97 kPa and increase to 24 kN/m^3 as shown in Table 2. This shows that the amount of water pressure from the recharge area that supply the slope body can reduce the strength and the slope factor of safety. This is also proven from the photogrammetry analysis with result Circular-Earth Flow complex landslide type that occurred in the mine waste dump slope of the research area. The material cohesion (c) is reduced by 9.54 % and increase the material unit weight value by 3.52 %.

Table 2. Slope material after slope failure has occurred (FoS = 1) from back analysis.

Material Name	Unit Weight (kN/m^3)	Strength Type	Cohesion (kPa)	Phi (deg)
Waste Dump (Material Timbunan)	24	Undrained	88.97	-
Foundation Material (Material Insitu)	21.93	Mohr-Coulomb	779	28.65

Slope material is very sensitive to the influence of groundwater or pore water pressure, considering that the material in the mine waste dump is dominated by soil (clay grain sized). The effect of groundwater which increases the unit weight will increase the weight of the waste dump on the slope body plus the effect of groundwater will reduce the binding power between particles (cohesion), so that the strength of the slope will be reduced and failure occurs.

Structural control like joint and fault are not found in the field, this is because the slope material is soil, not rock. The authors suggest that when handling mining slopes which are predominantly soil material, therefore the handling can be in the form of reducing the influence of groundwater on the slopes and good material compaction.

REFERENCES

Agam, M.W., Hashim, M.H.M., Murad, M.I., & Zabidi, H. 2016. Slope Sensitivity Analysis using Spencer's Method in Comparison with General Limit Equilibrium Method. *Procedia Chemistry 19 (2016)* Page 651 – 658.

Bishop, A. W. 1955. The use of the slip circle in the stability analysis of slopes. *Geotechnique 5, No. 1,* 7–17.

Fellenius, W., 1936. Calculation of the Stability of Earth Dams. *Proceedings of the Second Congress of Large Dams, Vol. 4,* pp. 445–463.

Janbu, N. 1954. Applications of Composite Slip Surfaces for Stability Analysis. *In Proceedings of the European Conference on the Stability of Earth Slopes, Stockholm, Vol. 3, p.* 39–43.
Read, J. & Stacey, P. 2009. Guidelines for open pit slope design. *Collingwood, Vic, CSIRO Publishing.* Australia.
Spencer, E. 1967. A Method of Analysis of Embankments assuming Parallel Inter-slice Forces. *Geotechnique, Vol 17 (1)*, pp. 11–26.
Supriatna, S., Sukardi & Rustandi, E. 1995. Geological Map of the Samarinda Quadrangle, Kalimantan. *Pusat Penelitian dan Pengembangan Geologi, Indonesia.*
Varnes, D.J. 1978. Slope Movement types and Processes. *Special Report Page 68–76.* Washington D.C.
Zhang, J., Tang, W.H., & Zhang, L.H. Efficient Probabilistic Back-Analysis of Slope Stability Model Parameters. *Journal of Geotechnical and Geoenvironmental Engineering Volume 136 Issue 1 - January* 2010

The Evolution of Geotech - 25 Years of Innovation – Hammah et al. (Eds)

Reliability analysis of soil nail walls using SRV and RLEM approaches

Seyed Masoud Asadollahi
Faculty of Civil Engineering, University of Tehran, Tehran, Iran

Moslem Rezvani
Msc, Geotechnical expert, Rayam Geotechnics, Iran

Ali Fakher
Faculty of Civil Engineering, University of Tehran, Tehran, Iran

Sina Javankhoshdel
Rocscience Inc. Toronto, ON, Canada

ABSTRACT: In geotechnical engineering, most of the parameters have an inherent variability which should be considered in the analysis and design to attain more reliable results. The object of this study is to investigate the influence of the variability of soil strength parameters on the variability of the factor of safety of soil nail walls using the probabilistic option in the Slide2 software. In this study, Single Random Variable approach (SRV) and spatial variability analysis have been adopted to carry out the probabilistic analysis. Moreover, the effect of soil nail bond strength variability (while soil strength parameters are constant), which causes the variability of the pullout resistance, on the variability of factor of safety is investigated. The results show that considering variability of parameters such as soil strength parameters and bond strength has a remarkable effect on the wall factor of safety distribution and therefore designers should not solely rely on deterministic analysis results. Also, it is concluded that spatial variability analysis shows more realistic results compared to SRV analysis.

1 INTRODUCTION

In recent years, deep excavations have become an inseparable part of constructions in urban areas due to the growth of population and land restrictions. Various stabilization methods have been proposed so far and among these, soil nailing system is one of the most effective and common ways to stabilize vertical or near vertical slopes.

Soil nail wall (SNW) is an in-situ reinforcement technique which is extensively used by practitioners to stabilize man-made or natural slopes due to its speedy construction, economical issues and well experiences from previous usages of this method in attaining the adequate margin of safety (Lin et al. 2017).

The global stability of SNWs is firstly evaluated in a Limit Equilibrium Method (LEM) software, which is one of the most accepted methods used to assess the stability of slopes, by assuming series of potential slip surfaces and calculating factor of safety (FoS) on each slip surface (Rawat et al. 2016)

Formerly, the FoS was determined in a deterministic manner by considering only a single value for model input parameters. Over time it turned out that most of the geotechnical

DOI: 10.1201/9781003188339-52

parameters have an inherent uncertainty and by considering them the results would be on the safe side. Therefore, the single random variable (SRV) probabilistic analysis became popular among practitioners. Phoon (2005) and Zhang (2015) have investigated the importance of taking model uncertainty into account.

The SRV probabilistic analysis has a major flaw which ignores the uncertainty and differing of parameters across locations in a single model. To resolve this problem, spatial variability analysis emerged and has been widely used by researchers (Javankhoshdel et al. 2017)

Zevgolis (2018) has evaluated the effect of soil parameters uncertainty on the stability of SNWs. Although several research have been done on probabilistic analysis of SNWs, the effect of different probabilistic analysis approaches on FoS has been barely investigated.

The aim of this study is to evaluate safety level of a SNW (global stability) by considering soil and nail strength parameters (e.g. cohesion, friction angle and bond strength) as variables under different design scenarios (e.g. SRV and Spatial variability analysis, uniform and non-uniform soil nail patterns and different grout injection pressure) to see how the parameters uncertainty will affect the margin of safety (minimum FoS value for temporary walls according to the standards) of the excavated slope.

2 METHODOLGY AND MATERIAL PROPERTIES

In order to investigate the effect of considering different variables under different design scenarios in SNWs, a simple wall with 10m height, vertical face and horizontal backfill was modeled in the Slide2 limit equilibrium commercial software (Rocscience 2020). Based on the Federal Highway Administration (FHWA) guidance, which is used as the design manual to design and construct SNWs, the typical height of SNWs ranges from 4 to 12m. Also, it has proposed that the soil nails should be installed above water table, therefore no water table was considered in the modelling process.

According to FHWA, the SNWs must be designed according to the allowable stress design (ASD) to satisfy the global minimum FoS equals to 1.35 and 1.5 for temporary and permanent walls, respectively. Yousef et al. (2017) suggested that the non-uniform soil nail patterns may be more beneficial in some cases, therefore as it is shown in Figure 1 both uniform and non-uniform soil nail length patterns were considered in this study as two scenario models to make a comparison between them.

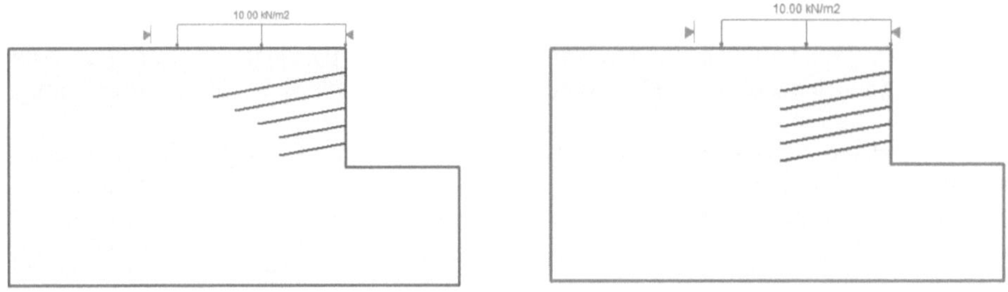

(a) (b)

Figure 1. Soil nail wall models: (a) non-uniform soil nail pattern, (b) non-uniform soil nail pattern.

For this study, a sandy soil with Mohr-Coulomb material model was chosen. It is worth mentioning that soil nail parameters were adopted in a way to attain minimum FoS = 1.35 of global stability in the deterministic analysis. Table 1 shows the design values used in this study.

Table 1. Model design values.

Parameters	Symbol	Value	Unit
Geotechnical Parameters			
Cohesion	*c*	7	kN/m^2
Friction angle	*φ*	29	°
Unit weight	*γ*	18	kN/m^3
Soil Nail Parameters			
Yield strength of nail	*fy*	400	N/mm^2
Nail spacing (horizontal × vertical)	*Sh*× *Sv*	1.5 × 1.5	m
Nail inclination	*I*	10	°
Ultimate bond strength (grout injection: gravity)	*qu*	300	kN/m^2
Ultimate bond strength (grout injection: under pressure)	*qu*	150	kN/m^2

FHWA has proposed values for the ultimate bond strength, which is one of the main factors that controls geotechnical failure of a soil nail, depending on the soil type and nail installation methods. On account of the fact that gravity injection can directly affect the ultimate bond strength, this study takes the design values of 150, 300 for gravity and under pressure grout injection respectively based on the FHWA recommendations.

3 PROBABILISTIC ANALYSIS WITH VARIABLE SOIL STRENGTH

Single random variable (SRV) and spatial variability analysis are two fundamental approaches which are mainly used to explore the uncertainty of parameters. In SRV method the value of soil property is randomly selected from the probability distribution of random variable, but only a single value allocates to the entire model in each iteration (Li 2016).

Spatial variability analysis takes the fluctuation of soil parameters in space into account. In this method each shear strength parameter (cohesion and friction angle) is generated using the local average subdivision method (LAS) developed by Vanmarcke (1997) and mapped onto the finite element mesh. Each node has different values of the soil property assigned to it, but nodes close to each other are correlated using horizontal and vertical correlation lengths (Javankhoshdel 2014) The spatial variability correlation is described by a parameter named "correlation length (θ)", which is a measure of the distance within which the properties are significantly correlated and there are no rapid changes in soil parameters within that distance. Higher correlation lengths imply the homogeneity of material while lower values imply the heterogeneity of soil material.

The probabilistic analysis was carried out in both approaches and for different scenarios with uniform or non-uniform soil nail patterns. Table 2 shows the soil strength variable parameters and their relative probabilistic parameters.

Table 2. Probabilistic design parameters.

Symbol	Mean Value	COV (used in present study)	COV (reported in literature)	Source
c	7 kPa	0.3	0.1-0.55	Phoon and Kulhawy (1999)
			0.3-0.5	Schneider and Schneider (2013)
φ	29°	0.15	0.05-0.15	Phoon and Kulhawy (1999)
			0.02-0.13	Duncan (2000)

Nguyen and Chowdhury (1985) have proposed that correlations between random values of shear strength parameters can influence the estimated probability of failure of slopes. These correlations are quantified by the cross-correlation coefficient (ρ or CC); therefore, in both SRV and random field probabilistic analysis different cross-correlation coefficient values were examined. In order to investigate the number of surfaces with FoS less than 1.35, the probabilistic analysis was carried out with 1000 number of samples and by using Latin-Hypercube sampling method due to its accuracy and efficiency.

Figure 2 shows the result of SRV probabilistic analysis for both uniform and non-uniform soil nail patterns under different cross-correlation coefficients.

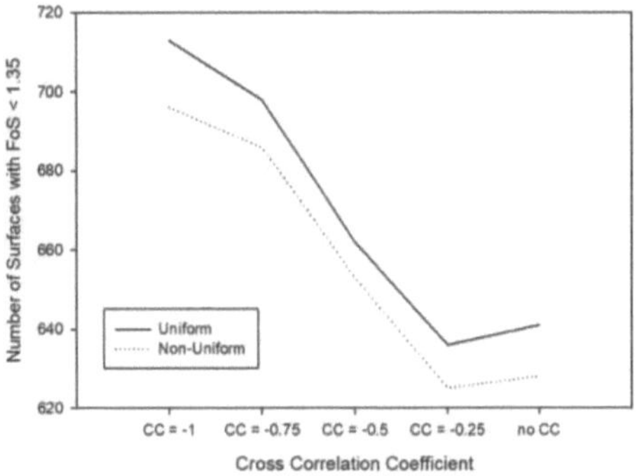

Figure 2. SRV analysis for both uniform and non-uniform soil nail patterns.

It can be noticed that more than half of the samples had FoS less than 1.35. Also, the difference between the results of uniform and non-uniform soil nail patterns was negligible. It can be seen in this figure that, by increasing cross-correlation coefficient from -1 to -0.25, the number of surfaces with FoS less than 1.35 decreases.

The results of the combination of the random field analysis and LEM approach known as random limit equilibrium method (RLEM) for both uniform and non-uniform soil nail patterns with different correlation lengths (CL) and cross-correlation coefficients are shown in Figure 3.

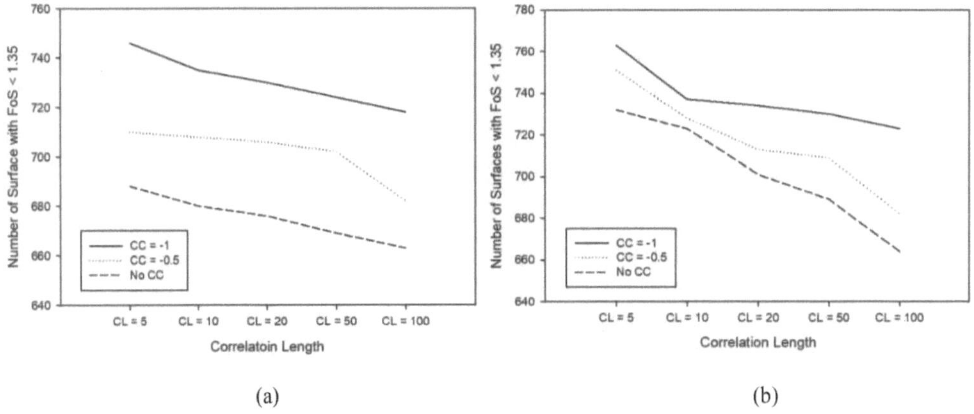

Figure 3. Random limit equilibrium method: (a) non-uniform soil nail pattern, (b) non-uniform soil nail pattern.

To investigate the effect of correlation length, different correlation lengths vary from 5m to 100m were adopted for RLEM analysis. As it is shown in Figures 3a and 3b, increasing the spatial correlation length decreases the number of surfaces with FoS below 1.35.

It is worth mentioning that in the RLEM analysis, the number of surfaces with FoS less than 1.35 was higher than SRV method in both uniform and non-uniform patterns, and increasing the spatial correlation length led to more surfaces with FoS less than 1.35 which is consistent with SRV analysis.

The reason for these behaviors for the SRV analysis with the cross-correlated cases and also the RLEM analysis with different spatial correlation length was due to the fact that increasing cross-correlation coefficient and also increasing the spatial correlation length both increase the level of uncertainty (make the FoS distribution wider). However, since the deterministic FoS is larger than the mean FoS (FoS = 1.35 is on the right-hand side of the mean FoS in Figure 4), widening the FoS distribution curve causes to have less cases on the left-hand side of the FoS = 1.35 which is shown with a vertical dashed line (See Figure 4)

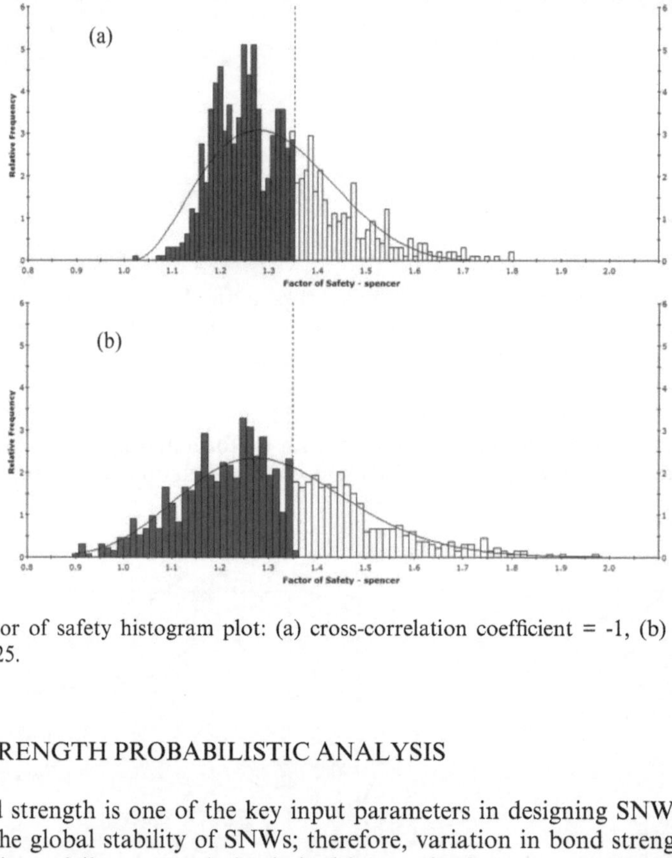

Figure 4. Factor of safety histogram plot: (a) cross-correlation coefficient = -1, (b) cross-correlation coefficient = -0.25.

4 BOND STRENGTH PROBABILISTIC ANALYSIS

Ultimate bond strength is one of the key input parameters in designing SNWs which is very important in the global stability of SNWs; therefore, variation in bond strength may lead to some instabilities or failure to reach the desired factor of safety.

In order to estimate the ultimate bond strength, two ways are adopted by designers. The first which is mostly common in the UK and North America is using tables which have recommended some values based on the soil type and the nail installation method. FHWA guideline has provided a table which was used as the reference for this study. The second is using closed-form numerical equations which incorporate nail geometry, surrounding strength parameters and confining stress level (Liu 2018).

Numerous factors including, ground conditions (e.g. soil type, surcharge load, degree of saturation), drilling method and grouting pressure directly affect the ultimate bond stress, and considering this parameter as a random variable is crucial for design purposes. Table 3 shows the adopted probabilistic values for bond strength probabilistic analysis.

Table 3. Ultimate bond strength probabilistic parameters.

Grouting Pressure	Mean Value	COV (used in present study)	COV (reported in literature)	Source
Gravity	150 kPa	0.15	0.152	Lazarte (2011)
Under Pressure	300 kPa	0.15	0.182	Devries (2013)
			0.23	Lin (2017)

In this analysis, the soil strength parameters were considered to be constant and equal to their mean values, also the sampling method and number of samples were selected the same as soil strength probabilistic analysis (section 3). The aim of this probabilistic analysis is to study the effect of the variation of ultimate bond strength under different scenarios including grouting pressure changing for both uniform and non-uniform soil nail patterns. Figure 5 shows the result of the analysis.

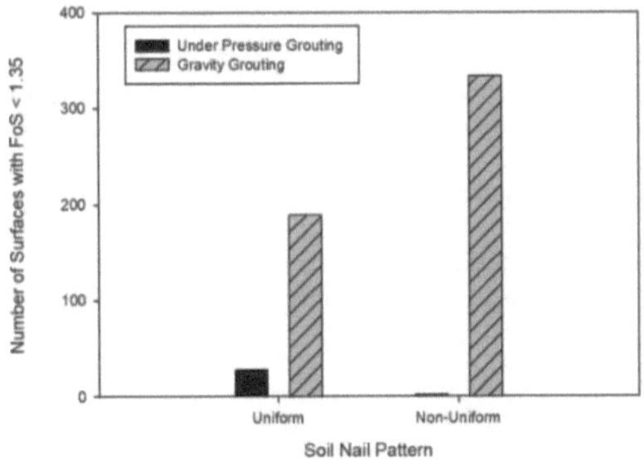

Figure 5. Bond strength probabilistic analysis.

The obtained results implies that in gravity grouting the variation of ultimate bond strength will lead to more surfaces with FoS less than 1.35 than under pressure grouting. The reason is that by decreasing the ultimate bond strength, the bond strength section in nail force diagram will be increased and since the force applied to a slip surface is determined by the point of intersection of force diagram and soil nail, less force may be applied to some slip surfaces related to under pressure grouting. In addition, the results showed that soil nail pattern does not directly influence the results.

5 CONCLUSION

In the current study, the variation of FoS against global stability analysis relative to the design value in temporary SNWs (FoS=1.35) under different scenarios was investigated. For this purpose, soil strength parameters (cohesion and friction angle) and ultimate bond strength were considered as variables in probabilistic analyses.

For soil strength probabilistic analysis, two different probabilistic methods (SRV and random field) with different cross-correlation coefficients and spatial correlation lengths were adopted. The results showed that random field analysis is more realistic and leads to more surfaces with FoS less than 1.35 than SRV analysis. Also, by increasing cross correlation coefficient and spatial correlation length, the number of surfaces with FoS less than 1.35 will decrease.

In bond strength probabilistic analysis gravity grouting and under pressure grouting were considered as two scenarios and it was concluded that gravity grouting is more prone in changing of ultimate bond strength. Therefore, to be on the safe side and in order to reduce the effect of changes in the ultimate bond strength, it is recommended to do grouting under pressure.

Above all, by conducting probabilistic analysis more than half of the samples were under design FoS which shows the importance of the probabilistic analysis.

REFERENCES

Duncan JM. Factors of safety and reliability in geotechnical engineering. Journal of geotechnical and geoenvironmental engineering. 2000 Apr;126(4):307–16

Devries B. LRFD Pullout Resistance Factor Calibration for Soil Nails Incorporating Survival Analysis and PLAXIS 2D.

Javankhoshdel, S. and Bathurst, R.J., 2014. Simplified probabilistic slope stability design charts for cohesive and cohesive-frictional (c-ϕ) soils. Canadian Geotechnical Journal, 51(9),pp.1033–1045.

Javankhoshdel, S. and Bathurst, R.J., 2017. Deterministic and probabilistic failure analysis of simple geosynthetic reinforced soil slopes. Geosynthetics International, 24(1),pp.14–29

Lazarte CA. Proposed specifications for LRFD soil-nailing design and construction. Transportation Research Board; 2011.

Lazarte CA, Robinson H, Gomez JE, Baxter A, Cadden A, Berg RR, Berg RR. Geotechnical engineering circular No. 7 soil nail walls-reference manual. National Highway Institute (US); 2015 Feb 1.

Li XY, Zhang LM, Gao L, Zhu H. Simplified slope reliability analysis considering spatial soil variability. Engineering Geology. 2017 Jan 12;216:90–7.

Lin P, Liu J, Yuan XX. Reliability analysis of soil nail walls against external failures in layered ground. Journal of Geotechnical and Geoenvironmental Engineering. 2017 Jan 1;143(1):04016077.

Liu H, Tang L, Lin P. Estimation of ultimate bond strength for soil nails in clayey soils using maximum likelihood method. Georisk: Assessment and Management of Risk for Engineered Systems and Geohazards. 2018 Jul 3;12(3):190–202.

Nguyen VU, Chowdhury RN. Simulation for risk analysis with correlated variables. Geotechnique. 1985 Mar;35(1):47–58.

Phoon KK, Kulhawy FH. Characterization of geotechnical variability. Canadian geotechnical journal. 1999 Nov 22;36(4):612–24.

Phoon KK, Kulhawy FH. Characterisation of model uncertainties for laterally loaded rigid drilled shafts. Geotechnique. 2005 Feb;55(1):45–54.

Rawat S, Gupta AK. Analysis of a nailed soil slope using limit equilibrium and finite element methods. International Journal of Geosynthetics and Ground Engineering. 2016 Dec;2(4):1–23.

RocScience Inc. 2020. Slide Version 2020 – 2D and 3D Limit Equilibrium Slope Stability Analysis. www.rocscience.com, Toronto, Ontario, Canada.

Schneider HR, Schneider MA. Dealing with uncertainties in EC7 with emphasis on determination of characteristic soil properties. Modern Geotechnical Codes of Practice. 2013:87–110.

Vanmarcke EH. Probabilistic modeling of soil profiles. Journal of the geotechnical engineering division. 1977 Nov;103(11):1227–46.

Youssef YG, Abdelrahman GE, Muhammed NE. PARAMETRIC STUDY FOR NON-UNIFORM SOILNAIL WALL.

Zevgolis IE, Daffas ZA. System reliability assessment of soil nail walls. Computers and Geotechnics. 2018 Jun 1;98:232–42.

Zhang DM, Phoon KK, Huang HW, Hu QF. Characterization of model uncertainty for cantilever deflections in undrained clay. Journal of Geotechnical and Geoenvironmental Engineering. 2015 Jan 1;141(1):04014088

The Evolution of Geotech - 25 Years of Innovation – Hammah et al. (Eds)

Probabilistic slope stability analysis of a case study using random limit equilibrium method and surface altering optimization

Sina Javankhoshdel, Brigid Cami, Terence Ma & Thamer Yacoub
Rocscience Inc., Toronto, ON, Canada

Reza Jamshidi Chenari
Geo-Engineering Center at Queen's, Kingston, Canada

ABSTRACT: Probabilistic analyses are often employed in slope stability analyses to account for the uncertainty associated with local variations in subsurface conditions. Practically, the consideration of spatially and stochastically varying soil properties is commonly achieved in combination with limit equilibrium methods to calculate the factor of safety. However, such analyses require advanced searching methods to locate critical slip surfaces with the minimum factor of safety. Surface altering optimization (SAO) is a novel searching method which minimizes the factor of safety by altering the geometry of a given slip surface using spline curves in 2D. By altering the points on the spline curves, the slip surface can be fine-tuned to trace the critical failure path in a slope. It is a local search algorithm that when combined with a coarser global search method, forms a powerful hybrid optimization technique to identify the critical slip surface. In this paper, a case study of the James Bay Dyke is presented whereby a probabilistic analysis is conducted on a slope with spatially varying shear strength. It was found that the use of random limit equilibrium method (RLEM) and SAO combined with a global search, provides accurate results while curtailing computational effort. It was also found that longer correlation lengths (coarser sampling resolution) can result in large regions of reduced material strength, leading to multiple failure modes and shorter slip surfaces. The results, which highlight multiple failure modes, were verified using the shear strength reduction (SSR) method.

1 INTRODUCTION

1.1 *2D Surface altering optimization*

The search for a critical slip surface is a global optimization problem in which the objective is to minimize factor of safety (F_S) through changing geometry of slip surfaces. Various search algorithms have been proposed in 2D slope stability applications.

Surface Altering Optimization (SAO) is a new technique to minimize factor of safety for a given slip surface by modifying its geometry. This method is a local optimization method. Although SAO can be used independently to find the critical slip surface in slope stability analysis, due to its local nature, it serves best when combined with a global search method to calculate factor of safety. In this way, the burden of finding an approximate geometry and location of the failure surface with minimum factor of safety is on the global search method, after which SAO is used to modify the geometry of that surface to further minimize the factor of safety.

DOI: 10.1201/9781003188339-53

1.2 *Spatial variability*

Probabilistic stability analysis results considering spatial variability of soil properties and using limit equilibrium method (LEM) have been reported in studies by Ji et al. (2012), Tabbaroki et al. (2013), Li et al. (2014), Javankhoshdel and Bathurst (2014) and Javankhoshdel et al. (2017).

Javankhoshdel et al. (2017) used a circular slip limit equilibrium method and random field theory called Random Limit Equilibrium Method (RLEM) to investigate the influence of spatial variability of soil properties on probability of failure. In RLEM, a given slope model containing soil layers is split into spatial regions in which the soil material parameters (e.g. cohesion, friction angle, etc.) in each region are individually sampled from a probabilistic distribution (e.g. Gaussian).

Cami et al. (2018) provided the results of comparison between the circular and non-circular RLEM approach and RFEM analysis. Mafi et al. (2020) conducted a similar comparison between a non-circular RLEM method with Monte-Carlo Optimization (MCO).

In this study, random and spatial variability of the soil shear strength (instead of cohesion and friction angle separately) have been investigated. The results of probabilistic analysis of a case study of a dyke located at James Bay in northern Canada are provided using the non-circular RLEM approach for different values of spatial correlation length. To confirm the factor of safety values and the location of the critical slip surfaces for the deterministic part of the analysis, these results have been compared to the same analysis using the more rigorous Finite Element Analysis.

2 SURFACE ALTERING OPTIMIZATION

2.1 *Structure of SAO*

Surface Altering is based on a sequence of transformations applied to the geometry of an input slip surface. Each SAO step is solved using the Bound Optimization BY Quadratic Approximation (BOBYQA) developed by Powell (2009). BOBYQA is a constrained derivative-free optimization method based on trust-regions (Wright and Nocedal 2006).

In two-dimensional slope stability analysis, a non-circular surface in its simplest form can be described as a linear spline curve. Coordinate values of control points will form the optimization input. As an example, Figure 1 illustrates a surface with 7 control points, yielding to 14 input variables to define x and y coordinates of the 2D surface. The geometry of the surface can be altered by modifying these coordinates. SAO offers a systematic set of steps to perform this alteration to minimize factor of safety such that it satisfies geometrical convexity and maintains the non-overlapping sequence of control points. These steps are repeated in multiple iterations until convergence criteria is met. A key consideration in SAO is to realize the surface geometry as a whole, such that changing coordinates of one control point is reflected on the other points to keep the surface convex and in order.

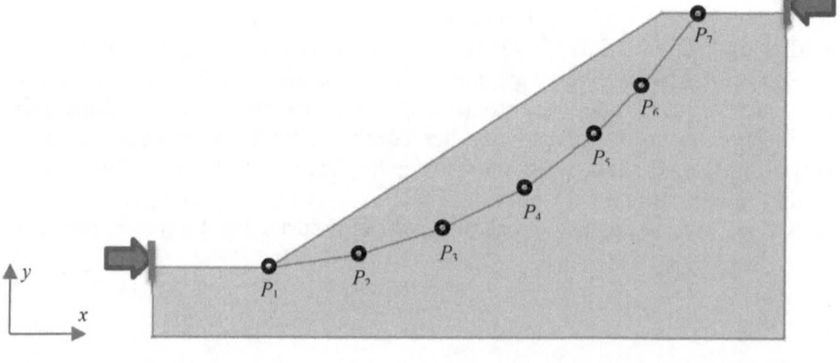

Figure 1. A linear spline 2D slip surface.

3 EXAMPLE JAMES BAY DYKE

The James Bay dyke presented by El-Ramly et al. (2001) was selected to provide baseline geometry and soil properties for this example. Soil properties, and the statistical properties of the soil parameters presented in this paper are the same as El-Ramly et al. (2001).

The metaheuristic Cuckoo Search method together with SAO approach was used in the deterministic and the RLEM analysis for this example.

The model and baseline material parameters used in the current example are shown in Figure 2.

Material Name	Color	Unit Weight (kN/m3)	Strength Type	Cohesion (kPa)	Phi (deg)	Cohesion Type	Allow Sliding	Water Surface	Ru
Embankment		20	Mohr-Coulomb	0	30			None	0
Clay Crust		18.8	Undrained	43		Constant		None	0
Marine Clay		18.8	Undrained	34.5		Constant		None	0
Lacustrine Clay		20.3	Undrained	31.2		Constant		None	0
Till		20	Infinite strength				Yes	None	0

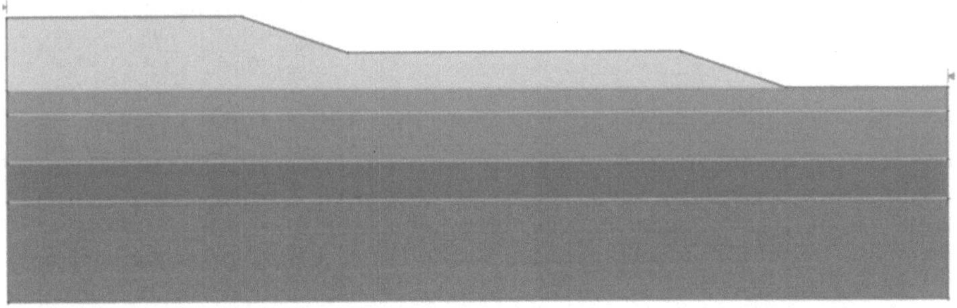

Figure 2. The James Bay dyke model and the material properties used in this example.

3.1 *Deterministic analysis*

In a deterministic slope stability analysis, the material parameters shown in Figure 2 are assumed to be constant values that do not vary within each material unit.

The critical slip surface computed via deterministic analyses using SAO is presented in Figure 3. The Spencer factor of safety for the case with SAO technique was $F_S = 1.17$.

A deterministic Shear Strength Reduction (SSR) analysis was also conducted to verify the LEM results. Figure 4 shows the maximum computed shear strain contours corresponding to the model, resulting in a shear strength reduction factor (SRF) of 1.22, which is in a good agreement with the results of LEM with optimization techniques. The surface shown in Figure 4 is also consistent with the critical non-circular LEM slip surface. The advantage of LEM method using SAO optimization compared to the FEM analysis is the computational time. FEM analysis uses SSR approach which is a lengthy process, while combination of LEM and SAO is much faster. This is especially very important when carrying out probabilistic analysis in which thousands of realizations are required.

3.2 *RLEM analysis*

Table 1 shows the random variable parameters for the shear strength of each material used in this example. Normal (Gaussian) distributions are assumed for all random variables. The clay crust and till materials were assumed to have the constant material properties defined in Figure 2.

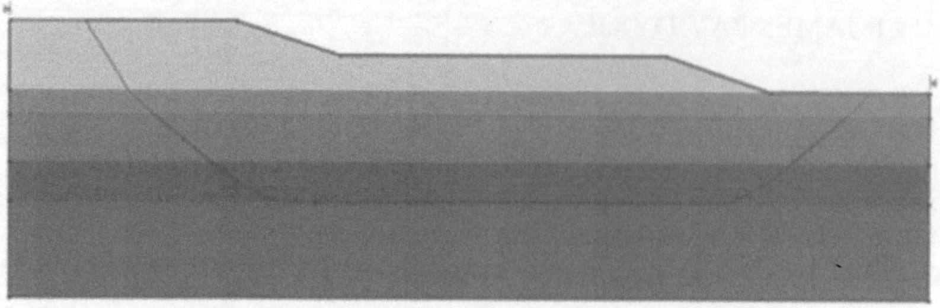

Figure 3. Critical slip surface geometry from deterministic analyses using SAO. $F_S = 1.17$.

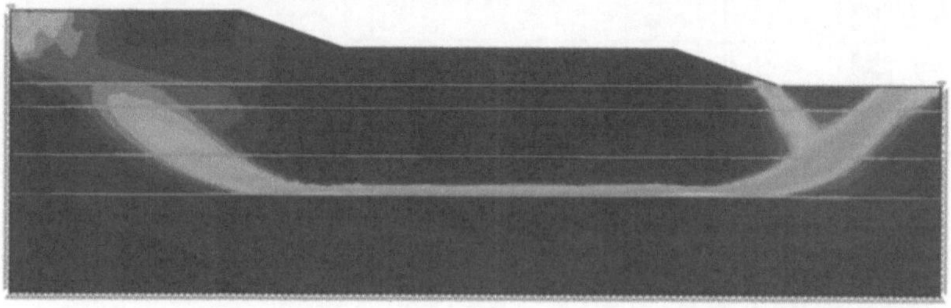

Figure 4. Maximum shear strain contours from an example SSR analysis. SRF = 1.22.

Table 2 summarizes the probability of failure from non-circular RLEM analyses with SAO algorithm. In these analyses, the soil properties were expressed as isotropic random fields with the spatial variability in the horizontal and vertical directions. Three different cases of spatial correlation lengths (5, 15, and 30 m) are considered. For each case, 10,000 Latin Hypercube simulations were used.

Figure 5 shows the failure surface bands generated during the RLEM analyses for the three different spatial correlation length cases. The deterministic failure surface with $F_S = 1.17$ presented in Figure 3 is also shown. It can be seen in these figures and in the results in Table 2, that as the spatial correlation length increases, probability of failure increases and there are more failure surfaces with shorter lengths appearing in the middle of the model. For the case where the spatial correlation length is highest, as shown in Figure 5c, it is more likely for very large regions of soil to sample weak parameters, which makes failure through the corresponding regions possible. In contrast, for the smallest spatial correlation length shown in Figure 5a, local reductions to the material strength are likely to be offset by stronger neighbouring regions, making failure through the middle of the model less probable.

Table 1. Random variable parameters for shear strength (normal distributions are used with all variables).

Material	Property	COV (Coefficient of Variation)
Embankment	shear strength	0.1
Marine Clay	shear strength	0.5
Lacustrine Clay	shear strength	0.3

Figure 6 shows two random field cases for shear strength with spatial correlation lengths of 5m for two different samples and their corresponding factors of safety. In this figure the shear strength of the soil is plotted in colour (note that the top two layers and bottom white layer

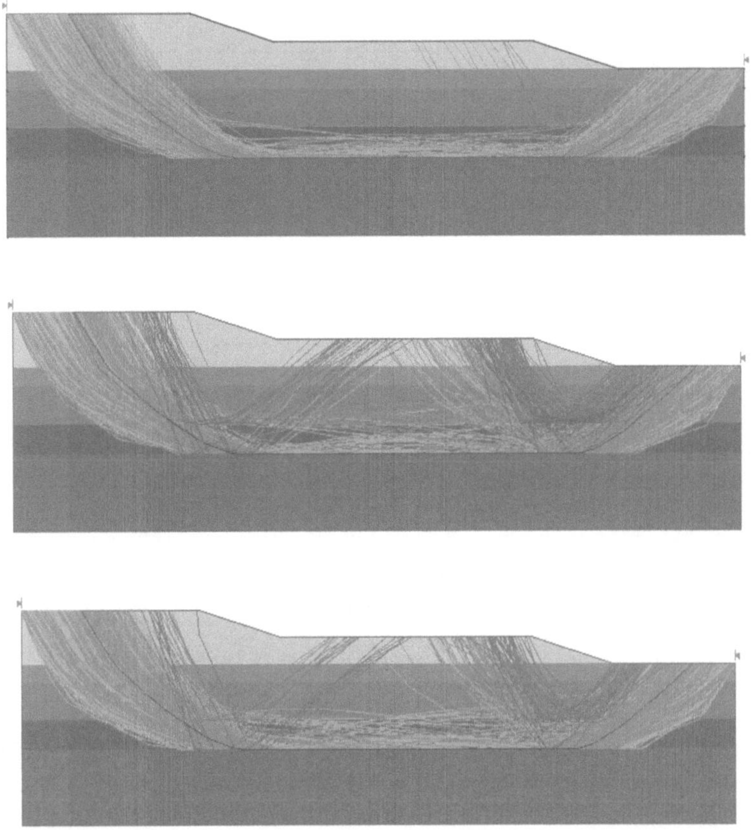

Figure 5. Envelope of critical slip surfaces for probabilistic analyses using RLEM with the deterministic FS = 1.17 and spatial correlation length of a) 5 m, b) 15 m, c) correlation length of 30 m.

Table 2. RLEM analyses results.

Parameter\Opt.	RLEM 5 m	RLEM 15 m	RLEM 30 m
Deterministic F_S^*	1.17	1.17	1.17
Mean F_S^*	1.08968	1.052	1.04
P_f (%)	9.23	23.35	29.66

* The factor of safety was calculated using the Spencer method

are deterministic). Figure 6a shows the shorter failure surface with the corresponding F_S value of 0.905, while Figure 6b is the case with the longer slip surface and Fs = 1.006. It can be seen in Figure 6 that the longer slip surface corresponds to a higher factor of safety. Thus, for the cases with larger spatial correlation lengths there are more short failure surfaces, and hence more failure surfaces, resulting in a higher probability of failure.

Another interesting observation in Figure 6 is that the failure surfaces in both Figure 6a and Figure 6b passed through the weakest material (blue region). This observation shows the strength of the combination of Cuckoo search and SAO to find the weakest failure path in the RLEM analysis with spatially variable soil layers.

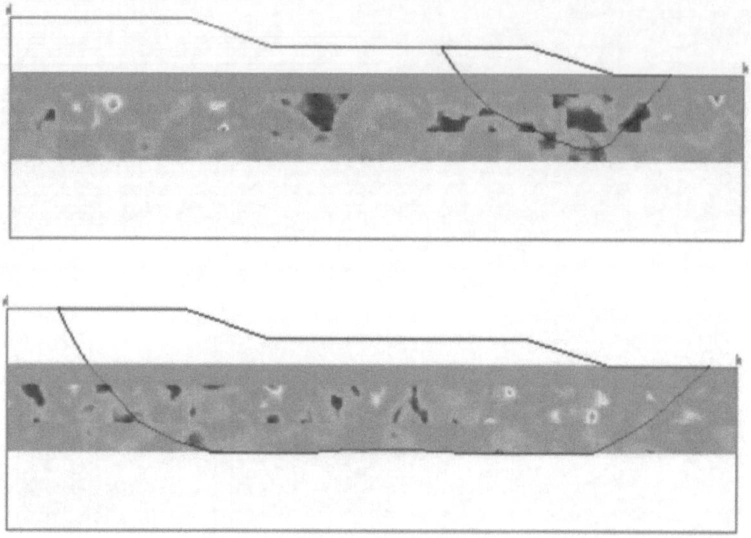

Figure 6. RLEM analysis for a) long spatial field (deterministic $F_S = 0.905$) with correlation length of 30 m, b) for short spatial field ($F_S = 1.006$) with correlation length of 5 m.

4 CONCLUSIONS

This study has demonstrated that the accurate identification of critical slip surfaces can be achieved via combination of the Random Limit Equilibrium Method (RLEM) for spatially and randomly varying material properties with Surface Altering Optimization (SAO). The James Bay Dyke case study was used as a numerical example, whereby multiple failure modes were identified by the RLEM. For longer correlation lengths (coarse sampling resolution), the multiple failure modes were more likely due to reductions in strength within large regions of the model.

REFERENCES

Cami, B., Javankhoshdel, S., Lam, J., Bathurst, R.J. and Yacoub, T. 2017. Probabilistic analysis of a tailings dam using 2D composite circular and non-circular deterministic analysis, SRV approach, and RLEM. 70th Canadian Geotechnical Conference, Ottawa, Ontario, 7 p.

El-Ramly, H. 2001. Probabilistic analyses of landslide hazards and risks: Bridging theory and practice. Ph.D. thesis, University of Alberta, Edmonton, Alta.

Javankhoshdel, S. and Bathurst, R.J. 2014. Simplified probabilistic slope stability design charts for cohesive and c-phi soils. Canadian Geotechnical Journal 51(9):1033–1045

Javankhoshdel, S., Luo, N. and Bathurst, R.J. 2017. Probabilistic analysis of simple slopes with cohesive soil strength using RLEM and RFEM. Georisk (online).

Ji, J., Liao, H.J., Low, B.K., 2012. Modeling 2D spatial variation in slope reliability analysis using interpolated autocorrelations. Computer and Geotechnics, 40: 135–146.

Li, D.Q., Qi, X.H., Phoon, K.K., Zhang, L.M., and Zhou, C.B. 2014. Effect of spatially variable shear strength parameters with linearly increasing mean trend on reliability of infinite slopes. Structural safety, 49: 45–55.

Mafi, R., Javankhoshdel, S., Cami, B., Jamshidi Chenari, R. and Gandomi, A.H., 2020. Surface altering optimisation in slope stability analysis with non-circular failure for random limit equilibrium method. *Georisk: Assessment and Management of Risk for Engineered Systems and Geohazards*, pp.1–27.

Powell, M.J., "The BOBYQA algorithm for bound constrained optimization without derivatives," University of Cambridge, Cambridge, 2009.

Tabarroki, M., Ahmad, F., Banaki, R., Jha, S. K., & Ching, J. (2013). Determining the factors of safety of spatially variable slopes modeled by random fields. Journal of Geotechnical and Geoenvironmental Engineering, 139(12), 2082–2095.

Wright S.J., and Nocedal, J. 2006. Numerical Optimization, Springer New York.

The Evolution of Geotech - 25 Years of Innovation – Hammah et al. (Eds)

Simulating the topography induced stresses using 3D numerical modeling

W. Ali
School of Civil Engineering, Universiti Teknologi Malaysia, Johor Bahru, Malaysia
Department of Mining Engineering, Balochistan University of Information Technology, Engineering and Management Sciences, Quetta, Pakistan

R.A. Abdullah
School of Civil Engineering, Universiti Teknologi Malaysia, Johor Bahru, Malaysia

ABSTRACT: The stress induced instabilities are commonly observed in tunneling projects. High in situ stresses can pose challenging and potentially dangerous ground conditions for the tunneling crew and can lag the project completion. There are several factors which contribute to the in situ stress at a site. In areas of very rugged topography, the in situ stress is often subjected to topographic perturbations. In this study the topography along the headrace tunnel in the Neelum Jhelum Hydropower project (Pakistan) is selected for analysis of the topography induced stresses. Digital elevation model of the project site was generated. The 3D real topography model was then run in RS3. The principal stresses magnitudes and orientations were determined by solving the modeled stress tensors. Then the tunnel was simulated under the prevailing rock and stress conditions. Analysis of the results showed failure zone depths consistent with the actual observations in the tunnel.

1 INTRODUCTION

Estimating the magnitude and orientation of in situ stresses is an essential part of all underground construction projects. Knowing the material properties and in situ stress level can be very beneficial to the engineer both in the design and construction stages. Stress-induced failures appear in the form of squeezing in weak rocks (Shrestha, 2006, Singh et al., 2007) and spalling and rock bursting (Broch and Sørheim, 1984, Kaiser et al., 1996) in brittle rocks.

The in situ state of stress is under the influence of several factors such as gravity, plate tectonics, topography, erosion, rock fabric, and glaciation (Amadei and Stephansson, 1997). Among these factors, the influence of current topography and sufficient knowledge of the rock material properties can be easily used in three-dimensional numerical models to get a reasonable estimate of the in situ stress at a certain depth. In this study, the Neelum Jhelum hydropower project in the Azad Kashmir region of Pakistan is selected as the project area. This area is characterized by very rugged topography with steep mountains and deep valleys. The twin TBM 8.5 m diameter headrace tunnels are crossing through the mountain with a maximum overburden of 1870 m. Frequent rock bursts were observed during the excavation of the headrace tunnels (Naji et al., 2019).

In this study digital elevation model of the project site is generated to simulate the in situ stress along the tunnel alignment using 3D finite element modeling in RS3. No horizontal displacement or strain was applied to the model, i.e., field stress type was set to gravity. After computing, the stress tensors were obtained by taking line query along the tunnel axis. The

DOI: 10.1201/9781003188339-54

magnitude and orientations of the principal stresses were obtained by solving the stress tensors for an 8000 m length of the tunnel. However, for comparing the modeled stresses with the actual conditions, a 1000 m long section of the headrace tunnel where very frequent rock bursting was recorded was selected. For this purpose, a separate model was generated to simulate the tunnel excavation process in RS3. The average values of the magnitude and orientation of principal stresses derived from the gravitational model were used as constant loading in this model. The vulnerability of the rock to brittle failure and rock bursting was established by comparing the rock material strength and modeled stress level with brittle failure criteria developed by Cai and Kaiser (2014), (Diederichs et al., 2004).

2 FINITE ELEMENT IN SITU STRESS MODELING

2.1 *Intact rock and rock mass properties*

The project site geology belongs to the Murree Formation, composed mainly of molassic sedimentary rocks of Paleocene and Eocene ages. Geographically the project site is part of the Sub Himalayan range. The main lithology of the Murree Formation is composed of strong grayish sandstone, reddish-brown sandstone and siltstone, reddish-brown mudstone, and rare shale. However, alternating beds of sandstone, siltstones, and mudstones were frequently observed during the headrace tunnel excavation. Five rock units were defined for the tunneling and underground works namely, Sandstone-1 (SS-1), Sandstone-2 (SS-2), Siltstone, Mudstone, and Shale. Neelum Jhelum Consultants conducted detailed laboratory testing. In this study, we used these laboratory test results as input into the RocData version 5.009 (Rocscience, 2020a) to determine the corresponding rock mass parameters (see Table 1). The output from RocData software is listed in Table 2.

Table 1. Rock material properties used as input into the RocData software.

Parameters	Mean values
Uniaxial compressive strength (MPa)	60.25
Geological Strength Index	53.75
Hoek-Brown constant "mi"	12.5
Young's Modulus (GPa)	21.6
Poisson's Ratio	0.27
Unit weight (MN/m^3)	0.027
Disturbance factor	0

Table 2. The output from RocData software.

Parameters	Mean values
Hoek-Brown constant mb	2.418
Hoek-Brown constant s	0.006
Hoek-Brown constant a	0.504
Uniaxial compressive strength of rock mass (Mpa)	4.58
Rock mass deformation modulus (GPa)	8.37

2.2 *Model geometry setup*

To simulate the influence of mountain topography on the in situ stress along the tunnel axis, a digital elevation model (DEM) of the project site is used in this study. The digital elevation data was downloaded from the Earthexplorer (USGS, 2020). The data were further processed in ArcMap 10.3 to get the coordinates and elevations for our desired model boundary. The coordinates and elevation data were then used as input in Slide3 (Rocscience, 2020c), and thus a 3D surface of the project site was obtained. We selected an 8000X8000 m (length X width) area as our model boundary. The boundary extent is symmetric on both sides of the tunnel. The vertical dimensions of the model vary because of the uneven topography. The vertical boundaries are extended by 2000m below the tunnel axis elevation so that the boundary conditions may not affect the stress evaluations at the tunnel level (Figueiredo et al., 2014) (see Figure 1 in section 2.3).

2.3 *In situ stress simulation*

The model boundary generated in Slide 3 was imported into the RS3 (Rocscience, 2020b) for finite element analysis. The model geometry was discretized by using 3.6 million 4-noded tetrahedra elements. A 100 m uniform element size was used for the boundary lying above the tunnel axis. The element size is then gradually increased to 200 m and 400 m below the tunnel alignment as shown in Figure 1. The generalized Hoek and Brown failure criterion (Hoek et al., 2002) with elastic material type was selected. After computing, a line query was taken along the tunnel axis to determine the state of stress before the excavation of the tunnel. Stress tensors were obtained at every 20 m along the tunnel axis. The stress tensors were solved in Mathematica (Wolfram Research, 2021) and the corresponding eigenvalues and eigenvectors were obtained. Thus, the magnitude and orientation of major (σ_1), intermediate (σ_2), and minor (σ_3) principal stresses were determined (See Table 3). The mean values of the principal stresses magnitude and orientations with the corresponding standard devaiton are listed in Table 3.

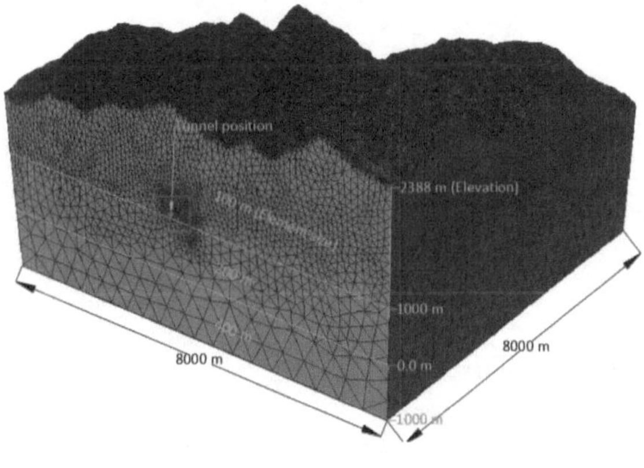

Figure 1. Model geometry and discretization.

Table 3. Modeled magnitude and orientation of principal stresses.

Principal stress	Magnitude (MPa) [Mean ± St. dev]	Plunge (°) [Mean ± St. dev]	Trend (°) [Mean ± St. dev]
Major principal stress	40.35 ± 0.19	3 ± 0.46	120 ± 1.65
Intermediate principal stress	35.55 ± 0.26	63.8 ± 1.54	36.2 ± 3.02
Minor principal stress	32.11 ± 0.18	26 ± 1.52	208.7 ± 1.52

2.4 *Simulating the influence of in situ stress on the tunnel*

To simulate the tunnel excavation and analyze its stability under the prevailing rock and stress conditions, a section of the headrace tunnel where frequent rock bursting was recorded was selected. The width, height, and length of the model boundary were set to 120, 120 and 70m respectively (see Figure 2). A 30m length of the tunnel was extruded in 20 stages. The model was discretized using 4-noded tetrahedra elements using graded mesh. A mesh refinement region of 1.5 times the tunnel diameter was selected. A 0.5 m uniform element size was used to discretize the refined region. Massive Sandstone-1 was the dominant rock encountered in this section with a GSI value of nearly 100 and a UCS of 86 MPa. Elastic analysis with Hoek-Brown brittle parameter m = 0 and s = 0.11 as per recommendations of Martin et al. (1999) was conducted. This approach was also recommended by Kaiser (2016) for GSI>65. The mean values of principal stress magnitudes and orientations determined from the gravitational model listed in Table 3 were used as constant loading in this model.

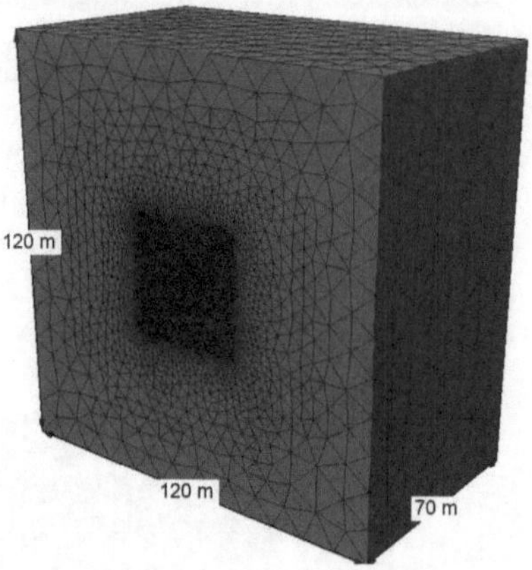

Figure 2. Model geometry for simulating the influence of in situ stress on the tunnel.

3 RESULTS AND DISCUSSION

The magnitude and orientation of principal stresses are plotted against the tunnel chainage in Figure 3 and Figure 4, respectively. A gradual decrease in principal stresses with the decrease in the overburden is evident from Figure 3. This change in magnitude is logical. However, the plunge and trend angles of principal stresses are not uniform. An abrupt change in the orientation of principal stresses can be observed in Figure 4. This change in orientation is attributed to the more uneven topography. The topography is nearly uniform on both sides of the tunnel up to approximately tunnel chainage 6680 to 8000m. In this section of the tunnel, the orientation of principal stresses is also uniform. However, beyond 8000m, the tunnel is passing below the valley bottom. Although the overburden was still in the range of 800-1600m, it was revealed that it is still under the influence of ridges on both sides of the tunnel.

To determine the over-stressing surrounding the tunnel, the Strength Factor (SF), which is the ratio of rock mass strength to induced stresses, is used. The SF < 1 implies the rock is failed or overstressed. In the Neelum Jhelum TBM tunnels, most of the rock bursts occurred approximately 4-5m behind the TBM shield (Jack˙Mierzejewski and Ashcroft, 2017).

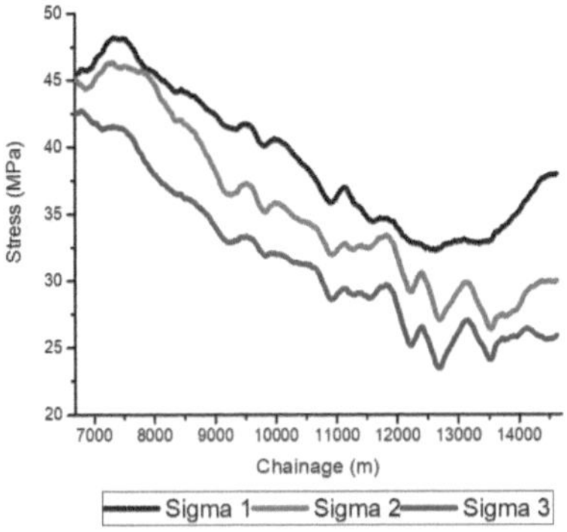

Figure 3. Modeled magnitudes of principal stresses.

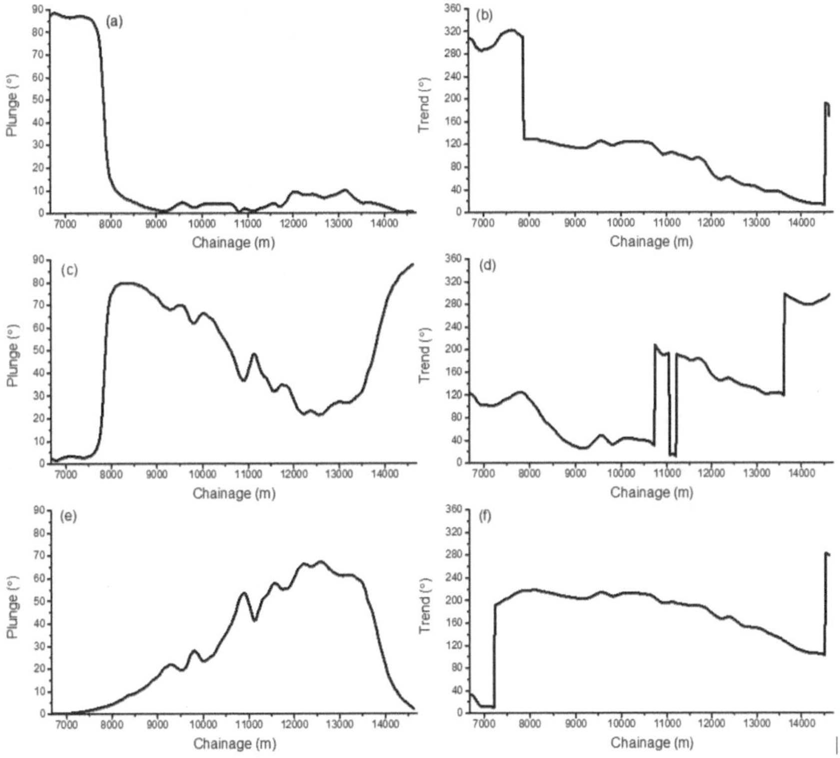

Figure 4. The orientation of principal stress derived from modeled stress tensors.

Contours of SF 5 m behind the tunnel face are shown in Figure 5a. Large zones of overstressing can be observed. The modeled depth of failed rock in the crown/invert and side walls is 2.59 m and 1.96 m, respectively. Since spalling is the development of extension fractures near the boundary of an excavation which is under compression (Diederichs et al., 2010), the larger failure depth shown in Figure 5a can practically be associated with the strain burst or rockburst. Also, the model results are nearly consistent with the actual damage depth in the tunnel section, which was subjected to rock bursting, as shown in Figure 5b.

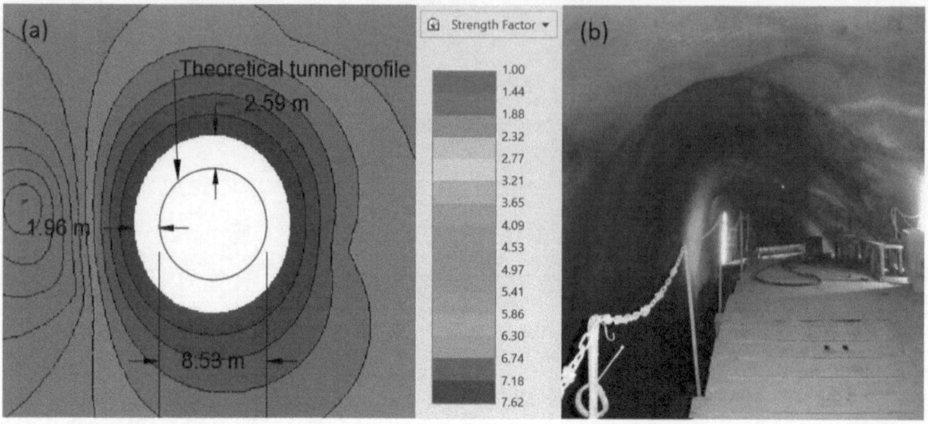

Figure 5. (a) Modeled tunnel profile showing the depth of failure depth, (b) TBM tunnel subjected to rock bursting (Bawden, 2015).

4 CONCLUSIONS

The following conclusions were drawn from the numerical analysis;

1. The in situ states of stress are significantly perturbed in regions with uneven topography. Particularly the orientation of principal stresses is more susceptible to change due to uneven topography.
2. The digital elevation models, in combination with 3D numerical modeling, presents a decent method to estimate the topographic perturbations of the in situ stresses in a project site.
3. The in situ stress during the design stage of an underground excavation can be better characterized by an integrated method of in situ stress measurement and numerical modeling of the 3D real topography.

ACKNOWLEDGEMENT

The authors would like to acknowledge for the financial funding from Higher Education Commission of Pakistan and Collaborative Research Grant National, Universiti Teknologi Malaysia (PY/2019/03147).

REFERENCES

Amadei, B. & Stephansson, O. 1997. *Rock stress and its measurement*, Springer Science & Business Media.

Bawden, W. 2015. Neelum Jhelum Hydroelectric Project Rockburst Investigation Report.

Broch, E. & SØRHEIM, S. 1984. Experiences from the planning, construction and supporting of a road tunnel subjected to heavy rockbursting. *Rock Mechanics and Rock Engineering*, 17, 15–35.

CAI, M. & KAISER, P. 2014. In-situ rock spalling strength near excavation boundaries. *Rock mechanics and rock engineering*, 47, 659–675.

Diederichs, M., Carter, T. & Martin, C. Practical rock spall prediction in tunnels. Proceedings of ITA World Tunnel Congress, 2010. 1e8.

Diederichs, M., Kaiser, P. & Eberhardt, E. 2004. Damage initiation and propagation in hard rock during tunnelling and the influence of near-face stress rotation. *International Journal of Rock Mechanics and Mining Sciences*, 41, 785–812.

Figueiredo, B., Cornet, F., Lamas, L. & Muralha, J. 2014. Determination of the stress field in a mountainous granite rock mass. *International Journal of Rock Mechanics and Mining Sciences*, 72, 37–48.

Hoek, E., Carranza-Torres, C. & Corkum, B. 2002. Hoek-Brown failure criterion 2002 edition. *Proceedings of NARMS-Tac*, 1, 267–273.
Jack Mierzejewski, G. & Ashcroft, B. Short-term rockburst prediction in tbm tunnels. Worl Tunnel Congress, 2017. 10.
Kaiser, P. 2016. Challenges in rock mass strength determination for design of deep underground excavations. *ISRM on-line lecture (45 min)* https://www.isrm.net/gca.
Kaiser, P. K., Mccreath, D. & Tannant, D. 1996. *Canadian rockburst support handbook*, Geomechanics Research Center.
Martin, C., Kaiser, P. & Mccreath, D. 1999. Hoek-Brown parameters for predicting the depth of brittle failure around tunnels. *Canadian Geotechnical Journal*, 36, 136–151.
Naji, A. M., Emad, M. Z., Rehman, H. & Yoo, H. 2019. Geological and geomechanical heterogeneity in deep hydropower tunnels: A rock burst failure case study. *Tunnelling and Underground Space Technology*, 84, 507–521.
Rocscience 2020a. RocData. 5.009 ed. Canada.
Rocscience 2020b. RS3. 4.01 ed.
Rocscience 2020c. Slide3. 3.006 ed.
Shrestha, G. L. 2006. Stress induced problems in Himalayan tunnels with special reference to squeezing.
Singh, M., Singh, B. & Choudhari, J. 2007. Critical strain and squeezing of rock mass in tunnels. *Tunnelling and underground space technology*, 22, 343–350.
USGS. 2020. *EarthExplorer* [Online]. Available: https://earthexplorer.usgs.gov/ [Accessed Novemeber 18, 2020 2020].
Wolfram Research, I. 2021. *Mathematica Online* [Online]. Wolfram Research, Inc. Available: https://www.wolfram.com/mathematica [Accessed December, 2020].

Session 12 - Numerical slope stability analysis II

The Evolution of Geotech - 25 Years of Innovation – Hammah et al. (Eds)

Consistent evaluation of slope safety factors between different strength criteria

Johan Clausen
Department of Civil and Architectural Engineering, Aarhus University, Denmark

ABSTRACT: Traditionally, safety factors in geotechnical engineering are defined by the largest possible reduction in shear strength at incipient failure. In numerical, elasto-plastic finite element calculations this definition is straight-forward to implement for the standard Mohr-Coulomb model. However, for models defined by non-linear Mohr envelopes it is not straight-forward to implement the standard definition of the safety factor. The problem being that material models in numerical codes are usually defined in cartesian stress space rather and by Mohr envelopes. This paper presents a consistent implementation of slope safety factors for elasto-plastic simulations using non-linear Mohr envelopes. The presented method renders comparisons of safety factors across different material models meaningful, and also allows for having different non-linear material models in the same calculation domain when performing shear strength reductions for obtaining the safety factor.

1 INTRODUCTION

The evaluation of slope safety is usually performed by calculating the safety factor. Several safety factor definitions exist. An example is a factor defined by the ratio of the stabilizing moment to driving moment of an assumed circular failure mode. In finite element calculations of slope safety factors, the most common definition of slope safety is that of

$$F = \frac{\text{Actual shear strength}}{\text{Reduced shear strength at incipient failure}} \tag{1}$$

For the classical Mohr-Coulomb material model this corresponds to the so-called phi-c reduction method, Zienkiewicz et al. (1975). Graphically, this corresponds to lowering the Mohr-Coulomb strength envelope in $\sigma - \tau$ space, as seen in Figure 1. Here σ and τ are the normal and shear stress, respectively, on a given plane. The procedure for determining the safety factor, F, in elasto-plastic finite element simulations is given in Table 1. For the Mohr-Coulomb model this method presents no difficulties, as this model is easily translated from $\sigma - \tau$ space,

$$S_{\mathrm{MC}} = c - \sigma \tan \varphi \tag{2}$$

into cartesian stresses,

$$f_{\mathrm{MC}} = k\sigma_1 - \sigma_3 - \sigma_c, \text{ with } k = \frac{1+\sin\varphi}{1-\sin\varphi}, \ \sigma_c = 2c\sqrt{k} \tag{3}$$

DOI: 10.1201/9781003188339-55

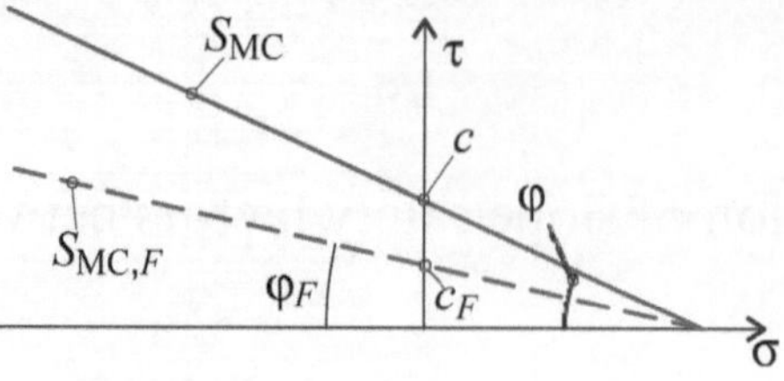

Figure 1. The original Mohr-Coulomb envelope and the resulting envelope after applying the safety factor, *F*.

Table 1. Procedure for calculation of slope safety factor.

1.	The current safety factor F_i is chosen, $Fi > F_{i-1}$
2.	Reduced shear strength is established from F_i
3.	Equilibrium iterations are performed:
	IF global equilibrium can be satisfied go to step 1.
	ELSE
	Take the factor of safety as $F=Fi$
	END IF

The constant friction angle is denoted φ and the cohesion *c*. The reduced strength parameters needed according to the method outlined in Table 1 is then found from Equation 2 as

$$\tan \varphi_i = \frac{\tan \varphi}{F_i} \text{ and } c_i = \frac{c}{F_i} \tag{4}$$

Experimental evidence suggests that the behaviour of frictional soils is better captured by models exhibiting stress dependent friction angle, especially in the low stress regime, Jiang et al. (2003); Baker (2004); Ottosen and Ristinmaa (2005); Anyaegbunam (2015). The low stress regime is especially relevant in slope safety calculations, due to the often shallow failures. Due to these reasons, a number of material models with stress dependent friction angles have been developed in cartesian stresses, Hoek and Brown (1980, 1997, 2019); Krabbenhøft et al. (2016, 2012); Lees and Clausen (2020). Models formulated in cartesian stresses are attractive from the point of finite element implementation. In cartesian stresses, however, the direct definition of the factor of safety as a reduction of the shear strength vanishes. The simple slope safety factor definition can be retained if non-linear Mohr envelopes are used. Several forms have been proposed in the literature, e.g. Jiang et al. (2003); Baker (2004); Hoek and Brown (1997); Anyaegbunam (2015); Clausen (2020), of which two will be used for demonstration in Section 2 of the present paper. The problem with material models formulated as non-linear Mohr envelopes is that they can not, in general, be translated into cartesian stress space which is needed for standard elasto-plastic finite element implementation.

1.1 *Elasto-plastic finite element calculations using mohr envelopes*

Previously, calculations with strength behaviour defined by Mohr envelopes, were restricted to bearing capacities only, i.e. settlements were not calculated. On the other hand, slope safety factor assessment with non-linear models based on yield functions in cartesian coordinates, e.g. the Hoek-Brown model, are not based on shear strength but rather on reduction on some chosen material parameters. This method renders comparison of safety factor across different material models meaningless. In a recent paper, Clausen (2020), a method for combining strength defined by Mohr envelopes and standard elasto-plastic finite element methodology was presented. This allows for applying Mohr envelopes on equal basis with strength defined by yield functions, even within the same calculational model. As mentioned in the previous section, this has the great advantage of rendering slope safety of different material models directly comparable, again also within the same calculational model, if several soil, or rock mass, types are present in the domain. In the following sections two non-linear Mohr envelopes are presented which will be used for slope safety factor calculation by reducing the shear strength in a standard elasto-plastic setting, as given in Table 1.

2 MATERIAL MODELS

In the next section, strength reduction is performed using the standard elasto-plastic Mohr-Coulomb model, along with two different non-linear Mohr envelopes, which will be presented in the following. In the calculations used in this paper, associated plasticity is assumed, but non-linearly associated plasticity is included in the formulation, see Clausen (2020).

2.1 *The Mohr power envelope*

Several references present Mohr envelopes defined by a power function of the normal stress, σ. Jiang et al. (2003); Baker (2004); Zhang and Yang (2019). Another example is the envelope presented in Hoek and Brown (1997),

$$S(\sigma) = A\sigma_c \frac{\sigma_a - \sigma^B}{\sigma_c} \tag{5}$$

In Anyaegbunam (2015), a non-linear failure envelope is presented, that is a generalization of the envelopes mentioned in the previous references, with a particular simple formulation;

$$\tau = S(\sigma) = (a - b\sigma)^n \tag{6}$$

where a is a cohesion-like parameter and b is a friction-like parameter. The curvature is controlled by the exponent n. The other power envelopes from the mentioned references can all be reformatted into the general form of Eq. (6). An example curve is shown in Figure 3 based on parameters given i Table 2. One important aspect of this model, is the unorthodox parameter units. As seen in Table 2, the units scale with the exponent, n.

It should be noted that the envelope slope tends to zero for increasing compressive stresses. This is illustrated in Figure 4, where the stress dependent friction angle, $\phi(\sigma)$, is shown for the three models.

2.1.1 *Strength reduction with the power envelope*

The reduction factor is multiplied onto the strength envelope of Eq. (6),

$$S_F(\sigma) = \frac{1}{F}(a - b\sigma)^n = \frac{a}{F^{1/n}} - \frac{b}{F^{1/n}}\sigma^n \tag{7}$$

leading to the reduced parameters

$$a_F = \frac{a}{F^{1/n}} \text{ and } b_F = \frac{b}{F^{1/n}} \tag{8}$$

It is noted that the curvature parameter n retains its initial value.

2.2 *An exponential Mohr envelope*

In the power envelope of the previous section, the values of the friction angle at extreme normal stresses are fixed at $\phi(\sigma \to -\infty) = 0$ and $\phi(\sigma \to \sigma_a) = 90°$, where σ_a is the apex stress, i.e.

$\tau = S(\sigma_a) = 0$ This is different in the exponential model which is defined by two friction angles

and two cohesion values

$$S(\sigma) = c_\infty - \sigma\mu_\infty - ae^{b\sigma} \tag{9}$$

where c_∞ is the asymptote intersection with the τ axis (the asymptotic cohesion) and μ_∞ is given by the asymptotic friction angle,

$$\mu_\infty = \tan\varphi_\infty \tag{10}$$

The material parameters a and b are found as

$$a = c_\infty - c_0 \tag{11}$$

and

$$b = \frac{\tan\varphi_0 - \tan\varphi_\infty}{c_\infty - c_0} = \frac{\tan\varphi_0 - \mu_\infty}{a} \tag{12}$$

where c_0 is the envelope intersection with the τ axis and φ_0 is the inclination at this intersection,

i.e. the instant friction angle at $\sigma = 0$.

The envelope is inspired by the non-linear yield criterion of Krabbenhøft et al. (2012) and the

GSK yield criterion of Krabbenhøft et al. (2016).

An example of the envelope can be seen in Figure 3 and the friction angle dependency on the normal stress is exemplified in Figure 4.

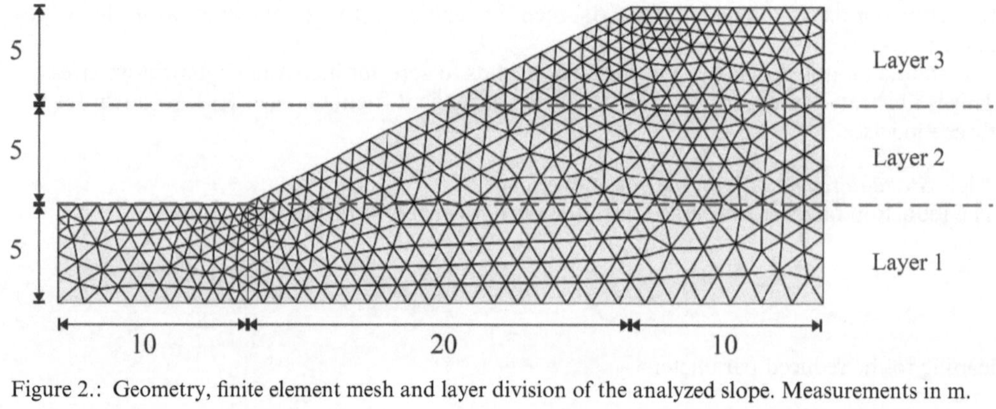

Figure 2.: Geometry, finite element mesh and layer division of the analyzed slope. Measurements in m.

2.2.1 *Strength reduction with the exponential envelope*

The reduction factor is applied to the strength envelope of Eq. (9),

$$S_F(\sigma) = \frac{c_\infty - \sigma\mu_\infty - ae^{b\sigma}}{F} = \frac{c_\infty}{F} - \frac{\sigma\mu_\infty}{F} - \frac{ae^{b\sigma}}{F} \tag{13}$$

leading to the reduced parameters

$$c_\infty, F = \frac{c_\infty}{F}, \ \mu_\infty, F = \frac{\mu_\infty}{F} \text{ and } a_F = \frac{a}{F} \tag{14}$$

It is noted that the curvature parameter b retains its initial value.

3 COMPUTATIONAL EXAMPLES

A computational example concerns the evaluation of the slope safety factor of the slope shown in Figure 2. The slope is discretised using standard LST elements, i.e. six noded triangular elements with quadratic shape functions. In total, the model consists of 902 elements with 3818 degrees of freedom. The vertical boundaries are supported in the horizontal direction and free vertically. The bottom horizontal boundary is free to move horizontally and supported in the vertical direction. The slope consists of three layers of materials, where each layer will consist of one of the materials mentioned above, including a standard Mohr-Coulomb material. For reference the homogeneous slope with all three layers consisting of the same material is also evaluated.

The meshing does not follow the layer boundaries for ease of changing the layer locations. Instead, each integration point of the elements are evaluated according to the layer it belongs to.

The material parameters are given in Table 2 and the corresponding Mohr envelopes are shown in Figure 3 and the friction angle dependency in Figure 4.

Different combinations of the materials of each layer is examined. The combinations and their designations are shown in Table 3.

Table 2. Material parameters in the numerical examples.

Power	Exponential	Mohr-Coulomb
$E = 600$ MPa	$E = 500$ MPa	$E = 400$ MPa
$v = 0.3$	$v = 0.25$	$v = 0.2$
$n = 0.684$	$\varphi = 45°$	$\varphi = 30°$
$a = 57.30\text{kPa}^{\frac{1}{n}}$	$c_0 = 18$ kPa	$c = 20$ kPa
$b = 12.34\text{kPa}^{\frac{1}{n}-1}$	$\varphi_\infty = 30°$	
	$c_\infty = 40$ kPa	

Selfweight for all materials is $\gamma = 10\,\text{kN/m}^3$

For each material combination of the slope, a calculation of the safety factor is carried out according to the procedure outlined in Table 1, and the results are shown in Table 3. The colour surface plots in the table display the incremental maximum shear strain in the final load step at incipient failure, and so is an indication of the failure mechanism.

Of the homogeneous slopes the Mohr-Coulomb material is seen to have the lowest factor of safety.

It can be seen that safety factor and the location of the slip surface changes with the different materials in the layers. The combinations with Mohr-Coulomb as the lower or middle layer tend to have the lowest factor of safety. When the lower layer consist of the Mohr-Coulomb material the slip surface extends deeper compared to the overall tendency.

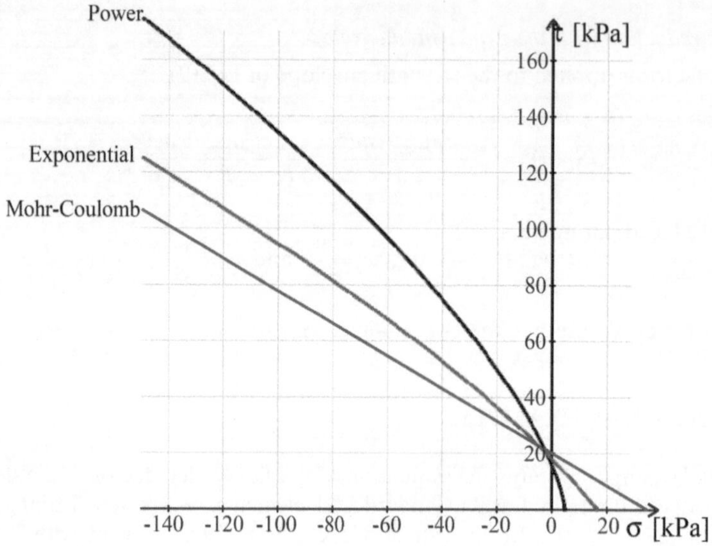

Figure 3. The three Mohr envelopes used in the computational examples. Material parameters are shown i Table 2.

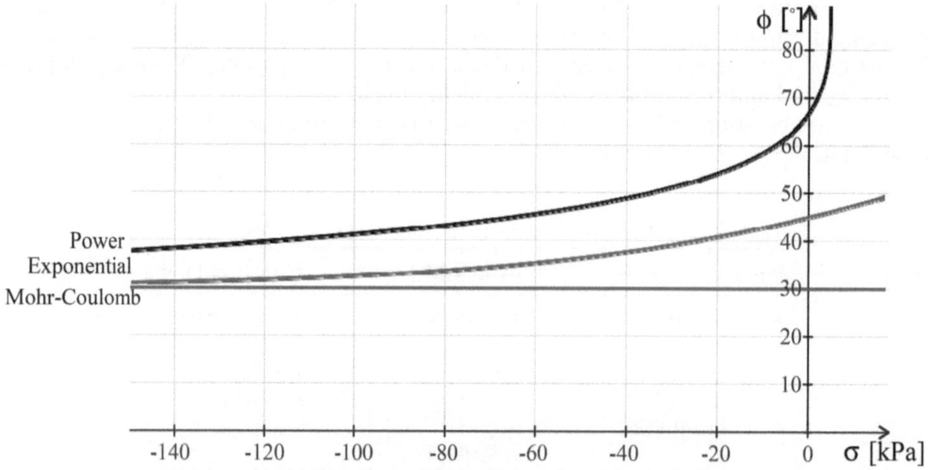

Figure 4. Stress dependent friction angle of the three models.

4 CONCLUSIONS

A method is presented that makes it possible to calculated safety factors based on consistent reduction of shear strength across different models with non-linear failure envelopes. With standard elasto-plastic material models formulated in cartesian stresses or their invariants comparison of safety factors across different material models is meaningless, but when safety factors are calculated on the basis of Mohr strength envelopes, the comparison can be made. A computational example on a slope composed of different material models demonstrates the use.

Table 3. Slope material composition and slope safety factor results. Surface plots represents maximum shear strain increment at last accepted load step.

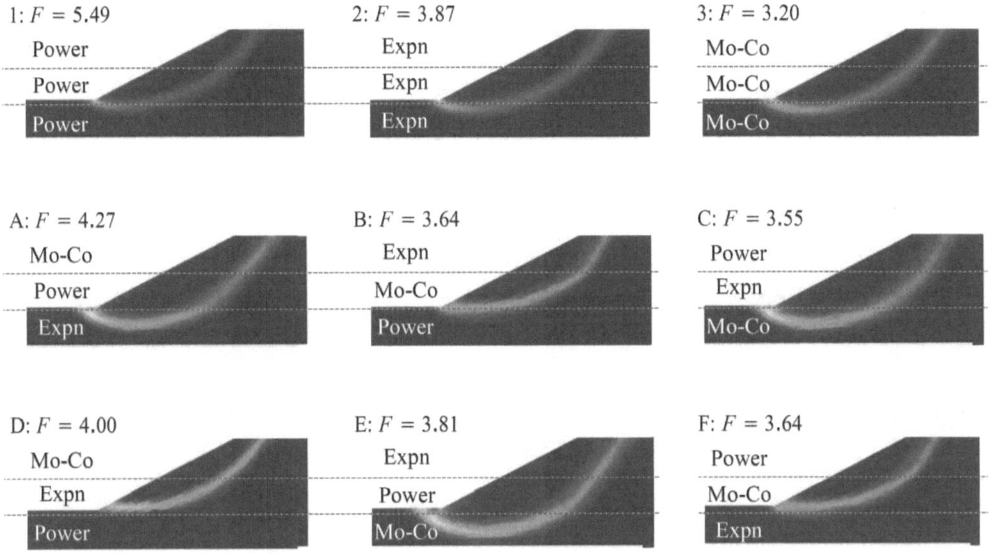

Expn: Exponential. Mo-Co: Mohr-Coulomb

REFERENCES

Anyaegbunam, A. J. (2015). Nonlinear power-type failure laws for geomaterials: Synthesis from triaxial data, properties and applications. *International Journal of Geomechanics*, 15(1).

Baker, R. (2004). Nonlinear Mohr-envelopes based on triaxial data. *Journal of Geotechnical and Geoenvironmental Engineering*, 130(5):498–506.

Clausen, J. (2020). Mohr envelope elastoplasticity. *Journal of Engineering Mechanics*, 146(5):15.

Hoek, E. and Brown, E. (2019). The Hoek–Brown failure criterion and GSI – 2018 edition. *Journal of Rock Mechanics and Geotechnical Engineering*, 11(3):445–463.

Hoek, E. and Brown, E. T. (1980). Empirical strength criterion for rock masses. *Journal of Geotechnical Engineering*, 106:1013–1035.

Hoek, E. and Brown, E. T. (1997). Practical estimates of rock mass strength. *International Journal of Rock Mechanics and Mining Sciences*, 34(8):1165–1186.

Jiang, J. C., Baker, R., and Yamagami, T. (2003). The effect of strength envelope nonlinearity on slope stability computations. *Canadian Geotechnical Journal*, 40(2):308–325.

Krabbenhøft, K., Lyamin, A., and Krabbenhøft, J. (2016). *OptumG2: Materials*. Optum Computational Engineering.

Krabbenhøft, S., Clausen, J., and Damkilde, L. (2012). The bearing capacity of circular footings in sand: Comparison between model tests and numerical simulations based on a nonlinear Mohr failure envelope. *Advances in Civil Engineering*, 2012.

Lees, A. S. and Clausen, J. (2020). Strength envelope of granular soil stabilized by multi-axial geogrid in large triaxial tests. *Canadian Geotechnical Journal*, 57(3):448–452.

Ottosen, N. S. and Ristinmaa, M. (2005). *The Mechanics of Constitutive Modelling*. Elsevier.

Zhang, R. and Yang, X. (2019). Limit analysis of anchor trapdoor embedded in nonhomogeneous and nonlinear soils. *International Journal of Geomechanics*, 19(8).

Zienkiewicz, O., Humpheson, C., and Lewis, R. (1975). Associated and non-associated visco-plasticity and plasticity in soil mechanics. *Ge'otechnique*, 25:671–689.

The Evolution of Geotech - 25 Years of Innovation – Hammah et al. (Eds)

Global stability assessment of open pit slopes using LEM and FEM: A comparison between the factor of safety and strength reduction factor

J. Goodale, E.I.T.
Teck Coal Limited, Elkford BC, Canada

B. Fisher, Ph.D., P.Eng.
Fisher Rock Engineering LLC, Radford VA, USA

ABSTRACT: A geotechnical stability assessment was conducted for a series of pit walls at an open pit coal mine in southeastern British Columbia. Cross sections were modeled with limit equilibrium (LEM) and finite element (FEM) software packages. Given the height of the pit walls and potentially high-induced stress regimes, the shear strength reduction method was applied to consider stress relaxation and plastic deformation of the rock mass during pit development. An iterative process was used to first establish a potential failure mechanism with FEM, then repeat it using LEM. During this evaluation, certain assumptions were required in LEM to achieve acceptable agreement between LEM and FEM critical failure mechanisms and stability factors.

1 INTRODUCTION

The design work for an open pit coal mine evaluated the geotechnical slope stability of footwalls and endwalls. The overall slope heights range from 400 to 600 meters, with inter-ramp angles between 30 and 50 degrees. Cross sections were modeled in SLIDE2018 and RS2 using Rocscience two-dimensional LEM and FEM software packages. The analyses were preliminary in nature, therefore LEM was chosen as the primary method for assessing the global stability of the pit walls. This allowed for expedited sensitivity analysis against various geotechnical assumptions. FEM was also employed to consider stress relaxation and plastic deformation of the rock mass during the phased pit development. An iterative process was used to first establish the critical failure mechanism (CFM) using FEM, and then attempt to replicate it in LEM. After establishing a repeatable CFM, the stability factors (factor of safety (FOS) and critical strength reduction factor (SRF)) were compared.

2 ENGINEERING GEOLOGY

The geology consists of “Fair to Good” (Bieniawski, 1989), medium-weak to very strong, interbedded sedimentary rock. Lithologies range from fine-grained mudstone and siltstone, to coarser grained sandstone, with coal seams and stringers throughout. Bedding has experienced flexural slip, and is expected to be persistent with orthogonal tension joints that formed during uplift and erosion. A regional thrust fault with several well-defined splays creates repetition and offset in the coal seams.

DOI: 10.1201/9781003188339-56

The final wall conditions are expected to be largely drained. Phreatic surfaces specific to each cross section and pit phase were incorporated into the models.

3 PIT WALL FAILURE MECHANISMS

3.1 *Footwalls*

Footwalls strike parallel to bedding and incorporate geotechnical step outs. In coal mines, footwalls are most commonly susceptible to translational failures caused by daylighting bedding in areas of pinch-outs and unexpected faulting. Unless identified during geotechnical characterization, the potential for translational failures is closely monitored during pit excavation and is not usually addressed during preliminary analyses. Instead bi-planar and ploughing failure mechanisms are considered during design. The bi-planar failure mechanism is typically more critical, and requires daylighting structure or shearing through the rock mass to allow kinematic release at the toe. Driving forces are transferred from the upper part of the wall (active zone) to the lower part of the wall (passive zone). The deformation zone between the active zone and passive zone is referred to as the transition or Prandtl zone, and has been discussed in detail in Kvapil and Clews (1971), Stead and Eberhardt (1997), Stead et al (2006), Fisher (2009), and Havaej et al. (2014). Figure 1 gives an illustration of the Prandtl zone between the active and passive blocks.

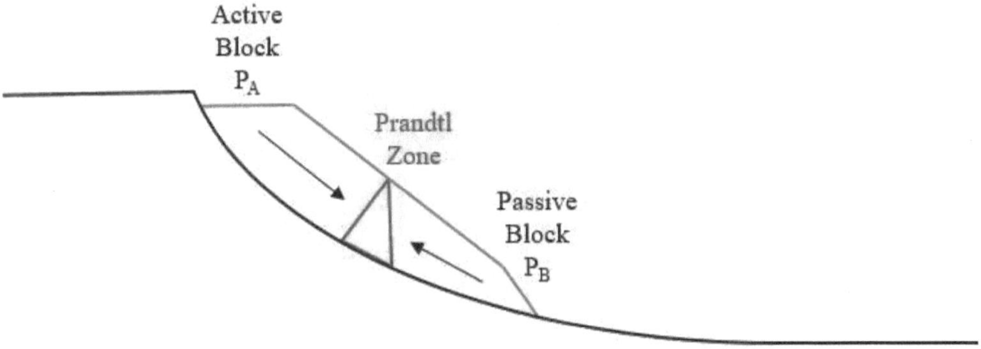

Figure 1. Active-passive failure surface after Kvapil and Clews (1971) and Stead et al. (2006).

3.2 *Endwalls*

Endwalls are excavated perpendicular to bedding strike, and often include access ramps. Crosssections typically show near-horizontal bedding, with coal seams that bound packages of rock mass of differing sedimentary lithologies and rock mass quality. CFMs include shearing through the rock mass starting at the pit wall crest (often requiring a tension crack in LEM), followed by toe breakout along near-horizontal coal seams. This failure mechanism is illustrated later.

4 SLOPE STABILITY METHODOLOGY

4.1 *Simulating stress relaxation in LEM*

Blasting and excavation during pit development can progressively lead to mobilization of peak rock mass shear strength followed by a decrease to post-peak strength. The degradation of the rock mass is most pronounced at the wall face, and decreases exponentially with depth. Given the height of the pit walls and potentially high induced stress regimes, the FEM models were integral in establishing the extent of stress-induced plastic deformation of the rock mass.

LEM is not readily capable of assessing this reduction in shear strength. To develop a direct comparison in LEM, a certain percentage of plastic failure was assumed to model stress relaxation and a transition from peak to post-peak strengths using the Rose et al. (2018) and Guzman and Perez (2015) methods. Figure 2 illustrates the zonation of rock mass stress relaxation in a staged FEM model. The transition from peak to post-peak strength is denoted by plasticity indicators.

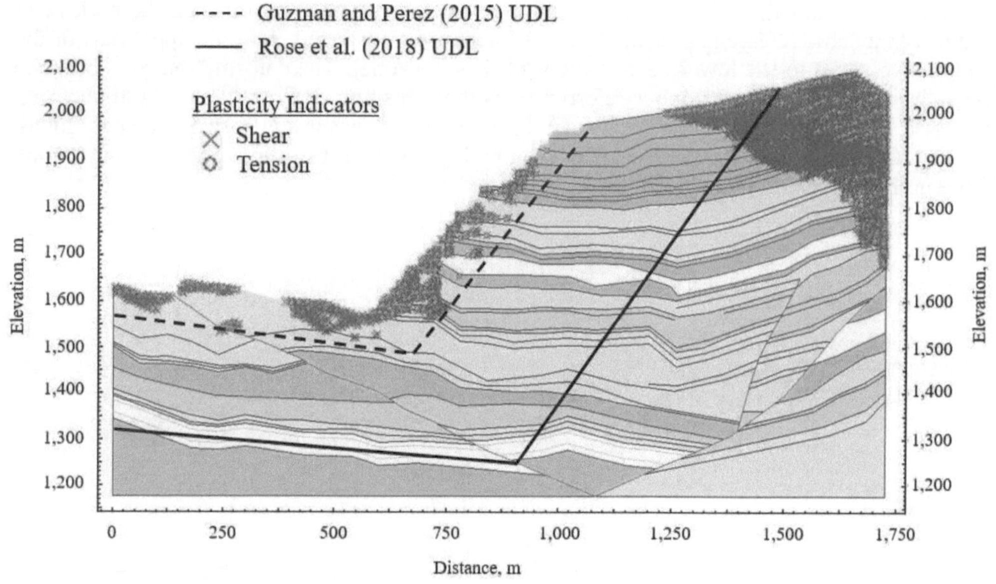

Figure 2. Undisturbed limit (UDL) suggested by Guzman and Perez (2015) vs. Rose et al. (2018).

Both methods assume that the Hoek-Brown failure criterion Disturbance "D" factor (Hoek et al., 2002) at the pit face should be equal to 1.0, and that this factor decreases exponentially with depth until reaching zero. The depth of disturbance suggested by Guzman and Perez (2015) is one-third the wall height, while Rose et al. (2018) state that a depth equal to the wall height is more applicable. Figure 2 shows that for this particular case, a disturbance zone of about one-third the wall height would be appropriate. However, our experience is that the actual disturbance zone is related to a number of factors, also discussed by Rose et al. (2018). For this particular study, it was found that a linear decrease in the D factor from the pit face to a depth equal to the wall height was conservative. This was determined by comparison of stability factors between LEM and FEM, and more importantly by comparison of CFMs (as discussed below).

Geotechnical engineers have been struggling with the proper application of the D factor for nearly two decades (Hoek et al. 2002 & Hoek, 2012). Rose et al. (2018) implies that a D factor of 1.0 represents post-peak rock mass strengths. This could provide a rationale for including blast damage and stress relaxation simultaneously in LEM.

4.2 *Establishing critical failure mechanisms*

Failure mechanisms within the pit walls are typically complex and involve shearing through the rock mass and translational sliding along geologic structure. There are several slip surface search routines available in SLIDE2018; only non-circular methods are applicable to this analysis. Of those non-circular methods, and assuming no prior knowledge of potential CFMs, our experience has been that the Cuckoo and Particle Swarm search methods are capable of predicting these complex failure mechanisms. However, these search methods are time consuming.

This analysis (and also experience with previous studies) showed that utilizing a methodology where FEM is employed to establish the CFM followed by targeting or 'seeding' the slip surface in SLIDE2018 consistently results in the identification of the critical (or near critical) slip surface and near lower-bound FOS. This allows for a targeted LEM approach with the collateral benefit of utilizing less time-consuming slip surface algorithms such as the Block or Path Search. This intermediate step was then followed by establishing the lower-bound FOS utilizing the more time-intensive Cuckoo and Particle Swarm Search methods. An example of the comparison of LEM and FEM failure mechanisms and stability factors is provided in Figure 3.

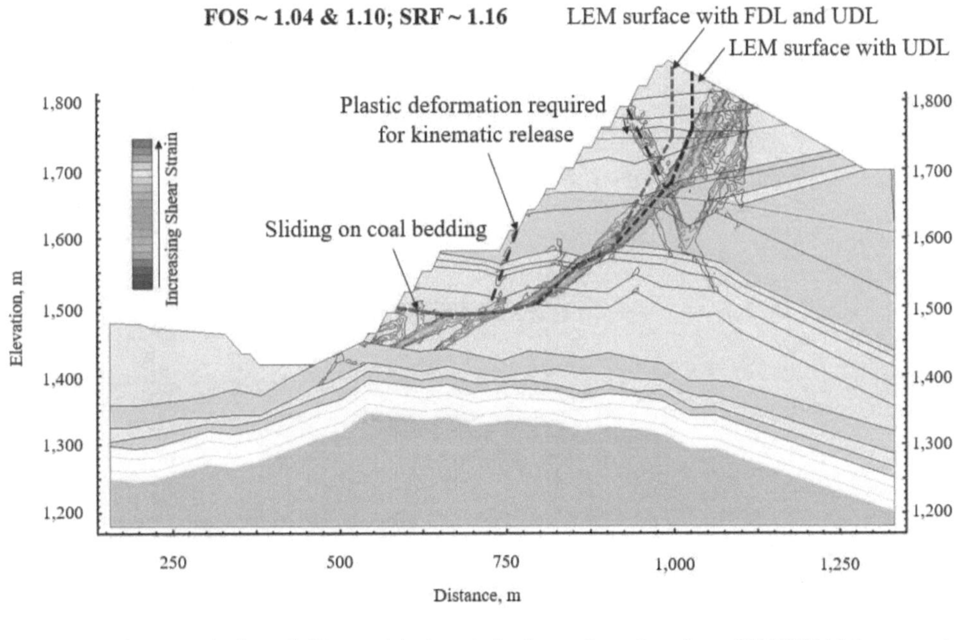

Figure 3. Shear strain from RS2 vs. critical optimized search surface from SLIDE2018 for an endwall model.

Two LEM critical slip surfaces are provided in Figure 3 while the critical slip surface from the FEM is outlined using shear strain contours. Tension cracks were routinely added to the LEM models to avoid computational errors. The first LEM slip surface (red with FOS of 1.04) is the result of using a "Fully Disturbed Limit" (FDL) as recommended by Rose et al. 2018. The FDL is a percentage of the wall height from the face that is considered fully disturbed with a D factor of 1.0. The second slip surface (black with FOS of 1.10) excludes the FDL but like the first slip surface, includes an Undisturbed Limit (UDL) located at a distance equal to the pit wall height.

Differences in the location of the LEM critical slip surface and FOS are the direct result of the amount of and decrease of the D factor assumed for the LEM models. One of the hazards of underestimating the percent of plastic failure within the LEM is the potential for the model to not reflect the CFM, highlighting the importance of this step.

Finally, zones of plastic deformation required for kinematic release of the pit wall are also emphasized in Figure 3. These are the Prandtl zones that are discussed above and will be further illustrated in the discussions to follow.

5 RECOGNITION AND RECONCILIATION OF PRANDTL ZONES

Figure 4 gives an example of a footwall where the difference between LEM and FEM stability factors is 11%. In LEM, the CFM is sliding along a weak coal seam which daylights at the toe of the wall. The same coal seam and sliding mechanism were identified in FEM, however FEM requires deformation of the rock mass to enable kinematic release of the block. This deformation is identified using the shear and tension plasticity indicators which extend between the coal seam and the pit face. By accounting for shear through the rock mass, the FEM returns a higher stability factor than the LEM which has primarily considered the ratio of resisting to driving forces along the coal seam while generating the FOS. The Sarma (1973) method of LEM could effectively consider plastic deformation of the wall required for kinematic release, however this method was not considered during the stability evaluations discussed herein.

Note that the footwall design in Figure 4 daylights bedding and was subsequently re-designed before continuing the stability analysis.

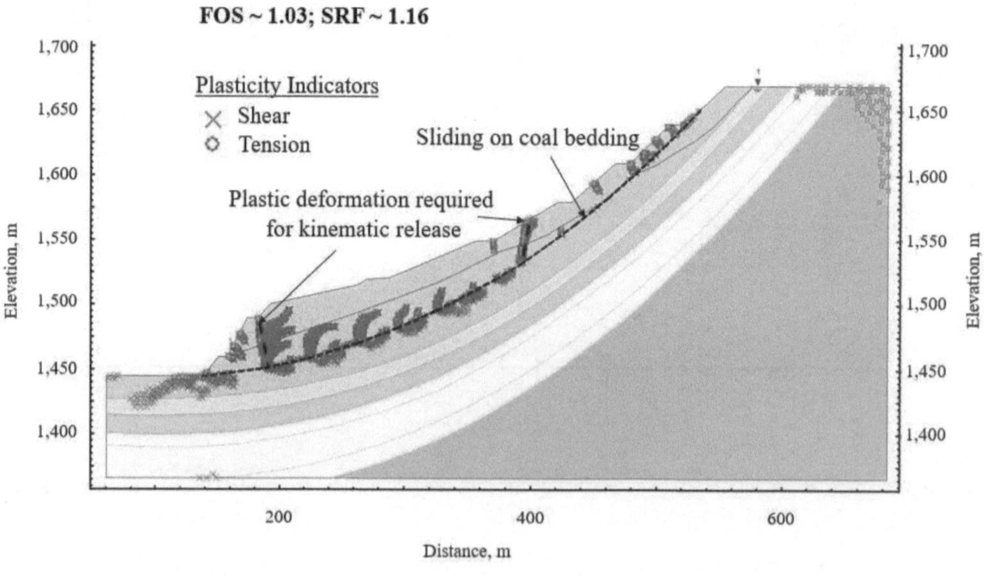

Figure 4. Shear strain from RS2 vs. critical optimized search surface from SLIDE2018 for a footwall model.

6 SUMMARY OF RESULTS AND CONCLUSIONS

When comparing the results of FEM and LEM analyses for pit walls in coal measure rocks, there are numerous factors to consider.

1. The CFM for endwalls is typically shearing through the rock mass near the crest (tension cracks are often observed) with toe breakout along coal seams.
2. The CFM for footwalls is typically sliding along daylighting structure. However, potential for pinch-outs and faulting that could offset bedding is one of the focuses of exploration investigations. Where these conditions are not identified, the CFM is often bi-planar. This requires sliding along a coal seam with toe breakout along adverse structure or through the rock mass.
3. A direct comparison of LEM and FEM requires the inclusion of stress relaxation in the LEM model. The Guzman and Perez (2015) and Rose et al. (2018) methods are good starting points. Similar results could be developed by completing wall-specific FEM modeling coupled with comparison of critical slip surface locations and stability factors.
4. First iterations of LEM analysis applied time-intensive Cuckoo and Particle Swarm search methods. Computation time was significantly reduced by using the shear strain contours from FEM analyses to apply a seeded search method instead (i.e., Block or Path Search). Aside from the time advantage, the seeded failure mechanism could return a near lower-bound (and sometimes lower-bound) stability factor. This highlights the importance of FEM in identifying the CFM, as well as leveraging FEM results to increase productivity.
5. LEM may not accurately incorporate internal shearing through the rock mass (i.e., Prandtl zones) and this needs to be considered when comparing results of the FEM/LEM models.

Interim stability factors from relevant models in this case study are summarized in Table 1.

Table 1. Comparison of FOS and SRF for endwalls and footwalls.

Type	FOS	SRF	Difference	Critical Failure Mechanism
Endwall	1.03	1.16	11%	Shearing through rock mass and sliding along coal seam; consistent
Endwall	2.02	2.18	7%	failure mechanisms with differing stress relaxation assumptions
Endwall	1.91	2.14	11%	between models.
Footwall	1.04	1.16	10%	Sliding along coal seam vs. sliding along coal seam with kinematic
Footwall	2.06	2.42	15%	releases required through Prandtl Zone.

Note: These models were taken from intermediate iterations of the pit wall and for illustration purposes only. They are not representative of the accepted pit design.

In all cases, and given the methods discussed herein, the stability factors from LEM are lower than those from FEM. This outcome is based on assumptions included in the geotechnical stability evaluations based on informed decisions and a thorough understanding of the conclusions discussed above. Ultimately, the LEM results were relied upon for this level of assessment, with the understanding that pit wall optimization will be conducted in future geotechnical stability assessments.

REFERENCES

Bieniawski, Z.T. 1989. *Engineering Rock Mass Classification*: 251. New York: Wiley.

Fisher, B. 2009. Improved characterization and analysis of bi-planar dip slope failures to limit model and parameter uncertainty in the determination of setback distances. University of British Columbia.

Guzmán, R.S. and Patricio G.P.P. 2015. Towards a Mechanically Based Definition of the Disturbance Factor Using the "Slope Model" Lattice Code. *ISRM Regional Symposium - 8th South American Congress on Rock Mechanics*. Buenos Aires, Argentina.

Havaej, M., Stead, D., Eberhardt, E. and Fisher, B. 2014. Characterization of bi-planar and ploughing failure mechanisms in footwall slopes using numerical modelling. *Engineering Geology*: 178.

Hoek E., Carranza-Torres C.T. and Corkum B. 2002. Hoek-Brown failure criterion-2002 edition. *Proceedings of the fifth North American rock mechanics symposium*, 1: 267–73. Toronto, Canada.

Hoek, E. 2012. Blast Damage Factor D. *Technical note for RocNews, Winter 2012 Issue*.

Kvapil, R. and Clews, K.M. 1971. An examination of the Prandtl mechanism in large dimension slope failures. *Transactions of the Institution of Mining & Metallurgy*: A1–A5.

Rose, N., Scholz, M., Burden, J., King, M., Maggs, C. and Havaej, M. 2018. Quantifying Transitional Rock Mass Disturbance in Open Pit Slopes Related to Mining Excavation.

Sarma, S.K. 1973. Stability analysis of embankments and slopes. *Géotechnique*, 23(3): 423–433.

Stead, D. and Eberhardt, E. 1997. *Developments in the analysis of footwall slopes in surface coal mining. Engineering Geology*, 46(1): 41–61.

Stead, D., Eberhardt, E. and Coggan, J.S. 2006. Developments in the characterization of complex rock slope deformation and failure using numerical modeling techniques. *Engineering Geology*, 82(1-2): 217–235.

The Evolution of Geotech - 25 Years of Innovation – Hammah et al. (Eds)

A simple method to model buckling slope instability using continuum numerical models

R.N. Ridl & D.H. Bell
School of Earth and Environment, University of Canterbury, New Zealand

M.C. Villenueve
Subsurface Engineering, Montanuniversität Leoben, Austria

D.F. Macfarlane
AECOM New Zealand Ltd, Christchurch, New Zealand

ABSTRACT: Buckling instability is a mode of slope failure associated with stratified rock masses. This paper demonstrates that buckling instability at a large-scale (>50 m slope height) can be modelled as a continuum, whereby the inherent anisotropy from the stratified nature of the rock mass is incorporated as a discrete fracture network. Three case studies of increasing complexity were used to demonstrate the capability of the anisotropic continuum Finite Element Method (FEM) for back-analysing buckling deformation. The case studies range from a single staged excavation, through a progressively excavated open pit slope, and finally to sequential unloading corresponding to the evolution of a large natural valley. The results of this study indicate that FEM analysis is a simple, fast and effective method for providing a first indication of buckling slope instability for a large-scale slope. Parametric comparison of constitutive models indicates the Generalized Hoek-Brown failure criterion typically produces more representative results than the Mohr-Coulomb failure criterion. Limitations of using infinitesimal strain theory for analysing buckling deformation are discussed.

1 INTRODUCTION

Buckling instability is a mode of slope failure associated with stratified rock masses which has been identified in road cuttings, open pit mines and large scale natural slopes. Buckling failure could result in rapid movement or creeping deformations dependent on a multitude of site conditions including, but not limited to: rock type/structure, the geometric attributes of the slope, and in situ stresses.

This paper intends to show that continuum based Finite Element Method (FEM) analysis can be used to model buckling instability of a stratified rock mass. Three case studies are back analyzed: a simple single-staged theoretical example, a sequentially excavated 120 m deep open pit mine, and a large-scale > 600 m high natural slope. The objectives of this paper are as follows:

- Demonstrate that simplified FEM modelling can be used to simulate buckling slope instability.
- Present a method of incorporating anisotropic rock mass behavior into a continuum based model.

DOI: 10.1201/9781003188339-57

- Use a parametric approach to determine the most appropriate constitutive criteria for modelling buckling deformation using a continuum-based model.
- Identify limitations to using infinitesimal strain theory for modeling buckling deformation.

2 ANISOTROPIC ROCK MASSES

Buckling slope instability is typically associated with stratified rock masses (Cavers, 1981). Persistent, closely spaced stratifications or defects result in an anisotropic behavior where the strength is directly dependent on the orientation of the defects relative to the imposed load (Ramamurthy, 1993). The anisotropic behavior of a rock mass can be numerically simulated using an anisotropic continuum method (Saroglou *et al.*, 2018), which involves incorporating a fracture network that represents the strength of the repetitive stratifications. Multiple publications have used this method for analysis of slope instability of a blocky rock mass (Barla and Barla, 2000; Azadeh and Curran, 2008; Hammah, Yacoub and Corkum, 2008; Riahi, Hammah and Curran, 2010; Azami *et al.*, 2013; Pereira and Lana, 2013).

The method of analysis demonstrated in this paper uses an equivalent continuum model to simulate the anisotropic behavior using RS2 (Rocscience Inc., 2019). The equivalent continuum model integrates an isotropic continuum representing the rock mass between the persistent defects, and a discrete fracture network representing the anisotropic planes as illustrated in Figure 1.

A parametric approach was adopted for all case study analyses by using two constitutive failure criteria to model the isotropic continuum, namely Mohr-Coulomb (MC) and Generalized Hoek-Brown (GHB), as well as varying the range of material strength parameters (Table 1).

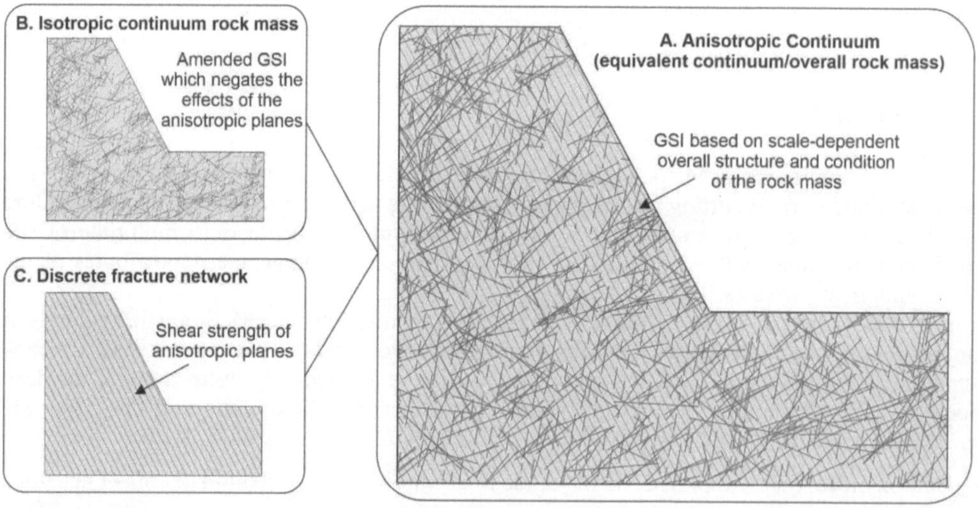

Figure 1. A) Anisotropic continuum incorporating: B) Isotropic continuum representing the rock mass comprising all other discontinuities (except for the major persistent stratifications), and C) Discrete fracture network representing persistent stratifications.

3 CASE STUDY 1: THE LIMIT EQUILIBRIUM EXAMPLE

Case Study 1 used Goodman's (1989) theoretical limit equilibrium example based on the Euler buckling beam solution for a buckling failure in the footwall of a coal mine. Input parameters provided by Goodman (1989) and derived by Ridl (2021) are given in Table 1. Joint stiffness parameters, namely normal shear stiffness (k_n) and shear stiffness (k_s), were derived using Equations 1 and 2.

$$k_n = \frac{E_i E_{rm}}{L(E_i - E_{rm})} \tag{1}$$

$$k_s = \frac{G_i G_{rm}}{L(G_i - G_{rm})} \tag{2}$$

where E_i = intact Young's Modulus; E_{rm} = rock mass deformation modulus; G_i = intact shear modulus; G_{rm} = rock mass shear modulus; and L = thickness of bedding/spacing of persistent defects.

The results of FEM modelling of Case Study 1 using GHB failure criterion achieved comparable results to Goodman (1989) worked example of a buckled length of 40 m. The GHB models produced a buckled length of slope varying between 40 m and 62 m for a range of material properties, whereas the MC models resulted in a shorter buckled length of <10 m (Figure 2).

Table 1. Case Study data including peak parameters with residual parameters in brackets.

			Case Study 1	Case Study 2	Case Study 3
		v	0.28[a]	0.28[d]	0.28[d]
	Material properties	γ (MN/m^3)	0.027[a]	0.027[b]	0.026[c]
		E_{rm} (GPa)	3[a]	9.2-15.6 (1.9-2.0)[d]	49-53 (12-27)[d]
		GSI	45-60 (25-27)[d]	50-60 (26-27)[d]	60-75[d]
		σ_{ci} (MPa)	50[a]	107[b]	146 (45)[c]
Isotropic continuum properties	GHB	m_i	13[d]	17[d]	11.8-12.3[d]
		D	0_d	0 (blast zone=0.7)[d]	0 (0.7)[d]
		ψ	6.5[d]	1.8[d]	0.7-1.6[d]
		ϕ (°)	46-50 (30.5-33.5)[d]	53-54 (46-48)[d]	30-50[c]
	MC	c (MPa)	0.9-1.2 (0.4-0.5)[d]	1.5-1.7 (0.85)[d]	13-30[c]
		σ_t (MPa)	0.69-0.87 (0)[d]	1.0-1.1 (0)[d]	0.5-1.0[c]
		ψ (°)	11.4-12.5[d]	13.3-13.6[d]	7-16[d]
Discrete fracture network (bedding planes)		ϕ (°)	10[a]	25[b]	21 (13.2)[c]
		c (MPa)	0[a]	0[b]	0.2 (0)[c]
		k_n (GPa/m)	8-10[d]	5-40[b,d]	2-13[c,d]
		k_s (GPa/m)	3.3-3.5[d]	5-17[b,d]	1-5.2[c,d]

[a](Goodman, 1989); [b](Wang, Cavers and Wong, 2004); [c](Tommasi *et al.*, 2009); [d]derived (Ridl, 2021)

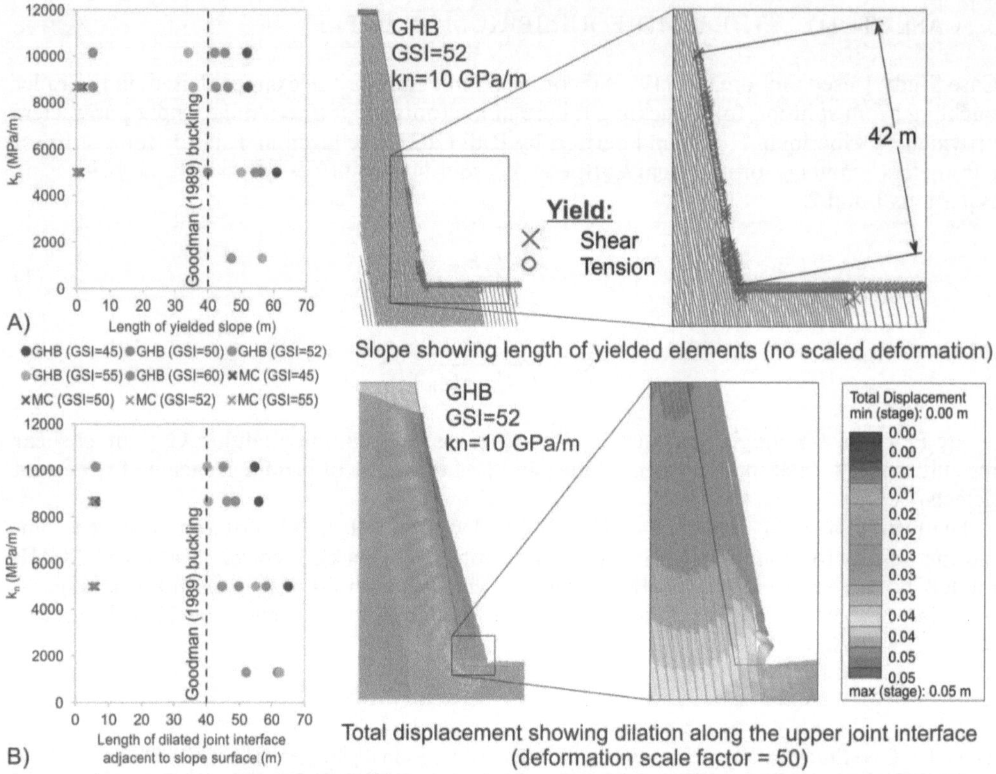

Figure 2. Length of yielded slope versus joint stiffness (kn) for parametric FEM models of Case Study 1.

4 CASE STUDY 2: QUINTETE OPEN PIT MINE

Case Study 2 involved back analysis of the 1997 Quintette Open Pit mine failure based on Wang, Cavers and Wong (2004) data. The failure involved a buckled footwall failure comprising the near surface rock ~1. 5m in thickness (two sandstone bedding planes) extending from the toe up slope to ~90 m in length.

The methodology adopted for FEM analysis of the failure involved sequential unloading using 9 stages to simulate progressive excavation from pre-mining level surface conditions to the vertical slope depth of ~120 m at failure. Excavation of the footwall was carried out in 15 m high batters at a constant dip angle equivalent to the dip of the bedding (44°). A veneer of 1.0-1.6 m thickness was adopted to account for blast damage (D = 0.7) along the pit bottom and slope face, respectively. A parametric approach was adopted with input parameters including GHB and MC failure criteria, and a range of material properties (Table 1).

The FEM model produced a buckling deformation characterized by an outwards bulge near the toe of the slope, yielded elements along the near-surface of the slope (Figure 3B), and dilated discrete interfaces (fracture network) near the toe of the slope. The results from the GHB models gave an average total yielded slope length of 75 m (range of 14 m-120 m), whereas the MC models resulted in an average total yielded slope length of 12 m for a range of 0 m-34 m (Figure 3A).

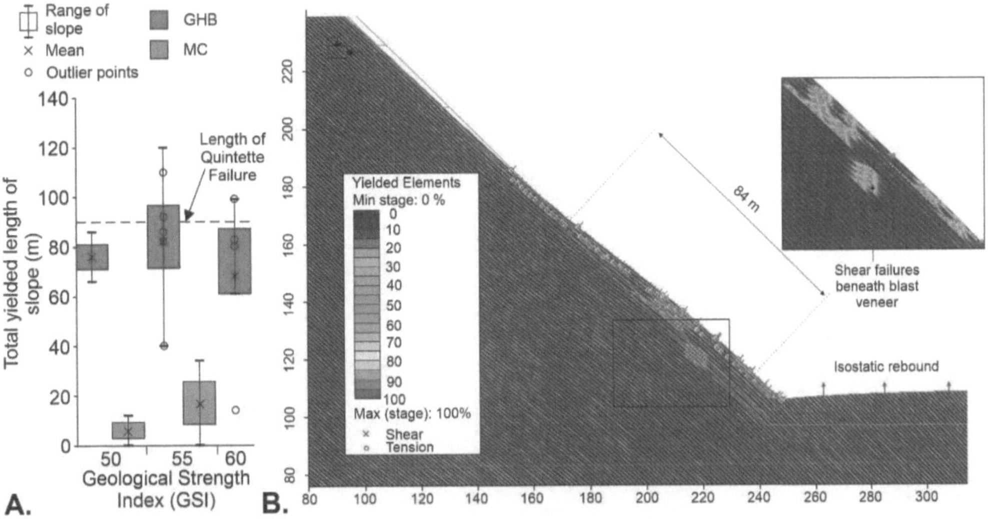

Figure 3. A) Summary of the total yielded length for parametric variations of Case Study 2. B) An example of buckling deformation derived from a GHB FEM model.

5 CASE STUDY 3: LAVINI DI MARCO LANDSLIDE

Case Study 3, which was the most complex analysis, involved the evaluation of buckling deformation associated with a large-scale natural slope. The analysis was based on work done by Tommasi *et al.* (2009) on the Lavini di Marco landslide along the flank of the Mount Zugna Torta, south of Rovereto in Trento, Italy. Tommasi *et al.* (2009) used Distinct Element Method (DEM) to model buckling deformation that developed at the slope break (change in slope angle) ~500 m below the slope crest. The buckling deformation involved the upper 3-6 m of moderately thinly to thickly bedded limestone, and was considered to have developed as a compressional feature associated with the progressive evolution of the landscape through uplift and erosion and glacial retreat.

Two geometric variations of the limestone structure were modelled based on the Tommasi *et al.* (2009) analysis, these being a) mesoscopic fold (FEM Model 1) and b) a concave change in the bedding plane angle (FEM Model 2) within the limestone (Figure 4). Sequential unloading along bedding planes was carried out in 50 stages (15 m excavation/stage) to simulate progressive development of the landscape through uplift and erosion of the slope.

The results of FEM Model 1 (mesoscopic fold) were similar to Tommasi *et al.* (2009) DEM results (Figure 4A,B,C), with a deformation sequence including initial convexity developing along the slope surface followed by shear and tensile failures on the downhill limb of the mesoscopic fold. Failure progressed laterally both upslope, downslope and with depth into the slope with subsequent unloading.

FEM Model 2 (concave bedding) did not however, produce observable buckling deformation comparable to the Tommasi *et al.* (2009) results (Figure 4D,E,F). The results do, however, show minor slope convexity and progressive plastic deformation developing downslope of the change in bedding angle. These results demonstrate the limitation of using infinitesimal strain theory to model large strain displacements.

6 DISCUSSION

RS2 is based on infinitesimal strain theory which considers that material properties are unchanged by deformation, and that displacements are generated as a function of rigid body

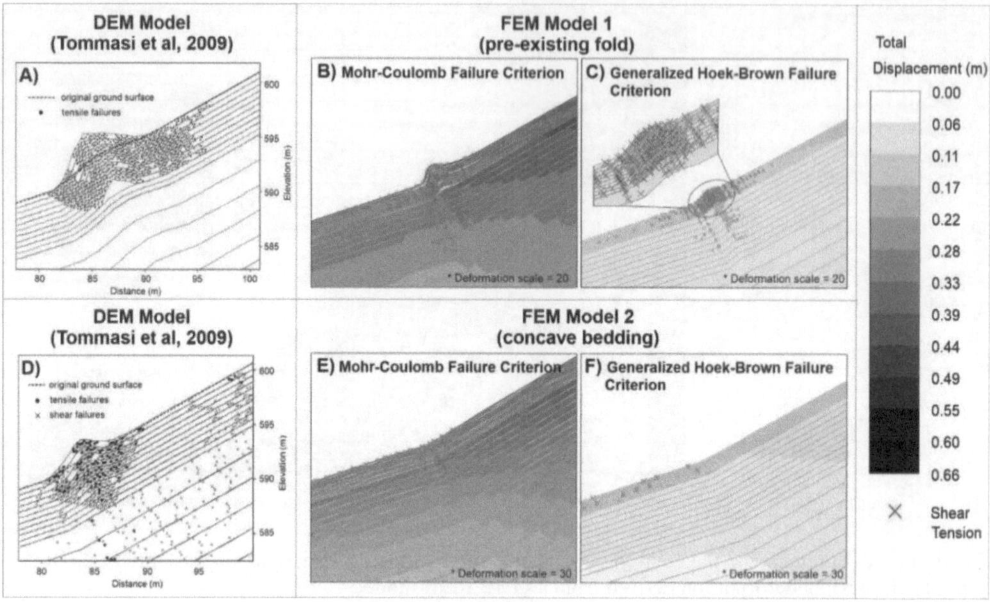

Figure 4. Comparisons of Tommasi *et al.* (2009) DEM models with the results of FEM Models 1 and 2.

motion. According to Chambon (2012) it is inappropriate to use infinitesimal strain theory to model buckling deformation because the magnitude of induced plastic indicators (displacement and rotation) associated with buckling deformation are similar to the element size.

The results of back analyzing three case studies of buckling deformation using 2-D FEM models show that although the deformation morphology of buckling using infinitesimal strain theory may not provide entirely accurate results in terms of the absolute magnitude of deformation, they can provide an initial indication of buckling deformation. Buckling deformation is evidenced by the development of slope convexity, displacement vectors indicating outwards bulging, joint dilation, tensile and shear failures along the upper slope (slope yielding), mobilization of plastic strain near the buckling zone and localized rotations of principal stresses.

Parametric analysis conducted by varying the constitutive models indicates that the GHB failure criterion typically produces deformation results more representative of the case study failures than the models using MC failure criterion. This is likely a function of the linearity of the MC failure envelope, as this does not accurately represent the non-linear rock mass behavior at low confining stresses. Buckling instabilities associated with the case studies are located proximal to

the slope surface, which is characterized by low confining stresses, making the model incorporating a GHB failure criterion appropriate for at least modelling preliminary deformation.

7 CONCLUSIONS

Three case studies of buckling slope instability with varying complexity ranging from a single-staged theoretical example to a large-scale natural slope were back analyzed using anisotropic continuum FEM models. The results indicate that buckling deformation comparable to the case studies are achievable using anisotropic continuum FEM models, and that models based on the GHB constitutive criterion are more representative of the case studies than the results from the models incorporating the MC failure criterion.

This paper shows that buckling deformation can be modelled using infinitesimal strain theory provided that limitations of the rigid body analysis are acknowledged. The major limitation of this analysis is the magnitude of displacements, which are not considered to be accurate due to use of engineering strain. Deformation indicators of buckling deformation include slope convexity, displacement vectors indicating outwards bulging, joint dilation, tensile and shear failures along the upper slope (slope yielding), mobilization of plastic strain near the buckling zone and localized rotations of principal stresses.

REFERENCES

Azadeh, R. and Curran, J. H. (2008) 'Application of Cosserat Continuum Approach in the Finite Element Shear Strength Reduction Analysis of Jointed Rock Slopes', in *12th International Conference on Computer Methods and Advances in Geomechanic*. Goa, India, pp. 110–118.

Azami, A. *et al.* (2013) 'A constitutive model for jointed rock mass', in *ISRM International Symposium - EUROCK 2013*. Wrocraw, Poland, pp. 385–390.

Barla, G. and Barla, M. (2000) 'Continuum and discontinuum modelling in tunnel engineering', *The Mining Geological Petroluem Engineering Bulletin*, 12, pp. 45–57.

Cavers, D. S. (1981) 'Simple methods to analyze buckling of rock slopes', *Rock Mechanics*, 14, pp. 87–104. doi: 10.1007/BF01239857.

Chambon, R. (2012) 'Small strain vs large strain formulation in computational mechanics', *Revue Française de Génie Civil*, 6(6), pp. 1037–1049. doi: 10.1080/12795119.2002.9692730.

Goodman, R. E. (1989) *Introduction to Rock Mechanics*. Second Edi. John Wiley & Sons.

Hammah, R., Yacoub, T. and Corkum, B. (2008) 'The practical modelling of discontinuous rock masses with finite element analysis', *42nd US Rock Mechanics Symposium and 2nd U.S. - Canada Rock Mechanics Symposium*, pp. 56–63.

Pereira, L. C. and Lana, M. S. (2013) 'Stress-Strain Analysis of Buckling Failure in Phyllite Slopes', *Geotechnical and Geological Engineering*, 31, pp. 297–314. doi: 10.1007/s10706-012-9556-8.

Ramamurthy, T. (1993) 'Strength, modulus responses of anisotropic rocks', in Hudson, J. A. (ed.) *Comprehensive Rock Engineering Volume 1: Principles, Practice & Projects*. Oxford: Pergamon Press, pp. 315–330.

Riahi, A., Hammah, E. R. and Curran, J. H. (2010) 'Limits of Applicability of the Finite Element Explicit Joint Model in the Analysis of Jointed Rock Problems', in *44th ARMA Conference*. Salt Lake City.

Ridl, R. N. (2021) *Evaluation of Buckling Deformation in the Schist of the Cromwell Gorge, New Zealand*. University of Canterbury.

Rocscience Inc. (2019) 'RS2 Version 10.0-2D Geotechnical Finite Element Analysis'. Toronto, ON, Canada.

Saroglou, C. *et al.* (2018) 'ARMR, a new classification system for the rating of anisotropic rock masses', *Bulletin of Engineering Geology and the Environment*. Bulletin of Engineering Geology and the Environment, (2010). doi: 10.1007/s10064-018-1369-4.

Tommasi, P. *et al.* (2009) 'Buckling of high natural slopes: The case of Lavini di Marco (Trento-Italy)', *Engineering Geology*. Elsevier B.V., 109(1–2), pp. 93–108. doi: 10.1016/j.enggeo.2009.02.002.

Wang, B., Cavers, D. S. and Wong, B. C. (2004) 'Surface buckling failure study and support design at the Quintette coal mine, Canada', in Lacerda, W. et al. (eds) *Landslides: Evaluation and Stabilization*. London: Taylor & Francis Group, pp. 475–480.

The Evolution of Geotech - 25 Years of Innovation – Hammah et al. (Eds)

Interaction of twin tunnels within weak rock masses and slopes

I. Vazaios
Arup, London, UK

N. Vlachopoulos
Royal Military College of Canada, Kingston, Canada

ABSTRACT: This paper investigates the influence of twin tunnels with respect to the twin-tunnel interaction and stability of tunnels running through slopes within weak rockmasses based on the slope angle and the quality of the in-situ material and rockmass strength. This work involved the use of 2D Rocscience numerical models of tunnel sequencing for numerical simulation of composite material behaviour and sequential tunnel deformation response. Validation data was obtained from the field for tunnels that were excavated as part of actual tunnels in Europe. The goal of the research is to highlight the use of numerical analyses tools and the selection of constitutive relationships for such analyses at the design stage whereby geotechnical risks can be determined prior to tunnel construction within slopes. Additionally, it is of interest the implicit influence of tunnel construction within slopes at a shallow overburden for different topography and rockmass quality cases with respect of slope stability and the potential shear failure mechanisms.

1 INTRODUCTION

As the population of cities around the world is growing, urban areas need to be developed and organized in dense and built-up environments. Underground construction projects have shown great improvements in many different aspects in the design and development of large cities in terms of safety, pollution, time management, scenery, and so on. However, such construction projects are now becoming bigger in an environment surrounded by critical infrastructure such as pipelines and heritage buildings, resulting in a complicated worksite full of risks. Furthermore, tunnels with various functionality might be located within close proximity of each other, as illustrated in Figure 1. Under these scenarios, important factors should be considered for design and construction of such tunnels based on their relative spatial configurations within their vicinity. One tunnel might intersect or overpass another one at an angle, as shown in Figure 1d. Currently, tunnels are frequently required to be parallel either in a horizontal (Figures 1a and 1b) or vertical (Figure 1c) configuration. Oftentimes, the alignment of a road or rail communication route through mountainous terrain (Figures 1e and 1f) will involve the design and construction of twin tunnels. This practice has come about to improve upon safety and efficiencies associated with route performance. In this way, one tunnel branch is constructed for each of the circulation directions. Another advantage of constructing twin tunnels is to reduce the size and, therefore, the geometrical and geomechanical challenges associated with constructing such large openings within weak rock masses and (or) unfavorable geotechnical conditions; for example, certain hydrological tunnels that have been constructed as twin tunnels rather than one tunnel with a larger diameter. Within a twin tunnel design, the method, type, and rate of construction of one of the bored tunnels can,

DOI: 10.1201/9781003188339-58

potentially, affect the behaviour and performance of the other parallel (adjacent or within close proximity) tunnel during the construction process. As with any tunnel construction, the excavation process affects the overall stress field around the tunnel. In this paper we will discuss about the twin tunnel interaction utilizing two specific case studies of twin tunnels within weak rock masses in Greece and investigate the interaction between twin tunnels using two dimensional (2-D) numerical analyses with the finite element (FE) code Phase2 (Rocscience 2012).

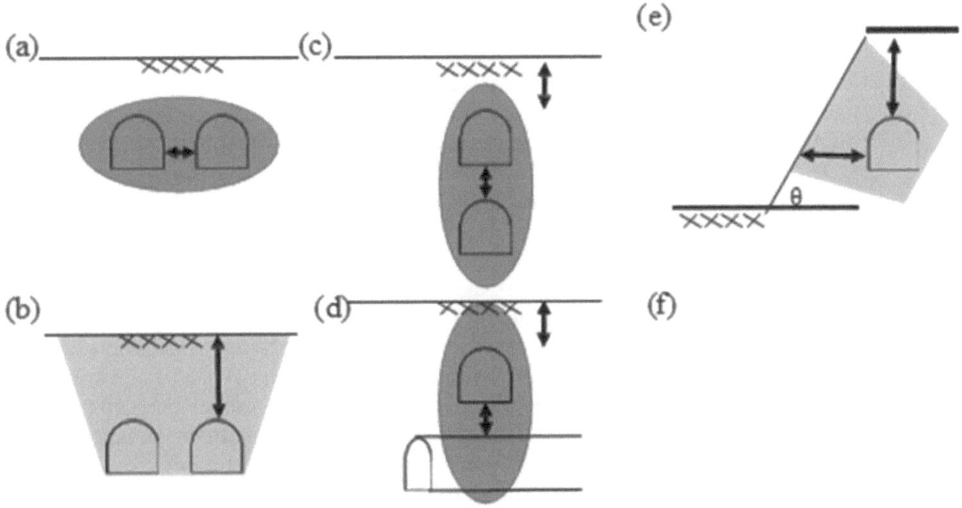

Figure 1. Interaction of tunnels with ground and other tunnels: (a) shallow twin parallel tunnels in a horizontal plane; (b) deep twin parallel tunnels in a horizontal plane; (c) twin tunnels parallel in a vertical plane; (d) a tunnel passing over another tunnel; (e) single tunnel near a slope; (f) twin parallel tunnels nearby a slope.

2 LITERATURE REVIEW

The interaction of twin tunnels has been investigated in literature using construction data as well as numerical and experimental methods mainly in terms of tunnel convergence and surface settlements. Terzaghi (1942) presented construction data from twin tunnels excavated within Chicago clay (with pillar width, $W = 0.425D$, where D is tunnel diameter) and Ward and Thomas (1965) from twin tunnels in London clay (pillar width $W = 0.60D$). Ghaboussi and Ranken (1977) performed 2-D numerical analyses and concluded that values of pillar width $>2D$ lead to negligible interaction of the two branches. Adachi et al. (1993) used the ratio Z/W (where Z is the vertical distance from the surface to the tunnel crown) to quantify the interaction between the two branches based on the results of 2-D model tests.

Addenbrooke and Potts (1996) investigated the eccentricity of the Gaussian settlements curve using 2-D finite element models for a circular tunnel section and the geotechnical conditions corresponding to the London clay formation. Addenbrooke and Potts (2001) further discussed the issue of twin tunnel interaction in London clay using 2-D numerical analyses for the case of side-by-side and piggyback geometries. Chang et al. (1996) proposed a simplified approach for the calculation of the factor of safety of the pillar between two tunnels, and provided a qualitative description of the methods that should be employed to decrease the interaction between the two tunnels and the potential failures. Kim (2004) studied the interaction between two tunnels via experiments in a plane strain tank and numerical analyses. Both approaches showed that the construction of the second tunnel led to additional downwards displacement to the crown of the existing tunnel and additional outwards displacement at the tunnel side (direction towards the new tunnel). Ng et al. (2004) investigated via 3-D coupled numerical analyses the influence of the lagging distance between the face of two tunnels on the induced settlements, convergence, excess porewater pressure, and internal forces developed in the tunnel lining (tunnel height H = 20 m, equivalent D=8.64 m, pillar width W = 1D). In a similar study, Do et al. (2016) found that the minimum normal force in the preceding tunnel liner occurred when both tunnels were excavated at the same time, while the normal force in the lining of the following tunnel reached a minimum when the preceding tunnel was far enough from the second tunnel.

Hage Chehade and Shahrour (2008) examined the effect of the relative spatial positioning of twin tunnels and construction procedures on the surface settlement using a 2-D FE analysis. Liu et al. (2009) studied the effect of construction of a second tunnel on the support elements of an existing tunnel in Sydney when the second tunnel axis is normal to the axis of the first tunnel. Chakeri et al. (2011) analysed the effect of shallow twin tunnel spacing and pillar width on the surface settlement and support response using 3-D continuum modelling.

2.1 *Monitoring of tunnels*

As with any design and construction method using the Observational Method, tunnel monitoring plays an essential role in assessing the behaviour of the ground due to excavation and the performance and suitability of the tunneling support scheme that has been designed and implemented. This is further reinforced by the International Tunnelling Association (AITES/ITA, 2011). As seen in Figure 2, a tunnel monitoring plan includes typical instrumentation equipment (Figure 2a) that is used to assess the performance of the tunnel support as well as the ground behaviour. Also included in Figure 2b are the typical locations of survey points that are utilized in order to determine the tunnel convergence of the tunnel cross-section at various intervals along the longitudinal profile of the tunnel alignment.

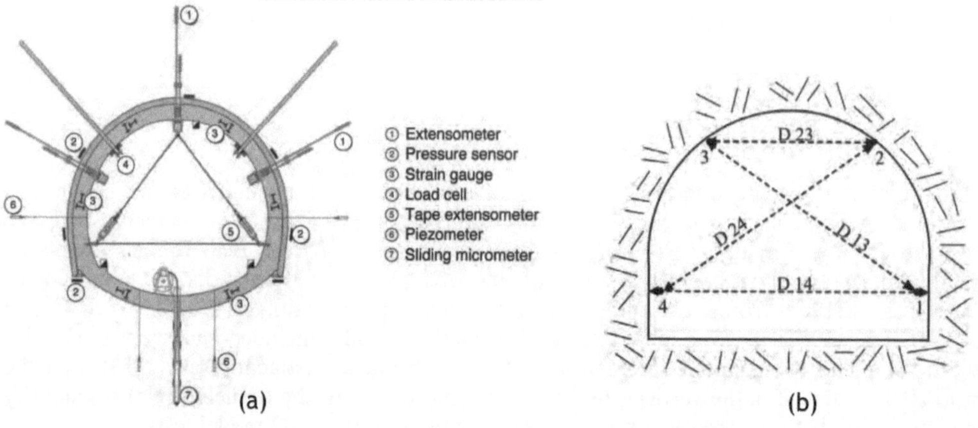

Figure 2. Tunnel Monitoring (a) Typical Instruments that are installed in order to assess tunnel behaviour, and (b) pin or survey point locations in order to determine tunnel convergence (AITES/ITA, 2011).

An example of the long-term monitoring of a surveyed cross section (similar to that shown in Figure 2b but for a circular bore) is shown in Figure 3 below. The survey data captures the vertical displacement at the locations (1 to 5) at the periphery of the tunnel opening. The tunnel was constructed using a top heading (a' phase) and bench (b' phase) and invert. Once the initial a' and b; phases plus the invert were constructed, the displacements stabilize at all locations. However, when the parallel tunnel tube (or twin tunnel) excavation passes by this location, it is evident that these exaction activities influence this initial bore or tube (as highlighted in the Green Section within Figure 3. In this way, the adjacent construction surely influences and affects the behaviour and performance of the other tunnel within its proximity. These type of data can be utilized in order to improve and assess support design as well as play a major role in calibrating and validating numerical models.

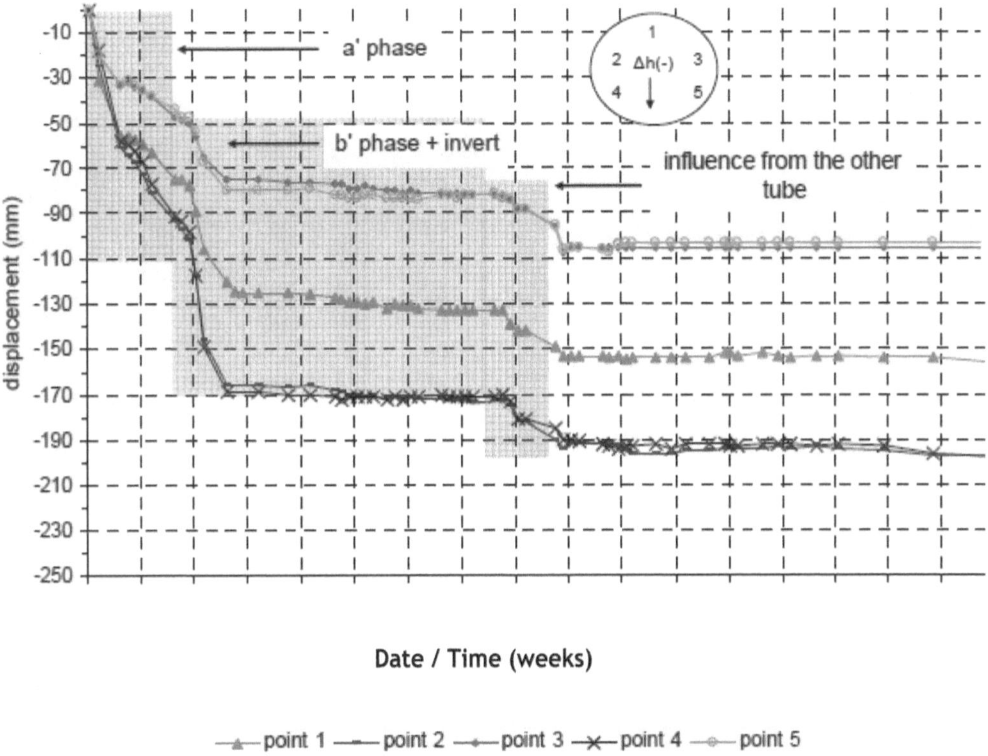

Figure 3. Vertical Displacements during excavation and primary support captured on-site during the construction of a circular tunnel with a twin bore excavated next to it (Vlachopoulos et al., 2013).

3 TWIN TUNNEL WITHIN WEAK ROCKMASSES

Within a twin tunnel design, the method, type and rate of construction of one tunnel can, potentially, affect the behaviour and performance of the other tunnel during the construction process. As with any tunnel construction, the excavation process affects the overall stress field around the tunnel. Depending on the geological conditions, this could create significant plasticity zones around the tunnel opening as summarized by Vlachopoulos and Diederichs (2009). The extent of the plasticity zone and re-arrangement of the stress field around the first tunnel will affect the conditions through which the second tunnel will pass parallel to the first tunnel. The construction of the second branch within such conditions may lead to the development of additional convergence, loads or even failures of the support shell of the first branch. Such interaction of twin tunnels has been recorded in many cases, however, many of the current design methodologies that are used have been formulated based on single tunnels and do not take into account the interaction between the two branches. Therefore, we investigate the interaction between twin tunnels using two dimensional (2D) numerical analyses with the finite element code Phase2.

3.1 *The case of the Driskos tunnel*

Based on the aforementioned, this paper focuses on the tunnelling completed in the Epirus and Western Macedonia regions of Northern Greece, as part of the massive Egnatia Odos Highway construction project. Specifically, the geological conditions that are considered in this paper come from those present at the Driskos Twin Tunnel Site (Figure 4). The tunnel construction at the Driskos site consists of twin, parallel tunnels that are approximately 4,570 m in length with a maximum overburden of 220 m.

Figure 4. Northern Tunnel Portals of Driskos Road Twin Tunnels denoting the spacing (S) or pillar width between the bores; the diameter of each bore is dented by D.

3.2 *Weak rock mechanics*

The Geological Strength Index (GSI) was used for this investigation (Hoek and Marinos, 2000). This classification system allows the estimation of the rock mass properties in varying geological conditions. The main criteria associated with the GSI classification system is a detailed engineering geology description of the rock mass that is qualitative in nature. This grew out of the notion that numbers on joints were largely meaningless for the weak and complex rock masses (Marinos et al., 2006). Accurate equivalent rock mass performance predictions for tunnels in such materials (including yield and residual strength as well as flow and dilation considerations) are complicated by other structural peculiarities (mixed face conditions, anisotropy due to the structural elements). The parameters/properties associated with the four weak rockmasses that were analyzed are seen in Table 1.

3.3 *The numerical model*

Within this investigation, Phase2 (Rocscience Inc., 2012) was used for the 2D numerical analysis. Phase2 uses an implicit Finite Element Method (FEM) which is widely used in the geotechnical and geological engineering industry in order to capture the behaviour of a tunnel (i.e. stress re-distributions and displacements) associated with tunnel excavation. The 2D numerical runs were limited to Elastic and Plastic models. The 2D model and associated geometry is shown in Figure 5.

In order to analyse the effects of twin tunnel excavation, the same baseline tunnel cross section of diameter of 11 meters was utilized in the modelling process. Models were created using two of the baseline cross sections positioned in the same vertical position varying in spacing from 0.1 to 10 diameters (1.1, 2.2, 2.75, 5.5, 8.25, 11, 16.5, 22, 33, 55, 110m) corresponding to percentages of the diameter (*D*). The goal of this portion of the analysis was to determine the

Table 1. Parameters used for 2D model comparisons.

Material/ Parameter	B1	C1	D1
p_o/σ_{crm}	8	6	4
σ_{ci} (MPa)	35	35	50
m_i	7	7	7
v	0.25	0.25	0.25
γ (MN/m^3)	0.026	0.026	0.026
E_i	19212	19249	27630
p_o (MPa)	28	28	28
GSI	35	45	48
m_b	0.687	0.982	1.093
s*	0.0007	0.0022	0.0031
a*	0.516	0.508	0.507
E_{rm} (MPa)	2183	4305	7500
σ_{crm} (MPa)*	3.5	4.7	7
c (MPa)*	1.100	1.753	2.145
ϕ	21.50	23.71	27.05

* Values calculated based on Hoek et al. 2002.

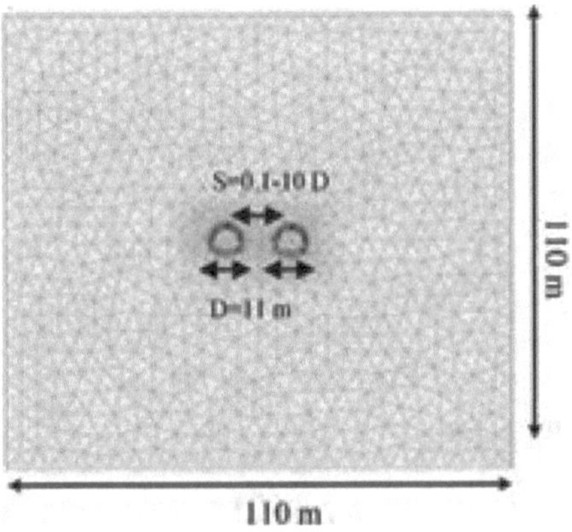

Figure 5. Selected geometries associated with (a) 2D (Phase2).

minimum spacing between tunnels in order to determine the uniquely independent behaviour of each tunnel excavation.

3.4 *Analysis results*

The results from the various model runs are shown in the following tables. Elastic and plastic unsupported analyses are summarized in Figures 6 and 7 respectively. Spacing models were run with tunnel to tunnel spacing ranging from 1.1m to 110m clear span between excavations. The results from the various model runs are shown in the following tables. Elastic and plastic unsupported analyses are summarized in Figures 6 and 7 respectively. Spacing models were run with tunnel to tunnel spacing ranging from 1.1m to 110m clear span between excavations.

The crown deflection profile decreases exponentially between 1.1 and 16.5 meters of spacing before exhibiting asymptotic behaviour. Minimal deflection change is recorded when spacing increases beyond 22 meters. A similar trend is observed for deflection in the floor. The results in Figures 6 and 7 depict the displacement relative to a nominally identical, single tunnel bore.

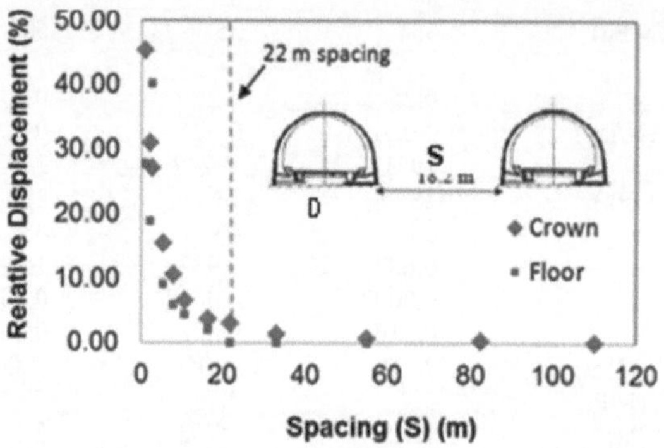

Figure 6. Relative Displacements of Twin Tunnels to Single Bore Displacements with varying spacing (pillar width) for the Elastic Case.

The plastic analysis exhibits the same trend as that of the elastic model. Similar to the elastic analysis, no significant changes in displacements (crown and floor) are observed beyond a clear spacing of 22 meters. (i.e. 2*D*-two diameter widths). In terms of design implications, the amount of relative displacement between a spacing of 1*D* and 2*D* for both the elastic and plastic case is approximately 5%.

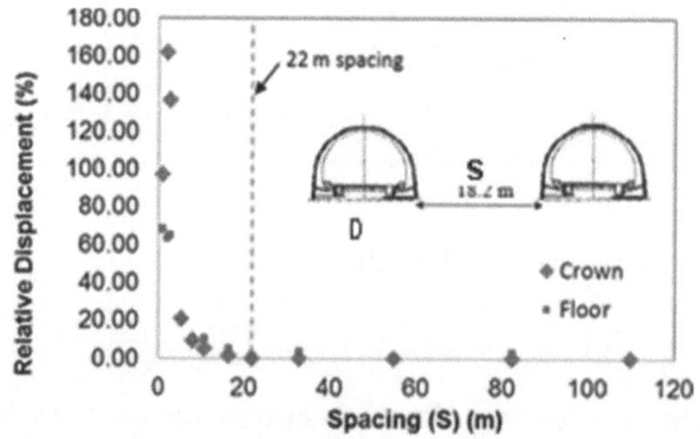

Figure 7. Relative Deflections of Twin Tunnels with varying Spacing for Plastic Case.

4 TWIN BORED TUNNELS WITHIN SLOPES OF WEAK ROCKMASSES

One of the major concerns in the design and construction of twin tunnels adjacent to mountainous slopes is that the slopes could be inherently unstable from the onset of the tunnel construction. Within twin tunnelling design, factors that must be taken into consideration include: the method, type and rate of construction as the behaviour and performance of one tunnel can certainly affect the behaviour and performance of the other tunnel, which when combined to an existing slope may result in rather challenging conditions. In the subsequent sections we are discussing the impact of twin tunnel boring on slope stability and the S3 Tunnel of the Egnatia Odos Project is presented.

4.1 *Numerical modelling*

The presence of the slope adds to the already complex conditions of the twin tunneling excavation, resulting to interaction phenomena which are rather critical both in the design and construction process of this kind of projects. Therefore, for the purposes of this paper, the finite element code RS2 (Rocscience 2015) was used in order to conduct two-dimensional (2D) numerical simulations for the parametric study. The parametric study involved varying the rockmass quality and the angle of the slope in a series of numerical analyses for a set distance between the twin tunnels.

4.2 *Model configuration*

Within this study, four different slope angle values of θ (15°, 25°, 35° and 45°) were investigated. In such slope stability analyses, it is crucial to perform a type of analysis which is going to yield a safety factor as a criterion of assessing the stability of the slope. In RS2, a slope stability analysis can be conducted by employing the Shear Strength Reduction (SSR) method. The main advantage of this method over traditional limit-equilibrium techniques is the non-requirement of erroneous assumptions concerning the shape or location of the failure surface. The numerical model consists of a slope having a height of H=200m, which is maintained constant for all four slope angle (θ) scenarios, two tunnels with a diameter of D=10m and a spacing of S=25m (i.e. 2.5D). Additionally, the tunnel overburden is within a range of h=25m to 30m, similar to the geometry that was employed for the S3 Tunnel of the Egnatia Odos. The remaining geometrical features of the numerical model as they were produced in RS2 are illustrated in Figure 8, with the model comprised of approximately 20,000 6-node, triangular elements of approximately the same size in order to ensure that the results would not be affected by the element size in each single case.

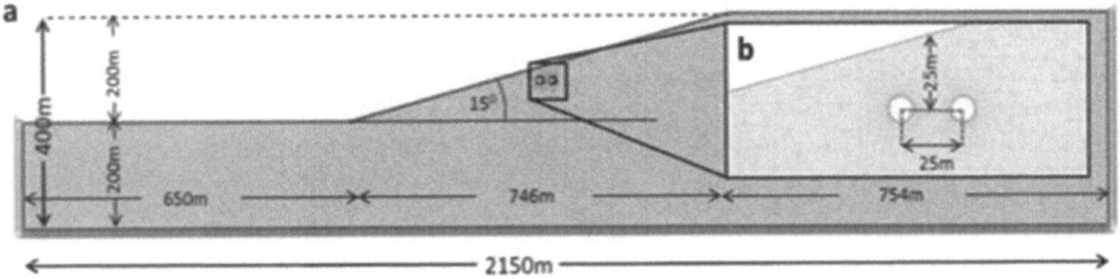

Figure 8. A. Numerical model configuration in RS2 b. Detail of the tunnel configuration within the numerical model; 15o slope angle, 25m overburden and 25m spacing between twin tunnels.

4.3 *Simulation process*

Due to the complexity of a twin tunnelling-slope stability problem, the stability of the slope must be investigated both, prior and post-construction of the tunnels. Therefore, the stability of the slope was examined prior to the excavation of the tunnels in order to determine the safety factor of the different slope angles and respective rockmass quality cases. Based on the SSR method, the Strength Reduction Factor (SRF), which is the equivalent of the safety factor associated with the traditional limit equilibrium methods, is estimated in order to determine whether a slope is stable or not. The estimated SRF for the slopes prior to the excavation of the tunnels is used as a reference point for the rest of the analyses (SRFref). Following the simulation of the slope as a single entity, six more steps were added to the simulation in order to account for the 3D effects of the excavation and the de-stressing taking place (i.e. taking into account the pre-convergence effects as the excavation face approaches a specific location) forming a simulation of seven stages in total.

Six different GSI values (10, 20, 30, 40, 50 and 60) were used in order to examine the impact of the rockmass quality on a twin tunnelling-slope stability problem. In the analyses performed, the rockmass is assumed to behave in an elasto-plastic manner according to the Hoek-Brown failure criterion which utilizes GSI as one of its input parameters. The material properties are summarized in Table 2.

Table 2. Material properties used in the numerical model. [1] Rockmass properties (Parametric study), [2] Rockmass properties for the S3 Tunnel (Koronakis et al. 2004), and [3] Liner properties.

Mechanical Properties	[1]	[2]	[3]
Young's modulus E (GPa)	0.30	0.30	30
Poisson's ratio v	0.3	0.3	0.2
Unit weight γ (kN/m^3)	26	26	25
Intact uniaxial compressive strength σci (MPa)	5.0	-	-
Constant mi	10	-	-
Friction angle φ (^{0})	-	27	
Cohesion c (kPa)	-	30.0	
In-situ stress ratio Ko	1.0	1.0	-
Constitutive model	Elasto-plastic	Elasto-plastic	Elastic

4.4 *Numerical analysis results*

Having discussed the basic configuration of the numerical model along with the major assumptions made, the results of the parametric analyses will be discussed in this section in order to denote how the slope stability was affected due to the twin tunnelling.

4.5 *Impact of twin tunnelling on the slope SRF*

To determine the stability of the slopes under investigation, a slope analysis was performed to estimate the SRF values pre-tunnelling construction and these results are illustrated in Figure 9a. For each analysis, all material properties remained the same except for the GSI value which was used to define the rockmass quality. In Figure 9b, the estimated SRF values after the excavation of the tunnels are illustrated. It can be observed that for slope angles equal to θ = 15°, 25° and 35° the estimated SRF values are approximately equal in both pre and post construction of the tunnels. However, for angle θ=45° and GSI values 40, 50 and 60 respectively the SRF values decrease when the tunnels are excavated. From Figure 9c, it can be inferred that for relatively shallow slope angles, the excavation of the tunnels does not significantly affect the stability of the slope. However, as the slope becomes steeper, the effect of the excavation of the tunnels becomes more significant regardless of the rockmass quality.

4.6 *S3 tunnel of Egnatia Odos*

The S3 Tunnel (Figure 10) is a shallow, twin bored tunnel, 230m long and 12m wide with a 30m axial length spacing between the two branches. Bored through a rock slope composed of sheared phyllite thrusting on top of crystalline limestone, significant stability issues were encountered during construction despite not having any signs of instability during the

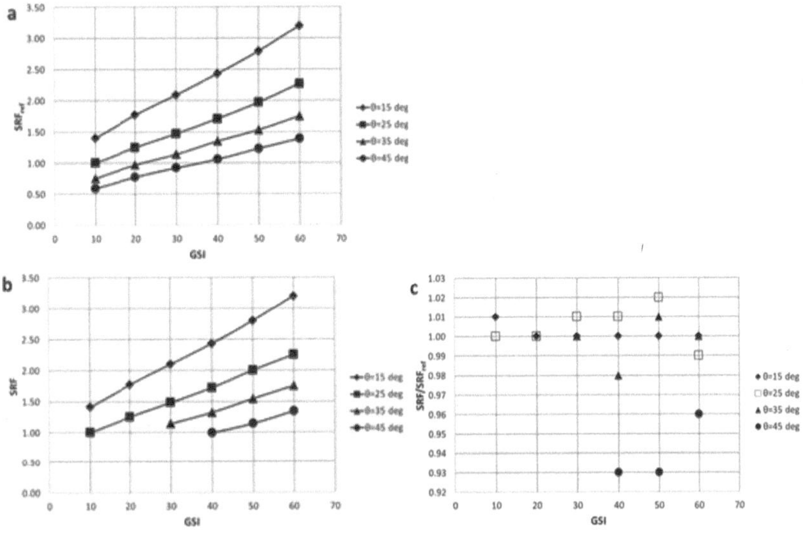

Figure 9. a. Strength Reduction Factors pre tunnelling (SRFref) and b. post tunnelling construction (SRF) for different angles and GSI values c. Normalized SRF/SRFref values.

geotechnical assessments (Koronakis et al. 2004). Due to the high degree of jointing in such rockmasses, failure does not occur on distinct structural features. Therefore, the medium can be treated as an equivalent continuum rather than a discontinuum and the assumption of a homogenous isotropic rockmass can be made.

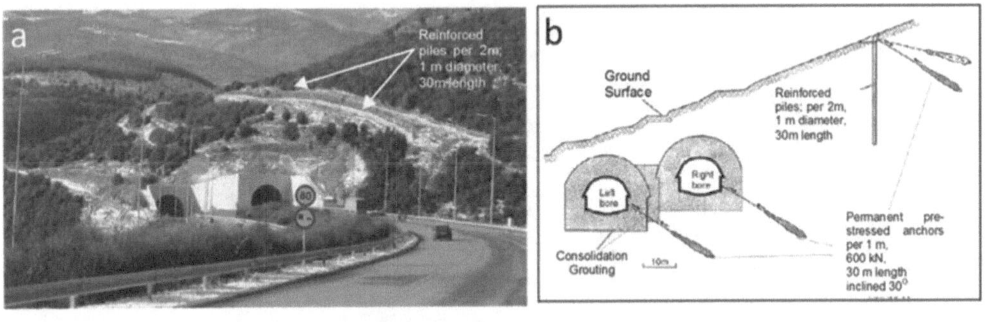

Figure 10. Support Measures for S3 Tunnel; (a) figure of support measures implemented (modified after Koronakis et al. 2004) and, (b) Photo of reinforced slope/twin tunnel portals.

To highlight the significance and impact of a tunnel constructed and passing through a slope in the case of the S3 Tunnel, a 2D numerical model was created by employing the techniques discussed in the previous sections. The slope angle in this case was θ= 210. The results of the analysis showed that the natural slope, prior to excavation of the tunnels, had an SRF value of 1.21, indicating that it was stable but its factor of safety is relatively low as a result of the low strength of the rockmass, despite the shallow angle of the slope. In Figures 11a and 11b, the total displacement and maximum shear strain contours of the slope are illustrated. However, post tunnel construction resulted in a lower SRF value of 1.15. This reduction is not that significant and it could be argued that this may be the result of the numerical analysis process as previously stated, as well as the result of the physical processes due to the excavation of the tunnels. Additionally, the potential shear failure surface is relocated as indicated in Figures 12a and 12b and passes through the tunnels, as a result of the loss of shear resistance due to the material removal and the change in the stress regime. This is a dissimilar potential failure mechanism than the one seen in Figure 11, making it apparent that there is indeed, an influence and interaction between the excavation of the tunnels and stability of the slope, it can be seen that the potential shear failure plane now passes through the excavated areas and passes deeper than the initial potential failure plane.

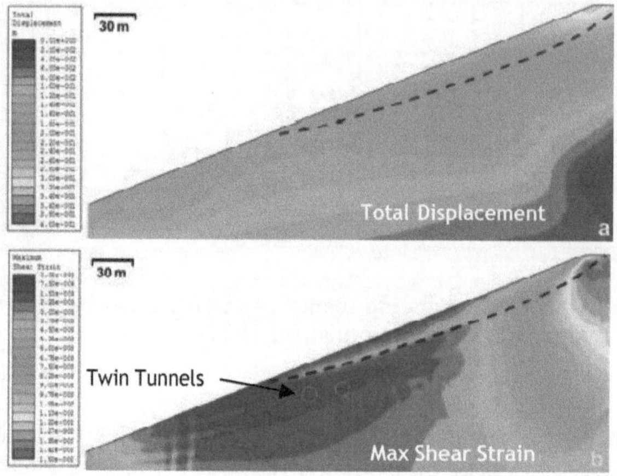

Figure 11. (a). Total displacement and (b). Maximum shear strain contours of the slope prior to the excavation of the tunnels. The potential shear failure surface is highlighted with a dashed line.

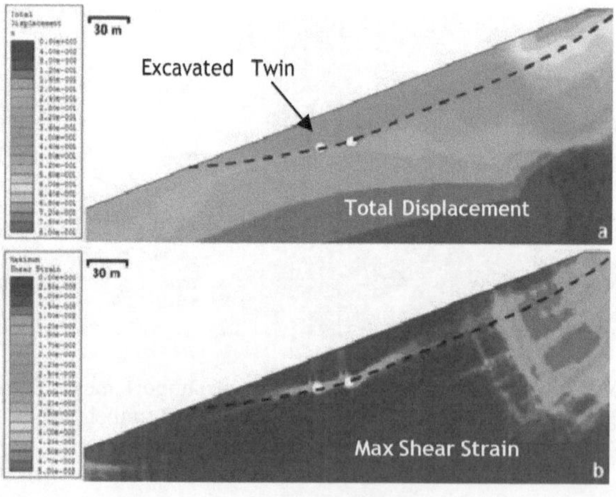

Figure 12. (a). Total displacement and (b) Maximum shear strain contours of the slope after the excavation of the tunnels. The potential shear failure surface is highlighted with a dashed line.

5 CONCLUSIONS

In this study, twin tunnelling interaction was investigated under different conditions. Conventional 2-D were created for multiple rock mass conditions, tunnel geometries. These numerical models serve as a baseline to examine the impact of rock mass conditions, the pillar width, and the presence of support on the interaction between the two tunnel branches. Results from these analyses suggest that the size and shape of the overall zone of plasticity around the tunnel dictate the rock mass and tunnel behaviour during excavation, and therefore the plastic zone around the tunnels needs to be taken into account for a slope–tunnel system as well, as it will dictate the amount of interaction between the two excavations. Furthermore, for twin tunnelling adjacent to rock slopes slope stability analysis was conducted by employing the novel SSR method. The main advantage of this method over traditional limit-equilibrium techniques is the nonrequirement of erroneous assumptions concerning the shape or location of the failure surface. These slope–tunnel models utilized the threshold value where the pillar width is assumed to be enough to ensure no interaction between the twin tunnels and thus isolates explicitly the effect of the excavation on the slope. The results indicated that by having the pillar width constant, the vertical translation of the tunnels (i.e., increase in the overburden) is actually more beneficial than that of a horizontal translation, resulting in a higher factor of safety. Therefore, redesigning an alignment with a vertical translation of the tunnels leads to better stress redistributions and slope stability conditions. This being said, it cannot be overstated how the in situ materials and their rock quality need to be specifically determined to ensure that a significant factor is not disregarded. As well, in a broader sense than included in these findings, specific guidelines and codes must be developed that take into consideration the influence of such tunnel construction activities within slopes.

ACKNOWLEDGEMENTS

The authors would like to thank the Natural Sciences and Engineering Council of Canada (NSERC), the Canadian Department of National Defence and The Royal Military College (RMC) Green Team.

REFERENCES

Adachi, T., Kimura, M., and Osada, H. 1993. Interaction between multi-tunnels under construction. In Proceedings of the 11th Southeast Asian Geotechnical Conference, Singapore, pp. 51–60.

Addenbrooke, T.I., and Potts, D.M. 1996. Twin tunnel construction—ground movements and lining behaviour. In Geotechnical Aspects of Underground Construction in Soft Ground, Edited by Mair and Taylor. Balkema, Rotterdam, pp. 441–446.

Addenbrooke, T.I., and Potts, D.M. 2001. Twin tunnel interaction: surface and subsurface effects. International Journal of Geomechanics, 1(2): 249–271. doi: 10.1061/(ASCE)1532-3641(2001)1:2(249).

AITES/ITA WG-2 – Research, 2011, Monitoring and Control in Tunnel Construction. International Tunnelling and Underground Space Association (ITA) report No 009/2011. No. ISBN: 978–2–9700776–3–3.

Chakeri, H., Hasanpour, R., Hindistan, M.A., and Unver, B. 2011. Analysis of interaction between tunnels in soft ground by 3D numerical modeling. Bulletin of Engineering Geology and the Environment, 70(3): 439–448. doi:10. 1007/s10064-010-0333–8.

Chang, C.T., Lee, M.C., and Hou, P.C. 1996. Design of twin-tube tunnel through soft rock. In Proceedings of Geotechnical Aspects of Underground Construction in Soft Ground. Edited by Mair and Taylor. Balkema, Rotterdam, pp. 251–255.

Do, N.A., Dias, D., and Oreste, P. 2016. 3D numerical investigation of mechanized twin tunnels in soft ground-influence of lagging distance between two tunnel faces. Engineering Structures, 109: 117–125. doi:10.1016/j.engstruct. 2015.11.053.

Ghaboussi, J., and Ranken, R.E. 1977. Interaction between two parallel tunnels. International Journal for Numerical and Analytical Methods in Geomechanics, 1(1): 75–103. doi:10.1002/nag.1610010107.

Hage Chehade, F., and Shahrour, I. 2008. Numerical analysis of the interaction between twintunnels: influence of the relative position and construction procedure. Tunnelling and Underground Space Technology, 23(2): 210–214. doi:10.1016/j.tust.2007.03.004.

Hoek, E., and Marinos, P.G. 2000. Predicting tunnel squeezing problems in weak heterogeneous rock masses. Tunnels and Tunnelling International, 32(11): 45–51.
Hoek, E., Carranza-Torres, C., and Corkum, B. 2002. Hoek-Brown failure criterion-2002 edition. In Proceedings of North American Rock Mechanics Symposium-Tunnelling Association of Canada, pp. 267–273.
Kim, S.-H. 2004. Interaction behaviours between parallel tunnels in soft ground. Tunnelling and Underground Space Technology, 19(4–5): 448. doi:10.1016/j. tust.2004.02.058.
Koronakis, N., Kontothanassis, P., Kazilis, N., and Gikas, N. 2004. Stabilization measures for shallow tunnels with ongoing translational movements due to slope instability. Tunnelling and Underground Space Technology, 19(4–5): 495. doi:10.1016/j.tust.2004.02.093.
Liu, H.Y., Small, J.C., Carter, J.P., and Williams, D.J. 2009. Effects of tunnelling on existing support systems of perpendicularly crossing tunnels. Computers and Geotechnics, 36(5): 880–894. doi:10.1016/j.compgeo.2009.01.013.
Marinos, P.G., Hoek, E., Kazilis, N., Agistalis, G., Rahaniotis, N., and Marinos, V. 2006. The tunnels of Egnatia highway, Greece. Design in a variety of rock masses under difficult geological conditions. In Proceedings of the 5th Greek Geotechnical and Environmental Engineering Congress, pp. 517–524.
Ng, C.W.W., Lee, K.M., and Tang, D.K.W. 2004. Three-dimensional numerical investigations of new Austrian tunnelling method (NATM) twin tunnel interactions. Canadian Geotechnical Journal, 41(3): 523–539. doi:10.1139/t04-008.
Rocscience. 2012. Phase2.
Terzaghi, K. 1942. Earth Pressure and Shearing Resistance of Plastic Clay: A Symposium: Liner-Plate Tunnels on the Chicago (IL) Subway. In Proceedings of the American Society of Civil Engineers, pp. 862–899.
Vlachopoulos, N., Fortsakis, P and Oke,J. 2013. Investigation into the Influence of Parallel Excavation of Twin Bored Tunnels within Weak Rock Masses. GEO Montreal, Canadian Geotechnical Society, Oct 2013. Montreal, Quebec, Canada.
Vlachopoulos, N., and Diederichs, M.S. 2009. Improved longitudinal displacement profiles for convergence confinement analysis of deep tunnels. Rock Mechanics and Rock Engineering, 42(2): 131–146. doi:10.1007/s00603-009-0176-4.
Ward, W.H., and Thomas, H.S.H. 1965. The development of earth loading and deformation in tunnel linings in London Clay. Building Research Station, Ministry of Technology.

Session 13 - Tunnels & caverns II

The Evolution of Geotech - 25 Years of Innovation – Hammah et al. (Eds)

The probabilistic analysis of steep lakeside slopes; geotechnical-geological-hydrogeological constraints and numerical analysis, an example from Hungary

Fruzsina Kápolnainé Nagy-Göde & Ákos Török*
Department of Engineering Geology and Geotechnics, Budapest University of Technology and Economics, Budapest, Hungary

ABSTRACT: Steep and unstable slopes often border large water bodies such as lakes and sea. The present paper provides an overview of the stability of high banks that border the largest Central European lake, Lake Balaton in Hungary. The stability analysis of the presented slope was performed in different software - Plaxis, Geo5 and Slide2 - and their results were compared. The modelling revealed that the stability of the high lakeside banks does not reach the required safety. Two major identified failure scenarios are deep seating curved landslides and at the top of the slope failures due to erosional processes.

1 INTRODUCTION

In previous works, several approaches were compared using various numerical methods (Eberhardt 2003) and analytical tools (Utili and Costa 2015). The methods of calculating and detecting slope failures have been developed rapidly in recent years, mostly using Limit Equilibrium (LE) (Li et al. 2020) and Finite Element (FE) analysis (Matthews et al., 2014, Cami et al., 2018, Oberhollenzer et al. 2018,). Recently, techniques that record the surface topography by UAV (Bar et al. 2020, Giordan et al. 2020) or other remote sensing techniques (Jaboyedoff 2012, Menegoni et al. 2020) have also been applied.

In this study, traditional analysis of slopes is presented with the application of FE and LE methods. The studied slopes are located close to the lakeside of Lake Balaton in Hungary. The lake is characterized by rapid changes in coastal morphology and variations in water levels. Climatic differences and diurnal to annual changes in precipitations and human interaction also influence the stability.

In the studied area, which is an expanding tourist destination, the infrastructure and recreational environment require slope stability assessment and failure risk analysis. The result of such a stability test is presented in this paper. The study focuses on the long-term behaviour of slope and probabilistic analysis of failures. The results of numerical calculations by Plaxis, Geo5 and Slide2 of Rocscience suggest a risk of a landslide at the studied slopes. The paper also compares the applicability of these numerical tools in slope stability assessment. The modelled layers and geometry represent Lake Balaton's shores, but the obtained results can be used as an analogy for other coastal slopes. An additional outcome of this study is to maintain the appropriate safety; it is necessary to perform monitoring and maintenance work on these types of coastal slopes regularly.

*Corresponding author
DOI: 10.1201/9781003188339-59

2 SITE DESCRIPTION

The studied slope is located in the coastal zone of Lake Balaton. At the lake basin eastern side, n the border of Balatonakarattya and Balatonkenese in Hungary (Central Europe). In this area, the slope's stability has emphasized importance due to the rail traffic; a railway line is cut into the slope. Additionally, a resort area at the bottom and the top of the slope is endangered by potential landslides.

The geometry of the study area is about 90 meters high and 340 meters wide. The studied cross-sections were selected based on five core drilled boreholes (the deepest being 80 m long), one cone penetration test and one soil physical measurement (60 m long). Additional data sources were also used: core drillings of the larger area of 400 m x 800 m were taken into account in the model's construction. The modelled section's approximate position and the morphological parts of the slope are marked in yellow, while the groundwater is indicated in blue in Figure 1. The geomorphological evolution of the site is unique since the lakeside has retreated due to several phases of sliding.

Nevertheless, the area was built on centuries ago and is still rapidly developing, despite major failures. One well-documented failure was in 1914 when the railway line with a steam engine and train slipped into the lake. The recent movements also signify the vulnerability of slopes. The hydrogeology is unique since the hinterland feeds groundwater; additionally, there is a complex interaction of lower horizons of groundwater with lake water. The lake level has been fluctuating, in ten meters of amplitude according to the records of the past 500 years. The groundwater table has an indirect link with precipitation. These conditions lead to the presence of confined aquifers that reduce slope stability. A monitoring system is in operation: groundwater monitoring wells and long-term monitoring displacements (surveying and inclinometers).

Figure 1. Study area, morphology and groundwater.
(Photo: Diána Pesztericz)

The prevailing sediments of the banks are Miocene clay which alternates with fine sand layers and sand lenses. Silty intercalations also occur in this lithostratigraphic unit which is called Tihany Formation. The cover beds in some areas are loess. From the point of view of soil mechanics, the subsoil comprises varieties of clay, silt and silty sandy soils. There are no parallel layers or uniform layering due to historical and recent slides (Kézdi, 1952). The layers representing Tihany Formation were deposited in the foreground of the estuaries. These palaeo rivers have variable depositional environments, from deltas arriving to a palaeo lake, called Lake Pannon, to shallow coastal marine sediments. In addition to changes in palaeo lake levels, environmental changes and climatic cyclicity also contributed to the non-uniform lithology and stratal pattern of these beds. As a result, 15–30 meter changes in water level were detected, indicating the progression and regression of the lake (resulting in the cyclic displacement of its shoreline). This series of events continued until Lake Pannon moved to the south about eight million years ago, and suffusion has become rarer. The deposited sediments have different particle sizes, but fine sand prevails. Calm sedimentation was interrupted by short-term water agitation events, settling mm to cm-scale thin fine sand lenses (Sztanó et al. 2005). From a geological engineering point of view, this sediment pattern is unfavourable. Sand layers between the clay layers behave like thin aquifers and facilitate water movement. It leads to surface movements, especially when these sand beds are capped by loess. The geology and characteristics of the Pannonian Basin's sedimentary structures have been described in detail in Sztanó et al. (2013) and Koroknai et al. (2020).

Based on the core descriptions and soil mechanical parameters obtained from laboratory tests, a 'simplified' 8-layer-model was generated (Figure 2).

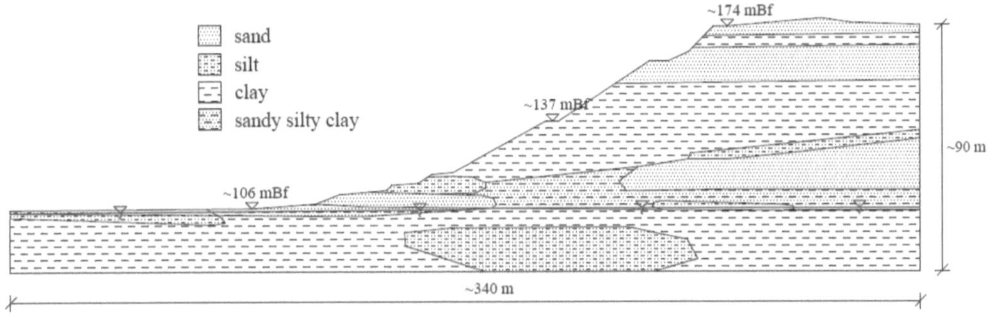

Figure 2. Model geometry, soil layers.

It has to be mentioned that the vast majority of past soil tests focused on soil classification rather than strength testing. As a result, there were not enough laboratory test results to only fit a function to the shear strength parameters. The input soil parameters of the numerical modelling were determined based on a combination of the limited results, and recommendations and former works (Table 1).

Table 1. Soil parameters applied in the model.

	g_{unsat} [kN/m^2]	g_{sat} [kN/m^2]	E [MPa]	n	c [kN/m^2]	ϕ [°]
Sa/1	17	18	6	0,35	5-8	24
Cl/1	19	20	10	0,3	50	22
Cl/2	20	21	11	0,3	60	26
Si/1	18	19	8	0,4	25	18
Sa/2	17	18	10	0,35	1	20
Sa-Si-Cl	20	21	10	0,4	30	25
Cl/3	21	22	16	0,3	80	34

3 METHODOLOGY

The stability calculations of the slope were performed using three different software applying three diverse calculation methods. In the first approach, a limit equilibrium deterministic method was applied using Geo5. It was followed by a finite equilibrium deterministic calculation method with the help of Plaxis. Finally, a probabilistic analysis of Rocscience Slide2 was undertaken. The study of a slope is based on the calculation of a safety factor and the delineation of the sliding surface.

In limit equilibrium methods, the equilibrium of a soil mass tending to slide down under gravity's influence is calculated (Duncan and Wright 2005, Utili and Crosta 2015). All these methods are based on the comparison of forces, moments, or stresses resisting the mass's movement with those that cause unstable motion (disturbing forces). Geo5 computes slope stability with a circular failure surfaces (for example, Bishop, Janbu or Spencer method) or polygonal slip surfaces (the Sarma method). Verification analyses are performed employing a classical approach (Sharma et al 2012). Two approaches to stability analysis are implemented in the program classical analysis according to the factor of safety and analysis following the theory of limit states (Görög and Török 2007). According to finesoftware (www2) in the limit state theory – what we used in Geo5 -the value of utilization Vu is calculated and then compared with the value of 100%. The value of utilization is given by:

$$V_u = \frac{M_a}{M_p} 100 < 100\%$$

where Ma is the sliding moment and Mp is the resisting moment.

As computer performance has improved, the application of FE in geotechnical analysis has become increasingly common. These methods have several advantages: to model slopes with a high degree of accuracy (complex geometry, sequences of loading, presence of material for reinforcement, the action of water, laws for complex soil behaviour) and to visualize better the deformations of soils (Matthews et al. 2014, Salunkhe et al. 2017). Plaxis software is widely used in modelling geotechnical problems, including slope stability issues (Abba 2014, Jacob et al. 2018). Plaxis can perform slope stability analysis by reducing the strength parameters of the soil. This process is called c-Phi reduction (for Mohr-Coulomb materials). In the approach, strength parameters are reduced during the calculations (Cheng and Huang 2005; Sun et al. 2017, Vinod et al 2020). The strength parameters, tan φ and cohesion, c, of the soil are successively reduced until the slope collapses. The total multiplier ΣMsf is defined as the ratio of the strength parameters entered as input values over the reduced ones. ΣMsf is set to 1 at the start of a calculation to set all material strengths to their unreduced values. The strength parameters of soil are thereby reduced systematically with an increment ΣMsf equal to 0.1 until failure. The safety factor SF is given by the ratio of the available strength over the strength at failure. It is equal to the value of ΣMsf at failure (Fawaz et al, 2015).

In Rocsience's Slide2 software, we performed a probabilistic analysis using Latin-Hypercube sampling method and global minimum analysis type. Slide2 has been widely used for probabilistic regional landslide hazard assessment (Hadjigeorgiou et al. 2006). The probabilistic Latin Hypercube sampling technique gives comparable results to the Monte Carlo technique but using fewer samples. Each analysis method in Slide2 (e.g., Bishop or Janbu) can result in a different Global Minimum slip surface. The probabilistic analysis is carried out independently on each Global Minimum slip surface that results from an analysis method. The approach assumes that the Probability of Failure calculated for the (deterministic) global minimum slip surface is representative of the Probability of Failure should all search surfaces be analyzed. In many cases, this is a valid or reasonable assumption (www1).

Random variables used in the calculation are shear strength parameters of soil layers: cohesion and friction angle. The vast majority of parameters used in geotechnical calculations follow a normal or lognormal distribution (Bond and Harris 2008). The normal distribution is commonly used for statistical analysis in geotechnical engineering. The distributions of many random variables conform to this distribution. It is generally used for probabilistic studies in geotechnical engineering unless there are good reasons for selecting a different distribution (Kádár and Nagy 2016). Based on these, we assumed a normal distribution of the parameters for the probability calculation. The values shown in Table 1 were taken as mean values. The minimum and maximum values were specified as relative values. It is recommended that the relative minimum and relative maximum values are equal to at least three times the standard deviation to ensure that a complete (non-truncated) normal distribution is defined, which has to be symmetric around the mean value (www1, Uzielli et al. 2006).

4 RESULTS AND DISCUSSION

The calculations with Geo5 software showed that utilization along local slip surfaces is 101-103%, while along global slip surfaces, it is 97-104%. These calculated values have already been determined, taking into account the expected safety of 1.35 by partial factors; thus, local stability is unacceptable, while the global stability is still just acceptable except for the Fellenius/Petterson method, which is also unacceptable.

Plaxis always detects the most probable failure according to the calculation method presented earlier. It means that local erosions are identified at the top of the slope, which is in good agreement with the site observations, i.e. erosion marks and small gullies that were found at the site. In this case, the calculated factor of safety was 1.18. This low FoS is linked to the presence of a sand layer with low cohesion. This layer is weak and easy to disintegrate. By increasing this cohesion value, we were able to study the global failure mechanism, which has a safety factor of 1.29. The geometry of the slip surface has a complex shape depending on the position of the weaker soil layers. Both safety values fall below 1.35 (cf. Eurocode 7).

Slide2 also showed that the Global Minimum slip surface to be the local surface (Figure 3).

We now present the results of the Bishop simplification, which was less in line with the results of the Plaxis software than the Janbu simplification. However, they also show a pretty good match. We can see that both deterministic and mean safety factors are 1.245. The Deterministic Safety Factor, FS (deterministic), is the safety factor calculated for the Global Minimum slip surface from the regular (non-probabilistic) slope stability analysis. The Mean Safety Factor is the mean (average) safety factor obtained from the Probabilistic Analysis. It is simply the average safety factor of all of the safety factors calculated for the Global Minimum slip surface. In general, the Mean Safety Factor should be close to the value of the deterministic safety factor, FS (deterministic). For a sufficiently large number of samples, the two values should be nearly equal. Equality was achieved, so the number of samples was sufficient.

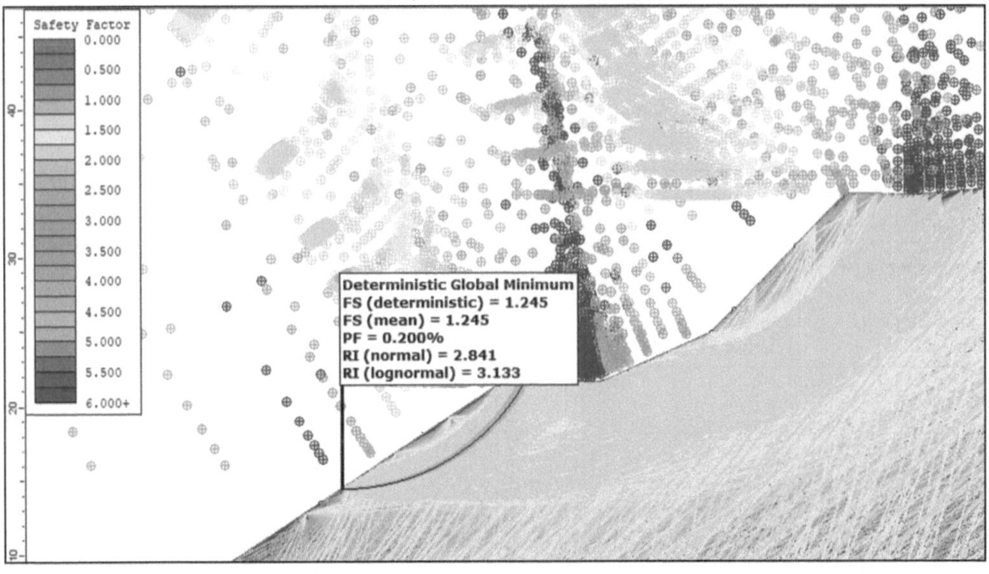

Figure 3. Global minimum slip surface.

Figure 4 shows slides with a safety factor of less than 1.4 (Bishop simplification). Here we note that, based on Janbu's simplification, the safety of the global slip surface with the lowest safety factor was only 1.27. So in Slide2, the safety factor does not reach the 1.35 values expected according to Eurocode7 for either local or global stability. The shapes and positions of the slip surfaces in the three software have similarities, but the essential differences are presented below.

Figure 4. The soil layer model (colour codes: red-clay, green-silt, yellow-sand, blue-ground water table) showing the Bishop simplified slip surfaces (FoS_{max} is 1.4).

We examined how random variables affect the value of the safety factor. The correlation coefficient of random variables is summarized in Table 2. The model is most sensitive to the parameters of the upper sand layers. In addition to the 88% (cohesion) and 45% (friction angle) correlation of these soil parameters (which are assigned to two layers in the model geometry), the influence of the other layer parameters below 6% is low. Statistical analyses of failure modes and failure criteria suggest that the applied software overlap, and sensitivity analysis identifies the dependence of failure on model parameters.

Table 2. Correlation coefficient of random variables.

soil layers/ correlation coefficient of random variables		Sa/1	Cl/1	Cl/2	Si/1	Sa/2	Sa-Si-Cl	Cl/3
Bishop simplified	**cohesion**	**0.88**	0.03	0.04	0.03	0.03	0.04	0.02
	friction angle	**0.44**	0.03	0.04	0.02	0.03	**0.06**	0.02
Janbu simplified	**cohesion**	**0.87**	0.03	**0.05**	0.03	0.03	0.04	0.02
	friction angle	**0.45**	0.03	0.04	0.02	0.03	**0.06**	0.03

The RI and PF values are given in Figure 3. A Reliability Index of at least 3 is usually recommended as a minimal assurance of a safe slope design. Calculating the reliability index (Figure 4), the best fit is the normal distribution, and the value of the Reliability Index is 2.84. Thus, the studied slope is not considered safe based on the probability calculation.

The Probability of Failure is simply equal to the number of analyses where the FS is less than 1, divided by the total Number of Samples. Using the worst possible combination of input parameters resulted in a probability of slope failure of 4.28%. In this case, approximately 43 out of 1000 cases have an FS <1.0. The distribution function (Cumulative plot) represents the cumulative probability that a randomly varying value is less than or equal to a given value. Where the distribution function has an inflexion point, it is the probability of failure (Figure 5). The minimum and maximum values of FS are given in Figure 4. The safety values of Bishop simplification are between 0.994 and 1.516. The probability of failure is 0.2%.

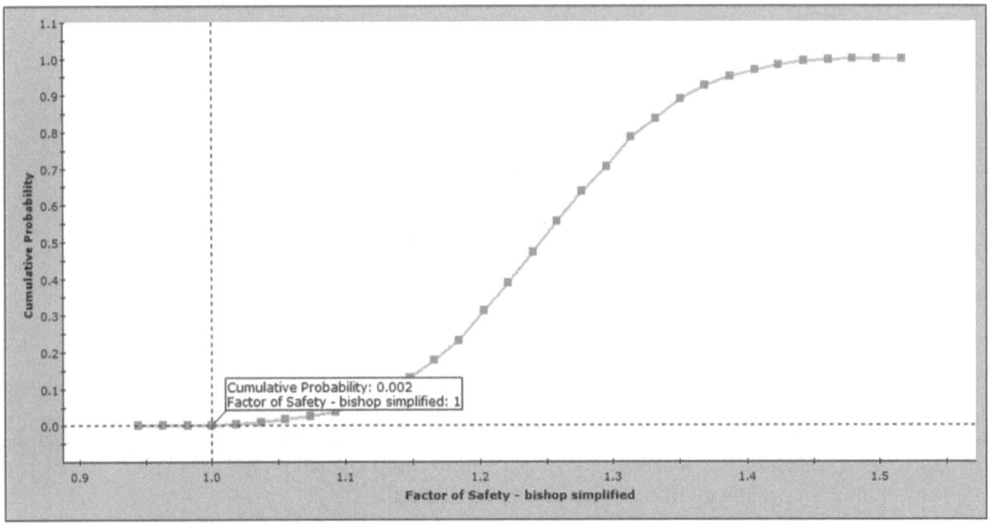

SAMPLED: mean=1.245 s.d.=0.08633 min=0.9508 max=1.504 (PF=0.200% RI=2.84090, best fit=Normal distribution)

Figure 5. Cumulative plot of FS (Bishop simplified).

The three software, Plaxis, Geo5 and Slide2, have shown that local failure at the top of the slope is more likely. Factor of safety calculated on the global slip surface is also below 1.35 and is thus not considered safe according to Eurocode7. The main results are the same, but the slip surfaces and safety factors are very different. Using Plaxis software, we detected a local failure mechanism with a safety factor of 1.18 and a global one with a value of 1.29. Due to the finite elements, the geometry of these slides is very complex. It is suggested that this model is the closest to reality, as it follows the geometry of soil layers. In this model, the sensitivity of weak soil lenses is also marked well. In Geo5, circle and polygonal slip surfaces have been obtained, and the user can edit even the geometry of these by handmade optimization. Another advantage of Geo5 is that the program performs several classical calculations for a given slip surface. Consequently, the obtained result is not a single value but a range of possible failures for any slip surface. The probability-based calculation of Slide2 is the most sensitive in terms of the soil parameters from the studied three calculations. These parameters are given on a statistical basis; hence the diversity of soil physical parameters are taken into account. It offers a strong advantage compared to the previous two software. As a result, there are many more variations on the slip surface geometry, but these are all circular surfaces. The obtained failure geometries are not as complex. Meanwhile, they represent weak points on the slope that the previous two software did not detect. The software draws a slip surface for each calculated parameter variation and specifies the safety factor. Only a few of these features are highlighted in Figure 6.

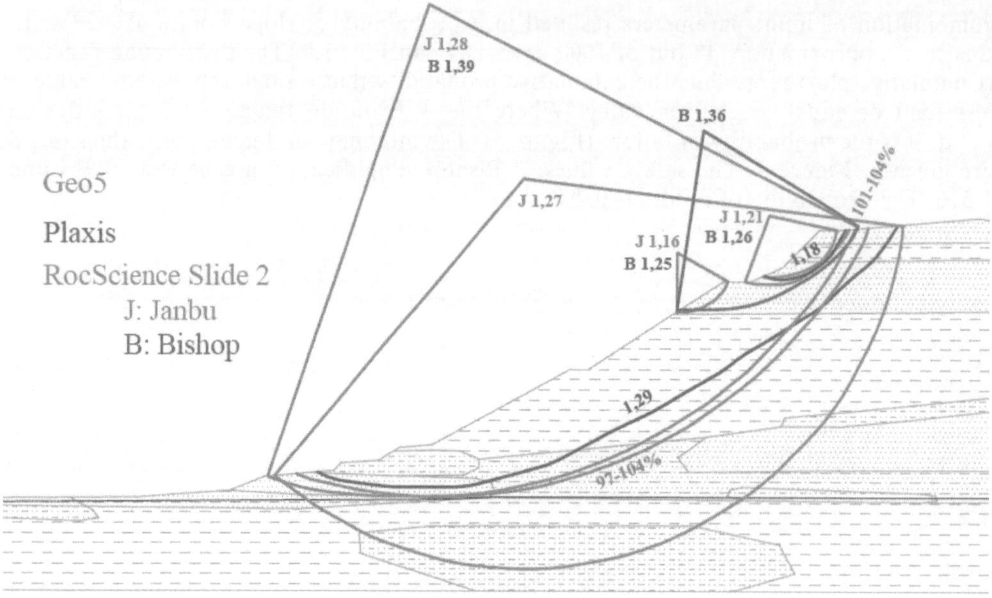

Figure 6. The results of the three software, different slip surfaces and safety factors.

Our results are in good agreement with that of Arun et al. (2020). In a thorough comparative study on the stability of Kuranchery slopes, they found that Plaxis 2D provided more precise results than Geo5. Plaxis 2D had higher values than Geo5.

The previous application of Slide2 has also proved that, for probabilistic regional landslide hazard assessment, the software is a handy tool (Acciaro 2018). This study is in line with the one of Hadjigeorgiou et al. (2006). In that paper, Bishop's Simplified Limit Equilibrium Method was used to assess a road embankment failure in Southwest Cyprus, presenting various slope failure scenarios. The current study also suggests that, similar to Islam et al. (2014), SLIDE2 is the most comprehensive slope stability analysis software with a Limit Equilibrium Method.

In light of these results, the slope is unstable. Thus it is suggested to monitor the displacement regularly (inclinometers, surveying) and set up a warning system. Periodical inspection of the drainage systems and the vegetated slope is also recommended.

5 CONCLUSIONS

This case study shows that modelling of slope stability of lakeside slopes is feasible using Plaxis, Geo5, and Slide2 software. Simple model calculations presented in this paper indicated significant differences in the results. Different failure surfaces were identified using the same soil layering, bed geometry and physical parameters in the models. The safety factor of local failure is 1,18 in Plaxis while using Slide the obtained values are between 1.16 and 1.36. According to the calculations with Geo5 101-104% utilization of slopes were found. The safety factor of global failure is 1.29 in Plaxis. The calculated minimum SF is 1.27, according to Slide. Meanwhile, Geo5 software has assigned 97-104% utilization. Slide 2 showed that the reliability index is below 3, which implies that the slope is not stable, not safe. The probability of slope failure is 0,2%.

In this study, we dealt with simplified stability. An even more accurate picture of the stability of the slope can be obtained by performing further studies taking into account differences in pore water pressures and the dynamic load of rail traffic. Considering the results, the monitoring system's application is essential to ensure recording the continuous displacement of the slope and set up a warning system.

ACKNOWLEDGEMENTS

The research reported in this paper has been supported by the National Research Development and Innovation Office of Hungary (Grant No. TKP2020 BME-IKA-VIZ).

REFERENCES

Acciaro M D, 2018. Probabilistic regional landslide hazard assessment for the Enguri dam (Jivari, Georgia) Open Access Master's Thesis, Michigan Technological University, 2018. https://doi.org/10.37099/mtu.dc.etdr/640

Arun K U, Jisna P, Simon R, Mathews O A, Anju E M, 2020. A Comparison Study on Stability of Kuranchery Slopes Using GEO5 and PLAXIS 2D Software, International Journal of Research in Engineering, Science and Management, Volume-3, Issue-3, 4 p.

Bar N, Kostadinovski M, Tucker M, Byng G, Rachmatullah R, Maldonado A, Pötsch M, Gaich A, McQuillan A, Yacoub T 2020. Rapid and robust slope failure appraisal using aerial photogrammetry and 3D slope stability models. International Journal of Mining Science and Technology, 30, 651–658.

Bond A, Harris A 2008. Decoding Eurocode 7. Taylor & Francis, London, 616 p.

Cami B, Javankhoshdel S, Bathurst R J, Yacoub T, 2018. Influence of mesh size, number of slices, and number of simulations in probabilistic analysis of slopes considering 2D spatial variability of soil properties, IFCEE 2018, Advances in Geomaterial Modeling and Site Characterization, GSP 295, DOI: 10.1061/9780784481585.019

Duncan J M, Wright SG, 2005 Mechanics of limit equilibrium procedures. Soil Strength and Slope Stability. John Wiley & Sons, Inc, New Jersey.

Eberhardt E 2003. Rock Slope Stability Analysis – Utilization of Advanced Numerical Techniques. University of British Columbia, Canada, Short Course note, 43p.

Fawaz A, Farah E, Hagechehade F 2014. Slope stability analysis using numerical modelling, American Journal of Civil Engineering, 2(3), 60–67.

Giordan, D.; Adams, M.; Aicardi, I.; Alicandro, M.; Allasia, P.; Baldo, M.; De Berardinis, P.; Dominici, D.; Godone, D.; Hobbs, P.; et al. 2020. The use of unmanned aerial vehicles (UAVs) for engineering geology applications. Bull. Eng. Geol. Environ. 79, 3437–3481.

Görög P, Török Á 2007. Slope stability assessment of weathered clay by using field data and computer modelling: a case study from Budapest. Nat. Hazards Earth Syst. Sci., 7, 417–422

Hadjigeorgiou J, Kyriakou E, Papanastasiou P, 2006. A road embankment failure near Pentalia in Southwest Cyprus, The South African Institute of Mining and Metallurgy, International Symposium on Stability of Rock Slopes in Open Pit Mining and Civil Engineering, 10 p

Islam S, Hussian A, Khan Y A, Chowdhury M A I, Haque B, 2014, Slope Stability Problem in the Chittagong City, Bangladesh, Journal of Geotechnical Engineering ISSN: 2394–1987 1/3, 13–25.

Jaboyedoff, M.; Oppikofer, T.; Abellán, A.; Derron, M.H.; Loye, A.; Metzger, R.; Pedrazzini, A. 2012. Use of LIDAR in landslide investigations: A review. Nat. Hazards, 61, 5–28

Jacob A, Thomas A A, Aparna G A G, MP A, 2018. Slope stability analysis using Plaxis 2D. International Research Journal of Engineering and Technology (IRJET) 5/4, 3666–3668.

Kádár I, Nagy L 2017. Comparison of Different Standards Based on Computing the Probability of Failure of Flood Protection Dikes, Periodica Polytechnica Civil Engineering, 61(1), 146–153. 8 p.

Kézdi A 1952. A Balaton északkeleti peremén bekövetkező mozgások vizsgálata, Hidrológiai Közlöny, 32(11–12), 403–408 (in Hungarian).

Koroknai B, Wórum G, Tóth T, Koroknai Zs, Fekete-Németh, Kovács G, 2020. Geological deformations in the Pannonian Basin during the neotectonic phase: New insights from the latest regional mapping in Hungary, Earth-Science Reviews 211 (2020) 103411, 30 p.

Li S, Cami B, Javankhoshdel S, Corkum B, Yacoub T 2020. Considering Multiple Failure Modes in Limit Equilibrium Slope Stability Analysis: Two Methods, Calgary 2020, 6 p.

Menegoni N, Giordan D, Perotti C. 2020. Reliability and Uncertainties of the Analysis of an Unstable Rock Slope Performed on RPAS Digital Outcrop Models: The Case of the Gallivaggio Landslide (Western Alps, Italy). Remote Sens. 12, (10), 1635. https://doi.org/10.3390/rs12101635.

Matthews C, Farook Z, Helm P (2014) Slope stability analysis–limit equilibrium or the finite element method. Ground Engineering: 48 (5) 22–28.

Oberhollenzer S, Tschuchnigg F, Schweiger H.F 2018. Finite element analyses of slope stability problems using non-associated plasticity. Journal of Rock Mechanics and Geotechnical Engineering, 10 (6), 1091–1101,

Salunkhe D P, Chvan G, Bartakke R P, Kothavale P R 2017. An Overview on Methods for Slope Stability Analysis, International Journal of Engineering Research & Technology, 6 (3), 528–535.
Sharma R K, Kumar V, Sharma N, Rathore A, 2012. Slope Stability Analysis Using Software GEO5 and C Programming, International Conference on Chemical, Ecology and Environmental Sciences (ICEES'2012) Bangkok 2012, 5 p.
Sztanó O, Magyar I, Szónoky M, Lantos M, Müller P, Lenkey L, Katona L, Csillag G (2013) Tihany Formation in the surroundings of Lake Balaton: type locality, depositional settings and stratigraphy. Földtani Közlöny 143(1):445–468 (in Hungarian)
Sztanó O, Magyar I, Müller P, Katona L, Babinszki E, Magyari Á (2005) Sedimentary cycles near the coast of Lake Pannon, Late Miocene, Hungary. In: 12th RCMNS Congress: patterns and processes in the Neogene of the Mediterranean region Universität Wien, pp 227–230.
Utili S, Crosta G B, 2015. Analysis Tools for Mass Movement Assessment. In: Shroder J F, Davies T (eds): Landslide Hazards, Risks and Disasters, Elsevier, Amsterdam, 441–465.
Uzielli M, Lacasse S,. Nadim F, Phoon KK, 2006. Soil variability analysis for geotechnical practice, DOI: 10.1201/NOE0415426916.ch3.
www1:rocscience.com/help/slide2/slide_model/project_settings/Probabilistic_Analysis.htm
www2:https://www.finesoftware.eu/help/geo5/en/analysis-according-to-the-theory-of-limit-states-safety-factor–01

The Evolution of Geotech - 25 Years of Innovation – Hammah et al. (Eds)

Weathered rock and crack detection of tunnel excavation using image analysis

Intan Norsheira Yusoff
School of Civil Engineering, Universiti Sains Malaysia (USM), Engineering Campus, Pulau Pinang, Malaysia

Mohd Ashraf Mohamad Ismail
Associate Professor, School of Civil Engineering, Universiti Sains Malaysia (USM), Engineering Campus, Pulau Pinang, Malaysia

Hayato Tobe
Chief Research Engineer, Rock Mechanics and Hydro-geology Group, Kajima Technical Research Institute, Kajima Corporation

Takako Miyoshi
Research Engineer, Rock Mechanics and Hydro-geology Group, Kajima Technical Research Institute, Kajima Corporation

Kensuke Date
Chief Research Engineer, Kajima Technical Research Institute Singapore, Kajima Corporation

Yasuhiro Yokota
Senior Research Engineer, Kajima Technical Research Institute Singapore, Kajima Corporation

ABSTRACT: Geological mapping and rock mass classification are two main quest in tunnel excavation. Most of tunnel excavation investigation are determined manually by geologist. Geological mapping may be different with each geologist and caused human error. To improvise this existed method, Working Face software is used as method to analyze image. RGB image as input for image analysis. This software can determine the deterioration weathered and crack detection (such as faults, joints and crack) by using an RGB image as input. This method also calculates the percentage of rock weathering grade of tunnel face. The weathering grade is divided into 6 grades and each grade represent by different color. The results are related with JH classification. All data and image for this study are taken from granite rock of Pahang-Selangor Raw Water Transfer Tunnel. Numerical and quantitative analysis are carried out to predict the rock fall of tunnel face excavation. The relationship between percentage of weathering grade, rock fall and JH classification will be discussed in this paper. Results are compared with geological mapping by geologist to determine the effectiveness of this method. The aim of study is to make the process of tunnel excavation investigation much easier for engineering geologist to design tunnel support.

1 INTRODUCTION

Weathering is used as a preparatory step for soil or rock mass denudation in sectors such as civil engineering, mining, heritage protection and urban planning (Dearman et al., 1978). As

DOI: 10.1201/9781003188339-60

the key cause of soil or rock mass erosion and slope failure during its engineering lifespan, (Hack & Price, 1997) attributed future weathering after completion of a slope or engineering works. Understanding their weathering states is one of the most important challenges in studying slope stability, foundations, and rock excavation (Alavi Nezhad Khalil Abad et al., 2014). The degree of weathering is one of the most imperative parameters influencing rock engineering behaviours. The degree of colour and discolouration of the soil or rock mass is also a good weathering indicator.

Tropical rainforests are sensitive to heavy rainfall, so water is filled with cracks and other voids in the rocks and soils close to the ground surface (Ghiasi et al., 2009). Discontinuities are a main and important observable parameter to be evaluated for engineering construction in and on a rock mass. The common weathering signatures for physical weathering are the geometry of discontinuity and shear strength depending on the discontinuity condition. The main weathering discontinuity properties are discontinuity spacing, persistence, roughness (ISRM, 1978; Otoo, 2012), aperture, weathering products' presence and character (or infill materials) and variation in the strength of intact soil or rock from fresh to weathered mass (BS 5930:1999, 1999; Tating, 2015). Mechanical and integral discontinuities are two basic forms of discontinuity (Price et al., 2009). Integral discontinuities have not yet been opened by movement or weathering and have the same shear strength without impacting the intact rock strength as the underlying rock materials. Due to weathering processes that alter the characteristics of the discontinuity, they may change into mechanical discontinuities, while mechanical discontinuities have been opened up by motion, stress or weathering. These are the weakness planes where the shear force is slightly lower than the material of the surrounding rock (Fakunle, 2016).

In rock tunnel projects, the New Austrian Tunneling Method (NATM) is the most used construction method. Geological logging and rock mass classification are two main tasks before NATM tunnel excavation (T & Chang, 2005). According to (Lee, 2012), because of the increase in economic and social development, long tunnels are in high demand due to their different uses, such as transport and water supply. This paper has therefore come out with a method that will make it much easier for engineering geologists to work. A non-destructive analytical method is digital imaging. It can be said that digital cameras and image processing are capable of measuring large textured samples quantitatively in colour (Lebrun et al., 2004). The main objective of image processing is to extract specific characteristics from the collected tunnel excavation facing images (such as faults, joints and shear zones) (T & Chang, 2005). In both qualitative and quantitative outcomes relating to the JH classification, the results of the study will contribute to determining the weathering grade of the tunnel face.

Two of the several interrelated factors in the overall design of a serviceable and economical tunnel are the design of the support system. Inseparable factors are the type of support, the excavation process, and the characterization of the ground. A tunnel support system's basic functions are to maintain the tunnel stable and to make the opening usable (Deere et al., 1970). However, the particular functions of the support structures depend mostly on the purposes of the tunnel.

2 METHODOLOGY

2.1 *JH Classification*

In this research, to determine the rock mass class using geological mapping and geological documentation of tunnel face and the sidewalls of excavated tunnels, geologist used a rock mass classification by Japan Highway Public Corporation (JH Classification). From the geological map of tunnel face data sheets, the compressive strength, weathering, spacing of joints, joint conditions, and groundwater condition evaluated (Figure 1 (a)). Figure 1 (b) shows the total points from observation faces of different JH rock mass classes. The table also shows different rock types from hard rock to soft rock. The JH classification is a rock mass rating system that relies mainly on the following four general rock mass strength observation data:

(a)

Name of tunnel: Main Tunnel NATM 3								Date of observation: 14/3/11 Time: 20:10		
CH: 5+683.34 TD: 971.90		Distance from portal: -m						No. of section		Rock Class: CII- CI
Height of earth covering ~21.00m		Formation: GRANITE								
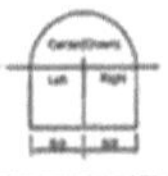		Description of other added support						A, B-Instrument:		
		Notes: Left Centre Right Calculation of Grade points (54 + (54x2) +54)/4= 54 Special Condition: Cutting face Collapse: Early-stage invert closure:								
Contents		Classification								
		1	2	3	4	5	6	Left	Centre	Right
A. Strength of intact rock material	Uniaxial comp. str	>100MPa	100~50 MPa	50~25 MPa	25~10 MPa	10~3 MPa	<3 MPa	3 → 2	3 → 2	3 → 2
	Point-load strength	>4 MPa	4~2 MPa	2~1 MPa	1~0.4 MPa	< 0.4 MPa	-			
	Strength judged by blow of hummer	NOT broken by strong blow of hummer	Broken by strong blow of hummer	Broken by normal blow of hummer	Broken by striking rocks each other	Broken easily by hand	Deformed by finger			
	Grade Point	32	26	19	13	6	0	24	24	23
B. Weathering/ Alteration	Degree of weathering	Fresh	Weathered along discontinuities	Weathered to the rock mass core	Sedimentary, Unconsolidated			3 → 2	3 → 2	3 → 2
	Hydrothermal alteration	No alteration	Partially altered and infilled with clay	Altered and weakened to the rock core	Heavily altered and become clayey or sedimentary					
	Grade Point	19	13	6	0			12	13	12
C. Spacing of discontinuities	Spacing of discontinuities	d ≥ 1m	1m>d ≥ 50cm	50cm >d ≥ 20cm	20>d ≥ 5cm	5cm > d		3 → 4	3 → 4	3 → 4
	RQD	>80	80~50	60~30	40~10	<20				
	Grade Point	24	18	12	6	0		5	8	6
D. Condition of discontinuities	Degree of opening	Fracture Totally-attached	Fracture Partly Opened	Fracture Mostly Opened	Fracture opened 1-5mm width	Fracture opened > 5mm width		3 → 2	3 → 2	3 → 2
	Infilled with	Nil	Nil	Nil	Clay (<5mm)	Clay (≥5mm)				
	Degree of Roughness	Course	Flat & Smooth	Partly Slickenside	Well-sharpened slickenside					
	Grade Point	25	19	12	6	0		14	13	14
E. Effect of discontinuity strike & dip orientation declination	Strike perpendicular to Tunnel axis	1.Drive with dip-Dip 45~90°	2.Drive with dip-Dip 20~45°	3.Drive with/against dip-Dip 0~20°	4.Drive against dip-Dip 20~45°	5.Drive with dip-Dip 45~90°		5	5	5
	Evaluation	Very favourable	Favourable	Normal	Unfavourable	Fair	A			
	Strike parallel to Tunnel axis	-	-	1. Dip 0~20°	2. Dip 20~45°	3. Dip 45~90°		-	-	-
	Evaluation	-	-	Normal	Unfavourable	Fair				

(b)

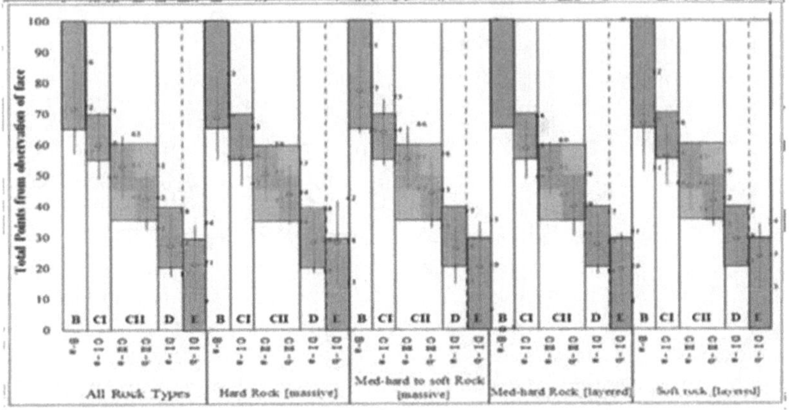

Figure 1. Geological documentation of JH Classification (b) JH rock mass classes (Azit & Ismail, 2014).

compressive strength, weathering, joint spacing, and joint condition. By adding the ratings of the parameters, the overall rating of the rock mass is calculated. The total rating (total point) is given a class that represents the rock mass quality. Then, the geological mapping will be compared with image analysis.

2.2 *L* a* b* coordinates*

A colorimeter is an instrument that mimics how people interpret colour. A colorimeter shines light down to the surface of the sample using an internal light source. It passes through three filters as the light reflects the device: red, green, and blue. These filters distil the values of tri-stimulus (RGB) that match how our eyes see colour. Even if two colours look the same to one-person, slight differences may be found when evaluated with a colour measurement instrument. For example, looking at the L*a*b* values for each apple, we can objectively determine that the apples don't match in colour. These values tell us that Apple 2 (sample) is lighter, less red, and more yellow than Apple 1 (standard) (Figure 2). It also shows that the rock sample is highly weathered with higher a* and b* value and vice versa. Sites investigation was conducted on granite rock surface area. A colorimeter can calculate the absorbency of light waves. The difference in the intensity of electromagnetic radiation in the spectrum's visible wavelength region after transmission or reflection by an object or solution is calculated during colour measurement (Roy Choudhury, 2015). In this research, FRU Colorimeter WR18 with 40mm aperture was specifically chosen because it can be used for large textured, uneven, and rough surfaces. Measurement results for colorimeter are expressed in brightness L*, redness a*, and yellowness b*. Deltas for L* (ΔL^*), a* (Δa^*) and b* (Δb^*) may be positive (+) or negative (-). However, the total difference, Delta E (ΔE^*) is always positive. The calculation for ΔE^* shown in Equation (1) Value of L, a* and b* of rock surface was recorded, transferred and saved into the computer.

$$Color\ difference,\ \Delta E^*_{ab} = \sqrt{\left(L^*_2 - L^*_1\right)^2 + \left(a^*_2 - a^*_1\right)^2 + \left(b^*_2 - b^*_1\right)^2} \qquad (1)$$

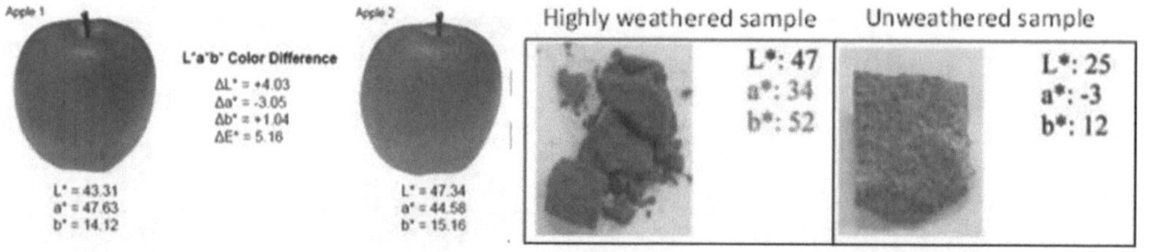

Figure 2. Example of L* a* b* colour difference.

2.3 *Image analysis of weathered rock and crack*

(Tobe et al., 2018) developed a technique based on image analysis for quantitatively evaluating the weathering grade on tunnel faces and applied this technique to an actual tunnelling site. The technique was implied in software named Workingface. The a* and b* value obtained from the colorimeter was used as a guide and range to get the best result to represent the real tunnel. Each RGB image was analyzed by using range value a* and b* from data collection to get the weathering grade results in qualitative and quantitative. The software also can detect cracks and discontinuities from the images. An image of the tunnel face is converted into a binary image that later being used to calculating intervals between cracks. Image analysis results of tunnel face will be discussed in Result & Discussion.

2.4 *Tunnel support design*

NATM has been developed for tunnelling in weak and squeezing ground. To limit displacements and re-establish equilibrium around the tunnel, it requires structural support for such ground. The two stages of rock support have been described by Rabcewícz and Golser (1973) as a part of the NATM 'the dual-lining support' (initial and final support) for tunnels was introduced. This is the idea of having the masses of rock surrounding the tunnel and deforming the initial support before installing the final or permanent support. The initial support always carried out as an outer lining designed during excavation to support the rocks. It mainly consists of shotcrete, systematically bolted and, if necessary, strengthened by additional steel ribs. Besides, in the very weak ground, a closing of the invert is carried out.

3 RESULTS & DISCUSSION

3.1 *JH classification and L* a* b* coordinate*

Geological mapping and geological documentation of tunnel face have been determined by using JH classification. Equation (2) shows how to calculate the total rating point for tunnel face. The bar chart in Figure 3(a) shows different class and their total rating. According to the bar chart, less weathered tunnel face has high total rating compare to weathered tunnel face.

Statistical values were calculated with all collected data of a* and b* in site investigation. The results used as input in Workingface software to analyzing RGB images. The graph of correlation between a* and b* (Figure 3(b)) shows that when the value of a* increase, b* also increase. This correlation interprets that with a high value of a* and b*, the rock surface is more weathered. It also shows that the $R^2 = 0.8641$ which has a very good correlation.

$$Total\ Rating\ Point = [Left + (Center x2) + Right]/4 \quad (2)$$

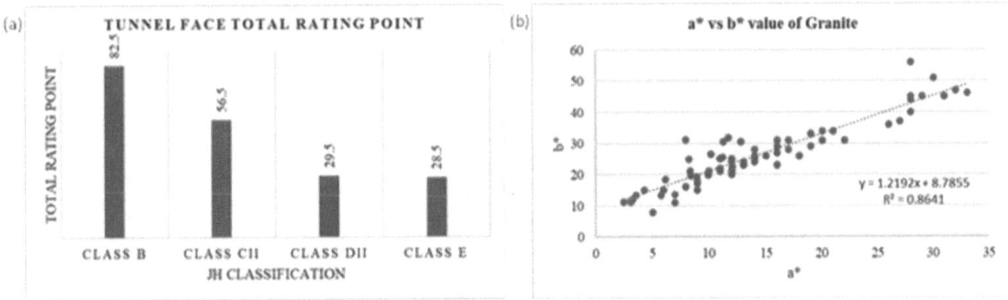

Figure 3. (a) Tunnel face total rating point. (b) Correlation graph of a* and b* value of Granite.

3.2 *Image analysis of weathered rock and crack*

Weathering grade of each tunnel was determined by analyzing RGB images of the tunnel face. In this study, different tunnel faces were analyzed to show the distinct results to compared and justify. Figure 4 shows class B tunnel face. The majority colour shows grade 1 and 2. The binary image shows high crack but it is sparsely distributed in crack interval image. Next tunnel face of class CII (Figure 5) shows majority colour of grade 2 and 3. Binary image and crack intervals show less crack Next results show image analysis of class DII (Figure 6) with majority colour of grade 6, 2 and 3. The cracks focused on highly weathered area which is grade 6. Lastly, image analysis of class E (Figure 7) with majority colour of grade 6, 3 and 4. This tunnel face has fault line and clearly shown in binary and crack intervals image. Results of image analysis can be compared qualitatively by looking at the colour domain in each tunnel face. All image analysis came out with the percentage results of each colour on each tunnel face. Results show in Table 1 helps engineering geologist to see which grade is the domain in tunnel face. Therefore, rock classification can be predicted from the percentage easily.

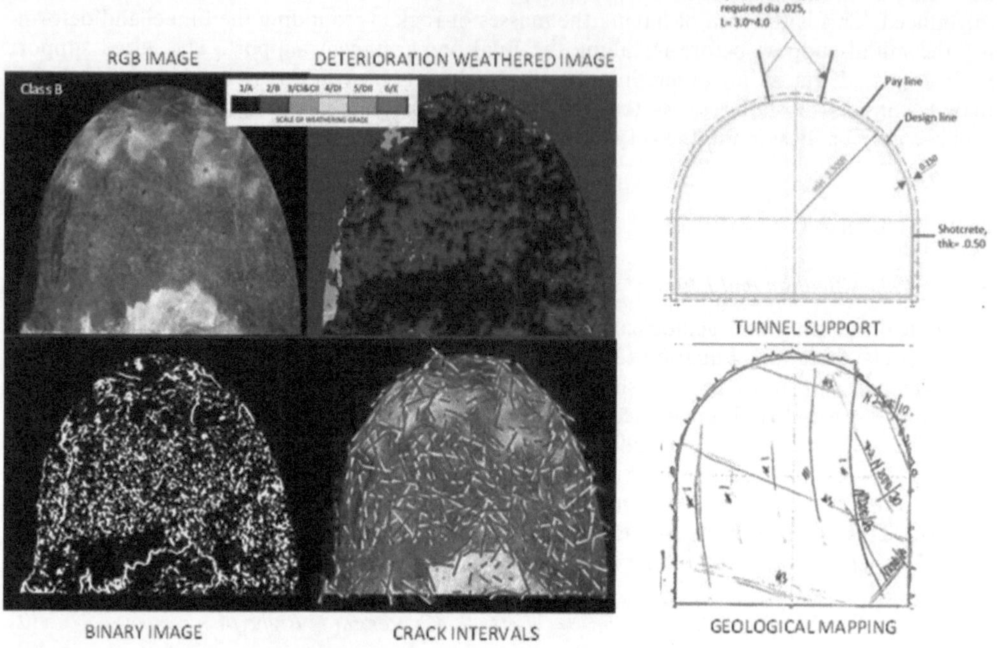

Figure 4. Image analysis of Class B & geological mapping.

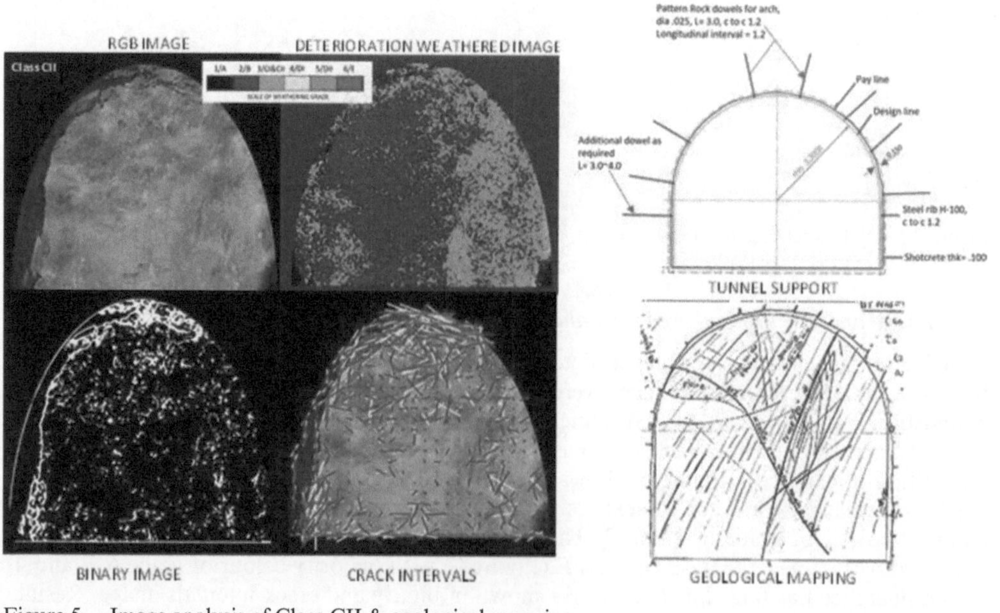

Figure 5. Image analysis of Class CII & geological mapping.

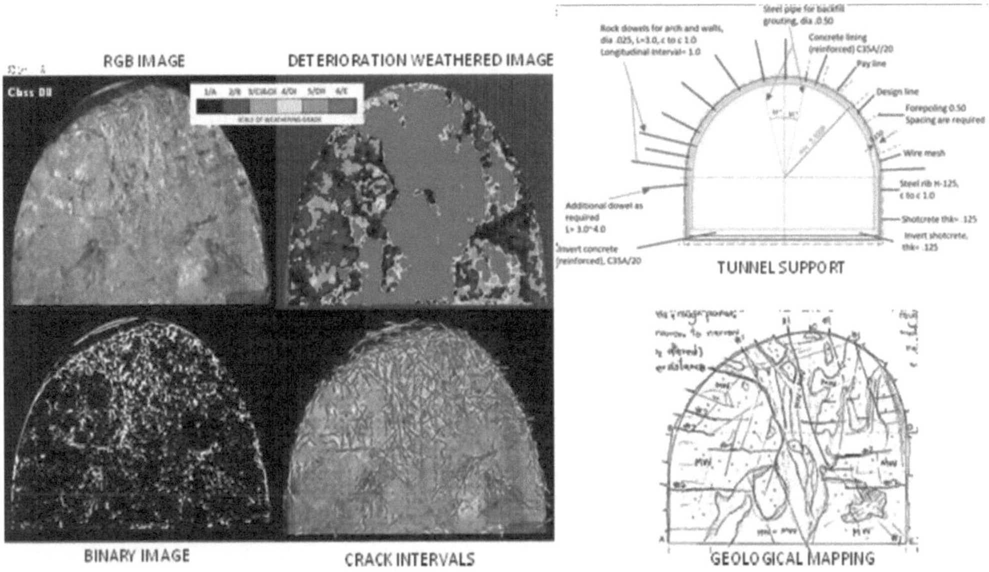

Figure 6. Image analysis of Class DII & geological mapping.

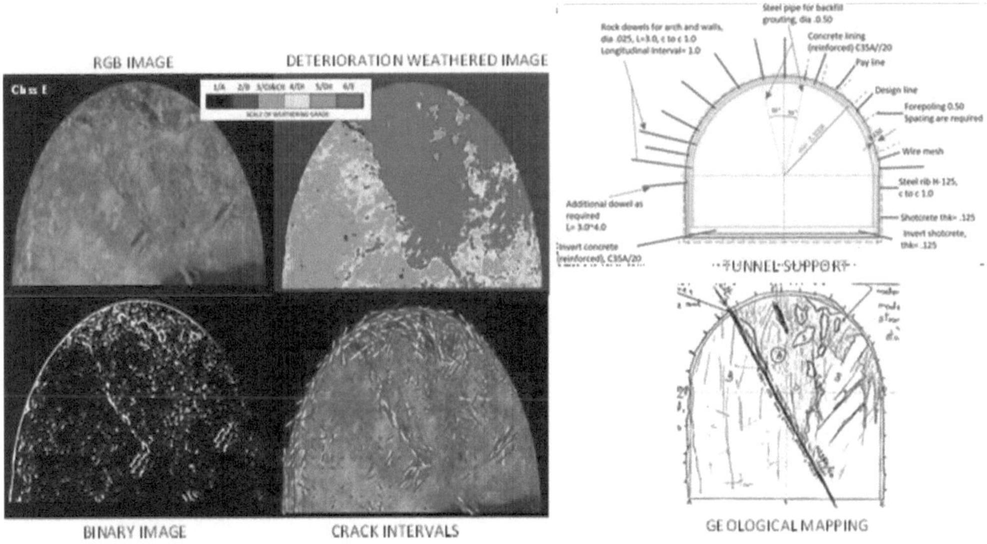

Figure 7. Image analysis of Class E & geological mapping.

Table 1. Weathering grade percentage of tunnel face.

Weathering grade	Tunnel Grade B	Tunnel Grade CII	Tunnel Grade DII	Tunnel Grade E
1	61	2	7	1
2	36	59	20	1
3	3	38	17	27
4	0	1	16	20
5	0	0	4	7
6	0	0	37	45

3.3 *Tunnel support*

This research allows an engineer to design temporary tunnel support easier by looking at the percentage of each tunnel face. Temporary support is described as any system designed and installed from the time it is first excavated up to the time a permanent lining is in place to support the perimeter of an underground opening. Figure 8 shows tunnel support required for each rock class for temporary support. Rock class a required only spot rock dowels with diameter of 0.025 m with a length of 3 to 4 m. While rock class B, shotcrete is needed with 0.05m thick along with spot rock dowels. Rock class CI is moderately weathered. It required 0.5m thick of shotcrete with an additional dowel. Next rock class CII, tunnel support required shotcrete with 1.0 thick, additional dowel and steel rib. Rock class DI required shotcrete with 1.25m thick, additional dowels, steel ribs and wire mesh. Lastly, for rock class DII and E, tunnel support required shotcrete, additional dowels, steel ribs, wire mesh and concrete for highly weathered rock.

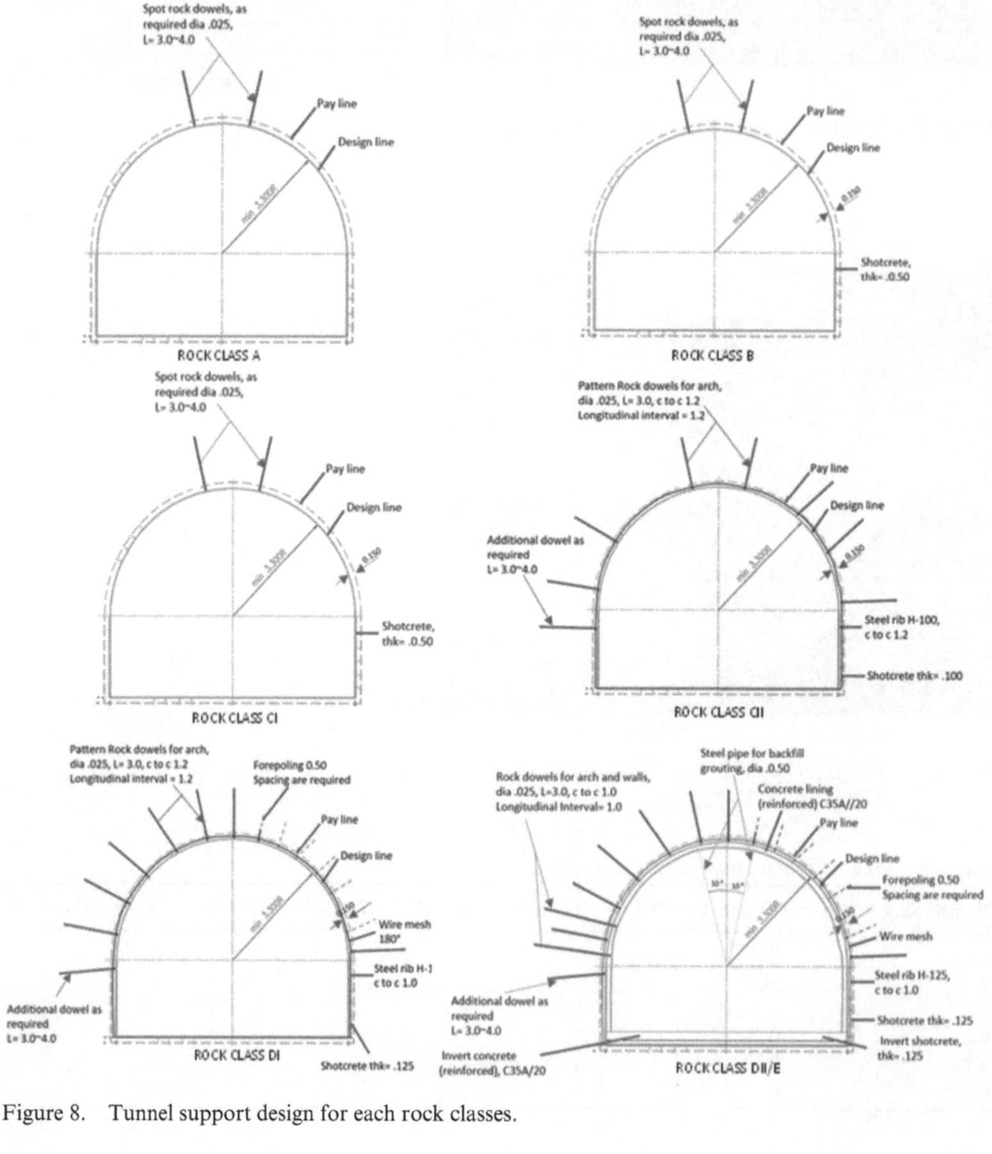

Figure 8. Tunnel support design for each rock classes.

4 CONCLUSION

It can be concluded that colour represents in tunnel face image analysis is important to determine the rock classification. This method also calculates the percentage of each weathering grade of the tunnel face. Besides weathering grade, this method also detects cracks of tunnel face as a binary image and crack interval image. It shows the distribution of cracks on tunnel face. Results from this research can be used for engineering geologist to design temporary tunnel support for first excavation much easier. However, this study's limitations require clear, bright and good RGB images to run in the software to get good and more accurate results.

REFERENCES

Alavi Nezhad Khalil Abad, S. V., Mohamad, E. T., & Komoo, I. (2014). Dominant weathering profiles of granite in southern Peninsular Malaysia. *Engineering Geology*, *183*, 208–215. https://doi.org/10.1016/j.enggeo.2014.10.019

Azit, R., & Ismail, M. A. M. (2014). Rock Mass Classification System Used for Pahang-Selangor Raw Water Transfer Tunnel. *InCIEC 2013*, 519–529. https://doi.org/10.1007/978-981-4585-02-6_45

Dearman, W. R., Baynes, F. J., & Irfan, T. Y. (1978). Engineering grading of weathered granite. *Engineering Geology*, *12*(C), 345–374. https://doi.org/10.1016/0013-7952(78)90018-2

Deere, D. U., Peck, R. B., Parker, H. W., Monsees, J. E., & Schmidt, B. (1970). Design of tunnel support systems. *Committee on Soil and Rock Properties*, *339*, 26–33. http://onlinepubs.trb.org/Onlinepubs/hrr/1970/339/339-003.pdf

Fakunle, A. A. (2016). *Detection of Weathering Signatures using Uav Photogrammetry in Comparison with Ground-Based Sensors*. 102. https://doi.org/10.2217/pmt.15.9

Ghiasi, V., Omar, H., & Huat, B. K. (2009). A study of the weathering of the Seremban granite. *Electronic Journal of Geotechnical Engineering, 14 D.*

Hack, R., & Price, D. (1997). Quantification of weathering. *Engineering Geology and the Environment. Proc. Symposium, Athens, 1997, Vol.1, June*, 145–150.

Lebrun, V., Toussaint, C., & Pirard, E. (2004). Monitoring color alteration of ornamental flagstones using digital image analysis. *Dimension Stone*, 139–145.

Lee, S. C. (2012). *The Evaluation of Tunnel Behaviors under High Rock Stress Using Numerical Analysis Method.*

Roy Choudhury, A. K. (2015). Visual measures of colour. In *Principles of Colour and Appearance Measurement* (Vol. 2). Woodhead Publishing Limited. https://doi.org/10.1533/9781782423881.1

T, S. L., & Chang, S. (2005). *Digital image processing based approach for tunnel excavation faces. 14*, 750–765. https://doi.org/10.1016/j.autcon.2005.02.004

Tobe, H., Miyajima, Y., Shirasagi, S., Yamamoto, T., & Kawabata, J. (2018). A rapid image analyzing method for determining crack distribution and interval on tunnel faces. *ISRM International Symposium - 10th Asian Rock Mechanics Symposium, ARMS* 2018.

The Evolution of Geotech - 25 Years of Innovation – Hammah et al. (Eds)

Numerical investigation of the stability of tunnel excavation faces in deep tunnels

D. Georgiou
MSc Mining Engineer, National Technical University of Athens, Greece

M. Kavvadas
Professor, National Technical University of Athens, Greece

A. Kalos
PhD Civil Engineer, National Technical University of Athens, Greece

ABSTRACT: This paper studies the effect of rockmass conditions, in the tunnel face deformation (extrusion) for deep tunnels (overburden height $H \geq 100$m) via three-dimensional (3D) numerical analyses in the commercial Finite Element Code RS3 Rocscience. The numerical investigation focuses on tunnel face stability in cases of unsupported tunnel face and reinforced tunnel face by fiberglass dowels, through the tunnel face deformation. The paper produces a non – dimensional ground factor (Λo) working on the rockmass conditions as a function of the tunnel face extrusion (Ωh). In cases of reinforced tunnel face fronts by fiberglass dowels, the reduction of the tunnel face extrusion (ΔΩh) is correlated with the tunnel face support characteristics.

1 INTRODUCTION

One of the most significant parameters for the tunnel construction safety, is the tunnel face stability due to the direct contact with the rockmass. In cases of good ground conditions (on high GSI values), tunnel face failure manifests by means of wedge failures as dense rockmass joint systems form segments, that can no longer sustain their own weight without failing. Failure on weak rocmasses (on low GSI values) manifests in a more classical ductile failure mode due to inelastic straining-induced strength damage. *Lunardi P. (2008)* describes three main modes of tunnel face behavior based on the ground response, where in cases of poor ground conditions and high overload the geomaterial sustains significant inelastic straining, by thus giving significant tunnel face deformations.

When the ground strength is lower than the vertical stress, the tunnel face collapses due to zero horizontal stresses on the tunnel face at the time of tunnel excavation. Scoping to reduce the tunnel face deformation by increasing the tunnel face stability, the tunnel face can be reinforced by fiberglass dowels, applying an equivalent tunnel face pressure on the face front. Thus, the failure strength increases as the fiberglass dowels apply an equivalent tunnel face pressure on the face front similar to the minor horizontal stress (σ_3) in the classical triaxial shear test.

Tunnel face stability constitutes a topic gathering significant attention in the international literature due to its practical importance. Numerous analytical methods (e.g. Horn, 1961; Atkinson & Mair, 1981; Panet, 1995; Kovari & Anagnostou, 1996) compute the factor of safety of a suitably selected failure surface, based on stress equilibrium of a critical ground

DOI: 10.1201/9781003188339-61

wedge at the excavation face. Such methods usually have limited accuracy and applicability due to the simplifying assumptions they are based on, stemming from the selected wedge. Numerical analyses can also address the safety factor via the Strength Reduction Method (e.g. Zienkiewicz et al, 1975), by gradually reducing ground strength until the tunnel face fails (signified by divergence of the numerical integration scheme). In such analyses, the safety factor computes as the inverse of the strength reduction factor causing face instability, by thus producing useful design charts for calculation of the safety factor as a function of ground strength, tunnel depth and size (e.g. Kavvadas et al., 2009; Prountzopoulos, 2012).

The present investigation uses three – dimensional (3D) numerical analyses to study tunnel face stability. The numerical analyses focus on an ensemble of tunnel diameters (D), tunnel overburden heights (H) and rockmass conditions controlled by the Generalized Hoek & Brown (2002) failure criterion. In cases of tunnel face support simulation, the paper studies numerous densities and diameters of fiberglass dowels for the tunnel face reinforcement.

The paper correlates non – dimensional tunnel face extrusion (Ω_h) to the non – dimensional ground factor (Λ_o) working on all aforementioned parameters. Tunnel face reinforcement by fiberglass dowels is accounted for through the reduction in the tunnel face extrusion (Ω_h), between reinforced and unsupported tunnel face fronts. The reduction (Ω_h) is further correlated with the appalling tunnel face pressure parameters described via the non – dimensional factor (N_{FG}).

2 TUNNEL FACE REINFORCMENT BY FIBERGLASS DOWELS

Reinforcement of the tunnel face is usually performed by fiberglass dowels in conventional tunnel excavations. Fiberglass dowels are elongated elements with great tensile strength and elasticity. When the tunnel face is reinforced by that elements, the tunnel face deformation is limited, due to the passive behavior of the dowels which are anchored in front of the tunnel face (in the rockmass) and "hold" the tunnel face. Due to the tunnel face reinforcement an equivalent pressure (P) can be applied in the tunnel face, where the amount of it, depends on the tunnel face reinforcement density and the braking load of each dowel. The most common tunnel face reinforcement densities by fiberglass dowels (FG) range between $1FG/1m^2$ to $1FG/4m^2$. It is preferable to use fiberglass dowels than steels dowels in tunnel construction, as they are easier to handle/cut during the tunnel face advance. Figure 1 shows a typical case of tunnel face reinforcement by fiberglass dowels in the construction of Athens Metro (Nikaia underground station).

Figure 1. Reinforced tunnel face by fiberglass dowels in the construction of Athens Metro.

3 NUMERICAL INVESTIGATION OF THE TUNNEL FACE STABILTY

3.1 *Numerical model*

The numerical investigation of two non–circular tunnels of equivalent tunnel diameter $D = 6 \div 10m$, shown in Figure 2.a, was performed in the Commercial Finite Element Code RS3 Rocscience *v. 4.013*. In both cases, the tunnel face advances by 1m in each excavation step, as the tunnel is excavated in a single phase (full – face excavation). Tunnel support measures include a primary lining of combined shotcrete and rockbolts installed one excavation step back from the tunnel face. In reinforced tunnel face fronts by fiberglass dowels, the parametric investigation works on two reinforcement densities (1FG/1 ÷ $2m^2$) with an overlap length of 10*m*, aiming to minimize its (the overlap length) effect on the tunnel face stability.

The overburden height (*H*), measured from the tunnel axis to the ground surface, varies from $H = 100$ to 200*m*. The analyses work on constant H/D (=10, 15, 20) ratios, as the overburden height is normalized with the tunnel face diameter ($D = 6 \div 10$m). The degraded rockmass analysed herein justifies use of Finite Element numerical tools; thus, the rockmass is discretized via four-noded tetrahedral finite elements. The lateral and bottom boundaries are placed far from the tunnel to ensure minimal boundary effects. To further eliminate possible additional boundary effects stemming from the tunnel end, results at the excavation face were processed when the excavation had advanced to a length $L = 5D$ from the start, leaving a total unexcavated length of $3D$ in front of the tunnel face. The model extends $4D$ above and $3D$ below the tunnel center as it is shown in Figure 2.b. In order to simulate high overburden loads (higher than $4D$), the additional height was simulated by means of a uniform additional pressure applied on the top surface.

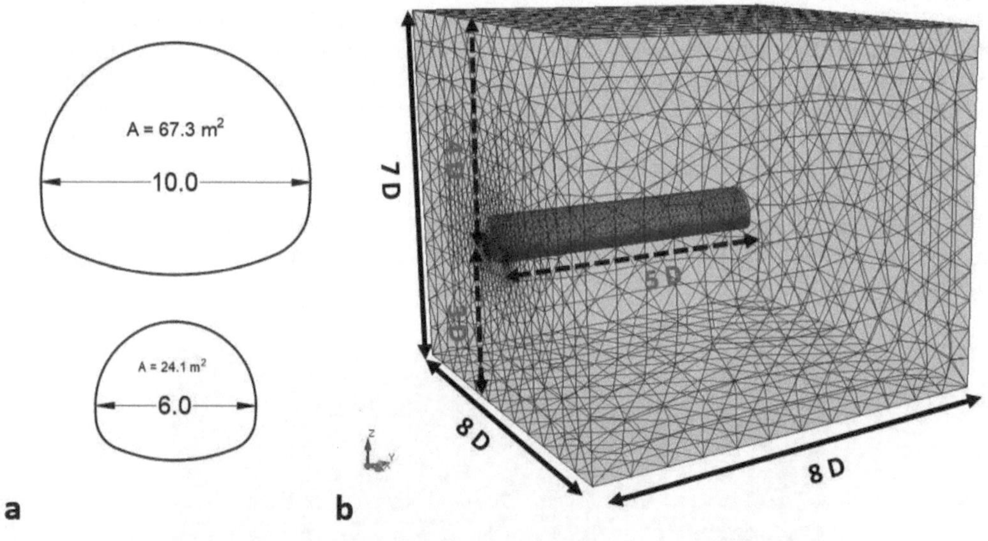

Figure 2. a) Cross section of the two tunnel types with equivalent tunnel diameters D = 6 ÷ 10m.
b) Typical Finite element mesh used in the numerical analyses. The case shown corresponds to tunnel width D = 10m.

3.2 *Rockmass properties*

The numerical study works on a degraded (GSI = 15 ÷ 45) rocksmass with a unit weight $\gamma = 25kN/m^3$, and a lateral stress coefficient at rest $K_o = 0.5$ and 1.0. The analyses work on Generalised Hoek-Brown failure criterion (Hoek et al, 2002), which builds on the premise of a linear elastic – perfectly plastic for the rockmass. The intact rock properties are listed here below:

- Young's Modulus: E_i = 2000 MPa;
- Poisson's ratio: v = 0.33;
- intact rock strength: σ_{ci} = 10 MPa

Table 1 lists the full set of rockmass mechanical parameters used in the parametric analyses. The rockmass strength "σ_{cm}"and modulus "E_m" are calculated via the following empirical formulae (Litsas et al., 2017; Hoek & Diederichs, 2006):

$$\sigma_{cm} = 0.02\sigma_{ci}\exp\left(\frac{GSI}{25.5}\right)\text{and } E_m = E_i\left[0.02 + \frac{1}{1+\exp[60 - GSI/11]}\right] \quad (1)$$

Table 1. Rockmass mechanical properties based on the Generalised Hoek-Brown failure criterion.

GSI	σ_{cm} (MPa)	m_b	s	a	E_i (MPa)	E_m (MPa)
15	0.36	0.48	$7.91\cdot10^{-5}$	0.561	2000	72.9
25	0.53	0.687	$2.4\cdot10^{-4}$	0.531	2000	119.7
35	0.79	0.981	$7.3\cdot10^{-4}$	0.516	2000	226.8
45	1.17	1.403	$2.0\cdot10^{-3}$	0.508	2000	447.3

3.3 *Tunnel face pre-support measures*

Tunnel support measures include a primary lining of combined shotcrete and rockbolts installed one excavation step back from the tunnel face. The primary lining is placed in all the examined cases, both unsupported and pre-supported tunnel face fronts. The rockbolts are placed around the tunnel on a radial distance of 1.5*m* and spanned axially (along the tunnel axis) by the same distance (1.5*m*). Figure 3 shows the simulated tunnel lining (the shotcrete is shown purple, while the rockbolts are coloured red) and face pre-support measures working via fiberglass dowels are shown in blue. The numerical analyses work on the following support measure behaviour premise : (a) linear elastic shotcrete and (b) elastic – perfectly plastic (until breaking load) rockbolts.

Reinforced tunnel face fronts by fiberglass dowels work on fully bonded bolts placed in the tunnel face at a density ranging from 1FG/1m^2 to 1FG/2m^2. The fiberglass nails are modelled as elastic – perfectly plastic truss elements working on a diameter (*d*), calculated to give the same cross-sectional area as the actual rectangular cross-section of the dowels. At the final simulation step, the fiberglass dowels extend 10*m* ahead of the tunnel face to minimized the overlap effect in the tunnel face deformation. Table 2 presents the geometry and relative mechanical properties of the tunnel support measures.

4 UNSUPPORTED TUNNEL FACE RESULTS

This section studies the tunnel face response in cases of unsupported tunnel faces, for different ground conditions and overburden heights. The maximal tunnel deformation (U_h) in the middle of the tunnel face was calculated via numerical analyses; it can be expressed as a function of the non – dimensional "average tunnel face extrusion" (Ω_h), as follows:

$$\Omega_h = \left(\pi\frac{U_h}{D}\right) \times \left[\frac{E_m}{p_\circ}\right] \quad (2)$$

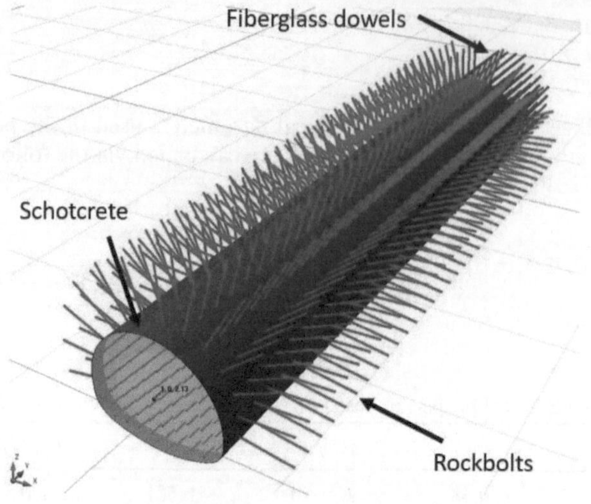

Figure 3. Simulated tunnel lining and tunnel face pre – support measures.

Table 2. Tunnel support measures' mechanical properties.

Property	Shotcrete	Rockbolts	Fiberglass Dowels
Elastic modulus (E)	$15GPa$	$200GPa$	$40GPa$
Poisson's ratio (v)	0.2	0.2	0.2
Thickness (t)	$30cm$	-	-
Breaking load (f_Y)	-	$245kN$	$200kN$
Diameter (d)	-	$25mm$	$13.5mm$
Type	-	Fully Bonded	Fully Bonded

where"D" is the tunnel diameter,"E_m" is the rockmass elastic modulus and"$p_\circ$" is the mean geostatic field stress at the tunnel axis. Figure 4 shows the average face extrusion"Ω_h" against the ground factor"$\Lambda_\circ$", a dimensionless parameter aggregating the main problem controlling factors (ground strength"σ_{cm}", tunnel depth"H" and diameter"D") aiming to minimise the scatter of the plotted pairs $(\Omega_h, \Lambda_\circ)$. The ground factor"$\Lambda_\circ$" is a modification of the classical"ground strength over geostatic stress" ratio often used to describe tunnel behavior (e.g. Hoek, 2000) and is expressed via the following formula:

$$\Lambda_\circ = 3.8\left(\frac{\sigma_{cm}}{\gamma \times H \times \sqrt{1 + (2/3) \times K_\circ}}\right) \times \left(\frac{H}{D}\right)^{0.35} \tag{3}$$

The best-fit estimate correlating the average tunnel face extrusion (Ω_h) with the non-dimensional ground factor $(\Lambda_\circ)$ is shown in the black solid line in Figure 4, and can be expressed as follows:

$$\Omega_h = 1.4 \times \Lambda_\circ^{-1.2} \tag{4}$$

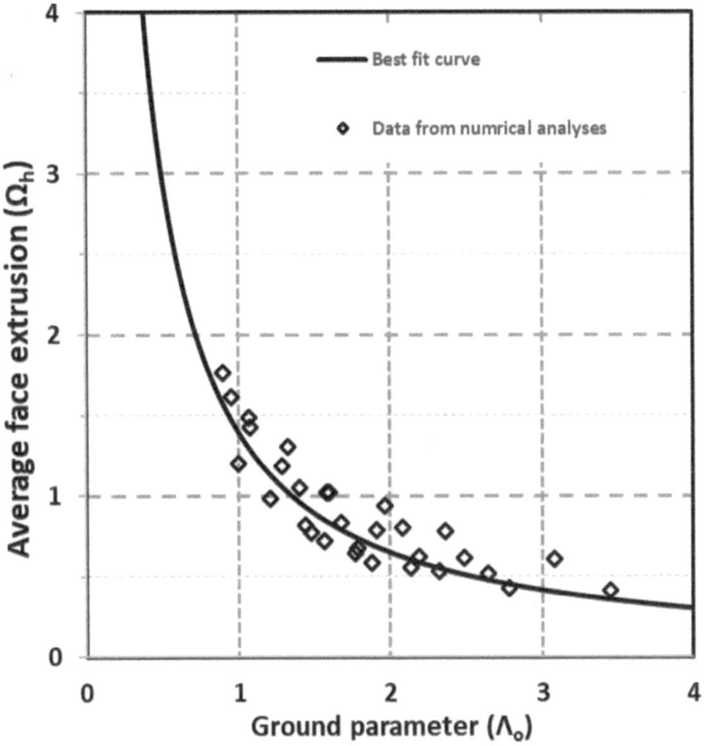

Figure 4. Numerical results of face extrusion (Ω_h) as a function of the ground parameter Λ_o shown in rhombi, compared to the best-fit estimate depicted via the black solid line working on Equation 4.

5 REINFORCED TUNNEL FACE RESULTS

The numerical study extends to reinforced tunnel face fronts by fiberglass dowels. This section studies the reduction of the tunnel face extrusion ($\Delta\Omega_h$) due to the tunnel face reinforcement. The numerical analyses work on two fiberglass dowels densities $0.5FG/m^2$ and $1FG/m^2$.

The parametric investigation shows that the reduction of the tunnel face extrusion ($\Delta\Omega_h$) due to the tunnel face reinforcement depends on the equivalent tunnel face pressure (PF), applied through the fiberglass dowels, and the rockmass strength (σ_{cm}). The tunnel face pressure (P_F) can be expressed as follows:

$$P_F = \frac{n \times F_u}{A_{Face}} \tag{5}$$

where "n" is the number of fiberglass dowels in the tunnel face,"F_u" is the ultimate breaking load of each nail and "A_{Face}" is the net tunnel face reinforced area. To best describe the effect of tunnel face reinforcement, this paper proposes a new dimensionless tunnel face pressure factor "N_{FG}" given through the following formula:

$$N_{FG} = \left(\frac{P_F}{\sigma_{cm}}\right)^{1.35} \times \left(\frac{H}{D}\right)^{0.3} \tag{6}$$

The reduction of the tunnel face extrusion ($\Delta\Omega_h$) between reinforced tunnel faces fronts (Ω_h) and their unsupported tunnel face counterparts (Ω_{h0}) is shown in Figure 5, as a function of the proposed dimensionless tunnel face pressure factor "N_{FG}".

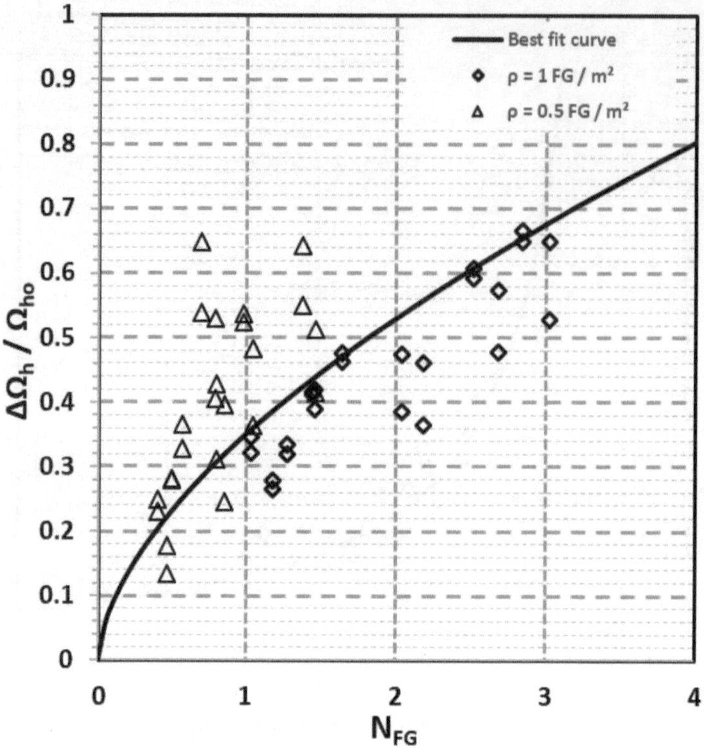

Figure 5. Reduction of the tunnel face extrusion ($\Delta\Omega_h$) due to the tunnel face reinforcement by fiberglass dowels normalized by the unsupported tunnel face extrusion (Ω_{h0}) versus the tunnel face pressure factor (NFG). The data points for the dense and coarse Fiberglass dowels mesh are shown in rhombi and triangles respectively. The black solid line gives the best-fit estimate working on Equation 7.

The best-fit estimate correlating the reduction of the tunnel face extrusion ($\Delta\Omega_h$), normalized by the unsupported tunnel face extrusion (Ω_{h0}), with the dimensionless tunnel face pressure factor (N_{FG}) can be expressed as follows:

$$\frac{\Delta\Omega_h}{\Omega_{h0}} = 0.35 \times N_{FG}0.6 \tag{7}$$

6 CONCLUSIONS

The numerical investigation proves that tunnel face stability in deep tunnels depends not only on the rockmass joint sets, but also on the rockmass strength and stiffness. In cases of weak rockmasses (low GSI values), the ground yields as its response proves similar to that of stiff soils. Unsupported tunnel face fronts tend to deform rapidly, by thus giving tunnel face extrusions that can potentially endanger face stability. Tunnel face stability also depends on the rockmass strength (σ_{cm}) and the horizontal geostatic stress coefficient (K_0); tunnels on $K_o = 1$ the tunnel give higher face extrusions than tunnels on $K_o = 0.5$. The paper produces a non – dimensional ground factor (Λ_o) working on the rockmass conditions as a function of the tunnel face extrusion (Ω_h) . The tunnel face reinforcement by fiberglass dowels reduces significantly the tunnel face extrusion. As the reinforcement density increases, the reduction of the tunnel face extrusion inflates, especially in cases of poor rockmasses. In such tunnel face fronts, the reduction of the tunnel face extrusion ($\Delta\Omega_h$) is correlated with the tunnel face support characteristics.

ACKNOWLEDGMENTS

Mr. Georgiou's present research was funded by the Onassis and Eugenides Foundations.

REFERENCES

Anagnostou, Georg, and Kalman Kovári. "Face stability conditions with earth-pressure-balanced shields." Tunnelling and underground space technology 11.2 (1996): 165–173.

Atkinson, J. H., and R. J. Mair. "Soil mechanics aspects of soft ground tunnelling." Ground Engineering 14.5 (1981).

Hoek, E. Big Tunnels in Bad Rock. 2000 Terzaghi Lecture. ASCE Journal of Geotechnical and Geoenvironmental Engineering, Vol. 127, No. 9. September 2001, pages 726–740.

Hoek E, Carranza-Torres C, Corkum B. Hoek-Brown criterion, 2002 edition. In: Hammah R, Bawden W, Curran J, Telesnicki M, editors. Mining and tunnelling innovation and opportunity, proceedings of the 5th North American rock mechanics symposium and 17th tunnelling association of Canada conference, Toronto, Canada, p. 267–73.

Hoek, E. and Diederichs, M.S. (2006). Empirical estimation of rock mass modulus. International Journal of Rock Mechanics and Mining Sciences, 43, 203–215

Zienkiewicz O C, Humpheson C, Lewis R W. Associated and non-associated visco-plasticity in soil mechanics. Geotechnique, 1975, 25 (4): 671–689

Kavvadas, M., G. Prountzopoulos, and K. Tzivakos. "Prediction of face stability in unsupported tunnels using 3d finite element analyses." Proc. 2nd International Conference on Computational Methods in Tunnelling EURO. TUN. 2009.

Litsas D., Sitarenios P. & Kavvadas M. "Effect of face support pressure on tunnelling induced ground movement", Proc. World Tunnel Congress (WTC 2017), Bergen, Norway, June 2017.

Lunardi, P. (2008). Design and construction of tunnels: Analysis of Controlled Deformations in Rock and Soils (ADECO-RS). Springer Science & Business Media.

Marinos, Paul, and Evert Hoek."GSI: a geologically friendly tool for rock mass strength estimation." ISRM international symposium. International Society for Rock Mechanics and Rock Engineering, 2000.

Prountzopoulos, G."Investigation of the excavation face stability in shallow tunnels." *National Technical University, Geotechnical Division, Athens (in Greek)* (2012).

The Evolution of Geotech - 25 Years of Innovation – Hammah et al. (Eds)

A numerical investigation on the influence of rockmass parameters and yield mechanics in pillar design

E.J. Dressel & M.S. Diederichs
Queen's Geomechanics and Geohazards Group, Queen's University, Kingston, Ontario, Canada

ABSTRACT: Pillar stability guidelines have been presented by various authors in the past, based on a combination of empirical (mostly visual) observations from mining applications and simple numerical elastic stress indices or data from continuum plasticity analysis. There have been numerous developments in the past several decades in mechanistic classification (brittle spalling, effective continuum rockmass, structural control) and mechanism-appropriate modelling and design metrics. The design of pillar dimensions and spacing is a key factor in construction projects beyond mining including the engineering for deep geological repositories, hydroelectric power cavern complexes, underground storage, quarrying and other applications. These applications are situated in a variety of rockmass environments and at stress regimes from shallow to deep. In such applications, the geometry of the pillars and surrounding excavations deviates from the classic mine pillar scenario. This paper redefines design guidelines for pillar dimensioning based on numerical simulation results. It includes the consideration of rockmass deformation and yield (rockmass strength approach), brittle strength and damage behavior where appropriate. The investigation includes 3D FEM models with equivalent continuum networks.

1 MODERN PILLAR DESIGN

A majority of previous research on pillar stability and strength has been performed on coal or sedimentary pillars, while little research has been performed on the strength of hard-rock pillars aside from Lunder and Pakalnis (1997), Von Kimmelman et al. (1994), Hudyma (1998), Hedley and Grant (1972), Krauland and Soder (1987), Sjoberg (1992), and Brady (1977). Previous studies have attempted to develop failure criteria for pillars using linear-elastic approaches. Using best-fit lines, and limited elastic modelling, a number of failure criteria were created. These previous studies conducted on hard-rock pillars may contain some 2D or even 3D simple elastic modelling, but there is a lack of large-scale plastic analysis conducted to predict pillar strength. With modern computing and more robust programs such as Rocscience RS3, plastic analysis is now more accessible and feasible and can be applied to pillar design (Rocscience 2021).

Lunder and Pakalnis (1997) developed a database of mining pillars from four main sources. Von Kimmelman et al. (1994) collected data from a Botswanan mine, using 47 pillars with an average unconfined compressive strength (UCS) of 94. Stress was calculated using 2D elastic displacement discontinuity methods. Hudyma (1998) collected data from 47 pillars from 13 Canadian mining operations to develop a design guide for rib pillars. UCS values of the Hudyma (1998) dataset range from 70 to 316 MPa.

DOI: 10.1201/9781003188339-62

Stress was calculated using 2D fictitious force and displacement discontinuity boundary element methods. Hedley and Grant (1972) collected data from 28 pillars, with UCS ranging from 210 to 275 MPa, in the Elliot Lake mining district in Canada. Pillar stress and strength was calculated based on Salamon and Munro (1967) using the "Size Effect Formula". Westmin Resources Ltd. and CANMET present data from 72 pillars Westmin Resources Ltd. Myra Falls Operations. They collected 32 pillars with a UCS of 172 MPa. Stress was calculated using simple 3D boundary element techniques calibrated to mine conditions. Some additional pillar data was collected and used from Krauland and Soder (1987), Sjoberg (1992), and Brady (1977). The pillar stability condition was classified for all data as shown in Figure 1. The majority of pillars in the database had an RMR between 65 and 80.

The pillars in the database were classified based on the Westmin Resources Ltd. pillar classification seen in Figure 1. Pillars classified as 1 are stable, with no damage, 2-4 are unstable with progressive fracturing and 5 is failed through the pillar core, and blocks falling out. Lunder and Pakalnis (1997) developed a set of empirical strength criteria, as seen in Equations 1-3, from the pillar database, which is widely used today to design mine pillars:

$$P_S = (K \times UCS) \times (C_1 + C_2 \times \kappa) \tag{1}$$

$$C_{pav} = 0.446 \times \log\left(\frac{W}{h} + 0.75\right)^{\frac{1.4}{\frac{W}{h}}} \tag{2}$$

$$\kappa = \tan\left(\cos^{-1}\left(\frac{1-C_{pav}}{1+C_{pav}}\right)\right) \tag{3}$$

where Ps = Pillar Strength (MPa); K = Rockmass strength size factor = 0.44; UCS = Uniaxial compressive strength of intact material (MPa); C_1 = 0.68, C_2 = 0.52; κ = Mine pillar friction term; C_{pav} = Average pillar confinement; w = Pillar width (m); h = Pillar height (m).

Martin and Maybee (2000) suggest that the Lunder and Pakalnis (1997) empirical chart fails to properly include confinement as the fit is very similar to other strength formulas. There should be an exponential increase due to confinement at larger W:H ratios, which is not seen in the empirical curve fit presented by Lunder and Pakalnis (1997). To investigate, Martin and Maybee (2000) conducted a study analyzing hard rock pillars in a South African mine as well as Canadian hard rock pillars using the generalized Hoek-Brown failure criterion. The analysis included a simple elastic analysis to compare strength factors, not failure mechanisms of pillars.

Figure 1 contains two new stability envelopes derived for this paper through discriminant analysis. The solid curve defines a limit above which pillar core yield occurs. Above the dashed line, any kind of pillar wall instability is observed. The curves developed are described by Equation 4.

$$SL = \frac{AR^B}{R^B + C^B} + D + ER \tag{4}$$

where SL = Stress limit (Pillar vertical stress /UCS); R = Pillar width to height ratio; Constants A thru E are shown in Figure 1.

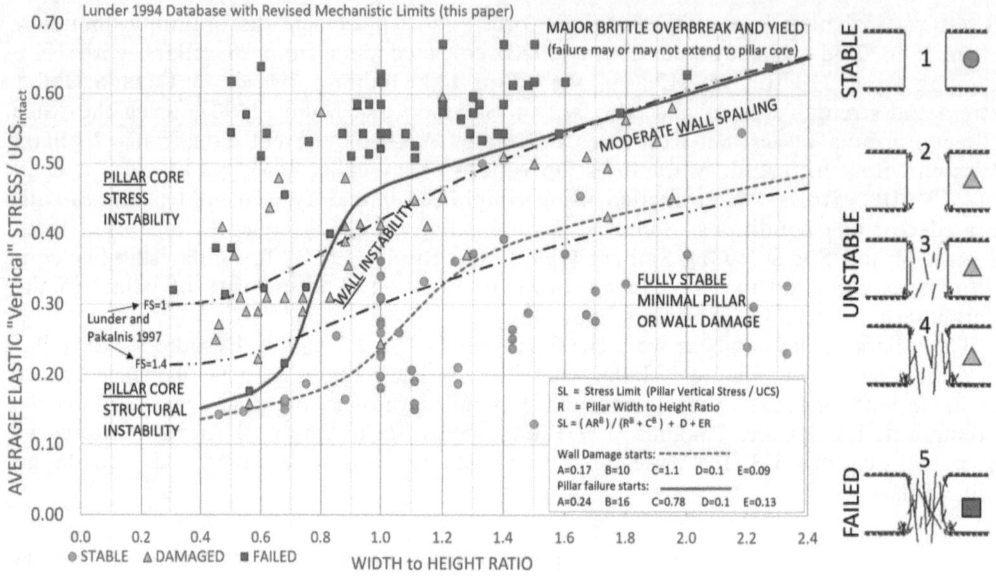

Figure 1. Pillar stability data from Lunder (1994) combined with empirical stability limits by Lunder and Pakalnis (1997). New mechanistic discriminant limits (red solid= failed/unstable and blue dashed =unstable/stable) have been added for this paper based on the case data and Equation 4.

2 METHODS USED TO ESTIMATE ROCKMASS RESPONSE

To examine plastic pillar response, the Generalized Hoek-Brown (H-B) criterion, as seen in Equation 5, is selected as it is very commonly and successfully used in industry (Hoek et al. 2002):

$$\sigma_1 = \sigma_3 + \sigma_{ci}\left(m_b \frac{\sigma^3}{\sigma_{ci}} + S\right)a \tag{5}$$

where σ_1= Major principal stress (MPa); σ_3= Minor principal stress (MPa); σ_{ci}= Uniaxial compressive strength of intact material; m_b = Frictional component or m from Hoek-Brown criterial; a,s = empirically derived constants based on rockmass response (Hoek et al 2002).

The Geological Strength Index (GSI) is used as a means to reduce rockmass strength based on visual observations of jointed rockmasses in the field. GSI (Marinos and Hoek, 2000) allows for non-intact rockmasses to be assessed as a whole rockmass rather than requiring many discrete joints in models. Parameters m_b, s and a can all be calculated for rockmass response using GSI as outlined in Equations 6-8, so long as GSI > 25, from Hoek et al. (2002).

$$m_b = m_l * \exp\left(\frac{GSI - 100}{28}\right) \tag{6}$$

$$S = \exp\left(\frac{GSI - 100}{9}\right) \tag{7}$$

$$a = 0.5 \tag{8}$$

where m_i= Hoek-Brown frictional component of intact rockmass.

Martin and Maybee (2000) assessed a typical range of Canadian hard rock pillar GSI. Of 177 pillars analyzed, the mean GSI assessed was 67, with 60% of pillars falling in the 60-80 GSI range, or 38 pillars from 60-70 GSI and 68 pillars from 70-80 GSI. These GSI values are comparable to the Lunder and Pakalnis (1997) database, as most pillars are located in Canadian hardrock mines. The ideal sample pillar is ~70 GSI to compare to the Lunder-Pakalnis database, as it is roughly the average pillar GSI and it falls within the mode of Canadian hard-rock pillars.

2.1 *Damage Initiation Spalling Limit Criterion (DISL)*

The damage initiation spalling limit (DISL) was developed (Diederichs 2007) to explain behavior of hard, brittle rocks. Using a cohesion weakening, friction strengthening parameters, the DISL criterion is able to predict an accurate rockmass response for hard rock.

As outlined in Diederichs (2007) brittle rock parameters can be calculated using *UCS*, *T*, m_i and crack initiation stress (*CI*). The parameter *a* is set to 0.25 pre-failure (instead of 0.5 in the HB equation) to represent systemic crack accumulation, with *s* and *m* seen Equations 9 and 10.

$$S = \left(\frac{CI}{UCS}\right)^{\frac{1}{a}} \tag{9}$$

$$m = S\left(\frac{UCS}{|T|}\right) \tag{10}$$

For post damage, *a* is set to 0.75, s is reduced to near zero and m ranges between 6 and 10. This formulation allows for mostly cohesive crack initiation followed by a confinement sensitivity that is part frictional and part due to the mechanics of crack interaction and propagation.

3 DEVELOPMENT OF PILLAR MODELS

To evaluate a realistic application of pillars in a non-mining environment, a power cavern with an accompanying transformer cavern connected by four cross tunnels is simulated. The 3D finite element analysis software RS3 allows a plastic response to be observed in multiple pillars and allows for 3D confinement to affect the pillar response. Over 600 pillars are modelled. Using RS3, the geometry is developed and illustrated in Figure 2. The cross tunnels create three pillars per model for analysis in hard rock conditions. A uniform secondary and tertiary mesh refinement are used in the final model mesh around the pillars, as well as the caverns.

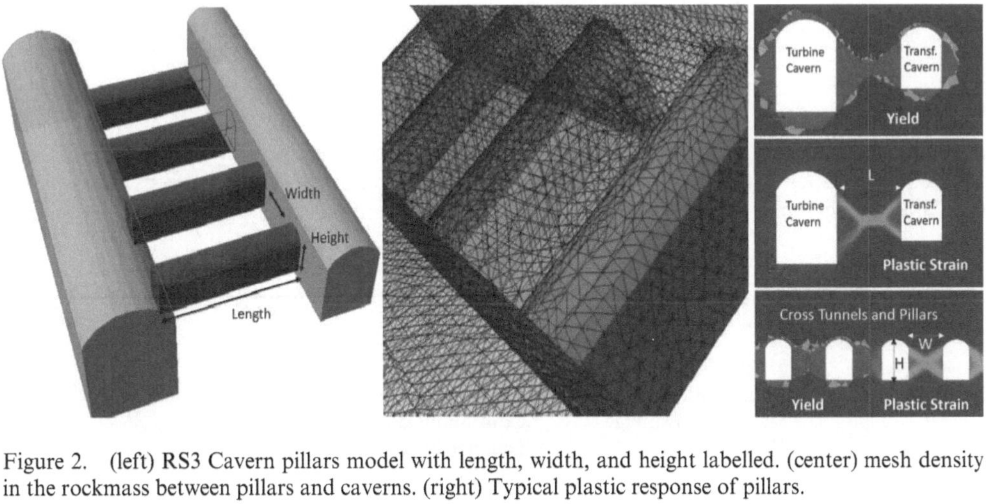

Figure 2. (left) RS3 Cavern pillars model with length, width, and height labelled. (center) mesh density in the rockmass between pillars and caverns. (right) Typical plastic response of pillars.

Table 1. Hoe k-Brown model parameters used for GSI 60, 70 and DISL models.

	GSI 60		GSI 70		DISL Brittle	
Parameter	Pre-Failure	Residual	Pre-Failure	Residual	Pre-Failure	Residual
m_i	10.6	-	10.6	-	10.6	-
m_b	2.5403	1.7774	3.6307	2.5403	0.32	8
s	0.01174	0.003865	0.03567	0.01174	0.0299	0.0054
a	0.5028	0.5057	0.5014	0.5028	0.25	0.75

Cobourg Limestone from Ontario was selected as an analog for a low-mi rock-type to investigate. The rock has a UCS of 113 MPa, a Young's Modulus of 39 GPa, and a Poisson's Ratio of 0.3. Hoek-Brown parameters as calculated from Equations 6-8 for GSI 70 and GSI 60 models and are listed in Table 1. Residual (post-yield) parameters for each GSI are calculated using a GSI 10 below the pre-failure GSI. Additionally, brittle parameters as per the Diederichs (2007) DISL criterion are generated using Equation 8 and 9 and are listed in Table 1.

Figure 2 displays and defines the dimensions of the pillars generated. Pillars generated range from a w:h ratio of 0.5 to 2.0 and a vertical stress ratio (σ_v/UCS) of 0.1 to 0.6, while maintaining a constant l:w ratio of 2.5. Stress ratio was calculated by simulating the pillar elastically and taking the average stress across the pillar center. A k stress factor (horizontal/vertical stress) of 1 is used in the models to eliminate variable in-situ stress effects.

4 MODELLING RESULTS

Each pillar in a model has a different average stress, which is measured elastically and plotted according to plastic response. Three pillars per model are generated and results are outlined in Table 2. Fewer brittle pillars are modelled because the lower bound for undamaged is a 0.2 vertical stress ratio, whereas GSI 70 and 60 is undamaged at 0.1 and <0.1 vertical stress ratio.

Table 2. Resultant pillar model generated by classification.

Classification	GSI 60	GSI 70	DISL Brittle
1	0	50	49
2	45	49	28
3	91	100	50
4	3	7	10
5	83	46	58
Total Pillars	222	252	195

4.1 *Defining pillar failure*

A system for defining pillar condition was developed as per Lunder and Pakalnis (1997) but adapted for plastic 3D pillar models as shown below in Table 3. The FEM analysis results from this study are shown in Figure 3. "Sigma1" is the average vertical elastic stress across mid pillar.

Table 3. Pillar classification definition.

Classification	Name	Definition
1	Stable	Undamaged
2	Minor skin damage	Limited skin damage
3	Significant Damage	Continuous damage in pillar wall (Hourglass)
4	Yielding	Damage impacting core
5	Failed	Full Yield throughout pillar – Continuous plastic strain

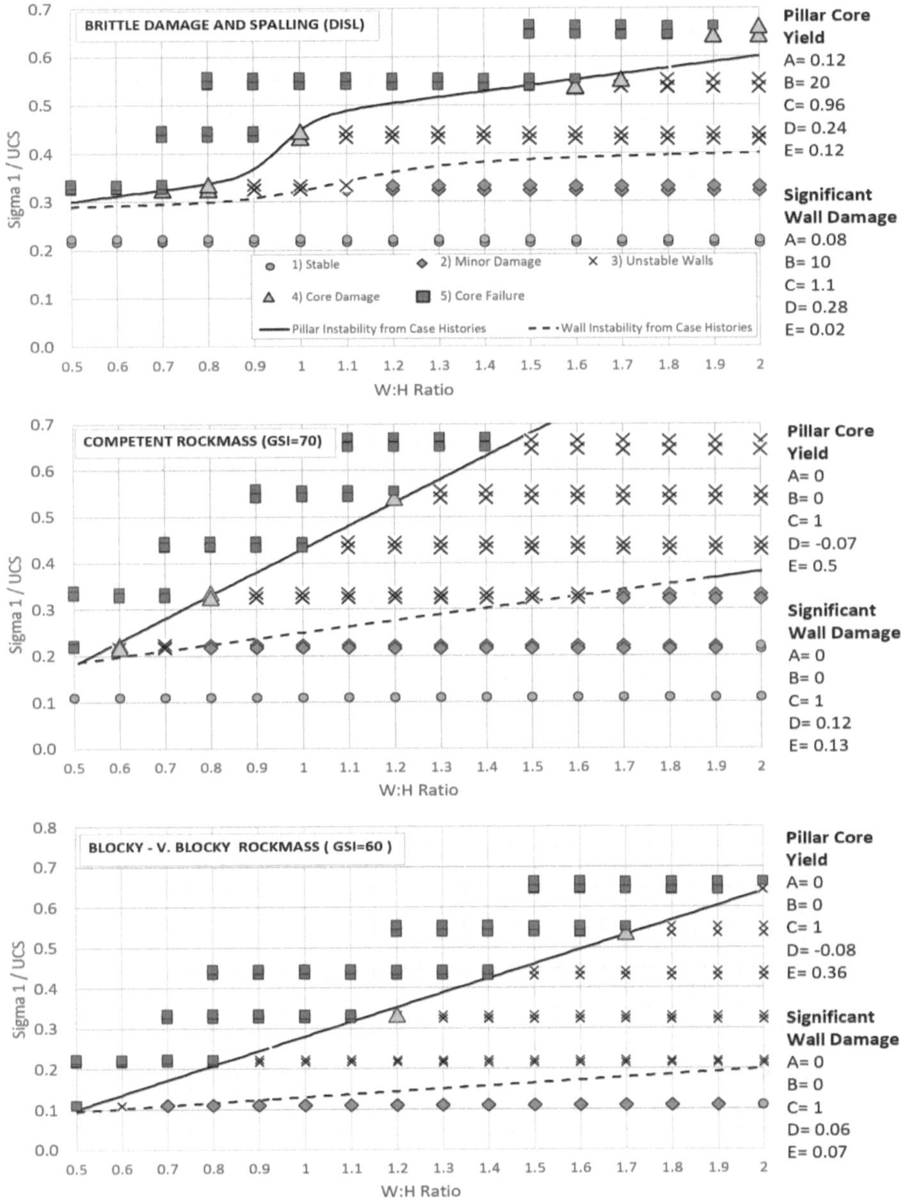

Figure 3. (Top) Model data for GSI 70 with legend for all plots, (Mid) Model data for GSI 60, (Bottom) Model data for Brittle pillar parameters. On the right are curve fit parameters for Equation 1. New discriminant envelopes (Eq. 4) are shown (Failed/Unstable division and Unstable/Stable division) for each case.

Rockmass failure (GSI-based models) show a near linear stability response with increasing W: H ratios. Low GSI leads to lower maximum stress ratios before instability and failure as expected. For brittle pillars, the brittle behaviour allows higher stability at lower stresses as cracks cannot propagate as easily within the brittle rockmass. However, as stresses increase, failure is more early-onset for brittle behavior when compared with GSI 70 failure, which is more gradual, and failure does not occur at high stresses for pillars with >1.5 w:h ratio. The brittle pillar model data follow the Lunder-Pakalnis data, suggesting that the pillars observed are brittle in behavior.

5 CONCLUSIONS

The new pillar failure criterion redefines pillar failure as not a safety factor, but a method on which to design and classify the failure of a pillar. Overall, pillar failure has begun to be reclassified, as pillar failure does not just occur with crack development. Pillar yield does not directly mean the pillar has failed. The pillar pre-peak and post-peak stresses must be investigated to further define pillar failure. Using shear strains as an indicator for pillar failure, the analysis of a pillar goes beyond surface deformities, and assess the mechanism behind the pillar failure, which in the case of hard rock pillars is dominantly brittle spalling.

The strength of a pillar is heavily dependent on both the w:h ratio of the pillar, as well as the rockmass strength. Previous attempts to define pillar strength have lacked in that there was no plastic analysis performed, and the fit was very linear for a non-linear relationship. Using plastic analyses, the brittle pillar failure criterion is better defined, and matches the empirical results. The rockmass strength curves are a tool to begin design based on GSI and elastic stresses, but pillar failure must be assessed using plastic modelling to validate pillar strength for a given rockmass strength.

REFERENCES

Brady, B.H.G. 1977. An analysis of rock behaviour in an experimental stoping block at the Mount Isa Mine, Queensland, Australia. *Intl. J. of Rock Mechanics Mining Science and Geomechanics*, 14:59–66.

Diederichs, M.S. 2007. Mechanistic interpretation and practical application of damage and spalling prediction criteria for deep tunneling. *Can. Geotech. J.* 44: 1082–1116.

Hedley D.G.F., Grant, F. 1972. Stope-and-pillar design for the Elliot Lake Uranium Mines. *Bulletin Can Inst Min Metall* 65:37–44.

Hudyma, M.R., 1988. *Rib pillar design in open stope mining*. M.A.Sc. thesis, The University of British Columbia.

Hoek, E., Carranza-Torres, C. & Corkum, B. 2002. Hoek-Brown Failure Criterion – 2002 Edition. *Proceedings of the 5th North American Rock Mechanics Symposium, Toronto, Canada*, vol. 1, pp 67–273.

Krauland, N., Soder P.E. 1987. Determining pillar strength from pillar failure observations. *Engineering and Mining Journal* 8:34–40.

Lunder, P.J. 1994. Hard rock pillar strength estimation: An applied empirical approach. *M.A.S.c thesis*, The University of British Columbia.

Lunder, P.J. & Pakalnis, R.C. 1997. Determination of the strength of hard-rock mine pillars. *Bull Can Inst Min Metall* 90:51–5.

Marinos, P., & Hoek, E. 2000. GSI: A Geologically Friendly Tool for Rock Mass Strength Estimation. *In Proc. GeoEng2000 Conference*. Melbourne. pp. 1422–1442.

Martin, C.D., Kaiser P.K., & McCreath, D.R. 1999. Hoek-Brown parameters for predicting the depth of brittle failure around tunnels. *Can Geotech* 36:136–51.

Martin, C.D. & Maybee, W.G. 2000. The strength of hard-rock pillars. *International Journal of Rock Mechanics & Mining Sciences* 37: 1239–1249.

Rocscience. 2021. RS3 14.013. Toronto, Ontario, Canada

Salamon, M.D.G., & Munro, A.H. 1967. A study of the strength of coal pillars. *J. of SAIMM* 67: 55-67. Sjoberg, J.1992. Failure modes and pillar behaviour in the Zinkgruvan Mine. *Proceedings of the 33rd U.S Symposium on Rock Mechanics*. Edited by Tillerson & Wawersik.

Von Kimmelmann M.R., Hyde, B. & Madgwick R.J. 1984. The use of computer applications at BCL Limited in planning pillar extraction and design of mining layouts. *Proceedings of ISRM Symposium: Design and Performance of Underground Excavations*. London: p. 53–63.

Session 14 - Stability analysis of mining slopes III

The Evolution of Geotech - 25 Years of Innovation – Hammah et al. (Eds)

Continuous wall stabilization of impoundment slopes

W. Carswell, L.C. Barlow & D.R. Siebert
Haley & Aldrich, Inc.

ABSTRACT: Final closure was proposed for three impoundments containing iron oxide (IOX) at a 500-acre site in the southeastern United States. The closure plan included placing 490,000 cubic yards of additional IOX material within the impoundments, reaching heights 10 to 28 ft on top of the existing embankments. Stability models of the raised impoundment indicated a factor of safety less than required without mitigation measures. Reinforced concrete diaphragm (slurry) walls were proposed as a multi-purpose solution, providing contaminated groundwater cut-off and slope stabilization. Preliminary research indicated that many approaches and solutions were available for pile-stabilized slopes but little design guidance existed for the design of continuous wall-stabilized slopes. The final design used a combined approach aided by the Rocscience products Slide2 (for slope stability) and RSPile (for design of the slope-supporting slurry wall). This paper discusses the approach and methodology adopted for the analysis and design of the slope supporting slurry wall and will conclude with recommendations for further study.

1 INTRODUCTION

Poor ground conditions posed a challenge to meeting closure stability requirements for three existing impoundments (factor of safety [FS] = 1.30, circular failure surface). Weak native clay underlies the site and impounded iron-oxide rich silt (IOX) has remained saturated since deposition. Undrained shear strength of the clay and IOX is low (300 to 1000 psf), the thickness of native clay ranges from 20 to 50 ft across the site. Proposed final closure included adding 490,000 cubic yards of material within the impoundments below a cover system, increasing the final grades 10 to 28 ft.

To satisfy stability requirements of the impoundments, a structurally reinforced concrete containment wall (slurry wall) was designed. The wall alignment was designed parallel to the embankment slope, offset 20 to 50 ft from the toe. The structural wall extended through the weak native clay into competent sand. Calculating the required embedment depth into sand and the structural demand (shear and moment) on the wall were critical elements to successfully achieving an adequate and efficient design. The final structural wall design used a combined approach aided by the Rocscience products Slide2 for slope stability and RSPile for development of forces in the slurry wall.

2 PROJECT BACKGROUND

IOX material was placed over the past 30 years in a complex of three impoundments. The proposed impoundment closure includes adding 10 to 28 ft of material above the embankment crest. Achieving the required FS for stability in the final condition was challenging due to weak native clay and slurry deposited impounded IOX material.

DOI: 10.1201/9781003188339-63

The soil profile consisted of fill soils and IOX material underlain by soft clay of the Pliocene-Pleistocene age (20 and 55 ft below grade). This clay unit acts as a confining aquitard, in-situ and laboratory testing indicates the consolidation state is approximately normally consolidated outside the embankments and potentially under-consolidated beneath the impoundments. Beneath the clay is a Miocene age silty sand, encountered approximately 55 to 105 ft below grade.

Water levels at the site were typically observed between 2 and 8 feet below grade outside the impoundments. Water within the impoundments was found near or above the surface, depending on seasonal precipitation and location.

Table 1 below presents soil properties as determined from SPT, CPT, in-situ strength testing, lab data from undisturbed samples, and index testing.

Table 1. Soil Properties.

	Temporary Condition (Undrained)		Final Condition (Drained)	
Soil Type	Undrained Shear Strength (psf)	Friction angle (deg.)	Cohesion (psf)	Friction angle (deg.)
IOX	800	–	0	27
Embankment Fill	0	34	0	34
Miocene Silty-Sand	0	32	0	32
Native Clay (Outside Impoundment)	$0.075\sigma_v$' + 350 psf	–	120	17
Native Clay (Inside Impoundment)	$0.124\sigma_v$' + 485 psf	–	120	17

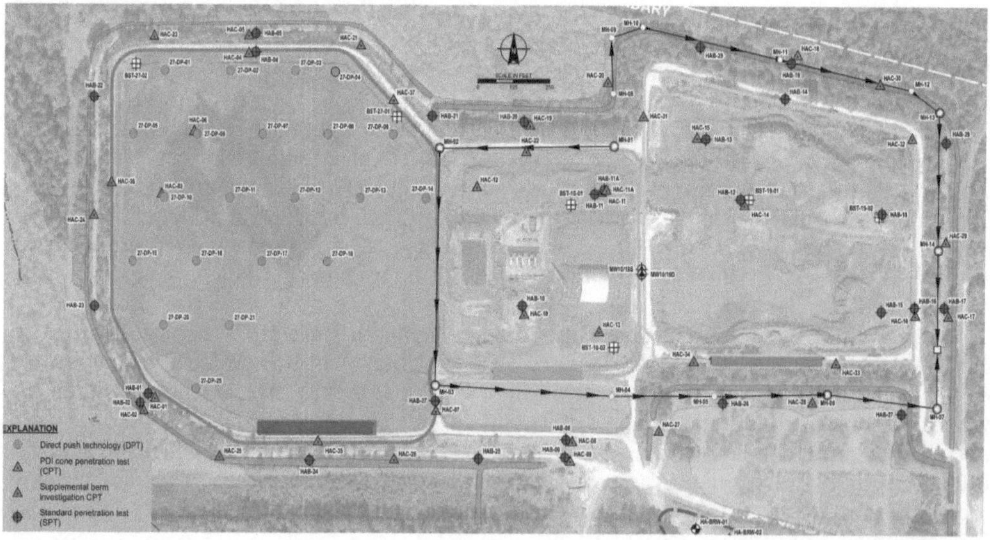

Figure 1. Project plan.

3 SLURRY WALL DESIGN

The impoundment was analyzed at several sections to account for variability in stratigraphy (clay thickness) and impoundment height. When analyzed using the two-dimensional slope stability program Slide2, several sections did not meet the required minimum FS. The proposed solution was to install a 4-ft thick, 5,000 psi reinforced concrete diaphragm wall constructed using the slurry method to support the slopes and increase global stability of the system, with the additional benefit of providing groundwater cut-off for containment.

While literature exists describing methodologies for analyzing and designing pile-supported slopes (e.g. Hassiotis & Chameau, 1984; Loukidis & Vavourakis 2014), little guidance was found on how to define the design loads for a continuous wall to reach achieve a desired FS.

Two-dimensional finite element modeling was considered to define design loads and associated "factor of safety" via strength reduction methods, as the features of the project would be conducive to plane strain analysis; however, due to the rigorous level of effort required to reproduce multiple slope stability analyses with a finite element modeling program, a combined approach of Slide2 and RSPile successfully calculated design loads and member reactions.

The combination of Slide2 and RSPile allows the user to account for the strength and stiffness of the supporting element (in this case, a reinforced concrete diaphragm wall). The procedure entailed the following steps:

- A 1.00-in. (25.4 mm) movement criteria was selected as the allowable movement of the support element at the depth of the critical slip circle.
- Soil properties were defined in RSPile corresponding to the soil strata in the Slide2 model (i.e., definition of p-y and t-z curves based on the Mohr-Coulomb properties used in Slide2), assuming effective stresses approximately equivalent to the middle of each soil stratum.
- The 4-ft thick concrete slurry wall properties were defined in RSPile, assuming a 1-ft wide element.
- The pile and soil properties from RSPile were linked to the slope-supporting element in Slide2, assuming a 1-ft spacing.
- Perform a slope stability analysis in Slide2 (Spencer method).
- Identify the resultant force from the slip circle with the minimum FS, the depth at which the slip circle intersected the support element, and the angle at the base of the slice where the support element was located (refer to Figure 2; in this example, the resultant was 102 kips per linear foot with a base angle of -13 degrees acting approximately 49 ft below ground surface).
- For verification, the allowable support element movement (1.00 in.) was decomposed into vertical and lateral components based on the intersecting angle of the slip circle at the element location. This displacement was imposed on the reinforced concrete element in the RSPile model at the same depth of the critical slip circle intersection in the Slide2 model.
- The resultant of the shear and axial load in the RSPile support element at the enforced displacement depth was compared to the resultant value from the Slide2 model (refer to Figure 3; the shear, axial load, and resultant were approximately 21.1 kips, 95.5 kips, and 97.8 kips respectively).
- The maximum loads from the RSPile verification analysis were then used to design the reinforcement of the supporting concrete slurry wall.

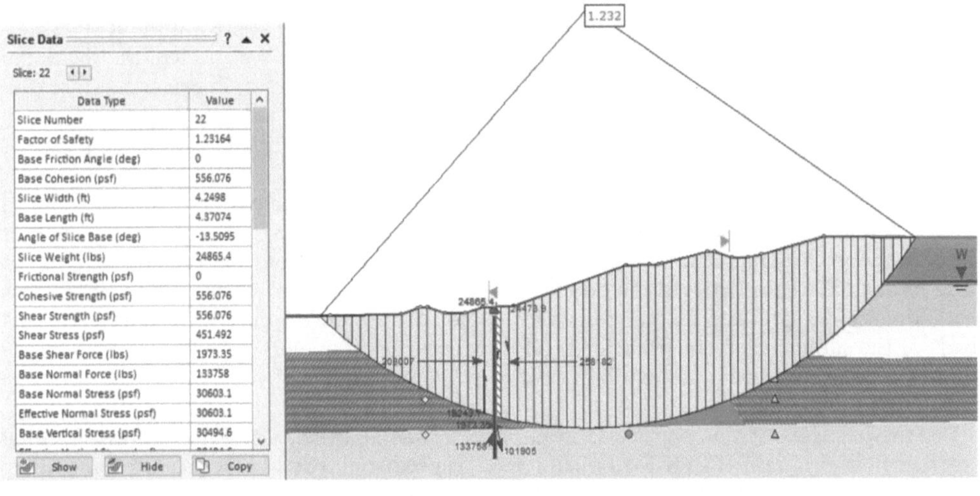

Figure 2. Example Slide2 slice results with support wall (results in units of lbs., ft).

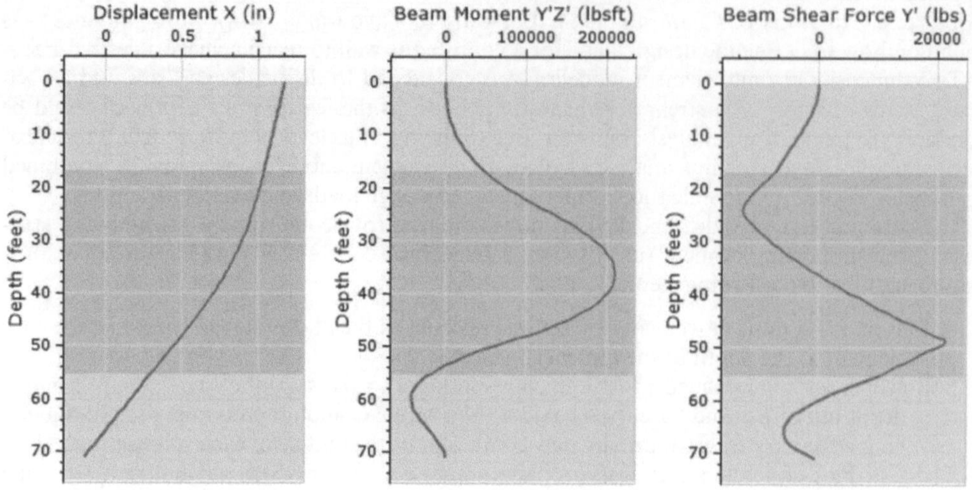

Figure 3. Typical RSPile Results.

4 DISCUSSION

While the combined approach of Slide2 and RSPile described above provided design loads with depth, there were concerns regarding the potential overlap or "shadowing" of passive wedges when assuming unit spacing and thickness of a pile element (in lieu of a continuous wall). Analysis of different spacing indicated that the FS was sensitive to spacing, but that the assumed resistance per "pile" was less sensitive. To ensure that the FS for wall-supported sections were acceptable for project conditions, additional Slide2 analyses were performed assuming an infinitely rigid, continuous support wall element (FOS = 1.30). Several different methods were employed to explore and evaluate the resistance of the support element and the associated design loads (refer to Table 2).

The simplest method used was a beam model which was pinned at the base (at the top of the competent sand layer) and rigidly connected at the top (at the approximate elevation where the slip circle intersected the support wall). The top of the beam element was displaced 1.00 in. using the same decomposed vertical and lateral displacements used in the RSPile verification analysis. Because the contribution of the soil was completely neglected, the shear and moment results were assumed to be a highly conservative upper bound. Because the structural axial load capacity of the reinforced concrete wall was much higher than the axial results from any of the methods used, the shear and moment results were of more interest for the purposes of reinforcement design.

A shear equilibrium method was also examined, where the total required resistance assumed to be taken by the support wall (in order to achieve the desired FS) was equal to the sum of the driving forces along the critical failure surface of a section without the support wall, multiplied by the difference between the target FS and the FS of the "no-wall" section. While this methodology did produce a total resistance value, it was unclear how to estimate internal loads of the support element from the total resistance; additionally, the overall methodology did not account for strain compatibility between the movement of the support wall and the surrounding soils.

For the purposes of this paper, the example case was analyzed using the two-dimensional finite element program PLAXIS as a source of comparison. The concrete diaphragm slurry wall was modeled using a "wished-in-place" beam element, used similar soil properties as the Slide2 model, and resulted in a FS that was very similar to Slide2 (1.23 from Slide2 vs. 1.28

Table 2. Comparison of Results (per Linear Foot of Impoundment).

	RSPile verification	Beam model (fixed-pinned)	Shear equilibrium	PLAXIS strength reduction method	Slide2 (rigid wall)
FS	1.23	–	1.01	1.28	1.30
Total resistance	102 kips	–	75.8 kips	–	–
Moment	212 kip-ft	1700 kip-ft	–	121 kip-ft	–
Shear	21.1 kips	81.1 kips	–	8.0 kips	–
Axial	95.5 kips	41.2 kips	–	13.8 kips	–
Displacement	1.00 in.	1.00 in.	–	–	–

from PLAXIS). The wall loads presented in Table 2 were obtained from the "wished-in-place" step prior to performing the safety analysis (strength reduction method for FS).

The output of the finite element model is believed to be the most accurate of the methods presented for estimating wall loads because the model included a continuous wall, produced a factor of safety consistent with Slide2, and provided wall loads directly from the output without additional verification cases in other programs. It should be noted that the factor of safety analysis in the finite element model used strength reduction of the soil materials and did not include the flexibility of the wall. The method used for design was conservative relative to the finite element model results, with shear and moment loads ranging from two to three times higher than the results from the finite element model.

The Slide2 and RSPile method (as well as the finite element method) resulted in FS less than 1.30; however, when the continuous wall was modeled in Slide2 as an infinitely rigid element, the critical slip surface passed underneath the toe of the wall and therefore total resistance against the wall was zero (and FS = 1.30).

5 CONCLUSIONS AND RECOMMENDATIONS

Soft soils, both in and below the impoundments, resulted in slope stability factors of safety below acceptable levels and thus required additional means of slope support. Design of a reinforced concrete diaphragm wall to increase the FS and facilitate the closure of the impoundments containing iron oxide was successfully performed using a combination of Slide and RSPile and compared to other methods (beam model and finite element analysis).

Comparison of methods provided a useful way to bound the problem and ensure results were reasonable. All methodologies presented imperfect solutions, requiring engineering judgment to understand the limitations of each methodology relative to critical design requirements. For the case study presented, the combined methodology approach was used to evaluate slope stability and support wall design loads. Use of Slide2 and RSPile for design of a continuous stability wall resulted in design wall loads that were more conservative than a finite element evaluation of slope stability (which did not consider wall flexibility). Further exploration of the shear equilibrium method to provide wall loads could be performed to provide another method of comparison.

REFERENCES

Hassiotis S. & Chameau, J.L. (1984). "Stabilization of slopes using piles." Joint Highway Research Project, JHRP-84-8, Indiana Department of Transportation.

Loukidis, D. & Vavourakis, V. (2014). "Limit lateral resistance of vertical piles in plane strain." Numerical Methods in Geotechnical Engineering – Hicks, Brinkgreve & Rohe (Eds). Taylor & Francis Group, London.

The Evolution of Geotech - 25 Years of Innovation – Hammah et al. (Eds)

Landslide hazard and risk level assessment of quarried slopes in Lebanon using drone imagery

R. Kaafarani & G. Abou Jaoude
Lebanese American University, Byblos, Lebanon

ABSTRACT: Human activities are continuously altering the steepness of natural slopes, making them vulnerable to slide under natural triggering factors like earthquakes and rainfall. Quarrying activities in particular, contribute largely to slope failures worldwide, especially when unorganized and chaotic. In Lebanon, quarries are scattered randomly across the country and lack proper urban planning and management. Regional scale maps were recently generated in a Geographic Information System (GIS) platform to identify hazard and risk levels of co-seismic and rainfall induced landslides. However, their applicability to quarried slopes was not tested. In this research, we present a methodology to assess, at a site specific level, the hazard and risk levels of quarried slopes under three conditions: dry, heavy rainfall, and seismic events. The aim is to also ascertain the degree of accuracy of the regional scale maps in predicting landslides in quarried areas. A limestone quarry in Bafliye, Southern Lebanon, was mapped with a DJI Phantom 4 V2.0 drone, recreated as a 3D scene in Pix4D Mapper, assessed kinematically in Rocscience DIPS 7.0, and analyzed using limit equilibrium and numerical modelling techniques using both Rocscience SWEDGE 7.0 and RS2 11.0. At failure, the maximum runout distance and the corresponding angle of reach were determined in Rocscience RocFall 8.0. The studied slope failed under both seismic events with 10% exceedance probability in 50 years, and rainfall events with 10-year return period. The runout distance was relatively small, yielding a low risk failure. The safety factors matched between the site specific and the regional scale analyses, while risk levels did not. This indicated that the regional scale analysis is adequate in predicting only hazard levels at quarried sites.

1 INTRODUCTION

Human activities are largely contributing in increased hazards of landslide occurrence worldwide. Quarrying is one of the major triggers for landslide occurrences, with 38.6% of non-seismic and non-rainfall triggered landslides being caused by mining activities (Froude et al. 2018).

In Lebanon, quarries have been significantly contributing to the economy after the end of the civil war in 1990 (Verdeil et al., 2007). Due to the increased demand on construction materials, quarries spread chaotically over the country, covered approximately 5267 ha, and extended to urban areas (Darwish et al., 2010).

Even with the persistence of the quarries problem in Lebanon, no researches on quarried slopes was done, especially under triggering events. Quarries were only studied for their impact on the environment (Darwish et al., 2010). Researchers rather focused on co-seismic stability assessments of individual natural slopes given the rugged topography of Lebanon and its location in an active seismic zone (Fawaz et al., 2014). Grant et al. (2016) and Kaafarani et al. (2019) assessed slope stability under earthquake and rainfall events on a regional scale using a multi-modal approach in a GIS framework, based on

DOI: 10.1201/9781003188339-64

the geologic map of Lebanon, a 15-m digital elevation map (DEM), earthquake peak ground acceleration maps and rainfall data.

Assessment of quarried slopes stability in the literature has a main focus on studying quarries that have already failed or show signs of failure. The aim is to usually give safety recommendations for quarrying operations (Karaman et al., 2013; Costa et al., 1999). Few researchers presented new mapping or analysis techniques (Wang et al., 2019), and no researcher presented a coherent methodology to assess the stability of quarried slopes.

In this paper, a site specific analysis methodology is presented and tested on a quarried site in Bafliye, South Lebanon. The stability of the slope is assessed using kinematic analysis, limit equilibrium methods (LEM) and finite element methods (RS2), and the corresponding runout distance and reach angles are determined. The purpose is to also assess the accuracy of the multi-model approach to predict hazard and risk levels on quarried slopes.

2 SITE DESCRIPTION

The selected site is located in Bafliye, South Lebanon, is 58.5m high, 250m wide, 70° steep and dips at 355°.

The geology of the site is limestone having a unit weight of 23 kN/m^3 (Dubertret, 1945). Strength parameters are determined from RocData. The site is subject to a PGA of 0.25g for an earthquake having a 10% probability of exceedance in 50 years (Huijer et al., 2011). The rainfall intensity is 103 mm/day causing a 17% saturation for a storm having a 10 year return period (Kaafarani et al., 2019).

3 METHODOLOGY

3.1 *Mapping and reconstructing the quarry*

The quarry was mapped with a DJI Phantom 4 drone utilizing the double grid mission in Pix4D Capture. An overlap of 90% was maintained between consecutive images to ensure the capture of minimal details. The camera was inclined at 60° from horizontal. The total covered area was 0.1 km2 with 210 geo-located images.

The quarry was reconstructed using Pix4D Mapper (Figure 1), with a minimum number of matches of 3 to reduce noise in the model.

Features other than the quarry face were eliminated using CloudCompare. The rightmost section of the quarry was highly jointed compared to the rest of the quarry face, and hence was extracted as a separate point cloud to be analyzed.

Discontinuity Set Extractor (DSE), a MATLAB plugin, was then used to detect discontinuities semi-automatically. DSE requires statistical input such as the nearest neighbor coefficient, number of bins, etc. to determine discontinuity planes and group them into sets. Their values were adopted from Riquelme et al., 2014. The detected joint planes were validated visually, and only points tracing actual discontinuities were kept.

Figure 1. 3D model in Pix4D Mapper.

3.2 *Kinematic analysis*

The kinematic stability of the quarry was assessed using DIPS. Equal angle lower hemisphere projection were used to represent the joints. The susceptibility to flexural toppling, planar sliding, wedge sliding, oblique and direct toppling was assessed and the probability for each mode of failure was obtained. A lateral limit of 20° was used in planar sliding and 30° was used in flexural toppling (Goodman, 1980; Hudson et al., 1997).

3.3 *Slope stability assessment using LEM and FEM*

The mode(s) with the highest probability were assessed using mode specific software. In this case, wedge sliding had the highest probability, and hence, the slope was analyzed using LEM in SWEDGE. The deterministic option was used since the intersecting joints are known. The Barton-Bandis criterion was adopted with a residual friction angle of 35°, joint roughness coefficient of 10 and joint compressive strength of 51.74 MPa, which were determined from RocData.

Slope stability was also assessed using FEM in RS2. Hoek-Brown criteria was selected to model intact rock, and the Barton-Bandis criterion was used to model discontinuities.

Intersecting joint sets were modelled using the joint network option, and the parallel deterministic joint model. However, since the orientation of the joint sets is in 3D, a trace plane with dip direction = 85° was defined.

Discontinuities length and spacing were measured in CloudCompare, and discontinuities were assumed to be fully persistent.

In both analyses, the slope was also assessed under dry, seismic and rainfall conditions; however, in the FEM case the actual rainfall intensity was used instead of an approximated saturation.

3.4 *Runout distance and angle of reach*

The runout distance of failing rock blocks was determined in RocFall. The seeder was set at the highest point of the slope, and the initial velocity was set to zero as the block will detach and fail under its weight when triggered. Weights of rock blocks were determined from SWEDGE, and restitution and friction coefficients were assigned as per Table 1. The slope roughness was directly measured on the 3D model. Then, the angle of reach was calculated, and corresponding risk levels were assigned.

Table 1. Input used in RocFall

Parameter	Restitution		Friction		Roughness	
	Normal	Tangential	Dynamic	Rolling	Spacing (m)	Amplitude (m)
Value	0.32	0.71	0.42	0.40	0.50	0.01

3.5 *Comparison with regional scale assessment*

The Bafliye slope was not captured given the coarseness of the DEM, and was stable in the regional maps under both rainfall and seismic events.

To compare the site specific analysis to the regional scale assessment, a quarry in Halate, similar in geology, geometry, PGA, and rainfall intensity was selected.

4 RESULTS AND DISCUSSION

4.1 *Site specific analysis*

4.1.1 *Kinematic analysis*

748 joints were identified and grouped into 10 joint sets (Table 2), and were visually validated as tracing actual discontinuities on the slope face.

Table 2. Joint setsdetermined in DSE.

Joint Set	1	2	3	4	5	6	7	8	9	10
Dip (°)	3.86	56.72	56.72	44.14	58.53	76.34	91.36	91.36	90.61	89.71
Dip Direction (°)	315	41.42	318.58	356.63	347.74	1.74	1.33	91.33	105.02	158.96

Kinematic analysis showed that 22.33% (mainly sets 4 and 5) of the discontinuities are susceptible to planar sliding. This result is expected since their dip directions are almost equal to the slope dip direction making them daylighting and prone to slide (Figure 2a). These sets are intersecting, and hence will be more likely to slide as wedges.

22.06% of joints (sets 7 and 10) are prone to flexural toppling (Figure 2b). By inspecting the 3D model, these sets are vertical, short, result of excavation, and focused on the lower part of the slope. They are not persistent into the slope and thus will not topple, hence this mode was not analyzed further.

Wedge sliding had the highest probability (46.10%), which was confirmed visually on site where many wedges were detected. Figure 2c shows planes intersections vulnerable to wedge

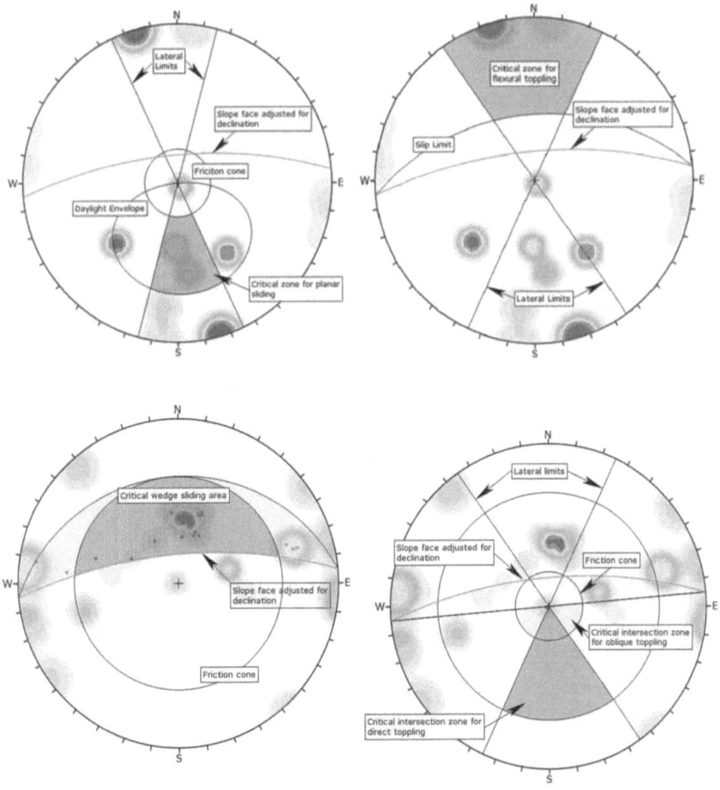

Figure 2. (a) Planar sliding (b) Flexural toppling (c) Wedge sliding (d) Direct and oblique toppling.

sliding in the crescent shaped area. The analysis of direct and oblique toppling resulted in no chance of occurring and thus are not analyzed further (Figure 2d).

4.1.2 *LEM and FEM analyses*

The wedge formed by sets 3 and 5 yielded the lowest safety factor (1.42) under dry conditions (Figure 3), indicating a stable slope, which is compliable with observations made on site. Under rainfall, this factor decreases to 1.41 on the same wedge, further indicating a stable slope. This decrease in the safety factor is expected, since water decreases friction between joint asperities, hence increasing the susceptibility to sliding. The same wedge failed under seismic loading (safety factor = 0.86).

The lowest factor of safety of 1.41 occurs on the wedge formed by sets 3 and 5 (Figure 4) under dry conditions, which is similar to the one obtained in LEM. However, under rainfall conditions sets 4 and 5 are more critical with a safety factor of 1.19. This variation is anticipated, given that the actual rainfall intensity is used instead of an approximated saturation level. The wedge formed by sets 10 and 4 will fail under seismic loading (F.S=0.97), which is inconsistent with SWEDGE. This discrepancy is predicted given that in RS2, the user has more control over joints properties such that persistence, length and spacing.

4.1.3 *Runout and reach angle*

The maximum runout distance reached by failing wedges was 7.76m from the slope face, corresponding to a reach angle of 9.5°, and hence yielding a low risk failure.

4.2 *Regional scale analysis*

The Bafliye quarry was stable under potential natural events with a steepness of 23° (coarse DEM).

The Halate was detected by the regional analysis as failing in wedge sliding. The safety factor under rainfall is 1.02 which is lower than the one obtained from site specific analysis (1.41) (difference of 14.3 %). Under seismic events, the regional maps predict critical displacements, rather than safety factors. At this quarry, the critical displacement is 0.29 cm compared to 0.3 cm in site specific analysis (difference of 3.33%).

In regional analysis, the reach angle was determined as 34°, hence predicting a high risk failure compared to a low risk level in site specific analysis.

This divergence of results is foreseen given that RocFall takes into account the roughness of the slope and lying ground, which increases the friction between these surfaces and sliding wedges, causing energy losses and therefore stopping them at closer distances.

The regional scale assessment neglects the presence, effects and interactions of discontinuities and is based on Mohr-Coulomb criteria, whereas the site specific assessment takes these factors into account, and uses more reliable models for the analysis of rock material.

Figure 3. Safety factor using LEM.

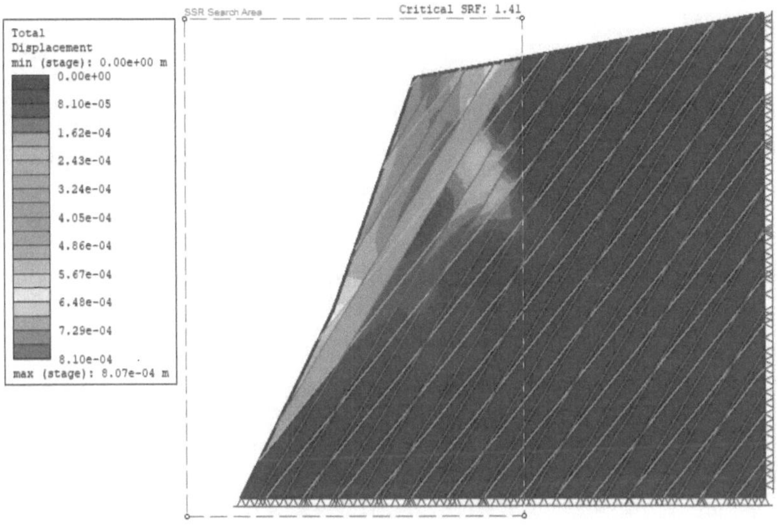

Figure 4. Safety factor using FEM.

Nevertheless, the regional assessment when capturing quarried slopes, provides a good approximation of possible hazard levels.

5 CONCLUSION

This research presented a coherent methodology for site specific assessment of limestone quarries. It also assessed the ability of the regional scale model to predict hazard and risk levels at quarried slopes. Results came in good agreement between both analyses only in terms of hazard. The findings imply that separate regional maps should be developed for quarried sites in Lebanon to better perceive the potential landslide hazards and risks in quarries, which in turn will impact urban planning in the country.

REFERENCES

Darwish, T., Khater, C., Jomaa, I., Stehouwer, R., Shaban, A., Hamze, M. (2010). Environmental impact of quarries on natural resources in Lebanon. Land Degrad. Develop. 22, 345–358.

Fawaz, A., Farah, E., Hagechehade, F. (2014). Slope Stability Analysis Using Numerical Modelling. American Journal of Civil Engineering. Vol. 2, No. 3, 2014, pp. 60–67.

Froude, M., Petley, D. (2018). Global fatal landslide occurrence from 2004 to 2016. Nat. Hazards Earth Syst. Sci., 18, 2161–2181.

Goodman, R.E. (1980). Introduction to Rock Mechanics (Chapter 8), Toronto: John Wiley, pp 254–287.

Grant, A., Wartman, J., Abou-Jaoude, G. (2016). Multimodal method for coseismic landslide hazard as-

Hudson, J.A., Harrison, J.P. (1997). Engineering Rock Mechanics – An Introduction to the Principles, Pergamon Press.

Kaafarani, R., Abou Jaoude, G., Wartman, J., Tawk, M. (2019). Landslide susceptibility mapping based on triggering factors using a multi-modal approach. Proceedings of International Conference of Engineering, 3-5 April, 2019. Beirut, Lebanon.

Riquelme, A., Abellán, A., Tomás, R., Jaboyedoff, M. (2014), A new approach for semi-automatic rock mass joints recognition from 3D point clouds, Computers and Geosciences. Vol. 68, pp. 38–52.

Verdeil E, Faour G, Velut S. (2007). Atlas du Liban territoires et societes. Coll. Etudes contemporaines.

Wang, S., Zhang, Z., Wang, C., Zhu, C., Ren, Y. (2019). Multistep rocky slope stability analysis based on unmanned aerial vehicle photogrammetry. Environ Earth Sci 78, 260. doi:10.1007/s12665-019-8145-z.

The Evolution of Geotech - 25 Years of Innovation – Hammah et al. (Eds)

Back-analysis of ductile slope failure mechanisms and validation with aerial photogrammetry, InSAR and GbRAR to proactively manage economic risks to protect the mine plan

N. Bar, M. Arrieta, A. Espino, C. Diaz, L.A. Mosquea & B. Mojica
Barrick Gold Corporation, Santo Domingo, Dominican Republic

A. McQuillan
Rocscience Inc, Gold Coast, Australia

G. Baldeon & G. Falorni
TRE ALTAMIRA Inc, Vancouver, Canada

ABSTRACT: Slope stability modeling techniques and deformation monitoring technology has significantly improved in the last decades. Integration between modelling and monitoring is becoming seamless enabling practicing geotechnical engineers to validate their stability models with actual slope performance data. This paper presents a case study demonstrating the use of aerial photogrammetric settlement analysis, satellite and ground-based radar data from a ductile slope failure mechanism on an overburden stockpile to validate a geotechnical model applied in 3D limit equilibrium analysis. Following the calibration of 3D models, stockpile design options are reviewed to minimize instability risks to ensure safety and protect the mine plan.

1 INTRODUCTION

For the last two to three decades, slope stability modelling by practicing engineers has mostly been limited to 2D cross-sections. Since 2016, the routine use of 3D slope stability assessments using three-dimensional limit equilibrium (LEA) and finite element analysis (FEA) models has rapidly become common practice (Bar & Weekes, 2017; Bar et al. 2019). Most rock masses, particularly those in large open pit mines, have complex geological conditions with anisotropic or sheared host rocks bounded by major geological structures in a variety of orientations. Such ground conditions and/or curved pit geometries cannot be adequately modelled using two-dimensional cross-sections (Bar & McQuillan, 2018). Comparably, waste rock dumps and stockpiles are typically constructed on geometrically-complex, three-dimensional terrain, which also warrant the use of 3D analysis for understanding realistic failure mechanisms (Semi et al. 2019; Bar et al. 2020).

In the last 15 years, tactical deformation monitoring technology has also improved in terms of data acquisition and processing speed, as well as reliability with real-time radar systems now capable of providing updates every minute across an entire slope. This is a step-change in safety risk management compared with manual surveying of individual survey prisms as little as 15 years prior. Radars are routinely geo-referenced to the same coordinate system used in 3D slope stability models and integration between the models and monitoring data has become seamless (McQuillan et al. 2020).

DOI: 10.1201/9781003188339-65

Strategic deformation monitoring has become simpler and easier with the commercial use of satellite-based InSAR (interferometric synthetic aperture radar) capable of covering kilometers of area on a fortnightly (or faster) frequency being applied to mine operations in the last 5 years.

Safety risks associated with ground failure at Pueblo Viejo mining operation are managed through multiple control measures including geotechnical design, monitoring and response, design execution and operating practices, as well as dewatering and depressurization. The use of multiple control measures provides redundancy in the event that one fails or becomes ineffective, the others continue to function and appropriately manage risks.

1.1 *Geographic and climatic setting*

Pueblo Viejo mine is located approximately 80 kilometers north of Santo Domingo, Dominican Republic, on the island of Hispaniola in the Caribbean archipelago as shown in Figure 1.

The mine is located at an altitude of approximately 300 to 500m above sea level. The Pueblo Viejo district has a tropical climate with the nearby cities of Cotuí and Bonao receiving 1843mm and 1956mm of average annual rainfall, respectively. Rainfall occurs all year round, although the wet seasons (May to November) receives almost twice the rainfall compared to the dry.

Figure 1. Location of Pueblo Viejo mine, Dominican Republic.

1.2 *Geological and geomorphological setting*

Gold mineralization in the Pueblo Viejo district, Dominican Republic, is spatially and temporally related to a series of Early Cretaceous volcanic domes (Nelson, 2000). Two principal deposits (Moore and Monte Negro) and a number of smaller deposits including Cumba, Mejita, Banco V, Arroyo Hondo I and II have contributed ore since mining commenced in 1975.

Mineralization is hosted in the Los Ranchos Formation, which is conformably overlain by the Hatillo Limestone (Torro et al. 2017b; Nelson et al. 2020). Near surface materials are often highly weathered and saprolitic in nature. Thick sequences of saprolites or residual soils up to 20m are quite common in the Caribbean (Larsen & Simon, 1990; Torro et al. 2017a). In the nearby Maimon region, oxidized zones, or the depth of weathering can exceed 50m (Andreu et al. 2015).

Designing slopes or foundations in saprolites and highly weathered rock present engineering challenges, including (Bar et al. 2016):

- Shearing and rotational sliding can occur in low strength material.
- Sliding (planar and wedge) can occur along relic geologic structures, which may be inconspicuous in saprolites or highly weathered rock masses.

Modes of instability in saprolitic or highly weathered rock masses often include a combination of shearing through weak material and sliding on relic geological structures. Such structures can be very difficult or impossible to detect through routine site investigations.

2 HONDO OVERBURDEN STOCKPILE OVERVIEW

Overburden removed from the principal deposits at Moore and Monte Negro is stored at Hondo Stockpile. Due to the hilly terrain surrounding the mine and environmental regulations that prevent untreated surface water discharge offsite, the operation is constrained by space to develop overburden and low-grade ore stockpiles. That is, the life-of-mine (LoM) plans and schedules are constrained by the capacity and integrity of Hondo overburden stockpile. Safety control measures used at Hondo stockpile are:

- Geotechnical design – routine updating of stockpile design based on performance.
- Monitoring and response – visual inspections and deformation monitoring.
- Design execution and operating practices – conformance to design, surface water management and geohazard exclusion areas.

2.1 *Stockpile design and construction sequence*

Hondo stockpile construction requires valley fill from east to west across Arroyo Hondo in three key stages as shown in Figure 2:

- Phase 1: Top-down construction on steep foundation on the upper, eastern valley slope.
- Phase 2: Bottom-up construction, buttressing Phase 1 on moderately steep foundation on the lower eastern valley slope.
- Phase 3: Valley is filled to the western slopes, providing natural confinement.

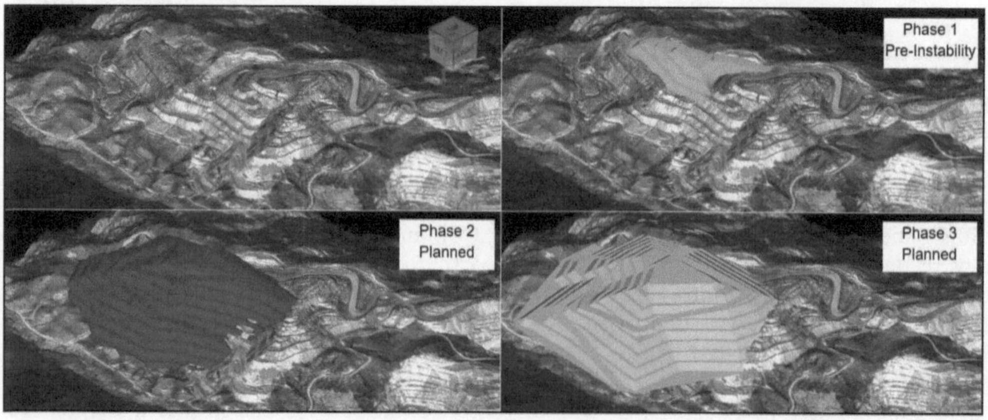

Figure 2. Hondo Stockpile Design and Sequence (Clockwise from Top-Left: As-Constructed Stockpile in June 2020; Phase Pre-Instability; Planned Phase 3 Design; Planned Phase 2 Design.

2.2 *Foundation topography and characteristics*

The initial stages of Hondo stockpile are founded on a moderately steep slope, ranging from 18 to 25° in Phase 1 and <10 to 18° in Phase 2.

The foundation slopes comprise saprolites and highly weathered, sedimentary rocks. These host relic, bedding planes and shears sub-parallel to the bedding, and exhibit anisotropic behavior. Prior to construction of Phases 1 and 2, the foundation was prepared with the removal of vegetation and topsoil.

Site investigations were carried out for Phase 3 of stockpile and included three diamond cored boreholes, in-situ dynamic standard penetration tests (SPT), and laboratory testing of remolded soil samples from within the saprolitic and highly weathered foundation. Table 1 provides a summary of typical foundation material characteristics. The USCS (Unified Soil Classification System) descriptions and Atterberg Limits results were obtained only for the soil-like material clasts. Similar foundation materials were observed in exposed foundation slopes in Phases 1 & 2.

Table 1. Foundation Material Characteristics.

	Depth (m)		Atterberg Limits				SPT
Material Name	From	To	LL%	PL%	PI%	USCS	N_{value}
Saprolite with Sandy Clay	0.00	8.00	35.6	24.1	11.5	CL	18 - 30
Saprolite with Highly Plastic Clay	8.00	14.00	45.2	12.1	33.1	CL	30 - 60
Highly Weathered Rock	14.00	20.00+	-	-	-	-	>70

* *LL: liquid limit, PL: plastic limit, PI: plasticity index*

The foundation material has low permeability (kv ≈ 10^{-8} to kh ≈ 10^{-5} m/s), and for the purpose of stability analyses, is generally considered saturated.

2.3 *Rock fill characteristics*

Rock fill strength was estimated using the non-linear Barton-Kjaernsli method for crushed rock (Barton & Kjaernsli, 1981; Barton, 2008):

- Strength, S, estimated to range from 6 to 7.
- Roughness, R, estimated to range from 5 to 7.
- Friction, Φr, estimated at 25° on average considering historic laboratory testing.

Figure 3 shows the comparison of strength estimation against Mohr-Coulomb linear failure criteria. The Barton-Kjaernsli method indicates the presence of some apparent cohesion for normal stresses above 200 kPa (0.2 MPa). These are supported by field observations whereby tipheads or individual lifts can locally be constructed steeply with face angles, in some instances, exceeding 40°). Based on this assessment, Mohr-Coulomb strength parameters: cohesion, c' of 5 kPa and an effective friction angle, Φ' of 36° were adopted.

2.4 *Water management*

Surface water runoff is strictly managed at Hondo stockpile to prevent acid rock drainage (ARD) into the natural waterways. This is achieved through a series of detention basins at the base of the stockpile and surface water drainage systems that comes from the north-western sector of the mine to Acid Rock Drainage Dam 1. Water is

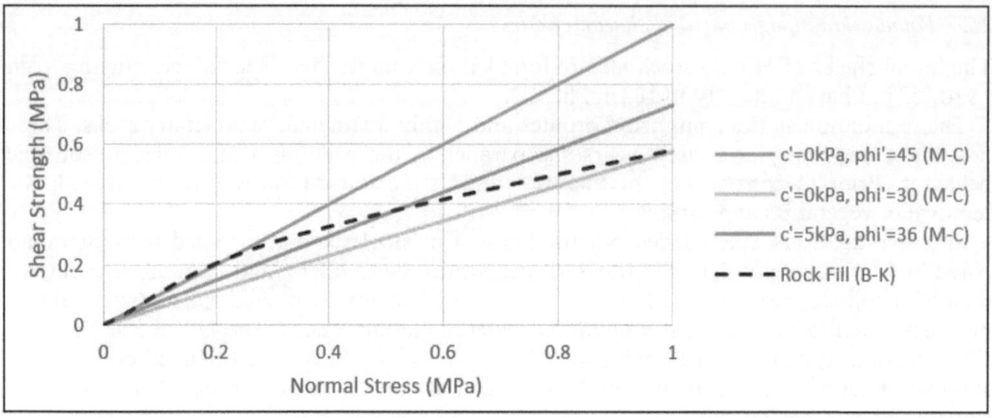

Figure 3. Rock fill shear strength comparison (M-C: Mohr-Coulomb; B-K: Barton-Kjaernsli).

subsequently treated to remove effluents and balance pH in order to comply with national and international environmental standards.

3 HONDO STOCKPILE INSTABILITY AND DEFORMATION

3.1 *Aerial reconnaissance and photogrammetry*

Aerial reconnaissance of Hondo stockpile is undertaken using a multi-rotor UAV (unmanned aerial vehicle) every two days with high-resolution photography to understand dump advance and to identify any visible changes to conditions.

UAV photogrammetry is undertaken using a fixed-wing UAV every two weeks to measure dump advance, and for Hondo stockpile in particular, to identify signs of instability, and monitor deformations as shown in Figures 4 and 5.

Figure 4. Instability outline formed by subsidence and cracks, which are mapped in the 3D photogrammetry model using *ShapeMetriX UAV* software. Isometric View (Looking East).

Photogrammetry enables the generation of 3D models from a series of overlapping photographs. The accuracy or precision of the 3D model is dependent on the resolution of the mesh generated and the accuracy of the ground control points. In Figure 4, the accuracy of the 3D models is approximately 10 cm, or 100 mm; i.e., deformations less than 100 mm is not detected. One-dimensional (1D) settlement analysis demonstrates downward movement at the crest (top) of the stockpile and upward (i.e. out-of-slope) movement at the base.

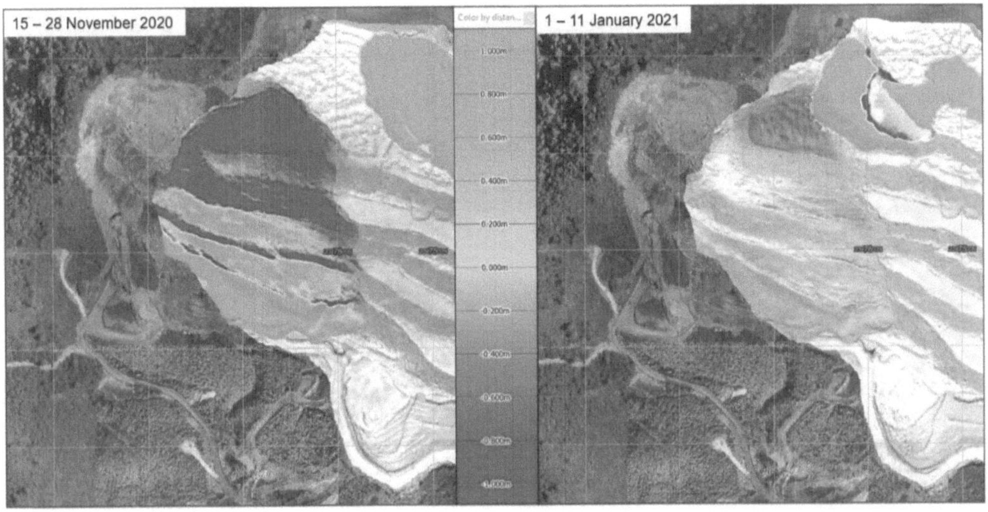

Figure 5. Settlement Analysis (Vertical Displacements) completed using UAV Photogrammetry. Left: Instability visible through settlement on the crest, and outward movement at the base of the stockpile. Right: Instability reactivation with additional loading at the top of the stockpile. Plan View (East-up-the-page).

3.2 *GbRAR*

Following the detection of deformation through UAV photogrammetry, the stockpile area was treated as a hazardous area. Deformations at Hondo stockpile were monitored using a *Reutech Mining MSR*: a ground-based real aperture radar (GbRAR) system. Physical barriers were used to prevent unauthorized access to the at-risk area beneath the stockpile and at the top.

The GbRAR was located approximately 120 m from the base of the stockpile and monitored deformations for several weeks. Its location at the base of the stockpile was sub-parallel to the direction of movement (later confirmed by 3D prism monitoring), enabling effective, high resolution monitoring using one-dimensional LoS (line-of-sight) radar.

Cumulatively, in the order of 15 m (15,000 mm) of LoS deformation was observed. During the peak of the movement, velocities of up to 50 mm per hour were observed (Figure 6), which occurred after several days of heavy rainfall. Velocities progressively reduced to less than 5 mm per hour. Velocities further reduced thereafter, with the exception of a short period during which the instability was reactivated with additional loading (Figure 5).

3.3 *InSAR*

InSAR bulletins are produced every 11 days following ascending orbit imagery acquired by the TerraSAR-X satellite (Figure 7).

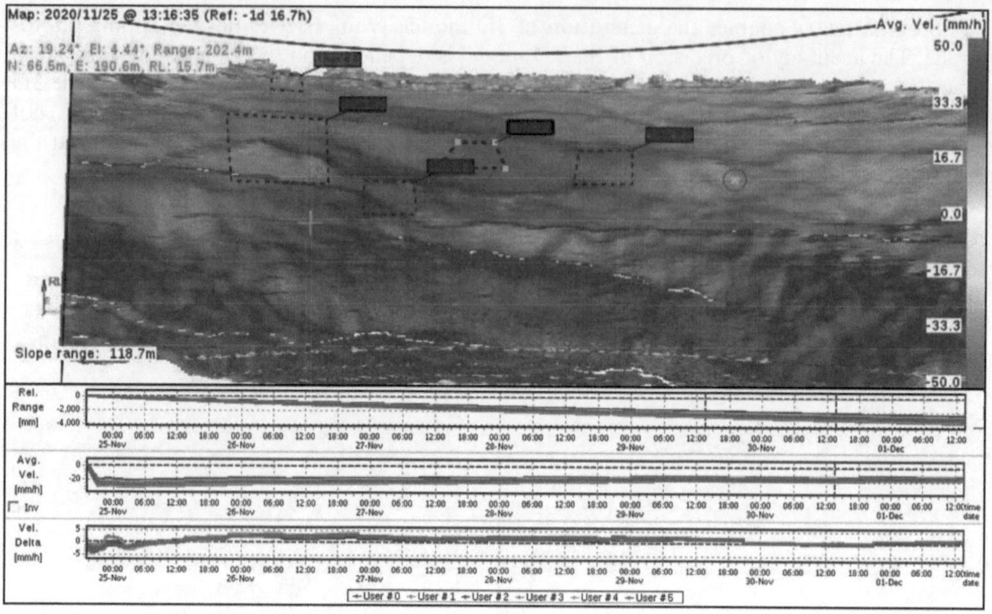

Figure 6. GbRAR Average Velocity Data illustrating the deforming portion of the Stockpile with a clear limit of deformation on the right hand side of the image, which is consistent with the UAV Photogrammetry.

InSAR bulletins started identifying displacement over geotechnical hazard 183 since late November 2020. The slope orientation combined with the satellite orbit and line-of -sight viewing angle indicate a possible component of westward movement. The limited surface disturbance, gradual onset of the movement and 11-day satellite revisit frequency provided an optimal setting for the application of InSAR and allowed the early detection of the incipient movement. An increasingly larger area is captured in the southern portion of the Hondo Dump until mid-November and a new area of movement to the northwest starts being captured at the same time. Displacement in this new area has an opposite direction (i.e. towards the satellite) and may represent movement at the base of the stockpile. It appears likely that the gradually increasing InSAR coverage in the area corresponds to an equally gradual decrease in ground displacement rates in the area.

The InSAR results over geotechnical hazard 168 (i.e. area of instability) are less clear, with limited areas below the detection threshold (<10 mm during an 11-day period), alternated with areas devoid of information since September 2020. Amplitude change detection analyses indicated significant changes to the ground surface, which were caused primarily by large and rapid slope deformation as shown in Figures 5 and 6.

4 SLOPE STABILITY MODELS

Due to the complex nature of the foundation terrain, slope stability models needed to be three dimensional (3D) in nature. Slope stability models were developed in 3D limit equilibrium analysis software *Slide3* and validated in finite element analysis software *RS3*, by Rocscience Inc.

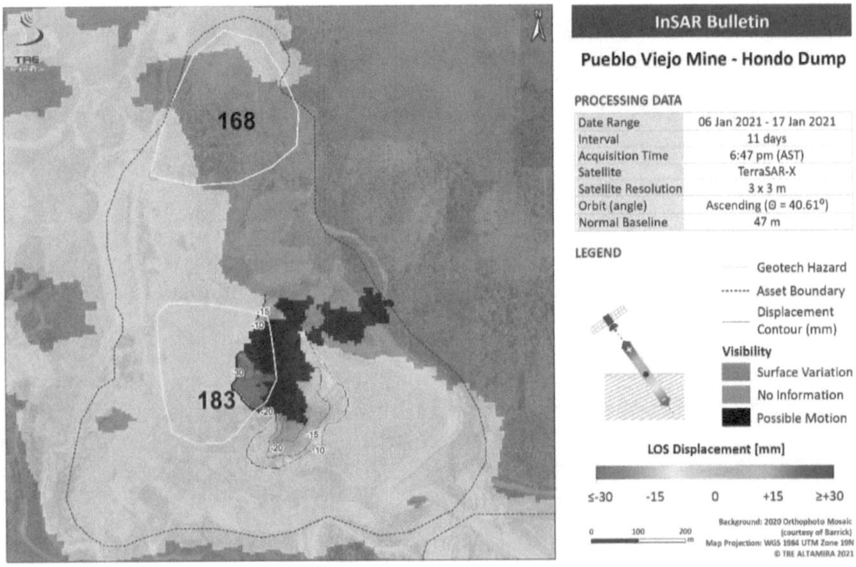

Figure 7. InSAR Bulletin identifying up to 35mm of LoS displacement over 11-day period. Geotechnical hazard 168 refers to the area of instability. North-up-the-page.

4.1 *Failure mechanism back-analysis*

The geotechnical model was developed considering a foundation with two types of saprolites, assumed to be conformably overlaying highly weathered rock (Table 2). Rock fill was placed over the foundation.

Based on field observations, the phreatic surface appears to follow the topographic surface. Pore pressures were simulated using Hu coefficients to adjust pressure heads:

- Foundation material (saprolites) have low permeability & are saturated: Hu=0.95-1.00.
- Dumped material (rock fill) has variable permeability, is free-draining in some areas, and saturated in others. H_u=0.50-0.75.

3D LEA simulations were completed with a focus on:

- Obtaining a defensible failure mechanism that has a geometric resemblance to observed deformations and physical signs of instability.
- Has a factor of safety (FoS) close to, but below equilibrium.
- Validating foundation material properties with the Mohr-Coulomb failure criterion.

Figure 8 illustrates the back-analysis result after several iterations. The failure mechanism attained involved shearing through the foundation with an imperfect geometrical match to the

Table 2. Back-Analyzed Material Properties.

	Depth*	(m)	γ	c'	Φ'		
Material Name	From	To	kN/m^3	kPa	°	H_u	Color
Rock Fill	-100.00	0.00	20	5	36	0.50	Grey-White
Saprolite with Sandy Clay	0.00	8.00	19	50	22	0.95	Orange
Saprolite with Highly Plastic Clay	8.00	14.00	19	25	22	0.95	Light Green
Highly Weathered Rock	14.00	20.00+	21	50	30	1.00	Dark Green

**Depth is measured downwards from foundation topographic surface. Negative values represent material deposited above. γ refers to unit weight.*

actual instability (subtle variation). It should be noted that the foundation topography used was over 20 years old, and is likely to be the cause of the variation.

The 3D LEA model region of focus was expanded to the entire Hondo stockpile using the input parameters from Table 2. An additional low FoS area was identified to the south of the area of instability (Figure 9).

4.2 *Model validation – Confirmatory finite element analysis*

The *Slide3* model was imported into 3D FEA software, *RS3*, to validate the results calculated using 3D LEA, as is common practice using 2D cross-section (Hammah et al. 2005).

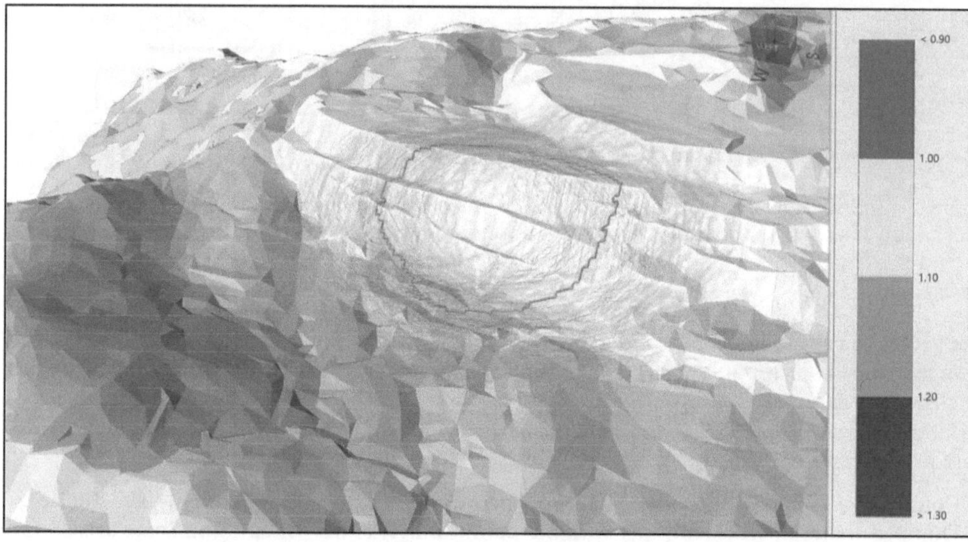

Figure 8. 3D LEA back-analysis of Hondo Phase 1 stockpile instability illustrating slip surfaces with FoS<1.2 and minimum FoS<1.0. Slip surface model involves shearing through foundation but has an imperfect geometric match to the actual instability. Isometric View (Looking East).

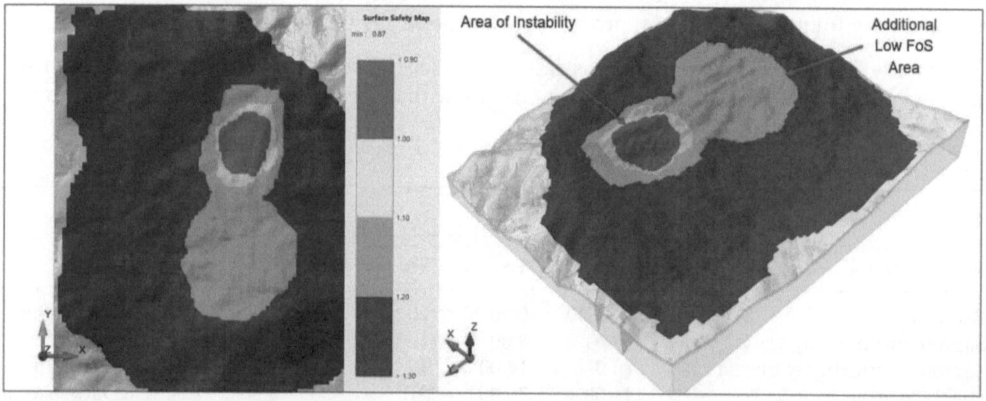

Figure 9. Extended 3D LEA of Hondo stockpile showing FoS map. Slip surfaces modelled involve shearing through foundation. Left: Plan View (north-up-the-page); Right: Isometric View (Looking South East).

Material properties were modelled as elastic. In such analysis, stress results are independent of the modulus of elasticity applied, but strain results are directly proportional to the value of the modulus (Stacey et al. 2003). As such, RS3 strain results were interpreted as relative indicators of deformation only, where the primary purpose of running finite element models was to validate likely zones of movement (i.e., to answer the question: did low FoS zones calculated in Slide3 correlate with high strain and high deformation zones calculated in RS3?).

A graded 4-noded tetrahedra mesh was applied to RS3 models. 3D FEA supported 3D LEA results where high concentrations of strain were modelled at the foundation of the stockpile, supporting the failure mechanism of shearing through the stockpile foundation, Figures 10 and 11.

Larger values of deformation were recorded in low FoS zones, validating 3D LEA results, Figure 12 . The validation is qualitative only, i.e., model stiffness calibration to measured displacements has not been completed.

3D FEA indicated that larger slope displacement may occur in the southern part of the stockpile, outside the area of instability. These results are also qualitatively supported through satellite InSAR LoS deformation (Figure 7).

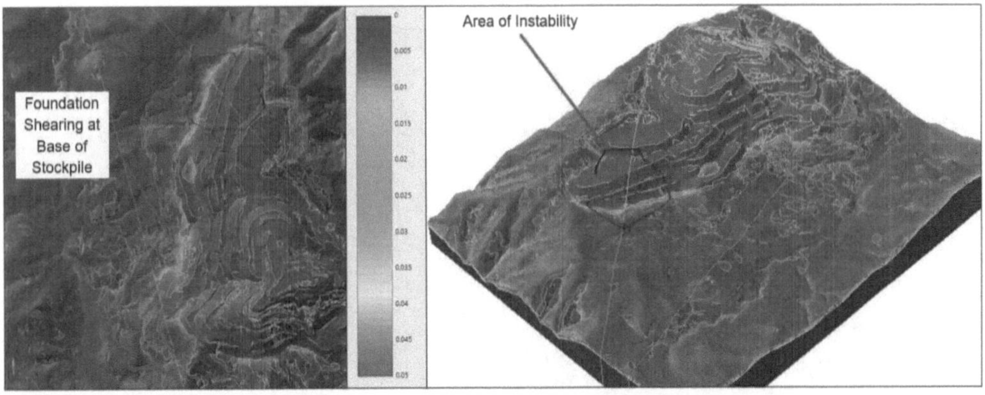

Figure 10. 3D FEA results of Hondo stockpile showing contours of maximum shear strain concentrated at the stockpile foundation up to a distance of approximately 20m beyond the base of the stockpile. Left: Plan view (north-up-the-page); Right: Isometric View (Looking South East).

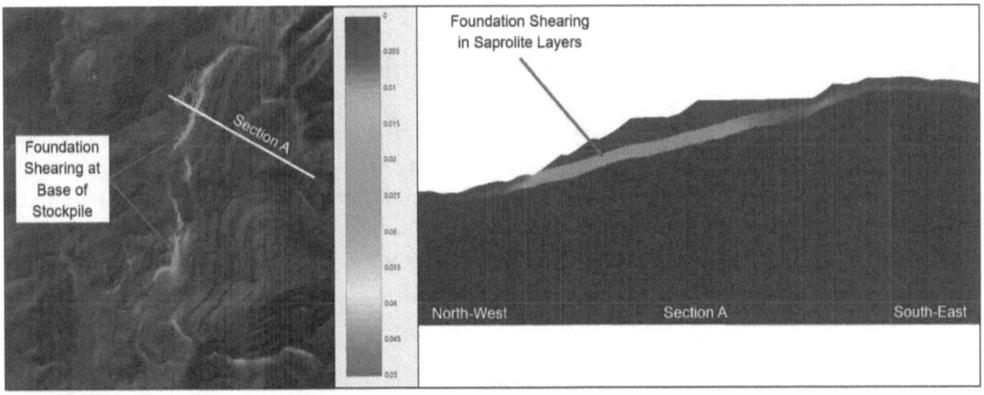

Figure 11. 3D FEA results of Hondo stockpile showing contours of maximum shear strain concentrated at the stockpile foundation. Left: Plan view (north-up-the-page) showing location of 2D section; Right: 2D section through stockpile with shearing in saprolite layers in the foundation.

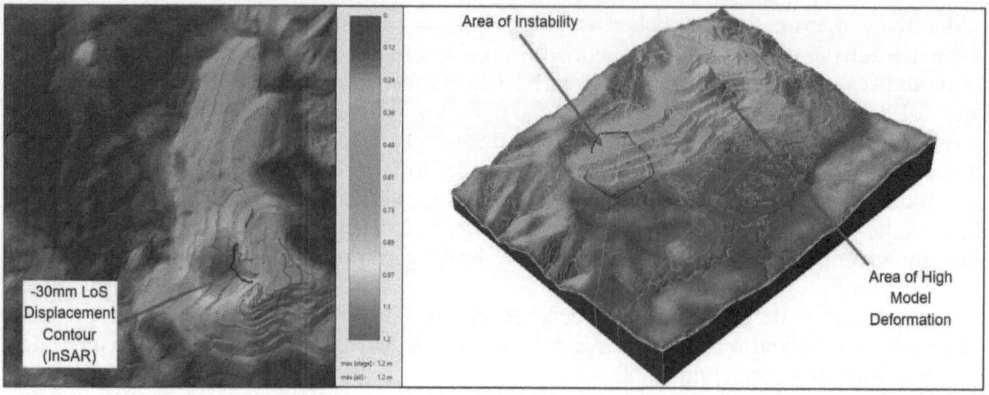

Figure 12. Plan view (north-up-the-page) 3D FEA results of Hondo stockpile showing predicted displacement (model stiffness not calibrated) and measured InSAR LoS displacement (-30mm displacement in 11 day period represented as black counter lines was previously shown in Figure 7).

Figure 13. Plan view (north-up-the-page): Left: Lowest FoS slip surface contoured with dip of column base (i.e. dip of slip surface); Middle: Modelled lowest FoS slip surface (black outline) overlain by relative range deformation from GbRAR; Right: Isometric view (Looking South East).

4.3 *Qualitative model validation – GbRAR monitoring data*

Relative range displacement (one-dimensional LoS displacement) data from GbRAR was imported into *Slide3* to qualitatively validate the 3D LEA model. Figure 13 shows a comparison of 3D LEA model results, which indicate a shallow slip surface moving westward at the base of the stockpile (in the saprolitic foundation). GbRAR relative range data shows the highest deformations in the same area at the base of the slope. The deformation monitoring agrees with the location of the lowest FoS slip surface and the movement direction.

4.4 *Forward prediction and stockpile redesign – phase 2*

The planned stockpile design was analyzed using 3D LEA and the back-analyzed model input parameters to predict the behavior of Phase 2. The analysis was undertaken considering sequential lifts with bottom-up construction as illustrated in Figure 14.

The 3D LEA results indicated stable conditions with FoS exceeding 1.2 for lifting Phase 2 to 440mRL. Subsequent lifts had unacceptable FoS involving shearing of the saprolite foundation beneath the rock fill:

- Lift to 480mRL: marginally stable (FoS<1.05) with potential for 80m high instability.
- Lift to 520mRL to 560mRL: unstable overall slope (FoS<1.00). Impact to detention basin (pond) increasing with each lift.

Based on this understanding of expected behavior with respect to the sequence, Phase 2 of the stockpile was redesigned. The redesign was an iterative process between mine planning and geotechnical engineering to achieve an optimal outcome (Figure 15):

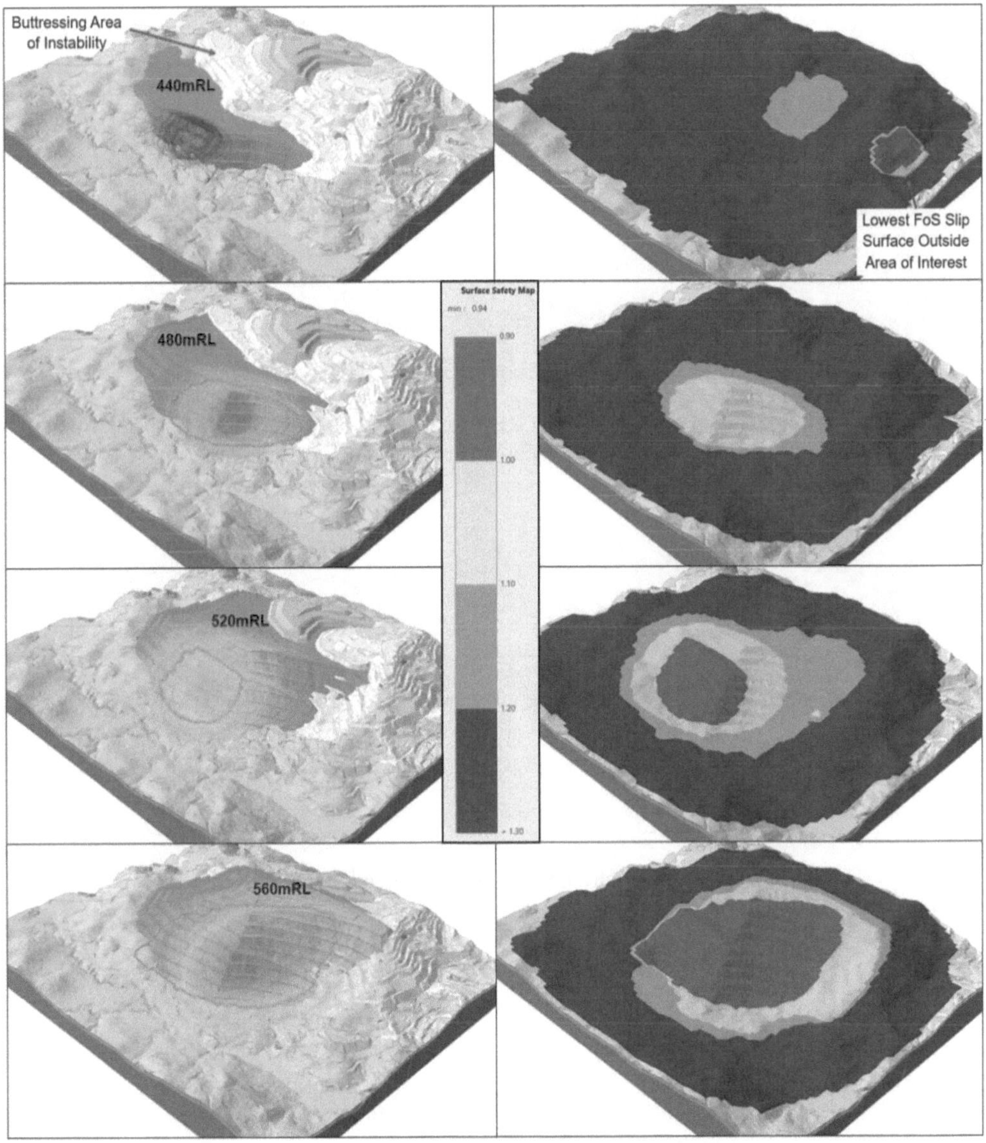

Figure 14. 3D LEA for planned Hondo Phase 2 stockpile design. Left: Slip surfaces involving shearing through foundation. Right: FoS map. Isometric View (Looking North East).

- Slope decoupling benches were included to reduce the overall slope angle:
- 30m wide on 440mRL, which will be utilized for subsurface deformation monitoring (inclinometers), pore pressure monitoring in the foundation (vibrating wire piezometers) and surface displacements (prisms). The decoupling bench will also be used for surface water drainage.
- 30m wide on 480mRL, which will also be utilized for surface water drainage and monitoring surface displacements.
- Maintaining a 50 m offset from the base of the stockpile to the detention basin considering the proximity of shearing from the stockpile loading of the foundation.

Figure 16 presents the overall slope stability analysis results for the redesigned stockpile constructed to full height using 3D LEA. The minimum FoS attained was 1.18 (approx. 1.20).

Phase 3 of the stockpile was not analyzed since it does not commence construction for several years. It shall be reviewed following additional drilling campaigns in Arroyo Hondo valley. It will also buttress Phase 2.

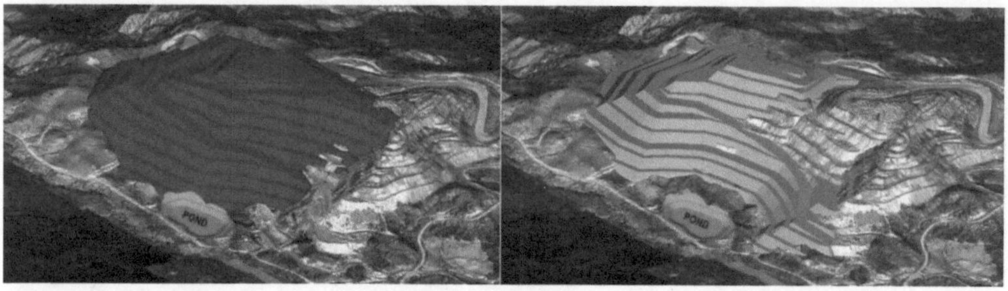

Figure 15. Hondo stockpile design Phase 2. Left: Planned (original). Right: Redesigned with slope decoupling benches on 440 and 480mRL and a 50m offset from the detention basin (pond) at the base. Isometric View (Looking North East).

Figure 16. 3D LEA for redesigned Hondo Phase 2 stockpile design. Left: Slip surfaces involving shearing through foundation. Right: FoS map. Isometric View (Looking North East).

5 KEY FINDINGS AND FUTURE WORK

Three-dimensional LEA and FEA were used in conjunction with deformation monitoring data to identify and validate realistic failure mechanisms at Hondo overburden stockpile. Through 3D analysis, a detailed understanding of ground failure risk was obtained:

- Spatially, since the analyses encompassed the entire stockpile (compared to a series of 2D sections).

- Consequentially, through estimation of potential failure volumes (i.e. understanding the impact of an instability or potentially unstable zone subject to shearing).
- Complimentarily, where a zone of higher deformation predicted by 3D FEA modelling prompted review and further monitoring of a previously unidentified moving zone in the stockpile.

The improved ability to understand ground failure risks facilitates the development of more effective control measures:

1. Geotechnical design, i.e., Phase 2 redesign, and planned deferral of material capacity into Phase 3. The design also provisions for effective surface water drainage.
2. Monitoring and response through the installation of strategic, subsurface monitoring in the foundation during early stages of construction.

The above mentioned risk management approach for Phase 2 of the stockpile is proactive and protects the mine plan for continuous availability of overburden storage capacity without disruption.

Safety risks will continue to be managed through appropriate design execution and operating practices. Surface monitoring through the use of aerial reconnaissance, photogrammetry and InSAR will continue. InSAR will be expanded to include dual orbits (ascending and descending) with dynamic monitoring (i.e. time series data) on a fortnightly basis for the early detection of potential instabilities and future model calibration. This will be complemented by subsurface monitoring mentioned earlier.

Future work comprises the periodic (quarterly to semi-annual) review of stockpile slope performance, which will also be an input into the detailed design for Phase 3, alongside additional site investigations.

ACKNOWLEDGEMENTS

Paul Ramos, Dionifel Ripoll, Silverio Nunez, Henry Rosario, Fernando Gonzalez, Yohan Mendez, Jose Hawil Felix, Brendon Douglas and Jose Recio from Barrick Gold Corporation are also thanked for their ongoing support with management of the stockpile. The authors additionally thank Daniel Ortiz and Ben Hamilton of TRE ALTAMIRA Inc for their technical support with InSAR.

REFERENCES

Andreu, E., Torro, L., Proenza, J.A., Domenech, C., Garcia-Casco, A., Villanova de Benavent, C., Chavez, C., Espaillat, J., Lewis, J.F. 2015. Weathering profile of the Cerro de Maimón VMS deposit (Dominican Republic): Textures, mineralogy, gossan evolution and mobility of gold and silver. *Ore Geology Reviews* 65: 165–179.

Bar, N., Barton, N.R. & Ryan, C.A. 2016. Application of the Q-slope method to highly weathered and saprolitic rocks in Far North Queensland. In *Proc. ISRM International Symposium: Eurock 2016, Cappadocia,* 29 – 31 August 2016, 585–590.

Bar, N. & McQuillan, A. 2018. 3D Limit Equilibrium Slope Stability Analysis for Anisotropic and Faulted Rock Masses in Australian Coal and Iron Ore Mines. In *Proc. 10th Asian Rock Mechanics Symposium, Singapore,* 29 October – 3 November 2018, 12p.

Bar, N. & Weekes, G. 2017. Directional shear strength models in 2D and 3D limit equilibrium analyses to assess the stability of anisotropic rock slopes in the Pilbara Region of Western Australia. *Australian Geomechanics* 52(4): 91–104.

Bar, N., Yacoub, T.E. & McQuillan, A. 2019. Analysis of a large open pit mine in Western Australia using finite element and limit equilibrium methods. In *Proc. ARMA 2019: 53rd US Rock Mechanics/Geomechanics Symposium, New York,* 23–26 June 2019. ARMA 19-A–30, 8p.

Bar, N., Semi, J., Koek, M., Owusu-Bempah, G., Day, A., Nicoll, S. & Bu, J. 2020. Practical waste rock dump and stockpile management in high rainfall and seismic regions of Papua New Guinea. In *Proc. Slope Stability 2020, Perth,* 12–14 May 2020, 117–128.

Barton, N.R. 2008. Shear Strength of Rockfill, Interfaces and Rock Joints, and their Points of Contact in Rock Dump Design. In *Proc. Rock Dumps 2008, Perth,* 5–6 March 2008, 18p.
Barton, N.R. & Kjaernsli, B. 1981. Shear strength of rockfill. *Journal of the Geotechnical Engineering Division* 107(7), 873–891.
Hammah, R., Yacoub, T.E., Corkum, B. & Curran, J. 2005. A comparison of finite element slope stability analysis with conventional limit-equilibrium investigation. In *Proc. 58th Canadian Geotechnical and 6th Joint IAH-CNS and CGS Groundwater Speciality Conferences, Saskatoon,* 4p.
Larson, M.C. & Simon, A. 1990. Landslide processes in saprolitic soils of a tropical rain forest, Puerto Rico. In *Proc. 12th Caribbean Geological Conference, St. Croix,* 7–11 August 1989, 217–222.
McQuillan, A., Yacoub, T.E., Bar, N., Coli, N., Leoni, L., Rea, S., Bu, J. 2020. Three-dimensional slope stability modelling and its interoperability with interferometric radar data to improve geotechnical design. In *Proc. Slope Stability 2020, Perth,* 12–14 May 2020, 1349–1357.
Nelson, C.E. 2000. Volcanic domes and gold mineralization in the Pueblo Viejo district, Dominican Republic. *Mineralium Deposita* 35: 511–525.
Nelson, C.E., Polanco, J., Macassi, A., Dominguez, H., Proenza, J., Torro, L., Rhys, D. & Iturralde-Vincent, M. 2020. The Hatillo Limestone, Pueblo Viejo district, Dominican Republic: Marginal reef or impermeable cap? *Boletin de la Sociedad Geologica Mexicana* 72(3): A011119.
Semi, J., Bar, N., Bu, J., Moore, P. & Kuira, P. 2019. Management of Rock Fill Dumps and Stockpiles in a High Rainfall and Seismically Active Region. In *Proc. 15th International Conference on Geotechnical Engineering, Lahore,* 5–7 December 2019, 383–390.
Stacey, T., Xianbin, Y., Armstrong, R., Keyter, G. 2003. New slope stability considerations for deep open pit mines. *Journal of The South African Institute of Mining and Metallurgy* 103(6), 373–390.
Torro, L., Proenza, J.A., Aiglsperger, T., Bover-Arnal, T., Villanova de Benavent, C., Rodriguez-Garcia, D., Ramirez, A., Rodriguez, J., Mosquea, L.A. & Salas, R. 2017a. Geological, geochemical and mineralogical characteristics of REE-bearingLas Mercedes bauxite deposit, Dominican Republic. *Ore Geology Reviews* 89: 114–131.
Torro, L., Proenza, J.A., Camprubi, A., Nelson, C.E., Dominguez, H., Carrasco, C., Reynoso-Villafana, R., Melgarejo, J.C. 2017b. Towards a unified genetic model for the Au-Ag-Cu Pueblo Viejo district, central Dominican Republic. *Ore Geology Reviews* 89: 463–494.

The Evolution of Geotech - 25 Years of Innovation – Hammah et al. (Eds)

Comparison between GSI-based implicit and explicit structure models

C.P. Fischer & M.S. Diederichs
Queen's University, Kingston, Ontario, Canada

ABSTRACT: The Geological Strength Index, GSI, is a rockmass classification system based on two fundamental parameters: the blockiness of the rockmass, and discontinuity condition. The Hoek-Brown failure criterion is used with GSI to estimate the strength and stiffness of jointed rockmasses. This strength estimation method considers the effect of isotropic structure implicitly in continuum rockmass analyses. Modern numerical tools allow the creation of models where dense networks of structure are represented explicitly, with assigned properties based on site investigation data. In this paper, equivalent explicit and implicit models of the rockmass around a tunnel are produced and compared to theoretical depth of yield and closure estimates, for multiple cells in the GSI chart. This verification provides greater confidence in traditional continuum modelling methods, which continue to have practical value for analyzing engineered excavations.

1 INTRODUCTION

The Geological Strength Index (GSI) is a rockmass classification scheme used worldwide in site characterization for tunnelling, mining, and open pit slope stability (Hoek et al. 1995). GSI classification combines two fundamental parameters: the blockiness of the rockmass, and discontinuity condition. Using GSI, rockmass classification can be incorporated directly into strength and stiffness parameters for use in equivalent continuum numerical modelling to predict rockmass behaviour. In such models, rockmass structure is accounted for implicitly. In modern numerical modelling tools, dense networks of discrete features like joints can be represented explicitly. Explicit models can demonstrate mechanisms of failure more clearly than implicit structure models.

1.1 *Rockmass classification and the Geological Strength Index*

The Hoek-Brown (1980) strength criterion combined with the GSI system (Hoek et al. 1995; Marinos & Hoek 2000) is widely used to estimate the strength and stiffness of intact and jointed rockmasses. The Generalized Hoek-Brown criterion (Hoek et al. 2002) incorporates GSI into the determination of the rockmass strength envelope as shown in Equations 1-4:

$$\sigma_1' = \sigma_3' + \sigma_{ci}\left(m_b\frac{\sigma_3'}{\sigma_{ci}'} + S\right)^a \tag{1}$$

$$m_b = m_i \exp\left(\frac{GSI - 100}{28 - 14D}\right) \tag{2}$$

$$S = \exp\left(\frac{GSI - 100}{9 - 3D}\right) \tag{3}$$

DOI: 10.1201/9781003188339-66

$$a = \frac{1}{2} + \frac{1}{6}\left(e^{-\frac{GSI}{15}} - e^{-\frac{20}{3}}\right) \tag{4}$$

where σ_1 and σ_3 are the major and minor principal stresses respectively, σ_{ci} is the unconfined compressive strength of the intact rock, m_b, s, and a are rockmass material constants, and D is a factor representing amount of blast damage (Hoek et al. 2002).

1.1.1 *Quantification of GSI*

Sonmez & Ulusay (1999) propose that the structure (vertical) axis of the GSI chart be quantified using the Structure Rating (SR), determined using the volumetric joint count outlined by Palmstrøm (1996). They also propose that the surface quality (horizontal) axis be quantified using the Surface Condition Rating (SCR). Similarly, Cai et al. (2004) use block volume (V_b) to quantify the structure axis, and the joint condition factor (J_c) to quantify the surface quality axis. Block volume is calculated based on joint spacing, orientation, number of sets, and persistence, and for practical purposes, can be estimated using $V_b = s_1 s_2 s_3$, for three joints sets of spacing s_i.

Hoek et al. (2013) use two linear scales to represent structure and surface quality axes, where the sum of the scales equals GSI. Scale A, for the structure axis, is related to Rock Quality Designation (RQD) by Deere et al. (1969) through the equation A=RQD/2. Scale B, for surface quality axis, is related to $JCond_{89}$ (from the 1989 RMR classification system by Bieniawski) through the equation B=1.5$JCond_{89}$. Additional guidance on quantification on a set-by-set basis is given by Day et al. 2019. The updated GSI chart by Hoek et al. (2013) is shown in Figure 1, and has been modified to include the cubic joint spacings and block volumes from Cai et al. (2004).

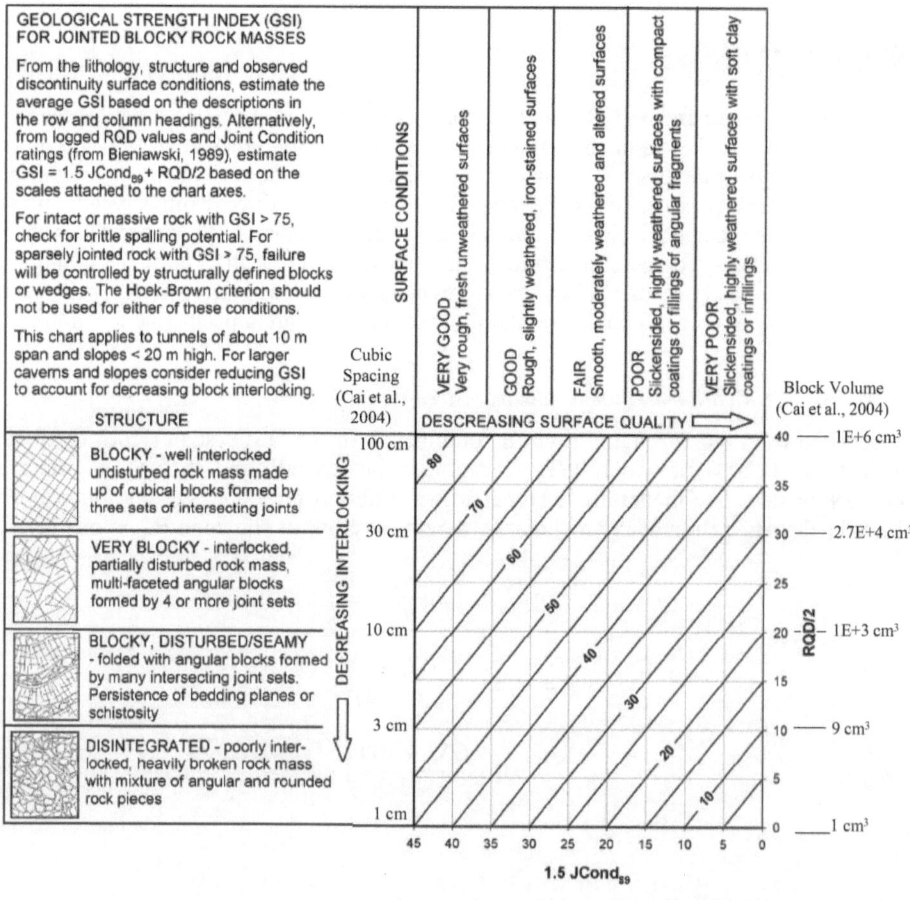

Figure 1. Quantified GSI chart after Hoek et al. (2013) with spacing included from Cai et al. (2004).

1.2 *Implicit and explicit numerical modelling*

GSI and the Generalized Hoek-Brown strength criterion remain common and effective methods to assess rockmass behaviour through equivalent continuum numerical modelling. However, explicit structure models can demonstrate mechanisms of yield and failure that may not be clearly shown in implicit models. To define explicit isotropic structure, deterministic or statistical geometric properties (orientation, density, and persistence) and mechanical properties (joint strength and stiffness) must be defined. If assigned properties are valid and representative, explicit and implicit models can be compared to verify implicit GSI-based results.

2 NUMERICAL MODELLING SCOPE AND INPUTS

In this study, equivalent implicit and explicit models are developed for a 10 m diameter tunnel and compared for five GSI ratings in the author-defined zone of applicability (scope) on the GSI chart. The Finite Element Method (FEM) program RS2 (Rocscience 2020a) is used. The scope is shown in Figure 2, where each node on the chart represents an explicit model developed. The upper limit of the scope zone corresponds to the transition to massive or discontinuously jointed rock, which does not satisfy fundamental GSI and Generalized Hoek-Brown assumptions. The lower limit of the scope zone corresponds to excessively high requirements for computational ability to mesh and run discrete models.

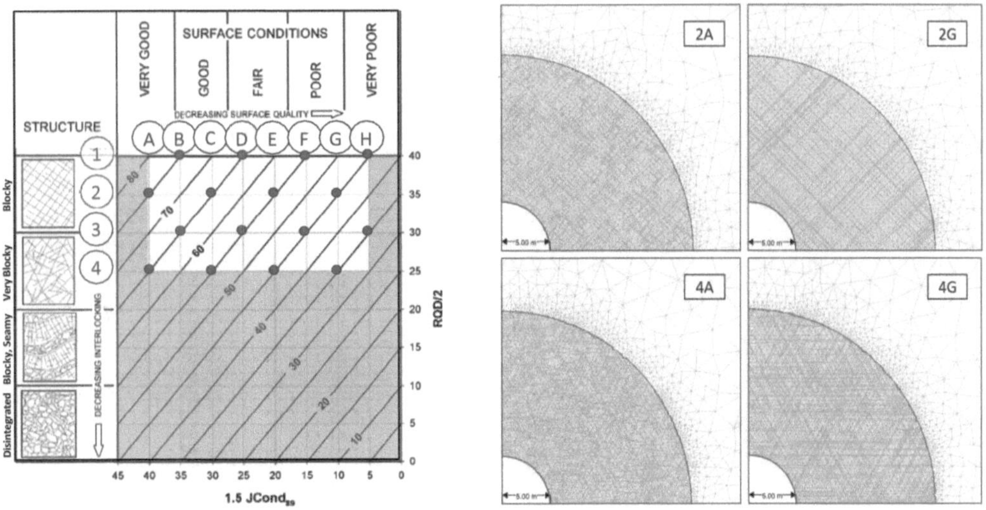

Figure 2. Left: Author-defined zone of applicability (scope) for the study. Right: Sample model geometry including joint networks, zoning, and mesh for models 2A, 2G, 4A, and 4G.

As seen in Figure 2, explicit models are created using a 40 m diameter zone of explicitly jointed rockmass surrounding the tunnel. Outside the explicitly jointed rockmass boundary is an equivalent continuum material extending to the pinned external boundaries a factor of 24 times the tunnel radius away. All rockmasses are represented plastically using the Generalized Hoek-Brown criterion, where residual strength equals peak strength for ease in model comparison.

The intact rock properties include a Uniaxial Compressive Strength (UCS) of 100 MPa, a Young's Modulus of 35.0 GPa, a unit weight of 0.027 MN/m^3, an m_i of 10, and a Poisson's ratio of 0.25. No tensile cut off is employed. For all rockmass materials, rockmass Young's Modulus is calculated using the GSI of the material and the Generalized Hoek-Diederichs method.

Two suites of models are run, each for different stress conditions shown in Table 1. For the first suite of models, the isotropic in-situ stress for each GSI rating is calculated to create a target rockmass UCS to in-situ stress ratio of 0.2. The rockmass UCS is calculated from the Generalized Hoek-Brown envelope. The second suite of models has a constant in-situ stress of 35 MPa.

Table 1. Two suites of stress conditions applied to each GSI rating of model.

Model GSI	75	65	55	45	35
In-situ stress for constant rockmass UCS to in-situ stress ratio of 0.2 (MPa)	124.4	71.0	40.2	22.4	12.0
Constant in-situ stress (MPa)	35.0	35.0	35.0	35.0	35.0

Baecher (1978) joint networks are used to create explicit jointing in RS2, which have a realistic appearance and input parameters with clear physical meaning. Networks are defined by joint density, 2D inclination, number of sets, and joint length, and are located through a Poisson point process. No statistical distributions are used to vary orientation or length. The number of joint sets in each model follows the qualitative descriptions of the structure axis of the GSI chart. One joint set is assumed to be out of plane for all models due to the 2D nature of the analysis.

To approximate joint density, the inverse of Cai et al.'s (2004) cubic joint spacing (in joints per meter) is applied to each joint set in the explicit models. The RS2 joint density setting "joint length per area" is used to account for longer joints traversing multiple unit areas. The joint geometry input parameters for each scope row of explicit models are summarized in Table 2.

Table 2. Joint network geometric modelling inputs.

Structure row from scope	Number of sets in 2D	Joint Spacing m	Joint density joints/m	Inclination increment degrees	Inclination 1 degrees	Inclination 2 degrees	Inclination 3 degrees
1	2	1.00	1.00	90	45	- 45	N/A
2	2	0.56	1.78	90	45	- 45	N/A
3	3	0.34	2.92	60	60	0	- 60
4	3	0.18	5.62	60	60	0	- 60

Between joints, intact rock is assumed to be GSI 85, which results in a decrease of the strength (to 44%) and stiffness from intact properties. The UCS of the intact rock achieved by applying GSI 85 is consistent with large-scale rockmass behaviour according to Hoek's scaling law for intact rock strength (Hoek & Brown 1980) and with observations of brittle wall damage in tunnels through massive rock at stresses of 33% to 50% of UCS (Martin 1997; Diederichs 2007).

To estimate joint length (persistence), the relation between the quantified surface quality axis and $JCond_{89}$ in Hoek et al. (2013) is used. $JCond_{89}$ is calculated for each explicit model, then related to joint length through a curve fit of the $JCond_{89}$ and discontinuity length information in Hoek et al. (2013), resulting in a range of joint lengths between 1 m and 25 m. Persistence is therefore attributed to the surface quality axis of the GSI chart.

Joint strength properties are defined using the Barton-Bandis failure criterion (Barton 1973; Barton & Choubey 1977; Barton & Bandis 1990). The upper limit (left boundary of the structure quality axis) for JCS is selected to be 44% of UCS, which is consistent with selecting GSI 85 for intact rock strength. The lower limit (right boundary of the structure quality axis) is selected to be 35%, which was obtained through end-member model calibration (models 1B and 3H). JRC's upper and lower limits are set to 20 and 2 respectively, which represents the range in joint undulation that could be observed at a large scale. The basic friction angle upper and lower limits are set to 33 and 27 degrees respectively, which is a minimal variation of friction angle consistent with large-scale joint strength. Joint normal (K_n) and shear (K_s) stiffnesses in MPa/m are estimated based on literature (Day 2017) and calibrated to implicit results for the end-member models. Values of JCS, JRC, and the basic friction angle for intermediate models are calculated linearly. K_n is estimated to have upper and lower limits of 250,000 MN/m and 25,000 MN/m respectively, while K_s is reasonably simplified to be 1/4 K_n. Intermediate K_n values are calculated through a non-linear progression.

3 RESULTS

To supplement the comparison of implicit GSI and explicit GSI model depth of yield and closure data, both datasets are compared to theoretical solution results obtained from RocSupport (Rocscience 2020b). The Carranza-Torres (2004) closed-form solution method and the Generalized Hoek-Diederichs modulus estimation method are used to calculate theoretical results. Results for wall displacement and depth of yield are demonstrated in Figure 3 and Figure 4 respectively, for the two stress conditions.

As shown in Figure 3 and Figure 4, the trend of the explicit model points is in general agreement with the theoretical and implicit estimates for closure and depth of yield at both stress levels investigated. This is expected, since some explicit model input parameters (joint stiffness and JCS) were calibrated to the end member models 3G and 1B.

Explicit data points demonstrate significant scatter and deviation from the expected implicit and theory values, even though the general trend of expected depth of yield and closure agrees. This means rockmasses of different blockiness and surface conditions that are classified as the same GSI will not exhibit the same depth of yield and displacement response. More massive models (structure codes 1 and 2) tend to have lower average depth of yield and closure than is predicted by theory, while blockier models tend to exhibit higher average depth of yield and closure. Explicit model closure deviates from theory in the constant in-situ stress results for models 3H and 4G, which behave more stiffly than theory expects. The average depth of yield in models 3H and 4G is less than theory for the constant in-situ stress models, while average depth of yield in the same models is higher than theory for the constant stress ratio models.

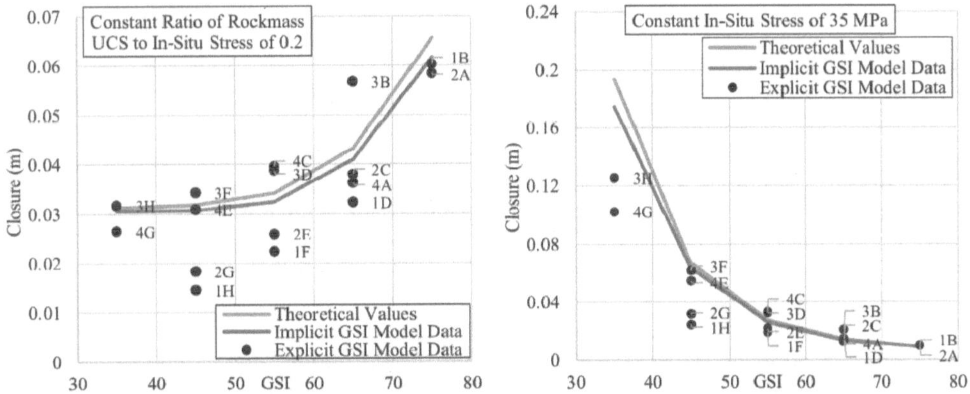

Figure 3. Tunnel closure in theory, implicit, and explicit GSI models. Left: Constant ratio of rockmass UCS to in-situ stress results. Right: Constant in-situ stress results.

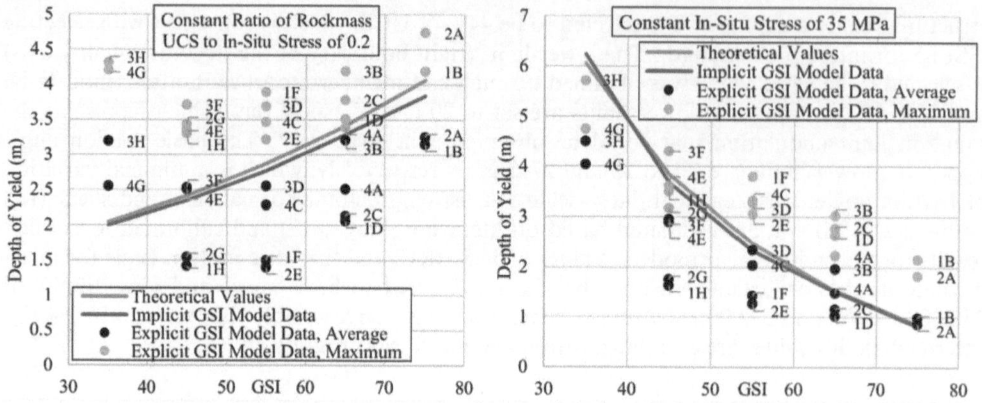

Figure 4. Average and maximum depth of yield in theory, implicit, and explicit GSI models. Left: Constant ratio of rockmass UCS to in-situ stress results. Right: Constant in-situ stress results.

The difference between the average and maximum depths of yield in Figure 4 is attributed to the anisotropy generated by the joint networks. This also sets up a variability in the local closure measured across differently oriented diameters (not shown here). Implications include the generation of liner moments even in a nominally isotropic stress field and GSI rockmass. Bolt loads would also vary. Investigations of these issues will be part of future work. In addition, massive explicit models (structure codes 1 and 2) visually showcase distinct wedges between failed joints.

4 CONCLUSIONS

Traditional continuum modelling using Hoek-Brown parameters based on intact rock properties and GSI classification is commonly used in rock excavation design. In modern numerical modelling software, dense networks of explicit structure can be represented with assigned properties from site investigation and laboratory testing data. If discontinuity properties are valid and representative, explicit models can be used to verify traditional modelling results and can show increased detail on rockmass yield mechanisms.

To develop 2D explicit structure networks, geometric inputs such as joint density, 2D inclination, number of sets, and joint length are required. Such parameters are determinable using GSI chart quantifications such as those by Hoek et al. (2013) and Cai et al. (2004), as performed in this study. Joint strength and deformation parameters must also be estimated reasonably, as results are sensitive to these parameters.

Explicit model closure and depth of yield data generally follow implicit and theoretical trends for both stress scenarios investigated in this study. Individual data points showcase significant scatter and deviation from theory values, meaning that rockmasses of different blockiness and surface conditions that nonetheless have the same GSI will not necessarily exhibit the same rockmass response, with implications for support design. More massive models demonstrate more differential displacement along the tunnel boundary due to the movement of different distinct rock wedges. More massive models also demonstrate greater differences between the average and maximum depths of yield, while more highly jointed rock converges and yields more isotropically.

Future work on this topic should involve developing explicit models with different joint orientations and different network realizations to minimize the impacts of specific networks on results. The impacts on support performance will be investigated, as will post-peak GSI recommendations.

REFERENCES

Baecher GB, Lanney NA and Einstein HH. 1978. Statistical Description of Rock Properties and Sampling. *Proceedings of the 18th U.S. Symposium on Rock Mechanics*, 5C1–8.

Barton, N.R. 1973. Review of a new shear-strength criterion for rock joints. *Eng. Geol.*, 7(4): 287–332.

Barton, N.R. & Bandis, S.C. 1990. Review of predictive capabilities of JRC-JCS model in engineering practice. *Rock Joints, Proc int symp on rock joints, Læn, Norway*, pp 603–610. Rotterdam: Balkema.

Barton, N.R. & Choubey, V. 1977. The shear strength of rock joints in theory and practice. *Rock Mechanics* 10(1): 1–54.

Barton, N.R., Lien, R., & Lunde, J. 1974. Engineering classification of rock masses for the design of tunnel support. *Rock Mechanics* 6(4): 189–236.

Bieniawski, Z.T. 1989. *Engineering Rock Mass Classifications: A Complete Manual for Engineers and Geologists in Mining, Civil, and Petroleum Engineering*. New York: John Wiley & Sons.

Cai, M., Kaiser, P.K., Uno, H., Tasaka, Y., & Minami, M. 2004. Estimation of rock mass deformation modulus and strength of jointed hard rock masses using the GSI system. *International Journal of Rock Mechanics and Mining Sciences* 41(1): 3–19.

Day, J.J. 2017. *The Influence of Healed Intrablock Rockmass Structure on the Behaviour of Deep Excavations in Complex Rockmasses*. PhD Thesis, Queen's University, Kingston, Canada.

Day, JJ., Diederichs, MS., & Hutchinson, DJ. 2019. A new Composite Geological Strength Index approach with application to healed intrablock rockmass structures. *Geotechnical and Geological Engineering* 37(6).

Diederichs, M.S. 2007. The 2003 Canadian Geotechnical Colloquium: Mechanistic interpretation and practical application of damage and spalling prediction criteria for deep tunnelling. *Canadian Geotechnical Journal* 44(9): 1082–1116.

Hoek, E. & Brown, E.T. 1980. *Underground Excavations in Rock*. Boca Raton: CRC Press.

Hoek, E., Carranza-Torres, C., & Corkum, B. 2002. Hoek-Brown Failure Criterion – 2002 Edition. *Proceedings of the 5th North American Rock Mechanics Symposium, Toronto, Canada*, vol. 1, pp 67–273.

Hoek, E., Carter, T.G., & Diederichs, M.S. 2013. Quantification of the Geological Strength Index Chart. Proceedings of the 47th US Rock Mechanics Symposium, ARMA, San Francisco, USA.

Hoek, E., Kaiser, P.K., & Bawden, W.F. 1995. *Support of Underground Excavations in Hard Rock*. Rotterdam: Balkema.

Marinos, P. & Hoek, E. 2000. GSI: A Geologically Friendly Tool for Rock Mass Strength Estimation. Proceedings of the ISRM International Symposium, ISRM, Melbourne, Australia, 2000.

Martin, C.D. 1997. Seventeenth Canadian Geotechnical Colloquium: The effect of cohesion loss and stress path on brittle rock strength. *Canadian Geotechnical Journal* 34(5): 698–725.

Palmstrøm, A. 1996. Characterizing rock masses by the RMi for use in practical rock engineering: Part 1: The development of the Rock Mass index (RMi). *Tunnelling & Undergr. Space Tech.* 11(2): 175–188.

Rocscience. 2020a. RS2, Version 11. Toronto, Ontario, Canada. Rocscience. 2020b. RocSupport, Version 5. Toronto, Ontario, Canada.

Sonmez, H. & Ulusay, R. 1999. Modifications to the geological strength index (GSI) and their applicability to stability of slopes. *Int. Journal of Rock Mechanics and Mining Sciences* 36(6): 743–760.

Session 15 - Spatial variability in probabilistic slope stability analysis

The Evolution of Geotech - 25 Years of Innovation – Hammah et al. (Eds)

Probabilistic analysis of geosynthetic reinforced slopes using 2D and 3D models

Pooya Dastpak
Ferdowsi University of Mashhad, Mashhad, Iran

Daniel Dias
University of Grenoble Alpes, Grenoble, France

Reza Jamshidi Chenari
Geo-Engineering Center at Queen's, Kingston, Canada

Brigid Cami & Sina Javankhoshdel
Rocscience Inc. Toronto, ON, Canada

ABSTRACT: Recent studies have provided deterministic and probabilistic design charts for reinforced slopes considering simple geometries and circular failures. The importance of the non-circular Limit Equilibrium Method (LEM) combined with optimization techniques to find the critical failure mechanism when considering the soil properties spatial variability influence has not previously been addressed. In this paper, 2D and 3D deterministic and probabilistic design charts for simple reinforced slopes are presented. The Slide2 and Slide3 software are used. For the probabilistic part, an efficient Response Surface Method is used rather than the methods that are commonly used in literature like the Latin Hypercube Sampling or Monte Carlo simulations. For the 2D case, the Random Limit Equilibrium (RLEM) approach is used to investigate the soil properties spatial variability influence on the Probability of Failure (PF).

1 INTRODUCTION

2D and 3D Limit Equilibrium Methods (LEM) can be used to determine the Factor of Safety (FOS) in slope stability analyses (Don-ping et al., 2019; Guo and Dias, 2020; Guo et al., 2020; Firincioglu and Ercanoglu, 2021; and many others). To this end, different conventional failure surfaces namely circular (Kitch, 1994), log-spiral (Leshchinsky and Boedeker, 1989), and two-part wedges (Bathurst and Jones, 2001) can be assumed to calculate FOS. In terms of reinforced slopes, these assumptions can result in quantitative differences in FOSs. Most recently, Cami et al. (2018) and Dastpak et al. (2019), took advantage of the advanced limit equilibrium-based method using global and local optimization techniques to accurately calculate the lowest FOS for a noncircular shape.

Kitch (1994), Kitch et al. (2011), and Javankhoshdel and Bathurst (2017) investigated simple geometry reinforced slopes using circular slip surfaces. Three different mechanisms were observed for the critical slip failure: internal, external, and transitional. Internal failure mechanism can be defined when the slip surface impacts all the reinforced layers and the failure region is located within the reinforced zone. In this case, for a given reinforcement tensile strength (T), a critical reinforcement length (L) can be found in which for

DOI: 10.1201/9781003188339-67

a longer L, the FOS remains constant, while the internal failure mechanism is preserved. While external mechanisms can be defined when the slip surface passes just beyond the reinforcement layer. Correspondingly, at a specific L, a critical T can be determined in which for a greater T, FOS stays unchanged. Therefore, Javankhoshdel and Bathurst (2017) presented the threshold values for both T and L for various slope angles. They pointed that the transitional zone is between the external and internal failure mechanisms which can be enlarged by increasing the slope angle. Transitional zone is a state in which at least one layer of the reinforcement layers is intersected by the slip surface. Dastpak et al. (2019) developed the threshold values presented by Javankhoshdel and Bathurst (2017) for non-circular slip surfaces using a global metaheuristic optimization method and a local optimization technique known as Cuckoo search and Surface Altering Optimization, respectively. It was shown that the transitional region enlarges significantly for non-circular slip surfaces.

Deterministic approaches for conventional slope stability analyses have drawbacks when it comes to nominal similar slopes with the same SOF. In such cases, although they can give equal FOS, they may lead to different Probabilities of Failure (PF) and due to the soil mass heterogeneity, every slope has consequently its unique PF. More recently, the soil property uncertainties effect on FOS has been investigated by a number of authors using a LEM approach (see Cho (2010); Javankhoshdel and Bathurst (2014); Guo and Dias (2020); amongst others). Similarly, Cami et al. (2018), Javankhoshdel et al. (2020), Mafi et al. (2020) and others, considered the spatial variability of soil properties to determine PF using the Random Limit Equilibrium Method (RLEM). As per Javankhoshdel and Bathurst (2017), similar design charts for either deterministic and probabilistic analyses were presented by Dastpak et al. (2019) for purely frictional simple reinforced slopes using noncircular slip surface. As indicated in the literature, none of the previous efforts have performed 3D reinforced stability analyses using simultaneous deterministic and probabilistic LEM approach. In this research, a comparison between deterministic and probabilistic 2D slope stability analyses and corresponding 3D analyses is presented. Eventually, several 2D RLEM analyses conducted for the worst case scenario and the results are compared with 3D simple probabilistic analyses.

2 PROBLEM DEFINITION AND GENERAL APPROACH

Figure 1 illustrates the 2D and 3D examples of slope with four layers of geosynthetics, $n = 4$. The slope height (H) is $H = 5\ m$ with an unit weight of $\gamma = 20kN/m^3$. The slope angle varied from $\beta = 45°$ to $90\ °$ and the extrusion depth (Le) varied from $Le = 10\ m$ to $120\ m$. The soil friction angle was ranged between $\varphi = 20°$ to $65°$.

In the current study by taking advantage of the Slide 2D and Slide 3D Software (Rocscience, 2020), the reinforced slope stability using LEM approach was analyzed. Initially, a sensitivity analysis is carried out in order to find the appropriate extrusion length (Le) for 3D cases. Then, the threshold values of T and L for a 3D 63°-reinforced slope is presented. To this end, the GLE-MP method incorporating a global metaheuristic optimization method, called Cuckoo search, and a local optimization technique (the Surface Altering Optimization) is used to determine FOS. Further, a response surface method coupled with the deterministic analyses was used to determine PF.

3 2D AND 3D DETERMINISTIC ANALYSES

As per Javankhoshdel and Bathurst (2017), T and L threshold values can be given in which for larger T and L, FOS values would remain unchanged. Therefore, for every reinforced slope the threshold values can be exclusive depending on the slope angle, soil friction angle and other parameters. In this research, T and L are assigned to be constant for all the reinforced layers. It is worth mentioning that the external failure mechanism cannot practically appear for 3D cases since the reinforcement layers must fail to cause the slip surface to pass beyond the

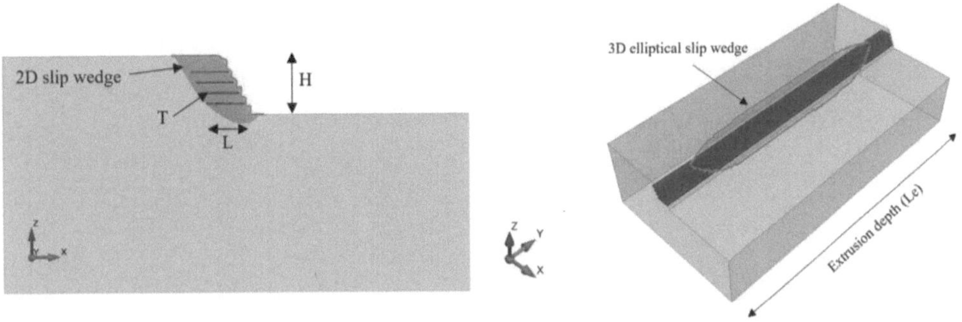

(a) (b)

Figure 1. Example of a simple geosynthetic reinforced slope; a) Slide 2D, b) Slide 3D.

reinforcement layers (refer to Figure 1). As seen in Figure 1, the slip wedge has an elliptical shape which may either dwindle or expand according to the T and L values. For example, for a given L, if T is low enough, the elliptical region expands and FOS increases; consequently, while the slip wedge still passes beyond the reinforcement layers (external failure). On the other hand, if T is high enough and L decreases, the elliptical slip wedge may taper off while FOS increases and the failure mechanism remains still external. Specific threshold values cannot be obtained for the external mechanism as well as for the 3D transitional zone. In other words, for a specific T and L which leads to an external 2D mechanism, the mechanism may be external, internal, or transitional for its 3D correspondent simulations with different FOS and extrusion lengths.

Figure 2 reveals the extrusion length effect on FOS for different slope angles (same conditions to hold for the 2D and 3D scenarios). The FOS values for the 3D simulations are higher than the 2D ones and it becomes more significant for lower extrusion length as well as for steeper slopes. Hence, the 2D modeling is deemed conservative. It should be noted that for higher FOS, e.g. FOS = 1.7, the 2D modeling becomes more conservative even for lower slope angles.

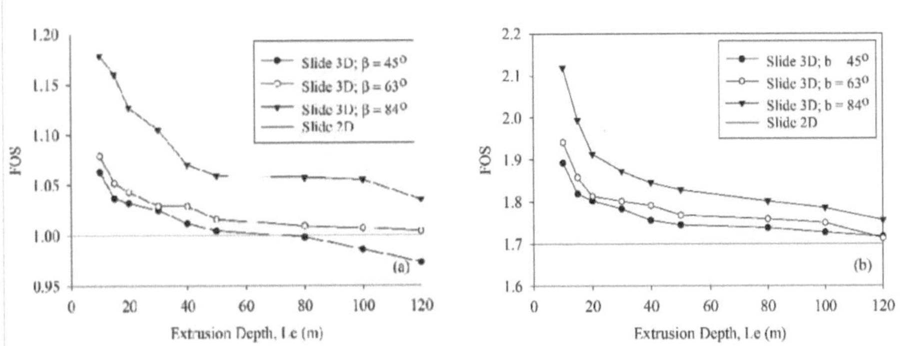

Figure 2. Variation of FOS versus extrusion length for different slope angles; a) $FOS_{2D} = 1$, b) $FOS_{2D} = 1.7$.

Figure 3a shows the T and L threshold values for the 2D and 3D cases (slope angle of 63°). The 3D values were obtained for Le = 120 m. Particular threshold values cannot be considered for the 3D external failure. Nevertheless, T and L threshold for the 3D internal failure mechanisms have the same trend as the 2D models. However, lower L and higher T are needed for the internal failure occurrence.

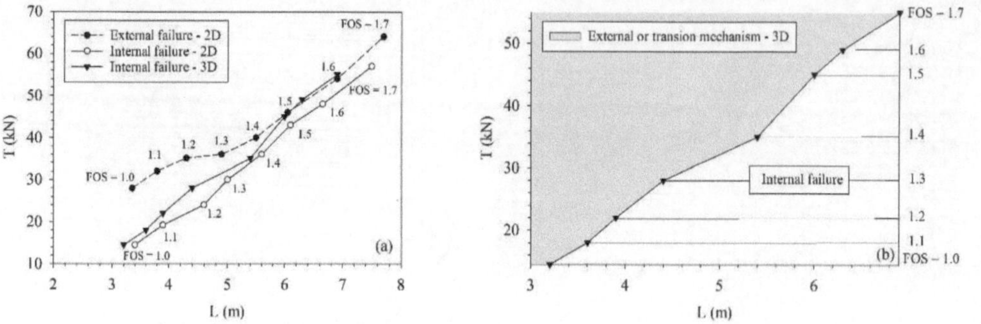

Figure 3. a) Threshold values of reinforcement length (L) and tensile strength (T) to generate only internal failure mechanisms and only external failure mechanisms, b) Contour plots for FOS values corresponding to internal failure (H = 5 m, n = 4, $\gamma = 20$ kN/m^3, $\varphi = 30°$).

Figure 3b shows the 3D FOS contour plots for the same example slope as in Figure 3. The grey region corresponds to either external or transitional failure mechanisms and the FOS contour cannot be exclusively showed. While, for the internal failure, the FOS contours can be defined. For a given T, if L is larger than its threshold value, the FOS remains unchanged.

4 PROBABILISTIC SLOPE STABILITY ANALYSIS

In order to calculate the slope PF, soil properties variability should be assigned. For this purpose, according to Phoon and Kulhawy (1999), the coefficient of variation for internal friction angle and unit weight is assumed as $COV_\varphi = 20\%$ and $COV_\gamma = 10\%$, respectively. The probability distributions were assumed to be lognormal for both γ and φ. Moreover, 15000 response surface simulations were sufficient to calculate PF with a global minimum analysis. It is worth noting that the global minimum (fixed) scheme calculates the deterministic FOS at first; then, it uses the corresponding deterministic slip surface for all the random variables simulations for the PF calculation. Moreover, the extrusion depth was assumed 120 m for all the 3D cases.

To present a probabilistic design chart for reinforced slopes, it is necessary to figure out the critical combination of L and T to merely create either external or internal failure mechanisms for a target FOS. As stated in the previous section, external thresholds cannot be exclusively estimated for 3D cases due to the elliptical slip wedge which must impact the reinforcement layers to pass beyond the reinforcement elements. Therefore, in this study, the probabilistic design charts are solely presented for the internal failure. Javankhoshdel and Bathurst (2017) calculated a factored friction angle (φ_f) according to Eq. 1 which normalizes φ with the corresponding FOS.

$$\varphi_f = \tan^{-1}\left(\frac{\tan\varphi}{FOS}\right) \quad (1)$$

Figure 4 represents the design chart for both 2D and 3D simulations using Single Random Variable (SRV) based on LEM. As indicated before, higher values of φ_f are associated with higher PF. This can be attributed to the fact that the internal failure PF largely depends on T. For those scenarios with greater φ_f, smaller T should interact to get the same FOS; thus, PF and φ_f have a direct relationship. Note that, as per Javankhoshdel and Bathurst (2017), the 1% PF criterion can be acceptable in the case of internal failure mechanism based on similarities to other high strength-soil structure systems, e.g. MSE walls. Regarding Figure 4, maximum PFs for 2D models are greater than 3D models which shows that the 2D designs are conservative. Likewise, maximum PF gently decreases for higher φ_f Also, for higher FOS, a small difference can be seen between the 2D and 3D PFs.

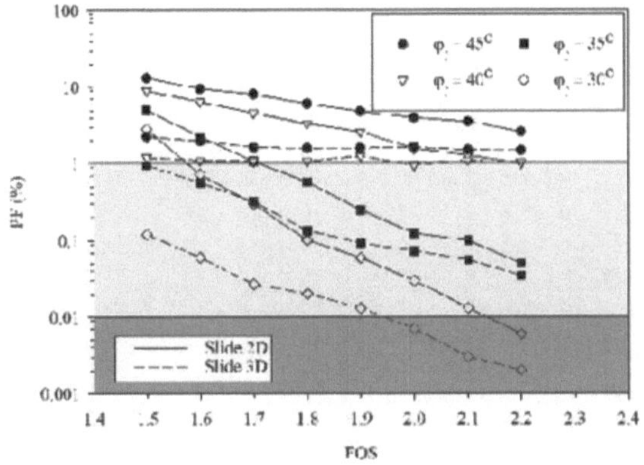

Figure 4. Maximum internal failure probability for reinforced soil slopes versus deterministic FOS for different values of factored friction angle φ_f, comparison between 2D and 3D LEM.

To add the spatial variability analysis in 2D models, the random field generation was used based on the Local Average Subdivision Method (LAS) proposed by Fenton and Vanmarcke (1990). Dastpak et al. (2019) reported that the worst case of vertical correlation length is 10 m for a simple reinforced slope. Here, as indicated in Figure 5, the horizontal and vertical spatial correlation lengths were considered to be infinite and equal to 10 m, respectively according to Dastpak et al. (2019).

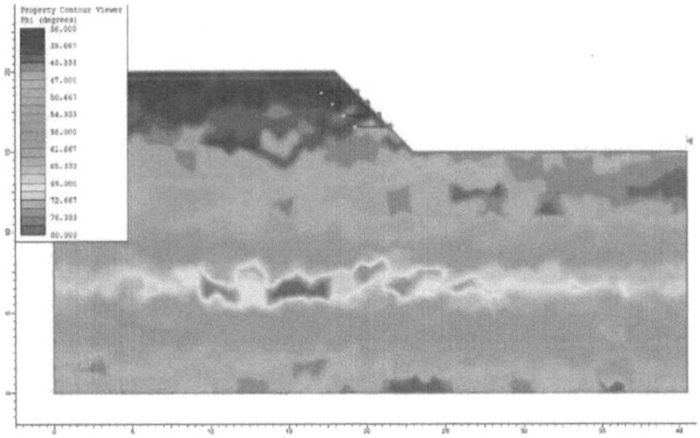

Figure 5. An example of random field generation for simple reinforced slope in Slide2 software; $\theta_H = \infty$, $\theta_v = 10$ m.

Figure 6 shows PF obtained by the 3D global minimum probabilistic analysis using SRV LEM and 2D spatial RLEM approach. As seen, 2D RLEM design is still conservative particularly for lower FOS. While, for higher φ_f and FOS, PF calculated by 3D SRV LEM is greater than the 2D spatial RLEM.

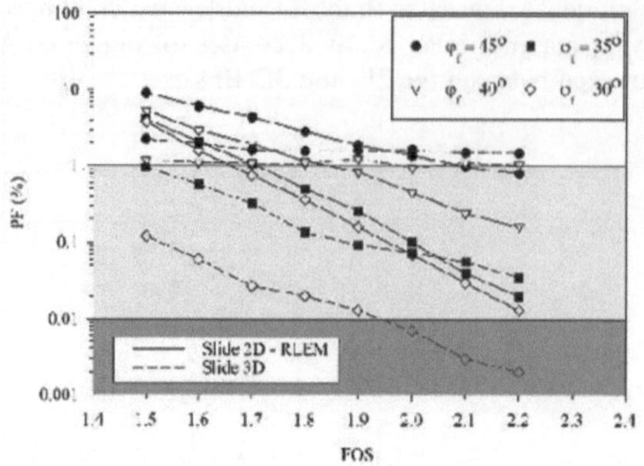

Figure 6. Maximum probability of internal failure for reinforced soil slopes versus deterministic FOS for different values of factored friction angle φ_f, a comparison between 2D RLEM and 3D LEM.

5 CONCLUSION

This study introduces the 3D deterministic and probabilistic Limit Equilibrium Method (LEM) for the study of geosynthetic reinforced slopes in purely frictional soils. Results show that in the 3D simulations, the extrusion depth (Le) plays an important role in the Factor of Safety (FOS) determination. Moreover, external failure mechanisms do not practically appear in 3D slope models since the reinforcement layer must be intersected by a slip surface passing beyond the reinforcement. Therefore, elliptical slip wedges do not have a constant shape. They are based on the tensile strength (T) and reinforcement length (L) as well as the soil strength. The elliptical slip wedges may be varied and not a unique threshold can be considered for the external failure mechanism. Indeed, with a given L and T in a 2D model that only generates an external failure with a specific FOS, the failure mode may be external, internal, or transitional for its corresponding 3D model with a different FOS. In general, 2D models are more conservative in comparison with 3D ones. In terms of the probabilistic analysis, the Probability of Failure (PF) tends to be smaller for 3D models even for 2D Random Limit Equilibrium Method (RLEM). In summary, using 2D LEM is highly recommended for low factored friction angles due to its very fast analyzing engine and simplicity rather than 3D simulations.

REFERENCES

Bathurst, R.J. and Jones, C.J.F.P., 2001. Earth retaining structures and reinforced slopes. In Geotechnical and geoenvironmental engineering handbook (pp. 501–537). Springer, Boston, MA.

Cami, B., Javankhoshdel, S., Yacoub, T. and Bathurst, R.J., 2018, November. 2D spatial variability analysis of Sugar Creek embankment: comparative study. In International Congress and Exhibition" Sustainable Civil Infrastructures: Innovative Infrastructure Geotechnology" (pp. 118–125). Springer, Cham.

Cho, S.E., 2010. Probabilistic assessment of slope stability that considers the spatial variability of soil properties. Journal of geotechnical and geoenvironmental engineering, 136(7), pp.975–984.

Dastpak, P., Mousavi, S., Chenari, R.J., Cami, B. and Javankhoshdel, S., General probabilistic analysis of simple reinforced slopes using RLEM approach. In GeoSt'Jonhs 2019, Canada, September 2019.

Dong-ping, D., Liang, L. and Lian-Heng, Z., 2017. LEM for stability analysis of 3D slopes with general-shaped slip surfaces. International Journal of Geomechanics, 17(10), p.06017017.

Fenton, G.A. and Vanmarcke, E.H., 1990. Simulation of random fields via local average subdivision. Journal of Engineering Mechanics, 116(8), pp. 1733–1749.

Firincioglu, B.S. and Ercanoglu, M., 2021. Insights and perspectives into the limit equilibrium method from 2D and 3D analyses. Engineering Geology, 281, p.105968.
Guo, X. and Dias, D., 2020. Kriging based reliability and sensitivity analysis–Application to the stability of an earth dam. Computers and Geotechnics, 120, p.103411.
Guo, X., Sun, Q., Dias, D. and Antoinet, E., 2020. Probabilistic assessment of an earth dam stability design using the adaptive polynomial chaos expansion. Bulletin of Engineering Geology and the Environment, 79, pp.4639–4655.
Javankhoshdel, S. and Bathurst, R.J., 2014. Simplified probabilistic slope stability design charts for cohesive and cohesive-frictional (c-ϕ) soils. Canadian Geotechnical Journal, 51(9), pp.1033–1045.
Javankhoshdel, S. and Bathurst, R.J., 2017. Deterministic and probabilistic failure analysis of simple geosynthetic reinforced soil slopes. Geosynthetics International, 24(1), pp.14–29.
Javankhoshdel, S., Cami, B., Chenari, R.J. and Dastpak, P., 2020. Probabilistic Analysis of Slopes with Linearly Increasing Undrained Shear Strength Using RLEM Approach. Transportation Infrastructure Geotechnology, pp.1–28.
Kitchl, W.A., Wright, S.G. and Gilbert, R.B., 1994. Probabilistic analysis of reinforced soil slopes. Engineering Mechanics, pp. 325–328.
Kitch, W.A., Gilbert, R.B. and Wright, S.G., 2011. Probabilistic assessment of commercial design guides for steep reinforced slopes: implications for design. In Geo-Risk 2011: Risk Assessment and Management (pp. 1055–1062).
Leshchinsky, D. and Boedeker, R.H., 1989. Geosynthetic reinforced soil structures. Journal of Geotechnical Engineering, 115(10), pp.1459–1478.
Mafi, R., Javankhoshdel, S., Cami, B., Jamshidi Chenari, R. and Gandomi, A.H., 2020. Surface altering optimisation in slope stability analysis with non-circular failure for random limit equilibrium method. Georisk: Assessment and Management of Risk for Engineered Systems and Geohazards, pp.1–27.
Phoon, KK and Kulhawy, FH, 1999. Characterization of geotechnical variability. Canadian geotechnical journal, 36 (4), pp. 612–624.
RocScience Inc. 2020. Slide Version 2020 – 2D and 3D Limit Equilibrium Slope Stability Analysis. www.rocscience.com, Toronto, Ontario, Canada.

The Evolution of Geotech - 25 Years of Innovation – Hammah et al. (Eds)

A stochastic spatial modeling approach for pit slope stability analysis using 3D limit equilibrium analysis

F. Cudjoe & K. Esmaeili
Lassonde Institute of Mining, University of Toronto, Canada

ABSTRACT: This paper presents a stochastic spatial modelling approach to evaluate the stability analysis of a pit slope in an open pit mine in Canada. More than 200 km of geotechnical borehole data drilled and logged in the mine area was used to develop 3D block models of Rock Mass Rating (RMR) and Uniaxial Compressive Strength (UCS), using stochastic Sequential Gaussian Simulation (SGS) method. The pit design was then excavated into the 3D RMR block models. The block models of the geotechnical attributes (UCS and RMR) were then embedded as input into discretized 3D limit equilibrium models created in SLIDE 3D to conduct a stochastic stability analysis of the pit slope. The modeling results were used to investigate possibilities for optimization of the pit slope design.

1 INTRODUCTION

A thorough understanding of the spatial variation of rock mass geomechanical properties (i.e. intact rock strength, and rock mass geomechanical quality) is essential for safe and more efficient pit slope design and operation (Eivazy et al. 2017). A detailed knowledge of rock mass gemechanical properties allows for more accurate identification of high-risk zones (low-quality rock mass zones) and can provide opportunities for steepening pit slope angles which can lead in significant economic benefits (Cudjoe and Esmaeili, 2020). Probabilistic techniques, like random field simulation have been employed to investigate the effect of rock mass heterogeneity on slope stability (Allan et al., 2012; Griffiths et al., 2009). A most important shortcoming of this approach is that it does not consider spatial pattern (anisotropy) of the geomechanical properties. Geostatistical estimation and simulation techniques have been implemented for spatial modeling of geomechanical attributes (Pinheiro et al., 2018; Eivazy et al., 2017; Mayer & Stead, 2017; Egaña & Ortiz, 2013; Stavropoulou et al., 2007). Among the simulation techniques, Sequential Gaussian Simulation (SGS) which is built on conditional simulation algorithm is the most valuable method in modelling the variability of geomechanical properties (Srivastava, 2013). Despite the challenges and difficulties associated with the geomechanical attributes (e.g. data scarcity and no-additivity of some data; discussed in Eivazy et al. (2016)), these spatial models have been implemented for rock slope stability evaluations using 2D numerical modeling techniques (Renani et al., 2019; Mayer & Stead, 2017; Marchesi et al., 2009). Despite these studies, limited investigations have been carried out to quantify the effect of spatial variability/heterogeneity of rock mass geomechanical properties on the stability of rock slopes in three dimensions.

In this paper, a stochastic spatial modelling approach was developed to evaluate the stability of pit slope in an open pit mine. Geostatistical SGS technique was implemented to develop 3D stochastic geomechanical models of rock masses present in the pit walls using core logging data from the mine. The block models of the geotechnical attributes (UCS and RMR) were then embedded as input into discritiezed 3D limit equilibrium numerical models using SLIDE

DOI: 10.1201/9781003188339-68

3D to conduct a stochastic slope stability analysis. The modeling results were used to determine the probability of pit wall failure.

2 GEOMECHANICAL DATA COLLECTION AND ANALYSIS

Geomechanical borehole datasets of an iron ore open pit mine were acquired through diamond drilling campaigns conducted since 1949. More than 200 km of geomechanical borehole data have been drilled and logged in the pit area. The core logging data consists of lithological and geomechanical rock mass properties, accounting for 869 drillholes, mostly ranging from 6.6 m to 981 m in length. The geomechanical properties of core samples have been logged based on RMR rock mass classification system (Bieniawski, 1976), which includes: Rock Quality Designation (RQD), joint surface condition, joint spacing, intact rock strength, water condition and the ratings for these parameters were summated as RMR_{76}. Based on the field data, the water condition was considered as dry. For the purpose of this study, slope stability investigations were conducted on the final pit limit. The final pit limit is 3700 m long, 1500 m wide and 600 m depth. Figure 1a shows the final pit limit along with drillholes logged according to lithology codes in the pit area. Figure 1b shows a cross section of the final pit limit with the drillholes displaying the RMR_{76} rock mass classes, logged along the holes. Figure 1c presents a North-South cross section of the mine's geological model with extremely folded and metamorphosed lithological units. These structural folds have created duplicate iron ore lenses with various thicknesses.

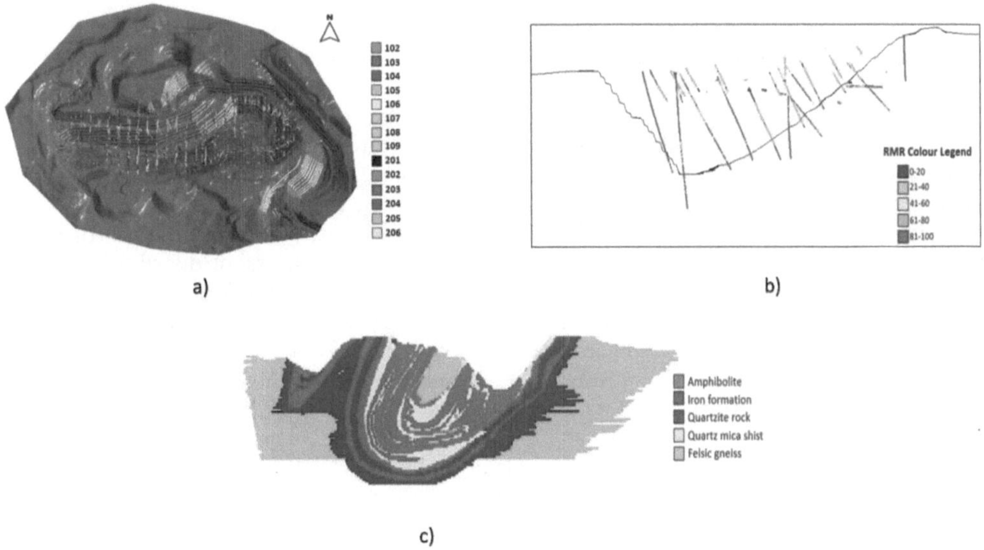

Figure 1. a) Drillholes logged according to lithology codes in the final pit limit, b) cross section of the final pit with the drillholes logged based on RMR_{76}, c) North-South cross section of the mine's geological model.

Exploratory statistical analysis was carried out on cleaned borehole dataset, consisting of 65,671 sample points with majority of the samples logged at 3 m interval length. Figure 2 (a) and (b) represent a comparison of RMR and UCS, respectively for the four major lithologies present in the pit wall in terms of box plots.

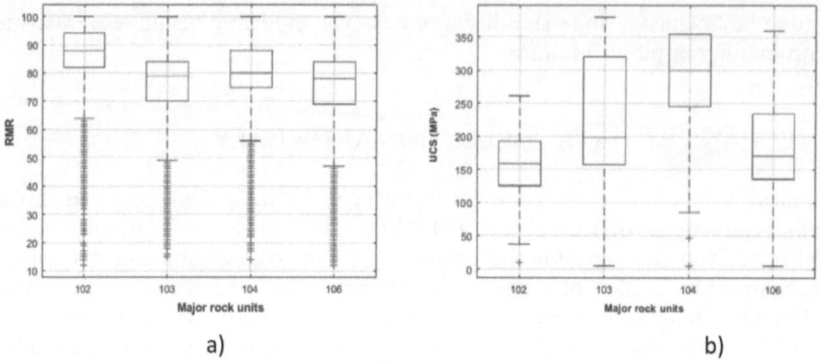

Figure 2. a) Box plot of RMR for major rock units in the pit, b) Box plot of UCS for major rock units in the pit the pit.

From these figures, Iron formation (103) shows the highest dispersion and range for UCS while Quartz mica schist (106) shows the highest dispersion for RMR. The mean RMR for Amphibolite (102) is the highest and the mean RMR for Quartz Mica Schist (106) is the lowest. Also, Amphibolite (102) shows the least dispersion for RMR and UCS geomechanical properties. Among these rock types, the Quartzite rock (104) which contains 0-40% mica is the strongest while Amphibolite (102) has the lowest strength in terms of UCS. However, in terms of RMR, Amphibolite (102) has the highest rock mass quality while Quartz mica schist (106) which contains 40-60% mica has the lowest rock mass quality.

3 3D SPATIAL MODELLING OF ROCK MASS GEOMECHANICAL PROPERTIES

Geotechnical borehole logging data was used to produce multiple equi-probable realizations of 3D RMR and UCS spatial models for the pit area using SGS technique. In this study, the spatial modelling technique was implemented to produce 3D RMR and UCS block models of the pit using Maptek Vulcan 10.1 software (Maptek, 2016). The SGS technique is based on multi-gaussian function which requires a gaussian distribution of input data. Hence, RMR and UCS samples were transformed into gaussian distribution to be compatible with the simulation approach. The normal score transformation process is essential to the overall spatial variability modelling because it reduces proportional effect which is intrinsic to regionalized variables mostly with skewed distribution. Variogram analysis including downhole variogram and experimental variogram modelling were carried out to quantify the spatial variability of normalized RMR and UCS values for each lithology group.

Twenty realizations of the 3D block models of RMR and UCS were generated with each realization consisting of over five million blocks at 10 m ×10 m × 14 m (east, north and elevation directions) block sizes. The block size matches the geological block model and the bench height of the mine. Figure 3 (a) and (b) present the North-South cross-section of one RMR and UCS realization, respectively. The statistics of the twenty block model realizations (e.g. histogram, Q-Q plot, variogram) were compared with the geomechanical field data to validate the results. According to Figure 3a, it can be observed that the quality of most of the rock masses range from good (61- 80/green) to very good (81-100/red) in terms of RMR_{76}. However, very poor-quality rocks (blue areas) are concentrated along the top-most boundary of the block model (2015 pit outline) which indicates that this area is likely to be critical for slope stability. This is an area where a large amount of weak mica-schist is exposed on the pit wall. Figure 3b generally reveals a high (100- 200 MPa) to very high (200-400 MPa) intact rock strength. However, blocks with very low intact rock strength (blue areas) are concentrated in the northern part along the 2015 pit outline.

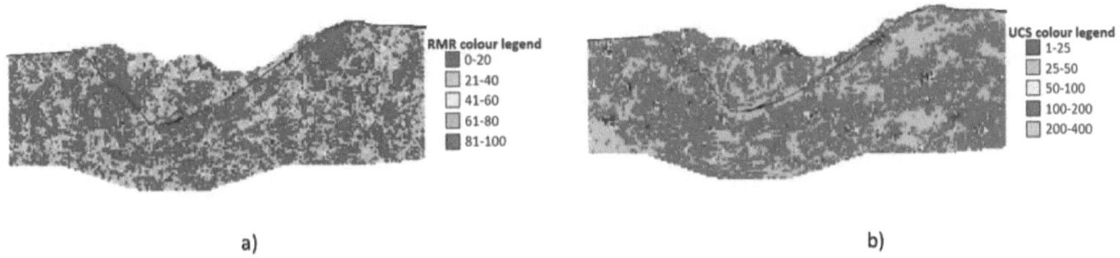

Figure 3. a) North-South cross section of one realization of RMR block model, b) North-South cross section of one realization of UCS block model.

4 SLOPE STABILITYANALYSIS USING 3D LIMIT EQUILIBRUIM ANALYSIS

In this study, 3D limit equilibrium modelling approach was applied to perform stability analysis of the pit slopes using SLIDE 3D v.2.015 (Rocscience, 2019). The block models of the geotechnical attributes (UCS and RMR) were embedded as input into a discretized 3D limit equilibrium (LE) model for the slope stability analysis. Overall, twenty heterogeneous slope models were developed for the final pit slope in SLIDE 3D, corresponding to twenty 3D block models of RMR and UCS. The 3D LE models were used to calculate the probability of failure for the pit slopes. The southeastern part of the final pit wall was considered for the slope stability analysis. This is because previous historical investigations revealed large-scale instabilities occurring in the south pit wall. The Mohr-Coulomb strength parameters, cohesion (c) and friction angle (φ) were determined for each individual block of the pit slope based on the Hoek-Brown parameters which were extracted from the developed 3D RMR and UCS block values. The Hoek-Brown failure criterion was initially used to determine the rock mass parameters from which the equivalent Mohr-Coulomb strength parameters were estimated. Table 1 presents some of the Hoek Brown parameters used for the modeling. The other parameters (e.g. GSI) were estimated for each discrete block. A disturbance factor (D) of 1, corresponding to production blasting, was assigned to blocks closer to the pit walls whiles a lower D factor of 0.85 was assigned to blocks farther away from the pit wall. The m_i values assigned to the four lithology domains are based on data from triaxial and UCS tests, conducted by consultants to characterize rock units at the mine site. The GSI index for each block was replaced by the equivalent RMR_{76} block values and the UCS block values were directly used as the uniaxial compressive strength of intact rock.

Table 1. Hoek Brown parameters for LE slope stability modelling.

Lithology Domain	m_i	Disturbance factor (D)		Unit weight (kN/m^3)
		Along pit walls (20 m depth)	Away from pit wall	
Iron Formation	25	1	0.85	28.0
Amphibolite	15	1	0.85	28.5
Quartzites	23	1	0.85	26.5
Quartzite Mica Schist	14	1	0.85	26.5

Figure 4a shows the geometry and slope dimensions of the pit south wall under study. The overall slope angle is relatively shallow (38°). Figure 4b and 4c present the spatial distribution of material properties (cohesion, internal friction angle) in a 2D cross-section model of the pit slope. The results indicate that cohesion distribution in the pit wall ranges from 3-36 MPa (with majority of values between 3-10 MPa). The distribution of internal friction angle is ranging from 4°-58° (with majority of values between 50°-58°).

The stability analysis was carried out using Janbu's simplified method in conjunction with the non-circular search path and cuckoo search method. The slope was considered to be dry since the pit slopes are geographically located in a permafrost zone coupled with an efficient mine water drainage system in summer.

The factor of safety (FOS) and the critical failure surfaces were determined for all the twenty heterogeneous slope models. The minimum FOS varies between 6.6 - 7.17 for all twenty models. Figure 5a and 5b present the failure surface and contour map of FOS projected on one 3D LE model, respectively. The failure surface covers almost the entire pit wall indicating a multi-bench failure surface. Since the FOS for all 20 models was above 1, it implies that the probability of failure (probability of FOS below 1) is almost zero. The failure volumes for the models range between 276.5Mm3 to 604.9Mm3. The variation of failure volumes for the LE models are attributed to variation in the slip surface caused by the heterogeneous nature of rock mass properties in each model. The higher volumes are associated to the failure surfaces which cover a wider area or extends farther behind the crest.

A sensitivity analysis was carried out to investigate the possibility of increasing the overall pit slope angle with all other parameters kept constant for optimal pit slope design. A small increment of the pit slope angle can have a significant impact on the operating cost of the mine through increased ore recovery and/or reduced stripping. The analysis revealed that as the slope angle of the south wall increases from 38° to 60°, the factor of safety reduces from ~7.0 to 3.8. This is attributed to the high rock mass qualities along the pit wall.

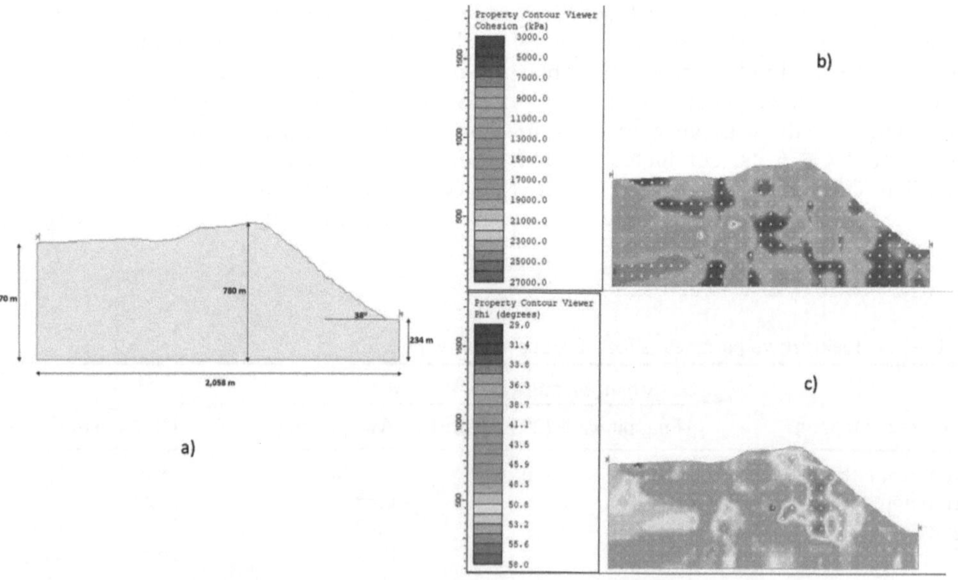

Figure 4. a) Slope geometry of the pit wall, b) spatial distribution of rock mass cohesion in a 2D cross-section of the pit slope, c) spatial distribution of friction angle in a 2D cross-section of the pit slope.

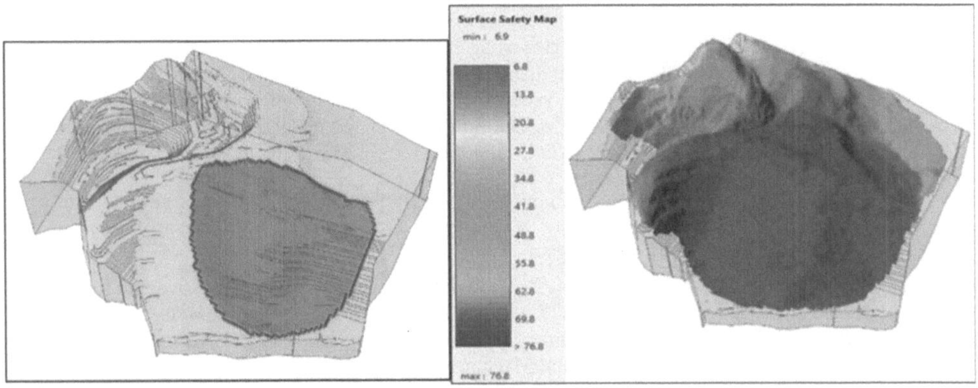

a) b)

Figure 5. a) 3D LE model showing the failure surface for a heterogeneous slope, b) 3D LE model showing FOS contour map for a heterogeneous slope.

5 CONCLUSIONS

A stochastic spatial modeling approach was applied to analyze the stability of a pit slope using 3D limit equilibrium analysis. The stability analysis accounted for the heterogeneity of rock masses in the mine area by embedding spatial geotechnical attributes (UCS and RMR) as input into discretized limit equilibrium models. The stochastic stability analysis revealed that the safety factors determined for the south pit slope is above the tolerable FOS for mine rock slopes under static condition. The proposed stochastic slope stability approach allows integrating the spatial variability of rock mass geomechanical properties in slope stability analysis. In addition, the limit equilibrium technique allows conducting slope stability analysis in a relatively fast and easy manner. The developed approach can be used to optimize pit slope design by investigating the possibility of increasing the pit slope angle.

The proposed approach takes into account the spatial variation of rock mechanical properties on the slope stability, however, due to the nature of geological structures and the high degree of folding in the lithological units in the mine, the mechanical properties of the rock units are anisotropic. The contact between Quartz mica schist and Iron formation in the south wall slope can be considered as an anisotropy surface and the variations of shear strength parameters can be investigated on the slope stability using Generalized Anisotropic Strength option in RS3 program. This is an ongoing study and the results will be validated based on pit wall monitoring data.

ACKNOWLEDGEMENTS

The authors would like to thank MITACS for the financial support. The authors also appreciate Rocscience for the educational license of SLIDE3.

REFERENCES

Allan, F. C., Yacoub, T. E. & Curran, J. H. 2012. On Using Spatial Methods for Heterogeneous Slope Stability Analysis. 46th US Rock Mechanics/Geomechanics Symposium. Chicago, IL, USA.

Bieniawski ZT. 1976. Rock mass classification in rock engineering. In: Proceedings of the symposium on exploration for rock engineering. 1–5 Nov 1976, Johannesburg, South Africa, pp 97–106.

Cudjoe, F., & Esmaeili, K. 2020. A 3D Spatial Model of Slope Mass Rating to Assess Potential Risk of Pit Wall Failure. 54th US Rock Mechanics/Geomechanics Symposium. American Rock Mechanics Association.

Egaña, M., & Ortiz, J.M. 2013. Assessment of RMR and its uncertainty by using geostatistical simulation in a mining project. Journal of GeoEngineering, 8(3), 83–90.

Eivazy, H., Esmaieli, K., & Jean, R. 2016. Challenges in modelling geomechanical heterogeneity of rock masses using geostatistical approaches. In Proceedings of 24th world mining congress (pp. 18–21).

Eivazy, H., Esmaieli, K., & Jean, R. 2017. Modelling Geomechanical Heterogeneity of Rock Masses Using Direct and Indirect Geostatistical Conditional Simulation Methods. Rock Mechanics and Rock Engineering, 50(12), 3175–3195.

Griffiths, D., Huang, J. & Fenton, G. 2009. Influence of Spatial Variability on Slopen Reliability Using 2-D Random Fields. J. Geotech. Geoenviron. Eng., 135, 1367–1378.

Maptek 2016. *Vulcan Version 10.1.*

Marchesi, VR., Da fontoura, SAB., Rubio, NPR. 2009. How 3D modeling can improve quality and reliability of geotechnical projects. In: EUROCK symposium, 29–31 Oct 2009, Cavtat, Croatia.

Mayer, J. M., & Stead, D. 2017. A comparison of traditional, step-path, and geostatistical techniques in thestability analysis of a large open pit. Rock Mechanics and Rock Engineering, 50(4), 927–949.

Pinheiro, M., Emery, X., Miranda, T., Lamas, L., & Espada, M. 2018. Modelling geotechnicaln heterogeneities using geostatistical simulation and finite differences analysis. Minerals, 8(2), 52.

Renani, H. R., Martin, C. D., Varona, P., & Lorig, L. 2019. Stability analysis of slopes with spatially variable strength properties. Rock Mechanics and Rock Engineering, 52(10), 3791–3808.

Rocscience Inc. 2019, SLIDE3 Version 2.015.

Srivastava, R. M. 2013. Geostatistics: a toolkit for data analysis, spatial prediction and risk management in the coal industry. *International Journal of Coal Geology*, *112*, 2–13.

Stavropoulou, M., Exadaktylos, G., & Saratsis, G. 2007. A combined three-dimensional geological-geostatistical-numerical model of underground excavations in rock. Rock mechanics and rock engineering, 40(3), 213–243.

The Evolution of Geotech - 25 Years of Innovation – Hammah et al. (Eds)

Developing spatially constrained Discrete Fracture Network (DFN) models for a stochastic pit slope stability analysis

Y. Zhao & K. Esmaeili
Lassonde Institute of Mining, University of Toronto, Canada

J. Ran
Kinross Gold Corporation, Canada

ABSTRACT: In this paper, data from an open pit mine in Western Africa were used to generate spatially constrained large-scale DFN models. Structural data from geotechnical boreholes and borehole tele-viewers, drilled and logged in the pit area, were used to develop stochastic 3D block models of volumetric fracture intensity (P_{32}) using Sequential Gaussian Simulation. A geocellular 3D DFN model with finite volume of cellular grid elements were then created and spatially constrained based on the 3D block models of P_{32}. This approach ensures that the fracture centers in the DFN generation are under the influence of cellular P_{32} values at corresponding locations, estimated based on the 3D block model of volumetric fracture intensity. The joint orientations were bootstrapped based on the joint orientations observed along the drillholes. The geocellular DFN models were further calibrated using joint trace mapping data, collected across the pit area using a drone-based aerial photogrammetry. The resulting DFN models are expected to accurately reflect the spatial variation of fracture geometry and intensity along the pit area. Finally, 2D cross sections of the generated DFN models were incorporated into 2D finite element models of the pit slope for a stochastic slope stability analysis. The results of the stochastic modeling were compared to a simplistic approach assuming an average rock mass structural condition.

1 INTRODUCTION

For moderate to strong intact rock masses, the stability of open pit slopes is mainly controlled by geological features. Therefore, detailed descriptions of rock structure distribution can provide accurate prediction of rock mass behaviors (Read and Stacey, 2009). Discrete Fracture Network (DFN), as a probabilistic simulation method, is considered the most practical tool for capturing rock mass structural heterogeneity and spatial variation of in-situ fracturing systems (Einstein et al., 1983). This is particularly significant for large-scale projects, where the spatial variation of discontinuities cannot be simply described as ubiquitous with deterministic spacing and orientation (Dershowitz and Einstein, 1988). With the introduction of DFN model and resulting non-persistent fractures, rock bridges are created, which are the crucial elements for large-scale open pit slope stability (Elmo et al., 2018). Step-path failure analysis has been identified as an essential technique for analyzing the interaction between intact rock bridges and open discontinuities by simulating the progression of fractures across rock slopes (Wong and Wu, 2014).

In this paper, we present a novel approach for large-scale conditional DFN modelling, which allows stochastic fractures to be spatially and geometrically constrained based on the fracture properties derived from core logging, borehole televiewer, and photogrammetry data.

DOI: 10.1201/9781003188339-69

The field structural data were used for developing fracture intensity spatial models using Sequential Gaussian Simulation (SGS) method. These spatial models along with photogrammetry of fracture traces on pit walls were used to develop stochastic and spatially constrained DFN models of the walls. These models were then incorporated into the Finite Element analysis for assessing pit slope stability using the shear-strength reduction method. The goal is to investigate the effect of fracture spatial distribution on open pit slope stability and to improve slope stability assessment with conditioned stochastic DFN models.

2 STRUCTURAL DATA COLLECTION AND ANALYSIS

This study was conducted at an open pit mine, located in Mauritania, Western Africa, lies within the Archean age Aouèounat greenstone belt, comprising mainly meta-volcanic and meta sedimentary sequences (Sims, 2019). Structural data used for DFN modeling were gathered across the pit, through various means as summarized here: core logging of 49 geotechnical drillholes with joint orientation and geomechnical properties; televiewer survey of 29 boreholes, containing joint orientation and spacing information; Unmanned Aerial Vehicle (UAV) photogrammetry of pit walls, including fracture orientation and trace length data (details presented in Bamford et al., 2020); and wireframes of major structures (i.e., faults, dykes) across the open pit.

Figure 1 presents a photography of the pit hangingwall (East wall), the locations of geotechnical and televiewer holes on the pit triangulation, and an example of the digital fracture tracing results using UAV mapping of the pit walls. The structural data were then processed to be used as primary input parameters, and DFN models were created using FracMan (FracMan, 2020). Fracture orientation is the key component for fracture set characterization. Stereographic analyses of fracture orientation data in DIPS demonstrated a controlling foliation set, trending N-S, together with three major joint sets, as summarized in Table 1. The distribution of fracture sizes (3D features) is often predicted using the distribution of fracture traces, the 2D measurements of the extent of structures on geological exposures, in the form of a power law distribution (Tonon, 2007; Rogers et al., 2010). The statistics of fracture trace length were also summarized in Table 1, for estimating fracture sizes distributions. Since foliation was considered ubiquitous and persistent, it was excluded from both analyses, for maintaining the statistical descriptions of major joint set properties. Fracture intensity describes the degree of fracturing within rock masses, using the notation Pij (Dershowitz and Herda, 1992). Although linear fracture intensity P10 [m-1] can be calculated directly from core logging, it is considered orientation-dependent of sampling region (Esmaieli et al. 2010), therefore volumetric fracture intensity, P_{32} [m2/m3] was used instead for DFN modelling. Conversion from P10 to P_{32} can be performed using the approach proposed by Wang (2006).

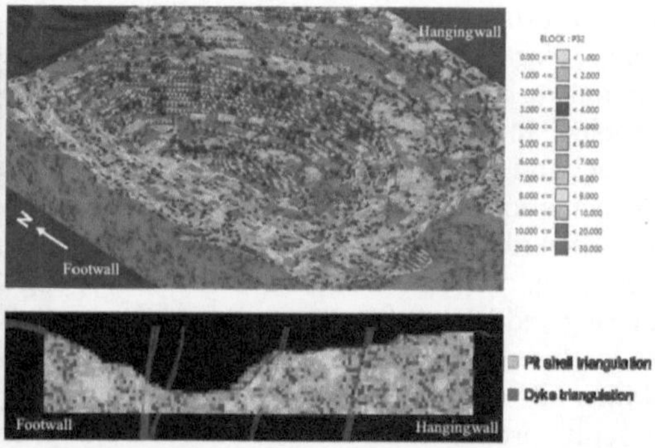

Figure 1. LEFT (A) Borehole locations plan view of the pit, (B) a view of the pit looking north-west (C) UAV-mapped tracing. Figure 2. RIGHT Simulated SGS P_{32} distribution within the pit limit.

Table 1. Weighted orientations of major joint sets and trace length statistics in the pit.

	Joint Set 1	Joint Set 2	Joint Set 3	Overall	Foliation
Fracture Orientation					
Dip Angle	88°	64°	53°	-	45°
Dip Direction	345°	245°	307°	-	76°
Fisher's κ	20.60	8.77	23.34	-	23.94
Trace Length Statistics					
Min (m)	1.00	1.00	1.00	1.00	-
Max (m)	10.57	6.01	13.21	13.21	-
Mean (m)	2.20	1.97	2.32	2.20	-
Std. Dev.	1.62	1.02	1.46	1.45	-

3 CONDITIONED DFN MODELLING AND CALIBRATION

Stability of the pit hangingwall is of primary importance for West Branch as this is the high wall that will be pushed back during the life of mine. Therefore, the ultimate hangingwall was used for this study. The ultimate pit is 2,700 m long, 1,500 m wide and 510 m deep. Recent studies suggested DFN model performance and accuracy can be significantly improved when using geocellular models, where fractures were geometrically and spatially conditioned to cell-based input data (Rogers et al., 2016). In this study, geocellular DFN models were developed, where the spatial model of P_{32} was used for intensity conditioning over fracture centers (Bym et al., 2014). For an improved model accuracy, further calibration was performed using high-quality fracture trace data, collected by UAV and remote sensing technology (Medinac & Esmaeili, 2020) as described below:

1. Collect the fracture traces on pit walls, excluding traces less than 1 m and foliations;
2. Discretize mapping area into cells of 10 m x 10 m (i.e., geocellular DFN grid size);
3. Sum up the trace lengths within one mapping cell and divide it by the area for P21 estimation;
4. Calculate simulated P21 on the corresponding DFN region using FracMan;
5. Compare the field and simulated P21 cell by cell and record the differences for quantifying model accuracy;
6. Trace back to P_{32} of the cells with large differences and modify them accordingly.

Table 2. Comparison of the primary input parameter statistics of the ultimate pit DFN model.

		Field Data	DFN Model
P_{32} (m^2/m^3)	Min	0.02	0.05
	Max	29.98	31.42
	Average	2.88	3.05
Orientation (Dip(°)/Dip direction(°))	Joint Set 1	88/175	88/176
	Joint Set 2	64/245	62/245
	Joint Set 3	73/307	73/308
	Foliation	45/076	45/076

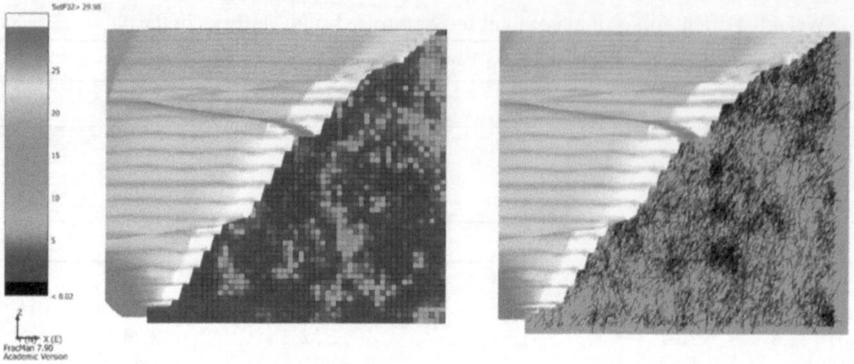

Figure 2. Block model of P_{32} (left) and fracture traces from DFN model (right) on a plane perpendicular to the ultimate pit shell limit.

After calibration, the number of over- and under-estimating P_{32} cells were significantly reduced by 80%. The statistics of fracture properties from survey and conditioned DFN models were compared for validation purpose, see Table 2. The characteristics of the DFN models are close to the field collected data. Moreover, there is a good matching between fracture clusters and high P_{32} regions on trace maps of DFN models and those of the 3D block models of fracture intensity (see Figure 2). Results suggested the simulation and calibration process were reliable and can accurately reflect the true spatial variation of fracture properties. The slight overestimation of P_{32} is hypothesized to be caused by the over-writing nature of fracture generation process, where the average target of P_{32} was ignored in order to match the high cellular P_{32}. In addition, having fractures crossing cell boundaries could also influence the estimation.

4 DFN-FEM MODELLING OF STEP-PATH FAILURE ANALYSIS

For inter-ramp to mine scale slope stability modelling, both geological structures and shear strength of intact rock bridges become equally significant. In this study, the stability of the ultimate pit hangingwall was assessed using the developed novel approach. The results were compared by a simplistic slope stability approach used by the mine's consultant. A step-path failure analysis, incorporating the effect of rock bridges, becomes significantly crucial. In this study, a finite element analysis was performed in RS2 (Rocscience 2020), using the Shear Strength Reduction (SSR) technique. Fracture traces extracted from conditioned DFN models were explicitly imported into a 2D finite element model of the pit slope. To improve the modeling efficiency, only fracture trace lengths longer than 10 m (bench height at mine) were incorporated into the RS2 model. A Mohr-Coulomb strength criterion was applied to the discontinuities, assuming joints were cohesionless at a friction angle of 30°. The geological units within pit slope were derived from past geotechnical investigations conducted by consulting firms. A sub-region of oxide/transition zone was considered on the upper 100 m of the model to represent the alteration zone near surface. All intact rock properties were modelled as being plastic, and Hoek-Brown shear strength parameters and equivalent Mohr-Coulomb shear strength parameters were calculated (see Table 3). A disturbance factor (D) of 0.7, representing disturbances induced by blasting damage, was considered. The geometry of the slope is shown in Figure 3.

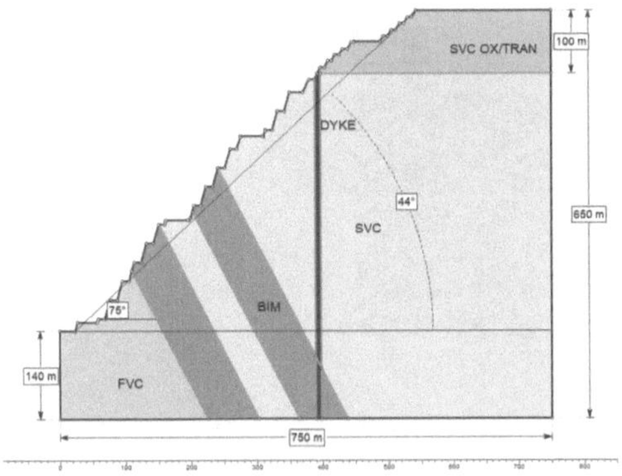

Figure 3. The ultimate pit slope geometry and rock types used for RS2 models.

Table 3. Summary of material properties used for RS2 modelling.

	SVC OX/TRAN	SVC	BIM	FVC	DYKE
Unit Weight (kN/m^3)	22.7	27.8	27.8	26.7	29.9
Poisson's Ratio	0.23	0.23	0.23	0.23	0.23
Young's Modulus (kPa)	1.9e+07	7.1e+07	7.1e+07	7.1e+07	7.1e+07
Peak Tensile Strength (kPa)	94	2472	2376	3847	1969
Peak Friction Angle (deg.)	33.30	42.67	46.43	48.78	51.25
Peak Cohesion (kPa)	95	2472	2498	4391	2454
Residual Tensile Strength (kPa)	0	0	0	0	0
Residual Friction Angle (deg.)	33.30	42.67	46.43	48.78	51.25
Residual Cohesion (kPa)	95	2472	2498	4391	2454
Dilation Angle (deg.)	9.99	28.16	30.64	29.57	33.82

Table 4. RS2 modelling results for the ultimate pit limit using conditioned DFN fractures.

	Critical SRF		Total Displacement	
Ultimate	Min	1.30	Min	0.01
	Max	1.45	Max	0.26
Pit limit	Mean	1.38	Mean	0.19
	Std. Dev.	0.06	Std. Dev.	0.06

Five iterations of DFN models were created and the stochastic fracture were incorporated into RS2 models using the ultimate pit slope geometry for SSR analyses. The modeling results including the critical SRF and total displacement are summarized in Table 4. The average critical SRF of the ultimate pit model was at 1.38 with an average maximum total displacement of 1.23 m, which was considered acceptable with SRF being higher than 1.3 (Read and Stacey, 2009). The maximum total displacement took place within the top region of the slope, and the main cause of instability was found to be lateral displacement (see Figure 4A). Results also suggested that fracture distribution was related to total displacement, where the slope toe region with significantly fewer fractures was more stable. It can be concluded that an increase of slope stability with a decrease of fracture intensities. While foliations were not affecting the slope performance due to their favorable orientation with respect to the pit hanging wall, large wedges were often enclosed by Joint Set 2 due to its distinct orientation. In addition, dyke was found causing hangingwall slope instability, where large displacements were often found near the dyke region. This is considered reasonable since the intrusion of dyke can cause regional fracturing.

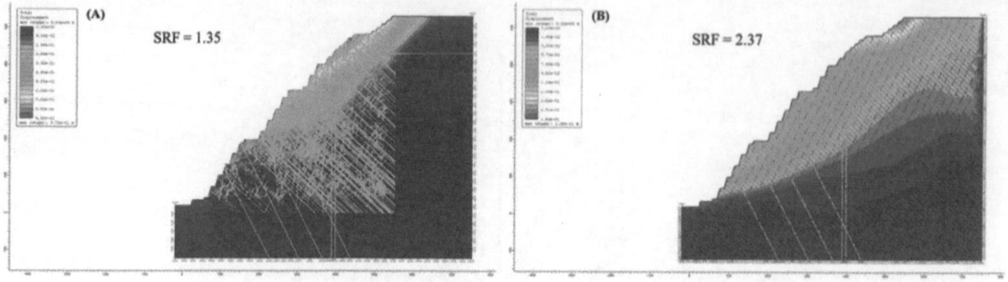

Figure 4. Total displacement comparison between (A) DFN fracture and (B) joint network RS2 models.

Nevertheless, the degree of improvement in slope stability assessment brought by DFN-FEM models were still unclear at this point. As a result, RS2 models with simple joint networks which were commonly used by geomechanical practitioners were created for the pit slope model. The foliation set and Joint Set 2 were included in this model, where the foliation assigned a dipping of 45° into the slope and considered fully persistent with a joint spacing of 5 m, and Joint Set 2 was assigned an average dip angle of 64° with 65% persistency with a joint spacing of 10 m (representing average structural data analysis for joint set 2), see Figure 4B. Same properties were assigned to the RS2 model with joint network. The modeling results indicate that the deterministic model with joint network has a considerably higher SRF (2.37) compared to the stochastic RS2 models. One significant difference is that the slope stability in joint network model was more reliant on material properties, where the maximum total displacement located merely within the oxide/transition zone. The instability within fresh zone was expressively overlooked, as no potential wedges were expected to be formed. In addition, because of the uniform and parallel distribution of joint sets, the joint network models failed to identify the impact of major structures as it was shown in the stochastic RS2 models.

5 CONCLUSION

The deficiency for creating reliable large-scale DFN models primarily comes from two aspects which are data uncertainty and model calibration. Since DFN models heavily rely on the statistics of structural dataset, it is crucial to have the input parameters as representative as possible. This paper presents the methodology of developing stochastic conditioned geocellular DFN models with appropriate calibration using field mapping data. By using spatial models of volumetric fracture intensity together with other fracture characteristics obtained from field data analysis, the generated DFN models best preserve the spatial variability of fractures across a large volume such as an open pit mine.

The DFN models can subsequently be linked to a 2D finite element model for a probabilistic step path slope stability analysis. The DFN-FEM slope stability analyses conducted in this study suggested that lateral displacement was the main cause of slope failure for the slope hangingwall and a key joint set was found to play an important role in formation of large-scale (multi-bench) wedges. Weak material strength and proximity to major structures were the two controlling factors for high fracture intensities, which can also affect hangingwall stability. Although foliation is persistent throughout the pit, it is not considered significant affecting slope stability of the hangingwall given its orientation. The conventional joint network models using average joint parameters resulted in over-estimation of the slope factor of safety. This can lead to the design of pit slopes with a higher risk of instability. Moreover, the lack of spatial and geometry variation of fractures made the identification of failure locations be merely dependent on material strengths. Therefore, the significance of using conditioned DFN models in large-scale slope stability assessments lies within the preservation of fractures distribution variability, which is critical for determination of failure pattern and mechanism. A more precise identification of slope failure location and mechanism can prevent over or under-estimation of slope stability, allowing designs to achieve the maximum economic returns with minimum likelihood of failure. However, it is essential that the modelling results be further validated using pit wall monitoring data, for making more robust conclusions regarding the overall pit wall stability.

ACKNOWLEDGEMENTS

The authors would like to thank the financial support by Kinross Gold and MITACS. The authors also appreciate the educational license of FracMan by Golder and RS2 by Rocscience.

REFERENCES

Bamford, T., Medinac, F., & Esmaeili, K. 2020. Continuous Monitoring and Improvement of the Blasting Process in Open Pit Mines Using Unmanned Aerial Vehicle Techniques. Remote Sensing, 12(17), 2801.

Bym, T., Fox, A. and Josephson, N. 2014. Development and demonstration of algorithms for conditional Simulation of Discrete Fracture Networks. International Discrete Fracture Network Engineering Conference proceeding. Vancouver, Canada. pp. 267.

Dershowitz, W.S. and Einstein, H.H., 1988. Characterizing rock joint geometry with joint system models. Rock mechanics and rock engineering, 21(1), pp.21–51.

Dershowitz, W.S. and Herda, H.H., 1992, January. Interpretation of fracture spacing and intensity. In The 33th us symposium on rock mechanics (USRMS). American Rock Mechanics Association.

Deutsch, C.V. and A.G. Journal. 1998. GSLIB: Geostatistical Software Library and User's Guide. New York: Oxford University Press.

Einstein, H. H., Veneziano, D., Baecher, G. B., & O'reilly, K. J. (1983). The effect of discontinuity persistence on rock slope stability. In International journal of rock mechanics and mining sciences & geomechanics abstracts (Vol. 20, No. 5, pp. 227-236). Pergamon.

Elmo, D., Donati, D. and Stead, D., 2018. Challenges in the characterisation of intact rock bridges in rock slopes. Engineering Geology, 245, pp. 81–96.

Esmaieli, K. Hadjigeorgiou, J. Grenon, M. 2010. Estimating geometrical and mechanical REV based on synthetic rock mass models at Brunswick Mine. Int. J. Rock Mech. Min. Sci., 47(6): 915–926.

FracMan. 2020. FracMan User Manual 7.70. Seattle, Washington, US: Golder Associates.

Medinac, F. and Esmaeili, K., 2020, May. Integrating unmanned aerial vehicle photogrammetry in design compliance audits and structural modelling of pit walls. In Proceedings of the 2020 International Symposium on Slope Stability in Open Pit Mining and Civil Engineering (pp. 1439–1454). Australian Centre for Geomechanics.

Read, J. and Stacey, P., 2009. Guidelines for open pit slope design. CSIRO Publishing. Rocscience 2020. RS2: 2D geotechnical finite element analysis.

Rogers, S., Elmo, D., Webb, D., & and Catalan, A. 2010. A discrete fracture network-based approach to defining in situ, primary and secondary fragmentation distributions for the Cadia East panel cave project. The Second International Seminar on Block and Sublevel Caving, 425-440.

Rogers, S. F., Elmo, D., Webb, G., & Moreno, C. G. 2016. DFN Modelling of major structural instabilities in a large open pit for end-of-life planning purposes. In 50th US Rock Mechanics/Geomechanics Symposium. American Rock Mechanics Association.

Sims, J. 2019. Tasiast Project, Mauritania. National Instrument 43-101 Technical Report. Kinross Gold Corporation.

Tonon, F. 2007. Determining fracture size probability distribution functions from trace length probability distribution functions. In 11th Congress of the International Society for Rock Mechanics. Taylor & Francis (pp. 215–222).

Wang, X., 2006. Stereological interpretation of rock fracture traces on borehole walls and other cylindrical surfaces (Doctoral dissertation, Virginia Tech).

Wong, L., Wu, Z. 2014. Application of the numerical manifold method to model progressive failure in rock slopes. Engineering Fracture Mechanics (119:1-20).

The Evolution of Geotech - 25 Years of Innovation – Hammah et al. (Eds)

Probabilistic analysis of an open pit mine slope in the Central African Copperbelt with spatially variable strengths

S.D. Cylwik, S.B. Cox & J.J. Potter
Call & Nicholas, Tucson, Arizona, USA

ABSTRACT: The highly variable nature of weak rock in weathered sedimentary deposits poses significant challenges to open pit and underground mine design. Traditional probabilistic analysis fails to consider all potential mechanisms of instability that may influence slope stability. An approach with spatially variable strengths allows the natural variability of the shear strength properties within the rock mass to be simulated. This paper examines the congruence of geotechnical block modeling and spatial variability analysis to estimate the probability of failure of a highly variable weathered open pit slope in the Central African Copperbelt. The input parameters for the random spatial field simulations are estimated from variography of composited drill hole data and univariate statistics of a 3D block model of rock-mass shear strength. The Rocscience software Slide2 is used to perform the random field simulations and analysis. It is also demonstrated that the total variance can be reduced by the small-scale variability (nugget effect variance) for spatially averaged shear strengths. Typical rock-mass spatial parameters from other projects are summarized.

1 INTRODUCTION

Factor of safety (FOS) is not a consistent measure of the risk of slope failure since it is possible for many levels of risk to exist for the same FOS value. The probability of failure (Pf) is a more explicit measure of risk than FOS, and is estimated by performing a probabilistic analysis considering input parameters as random variables. All geotechnical properties are spatially variable. If spatial variability is not considered in a probabilistic slope analysis, severe over-or under-estimation of the probability of failure can result (Cylwik et al., 2018). In addition to the mean and standard deviation of a parameter, an estimate of the correlation distance is also required to model spatial variability. It is often difficult to obtain the correlation distance of rock-mass strength parameters because they must be estimated with a transformation model and cannot be measured directly (as opposed to fine-grained soil). This publication presents the methodology that has been developed to perform spatially variable probabilistic slope stability analysis within oxide zone deposits in the Central African Copperbelt.

2 CENTRAL AFRICAN COPPERBELT REGIONAL GEOLOGY

The Central African Copperbelt (CACB) is a series of sediment-hosted, stratiform coppercobalt deposits. Most mining is currently focused on the oxide deposits, which typically range in depth from 50 to 300 meters below surface. Ore and waste rocks include dolomites, siltstones, limestones, and sandstones that are variably bedded, brecciated, and altered, as shown in Figure 1.

DOI: 10.1201/9781003188339-70

Slope stability is controlled by bedding orientation, faults, rock strength, and groundwater. Most deposits typically consist of a footwall comprised of RGS, and a hanging wall comprised of SDS and CMN rock types, separated by the RAT/RGS contact fault.

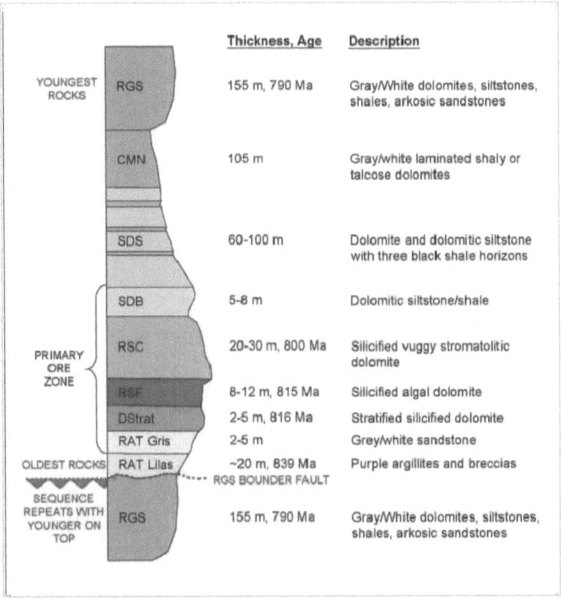

Figure 1. Stratigraphy of Mine Series rocks within the Central African Copperbelt (after Schuh et al., 2012).

3 ROCK-MASS CLASSIFICATION AND STRENGTH ESTIMATION IN WEAK OXIDE ROCKS

A geotechnical rock type (GTR) classification system has been developed specifically for the CACB to provide an index of rock quality using available data. This system focuses on the soft, transition zone rocks that are typically encountered during mining of the oxide portion of the deposits. In comparison, more commonly used classification systems focus on stronger rocks. A sulphide zone is present below the oxide zone and consists of hard rock. The sulphide zone is not mined in the current open pits, but will be part of future surface and/or underground mining operations. The GTR classification system utilized is based on hardness, talc content, and recovery, and is presented in Table 1. These parameters are interpolated into three-dimensional block models and are used to define geotechnical rock types and domains.

Table 1. GTR classification system.

Geotechnical Rock Type	Talc Index*	Calculated GTR**	Description
GTR-T	1-3	-1 - 6	High talc content
GTR0	0,4	≤ 0.5	Soil-like material
GTR1	0,4	≤ 1.5	Friable, weak rock
GTR2	0,4	≤ 2.5	Transition rock
GTR3-6***	0,4	≤ 3.5	Hard rock

* 0-no talc, 1-massive talc, 2/3-talc along bedding, 4-traces of talc.
** C_GTR = (1-%Recovered)*0 + (% ≤ S6)*-1 + (% ≤ R2)*1 + (% > R2)*Hardness.
*** GTR3-GTR6 grouped as one strength for analysis.

Different methods of strength estimation are utilized depending upon the character of the material:

1. Rock-mass strength using the CNI method (Read & Stacey, 2009) for hard materials (GTR2 to GTR6)
2. Joint strength plus 1.5 percent intact strength along bedding for analysis of anisotropic failures in hard rock (GTR2 to GTR6)
3. Direct shear strength and intact shear strength for soft rock, talc, RAT/RGS shear, and fault gouge (GTR0, GTR1, GTR-T, faults)

3.1 *Block model*

In order to estimate the variability in rock quality for a given deposit, a three-dimensional block model of GTR is created from available drill hole data.

3.1.1 *Domains*

Prior to estimation of GTR, the model area is divided into geotechnical domains based on rock type and depth from surface. Rock types are categorized into the following groups:

- SDS Group: SDB, SDS, and CMN formations
- RSC Group: RSF and RSC formations
- RGS Group: RAT and RGS formations

Rock quality generally improves with depth. Rock groups are further divided into zones of similar geotechnical characteristics based on depth from surface.

- Weathered zone (1) – well-developed saprolitic horizon consisting primarily of GTR00 material. The weathered zone typically extends between 5 and 60 meters below the surface.
- Intermediate zone (2) – variably weathered rocks that range in strength from soft, soillike material (GTR00 to GTR01) to harder, more competent rock (GTR02 to GTR03). This zone comprises the majority of the oxide pit slopes.
- Competent, lower zone (3) – predominantly hard rock (GTR03). The contact between this and the overlying intermediate zone is typically located at or just below the toe of the oxide pit slopes.

Additional domain boundaries are defined as needed based on field observations and/or analysis of drill hole data (e.g., around fault zones).

3.1.2 *Estimation of GTR*

Drill hole data are composited on fixed 1.5-meter intervals honoring logged or modeled rock type boundaries. Variograms of composited GTR data are generated for each domain to determine input parameters for ordinary kriging. Model values for the sample area are estimated using ordinary kriging with domain-appropriate (isotropic or bedding parallel) search criteria.

3.1.3 *Conversion of GTR to shear strength*

Cohesion and friction angle vales are estimated for each individual drill composite. This is accomplished by creating continuous functions that estimate shear strength in between the integer values of GTR, as shown in Figure 2.

4 PROBABILISTIC ANALYSIS WITH SPATIAL VARIABILITY

4.1 *Example cross section – FZ_2*

The example deposit trends roughly southwest to northeast and is geologically and structurally complex. The sedimentary sequence is repeated several times due to folding and thrusting along the RGS bounding thrust fault system.

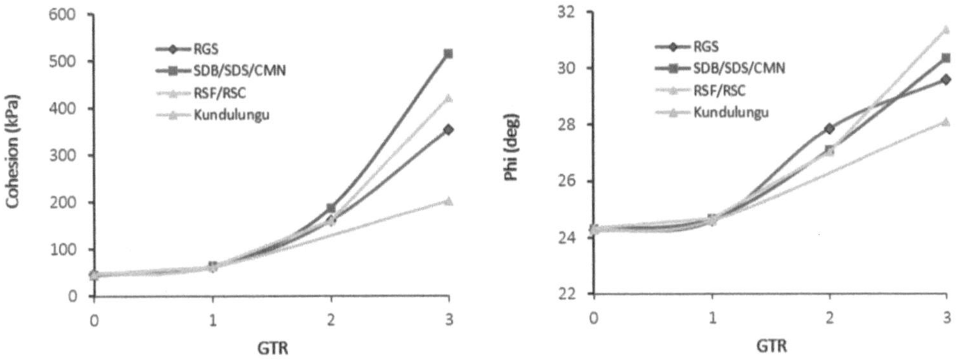

Figure 2. Example continuous functions to convert GTR to cohesion and friction angle.

Section FZ_2 in contains two fault blocks with near-vertical stratigraphy separated by the RGS bounding thrust fault, as shown at the bottom of the cross section in section 4.3. In the upper benches, the RGS bounding fault puts the Mines Series uncomformably against the CMN formation. Another Mines Series sequence is at the toe of the pit slope and is in fault contact with the RGS Breccia on the south wall.

4.2 *Input parameters*

Cohesion and friction angle are assumed to be spatially variable for this analysis. Unit weight is assumed to be constant. The following procedures are used to estimate the input properties:

- The mean value (μ) is estimated from the model blocks of cohesion and friction within each geotechnical domain, limited to within 25 meters in front of and 75 meters behind (into the wall) the pit slope design. The block model is the best mean estimator of the values of cohesion and friction angle within the potential zone of instability. The median value of the blocks is utilized due to the long tail in the distribution of strength values (Figure 3).
- The distribution type is estimated from the distribution shape of the drill hole composites of cohesion and friction angle. An example distribution of the SDB/SDS/CMN Rock type for weathered zone 2 is shown in Figure 3.
- The correlation coefficient between cohesion and friction angle is estimated form the cohesion and friction angle values of the drill hole composites.
- The long-scale variance (σV^{2}) is used as the input variance for the spatial analysis, and is estimated by fitting a variogram model to the experimental variogram of drill composites of cohesion and friction. Example variogram models are shown in Figure 3.
- The correlation distance (θ) is estimated by fitting a variogram model to the variogram of drill composites of cohesion and friction.
- The minimum strength considered in the simulations is that of GTR0 material, or cohesion of 48.2 kPa and friction angle of 24.3 degrees.
- *Important Considerations for Spatial Variability Input Parameters*
- Note that the variance and correlation distance of a parameter should not be estimated from interpolated data such as a block model, since block interpolation methods do not retain the variance of the raw data (they are typically best mean estimators).
- Only the long-scale variance of the data set must be considered for slope stability analysis with spatial variability. The nugget variance (σN_2), or "nugget effect,"

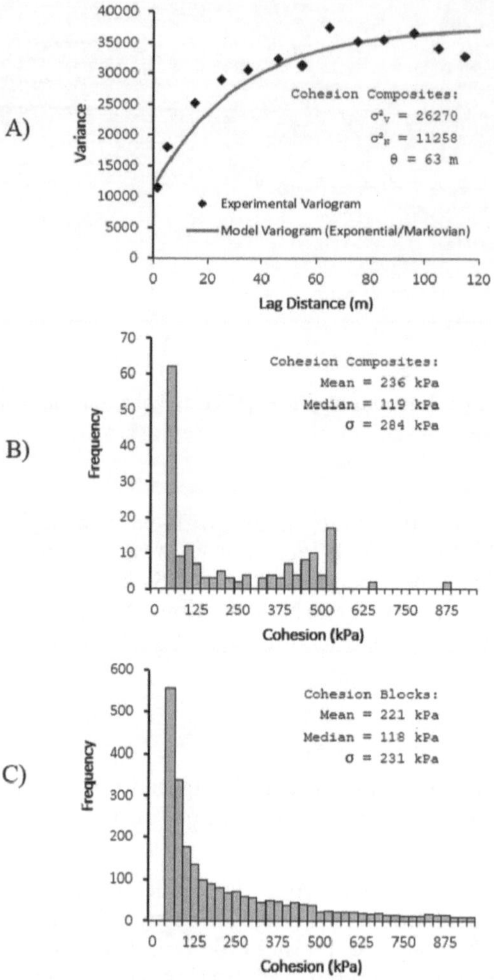

Figure 3. Section FZ_2, intermediate zone 2, SDS rock type, cohesion A) Variograms of drill hole composites, B) Histograms of drill hole composites, and C) Histograms of model blocks.

which is due to short-scale variation, measurement error, positional error, and/or inherent randomness, does not need to be considered. An example validating this concept is shown in Figure 4. Calculation of a variance reduction function (Cylwik et al. 2018, Vanmarcke 2010) shows that the nugget effect variance does not contribute to the variance of a spatial average. Slope stability is controlled by the spatial average strength along the slip surface, not by the strength at any one point along it, and therefore the nugget variance is not required.

- It is also important to ensure that the covariance functions used by the variogram fitting software and by the slope stability random field generator are equivalent. For this analysis, Hexagon MineSight software was used to perform the variography. For the exponential/Markovian covariance function, MineSight uses "$-3\tau/\theta$" as the exponent whereas Slide2 uses "$-2\tau/\theta$", and therefore correlation distances exported from Mine-Sight must be multiplied by two-thirds before being input into Slide2.

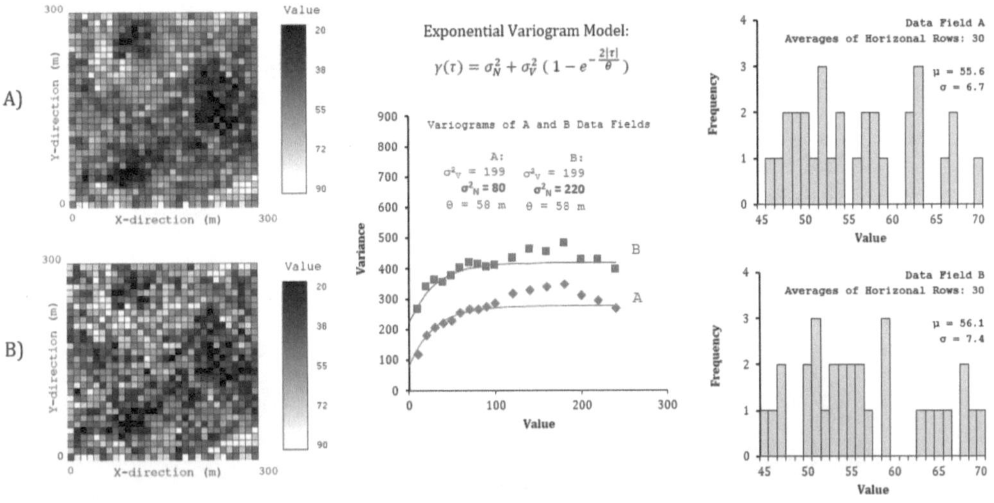

Figure 4. Comparison of two data fields to demonstrate that the nugget effect does not alter the spatial average of a data field.

4.3 *Results*

Figure 5 presents the results of the spatially variable probabilistic analysis for cross section FZ_2 with 1000 spatial simulations. Each of the 1000 simulations optimizes a global minimum critical slip surface that tends to pass through the weakest materials for that particular simulation. The deterministic global minimum FOS value is 1.404 (shown in red), the mean probabilistic FOS value is 1.119, and the Pf is 8.1 percent. The geology model is shown at the bottom of the section for reference and the slope is underlain by one of the 1000 spatial simulations. The 1000 optimized slip surfaces are colored by FOS value, and the distribution of FOS values is shown in the upper left. Many potential critical failure mechanisms are identified by the analysis. The majority of critical surfaces with low FOS values are located in the lower slope (indicating that this mechanism has the greatest Pf) and represent a failure mechanism different from the deterministic minimum surface.

4.4 *Typical spatial variability parameters for rock*

Correlation length and long-scale variance parameters for rock properties are often difficult to estimate due to lack of available data and also because of a scarcity of published values. Typical observations from the variograms of published data (Cylwik et al. 2018, Exadaktylos et al. 2008, Hsu and Nelson 2006) include:

- Parameters that measure rock-mass fracturing (e.g., RQD, GSI structure rating) typically have a very large correlation length, whereas fracture strength and intact strength typically have very small correlation lengths. RQD data correlation lengths can be as small as 15 meters or as large as 200 meters (with typical values being 20 to 80 meters).
- All variograms of rock-mass strength properties typically show a nugget effect. This may occur in rock and not as often in soil because of the discontinuous nature of rock.
- Typical ranges for spatial parameters of rock properties observed across many mine sites are provided in Table 2.

Table 2. Typical correlation length and nugget effect ranges for rock properties

Rock Property	Correlation Length	Nugget Effect
RQD, RMR	20-80 m	20-50%
UCS, $\sigma_{Tensile}$	< 10 m	50-100%

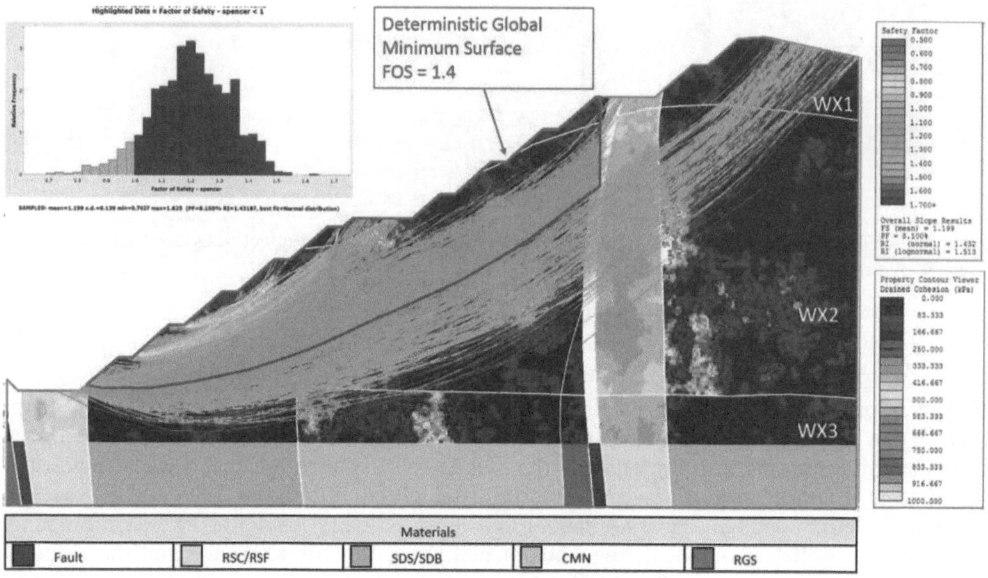

Figure 5. Results of spatially variable slope analysis of section FZ_2.

5 CONCLUSIONS

Probabilistic slope analysis allows for the risk between various mine plans to be directly and quantitatively compared. Spatial variability modeling in Slide2 allows for many random field simulations to be run to identify potential failure mechanisms that otherwise may not be considered in traditional analysis. It is often difficult to estimate spatial parameters of rock-mass strength properties, since rock-mass strength cannot be measured directly. In this paper, it was demonstrated that rock-mass strength can be estimated and assigned directly to the drill hole composites and spatial parameters may be estimated from them. A block model is used to obtain mean estimates of strength for each domain. All sources of potential uncertainty deserve consideration in a probabilistic analysis, including uncertainty of the statistical mean, transformation model, lithological boundaries, pore water pressure, etc. Potential failure mechanisms that involve structural components must also be considered and incorporated into rock slope analysis.

REFERENCES

Call, R.D., Cicchini, P.F., Ryan, T.M., & Barkley, R.C. 2001. Managing and Analyzing Overall Pit Slopes, in *Slope Stability in Surface Mining*, Hustrulid, W.A., McCarter, M.K., and Van Zyl, D.J.A. (eds), SME, pp. 39–46.

Cylwik, S.D., Beck, J.A., & Ryan, T.M. 2018. The Uncertainty of Rock Mass Shear Strength Estimates: How to Incorporate the Reduction in Variance Due to Spatial Averaging for Use in Probabilistic Analysis. In, *Slope Stability 2018*, April 10-13, 2018, Seville, Spain.

Exadaktylos, G. & Stavropoulou, M., Xiroudakis, G., de Broissia, M., Schwarz H. 2008. A spatial estimation model for continuous rock mass characterization from the specific energy of a TBM. *Rock Mech Rock Eng* 2008 41:797–834.

Hsu, S.C. & Nelson, P.P. 2006. Material Spatial Variability and Slope Stability for Weak Rock Masses. *Journal of Geotechnical and Geoenvironmental Engineering*. 132 (2), 183–193.

Read, J., & Stacey, P. 2009. *Guidelines for Open Pit Slope Design*, CSIRO Publishing, 510 p.

Schuh, W., Leveille, R.A., Fay, I., & North, R. 2012. Geology of the Tenke-Fungurume Sediment-Hosted Strata-Bound Copper-Cobalt District, Katanga, Democratic Republic of Congo, *Geology and Genesis of Major Copper Deposits and Districts of the World: A Tribute to Richard H. Sillitoe, Jeffrey W. Hedenquist, Michael Harris, Francisco Camus.*

Vanmarcke, E.H. 2010. *Random Fields: Analysis and Synthesis*, World Scientific Publishing, 350 p.

Session 16 - Numerical analysis of underground excavations

The Evolution of Geotech - 25 Years of Innovation – Hammah et al. (Eds)

Options for designing a hold-retain strategy for excavations in stratified rock

R.W. Seedsman
Byrnes Geotechnial Pty Ltd

ABSTRACT: Site-based geotechnical engineers have a limited number of numerical tools to apply to the design of ground support in stratified rock. A simplification of the discontinuous rock mass to be an equivalent transversely-isotropic continuum is necessary. The consequent inability to invoke plasticity is not a limitation as near the excavation boundary the rock mass is more likely to be behaving in a brittle manner. Compared to assuming isotropy, the transversely isotropic approach is less dependent on strength reduction factors and provides a better prediction of stress redistribution after the excavation is formed. Improvements in the understanding of shear stiffness of bedding and foliations are required.

1 INTRODUCTION

In excavations in bedded or laminated rock masses where the stresses are high relative to the strength of the rock substance there is the possibility of large falls of ground. Where the muck pile has been subsequently removed the shape of the fall activity is typically found to be parabolic and can extend to a height equal to or more than the excavation span (Figure 1). Such falls of ground are unacceptable, and a support strategy needs to be developed that prevents the collapse. For these high falls, the stress/strength relationship is such that reinforcement is not appropriate and a "hold and retain" or suspension strategy is required.

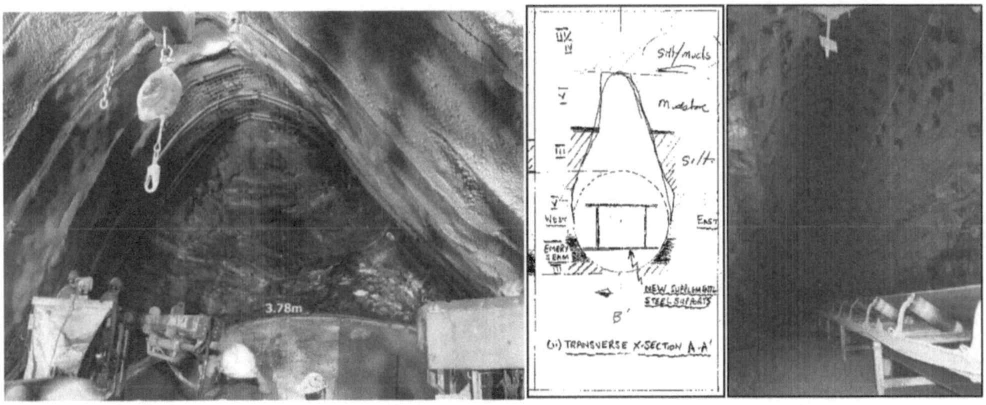

Figure 1. Typical shapes of fall cavities in sedimentary rocks (courtesy of M.A. Perras; centre – Seedsman, 2009; right – Payne, 2008).

DOI: 10.1201/9781003188339-71

A hold and retain design requires a prediction on the size and shape of the potential fall mass so that installed load capacity can be calculated. Since the support is installed prior to the development of the instability, the design also needs to consider the amount of dilation within the fall volume. In the last 2 decades there have been major changes in the understanding of failure near excavation boundaries in massive rocks leading to the recognition of the need to consider brittle failure. The resulting S shaped failure criterion has been applied successfully to rock masses when combined with the assumption of isotropy (Kaiser, 2019); this approach has not been as useful when applied to bedded rock masses (Perras et al, 2014; Seedsman, 2018a). One explanation lies in the assumption of isotropy when calculating the stresses around excavations in what may be better considered to be transversely isotropic.

There are several design tools available for the specification of a hold and retain strategy including empirical relationships, stress analyses ranging from simple boundary elements to finite elements in 2D or 3D, to the recent synthetic rock mass tools (Figure 2). Most mines will have good knowledge of the compressive and shear strength of the rock and some appreciation of the orientation and spacing the discontinuities. There would be lesser knowledge of the shear strength and particularly the stiffness of the discontinuities. Lambe (1973) suggested that the accuracy of a prediction in geomechanics depends on a balance between the data available and the method of analysis. This insight can also be applied to the confidence in ground support design – workplace safety and possible over-support. For bedded rock masses, this author considers that the current engineering geological knowledge does not justify analyses beyond 2D boundary or finite element elastic continuum methods.

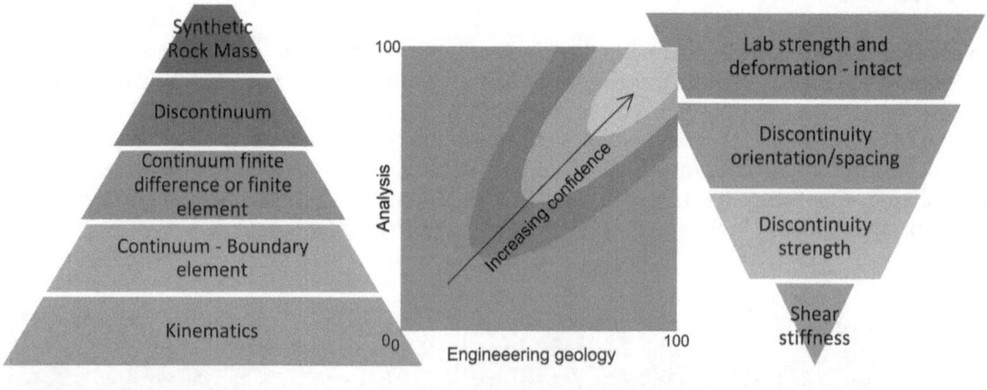

Figure 2. Hierarchies of accessibility of analytical tools and engineering geology knowledge on a mine site with an assessment of accuracy of prediction (after Lambe, 1973).

2 AN EQUIVALENT CONTINUUM FOR BEDDED OR LAYERED ROCK MASSES

2.1 *Engineering geology*

For convenience the rest of this paper will reference sedimentary rock, specifically clastic rocks, but the principles are considered to be transferable to foliated metamorphic rocks. The layers within a sedimentary rock mass have Class B anisotropy on a laboratory scale (Barla, 1974) but the features of interest for the rock mass are the discontinuities or partings. Bedding partings have high persistence (> 20 m) and are typically smooth and planar. The rock mass may be heterogeneous with layers of different strength but the partings mean that the rock mass is not isotropic. If the rock mass is to be modelled as a continuum, it will need to be transversely isotropic.

Of particular interest in this paper is the spacing of bedding partings. Bedding spacing is not directly relatable to grain size. Core segments from triple tube drilling programs are typically in lengths greater than 100 mm so the RQD of sedimentary rocks is often reported as 100. The Geological Strength Index is typically in the order of 35 (Hoek at al, 2005).

2.2 *Transverse isotropy*

Characterizing a transversely isotropic material requires 2 Youngs Modulus (E) values, 2 Poisson's Ratio (v) values, and a shear modulus (G). The modulus and Poisson's Ratio values can come from laboratory tests, and in the context of stratified rocks not much error if the values are taken as equal. If the numerical analyses are confined to stresses, the magnitudes of E and G are irrelevant, but the E/G ratio is critical.

By back analysis the author has found that E/G ratios of between 30 and 100 apply for bedded coal and thinly bedded sediments. However, if transverse isotropy is to be useful as a design tool there is a need to be able to relate this to engineering geology. The shear modulus of a stratified mass can be estimated as: $1/G= 2*(1+v)/E + 1/(Ks*S)$, where Ks is the shear stiffness of the joints (partings) and S is the spacing (Brady and Brown, 1985). Spacing can be derived from the engineering geology; but joint stiffness is not an easily measured or well-known parameter (Rocscience, 2021). A shear modulus can be obtained from a shear box test although the data is limited to relatively low normal loads (Bastola and Chugh, 2015; Bertuzzi, 2016). For excavation faces parallel to the parting plane, the normal stresses are low and the laboratory data may be applicable. Figure 3 shows how the shear modulus varies with parting spacing. The dependency of Ks on normal stress means that it is not allowable to apply the same values to other excavation faces where the geometry results in higher normal stresses.

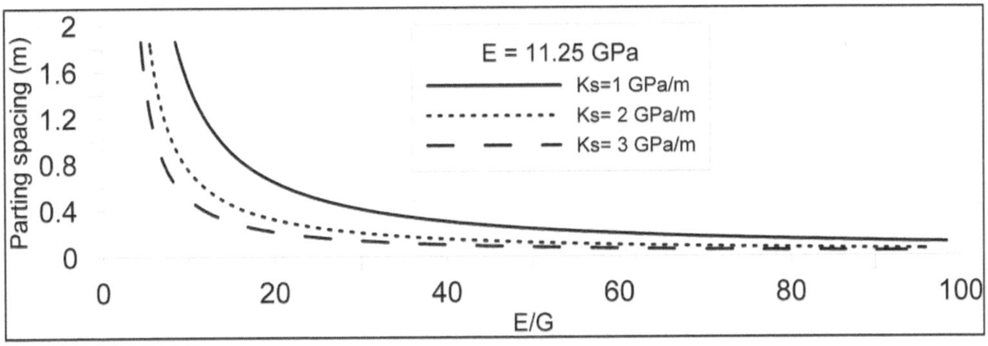

Figure 3. The relationship between parting spacing and the E/G ratio for 3 different shear stiffnesses.

2.3 *Corroboration*

In underground coal mines multiple roadways are often driven parallel to each other and separated by less than 50 m. It is well known that the ground conditions in the leading roadway are often markedly poorer than subsequent roadways – the first-driven roadway provides a stress shadow (Gale and Mathews, 1992). Seedsman (2018b) showed that that the lateral extent of the stress reductions (often more than 10 times the roadway height) cannot be explained if the rock mass is assumed to be isotropic but can be readily explained by a transversely isotropic continuum (Figure 4).

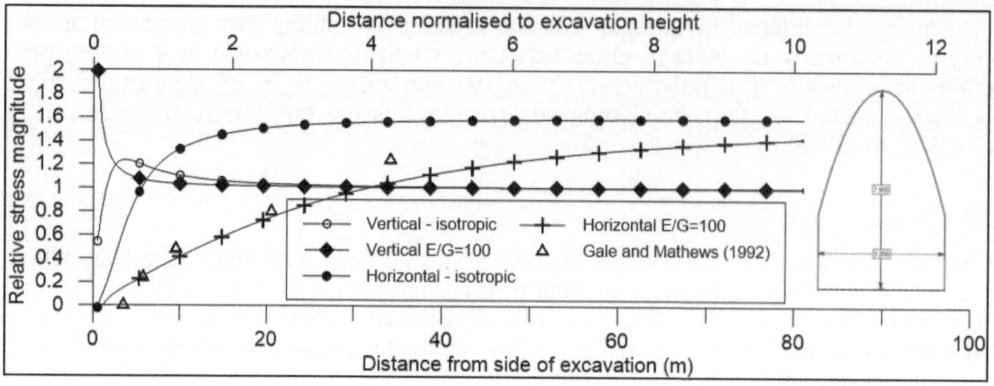

Figure 4. Transverse isotropy better models stress shadows than an assumption of isotropy.

3 FAILURE

Available failure criteria are Mohr Coulomb, Generalized Hoek Brown, and the S shaped brittle criterion (Kaiser, 2019); the latter incorporating a damage initiation of 1/3 of the uniaxial compressive strength (UCS) and a spalling limit of 7.5 for mudstone (Seedsman, 2018a). The choice of a suitable criterion should be based on how well the failure zone matches observations of a V-shaped notch geometry for both circular and rectangular excavations (Figure 1). Manipulating parameters will achieve same height of failure regardless of the criterion, but the notch geometry provides additional constraints. Adopting isotropy and homogeneity, circular excavations produced a clover leaf shape and rectangular excavations produced a bullet shape. Transverse isotropy resulted in the required notch geometry. In the following, the impact of the different failure criteria for a mudstone with closely spaced bedding (Table 1) is examined.

Adopting a spalling limit of 7.5, the failure height/excavation span ratio is 0.37 for a circle and 1.0 for a 2:1 rectangle (Figure 5). For circles, other failure criteria do not give the same notch shape:

- Using only the damage initiation component, the UCS needs to be reduced slightly to 48.7 MPa for circles and 40 MPa for rectangles.
- A similar shape as for damage initiation is produced with a cohesion of 4 MPa which implies a UCS of 12.6 MPa for circles; for rectangles the cohesion needs to be reduced to 3.32 MPa implying a UCS of 10.4 MPa.
- A V shaped failure zone closer to the base case is produced with a GSI of 62 for circles or 58 for rectangles. This is a misuse of the Generalized Hoek Brown which requires isotropy (Hoek and Brown, 1997); accepting this limitation the GSI of "seamy" rock masses with persistence of bedding should be in the order of 35-40 (Hoek et al, 2005).

Table 1. Engineering geology parameters – mudstone with closely spaced bedding partings.

UCS (MPa)	50	Poisson ratio	0.25
Triaxial and bedding friction angle (°)	25	Shear modulus (G) – transverse isotropy (MPa)	225
Triaxial cohesion (MPa)	13	E/G – transverse isotropy	50
Hoek Brown mb parameter	7	Depth (m)	300
Spalling limit	7.5	Horizontal/vertical stress ratio (k)	2.0
Modulus – E (GPa)	11.25		

Based on Figure 5, combined with transverse isotropy with the spalling limit strength criterion best reproduces the notch shape for both circular and rectangular excavations. Except for

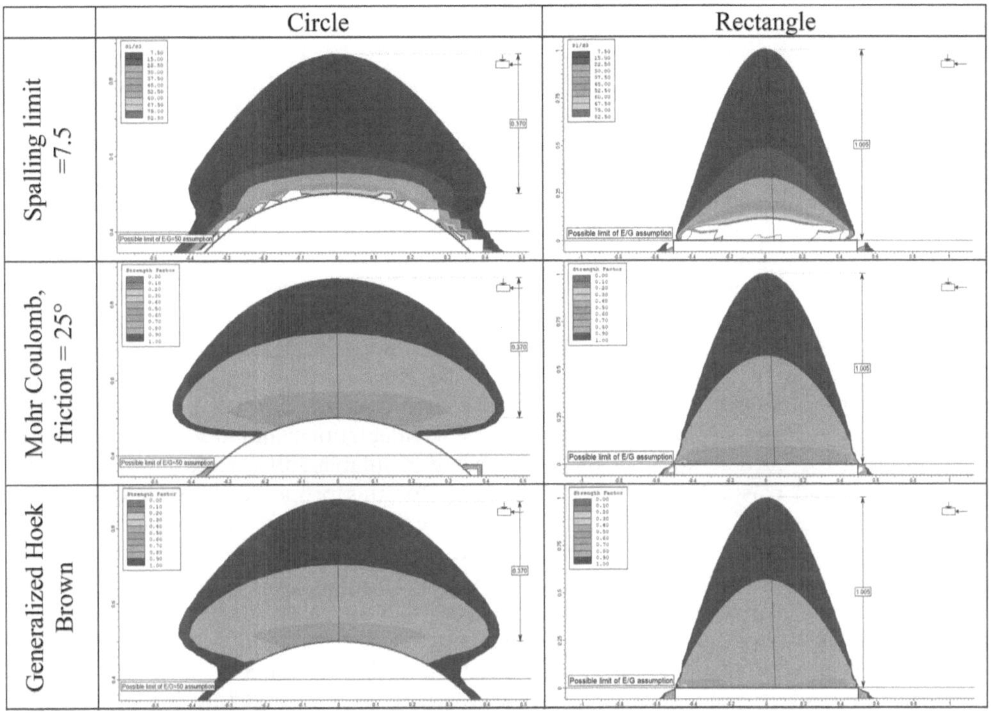

Figure 5. Failure shapes assuming transverse isotropy.

coal (Buzzi et al, 2013), measurement of the spalling limit parameter in the laboratory has not been reported. Values of 7.5 for mudstone and 10 for sandstones were obtained by back analysis (Seedsman 2018a) and it is noted that these are within the range recommended for hard rock by Diederichs (2007). The damage initiation component of the S shaped criterion is a similar concept to crack initiation which can be derived from laboratory tests (Nicksiar and Martin, 2013). The use of the S-shaped criterion also has advantages over the other criteria as it does not require strength reduction factors that cannot be related to either laboratory tests or geotechnical logging. The S shaped criterion has an additional advantage in that simple design charts can be prepared (Figure 6).

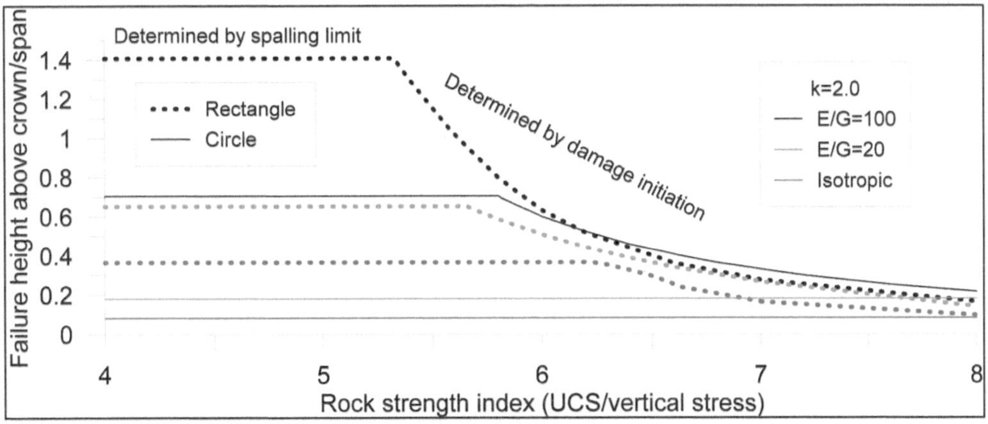

Figure 6. Design charts based on S shaped criterion for three E/G ratios (k=2.0).

4 HOLD AND RETAIN DESIGN

For the design of ground support the collapse zone can be obtained from a simple elastic failure analysis adopting transverse isotropy and brittle parameters. The failure zone is then considered to be free to collapse under gravity unless restrained with straps/mesh/membrane elements and held in place with tendons or an arch/cross member. The load on the holding elements is the dead-weight of the failure volume. The shape of the failure zones is required for the deadweight calculation and also the location of anchorages if suspension by tendons is adopted.

Ground support is installed prior to the failure developing – either because it is installed close to the working face where there is constraint from the face or, in a mining situation, in anticipation of a stress increase. The failing rock mass will impose deformations on the holding elements which need to be considered in the design. Given the assumption of brittle behavior, the stresses and deformations produced in either elastic or plastic analyses are artefacts of the numerical calculations and not real. It follows from this that the current inability of commercially available software to combine transverse isotropy and plasticity is not a disadvantage for this proposed design method. For isotropic rock masses Kaiser (2016) advocates the use of a linear bulking factor to estimate the expansion after failure. A default value of 3% is often used for initial excavations in massive hard rock. Seedsman (2019) proposed similar values for bedded rocks in coal mines. It is noted that these levels exceed the yield strain of steel so the design may need to consider ways of adsorbing the deformations by longer holding elements. Designs utilizing fully grouted tendons may need to consider slip elements so that the induced strain on the tendons is not localized. However, a corollary of invoking brittle failure is that the stresses normal to the grouted tendons that were present on installation are reduced to zero: hence the borehole will dilate and bond strength decrease (Hutchinson and Diederichs, 1996). It may be that any advantages of full column grouted tendons are negated once brittle failure is initiated; fully grouted systems still need a plate.

5 CONCLUSIONS

Combining transverse isotropy with brittle failure provides a failure criterion (TIB) which can be derived from engineering geology knowledge without the use of strength reduction factors. The input parameters are discontinuity spacing, shear stiffness, and the UCS of the rock substance. The spalling limit ranges between 7.5 and 10 for mudstones and sandstones as derived from back-analysis of high roof falls. Transverse isotropy results in higher predicted stresses than isotropy and this results in the ability to use intact strengths as opposed to strength reduction factors. For homogeneous rock masses, the TIB criterion can be implemented in a boundary element program: the maximum failure height for design is the one given by the spalling limit and for higher strength/stress ratios a lesser height is obtained from the damage initiation calculation.

Further development of the TIB criterion requires more research into shear stiffness. The empirical model of Bandis et al (1983) needs to be further developed and extended to higher normal stress so that there is greater confidence in the calculation of the E/G ratio from the spacing of discontinuities. A better understanding of how the spalling limit can be measured in the laboratory is also required.

REFERENCES

Bandis, S.C., Lumsden, A.C. and Barton, N.R. 1983. Fundamentals of Rock Joint Deformation. International Journal of Rock Mechanics and Mining Sciences, 20, 249–268.

Barla G 1974 Rock Anisotropy – Theory and Laboratory Testing. Chapter 8. Rock Mechanics Springer-Verlag Ed L Muller 131–169.

Bastola, S and Chugh, Y 2015 Shear strength and stiffness of bedding planes and discontinuities in the immediate roof rocks overlying the No 6 seam in Illinois. Proceedings 13th International Congress of Rock Mechanics.

Bertuzzi R. 2016. Strength and stiffness properties of defects within the Hawkesbury Sandstone and Ashfield Shale. Australian Geomechanics, 51(3): 93–101.

Brady, B.H.G. and Brown, E.T. 1985. Rock mechanics for underground mining. George Allen & Unwin, London.

Buzzi, O, Sieffert Y, Mendes J, Liu X, Giacomini A, Seedsman R. 2013 Strength of an Australian Coal Under Low Confinement. Rock Mech Rock Eng 2013. doi:10.1007/s00603-013-0493-5.

Diederichs, M. The 2003 Canadian geotechnical colloquium: Mechanistic interpretation and practical application of damage and spalling prediction criteria for deep tunnelling. Can Geotech J 2007. doi:10.1139/T07-033.

Gale, W.J. and Matthews, S.M. 1992. Stress control methods for optimised development and extraction operations. Report to National Energy Research Development, and Demonstration (NERD&D) Program, Project 1301.

Hoek, E. and Brown, E.T. 1997. Practical estimates or rock mass strength. Intl. J. Rock Mech. & Mining Sci. & Geomechanics Abstracts. 34(8),1165–1186.

Hoek E, Marinos P.G., Marinos V.P. Characterisation and engineering properties of tectonically undisturbed but lithologically varied sedimentary rock masses. Int J Rock Mech Min Sci 2005. DOI:10.1016/j.ijrmms.2004.09.015.

Hutchinson DJ, Diederichs MS. Cablebolting in Underground mines. BiTech Publishers, 1996.

Kaiser PK. 2016. Ground Support for Constructability of Deep Underground Excavations – Challenges of managing highly stressed ground in civil and mining projects. Sir Muir Wood lecture of International Tunneling Association at World Tunneling Congress, San Francisco, 33p.

Lambe TW. Predictions in soil engineering. Geotechnique 1973. doi:10.1680/geot.1973.23.2.151.

Nicksiar, M. and Martin, C.D. 2013. Crack initiation stress in low porosity crystalline and sedimentary rocks. Engineering Geology, 154, 64–76.

Payne, DA, 2008, Crinum Mine, 15 longwalls 40 million tonnes 45 roof falls - What did we learn? Proc. Coal 2008: Coal Operators' Conference, Wollongong. pp 22–43.

Pells, P.J.N. 1980. Geometric Design in Underground Openings for High Horizontal Stress Fields. 3rd ANZ geomechanics Conference, 2, 183–188.

Perras, M.A., Diederichs, M.S. and Besaw. 2014. Geological and geotechnical observations from the Niagara Tunnel project. Bull Eng Geol Environment, 73:1303–1323. doi 10.1007/s10064-014-0633-5

Rocscience, 2021. Estimating Joint Stiffness. https://www.rocscience.com/help/rs2/theory/estimating_joint_stiffness.htm?agt=index Accessed 28 January 2021

Seedsman, R.W. 2009. An update of conditions in the Donkin-Morien tunnels. Proceedings from the 3rd Canada-US Rock Mechanics Symposium, Toronto, Ontario, Canada.

Seedsman, R.W. 2018a The spalling limit in transversely isotropic materials, the prediction of the maximum height of failure, and the design of long-tendon roof support systems. Mining Technology 127:2, 65–74, DOI: 10.1080/14749009.2017.1325552.

Seedsman, R.W. 2018b. The redistribution of stresses around longwall extraction panels in bedded rock masses. https://ro.uow.edu.au/coal/691.

Seedsman, RW. 2019 Interpreting roof extensometry in coal mine roofs. Mining Technology, 128:3, 143–151, DOI: 10.1080/25726668.2019.1583458

The Evolution of Geotech - 25 Years of Innovation – Hammah et al. (Eds)

The fast multipole method for the computation of large-scale three-dimensional elastostatics boundary-element problems in underground excavations

H. Wahanik, S. Moallemi, J. Curran, T. Yacoub & B. Corkum
Rocscience Inc., Toronto, Ontario, Canada

ABSTRACT: We present the theoretical and computational implementation of an original version of the Fast Multipole Method (FMM) for solving large-scale problems in 3D elastostatics based on the indirect boundary integral fictitious formulation. The conventional boundary element method with N collocation nodes, translates into solving a dense and non-symmetric system of equations, with computational complexity $O(N^2)$. Additionally, for large models, storing the system's matrix in random-access memory (RAM) is intractable, requiring storing sections of the matrix in hard disk. This creates an important burden towards computational cost. The FMM is inspired on the intuitive idea of collecting influences (e.g., gravitational, electromagnetic, acoustic, elastic) from close sources into a single source, using approximations with the flavor of a Taylor expansion. We illustrate the power of the method applied in computing 3D elastostatics models for real-world excavations, running within Rocscience modelling software, where important speed-ups were observed for a wide range of models ranging from 100K up to 1M elements.

1 INTRODUCTION

Boundary value problems for the Laplace equation, Helmholtz equation, or 3D elastostatics problems, are conveniently expressed through systems of boundary integral equations defined on bounded regions in 2D or 3D space (see Cheng & Cheng 2005). A discretization of the boundary integral equations using line segments or triangular elements and a set of discrete collocation nodes, induces a linear system of algebraic equations represented by a dense and non-symmetric matrix. The preferred solver for such system is the iterative GMRES algorithm (Saad & Schultz 1986). In this algorithm, the main contributors to the computing cost are the calculation and storage of the matrix, and the evaluation of a matrix-vector product, both of complexity $O(N^2)$ (where N is the number of collocation nodes). Evidently for a large triangulated meshed boundary storing the system's matrix in RAM is intractable.

In the 1980's, Rokhlin and Greengard (Greengard 1988, Greengard & Rokhlin 1987, Rokhlin 1985) introduced the Fast Multipole Method (FMM), a powerful algorithm designed to reduce the CPU time in the solution of the aforementioned discrete algebraic system; the FMM only uses a fraction of RAM memory, and the computational complexity and storage requirements are reduced from $O(N^2)$ to $O(N)$. Since then, the FMM has since been widely adapted to accelerate the computation of many boundary integral problems in areas such as acoustic wave propagation, Stokes flow, and elastostatics (Liu 2009, Liu 2019) and many other problems where a summation kernel may be present. Additionally, it has been deployed in many commercial applications (e.g., Schoemann et al. 2020).

DOI: 10.1201/9781003188339-72

In our work, we use the FMM to solve three crucial parts of the 3D elastostatics problem: find the unknown fictitious stress intensities, recover the surface displacements across the surface boundary, and evaluate the stresses and displacements on the field points. We review the theory and implementation, and present the results of the numerical experiments, showcasing the massive power of the method when deployed on practical models of underground excavations.

2 THE BOUNDARY VALUE PROBLEM FOR 3D ELASTOSTATICS

Let u_{ij}, ε_{ij}, and σ_{ij}, denote the displacement, strain, and stress in a linearly elastic domain Ω, with boundary $\partial\Omega$, and outward normal with components n_i. The governing equations for the *equilibrium*, *strain-displacement*, and *stress-strain relationship* are given by:

$$\sigma_{ij,j} + f_i = 0, \tag{1}$$

$$\varepsilon_{ij} = \frac{1}{2}\left(u_{i,j} + u_{j,i} \right), \tag{2}$$

$$\sigma_{ij} = E_{ijkl}\, \varepsilon_{kl}, \tag{3}$$

where f_i is the body force and E_{ijkl} is the elastic modulus tensor given by

$$E_{ijkl} = \lambda\, \delta_{ij}\, \delta_{kl} + \mu \left(\delta_{ik}\, \delta_{jl} + \delta_{il}\, \delta_{jk} \right), \tag{4}$$

where λ and μ are the Lamé constants, which are functions of the Young modulus E and Poisson ratio ν (see Yoshida 2001, Liu 2009). In the equations above we use Einstein's summation notation, and the notation $()_{,k} = \partial/\partial k$. The boundary conditions are given by setting the displacements and the traction t_i (where $t_i = \sigma_{ij}n_j$) on the boundary $\partial\Omega$.

For solving the initial value problem described here, we use the indirect fictitious integral formulation, which is based on the integrals of the Green's kernels for displacements and traction. It states that for $\boldsymbol{x} \in \partial\Omega$, the displacements and stresses at $\boldsymbol{x}$ are due to the (unknown or fictitious) surface tractions $\phi_j(\cdot)$, and are given as:

$$u_i(\boldsymbol{x}) = \int_{\partial\Omega} U_{ij}(\boldsymbol{x},\boldsymbol{y})\phi_j(\boldsymbol{y})dS, \tag{5}$$

$$\sigma_{ij}(\boldsymbol{x}) = \int_{\partial\Omega} T_{jki}(\boldsymbol{x},\boldsymbol{y})\phi_j(\boldsymbol{y})dS, \tag{6}$$

where U_{ij} and T_{ijk} are the Green's functions or so-called fundamental solutions. The surface traction is recovered through $t_i(x) = \sigma_{ik}(x)n_k(x)$. The numerical discretization of (5) and (6) at a collocation node $\boldsymbol{x}^p$ and across a triangular element mesh gives:

$$u_i(\boldsymbol{x}^p) = \sum\nolimits_e \int_e U_{ij}(\boldsymbol{x}^p,\boldsymbol{y})dS\, \phi_j^e, \tag{7}$$

$$t_i(\boldsymbol{x}^p) = -\frac{1}{2}\,\delta_{ik}\,\phi_k(\boldsymbol{x}^p) + \sum\nolimits_e \int_e T_{jki}(\boldsymbol{x}^p,\boldsymbol{y})dS\, \phi_j^e\, n_k, \tag{8}$$

where the term $-\frac{1}{2}\,\delta_{ik}\,\phi_k$ is the residual term appearing due to a limit point approaching the non-excavated zone towards the boundary (Banerjee & Butterfield 1981). Equation (8) results in a linear system $Ax = b$, where the term b is given by the node-wise tractions originating from the surface stress field. Once a solution is found, the displacements are recovered using

Eq. (7). Extensive research has been carried to estimate numerically near-singular integrals as those appearing in Eqs. (7)-(8) (Vijayakumar & Cormack 1985, Vijayakumar & Curran 2007). The so-called node-centric approach (Vijayakumar et al. 2000) was pioneered to improve the accuracy and performance of the problem, and it is the flagship method currently used in the computation of 3D elastostatics deployed in Rocscience's EX3 software.

Evaluation and storage of the matrix A induced by (8) has complexity $O(N^2)$. For large models (above 100K surface elements for example) it becomes too large for temporary storage in RAM; alternatively, it must be stored in the hard disk; furthermore, the GMRES solver's complexity is $O(N^2)$. The FMM provides a solution to these issues. In the FMM, the matrix-vector multiplications required at each iteration of GMRES, is calculated through the direct evaluation of (8). Indeed, each single equation in the system corresponds to a sum of integrals ranging over the surface's elements, for every fixed collocation node $\boldsymbol{x}^p$. If the collocation node is close to a surface element $\boldsymbol{e}$, the integral is evaluated directly. On the contrary, we use the FMM's M2M expansion and the *pole* $\boldsymbol{y}_c$ to estimate such integral. This enables a far-field approximation: all element-to-element interactions are replaced by cell-node interactions, with cells belonging to a spatial hierarchy provided by an octree hierarchical data structure; the interactions are estimated through the M2M, M2L, and L2L expansions (Rayar 2005). Let us focus in the FMM algorithm in more detail.

3 THE FMM ALGORITHM

The Fast Multipole Method was introduced as a mean to accelerate sums of the type:

$$S(\boldsymbol{x}_i) = \sum_{j=0}^{N} K\left(\boldsymbol{x}_i, \boldsymbol{y}_j\right) m_j, \tag{9}$$

where $i \in \{1, \ldots, N\}$, by using a low-rank representation of the summation kernel K. This algorithm was chosen as one of top ten algorithms of the 20th century (Cipra 2000). An example where such a sum may appear, is in the estimation of the matrix-vector multiplications required by the iterative GMRES algorithm (for the solution of $Ax = b$). The FMM enables the evaluation of such sum without ever storing or calculating the matrix of the system explicitly. The basic tool on which the FMM relies on, is the expansion of the summation kernel K across a set of poles, similar to a Taylor expansion. Consider the ansatz of the kernel:

$$K(\boldsymbol{x}, \mathbf{y}) = \sum_i K_i^x(\boldsymbol{x}, \boldsymbol{y}_c)\, K_i^y(\boldsymbol{y}, \boldsymbol{y}_c), \tag{10}$$

with expansion pole $\boldsymbol{y}_c$ close to y, satisfying $|\boldsymbol{y} - \boldsymbol{y}_c| \ll |\boldsymbol{x} - \boldsymbol{y}|$. In the case of boundary-integral problems such as (7)-(8), the summation kernels are the integrals of the fundamental solutions (or so-called Green's functions) which we denote here as G. If such infinite expansion (10) exists for G, then the boundary integral equations (5) and (6) can be written in the form:

$$\int_e G(\boldsymbol{x}, \boldsymbol{y})\phi(\boldsymbol{y})\, dS = \sum_i G_i^x(\boldsymbol{x}, \boldsymbol{y}_c) \int_e G_i^y(\boldsymbol{y}, \boldsymbol{y}_c)\phi(\boldsymbol{y})\, dS, \tag{11}$$

or simply using the compact form:

$$\int_e G(\boldsymbol{x}, \boldsymbol{y})\phi(\boldsymbol{y})\, dS = \sum_i G_i^x(\boldsymbol{x}, \boldsymbol{y}_c)\, M_i(\boldsymbol{y}_c). \tag{12}$$

$M_i(\boldsymbol{y}_c)$ is the so-called *moment* at $\boldsymbol{y}_c$. Moreover, if the expansion pole $\boldsymbol{y}_c$ is shifted towards a new location $\boldsymbol{y}_{c'}$, we write a second ansatz:

$$M_i(y_{c'}) = \sum_{p=0}^{k} I_p(y_c, y_{c'})\, M_p(y_c). \tag{13}$$

If the expression (13) exists, then the expansion (12) using the pole $y_{c'}$ can be written in terms of the new pole y_c. Equation (13) is the so-called moment-to-moment expansion (M2M). In the moment-to-local expansion (M2L), we consider an expansion pole x_L close to x, i.e., satisfying $|x - x_L| \ll |x_L - y_c|$. We write a third ansatz,

$$G_i\left(x, y_c\right) = \sum_l I_l^x(x, x_L)\, O_l^y(x_L, y_c). \tag{14}$$

Upon substitution of the later equation in equation (12) and rearrangement of the sum we obtain:

$$\int_e G(x, y)\phi(y)\, dS = \sum_l I_l^x(x, x_L)\, L_l(x_L), \tag{15}$$

where:

$$L_l(x_L) = \sum_i O_l^y(x_L, y_c)\, M_i(y_c). \tag{16}$$

Equation (15) is called the *local expansion* around x_L; here the boundary integral equation is expressed in terms of the poles x_L and y_c. As expected, we write a fourth ansatz for the case when the local expansion pole x_L is shifted towards a new location $x_{L'}$:

$$L_l(x_{L'}) = \sum_{m=l}^{p} J_m(x_L, x_{L'})\, L_m(x_L), \tag{17}$$

which is the so-called local-to-local translation, L2L. If such set of expansions is possible, the implementation of the FMM algorithm is possible for the summations of type (8). In the case of the 3D elastostatics problem, the moments expansion (as Eq. (12) above) is given by:

$$\begin{aligned}\int_e T_{jki}(x,y)\phi_j(y)dS &= \tfrac{1}{8\pi\mu}\sum_{n=0}^{\infty}\sum_{m=-n}^{n}\left[F_{j,s,n,m}(x - y_c)\overline{M_{s,n,m,k,l,i}(y_c)}\right.\\ &\qquad + G_{j,n,m}(x - y_c)\overline{M_{s,n,m,k,l,i}(y_c)},\end{aligned} \tag{18}$$

where:

$$M_{s,n,m,k,l,i}(y_c) = E_{spki}\int_e \frac{\partial}{\partial y_p} R_{n,m}(y - y_c)\phi_j(y)\, dS\,, \tag{19}$$

$$M_{n,m,k,l,i}(y_c) = E_{spki}\int_e \frac{\partial}{\partial y_p}\left((y - y_c)_s\, R_{n,m}(y - y_c)\right)\phi_j(y)\, dS\,, \tag{20}$$

and the over-bar represents complex conjugation. Here $R_{n,m}$ is the solid complex-harmonic function (in complex-variable space) composed of Legendre polynomials (see Yoshida et al. 2001, Liu 2009). Note that we only consider the real-part of the expressions (18)-(20).

Moreover, when the expansion pole y_c is shifted towards $y_{c'}$, the moments (19)-(20) are provided by the M2M translation, and are given by:

$$M_{s,n,m,k,l,i}(y_{c'}) = \sum_{n'=0}^{n}\sum_{m'=-n'}^{n'} R_{n',m'}\left(y_c' - y_c\right) M_{s,\, n-n',\, m-m',k,l,i}(y_c)\,, \tag{21}$$

$$M_{n,m,k,l,i}(\boldsymbol{y}_{c'}) = \sum_{n'=0}^{n}\sum_{m'=-n'}^{n'} R_{n',\,m'}\left(\boldsymbol{y}_c' - \boldsymbol{y}_c\right)[M_{n-n',m-m',k,l,i}(\boldsymbol{y}_c) - (\boldsymbol{y}_{c'} - \boldsymbol{y}_c)_s\, M_{s,\,n-n',m-m',k,l,i}(\boldsymbol{y}_c)\,]. \tag{22}$$

The M2L expansion is given by:

$$L_{s,n,m,k,l,i}(\boldsymbol{x}_L) = (-1)^n \sum_{n'=0}^{y}\sum_{m'=-n'}^{n'} \overline{S_{n+n',\,m+m'}}(\boldsymbol{x}_L - \boldsymbol{y}_c) M_{s,\,n',m',k,l,i}(\boldsymbol{y}_c)\,, \tag{23}$$

$$L_{n,m,k,l,i}(\boldsymbol{x}_L) = (-1)^n \sum_{n'=0}^{y}\sum_{m'=-n'}^{n'} \overline{S_{n+n',\,m+m'}}(\boldsymbol{x}_L - \boldsymbol{y}_c)[M_{n',m',k,l,i}(\boldsymbol{y}_c) - (\boldsymbol{x}_L - \boldsymbol{y}_c)_s\, M_{s,\,n',m',k,l,i}(\boldsymbol{y}_c)]. \tag{24}$$

When the local pole $\boldsymbol{x}_L$ is shifted towards $\boldsymbol{x}_{L'}$, the L2L expansion is given by:

$$L_{s,n,m,k,l,i}(\boldsymbol{x}_{L'}) = \sum_{n'=n}^{\infty}\sum_{m'=-n'}^{n'} R_{n'-n,\,m'-m}(\boldsymbol{x}_{L'} - \boldsymbol{x}_L) L_{s,\,n',m',k,l,i}(\boldsymbol{x}_L)\,, \tag{25}$$

$$L_{n,m,k,l,i}(\boldsymbol{x}_{L'}) = \sum_{n'=n}^{\infty}\sum_{m'=-n'}^{n'} R_{n'-n,\,m'-m}(\boldsymbol{x}_{L'} - \boldsymbol{x}_L)\,[L_{n',m',k,l,i}(\boldsymbol{x}_L) - (\boldsymbol{x}_{L'} - \boldsymbol{x}_L)_s\, L_{s,\,n',m',k,l,i}(\boldsymbol{x}_L)]. \tag{26}$$

The local expansion (such as Eq. (15)) of the 3D elastostatics boundary integral equation around a pole $\boldsymbol{x}_L$ is given by:

$$\int_e T_{jki}(\boldsymbol{x},\boldsymbol{y})\phi_j(\boldsymbol{y})dS = \frac{1}{8\pi\mu}\sum_{n=0}^{\infty}\sum_{m=-n}^{n} [F^R_{j,s,n,m}(\boldsymbol{x} - \boldsymbol{y}_c) L_{s,n,m,k,l,i}(\boldsymbol{x}_L) + G^R_{j,n,m}(\boldsymbol{x} - \boldsymbol{y}_c) L_{n,m,k,l,i}(\boldsymbol{x}_L) \tag{27}$$

We have originally derived these expressions based on the derivation from Liu 2009. Let us expand briefly on the algorithm.

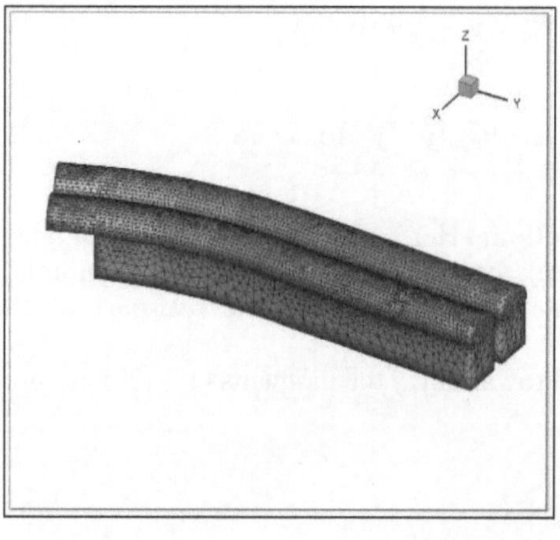

Figure 1. A meshed boundary of an underground excavation

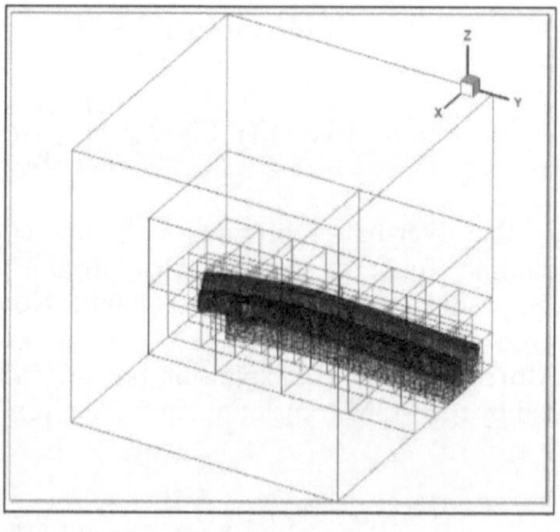

Figure 2. Octree hierarchical discretization

4 FMM ALGORITHM STEPS

4.1 *Spatial hierarchical discretization: Octree*

We start with the mesh of a surface boundary as shown in Figure 1. In the first step of the FMM we build T, an octree hierarchical discretization of the mesh in 3D space (see Meager 1982). Such octree is built using a recursive insertion algorithm: at step 0, a cube-shaped bounding box covering the entire mesh is defined; this is level 0; it is then subdivided in 8 child cube cells (level 1); through recursion, each element is located within the cells at each level and inserted in the leaf cells (a cell is a leaf if it has no child cells). Whenever reaching a maximum number of elements in a leaf cell a new set of child cells is introduced (and added to the next level); all elements contained in the cell are moved into the corresponding child cells. A clean-up algorithm eliminates empty leaf cells or empty intermediate cells. The resulting spatial hierarchical data structure is shown in Figure 2 for a 3D model of an underground excavation.

4.2 *Upward pass: Integration of leaf moments and M2M expansion*

We iterate throughout the tree in the *upwards* direction, starting from the lowest level (leaf level) up to level 2. At the leaf level, we use numerical integration to evaluate the integral moments (e.g., $M_i(y_c)$ in (11), or $M_{s,n,m,k,l,i}(y_c)$ in (16)), where y_c is the center of the leaf cell. At each level $\boldsymbol{l}$, ($\boldsymbol{l} > 2$), we collect the effects from the child cells by translating the moments towards the center node of the parent cell using the M2M translation. These moments are stored at each level of the octree.

4.3 *Downward pass: M2L and L2L*

We iterate through the tree *downwards*, starting from level 2. For every cell $\boldsymbol{C}$ at level $\boldsymbol{l}$ we use the M2L expansion (23)-(24) to translate the moments from cells in the interaction list of $\boldsymbol{C}$; the interaction list is defined by cells non-adjacent to $\boldsymbol{C}$ but whose parent is adjacent to $\boldsymbol{C}$'s parent. In the next step, the L2L translation, the local expansion coefficients are translated into $\boldsymbol{C}$'s child's using (25)-(26).

4.4 *Integral evaluation: Direct and far-field effects*

We now estimate (8), a sum of integrals from all elements of the mesh towards x_p; by construction, the collocation point x_p belongs to a leaf C of the octree T . To estimate such sum, for a fixed point x_p, we divide the set of elements on the mesh into two sets: the adjacent elements, i.e., those elements in the same leaf or belonging to adjacent leafs (i.e., that share a vertex with C), and its complementary set. The sum can then be written as:

$$t_i(\boldsymbol{x}_p) = -\frac{1}{2}\,\delta_{ik}\,\phi_k(\boldsymbol{x}^p) + \sum\nolimits_{adjacent}\int_e T_{jki}(\boldsymbol{x}^p,\boldsymbol{y})dS\phi_j^e\,n_k + \sum\nolimits_{far}\int_e T_{jki}(\boldsymbol{x}^p,\boldsymbol{y})dS\phi_j^e\,n_k. \quad (28)$$

The integral of an element in the adjacent leafs is calculated directly as in the usual boundary element method; the sum of effects from far-away cells is estimated using the local expansion from the center of the leaf towards the collocation node. We use the numerical method developed by Vijayakumar et al. 2000 for the evaluation of singular integrals that appear during the evaluation of the direct effects. The upward pass, downward pass, and the integral evaluation steps are run in each iteration of the GMRES algorithm to replace the costly matrix-vector multiplication (of complexity $O(N^2)$), by the FMM algorithm, of complexity $O(N)$.

4.5 *Expanded-FMM: Surface displacements and field points*

The convergence of the GMRES method accelerated by the FMM, provides the unknown fictitious stresses ϕ_j at each collocation node. Once such unknown values are found, it is possible to recover the surface displacements $u_i(x^p)$ (see Equation 7). This evaluation is also possible using the FMM algorithm: the 3 main steps carried out in the solver's stage are repeated across the same octree, but instead of using the expansions for the stiffness tensor T_{ijk}, we use the expansions for the displacement (for such expressions, please refer to Liu 2009. A single iteration of the FMM recovers the desired unknown values of u_i in linear time.

Moreover, as a direct consequence of the methodology, we calculate the stress and displacement at field points f^q throughout the domain. For doing so, a new custom octree is created, including both the surface mesh and the field-points. The FMM algorithm is

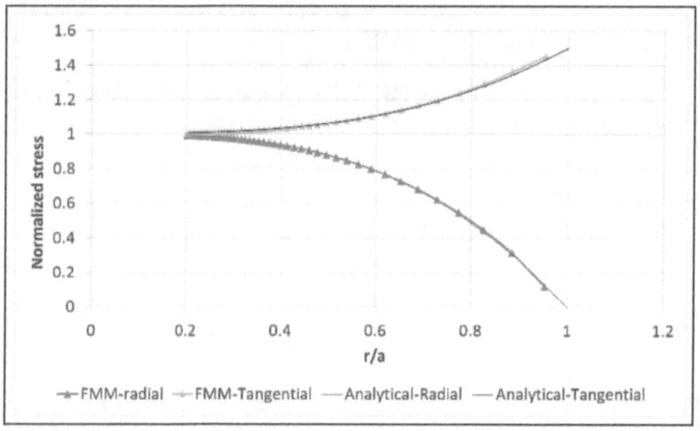

Figure 3. Normalized stress, FMM vs Analytical solution from Timoshenko (1970).

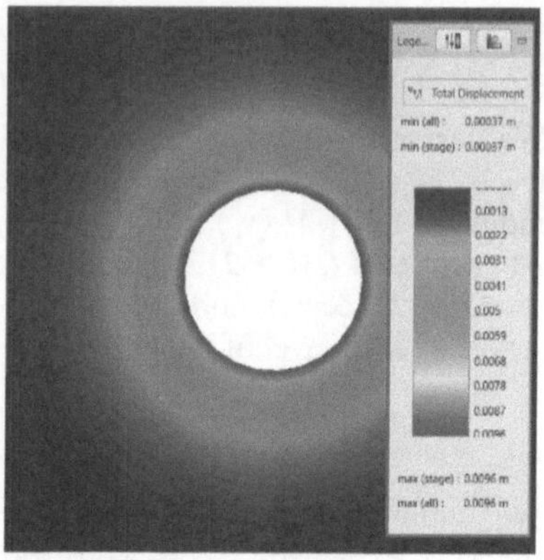

Figure 4. Total displacement outside excavated sphere.

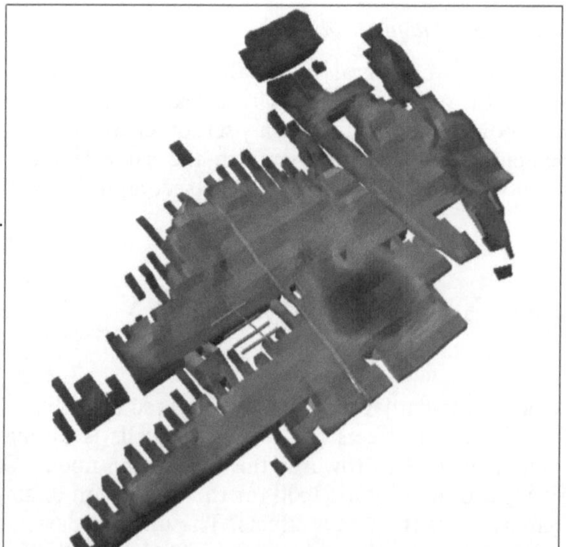

Figure 5. Displacement computed with FMM, real-life underground excavation subject to gravitational load, with unit weight of 21.4 KN/m3.

modified in the following manner: the upward pass is only enabled for branches that contain elements, while branches created exclusively for field-points are ignored. In the downward pass, exactly the opposite is done; effects are only translated downwards towards any branches containing field points. The set of field points is the new set of collocation nodes, implying that the direct integral evaluation is done over them; this includes both direct integrals for adjacent elements, and the local expansion estimating the far-field effects.

5 COMPUTATIONAL EXPERIMENTS

5.1 *Hardware*

Computations were performed in a Dell XPS laptop, with Intel Core i7 processor (2.3 GHz, 8 cores, with 16 Logical processors).

5.2 *Accuracy: Analytical expressions*

We compare the simulations with FMM using a model of an underground sphere under the hydrostatic stress field, in an infinite domain with radius = 1 m and constant field stress of 30 MPa, $E = 2000$ MPa, $v = 0.3$ (Timoshenko 1970). FMM produces accurate numerical results, as show in Figure 3 and Figure 4.

5.3 *Real-world excavations*

We tested the FMM in models of real-world excavations. In such models, the presence of small non-manifold features in the input geometry is common; for such cases, we have introduced a robust evaluation of the integrals in the presence of singularities. (see Figure 5).

5.4 *Time complexity: FMM vs classical solver*

To test the full power of the Fast Multipole Method, we generate a simple model of multiple spheres (Figure 6) and generate progressively finer meshes. The model is solved using the FMM, and the classical solver. The computational results are shown in Figure 6. The expense in computation increases linearly with the mesh size in the case of the FMM algorithm while increasing quadratically for the classical approach.

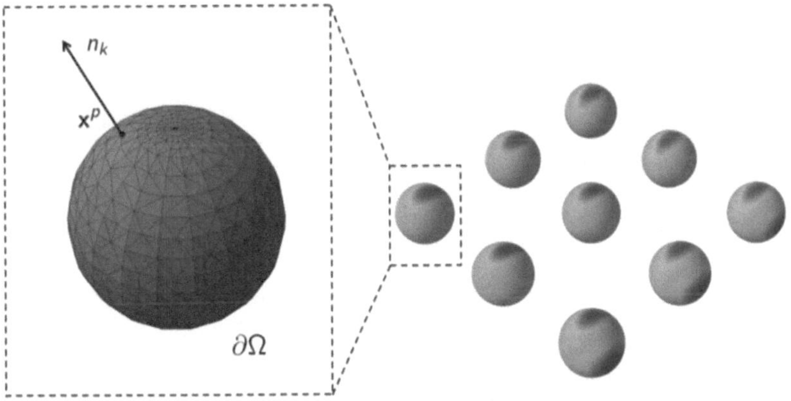

Figure 6. Spheres model with constant loading, equal to 10 MPa. The model shown here has 184K elements. Computing time: classic solver: 3.5h, FMM solver: 56.9 seconds.

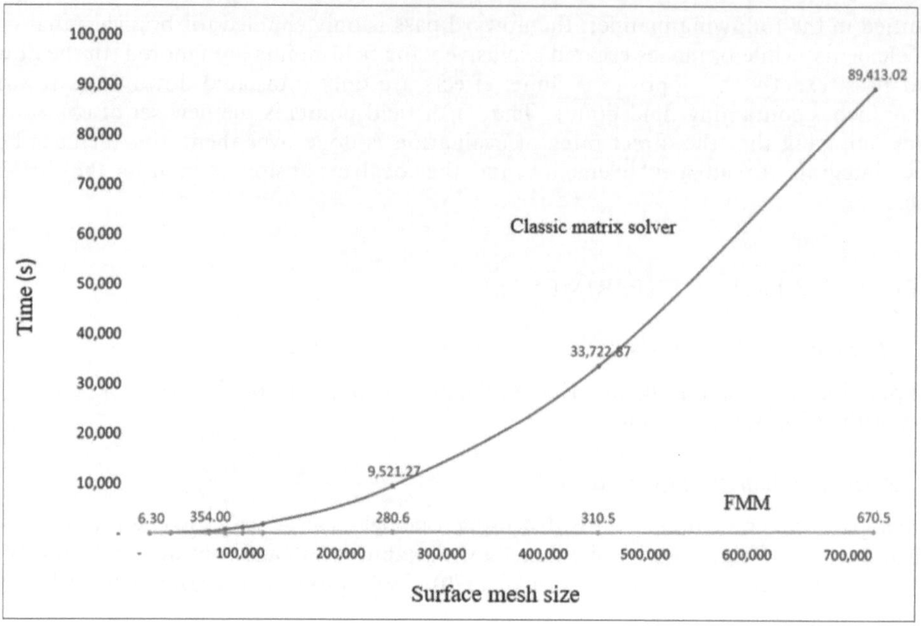

Figure 7. Computational experiments clearly show the trends of linear time complexity vs quadratic complexity in the FMM vs classical matrix solver approach.

6 CONCLUSIONS

We have deployed the FMM for single-material indirect fictitious 3D elastostatics problem, show-casing a massive performance, many times faster than the classical BEM approach. The algorithm has room for improved speed and generalization to more complex physical scenarios (multi-material non-excavated zones for instance). Moreover, initial experiments show that the FMM is multiple times faster than the FEM method in modelling 3D elastostatics variables on surface meshes above 100K elements.

ACKNOWLEDGEMENTS

We want to thank Rocscience management for their valuable support in the development of this project. Special thanks to Owen Westland for his help with parallelization of the code.

REFERENCES

Cheng, A.H.D. and Cheng, D.T., 2005. Heritage and early history of the boundary element method. *Engineering Analysis with Boundary Elements*, *29*(3), pp.268–302.
Cipra, B.A., 2000. The best of the 20th century: Editors name top 10 algorithms. *SIAM news*,*33*(4), pp.1–2.
Fong, W. and Darve, E., 2009. The black-box fast multipole method. *Journal of Comp. Physics*, *228*(23), pp.8712–8725.
Greengard, L., 1988. The rapid evaluation of potential fields in particle systems. MIT press.
Greengard, L. and Rokhlin, V., 1987. A fast algorithm for particle simulations. *Journal of computational physics*, *73*(2), pp.325–348.
Liu, Y., 2009. *Fast multipole boundary element method: theory and appl. in eng*. Cambridge univ. press.
Liu, Y., 2019. On the BEM for acoustic wave problems. *Eng. An. with Boundary Elements*, *107*, pp.53–62.
Meagher, D., 1982. Geometric modeling using octree encoding. *Computer Graphics*, *19*(2), pp.129–147.
Saad, Y. and Schultz, M.H., 1986. GMRES: A generalized minimal residual algorithm for solving nonsymmetric linear systems. *SIAM Journal on scientific and statistical computing*, *7*(3), pp.856–869.

Schoeman, M., Marchand, R., van Tonder, J., Jakobus, U., Aguilar, A., Longtin, K., Vogel, M. and Alwajeeh, T., 2020, July. New Features in Feko/WinProp 2019. In *2020 International Applied Computational Electromagnetics Society Symposium (ACES)* (pp. 1–2). IEEE.
Raykar, V.C., 2005. A short primer on the fast multipole method. *Available on the author's website, 1.*
Rokhlin, V., 1985. Rapid solution of integral equations of classical potential theory. *Journal of computational physics*, *60*(2), pp.187–207.
Timoshenko, S., 1970. Goodier. JN, Theory of Elasticity. *New. York McGraw—Hill*, *970*(4), pp.279–291.
Vijayakumar S, Cormack DE. A new concept in near-singular integral evaluation: the continuation approach. SIAM Journal on Applied Mathematics. 1989 Oct;49(5):1285–95.
Vijayakumar, S. and Curran, J.H., 2007. Influence of boundary curvature on tangential stresses for the displacement discontinuity method. *Engineering analysis with boundary elements*, *31*(3), pp.267–274.
Vijayakumar, S., Yacoub, T.E. and Curran, J.H., 2000. A node-centric indirect boundary element method: three-dimensional displacement discontinuities. *Computers & Structures*, *74*(6), pp.687–703.
Yoshida, K.I., Nishimura, N. and Kobayashi, S., 2001. Application of new fast multipole boundary integral equation method to crack problems in 3D. *Eng. Analysis with Boundary Elements*, *25*(4-5), pp.239–247.

The Evolution of Geotech - 25 Years of Innovation – Hammah et al. (Eds)

Life-of-mine deformation and stability assessment for Kibali conditions in Western 9000 series lodes

G.N. Kaleba
Kibali Gold Mine, DRC

ABSTRACT: Kibali Gold Mine embarked on a project to assess life-of-mine (LOM) mine-wide deformation and stability assessment for the Western 9000 series of stopes following the 2021 LOM plan. The exercise also aimed at identifying potential changes to the 9000 series lodes LOM design or sequence should the simulations flag high risks, including areas with elevated seismic potential.

1 INTRODUCTION

The Kibali Gold Mine (KGM) geomechanics team (a mine under the Barrick group) and KSCA Geomechanics Pty Ltd undertook a life-of-mine (LOM) deformation and stability assessment for the mine. The assessment built upon previous assessments of LOM deformation and stability work completed for Kibali in 2017, 2018, 2019 and 2020. The current paper describes the assessments, which were based on the results of mine-scale three-dimensional numerical modelling using the finite element method implemented in the Dassault System program, abaqus FEA.

The current study focused on the planned production from the 9000 series lodes. These stopes are shown on Figure 1 (which presents a perspective view of the mine).

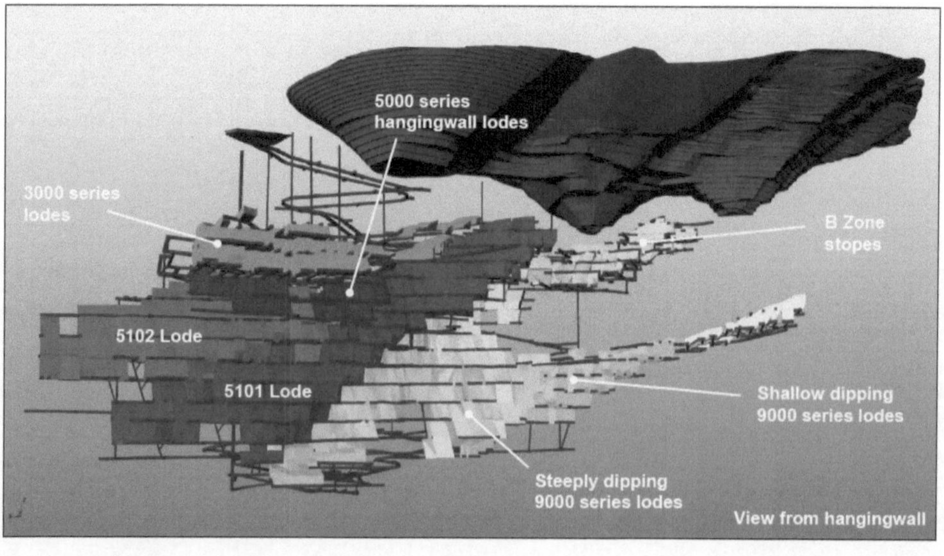

Figure 1. Perspective view showing the life-of-mine geometry.

DOI: 10.1201/9781003188339-73

The three stress measurements completed in 2020 indicate that the stress field is slightly higher than previously thought, although the difference is relatively small. At the maximum working depth of 750m, the major principal stress is now 41.9 MPa compared to the previous estimate of 39.8 MPa. This represents an increase of ~5% in the major principal stress.

In general, little deformation has been observed at the mine. It was therefore believed that the current model could be used to forecast conditions over the mine life with moderate reliability.

Mostly mining conditions at KGM have been good. However, more damage is forecast in the areas with major geological structures.

The main concerns at the mine have been the following:

1. Normal problems associated with mining secondary stopes. A variety of problems typically arise which result in reduced productivity, reduced recovery and higher dilution relative to the primary stopes. These mining problems are exacerbated where major structures are present and/or where there is rock mass yield.
2. The recovery factor of 96% that has, according to the NI 43-101 report, been applied to all stoping at Kibali, is rather optimistic. A recovery factor of 85% ± 5% is probably more realistic for the secondary stopes at Kibali.
3. Stress concentrations are forecast in some of the secondary pillars, in the close-out pillar, at the abutments and in the overlap between the 5101 and 9000 lodes. These stress concentrations are not particularly high, but they are expected to lead to some locally difficult mining conditions perhaps including larger and more frequent seismic events than recorded to date.
4. Damage is expected to develop in the extended crown spans of the shallow dipping 9000 stoping blocks as stoping proceeds. Reliable tight filling will be required to limit crown instability in these stoping blocks.
5. Although little seismicity has been recorded to date, and the LOM seismic potential appears to be generally low, the possibility of damaging seismicity cannot be ruled out.
6. Upgraded ground support may be needed in ore development where the capacity of the current support system, which comprises mesh and split sets, is exceeded.

2 MINING BACKGROUND

2.1 *9000 Series mining method*

The mine uses sublevel open stoping (SLOS). The stopes are mined in a bottom-up echelon retreat advanced face, which allows multiple mining fronts to be opened and offers greater flexibility. In plan view, the mining retreats towards the footwall. The geometry of the mine sequencing is shown on Figure 2.

When stopes are mined to a full height, they are completely backfilled before the next stope in the sequence begins production. The mine uses multi-levels to facilitate greater production rates.

2.2 *Structures*

Four main groups of structures have been interpreted at the mine and modelled. They are as follow:

1. Mine-scale geological structures at Kibali
2. Two primary joint sets
   a. Set 1 (Js1) – these structures dip at 60° to 89° towards ENE
   b. Set 2 (Js2) – these structures dip at ~80° towards SE. They strike approximately parallel to the stope hanging walls in the main SLOS zones. This is unfavourable for hanging wall stability.
3. Adverse lithological contacts. The dolerite contacts were modelled as structures in the present LOM model.

The discontinuities are described in Dempers and Seymour (2019) and shown in red on Figure 3 below.

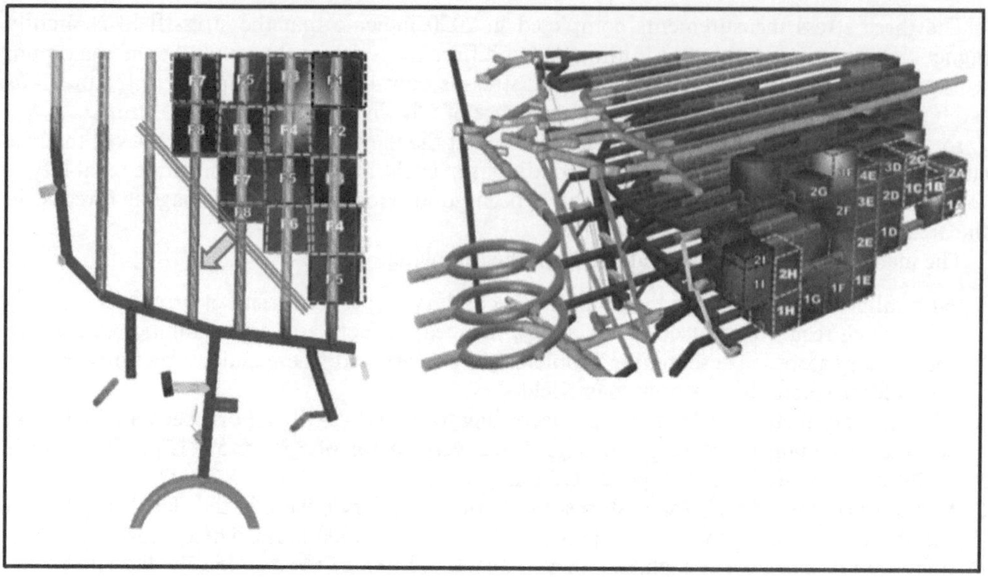

Figure 2. Views showing the transverse advance of the mining face and the bottom-up sequencing.

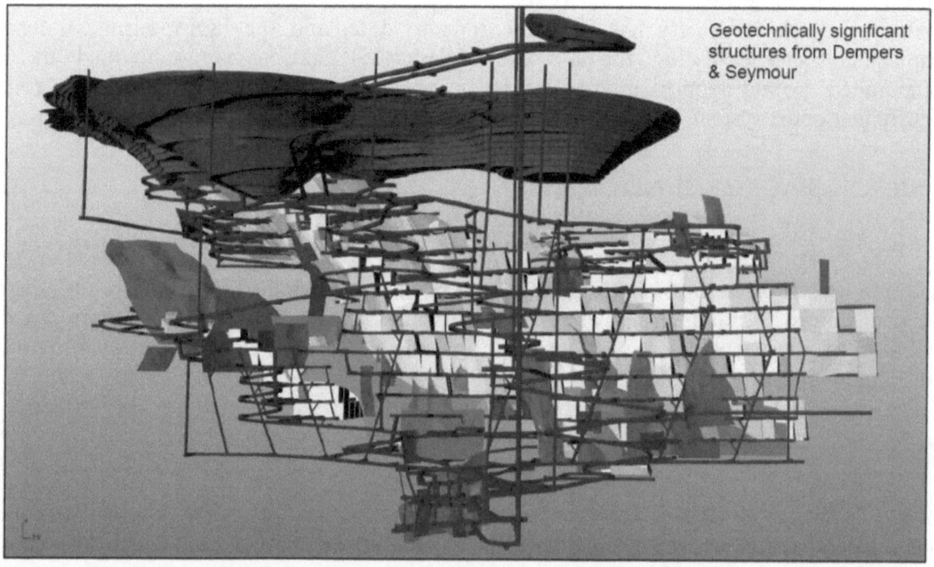

Figure 3. Structures explicitly modelled in the LOM deformation assessment project.

2.3 *Geotechnical domain assignment and rock mass conditions*

All the geotechnical domains, except for the schist units, were mapped into the FE model based on the lithology codes from the mining rock mass model (MRMM). The schist units were built explicitly from geometric shapes. The schist contacts are understood to be welded and were therefor not modelled as structural features.

The mean UCS and GSI values were used to define Generalized Hoek-Brown material properties, following the estimation scheme outlined by Beck et al. (2013).

A schematic of the geotechnical domains in the model is shown on Figure 4.

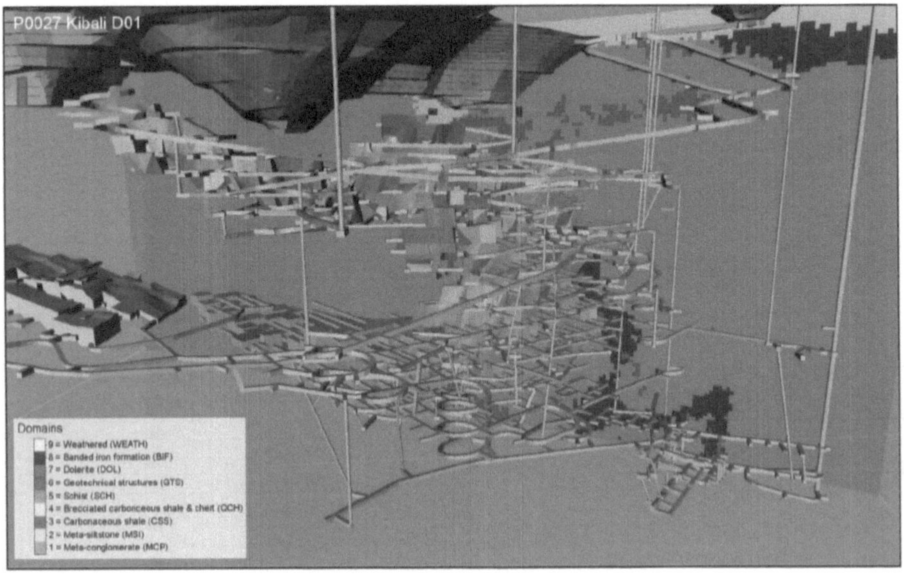

Figure 4. Perspective views showing geotechnical domains plotted on vertical cross-sections through the meshed LOM geometry. The cross-section plane is oriented at 110° to 290°.

2.4 *Mining geometry and sequence*

The geometry used for the model included the following:

1. The complete LOM development geometry. The slot drives were not included in the development geometry but were built into the FE mesh
2. The complete LOM stope geometry

The LOM sequence provided did not include the stoping sequence beyond 2030. Therefore a stoping sequence was assumed for stopes extending into 2032 Q3. This assumed stoping sequence mainly affected the north-eastern extension of the 5102 lode and the shallow-dipping 9000 lodes.

Stoping in the Western 9000 series lodes follows a continuous sequence. (Previously, a primary secondary stoping sequence was followed in the larger shallow-dipping orebody.)

2.5 *Material properties*

The strength and deformation properties assigned in the FE model are shown on Table 1 below.

2.6 *Stope filling methodology and fill properties*

In the FE model, stopes were excavated and filled by adjusting the Young's modulus of the stope volume over a simulation frame which has a duration of 3.0s. This method assumes tight filling of every stope. The properties of the fill were as follow:

- Young's modulus, Efill = 100 MPa.
- Poisson's ratio, νfill = 0.25.

In practice, the mine sometimes leaves stopes open for longer than modelled and does not al-ways achieve tight filling.

Table 1. Geotechnical properties of the rock mass units.

Number	Domain	Code	Density (kg/m³)	UCS (MPa)	GSI	Stage	Plastic strain	E (GPa)	ν	s	m	a	d	κ	Φ
1	Meta-conglomerate	MCP	2,920	122	80	Peak	ε_0 = 0.0%	25.7	0.27	6.33E-03	2.74	0.500	0.31	94	330
						Transition	ε_1 = 1.9%	25.7	0.27	6.33E-03	2.74	0.500	0.31	94	330
						Residual	ε_2 = 3.3%	12.9	0.27	2.02E-03	1.43	0.500	0.11	30	170
2	Meta-siltstone	MSI	2,830	129	86	Peak	ε_0 = 0.0%	27.4	0.28	8.35E-03	3.15	0.500	0.33	140	410
						Transition	ε_1 = 1.8%	27.4	0.28	8.35E-03	3.15	0.500	0.33	140	410
						Residual	ε_2 = 3.2%	13.7	0.28	2.32E-03	1.52	0.500	0.12	39	200
3	Carbonaceous shale	CSS	2,830	118	73	Peak	ε_0 = 0.0%	24.9	0.27	5.18E-03	2.40	0.500	0.30	72	280
						Transition	ε_1 = 2.0%	24.9	0.27	5.18E-03	2.40	0.500	0.30	72	280
						Residual	ε_2 = 3.4%	12.4	0.27	1.87E-03	1.38	0.500	0.11	26	160
4	Brecciated carbonaceous shale & chert QCH		3,020	120	71	Peak	ε_0 = 0.0%	25.3	0.27	5.40E-03	2.41	0.500	0.30	78	290
						Transition	ε_1 = 1.9%	25.3	0.27	5.40E-03	2.41	0.500	0.30	78	290
						Residual	ε_2 = 3.4%	12.6	0.27	1.94E-03	1.40	0.500	0.11	28	170
5	Schist	SCH	2,770	126	80	Peak	ε_0 = 0.0%	26.3	0.28	6.86E-03	2.82	0.500	0.32	110	360
						Transition	ε_1 = 1.9%	26.3	0.28	6.86E-03	2.82	0.500	0.32	110	360
						Residual	ε_2 = 3.3%	13.2	0.28	2.13E-03	1.46	0.500	0.11	34	180
6	Geotechnical structures	GTS	2,800	67	33	Peak	ε_0 = 0.0%	12.1	0.23	4.65E-04	0.61	0.500	0.15	2.1	41
						Transition	ε_1 = 2.8%	12.1	0.23	4.65E-04	0.61	0.500	0.15	2.1	41
						Residual	ε_2 = 4.9%	6.1	0.23	3.71E-04	0.67	0.500	0.05	1.7	45
7	Dolerite	DOL	2,900	131	80	Peak	ε_0 = 0.0%	27.4	0.28	7.85E-03	2.98	0.500	0.33	130	390
						Transition	ε_1 = 1.8%	27.4	0.28	7.85E-03	2.98	0.500	0.33	130	390
						Residual	ε_2 = 3.2%	13.7	0.28	2.32E-03	1.52	0.500	0.12	40	200
8	Banded iron formation	BIF	3,480	125	77	Peak	ε_0 = 0.0%	26.3	0.28	6.61E-03	2.71	0.500	0.32	100	340
						Transition	ε_1 = 1.9%	26.3	0.28	6.61E-03	2.71	0.500	0.32	100	340
						Residual	ε_2 = 3.3%	13.2	0.28	2.13E-03	1.46	0.500	0.11	33	180
9	Weathered	WEATH	2,600	32	44	Peak	ε_0 = 0.0%	6.5	0.22	2.77E-04	0.57	0.500	0.08	0.28	18
						Transition	ε_1 = 3.3%	6.5	0.22	2.77E-04	0.57	0.500	0.08	0.28	18
						Residual	ε_2 = 5.8%	3.2	0.22	9.07E-05	0.36	0.500	0.03	0.093	12
10	Faults Strong	FLT1	2,800	50	35	Peak	ε_0 = 0.0%	10.5	0.23	3.86E-04	0.58	0.500	0.13	0.96	29
						Transition	ε_1 = 3.0%	10.5	0.23	3.86E-04	0.58	0.500	0.13	0.96	29
						Residual	ε_2 = 5.2%	5.2	0.23	2.66E-04	0.58	0.500	0.05	0.66	29
11	Faults Moderate	FLT2	2,800	40	35	Peak	ε_0 = 0.0%	8.41	0.22	2.94E-04	0.52	0.500	0.10	0.47	21
						Transition	ε_1 = 3.1%	8.41	0.22	2.94E-04	0.52	0.500	0.10	0.47	21
						Residual	ε_2 = 5.5%	4.2	0.22	1.61E-04	0.46	0.500	0.04	0.26	19

Kibali - Material property set MCE

KIBALI_MATPROPS.xlsm

2.7 *Damage criteria*

The following criteria were used to assess damage and seismicity levels:

Table 2. Correlation of plastic strain values with the damage state of the rock.

Plastic strain	Damage state	Observed behaviour
>5%	Very significant	Gross distortion and comminution.
~3%	Significant	Extensive fracturing of intact rock.
~1.5%	Moderate	Constant load leads to increasing deformation.
~0.7%	Minor	No significant decrease in strength or stiffness.
<0.35%	None to very minor	Undisturbed in situ conditions.

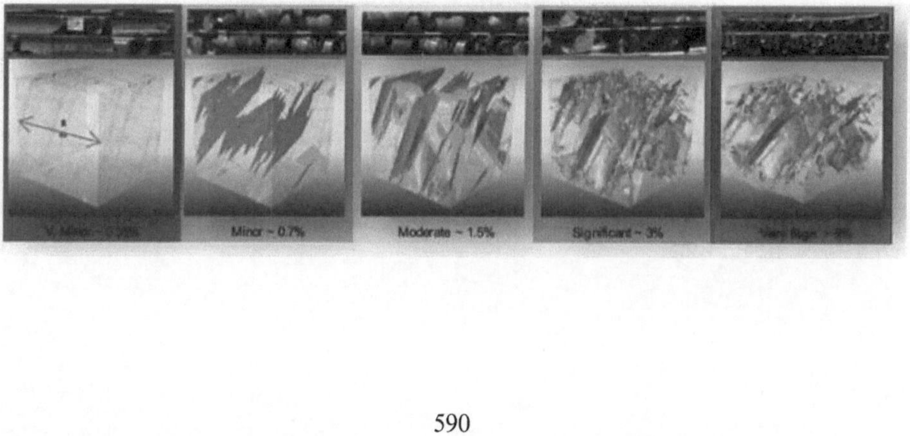

3 RESULTS AND DISCUSSION

This section summarizes the model results and the interpretation of the likely rock mass behaviour for the Kibali underground mine over the remaining mine life according to the current LOM plan and the current understanding of geotechnical conditions. The results are presented in terms of stress magnitudes, rock mass damage and RER (maximum instantaneous rate of Energy release).

3.1 *Conditions in Western 9000 series lodes*

Results for the western shallow dipping stoping blocks in the 9000 series lodes are shown in Figure 5.

The results show that

1. There will be little damage in early development (see Figure 5a).
2. Damage increases in the immediate hanging wall as the mined-out span increases (see Figure 5). This is called an extended crown span. Although the stopes are backfilled, damage starts to develop in this extended crown span because the backfill provides little support pressure. Provided effective tight filling can be achieved, the increasing damage does not indicate an increased frequency of hanging wall instability. However, ineffective tight filling would lead to more damage and more instability.
3. Generally low stress concentrations form at the stoping front (see Figure 5).

The generally low levels of forecast damage and moderate stress concentrations at the stoping front indicate that a more aggressive primary-secondary stoping sequence is prob-ably possible in these blocks, from a purely geotechnical perspective. However, the deci-sion on the sequence in this block of course depends on other factors and the possibility of a primary-secondary sequence is just raised as a possibility for consideration. In practice, the mine sometimes leaves stopes open for longer than modelled and does not al-ways achieve tight filling.

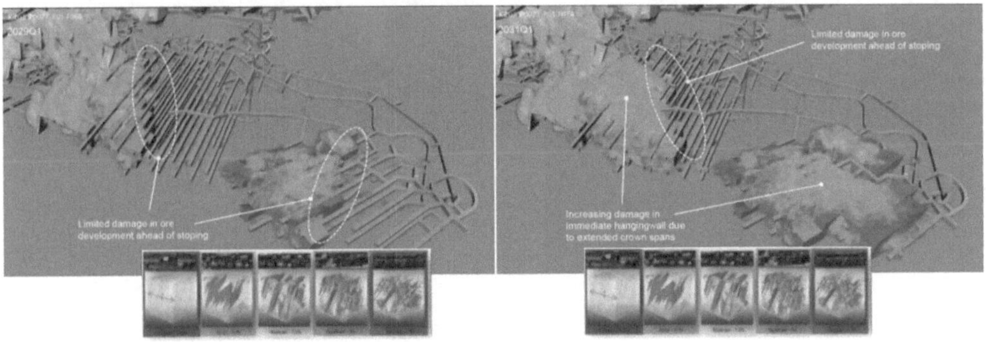

Figure 5. Views looking from the hanging wall showing forecast rock mass damage (red zones) in the Western 9000 series lodes. Here damage is plotted using 3D volume rendering.

3.2 *Forecasts of seismic potential*

Even without calibration, RER is still useful for evaluating relative seismic potential (e.g., periods of time and areas of the mine with elevated seismic risk). The modelling results, shown on Figure 7 below, indicated the following:

1. Early stoping is associated with widespread but generally low seismic potential.
2. The highest seismic potential is associated with stoping at the abutments and in the overlap region between the 5101 and 9000 lodes

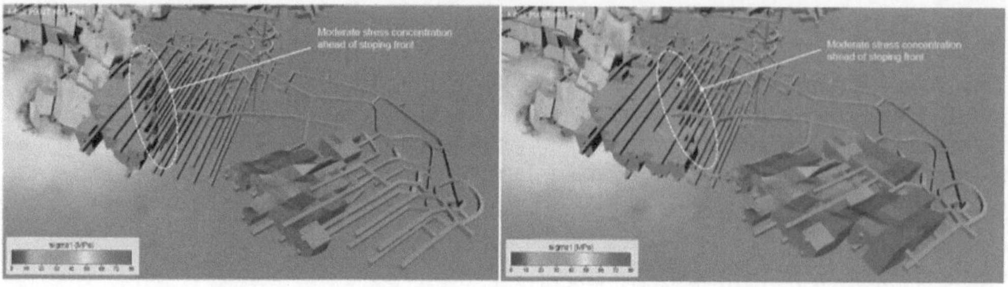

Figure 6. Views from the hanging wall showing forecast $\sigma 1$ magnitude in the Western 9000 series lodes. Here $\sigma 1$ is plotted using 3D volume rendering.

3. The seismic potential is lower when stoping in the upper levels later in the mine life. This is due to the naturally lower in situ stress environment in these levels due to the reduced mining depth.

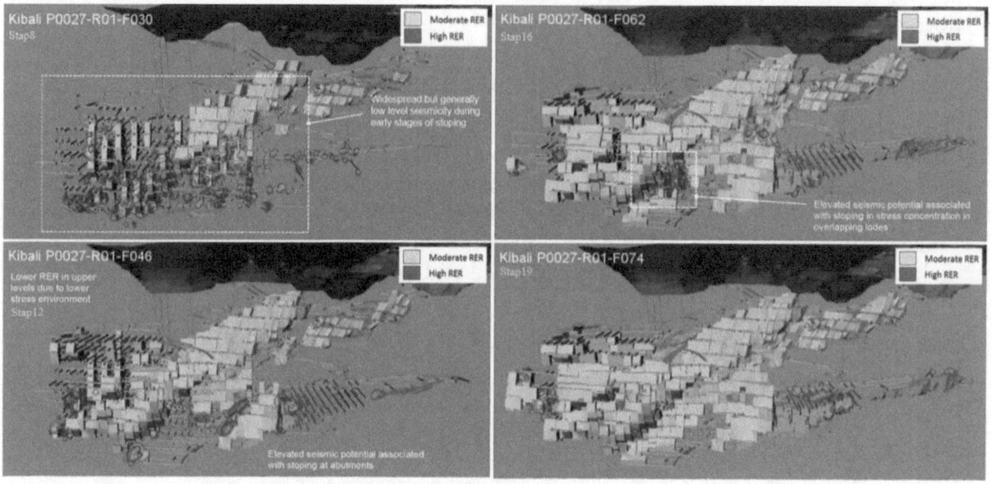

Figure 7. Views of RER for different stages of mining.

4 CONCLUSION

The geotechnical recommendations arising from this project are as follow:

1. In the main SLOS area, the mine should plan for reduced productivity and recovery rates, and higher dilution rates and mining costs in the secondary stopes, compared to the early primary stopes
2. Develop trigger action response plans (TARPs), including monitoring plans, to mitigate each of the risks associated with the secondary pillars and the close-out stopes
3. Ensure that the design allows for tight filling of the stopes in the shallow dipping 9000 series lodes
4. The as-built development geometry indicates two mining practices that must be reconsidered
   a. Early development of the crosscuts for the secondary stopes
   b. Driving the crosscuts beyond the hanging wall limit of each stope

5. Continue developing the structural model with new structural data and interpretations as they become available. The structural model needs to be incorporated into the LOM design and planning process.
6. Continue seismic monitoring and interpretation of source mechanisms, and damage mapping
7. Carefully monitor secondary stope performance to support the TARPs, purchase a laser scanner to measure deformation in development
8. Ensure that all the stopes under the extended crown 9000 series are tightly filled as soon as possible after completion
9. More aggressive primary-secondary sequencing, rather than the planned continuous sequence, may be possible in the shallow-dipping 9000 series lodes.

REFERENCES

1. KGM Gaby K & Applied Geomechanics Consulting. Review of 2020 life-of-mine plan & updated geotechnical conditions for Kibali. Final report to KSCA Geomechanics, 25th May 2020.
2. Barrick Gold Corporation. Technical report on the Kibali gold underground mine, DRC. NI 43–101 report (effective 31st December 2019), 15th January 2020.
3. Beck DA, Lilley CR, Reusch F, Levkovitch V & Flatten A. A preliminary, calibrated scheme for estimating rock mass properties for non-linear, discontinuum models. *Rock Characterisation, Modelling & Engineering Design Methods – 3rd ISRM SINOROCK Symposium, Eds. Xia-Ting Feng, John A. Hudson & Fei Tan, Shanghai, China, 18–20 June 2013.*
4. Beck Engineering. Global deformation modelling at Kibali. Final report to Randgold Resources,22nd January 2017.
5. Beck Engineering. Life-of-mine deformation & stability assessment for Kibali. Final report to Randgold Resources, 4th July 2018 (2018a).
6. Beck Engineering. Assessment of revised life-of-mine sequence for Kibali. Final letter to Randgold Resources, 4th July 2018 (2018b).
7. Beck Engineering. Life-of-mine deformation & stability assessment for Kibali. Final report to Barrick GoldCorporation, 13th May 2019.
8. KGM Geotech, Dempers & Seymour. Kibali Mine MRMM update. Final report to Barrick, June 2019.
9. Levkovitch V, Beck DA & Reusch F. Numerical simulation of the released energy in strain softening rock materials & its application in estimating seismic hazards in mines. *Proceedings 8th International Symposium on Rockbursts & Seismicity in Mines, 1–7 September 2013, Saint Petersburg, Russia.*
10. Western Australian School of Mines (WASM). Stress measurements from oriented core using the acoustic emission method – Kibali Gold Mine. Report to Randgold Resources, April 2011.
11. Western Australian School of Mines (WASM). Stress measurements from oriented core using the acoustic emission method – Kibali Gold Mine. Report to Dempers & Seymour, December 2012.
12. Western Australian School of Mines (WASM). Stress measurements from oriented core using the acoustic emission method – Kibali Gold Mine. Report to Kibali Gold Mines SPRL, May

The Evolution of Geotech - 25 Years of Innovation – Hammah et al. (Eds)

Numerical models for the design and construction of new underground structures at CERN (HL-LHC), Point 5

D. Merlini, M. Falanesca, F. Gianelli & G. Bella
Pini Group, Lugano, Switzerland

R. Schürch
Pini Group, Zurich, Switzerland

A. Luz Lopez-Hernandez
CERN, Genève, Switzerland

ABSTRACT: The Large Hadron Collider (LHC) is the latest, most powerful, world's largest underground particle accelerator realized on the CERN site. High-Luminosity LHC (HL-LHC) is a new project aimed to upgrade the LHC at Point 1 (ATLAS in Switzerland) and Point 5 (CMS in France) to enhance scientific progress. This paper describes the design and construction issues developed at Point 5 for the new underground structures near the LHC tunnel. The project requires new technical infrastructure: an additional shaft with a - 12 m-diameter and 60 m-height, cavern with 270 m^2 cross-section, approximately 500 meters of galleries connected to the LHC tunnel, vertical linkage cores and additional technical buildings at the surface. This site's geological ground model lies in an area covered by Quaternary moraine with two independent aquifers. The bedrock of Molasse comprises sub-horizontal lenses of heterogeneous sedimentary rock, that is known to locally retain hydrocarbons and to have a swelling behaviour. To investigate the heterogeneous behaviour of the rock mass composed of several layers with different strengths, numerical calculations have been performed, under a 2D plane strain condition with RS2 9.0 FEM-software. The purpose of using the software was to design both the rock-supports and the concrete inner lining for the galleries and the shaft. Data from a comprehensive monitoring system with pre-defined threshold values was compared to the 2D FEM results, confirming the importance of the observational method to verify the assumptions used in the numerical modelling.

The execution of the underground works started in April 2018. The excavation of the main underground works has been successfully completed without any critical impact on the nearby existing underground structures. The completion of the works is scheduled for September 2022.

Keywords: Finite element method, numerical analysis, rock mechanics, observational method

1 INTRODUCTION

The European Organization for Nuclear Research (CERN) is an organization with 23 member states worldwide, and its headquarter is based in Geneva. The Large Hadron Collider (LHC) is the world's largest underground particle accelerator placed on both sides of the

DOI: 10.1201/9781003188339-74

Swiss-French border. The data collected by this unique instrument in the world has allowed the discovery of the Higgs boson in 2012.

The High-Luminosity LHC is a new project aimed at enhancing the LHC experiments, in order to produce more data by increasing the number of particle collisions by a factor of 10. This project will be operational in 2026 and requires new technical infrastructure for the two main detectors, respectively, at Point 1 (ATLAS) and Point 5 (CMS), as shown in Figure 1.

At Point 5, the new HL-LHC underground structures are placed on the inner side of the existing LHC ring at an average distance of approximately 50 m from the LHC axis and located 7 m above the level of the existing LHC tunnel crown. The new underground structures consist of the following main objects: i) a new shaft PM57 12 m diameter and 60 m deep with at the base a service cavern US57/UW57 of excavation area 270 m^2, ii) a power converter gallery UR55 with an excavation area of approximately 60 m^2, iii) two pairs of service galleries UA57 /UA53 and UL57/UL53 with excavation areas of 45 m^2 and 20 m^2 respectively, iv) 16 vertical linkage cores to the existing LHC 1.7 m excavation diameter and 5m deep, v) 2 personnel escape galleries UPR53/UPR57 of excavation areas 25 m^2 (Figure 1, right). All the new underground structures are designed with a double lining system. For temporary support, a design working life of 10 years is required. For the final lining and the waterproofing system, a design working life of 100 years is needed according to the specification. Additional technical buildings at the surface are currently under construction.

The main challenges of the construction project (Canzoneri et al., 2019) were related to the following: i) the criteria against the vibrations and settlements induced from the excavation works on the existing structures where experiments were in progress; ii) the excavation inside a heterogeneous rock mass with known swelling behaviour and containing hydrocarbons.

This paper describes the key role of the RS2 9.0 software, which made it possible to overcome parts of these challenges.

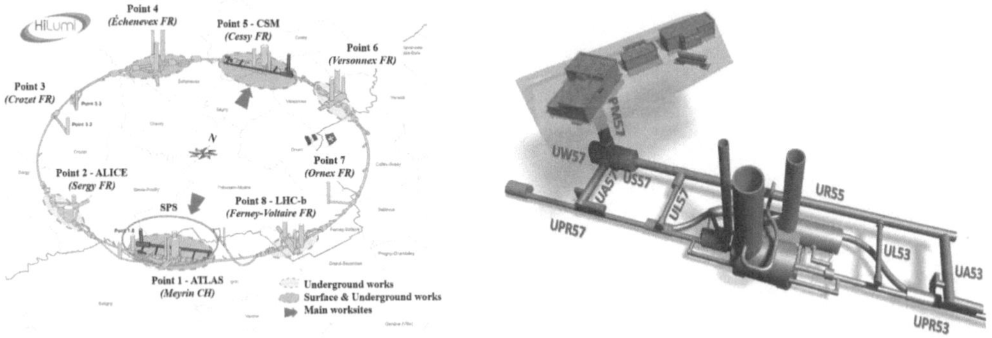

Figure 1. Left: HL-LHC project; Right: new (in blue) and existing (in grey) structures at Point 5 (courtesy CERN).

2 GEOLOGICAL CONDITIONS

The general geological profile at the site of Point 5 as shown in Figure 2 can be described, from surface downward, as follows: i) a limited thickness of fill put in place during the previous works comprising of colluvial soils; ii) fluvial-glacial soils resting on heterogeneous glacial (Wurmian) moraines, these deposits have an overall thickness ranging from approximately

30 m to 50 m; iii) underlying the moraine deposits, the rock formation of the red Molasse (Chattian age, Tertiary).

The new underground structures have an overburden of approximatively 60 m. These are excavated inside the Molasse unit, except the upper part of the PM57 shaft which extends to a depth of approximately 22 meters, crossing the soil units described in Table 1. Available ground investigation data from previous underground works showed that the Molasse in this area is a highly heterogeneous rock mass known to locally contain liquid and gaseous hydrocarbons (Kurzweil, 2004). The Molasse layers are usually 0.5 to 5 m thick, and their stiffness can vary significantly from one layer to another. Three main units can be distinguished: i) marls: the finest grain-size unit of the molasse, characterized by a high percentage of clay minerals, including the swelling minerals illite and smectite; ii) sandstones: characterized by coarser grain-sizes of silt and fine sand, in approximately similar percentages, cemented by calcareous cement; iii) transitional materials between marls and sandstones.

Taking into account the information from the previous underground excavations and the recent ground investigation, the different facies were modelled (Table 2) in the software RocLab 5.0. The in situ-stress was determined from the in-situ investigations where the K_0 varied between 1.25 and 2 for the rocks and the K_0 was 0.7 for the soils. In terms of hydrogeology, two aquifers were identified: i) upper aquifer, phreatic, located within the fluvio-glacial soils; ii) lower aquifer, within the underlying moraine. The two aquifers do not communicate, being separated by the less permeable layers, though tracing tests carried out in the Point 5 site showed local connections, natural or maybe due to anthropic activities. As the permeability of the Molasse obtained from in situ tests was very low ($k < 10^{-7}$ m/s), the rock mass was considered to be impermeable.

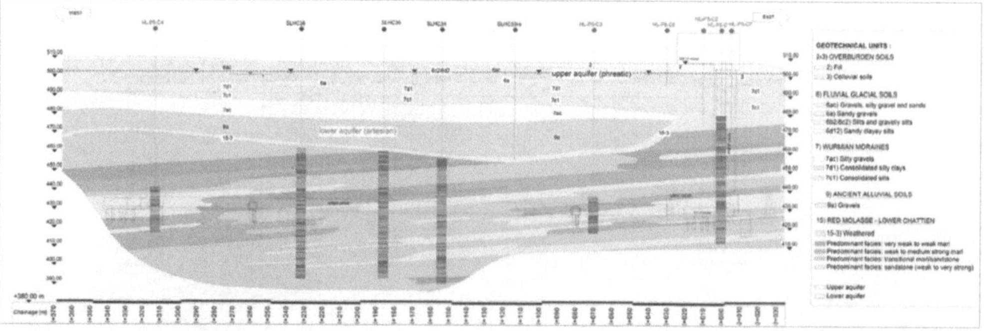

Figure 2. Geological profile at Point 5.

Table 1. Geotechnical parameters for soil units.

	γ [kN/m³]	OCR [-]	ϕ [°]	ψ [°]	c' [kPa]	cu [kPa]	E [MPa]	υ [-]	k [m/s]
Fill	21.0	-	28-30	8	0	-	40-120	0.3	-
Colluvial soil	19.5	-	27-29	7	5-10	> 40	>100	0.3	$4 \cdot 10^{-8} \div 8 \cdot 10^{-7}$
Silty gravel	23.5	-	32-36	12	0	-	200-450	0.3	$9 \cdot 10^{-4}$
Sandy silt	22.5	2	32-34	12	10-15	150-250	100-170	0.3	$9 \cdot 10^{-8} \div 1 \cdot 10^{-7}$
Silty clay	22.5	2	30-32	10	10-15	100-200	70-140	0.3	$(4 \div 9) \cdot 10^{-8}$

Table 2. Geotechnical parameters for rock units.

	γ [kN/m³]	ϕ [°]	ψ [°]	ϕ_{res} [°]	c [MPa]	c_{res} [MPa]	ϕ_{res} [°]	σ_c [MPa]	σ_t [MPa]	E_{mc} [MPa]	E_{md} [MPa]	υ [-]	k [m/s]
Very weak to weak marls	24.0	18.5	0	17.0	1.2	0.5	17.0	3.2	0.5	500	850	0.30	1·10-7
Weak to strong marls	24.5	30.0	10	30.0	2.6	1.3	30.0	9.4	1.0	1260	2500	0.30	1·10-7
Marl/ sandstone	25.0	38.0	18	34.0	4.8	1.6	34.0	20.0	1.8	3000	4500	0.30	1·10-7
Sandstone	23.0	41.0	21	34.0	2.1	1.0	34.0	9.1	0.6	1500	2500	0.30	1·10-7

3 THE ROLE OF NUMERICAL ANALYSES DURING THE DESIGN PHASE

In the first design phase, preliminary calculations were carried out using the software Roc-Support 4.0 for an initial assessment of the required rock supports and evaluating the stress-strain behaviour. The Convergence-Confinement analyses were performed according to the Duncan Fama (1993) solution based on the different values of the geomechanical parameters shown in Table 2. From these analyses, the behaviour of the rock masses was generally determined to be within the elastic range, generating a total maximum tunnel displacement without support of less than 1 cm. Only by considering a worst-case scenario with a section entirely within the unit of very weak marls were a total maximum displacement of 5 cm with a plastic radius of approximatively 3 m calculated. Analyses using the UnWedge 4.0 software were carried out for the stability analyses of the rock wedges to design the rock bolt pattern.

FE 2D analysis, under plane strain conditions, was performed using RS2 9.0 software to investigate the rock mass's heterogeneous behaviour composed of several layers with different strengths. The results of these analyses have: i) provided the design of the rocksupports; ii) assessed the potential impact on the existing underground structures; iii) provided the design of the permanent concrete inner lining.

Numerous calculation models were created to design the various standard cross-sections and rock support classes required according to the critical narrow tunnel geometric configurations and different scenarios for the location and thickness of the molasse layers. For all models, an elasto-perfectly plastic Mohr-Coulomb criterion was adopted to reproduce the soil behaviour; for the rock mass, the equivalent Mohr-Coulomb with peak and residual strength parameters were adopted (Table 2). The external dimensions of the FE models were chosen to minimize boundary effects. Boundary conditions consist of fixed horizontal/vertical displacements along x-direction (hinge) and fixed vertical displacements along y-direction (roller). The domain has been discretized through triangular meshes, increasing the discretization around the underground structures. To reproduce the excavation phases, the convergence-confinement method (Panet and Guenot, 1982) was adopted with the stress reduction λ evaluated based on the ground reaction curve and calibrated based on the monitoring results from the previous underground excavations. Bolts were modelled as a fully bounded element or Swellex type, and the shotcrete liner was assigned as a standard beam. The permanent concrete inner lining is also considered a standard beam but with a selected composite liner function.

Figure 3 shows the numerical model for evaluating the influences between the existing USC55 cavern (width of 19 m, height of 17m and length of 85m) and UR55 tunnel. The UR55 was modelled with support class 1 consisting of: friction bolts L = 4m $F_{tk} \geq$ 240kN (distance 1.5m/1.5m), 1st layer 5cm thickness of fibre reinforced shotcrete C20/25, wire mesh 1.89cm²/m and 2nd layer 10cm of shotcrete C20/25. For the USC55 cavern, rock supports were modelled using shotcrete C25/30 with total thickness of 30cm, rock bolts $F_{tk} \geq 600$ kN L = 9 m (transversal distance 0.85 - 1.5 m x longitudinal distance 1.5m) and rock bolts $F_{tk} \geq 400$ kN, L = 6.00 m (transversal distance 1.5 m x longitudinal distance 1.5 m). The inner lining of the US55 cavern comprises reinforced concrete C40/50 with a thickness of 57 cm. From the analysis of the results, the following main conclusions were identified: i) the behaviour of rock masses was mainly elastic with narrowed plastic zones that were not in contact with the nearby tunnels; ii) the increases in stresses and displacements of the rock masses around the existing tunnels were low; iii) the increase of displacements in the existing crown cavern were estimated to be a maximum of 3 mm and in the tunnel invert to be a maximum of 1 mm, all below threshold values for the ongoing CERN tests; iv) the increase in axial forces, bending moments and shear forces on the existing tunnel lining were within the ULS and SLS criteria.

Figure 4 shows the model used to evaluate the excavation of the new UA57 tunnel which is located at a minimum distance of approximately 5.5 m above CERN's main ring (R57). For the existing R57, the lining was modelled as a standard beam of concrete C25/30 without reinforcement with a thickness of 22 cm. For the UA57, the rock support class 2 was modelled comprising of: shotcrete C20/25 with total thickness of 25 cm, lattice girders 3G 70/20/26 with a spacing of 1.2 m and radial fully grouted bolts L = 4.00 m, $F_{tk} \geq 250$ kN with a spacing in plane of 1.50 and out of plane of 1.20 m. This model also confirmed a predominantly elastic rock mass behaviour with narrowed plastic zones that were not in contact with the nearby tunnels. The increase in displacements in the existing tunnel was estimated to be a maximum of 1 mm. From these analyses, it was concluded that no additional supports were necessary for the existing LHC.

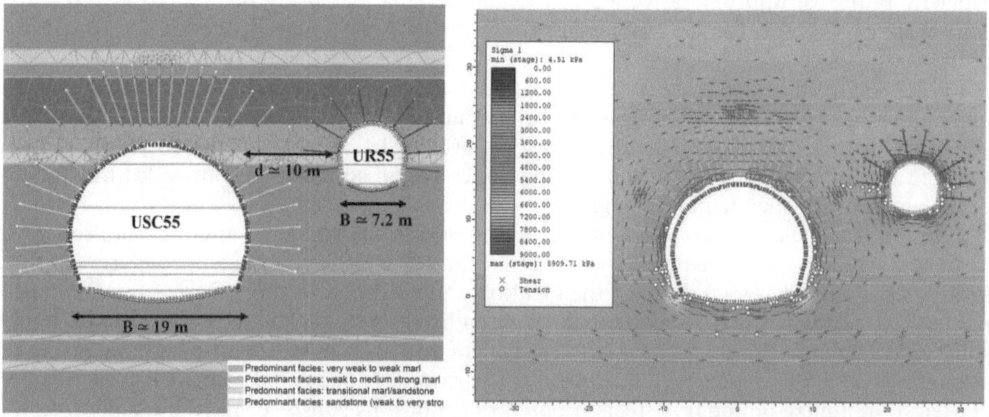

Figure 3. Model for the evaluation of UR55 tunnel excavation on the existing UCS55 cavern. Left: FE model geometry and rock units; Right: stress and yielded zones on the rock mass.

To evaluate the sections near the several crossing zones where 3D effects are critical, the same conservative geotechnical parameters were adopted as those used in the models analyzed using RS2 2D software. The tunnel inner lining was reinforced only in the crossing zones as determined from the results of the 2D numerical models. Stresses acting on the final supports were compared with those obtained from simplified beam-spring 2D models also taking into account the loads induced by temperature, swelling and creep effects according to the load combinations stated in Eurocode.

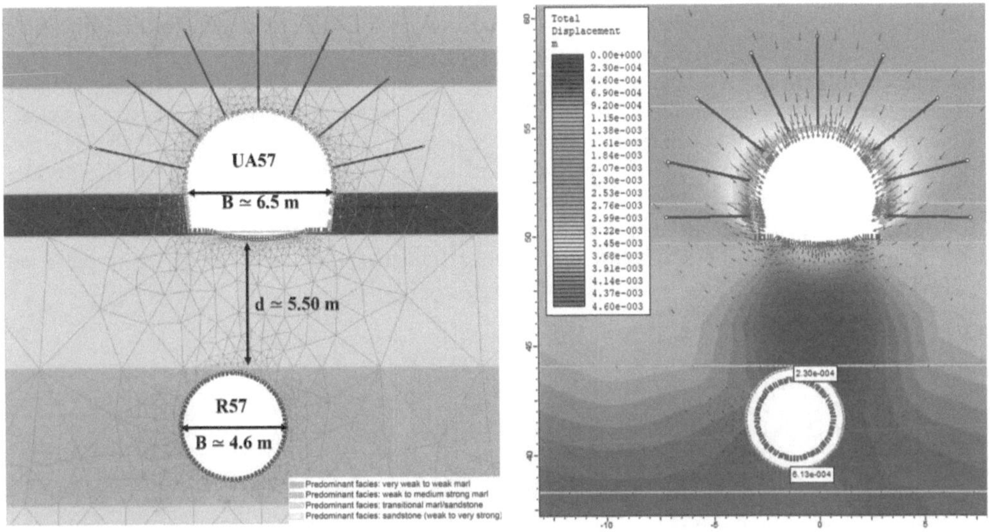

Figure 4. Model for the evaluation of UA57 tunnel excavation on the existing LHC. Left: FE model geometry and rock units; Right: total displacements resulting from the excavation of UA57.

4 THE ROLE OF NUMERICAL ANALYSES DURING THE CONSTRUCTION PHASE DATA

A comprehensive monitoring system with predefined threshold values was compared to the 2D FEM results, confirming the importance of the observational method to verify the assumptions used in the numerical modelling. Convergence monitored for the new underground structures were a maximum 1 cm and those for the existing structures, were on the scale of a millimeter.

Considering the actual ground conditions encountered during the excavation stage, some optimizations of the rock supports were possible. Figure 5 shows the back-analysis model for the UR55 rock support class 1 that was performed to assess the lower rock bolts' removal, thus allowing an increase in excavation rate and a reduction in project costs. The models' results showed that the structural safety was still within threshold values and that the effects on the nearby structures were similar to those in the original design solution.

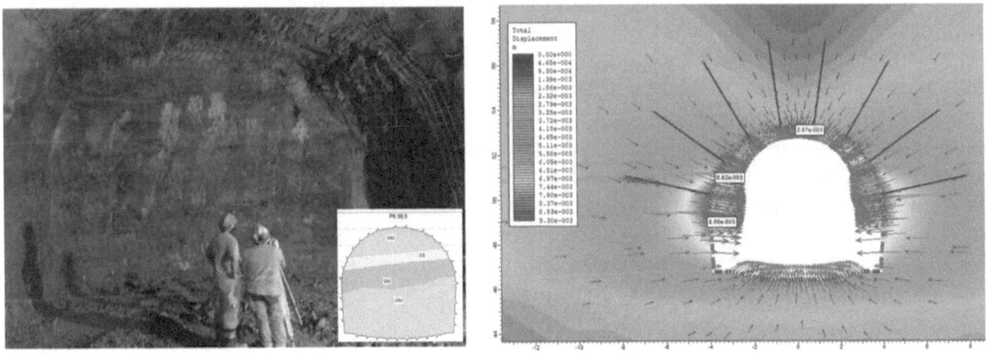

Figure 5. Left: convergence monitoring. Right: back-analysis for the optimization of rock supports.

Another model was done at the PM57 shaft, where it was also possible to optimize the reinforced concrete inner lining. Based on the actual ground conditions encountered and the monitoring data, FE axial-symmetric models were created, including all the effective excavation phases (Figure 6). The rock pressure along the depth of the shaft was compared to those estimated from the Convergence-Confinement analysis. The re-evaluation led to a reduction of the rock load acting on the final lining from 520 kPa to 400 kPa. Horizontal and vertical steel reinforcements were re-evaluated using Ø14 mm with 150 mm spacing saving a significant amount of steel (about 40%) while keeping the thickness of the inner lining (60 cm) and the type of concrete used (C35/45) unchanged.

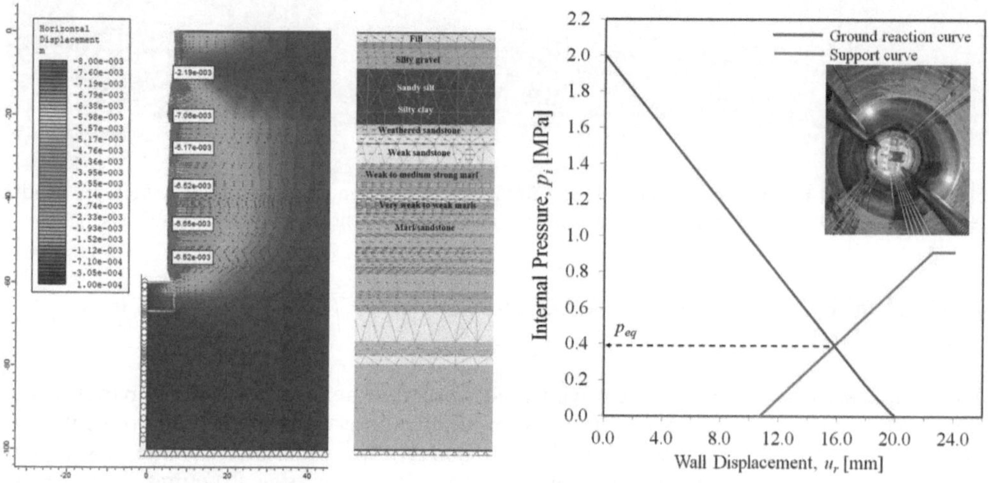

Figure 6. Back-analysis of Shaft PM57. Left: axial-symmetric FE model showing horizontal displacement calculated from the cross-check geology; Right: ground reaction and rock-support curves obtained from RocSupport 4.0.

5 CONCLUSIONS

In this paper, the benefits of using a high-performance and reliable calculation FEM software both in the design and the construction phases in conjunction with a comprehensive monitoring system installation are presented. Numerical modelling was a fundamental tool that simulates numerous excavation phases in complex ground conditions and evaluates the potential impact of the excavation works on nearby underground structures. The excavation of the main underground works was successfully completed without any critical impact recorded on the nearby existing underground structures. Comparisons with continuous monitoring data allowed for optimizing the overall design of supports, allowing an increase in excavation rate and a reduction in project costs. The execution of the underground works started in April 2018, and currently, the construction phase is on track with the contractual Construction Programme. The completion of the works is scheduled for September 2022.

ACKNOWLEDGEMENTS

The Authors would like to thank the CERN HL-LHC project for their support and permission to publish the data contained in this paper.

REFERENCES

Canzoneri A., Amiot J., Rozemberg F., Merlini D., Gianelli F., Como G., De Salvo F., Helou C., Luz Lopez- Hernandez A., Mattelaer P. (2019), *CERN (HL-LHC): New underground & surface structures at Point 1 & Point 5*, WTC 2019, Napoli - Italy.

Duncan Fama, M.E. (1993). *Numerical modelling of yield zones in weak rocks.* In Comprehensive rock engineering, (ed. J.A. Hudson) 2, 49–75. Pergamon, Oxford.

Kurzweil H. (2004), *The LHC-project of CERN – large caverns in soft rock.* In: WTC 2004, ITA, Singapore.

Panet M., Guenot A. (1982), *Analysis of convergence behind the face of a tunnel*, Tunneling 82, Bringhton.

Session 17 - Stability analysis of jointed rock slopes

The Evolution of Geotech - 25 Years of Innovation – Hammah et al. (Eds)

Modeling the effectiveness of a grout curtain for construction groundwater control in karst for a lock excavation

Vanessa C. Bateman
Chief of Civil Works Engineering, US Army Corps of Engineers, Engineering and Construction Division, HQUSACE, Washington, USA

ABSTRACT: Grout curtains have been widely used in the United States in order to mitigate or control seepage under and around a civil works structures throughout the 20th Century. Using a variety of methods, the purpose of a grout curtain is to provide a reduction in the seepage across the installation. Reduction in seepage can be particularly important in karst geology where the presence of solution widened openings in the rock that occur along bedding, joints and other features in the rock can sometimes be quite large. These features may be partially or completely infilled with clay and other materials. Grouting operations may not fully remove infilled materials and can leave weak points in the curtain that are exploited by the groundwater flow over time. Even with a well-developed ground investigation, it is impossible to completely define all the water pathways in the karst rock. Fortunately, modern computer codes supply a means to model the effectiveness of a grout curtain and to explore the consequences of a breach of that curtain in the subsurface. For this study, a lock excavation problem was set up in RS2 using the geological context from a lock excavation project that reflects the karst issues that have been encountered in multiple lock and dam projects in Tennessee and Kentucky. Two known potential breach points were established, and the effects were modeled. The results led to insights into the size of features that could cause constructability issues. In addition, the results are used to target specific zone of interest with an instrumentation and monitoring program.

1 INTRODUCTION

Seepage problems due to karst rock underneath dams and lock foundations have been well documented on many projects across the United States, requiring extensive ground modifications including repeated grout curtain installations and cutoff walls. This is particularly true of embankment dams built prior to the 1950s in Tennessee and Kentucky where the foundation designs did not adequately address karst seepage problems on projects such as Wolf Creek Dam (Zoccola, Haskins, & Jackson, 2006), Center Hill Dam (USACE, 2012) and the former Hales Barr Lock and Dam (Frink, 1946). However, the same karst terrane also presents challenges to lock construction as inflow into the excavation can cause both stability and dewatering issues. Both Chickamauga Lock (Tennessee Valley Authority, 1949) and Kentucky Lock (Tennessee Valley Authority, 1951) required grout curtain installation during original construction and the same is true of the new locks currently being constructed on both projects. Karst foundation issues are widely expected in both states where Ordovician to Mississippian aged limestone outcrops occur near the surface. Figure 1 below shows the distribution of carbonate rocks across the area with some selected locks and dams highlighted.

DOI: 10.1201/9781003188339-75

The Importance of Rock Structure and Infilling for Grout Curtain Installation in Karst

Where there are significant vertical discontinuities in the rock such as near vertical faults, joints, steeply inclined bedding or valley stress relief features, there may be significant development of vertical karst features such as those seen at Center Hill Dam (USACE, 2012) and beneath Kentucky Lock (Moneymaker & Rhodes, 1945). These can interact with near horizontal features such as bedding to produce the complicated karst development. Crawford gives an excellent explanation of the development of interconnected caves in the Cumberland Plateau and Eastern Highland Rim that shows how vertical features interact with near horizontal bedding to develop multi-level caves (Crawford, 1984). Waltham and Fookes show how this complexity can develop over time from "Juvenile" to "Complex karst" (Waltham & Fookes, 2003).

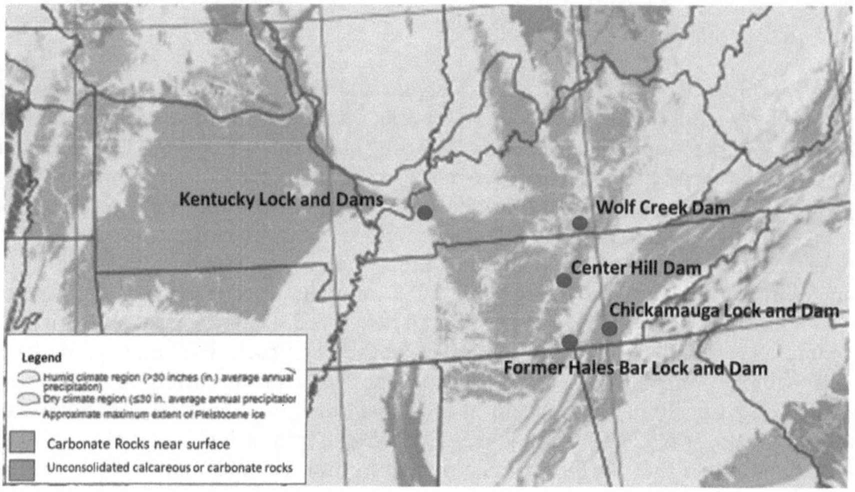

Figure 1. Distribution of Karst Terrane Across Kentucky and Tennessee with selected project shown (After (Weary & Doctor, 2014).

Further complicating matters, many of these karst features may be partially to completely infilled. Washing/grouting operations may not effectively remove this infilling leading to ineffective treatment or degradation of the curtain over time (USACE, 2017).

Many dam projects in karst (Center Hill, Wolf Creek, Hales Bar to name a few) have had repeated episodes of grout curtain installation over the lifetime of the project, leading to the conclusion by some that in this terrane it is difficult for grouting to be considered a "permanent solution." The inability of grouting to adequately address the clay infilling and vertical features below Hales Bar Dam was a large part of the decision to abandon the dam and construct Nickajack Dam as was the failure of multiple grouting programs (Tennessee Valley Authority, 1972).

2 GEOLOGY AND CONTEXT OF THE KENTUCKY LOCK SITE

To explore the effects weak points in a grout curtain used, a simplified model was set up in RS2 with the geologic context of the Kentucky Lock for an excavation. The context of this site was chosen because: 1) Kentucky Lock and Dam have previously explored significant vertical karst features (Moneymaker, 1941) and (Moneymaker & Rhodes, 1945), 2) the current construction of the new lock has exposed large areas of part of the lock foundation and 3) grouting was successfully used in a vertical feature below the upstream cofferdam to prevent excessive water infiltration and 4) the lock foundation and cofferdams are placed on one unit of limestone. Figure 2 below on the right shows the vertical features that were located beneath the upstream cofferdam and where conventional and jet grouting were successfully used to provide seepage. Grouting locations are shown as the small black circles (USACE, 2002).

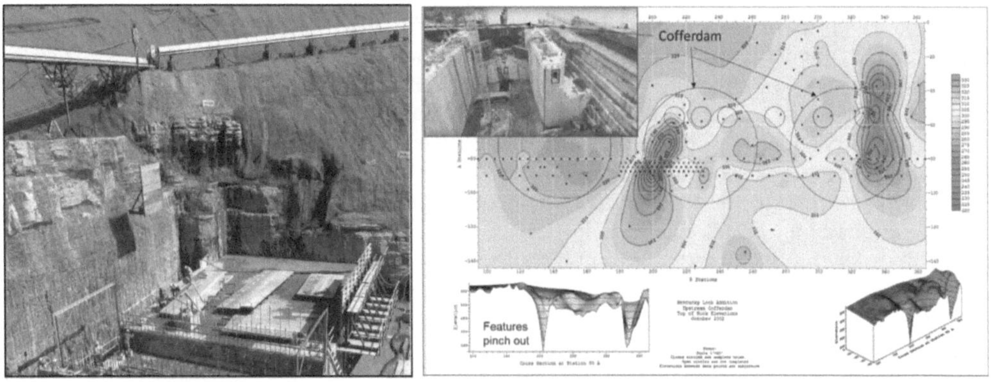

Figure 2. Right is Typical Ft Payne Formation at Kentucky Lock; Left is a Vertical Feature Beneath Upstream Cofferdam During Grouting.

The Mississippian age Fort Payne limestone has both significant vertical karst features and clear interconnection along bedding, particularly closer to the surface. Weathering in the epi-karst zone, as is typical, is more pronounced and the rock "tightens" with depth. Figure 2 below on the left shows this weathering pattern with several infilled bedding planes. A shallower vertical feature was intersected here as can be seen from the brown discoloration on the face which terminates at a bedding plane. The rock is nearly horizontally bedded.

3 SETTING UP THE SIMPLIFIED MODEL

For this initial study, the intent was to explore appropriate model parameters that could be used as a starting place for modeling on multiple projects. This model was set up with a theoretical lock excavation with a deep vertical karst feature along with two significant nearly horizontal karst features, one with little infilling and a deeper layer that was completely infilled with clay. For simplicity sake, a cliff was modeled on the right side, at the actual site, rock is not this high above the excavation. A similar, but not the same width was used for the excavation for installation of the lock and the water depth was increased.

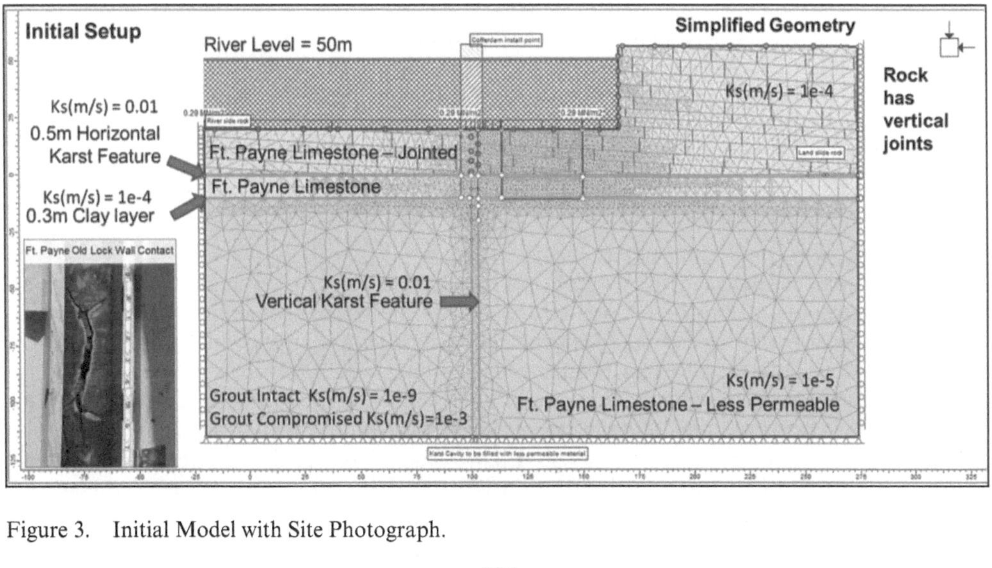

Figure 3. Initial Model with Site Photograph.

The upper layers of the Fort Payne limestone were modeled as jointed with a pattern like that evident in Figure 2 above. A higher hydraulic conductivity was used in this layer. This is a common pattern in Tennessee and Kentucky. A slight incline was added into the excavation from the river side as this situation occurred at both Hales Bar and Chickamauga Lock.

In order to produce a viable model, the vertical karst feature was modeled as a rectangle. A high static river level was used. As with the Kentucky Lock site, the vertical karst feature was placed underneath the to be constructed cofferdam, close to the wall of the excavation. The two intervening layers the 0.5m horizontal karst feature and the 0.3m clay layer were placed in order to test the grout curtain installed in the vertical karst feature. The upper layer was assumed to be only partially infilled and able to pass water freely, the lower layer was assumed to have the permeability of infilled clay.

4 EXECUTING THE EXCAVATION AND TESTING THE GROUT CURTAIN

The model was set up with 6 stages to simulate the grouting, initial dewatering and excavation of the lock. Stages 1-6 were used with variations to the grout curtain to model the effects of the following conditions: 1) Baseline – no grout curtain, 2) Grout curtain has a hole at the upper horizontal feature, 3) Grout curtain has compromised grout at the horizontal features, 4) Grout curtain is fully intact, and 5) Grout curtain is shorter, stopping 10m below grade of excavation. Sensitivity analysis was performed to test the models results with different parameters. Figure 4 below shows the stages used in the model.

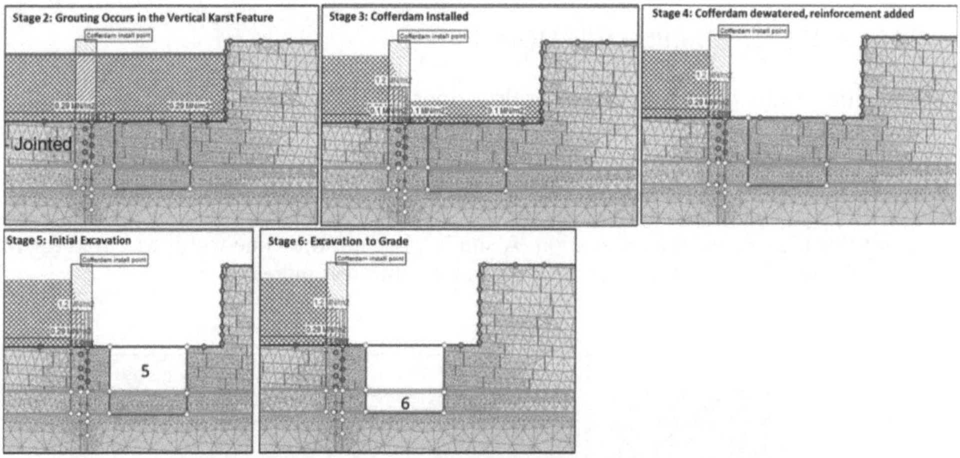

Figure 4. Stages set up in RS2 to model grout curtain and other dewatering measures.

Condition 1: Baseline - Initial Conditions

For the initial condition with no grouting, the excavation was estimated to fill with no pumping in around 14 hours. This was established as the worst-case scenario for the model. The vertical karst feature is assumed to be partially filled with a high permeability. Flow lines as expected flow downward along the feature, exiting at the first open horizontal feature. A quick estimate of the pump capacity to maintain the lock in a dewatered state with no additional measures was estimated and discharge points were established at multiple locations to estimate water flow from the various layers. Figure 5 below shows the flow lines and pressure head as well as locations of water infiltration into the excavation. An excavation length of 500m was assumed in order to estimate pumping capacity needed.

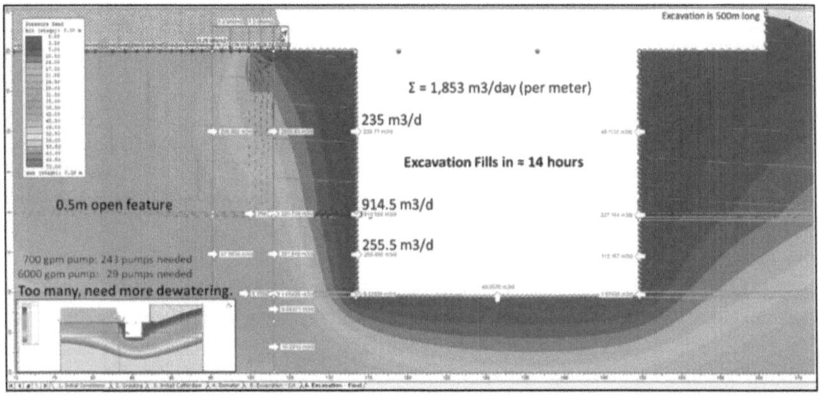

Figure 5. Initial Run with No Grouting in the Vertical Feature.

Condition 2: Grout Curtain has a Hole at the 0.5m feature.

Condition 2 assumes that grouting was performed in the feature, but that heavy flow of water along the 0.5m feature washes out the grout and the curtain is ineffective. While the excavation predicted to fill in 27 hours too many large capacity pumps would still be needed.

However, this 0.5m hole in the grout curtain produces a measurable effect that has implications for piezometer placement. The pressure head in the ground drops for a large area on the upstream size of the curtain. There is a corresponding drop on the downstream side, however, as the downstream (in excavation) side is free draining, this effect is not as noticeable. Placement of a piezometer within the 10m distance from the curtain shown in Figure 6 below both above and below the open feature would be well placed to identify a potential flaw developing over time in the grout curtain.

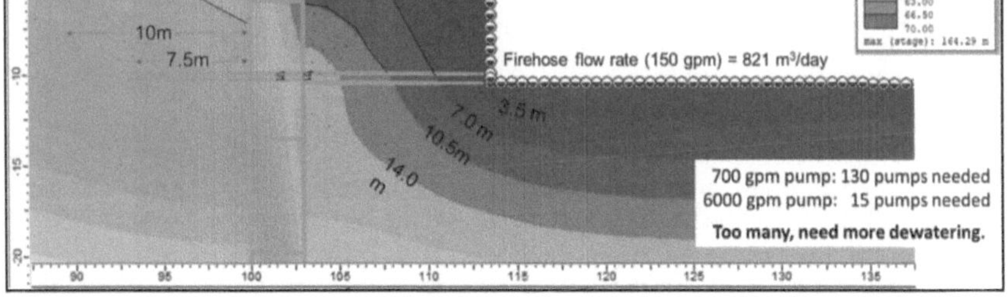

Figure 6. Condition 2 - Flaw in the Grout Curtain.

Based on this study, pressures in the piezometers placed on the upstream side would be expected to drop as the flaw developed. This is an important finding as piezometers placed to monitor flaws in grout curtains and barrier walls in Tennessee and Kentucky on many of the projects listed have historically been concentrated on the downstream side of the curtain or wall. A similar effect occurs if the bottom feature, is cleaned out from the grout curtain to the excavation.

Condition 3: Grout Curtain is compromised at the 0.5m feature, but partially effective.

Here the pressure distribution is much improved, and it takes 33 hours to fill the excavation. Under this condition, the groundwater coming into the land side is far more significant and the grout curtain can perform its function, even with compromised grout at the 0.5m horizontal feature.

Condition 4: Fully effective Grout Curtain.

For this condition, the filling without pumping takes 49 hours and only 8 large pumps are needed in the excavation. With a fully effective curtain, as with Condition 3, the groundwater infiltration from the land side becomes more significant to the problem and it becomes clear that under these groundwater conditions, dewatering wells from the land side may be needed to reduce the number of pumps that are required in the excavation. The flow on the river side of the excavation in the 0.5m feature reduces to 34 m3/day, down from 229 m3/day in condition 3.

Condition 5: Shorter grout curtain, ending 10m below the excavation grade.

Shortening of the grout curtain does result in an increase of seepage into the excavation, but it is a relatively small increase, indicating that a shorter curtain can accomplish the goal. Here it was assumed that the vertical feature "pinches out" as can be seen with features shown earlier in Figure 3. More discharge is expected through the floor and near the base of the excavation, but this only reduces the fill time of the excavation by a couple of hours.

5 CONCLUSION

While this model is not an exact replica of conditions at any of the projects mentioned, this idealized model demonstrates that this methodology can be adequately used at multiple projects with similar geologic conditions. The geology context of Kentucky Lock was used to provide a "reality check" for the parameters used. It demonstrates how karst conditions and grout curtains can be modeled with a Finite Element program for a practical solution that can be used to support design on a project site. This underused tool for grouting design can be used to explore not only the direct effects of installing a grout curtain but can also be used for exploring and understanding the consequences of inadequate or compromised grouting. This model, applied to the geometry and geology variations at other sites can be used to not only predict dewatering needed, but also can predict where to place instrumentation such as piezometers for the best chance of detection of developing problems. As grout curtains are known, particularly in karst, to often be less effective over time, this can inform long term monitoring at the site. This will be particularly significant for dam sites in karst where the seepage paths are longer and surface seepage warning signs may take more time to appear. Models such as this can be effectively used for predicting not only problematic conditions, but for helping to inform the overall long term dam safety monitoring.

REFERENCES

Crawford, N. (1984). Karst landform development along the Cumberland Plateau Escarpment of Tennessee. In R. Lefleur, *Groundwater as a Geomorphic Agent* (pp. 294–338). Boston, MA: Allen and Unwin.

Frink, J. (1946). The Foundation of Hales Barr Dam. *Economic Geology, 41*, 576–597. doi:10.2113/gsecongeo.41.6.576

Kellberg, J., & Simmons, M. (1977). Geology of the Cumberland River Basin and the Wolf Creek Damsite, Kentucky. *Environmenal and Engineering Geoscience, xiv*(4), 245–269. doi:10.2113/gseegeosci.xiv.4.245

Moneymaker, B. (1941). Subriver solution cavities in the Tennessee Valley. *Journal of Geology*, 74–76.

Moneymaker, B. (1948, February). Some Broad Aspects of Limestone Solution in the Tennessee Valley. *Transactions of the American Geophysical Union, 29*(1), 93–96. doi:10.1029/TR029i001p00093

Moneymaker, B., & Rhodes, R. (1945). Deep Solution Channel in Western Kentucky. *Geologic Society of Americal Bulletin, 56*(1), 39–44. doi:10.1130/0016-7606
Tennessee Valley Authority. (1949). *Geology and Foundation Treatment: Tennessee Valley Authority Projects.* Washington, DC: US Government Printing Office.
Tennessee Valley Authority. (1951). *The Kentucky Project: A Comprehensive Report on the Planning, Design, Construction and Initial Operations of the Kentucky Project.* Washington, DC: US Government Printing Office.
Tennessee Valley Authority. (1972). *The Nickajack Project: A Report on the Planning, Design, Construction, Initial Operations and Costs.* Knoxville, TN: Tennessee Valley Authority.
USACE. (2002). Kentucky Lock Upstream Cofferdam Installation Files. Nashville, TN: US Army Corps of Engineers.
USACE. (2012). *Supplement to Major Rehabilitation Evaluation Report.* Nashville, TN: US Army Corps of Engineers, Nashville District.
USACE. (2017). *EM 1110-23506: Grouting Technology.* Washington, DC: US Army Corps of Engineers.
Waltham, A., & Fookes, P. (2003). Engineering classification of karst ground conditions. *Quarterly Journal of Engineering Geology and Hydrogeology, 36,* 101–118.
Weary, D., & Docton, D. (2014). *Karst in the United States: A Digital Map Compilation and Database.* Reston, VA: US Geological Survey. Retrieved from https://pubs.usgs.gov/of/2014/1156/pdf/of2014-1156.pdf
Zoccola, M., Haskins, T., & Jackson, D. (2006). The Problem is the Solution: A History of Seepage, Piping, and Remediation in a Karst Foundatoin at Wolf Creek Dam. *U.S. Society on Dams (USSD) 26th Annual Meeting.* San Antonio, TX: USSD.

The Evolution of Geotech - 25 Years of Innovation – Hammah et al. (Eds)

Application of Discrete Fracture Networks (DFN`s) to the design of benches in an open pit mine in South Africa

V. Kuppusamy
[VK] *GeoPioneers Pty (Ltd), Johannesburg, South Africa*

ABSTRACT: The global mining market has become increasingly competitive and as a result design engineers are constantly striving to reduce the overall mining costs. In the open pit mining industry, this translates to an excavation of the steepest possible slope angles that would result in the lowest stripping ratio while ensuring maximum ore recovery. Achieving the steepest slope angles often starts with optimizing the bench design as this is the building block of any open pit geotechnical slope design. The DFN approach is perhaps one such method that has been identified to produce more realistic or optimal pit slope designs. This paper presents a case study in South Africa that illustrates the application of discrete fracture networks to the design of open pit benches. The results of this case study would suggest that the DFN approach using FracMan produces more optimal slope designs as compared to the traditional deterministic and pseudo-probabilistic methods (Kuppusamy, 2020).

1 INTRODUCTION

The commonly used phrase, "*we are only as strong as our weakest links*" adequately describes the role of rock mass structures in the stability of open pit benches. The methods applied to model these rock mass structures can result in markedly different stability analysis and bench configurations (Kuppusamy, 2020). Modelling these rock mass structures have always been challenging due to their high degree of variability and uncertainty. Therefore, methods that attempt to better address variability and uncertainty would produce more realistic representations of structures that are more readily observed in nature The DFN approach has perhaps had the most success with realistically modelling the heterogeneity of the rock mass structures because it allows for the characterization of individual fracture properties (e.g., orientation, spacing, size, aperture).

Furthermore, over the years with the advancement in technology, there has been many improvements to the structural data collection methods and instrumentation, which have subsequently increased the accuracy, quality, and availability of data relating to rock mass structures. Methods such as photogrammetry, acoustic and optical borehole televiewers, and high-resolution geophysics, are widely used in the mining industry (Rogers, et al., 2006). However, incorporating this structural data into the methodology commonly used for the design of rock slopes still remains a challenge for many practitioners. The kinematic analysis carried out in the conventional bench design approach includes many assumptions regarding the properties of these structures which may not be realistic. The discrete fracture network (DFN) approach offers practioners with an alternative tool that attempts to address some of these unrealistic assumptions and better utilize the structural data available today.

DOI: 10.1201/9781003188339-76

This paper describes a case study in South Africa that applies the DFN methodology to the design of open pit benches. It involved the use of available mine data to generate a three-dimensional DFN using FracMan (developed by Golder Associates) and then evaluating the associated wedge block stability results and its implications on the bench design process. For comparative purposes, the traditional wedge stability analyses consisting of the stereographic kinematic assessment and subsequent limit equilibrium analysis using SWedge was also included in this study.

2 FIELD DATA SUMMARY

2.1 *General site description*

This case study is based on data collected from an open pit phosphate bearing-ore mine located in Limpopo, South Africa. Figure 1 shows the pyroxenite open pit at the time of the site investigation which was approximately 200m deep and extended 1.5km long in the north-south direction and 850m wide in the east-west direction. This pyroxenite deposit forms part of the Phalaborwa Igneous Complex (PIC) which is considered the most important carbonatite complex in South Africa (Roux, et al., 1989). The Phalaborwa igneous complex is characterized by Cu-Zr mineralization that includes predominantly pyroxenites, foskorite and carbonatite.

Figure 1. Case study open pit pyroxenite mine in South Africa.

The pyroxenite open pit had been divided into pit design sectors based on the pit wall orientations and similar expected geotechnical conditions. Table 1 provides the average dip direction that the pit walls face in each pit sector.

Table 1. Open pit design sectors.

Pit Sector	North East Sector	East Sector	South Sector	South West Sector	West Sector	North West Sector
Average Wall Face Dip Direction	218°	295°	040°	352°	085°	147°

2.2 *Structural data collection and analyses*

The structural available consisted of only mapping data. Both scanline and window mapping surveys were carried out on the open pit bench exposures across various lithological units, namely, granite, dolerite, massive pyroxenite and feldspathic pyroxenite. The scanline surveys recorded the following geotechnical parameters according to standard ISRM methods (ISRM, 1978): discontinuity type (e.g., joint, bedding etc.); persistence; infilling; roughness; joint roughness coefficient (JRC) (Barton, 1987); termination index; aperture; rock type; and groundwater condition. Photogrammetry methods were used to digitally map structures within benches that were either inaccessible and/or considered too dangerous to complete physical bench mapping surveys. Figure 2 shows an example of the 3D outcrop model produced and digitally mapped using Sirovision.

Figure 2. Example of 3D outcrop model created and mapped using Sirovision.

All available structural orientation data relating to the massive pyroxenite (MP) lithological unit was processed and analyzed using statistical stereographic methods using ©Rocscience Dips program. This assessment together with an understanding of the local geological conditions allowed for the selection of discontinuity sets that were considered to be representative of the structural fabric of the rock masses prevalent in the pit walls. Figure 3 shows the stereonet showing the contours of the four discontinuity sets identified for the MP rock mass and Table 1 summaries the set orientation properties and histogram of discontinuity spacing and trace length data for each discontinuity set were fitted with various distributions types (e.g., exponential, lognormal, normal, inverse gaussian and gamma) and their equivalent parameters estimated accordingly. Furthermore, goodness-of-fit tests were performed to assess if the selected distributions adequately fit the data. Figure 4 shows an example of the histogram produced for the spacing data for set J1. The lognormal distribution was the best fit curve for the J1 set spacing data as indicated by visual estimation as well as the goodness-of-fit test results.

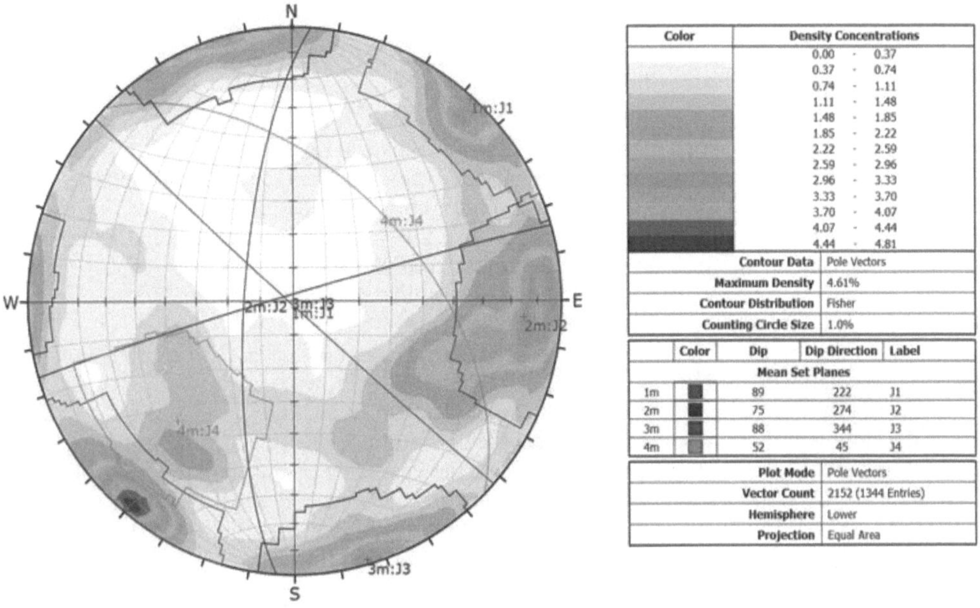

Figure 3. Stereographic projection showing contour concentration of major discontinuity sets.

Table 2. Summary of major discontinuity sets.

Set ID	Set Name	Dip (degrees)	Dip Direction (degrees)	Fisher`s K-value
1m	J1	89	222	35.1
2m	J2	75	274	25.8
3m	J3	88	344	28.9
4m	J4	52	045	23.1

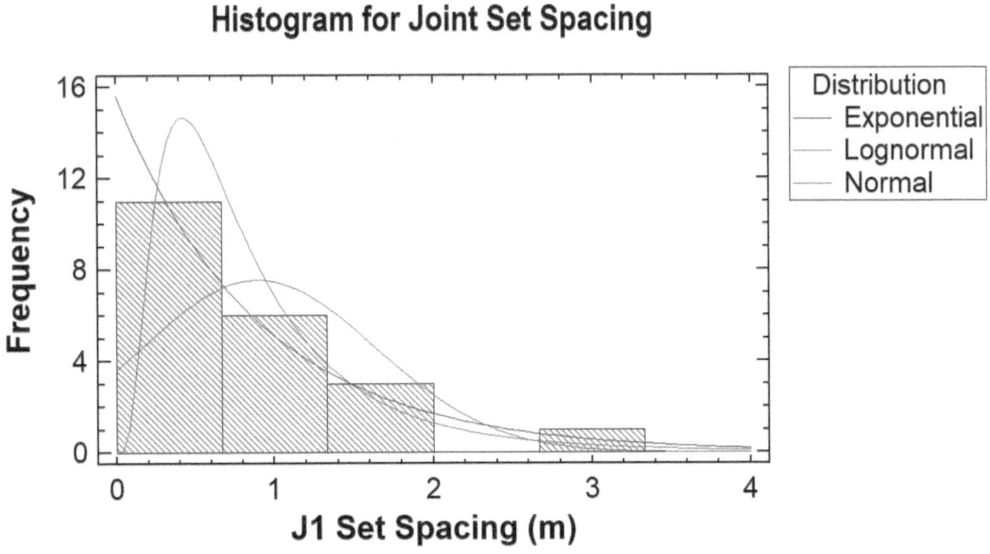

Figure 4. Example of histogram showing discontinuity spacing data.

3 DISCRETE FRACTURE NETWORK MODELLING

3.1 *DFN modelling methodology*

Discrete fracture network modelling is a methodology used to generate three-dimensional stochastic representations of rock mass structures from statistical distributions that describe their characteristics (e.g., orientation, spacing, size, aperture). The stochastic nature of the DFN modelling process is such that there are an infinite number of possible realizations of the 3D fracture system based on the sampled field data. The DFN model generated from this sampled field data (i.e., 1D/2D discontinuity data collected from variably oriented rock exposures and boreholes) share the sample statistics and allow for explicit modelling of an individual fracture or simplified fracture sets (Elmo, et al., 2015).

The principles of DFN modelling have been documented by several researchers such as (Dershowitz, et al., 1998), (Jing & Stephansson, 2007), (Dershowitz & Einstein, 1988), (Elmo, et al., 2015), (Grenon & Hadjigeorgiou, 2008) and (Rogers, et al., 2006).

The development of a DFN model requires the definition of several fracture properties summarized in Table 2. Primary properties represent the geometry of the fracture properties and is essential for all DFN modelling. The secondary properties are defined based on the DFN application (i.e., hydraulic or geomechanical). It should be noted that the term "fracture" and "discontinuity" are synonyms and are used interchangeably in this paper.

Table 3. Properties for defining a DFN model (Elmo, et al., 2015).

Primary	Secondary
Orientation distribution	Aperture distribution
Fracture length distribution	Fracture shear strength properties
Fracture intensity distribution	Fracture stiffness properties
Spatial variation	Transmissivity distribution
Termination percentage	Storativity distribution
	Termination Percentage

3.2 *Mine bench-scale DFN model*

Golder developed software FracMan generates stochastic simulations of fracture networks based on user-defined inputs that are generally obtained through the statistical analysis and interpretation of fracture data. For this case study, FracMan was used to generate the bench-scale DFN model using the geometric network construction method.

The DFN model inputs were derived using mine outcrop mapping data available. The P32 intensity (i.e., area of fractures per unit volume) was estimated using the P32 simulation method based on P21 (i.e., length of fractures per unit area) data derived from photogrammetric surveys as described by (Rogers, et al., 2017). Fracture size parameters were determined from trace length statistical distributions and converted to an equivalent fracture radius using analytical methods provided by (Zhang, et al., 2002). The lognormal normal distribution was the best fit for the trace length data for each discontinuity set.

The Fisher distribution was selected to model the orientation dispersion and the Enhanced Baecher model applied to the spatial distribution of fractures. Fracture termination was not assigned for this modelling exercise so that the fracture generation would not be conditioned to this. A four-sided polygon was selected to represent the fracture shape for the DFN model. The fracture networks were generated within an outer rock mass box model region (i.e., 45m x 45m x 22.5m) and clipped to the bench-scale model region (i.e., 30m x 30m x 15m) to reduce boundary effects (Kuppusamy, 2020). A total of five realizations were generated for the Mine X DFN bench-scale model as shown in Figure 5.

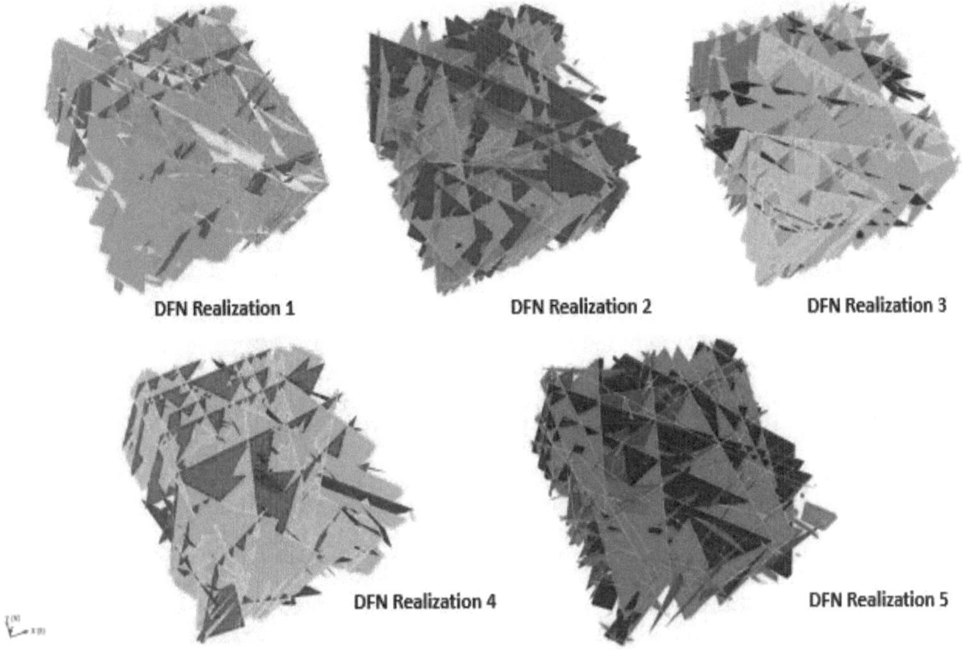

Figure 5. Case Study DFN model realizations *(Kuppusamy, 2020).*

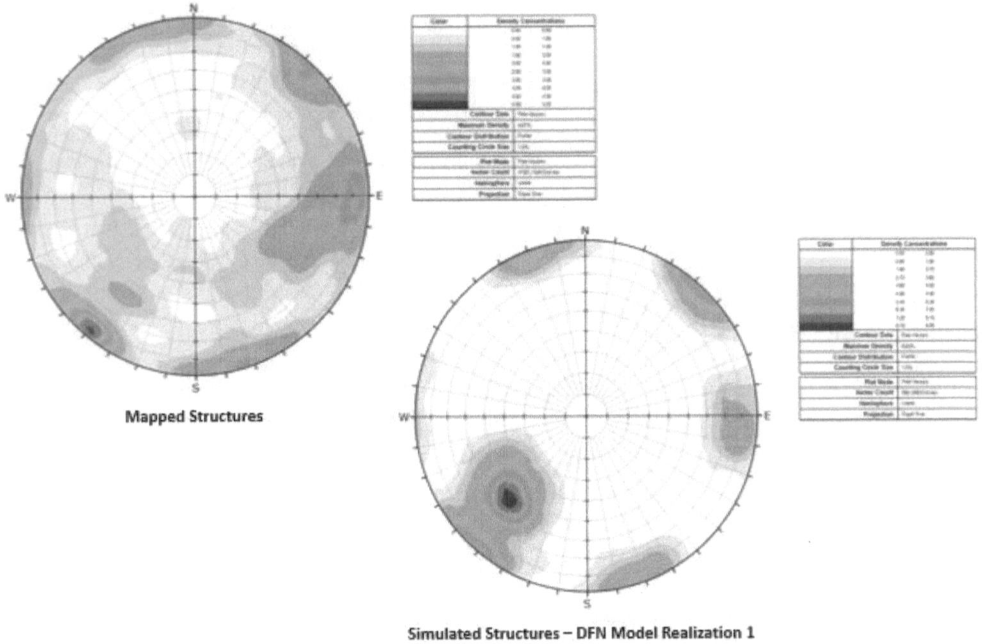

Figure 6. Model validation example: Comparison of mapped structures with simulated fractures from DFN model realization 1.

Model validation is critical in evaluating the suitability of the synthetic fracture network and its intended application. The DFN realizations generated with FracMan produce these statistical models that were validated by comparing model parameters (e.g., orientation and fracture intensity) with the field data and observations. In general, the DFN model validation process has shown a reasonable level of agreement between the DFN model and the mine field data. An example of the fracture orientation comparison between mapped and simulated fractures are shown below in Figure 6.

4 WEDGE STABILITY RESULTS

This case study focuses on the geometric characteristics related to wedge formation and stability and their subsequent bench design implications. It compares the results of the wedge stability as derived from the traditional method and the DFN approach. The wedge stability analyses were completed using these methods considering the following:

a. Bench face angle (BFA) of 90, 80 and 70 degrees.
b. Bench height of 15m; bench scale slope surfaces with slope face directions corresponding to the east, west, south, southwest, northeast, and northwest pit wall sectors of the mine.
c. Constant shear strength parameters (Mohr-Coulomb model, cohesion of 0kPa and friction angle of 30°).
d. Unit weight of 2700 kg/m^3.
e. No water pressure, seismicity, fracture stress or tension crack.

Selected results from this case study are provided in the sub-sections below with further details available in (Kuppusamy, 2020).

Table 4. Example of case study SWedge deterministic and probabilistic results.

		Wedge Information				Deterministic		Probability of Failure	
Pit Sector	Pit Wall Mean Dip Direction	Joint Set Combination		Trend (°)	Plunge (°)	Volume (m^3)	Factor of Safety	Factor of Safety <1.0	Factor of Safety <1.3
East	295°	J1	J2	309	72	130.2	0.37	0.75	0.81
		J1	J3	293	87	2.0	0.06	0.56	0.57
		J2	J3	261	75	102.1	0.15	0.89	0.90
		J2	J4	352	38	625.9	1.25	0.32	0.51
South	040°	J2	J4	352	38	3549.6	1.25	0.47	0.59
		J3	J4	72	49	3769.4	0.45	0.76	0.86
North East	218°	J1	J3	293	87	1.6	0.01	0.51	0.51
		J1	J4	132	04	1638.7	24.96	0.04	0.04
		J2	J3	261	75	31.5	0.20	0.76	0.77
North West	147°	J3	J4	072	49	84.9	0.75	0.49	0.60
West	085°	J3	J4	072	49	469.0	0.75	0.68	0.81
South West	352°	J1	J2	309	72	33.7	0.37	0.64	0.70
		J1	J3	293	87	3.7	0.02	0.56	0.56
		J2	J4	352	38	214.7	1.25	0.35	0.52
		J3	J4	072	49	108.0	0.45	0.73	0.81

4.1 *Traditional kinematic and wedge stability analysis*

The traditional method referred to in this paper involves a kinematic stereographic assessment and the subsequent limit equilibrium wedge stability assessment using ©Rocscience SWedge.

Both the deterministic and probabilistic analyses was carried out using SWedge for all pit design sectors. An example of the deterministic results considering a BFA of 90 degrees is summarized in Table 4. Figure 7 shows the example of the wedge formed in the east sector pit wall.

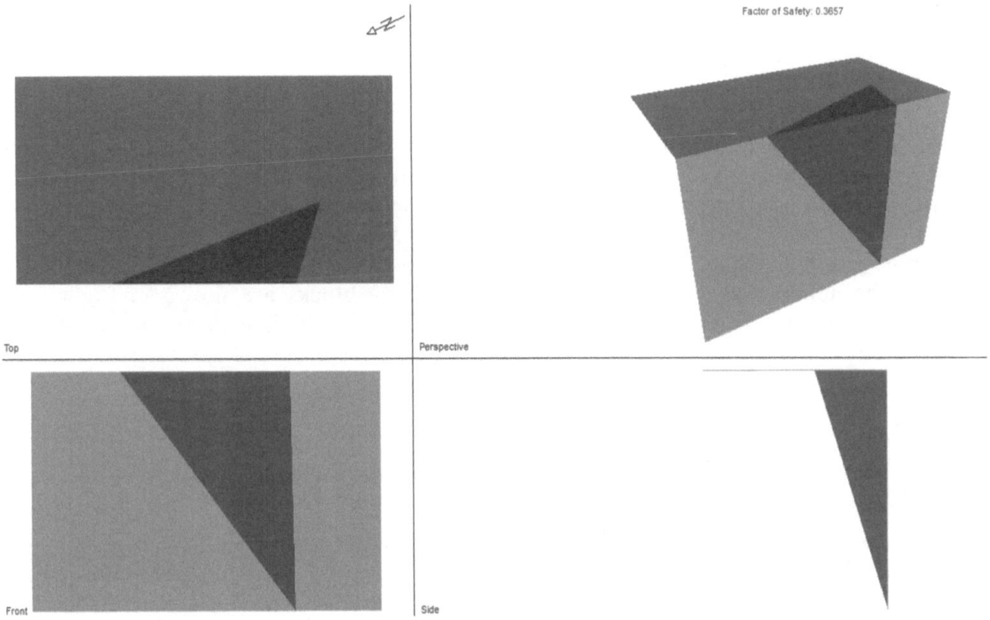

Figure 7. SWedge: Bench-scale wedge formed by J1 and J2 within the east pit wall (slope dip direction 295°).

Based on the SWedge deterministic results, the maximum sized unstable wedge (i.e., wedge volume of $3769.4m^3$) is expected to occur in the south pit sector walls. In general, decreasing the bench face angle reduces the volume of the maximum sized wedge and the number of potential wedges formed, and overall stability of the benches is improved. The equivalent probabilistic results for the south pit sector indicate the average maximum sized volume wedge formed by intersecting sets J3 and J4 is $1143.9m^3$.

4.2 *DFN rock wedge analysis*

FracMan software developed by Golder Associates utilizes an explicit block search algorithm, referred to as "Rock Wedge Analysis", to evaluate the rock block stability of surface or underground excavations (FracMan Technology Group, 2016). This allows FracMan to identify 3D blocks defined by the intersection of the underlying DFN model with a specified free surface (e.g., a rock slope face). The stability analysis in FracMan is based on (Goodman & Shi, 1985) block theory. Table 5 provides a summary of the FOS categories associated with the FracMan rock wedge analysis.

The rock wedge analysis was carried out for each DFN realization for each pit sector. Examples of the FracMan rock wedge results are shown in Figure 8 and Figure 9 (NB: Green blocks are stable and red blocks are unstable). As shown in Figure 8, the decrease in the BFA angles results in different stability analysis as

Table 5. FracMan rock wedge analysis factor of safety categories.

Factor of Safety (FOS)	Failure Mode Description
FOS = 100	Kinematically Inadmissible
FOS > 1.0	Stable Blocks
FOS < 1.0	Unstable Blocks
FOS = 0	Freefall Blocks

indicated by the different unstable wedges daylighting the slope. Based on all the DFN model realizations for all pit sectors rock wedge results, the maximum sized unstable block was identified in the southwest sector with a predicted block volume of 83.7 m^3.

Example graph showing the number of total blocks formed in the selected bench slope face and the number of stable, unstable, and freefalling blocks are shown in Figure 10. As shown in Figure 10, the highest number of unstable blocks is expected in the north-east pit sector.

4.3 *Comparison of traditional method and DFN approach*

Based on the wedge stability results from this case study, the following key observations were noted (Kuppusamy, 2020):

a. The SWedge predicted maximum volume for a single wedge is substantially larger than the corresponding scenario result using FracMan. For example, the south sector (90° BFA) was identified as having the largest maximum sized wedge volume (greater than 3000m^3), whilst the DFN model realizations predicted a maximum unstable block volume of ± 45m3. This is as expected given the continuous and infinite fracture length assumed in the traditional wedge stability analysis.
b. The smaller block volumes observed with the DFN approach could be attributed to the DFN models considering the distribution of fracture length and spacing. In addition, it allows for the intersections of multiple discontinuity sets (multiple wedges) whilst the traditional approach only considers the intersections of two discontinuity sets (i.e., single wedge).
c. The DFN method allows for both complex and simple block geometries to be formed.
d. Predicted failure volumes are not reduced with a decrease in bench face angles as there are cases where the DFN model has identified new wedges. This stochastic approach is more realistic as it better resembles the observations at the mine.
e. The traditional method is advantageous as it is relatively easy to use and understand, requires lower computational power and the results are obtained almost immediately whilst the DFN approach is comparatively more complicated and requires more model inputs and results are not as readily available. However, given the more realistic results produced and its potential impact on the mine`s economic performance, the additional effort is justifiable.
f. The stereographic projections are limited in that they consider only the angular relationships between lines and planes and do not represent the sizes and positions of the structures, unlike the FracMan rock wedge analysis.
g. The FracMan rock wedge algorithm results in a higher accuracy of block shape and volume, which is considered more appropriate for the kinematic assessment of block stability.
h. The DFN approach makes better use of all available structural data that technology has made easily available today.

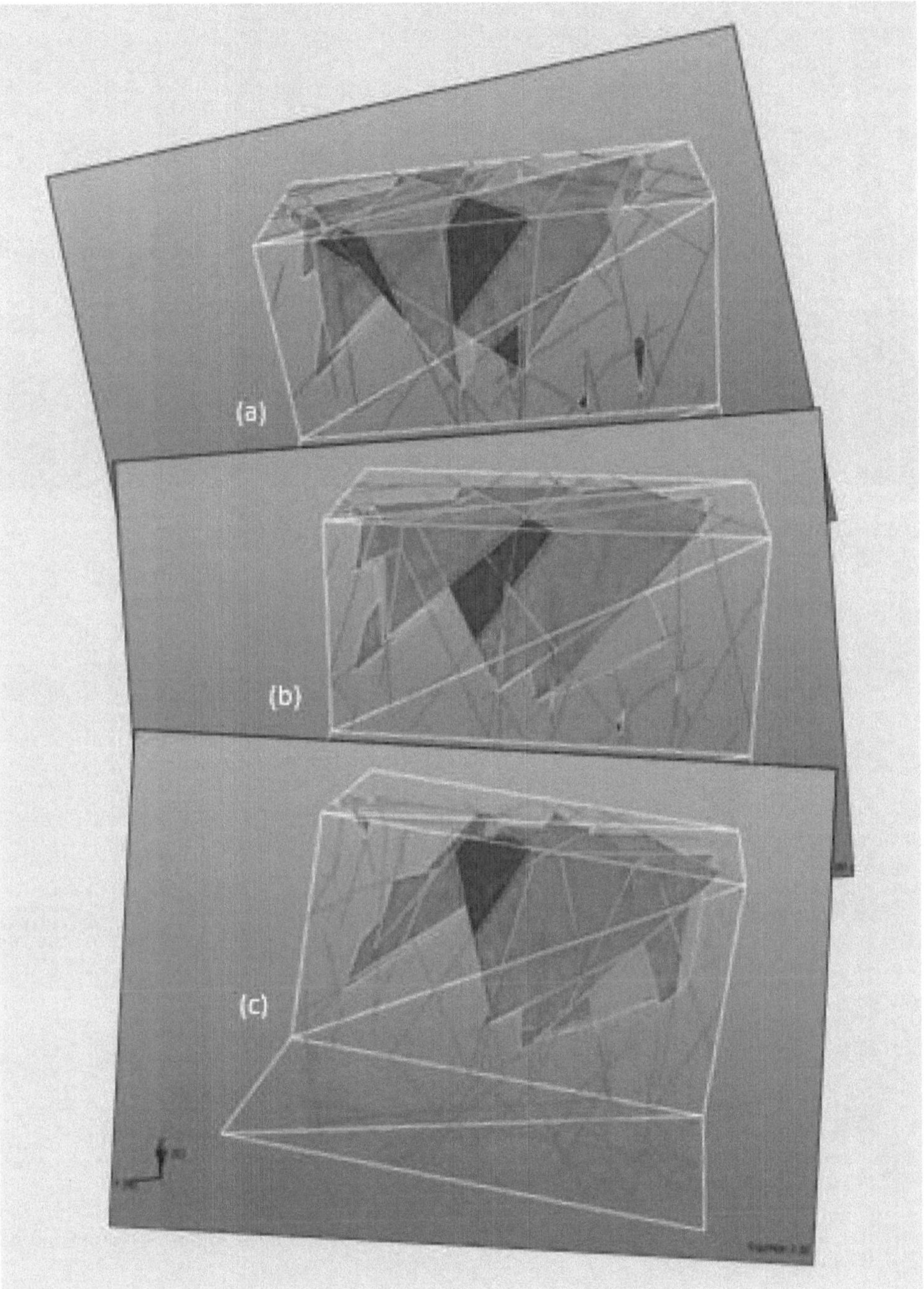

Figure 8. FracMan rock wedge result for DFN realization No. 5, East Pit Wall, (a) 90 ° BFA, (b) 80 ° BFA, (c) 70 ° BFA.

5 BENCH DESIGN CONSIDERATIONS

Bench design components (i.e., bench height, bench width and bench face angle) contribute to the inter-ramp geometry which comprises the overall slope design. Bench heights are generally fixed by the operational machinery; the bench face angles are selected by considering the

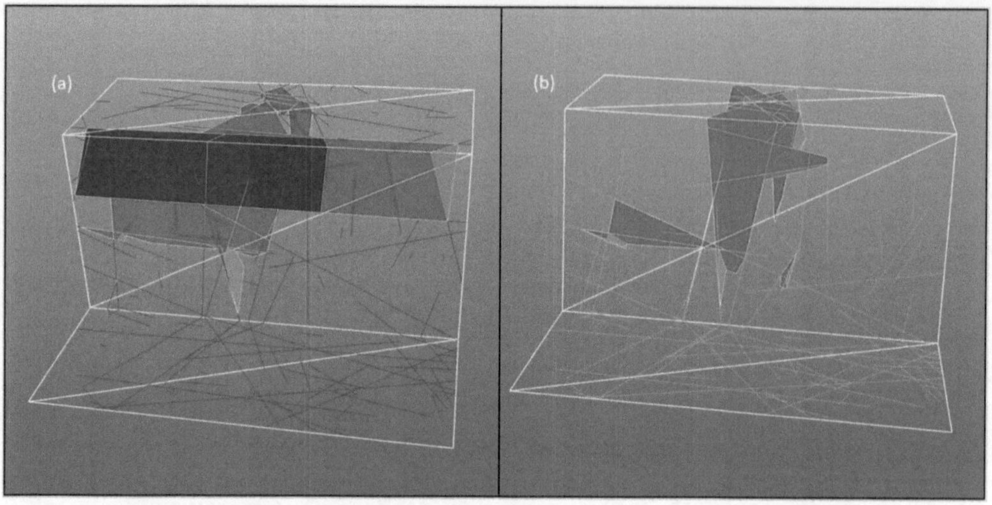

Figure 9. FracMan rock wedge result for DFN realization No. 2, Northeast Pit Wall, (a) 90 ° BFA, (b) 80 ° BFA.

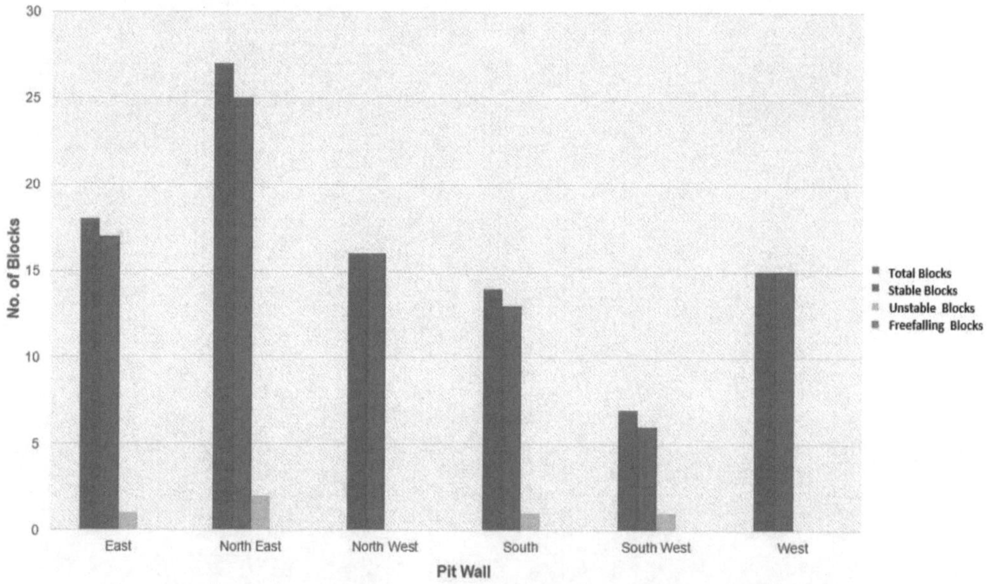

Figure 10. DFN realization 1 - Graph showing the no. of blocks identified within the various pit wall benches with a BFA of 90 degrees.

results of the kinematic failure modes in conjunction with the level of risk accepted by the operating mine. Bench/berm widths are design to accommodate potential failure material and compensate for planar back break to ensure adequate working space for mining machinery. This case study was focused on the wedge failure volumes and its influence of the design of bench widths and to some extent, bench faces angles.

This case study wedge stability results (i.e., SWedge deterministic, SWedge probabilistic and FracMan rock wedges results) were considered isolated and applied separately to determine bench design parameters. This is summarized in Table 6 and was based on the following (Kuppusamy, 2020):

a. From the SWedge deterministic results, wedges that exceed the accepted design criteria (i.e., FOS <1.1) were considered and an 80% design threshold was applied to these predicted wedge volumes.
b. From the SWedge probabilistic results, wedges that exceed the accepted design criteria (i.e., PoF [FOS<1] between 25-50%) and the 80% design threshold were applied to the corresponding wedge volumes predicted. Both the SWedge predicted minimum bench width and the (Haines, et al., 2006) berm width was provided.
c. FracMan rock wedge results considered the "worst case" scenario DFN realization with the 80% debris volume applied.
d. Based solely on wedges controlling the stability of benches and does not consider planar back break effects nor effective bench face angles.

Table 6. Case study bench design parameters estimated from wedge stability results.

		Pit Design Sector					
Slope Design Element	Wedge Stability Results	East Sector	South Sector	North East Sector	North West Sector	West Sector	South West Sector
		295°	040°	218°	147°	085°	352°
Bench Height	All	15m	15m	15m	15m	15m	15m
Bench Face Angle	SWedge Deterministic	90°	70°	90°	90°	80°	90°
	Swedge Probabilistic	90°	70°	90°	90°	80°	90°
	FracMan Rock Wedge	90°	90°	90°	90°	90°	90°
Minimum Bench Width	SWedge Deterministic	5.3m	10.6m	3.4m	4.7m	7.5m	5.1m
	SWedge Probabilistic	3.1m *5.0m	2.1m *10.6m	3.7m *3.4m	2.6m *4.7m	2.8m *7.5m	2.6m *3.4m
	FracMan Rock Wedge	7.3m	3.8m	4.1m	1.8m	3.3m	4.6m
Maximum Inter-ramp Angle	SWedge Deterministic	71°	43°	77°	73°	56°	71°
	SWedge Probabilistic	78° *72°	63° *43°	76° *77°	80° *73°	70° *56°	80° *77°
	FracMan Rock Wedge	64°	76°	75°	83°	78°	73°

*Based on (Haines, et al., 2006) bench width estimation method

As shown in Table 6, the wedge stability results derived from the different methodologies would have an influence on the recommended design parameters. This is most evident with the bench width estimation and related estimated maximum inter-ramp angle. The results of the DFN analysis suggests that the current bench design can be optimized.

6 CONCLUSIONS

The use of discrete fracture networks is becoming increasingly common in geomechanics, (Dershowitz, 1984), (Dershowitz & Einstein, 1988), (Grenon & Hadjigeorgiou, 2012), (Elmo, et al., 2015), (Rogers, et al., 2007), (Stacey, et al., 2014) and whilst this approach has been applied to large-scale slope stability problems, its application to smaller scale problems is comparatively more limited. One such application limitation is its use in the design of open pit benches.

This case study shows that the traditional methods applied to design open pit benches; whilst considered valuable, is better understood, not to mention, quick and easy to use; its conservative assumptions can have significant impacts on the mines` economic performance. The benchscale FracMan DFN model results from this case study suggests that the DFN approach would produce more optimal slope designs as compared to the traditional deterministic and pseudo-probabilistic methods. In light of the potential economic benefits, the effort (i.e., higher effort required with DFN methods as compared to traditional methods) versus the value (better economic performance and management of slopes) of applying the DFN method is justified for this case study. Therefore, practitioners should consider the use of DFN`s as a tool to optimize and produce more realistic slope design in the open pit mining industry.

It should be noted that although the DFN approach better represents rock mass structures and can produce more optimal bench designs, the full potential of a DFN model to capture stochastic variability can only achieved if the generated DFN geometry is statistically the same or at least similar (within some tolerance) to the actual fracture network it is meant to simulate. If this is not the case, then this would perhaps be an inefficient and complicated alternative approach to the traditional bench design approach (Poropat & Elmouttie, 2011).

REFERENCES

Barton, N., 1987. Discontinuities. In: F. G. Bell, ed. *Ground Engineers` Reference Book*. London: Butterworths.

Dershowitz, S. W., 1984. *Rock Joint Systems*, Cambridge, Massachusetts: Massachusetts Institute of Technology (unpublished PHD thesis).

Dershowitz, S. W. & Einstein, H. H., 1988. Characterizing Rock Joint Geometry with Joint System Models. *Rock Mechanics and Rock Engineering*, Volume 21, pp. 21–51.

Dershowitz, W. S. & Herda, H. H., 1992. *Interpretation of fracture spacing and intensity*. Santa Fe, New Mexico, 33rd US Rock Mechanics Symposium.

Dershowitz, W. S., Lee, G., Geier, J. & LaPointe, P. R., 1998. *FracMan: Interactive discrete feature data analysis, geometric modelling and exploration simulation - User Documentation*. Seattle, Golder Associates Inc..

Elmo, D., Rogers, S., Stead, D. & Eberhardt, E., 2014. Discrete Fracture Network approach to characterise rock mass fragmentation and implications for geomechanical upscaling. *Institute of Materials, Minerals and Mining and The AusIMM*, 123(3), pp. 149–161.

Elmo, D., Stead, D. & Rogers, S., 2015. *Guidelines for the Quantitative Description of Discontinuities for use in Discrete Fracture Network Modelling*. Vancouver, Canada, International Symposium on Rock Mechanics.

Elmouttie, M. K. & Poropat, G. V., 2015. *Uncertainty Propagation in Structural Modelling*. Vancouver, Canada, International Symposium on Rock Slope Stability in Open Pit Mining and Civil Engineering.

FracMan Technology Group, 2016. *FracMan Geomechanics Edition*. [Online] Available at: www.fracman.com [Accessed 2016].

Golder Associates, 2006. *FracMan - Interactive discrete fracture network (DFN) analysis for representing key discrete fractures in 3D space, geometric modelling and exploration simulation*, Seattle: Golder Associates Inc.

Goodman, R. E. & Shi, G. H., 1985. *Block Theory and its Application to Rock Engineering*. New Jersey: Prentice-Hall.

Grenon, M. & Hadjigeorgiou, J., 2008. A design methodology for rock slopes susceptible to wedge failure using fracture system modelling. *Engineering Geology*, Volume 96, pp. 78–93.

Grenon, M. & Hadjigeorgiou, J., 2012. Application of fracture system models in mining and civil rock engineering design. *International Journal of Mining, Reclamation, and Environment*, 26(1), pp. 55–73.

Haines, A., Voulgaris, P., Walker, D. & de Bruyn, I., 2006. Geotechnical design considerations for the proposed Oyu Tolgoi open pits, Southern Mongolia. *The South African Institute of Mining and Metallurgy International Symposium on Stability of Rock Slopes in Open Pit Mining and Civil Engineering*, Issue 44, pp. 133–154.

ISRM, 1978. Suggested Methods for the Quantitative Description of Discontinuities in Rock Masses. International Society for Rock Mechanics: Commission on Laboratory and Field Tests. *International Journal of Rock Mechanics and Mining Science & Geomechanics*, Volume 15, pp. 319–368.
Jing, L. & Stephansson, O., 2007. *Fundamentals of Discrete Element Methods for Rock Engineering: Theory and Applications.* Volume 85 ed. s.l.:Elsevier B.V.
Kuppusamy, V., 2020. *Application of Discrete Fracture Networks (DFN) to the Design of Open Pit Benches in Rock Slopes*, Johannesburg, South Africa: University of Witwatersrand (currently unpublished masters dissertation).
Poropat, G. V. & Elmouttie, M. K., 2011. *Uncertainty Propogation in Structural Modelling.* Vancouver Canada, Internation Symposium on Rock Slope Stability in Open Pit Mininh and Civil Engineering.
Rogers, S. F., Bewick, R. P., Codelco, A. B. & Gaudreau, D., 2017. *Intergrating photogrametry and discrete fracture network modelling for improved conditional simulation of underground wedge stability.* Perth, Austrailian Centre for Geomechanics.
Rogers, S. F., Kennard, D. K., Dershowitz, W. S. & Van As, A., 2007. *Characterising the In Situ Fragmentation of a Fractured Rock Mass Using a Discrete Fracture Network Approach.* Vancouver, Canada, ARMA, American Rock Mechanics Association.
Rogers, S. F., Moffitt, K. M. & Kennard, D. T., 2006. *Probabilistic Slope and Tunnel Block Stability Analysis using Realistic Fracture Network Models.* Colorado, American Rock Mechanics Association (ARMA), 41st U.S Symposium on Rock Mechanics (USRMS).
Roux, E. H. et al., 1989. Phosphate in South Africa. *Journal of South African Institute of Mining and Metallurgy*, 89(5), pp. 129–139.
Stacey, T. R., Armstrong, R. & Terbrugge, P. J., 2014. *Experience with the development and use of a simple DFN approach over a period of 30 years.* Vancouver, International Discrete Fracture Network Engineering Conference.
Zhang, L., 2017. *Engineering Properties of Rocks.* 2nd ed. Cambridge, USA: Elsevier Ltd..
Zhang, L., Einstein, H. H. & Dershowitz, W. S., 2002. Stereological relationship between trace length and size distribution of elliptical discontinuities. *Ge'otechnique*, 6(52), pp. 419–433.

The Evolution of Geotech - 25 Years of Innovation – Hammah et al. (Eds)

Comparison of 2D-3D limit equilibrium and finite element methods for the analysis of bi-planar rock slope failures

Y.U. Doğan
Geodestek Ltd. Sti., Ankara, Turkey

N.S. Isik
Gazi University Faculty of Technology Department of Civil Engineering, Ankara, Turkey

ABSTRACT: Rock slopes containing highly persistent discontinuities or discontinuity sets (i.e. bedding or foliation planes, faults) having parallel or sub-parallel inclination to the slope face are designated as footwall slopes. These slope types commonly form in weak, thinly bedded, orthogonally jointed, sedimentary rock environments. Although it has been suggested to use conventional kinematic rock slope stability techniques for footwall slopes, they are not readily analyzed using these methods since the discontinuities do not daylight on the slope face (due to having similar inclination as the slope face). In this study, slope stability analyses are conducted using 2D and 3D limit equilibrium (LE) and the finite element (FE) based method shear strength reduction technique for a theoretical fully and partially discontinuity controlled bi-planar failure mechanism. The advantages and disadvantages of using 2D-3D LE and FE-based methods in bi-planar failure analysis highlighted in the paper.

1 INTRODUCTION

Slopes containing highly persistent discontinuity or discontinuity sets having parallel or sub-parallel inclination to the slope are named as dip slopes or footwall slopes. Especially in open-pit coal mines in sedimentary rocks, dip slopes parallel to the stratum inclination can be formed as shown in Figure 1 (Alejano and Juncal, 2010). In such cases, using conventional kinematic rock slope stability techniques of planar, wedge and toppling failure analyses have been suggested. However, in dip slopes, discontinuities do not daylight on the slope face due to having similar inclination with the slope. As a result, dip slopes become difficult to evaluate using conventional kinematic methods (Hoek and Bray, 1974). Nevertheless, previous case studies on dip slopes have demonstrated the existence of different failure mechanisms (Serra de Renobales, 1987, Brawner et. al., 1971, Dawson et. al., 1983). These failure mechanisms do not necessarily have to occur on existing discontinuities but can also happen by crushing, shearing, or tensioning in the rock at the slope's toe.

The failure mechanisms observed in dip slopes are likely to be assessed using standard limit equilibrium (LE) based methods. However, considering that LE based methods require some assumptions to be made in terms of the geometry of the slip surface and interslice forces, it can be reasonable to evaluate such cases more reliably by using numerical approaches. In a study which examined the factors affecting the footwall slopes and the failure mechanisms in open-pit coal mines, Stead and Eberhardt (1997) suggested that the failure mechanisms can also be evaluated with numerical modeling methods.

Thanks to the advanced technologies and the latest developments in both software and hardware, numerical models can be readily applied to better understand the aforementioned failure

DOI: 10.1201/9781003188339-77

mechanisms and estimate the design-based factor of safety (FoS) values. Numerical analysis of failure mechanisms can be conducted and compared with other methods using a finite element (FE) based software like RS2 (Rocscience, 2020) together with a special technique called “shear strength reduction (SSR)” (Dawson et. al., 1999, Griffiths and Lane, 1999, Hammah et. al., 2004).

Within this general framework, we concisely reviewed the bi-planar failure mechanism, the most common failure mechanism that can be observed in dip slopes. Then, we performed 2D and 3D slope stability analysis on a theoretical dip slope using LE and FE-based methods and compared the obtained results in terms of FoS and failure mechanisms.

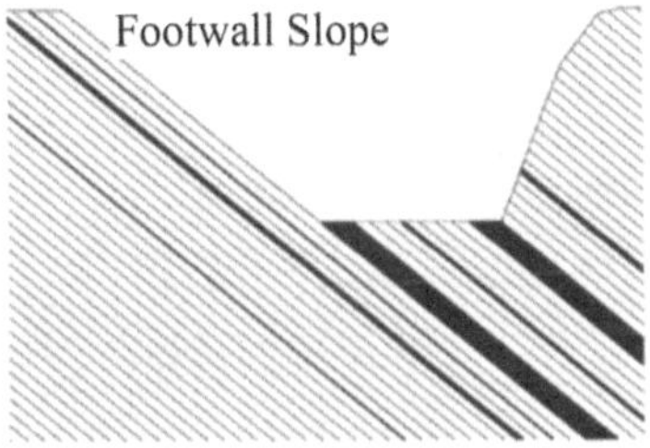

Figure 1. Typical footwall slope that failure mechanisms are expected to develop (modified after Alejano and Juncal, 2010).

2 BI-PLANAR FAILURE MECHANISM

In the literature, three different failure mechanisms exist for dip slopes, namely: i) buckling, ii) ploughing, and iii) bi-planar failure.These three different failure mechanisms are also studied under the following two sub-categories: i) fully discontinuity controlled, and ii) partially discontinuity controlled by the existence or non-existence of a secondary discontinuity. If a secondary unfavourably oriented discontinuity exists, the sliding mass will move only along the discontinuities and a fully discontinuity-controlled, failure mechanism will occur. On the other hand, if there is no unfavourably oriented secondary discontinuity, the discontinuity control will be partial, and the failure will occur due to shearing and/or crushing in the rock at the slope toe. Despite both fully and partially discontinuity-controlled failure mechanisms being similar in terms of geometry, fully discontinuity-controlled failure mechanisms are more likely to occur in practice.

Even though three different mechanisms can be observed in dip slope failures, only the bi-planar mechanism is reviewed in this study; buckling and ploughing failure mechanisms are mentioned briefly for the sake of completeness.

Bi-planar failure mechanisms consist of two blocks: one active and the other passive (Figure 2). As a result, this type of failure mechanism is also named active-passive slope failure in the literature. A load is generated on the passive block due to the slope down movement of the active block. This causes the passive block to lose its stability and failure occurs.

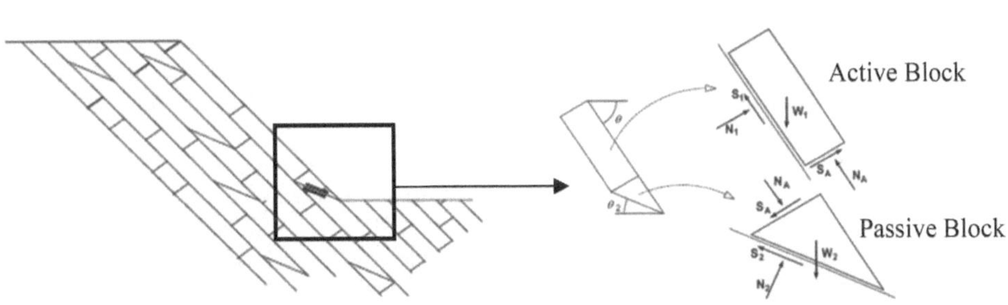

Figure 2. Geometric analysis of bi-planar failure mechanism (modified after Alejano et. al., 2011).

In literature, the bi-planar failure is usually examined in two different mechanisms according to the type of toe failure formed in the passive block. In the case of the existence of a secondary, shallow dipping discontinuity daylighting at the slope toe, the passive block will move along the shallow dipping discontinuity in conjunction with the basal plane (Figure 3a). Hence, a fully discontinuity-controlled failure mechanism will occur. Eberhardt and Stead (1998) have suggested that in cases where the shallow dipping discontinuity makes a 15° or less angle with the basal plane, the tension failure between the active and passive block and the deformation allows the bi-planar failure to kinematically take place. On the other hand, in the case where a secondary, shallow dipping discontinuity does not exist, the failure will occur by shearing and/or crushing in the rock due to the forces generated by active block. Thus, a partially discontinuity failure mechanism will be observed (Figure 3b).

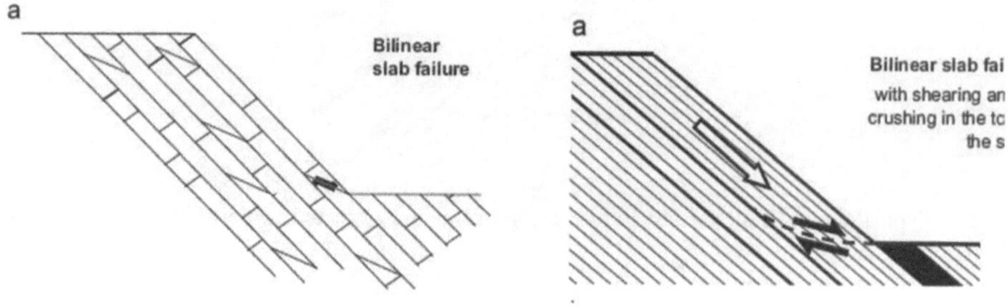

Figure 3. a) Fully discontinuity controlled bi-planar failure, b) Partially discontinuity controlled bi-planar failure (modified after Alejano et. al., 2011).

3 SLOPE STABILITY ANALYSES

Within this study's scope, a fully and partially discontinuity controlled bi-planar failure mechanisms were analyzed using 2D and 3D LE and FE methods to better understand these mechanisms. Then, the obtained results are compared. The slope geometry was taken from Dogan (2019)'s study by partly changing the geometry. The input parameters used in the models have been fictionally determined. The representative geometry of the slope is shown in Figure 4. Additionally, the stiffness and the strength parameters of sandstone material are presented in Table 1, while the stiffness and strength parameters of discontinuities are presented in Table 2. The material and discontinuity strength properties used in this study are defined in terms of the Mohr-Coulomb (MC) failure criterion.

2D and 3D LE analyses were conducted using Slide2 (Rocscience, 2020) and Slide3 (Rocscience, 2020), the advanced two-dimensional and three-dimensional LE-based slope stability analysis software. Additionally, the numerical analyses were performed using the 2D and 3D FE-based software RS2 (Rocscience, 2020) and RS3 (Rocscience, 2020). The numerical analyses require the post-peak strength of materials. Hence, in the slope stability analysis using the SSR approach, the materials are assumed to behave in an elastic-perfectly plastic manner. This allows the FE results to be compared to benchmark analyses with the LE method.

In the FE slope stability analysis using the SSR method, the initial user-defined strength parameters of soil, rock and discontinuities (i.e. cohesion, friction angle, tensile strength, etc.) are reduced by a factor, and then the analysis is conducted. If the results converge within predefined tolerance limits, the strength parameters are reduced by a larger factor and the analysis is conducted again. This procedure is repeated until the results do not converge within the tolerance limits (Zienkiewicz, 1975). The shear strength reduction factor at which results do not converge represents the critical strength reduction factor (SRF). The critical SRF obtained by the numerical analysis represents the FoS value expressed in LE analysis.

Unlike the LE method, two additional parameters are required to characterize the rock joint behavior in the numerical analysis. These parameters are defined as the joint normal (Jn) and joint shear (Js) stiffness, where Jn refers to the deformation of joint surfaces in axial direction subjected to normal stresses, while Js refers to the deformation of joint surfaces in the lateral direction due to shear stresses. Although there are empirical equations in the literature to estimate the Jn and Js parameters suggested by various researchers (Barton, 1972, Bandis, 1983), these equations require additional parameters that can be challenging to obtain. On the other hand, Jn and Js parameters have little effect on the critical SRF values, while they have a major impact on the estimated displacement values. Since their effect on critical SRF value is negligible and our intent in this study was not to obtain accurate displacement values, the Jn and Js parameters were used as the default values provided in the software for the sake of easiness.

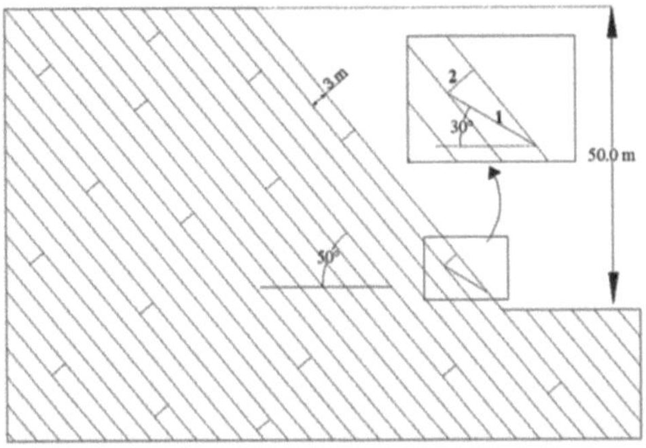

Figure 4. The geometry of the slope used in slope stability analysis.

Table 1. Stiffness and strength parameters of sandstone.

Material	c (kPa)	ϕ (°)	γ (kN/m^3)	Young Modulus (MPa)	Poisson Ratio
Sandstone	128	42	25	5000	0.32

Table 2. Stiffness and strength parameters of discontinuities.

Discontinuity	Dip (°)	c (kPa)	ϕ (°)	Jn (GPa/m)	Js (GPa/m)
Stratum	50	0	30	10	1
Joint 1	30	0	40	10	1
Joint 2	*Perpendicular to Stratum*	0	40	10	1

3.1 *Limit equilibrium method*

In the 2D LE approach, the sliding mass is usually divided into vertical slices. It is divided into vertical columns in the 3D LE approach, and the force and/or moment equilibrium equations for the sliding mass are solved. In the equations, the forces acting on the base of a slice or column are considered, together with the interslice forces. The cumulative ratio of the resisting forces to the driving forces defines the factor of safety (FoS) value.

In this study, two of the most common and conventional LE methods suggested by Spencer (1967) and Morgenstern-Price (1967) were used. The 3D LE slope stability analyses were conducted by extruding the 2D cross-section. The sliding mass was divided into 50 slices and 50 columns in 2D and 3D analyses, respectively. The failure surface type was assumed to be non-circular considering the nature of the failure mechanism. Additionally, for the investigation of non-circular slip surfaces, a very fast and efficient metaheuristic global optimization method developed by Yang and Deb (2010) called "Cuckoo search" was used together with a local search method called Surface Altering (SA) Optimization. This powerful tool provided in Slide2 and Slide3 uses the results of the primary search method to yield lower factors of safety by modifying the geometry of a given slip surface using control points (vertexes) along the surface.

The FoS values are estimated as 0.58 with Spencer method and 0.56 with Morgenstern-Price method for the 2D LEA of fully discontinuity controlled bi-planar failure mechanism. Moreover, as a result of the 2D LEA of partially discontinuity controlled bi-planar failure mechanism, the FoS values are obtained as 0.90 with Spencer method and 0.86 with Morgenstern-Price method. On the other hand, 3D LEA for both fully and partially discontinuity-controlled mechanisms yielded very similar FoS results which are withing the range of 2%. This minor discrepancy in the FoS values obtained by 2D and 3D LE is presumed to originate from the locations and dimensions of the columns in 3D LE. However, this assumption cannot be generalized and remains a subject for future research.

3.2 *Finite element method*

In this study, numerical slope stability analyses were conducted using the FE method in both 2D and 3D. The 3D FE slope stability analyses were conducted by extruding the 2D cross-section that was used in 2D FE analysis, as it was done in LE analyses.

The material and discontinuities were assumed to show plastic behavior and an elastic-perfectly plastic constitutive model was implemented for both material and discontinuities. The maximum required iteration number to reach convergence in numerical analysis was assumed as 1000, and the tolerance for convergence was set as 0.001, both in 2D and 3D analysis. Additionally, the coefficient of earth pressure at rest was selected as 1, in order to obtain comparable results to the LE method.

Boundary conditions or restraints are one of the most important variables that directly effects the results while conducting a 3D FEA. In 3D FE, the geometry should be long enough in the third dimension to avoid the confinement effects of boundary conditions on the results. Thus, 3D FE analyses were conducted on various geometries having different B/H ratios ranging B/H=1 to B/H=10, where H represents the height of the slope and B expresses the length of the geometry in the third dimension.

Figure 5 shows the critical SRF results for partially and fully discontinuity controlled bi-planar failure mechanisms obtained from 2D and 3D FE modelling. In 2D FE analysis, the critical SRF was 0.79 and 1.18 for fully and partially discontinuity-controlled mechanisms, respectively. Additionally, higher SRF values ranging between 10-32% were obtained for 3D FE modelling where the B/H ratios were between 1-3. On the other hand, in the 3D FE models where the length of the geometry in the third dimension was sufficient to prevent from boundary effects (B/H ratio over 4), both the 2D and 3D FEA yielded similar results.

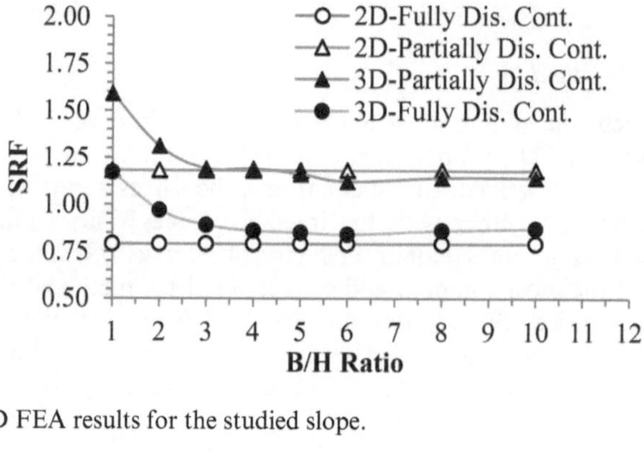

Figure 5. 2D and 3D FEA results for the studied slope.

4 CONCLUSIONS

This study helped to establish that both 2D and 3D LE and FE methods can well capture the bi-planar failure mechanisms analyzed by these methods. However, there are several advantages and disadvantages of using LE and FE methods for such cases and conducting 2D and 3D analyses.

In performing slope stability analysis using the LE method, arbitrary assumptions regarding the failure surface type and interslice forces are made to ensure static determinacy. The accuracy and reliability of the obtained results largely depend on how realistic and accurate these assumptions are. In failure mechanisms where the failure is fully controlled by structural elements, failure types and interslice forces cannot all be taken into account.

Although the FE approach has a few disadvantages, mainly regarding the additional required parameters and computation time, it does not employ initial assumptions on the failure type or forces acting on the sliding mass. As a result, together with the SSR technique, numerical analysis has the edge over the LE method where the failure mechanism is controlled by structural elements.

By the nature of a bi-planar failure mechanism, there needs to be a release joint perpendicular to the slope face for failure to occur. In 3D LE analysis where the geometry is generated by extruding a 2D cross-section and a release joint doesn't exist, the failure mechanism will be exactly the same as 2D and lead to very similar FoS values. Nonetheless, in 3D FE, the development of release joint can be well captured. However, the 3D FE results will produce results very similar to those of 2D FE analysis. However, FE models typically require significantly more computation time. 2D analysis is sufficient if the assessed geometry does not require 3D modeling.

REFERENCES

Alejano, L.R., Ferrero, A.M., Ramirez-Oyanguren, P., Alvarez Fernandez, M.I. 2011. "Comparison of limit equilibrium, numerical and physical models of wall slope stability". *International Journal of Rock Mechanics & Mining Sciences*, 48, 16–26.

Alejano, L.R., Juncal, A.S. 2010. "Stability analyses of footwall slopes in open-pit mining". *Dyna*, 77(161), 61–70.

Bandis, S.C., Lumsden, A.C., Barton, N.R. 1983. "Fundamentals of rock joint deformation". *International Journal of Rock Mechanics and Mining Science and Geomechanics Abstracts*, 20(6),249–268

Barton, N. 1972. "A model study of rock-joint deformation". *International Journal of Rock Mechanics and Mining Science and Geomechanics Abstracts*, 9(5), 579–602.

Dawson, E.M., Roth, W.H., Drescher, A. 1999. "Slope stability analysis by strength reduction". *Geotechnique*, 49(6), 835–840.

Dawson, R.F., Bagnall, A.S., Barron, K. 1983. "Rock anchor support systems at Smoky River Coal Limited". *Annual General Meeting of the Canadian Institute of Mining*. Grand Cache, Alberta, Canada.

Dogan, Y.U. 2019. Comparison of various methods for the rock slope stability analyses (91105), Master's Thesis, Gazi University Department of Civil Engineering.

Eberhardt, E., Stead, D. 1998. "Mechanisms of slope instability in thinly bedded surface mine slopes". *Proc: 8th International Congress of the International Association for Engineering Geology and the Environment*, Vancouver, BC, Canada, Sept. 21-25, 2018, 8(5), 3011–3018.

Griffiths, D.V., Lane, P.A. 1999. "Slope stability analysis by finite elements". *Geotechnique*, 49(3), 387–403.

Hammah, R.E., Curran, J.H., Yacoub, T.E., Corkum, B. 2004. "Stability analysis of rock slopes using the finite element method. *In Proceedings of the ISRM Regional Symposium EUROCK 2004 and the 53rd Geomechanics Colloquy*, Salzburg, Austria.

Hoek, E., Bray, J. 1974. *Rock Slope Engineering*. Institute of Mining and Metallurgy 1st ed. New York, USA, Spon Press.

Morgenstern, N. R., and Price, V. E. A numerical method for solving the equations of stability of general slip surfaces. Computer Journal, vol. 9, no. 4, s. 388–393, 1967.

RS2 v11.007. *2D finite element based software*. Rocscience Inc, Toronto, Canada, 2020.

RS3 v4.013. *3D finite element based software*. Rocscience Inc, Toronto, Canada, 2020.

Serra de Renobales, T. 1987. "Strata buckling in footwall slopes in coal mining". 6th *Congress of International Society of Rock Mechanics*. Montreal, Canada, 527–531.
Slide2 v9.012. *2D limit equilibrium based software*. Rocscience Inc, Toronto, Canada, 2020.
Slide3 v3.010. *3D limit equilibrium based software*. Rocscience Inc, Toronto, Canada, 2020.
Spencer, E. 1967. "A method of analysis of the stability of embankments assuming parallel interslice forces". *Geotechnique*, 17: 11–26.
Stead, D., Eberhardt, E. 1997. "Developments in the analysis of footwall slopes in surface coal mining". *Engineering Geology*, 46(1), 41–61.
Yang, X.S., Deb, S. 2010. "Cuckoo Search via Levey Flights". *2009 World Congress on Nature and Biologically Inspired Computing, NABIC 2009 – Proceedings*, 210–214.
Zienkiewicz, O.C., Humpheson, C., Lewis, R.W. 1975. "Associated and non-associated visco-plasticity and plasticity in soil mechanics". *Geotechnique*, 25(4), 671–689.

The Evolution of Geotech - 25 Years of Innovation – Hammah et al. (Eds)

On the use of Extended Finite Element Method (XFEM) for jointed rock slope problems

S. Moallemi, T. Yee, D. Qi, T. Yacoub, B. Corkum & J.H. Curran
Rocscience Inc, Toronto, Ontario, Canada

ABSTRACT: Finite element method as an advanced numerical technique has been used widely for modeling jointed rock mass, where both joint and intact rock would be modeled explicitly. One of the major challenges that is associated with this approach is conforming the mesh with the joints in complex geometries. As a solution, extended finite element method (XFEM) is introduced to trace the discontinuities implicitly where the discretization of the domain is independent of joints. In this paper, we expanded the application of XFEM in jointed rock slopes for any number of discontinuities in the domain and discussed the factors that can affect numerical instability in analysis.

1 INTRODUCTION

One of the most common problems in rock mechanics is modeling jointed rock slopes, where the position and stiffness of the joint has a major impact in the stability of the domain. In order to present the proper behavior of the discontinues medium, several studies have been done. Some are focused on a macroscopic description of jointed rocks by introducing constitutive models [1] and some are focused on modeling joints and rocks separately. The latter is the focus of this research. Goodman et al.[2] presented the mathematical description of jointed rocks in the context of finite element, in which 2D elements are used to discretize the intact rock and zero thickness elements are used to describe the opening/closure of the joints. This method has been used widely in rock engineering. One of the major problems of this approach can arise in discretization of the domain, when a number of closely spaced joints are presented. This approach can lead to produce sliver elements and ill conditioned system of equations. An alternative approach to skip the issue is using extended finite element method (XFEM). In this approach, the joint would be modeled implicitly, and mesh is not required to be conformed with the joint. This method has been proposed by Belytschko and Black [3] as a general mesh independent technique for modeling discontinuities in solids. After the initial proposal of XFEM, lots of research has been devoted to expand the applications of this approach. Moës et al. [4] generalized XFEM for modeling crack propagations in solids. Another advantage of this method is capturing the stress concentration without any mesh refinement around the crack tips [5].

The application of XFEM is not limited to solids. Retore et. al. [6] proposed the new enrichment for both solid and fluid, to model the fluid flow in the fracture and the porous medium. This approach was extended for modeling hydraulic fracturing in a saturated porous medium [7]. Deb and Das [8] proposed the applications of XFEM for jointed rock mass for simple geometries. Later, Moallemi et al. [9] used similar methodology for complex geometries with high number of joints in rock and compared computational time with explicit method.

DOI: 10.1201/9781003188339-78

In this paper, the mathematical framework for XFEM is expanded to capture the effect of intersecting joints more accurately. The governing equations for jointed rock would be explained and the weak form of equations based on XFEM are presented. At the end, two numerical examples on slope stability are given and the results are compared with explicit method.

2 MATHEMATICAL FORMULATION

2.1 *Governing equations*

In this section, the governing equation for a non-linear rock mass in 2D is reviewed. The domain Ω, as shown in Figure 1, contains discontinuities with the boundary of Γ_d, under the external traction $\boldsymbol{t}$ applied on Γ_t and internal body force $\boldsymbol{b}$ and prescribed displacements imposed on Γ_u.

A set of equations that are required to describe the domain are equilibrium equation and constitutive laws for the rock mass and the joint. The equilibrium can be written as:

$$div(\boldsymbol{\sigma}) + \boldsymbol{b} = \boldsymbol{0} \tag{1}$$

where, $\boldsymbol{\sigma}$ is the Cauchy stress. By assuming the small strain on the domain, the strain can be calculated as the symmetric gradient of displacement by $\boldsymbol{\varepsilon} = \nabla^s \boldsymbol{u}(\boldsymbol{x})$

There are different constitutive models to describe the behavior of rock mass such as Mohr-Coulomb and Hoek-Brown. In general, the relation between stress and strain is given by:

$$\dot{\boldsymbol{\sigma}} = D : \dot{\boldsymbol{\varepsilon}} \tag{2}$$

in which the dot above the stress and strain defines the rate of change of each and D defines the tangential operator, and its components are defined based on the constitutive model of rock. A similar idea can be applied to the joints in the rock mass. Regarding the stiffness and properties of the joint, the rate of traction ($\dot{\boldsymbol{t}}_{\Gamma_d}$) can vary based on the opening and closure of the discontinuity. Similar to equation (2), the constitutive model for the joint can be presented by:

$$\dot{\boldsymbol{t}}_{\Gamma_d} = \mathbf{K} : \dot{\boldsymbol{w}} \tag{3}$$

where $\mathbf{K}$ defines the stiffness operator of the joint and $\dot{\boldsymbol{w}}$ defines the rate of opening/closure along discontinuity.

The weak form of governing equation (1) can be found by multiplying the test function $\boldsymbol{\delta u}(\boldsymbol{x})$ and integrating over the domain, one can write:

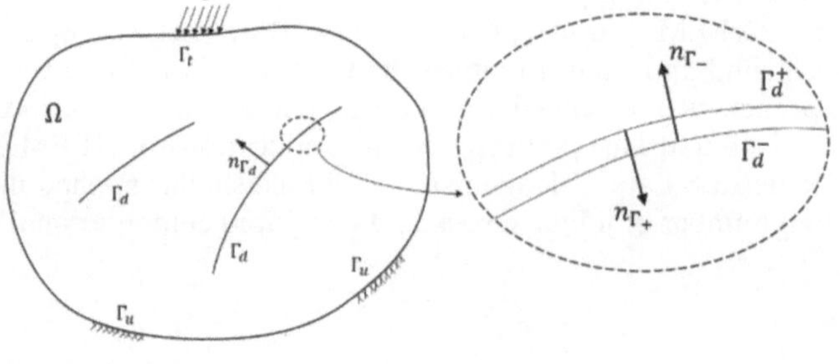

Figure 1. Scheme of a jointed rock mass.

$$\int_{\Omega} \boldsymbol{\delta u}\, div(\sigma) d\Omega + \int_{\Omega} \boldsymbol{\delta u}\, \boldsymbol{b}\, d\Omega = 0 \tag{4}$$

By expanding the above equation, we have:

$$\int_{\Omega} \sigma : \nabla \boldsymbol{\delta u}\, d\Omega - \int_{\Omega} div(\boldsymbol{\delta u}\, \sigma)\, d\Omega - \int_{\Omega} \boldsymbol{\delta u}\, \boldsymbol{b}\, d\Omega = \boldsymbol{0} \tag{5}$$

Using divergence theory, and assigning the integral to both sides of the joints, the above relation can be simplified as:

$$\int_{\Omega} \sigma : \nabla \boldsymbol{\delta u}\, d\Omega - \int_{\Gamma_d^+} \boldsymbol{\delta u}\, \sigma . \boldsymbol{n}_{\Gamma_+} d\Gamma_d^+ - \int_{\Gamma_d^-} \boldsymbol{\delta u}\, \sigma . \boldsymbol{n}_{\Gamma_-} d\Gamma_d^- - \int_{\Omega} \boldsymbol{\delta u}\, \boldsymbol{b}\, d\Omega - \int_{\Gamma_t} \boldsymbol{\delta u}\, \boldsymbol{t}\, d\Gamma_t = 0 \tag{6}$$

in which, $\boldsymbol{n}_{\Gamma_+}$ and $\boldsymbol{n}_{\Gamma_-}$ are the normal on the positive and negative side of the joint as shown in Figure 1. Since both normals on each side are parallel and in opposite direction, if we assume the normal direction of the joint is $\boldsymbol{n}$ and pointing to the positive side of the joint, equation (6) can be simplified to:

$$\int_{\Omega} \sigma : \nabla \boldsymbol{\delta u}\, d\Omega + \int_{\Gamma_d^+} [\![\boldsymbol{\delta u}]\!] \boldsymbol{\sigma} . \boldsymbol{n}_{\Gamma_d} d\Gamma_d - \int_{\Omega} \boldsymbol{\delta u}\, \boldsymbol{b}\, d\Omega - \int_{\Gamma_t} \boldsymbol{\delta u}\, \boldsymbol{t}\, d\Gamma_t = 0 \tag{7}$$

here, $[\![\boldsymbol{\delta u}]\!]$ defines the jump of $\boldsymbol{\delta u}$ on both sides of the joints, i.e. $[\![\boldsymbol{\delta u}]\!] = \boldsymbol{\delta u}^+ - \boldsymbol{\delta u}^-$ and $\boldsymbol{\sigma} . \boldsymbol{n}_{\Gamma_d}$ is the traction along the joint which is defined in equation (3).

2.2 *Discretization with XFEM*

In this section, the discretization of the weak form equation (7) in space with extended finite element method is discussed. In this method, nodes would be enriched based on the type of discontinuity. To capture the discontinuity in the domain, the Heaviside function ($H(\boldsymbol{x})$) is used, which is +1 on one side of the discontinuity and -1 on the other side. Therefore, if $\boldsymbol{u}(\boldsymbol{x})$ defines the displacement in domain, it can be approximated by

$$\boldsymbol{u}(\boldsymbol{x}) = \sum\nolimits_{i=\aleph} N_i(\boldsymbol{x}) \overline{u_i} + \sum\nolimits_{i=1:\aleph^d} N_i(\boldsymbol{x}) (H(\boldsymbol{x}) - H(\boldsymbol{x}_i)) \tilde{u}_i \tag{8}$$

Here, $\aleph$ and $\aleph^d$ are total number of nodes and the enriched nodes in domain, respectively, N_i denotes the standard shape function of ith node and $\bar{u}_i$ and $\tilde{u}_i$ represent the unknowns of the ith node. An example of nodal enrichment is presented in Figure 2. As it can be seen, any node whose support element is crossed by a joint would be enriched.

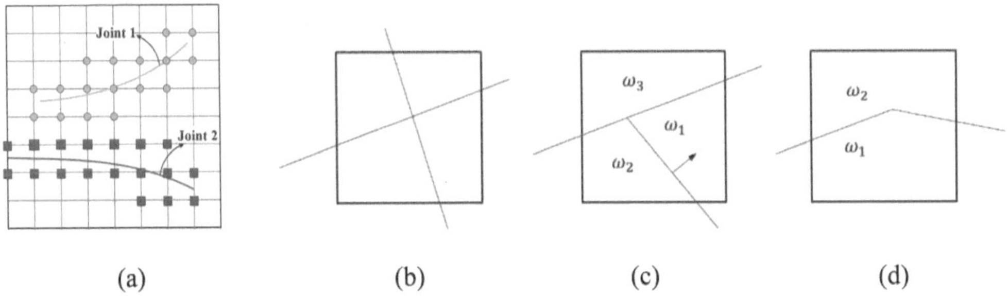

Figure 2. (a) nodal enrichment for elements crossed by joints, (b-d) different cases for crossing joints.

The approximation provided in equation (8) is proper when joints are not crossing each other. In the case of crossing, as discussed in Daux et al [11], an additional enrichment is required which is imposed by using junction function $J(\boldsymbol{x})$. Therfore, if $\aleph^j$ depicts all nodes whose support contains crossing joints, the approximation (8) can be updated as:

$$\boldsymbol{u}(\boldsymbol{x}) = \sum_{i=\aleph} N_i \overline{u_l} + \sum_{i=1:\aleph^d} N_i (H(\boldsymbol{x}) - H(\boldsymbol{x}_i)) \widetilde{u}_i + \sum_{i=1:\aleph^j} (J(\boldsymbol{x}) - J(\boldsymbol{x}_i)) \check{u}_i \tag{9}$$

Where $\check{u}_i$ is the additional unknown of the ith node due to assigning the junction function enrichment. In general, depending on the condition of crossing joints, three different forms for junction function is presented [11, 12]. In the first case, if joints are completely crossing each other (Figure 2-b), the junction is defined as the multiplication of Heaviside function for both joints, i.e. $J(\boldsymbol{x}) = H_1(\boldsymbol{x}).H_2(\boldsymbol{x})$. For the second case, as shown in Figure (2-c) when one joint is crossing both sides of an element (major joint), and the other joint (minor joint) is intersecting the major joint, it divides the area of an element into three zones. The junction function in this case, takes the value of +1 when point $x \in \omega_1$, if the point $x \in \omega_2$ the junction takes value of -1 and finally for any point belonging to ω_3, the junction would return the value of 0. For the last senario where 2 joints intersect at a point (Figure 2-d), the element would be divided into two areas, where $J(x) = +1$, when $x \in \omega_1$ and $J(x) = -1$ when $x \in \omega_2$.

In order to satisfy the stability of the solution for the second and third case of crossing joints, the Heaviside enrichment should be modified. In the second case, the Heaviside enrichment for all nodes inside the crossing element that are enriched with Heaviside functionwith the minor joint, would be removed. Similarly, for the third case, the Heaviside enrichment for both joints would be removed and only junction function would be used to enrich the nodes in the element. The details of the enrichment is explained by An et al. [12]. Taking that into account, by approximating the displacement $\boldsymbol{u}(\boldsymbol{x})$ by equation (9), the equilibrum equation given in equation (7) can be written in the matrix form by:

$$\int_{\Omega} \boldsymbol{B}^T \boldsymbol{\sigma} \, d\Omega + \int_{\Gamma_d} [\![\boldsymbol{N}]\!]^t \, \boldsymbol{t}_{\Gamma_d} \, d\Gamma_d - \int_{\Omega} \boldsymbol{N}^t \, \boldsymbol{b} \, d\Omega - \int_{\Gamma_t} \boldsymbol{N}^t \, \boldsymbol{t} \, d\Gamma_t = \boldsymbol{0} \tag{10}$$

in which, $\boldsymbol{B}$ is a matrix containing the derivative of shape functions, and $\boldsymbol{N}$ is the shape function in a vector form that includes standard shape functions for all nodes and the enhanced shape functions for the enriched nodes. Since equation (10) is non-linear, the linearization algorithm like Newtown-Raphson can be applied to the above equation to be solved iteratively. The details of these procedures can be found in the literature [9].

In order to take the integrals given in equation (10) over crossed element, the sub-triangulation is required to divide the element into continuous zones and then integration can be applied over the element. A sample of this sub-triangulation is shown in Figure (3-a). Noting that, in this work, three gauss points are assigned on each sub-element. For integration along the joints (Figure 3-b), two Gauss points are assigned on the joints inside the linear elements, and three Gauss points if the joints are inside higher-order elements.

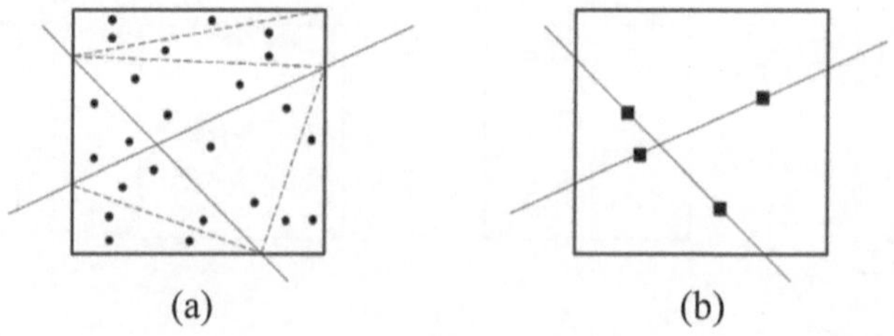

Figure 3. (a) sub-elements in an enriched element and the Gauss point position. (b) Gauss points assigned to the joints.

3 NUMERICAL EXAMPLES

In this section, two numerical examples of slope jointed rocks are solved by XFEM and the results are compared with explicit joint method.

3.1 *Slope with Voronoi joint network*

The stability of a jointed slope is investigated in this example. The geometry of the slope is provided in Figure 4. The dimension of slope is 80m×70m, the slope angle is 72° degrees and the Voronoi joint network is assigned to the slope. As it can be seen in Figure 4, by using the XFEM approach, the mesh is independent of the joint network.

The module of elasticity for rock is assumed to be 20,000 MPa and the Poisson's ratio is 0.3. Mohr-Coulomb constitutive model is applied for the rock with the frictional angle of 20° and cohesion of 0.5 MPa. The slope is subjected to gravitational load with the unit weight of $27KN/m^3$. The normal and tangential stiffness of joints are assumed to be 10,000 MPa/m and

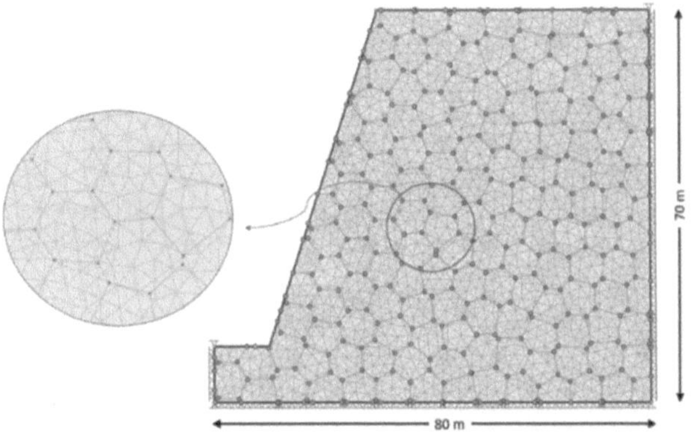

Figure 4. Geometry of the jointed rock slope.

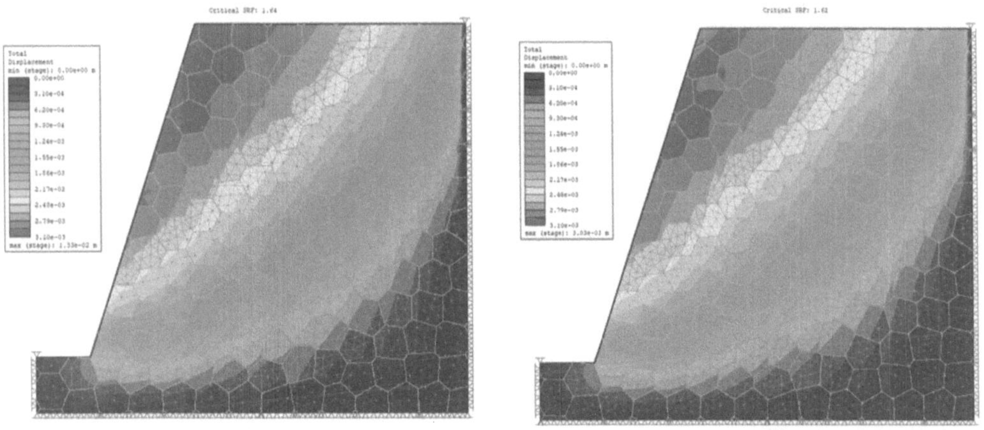

(a) (b)

Figure 5. Total displacement in domain before failure (SRF=1.6) for (a) XFEM (b) Explicit joint method.

1000 MPa/m, respectively. Similar to rock, the Mohr-Coulomb criterion is considered for failure criterion with the Cohesion of 0.5 MPa and 20° frictional angle.

To calculate the stability of the slope, the shear reduction analysis is done and compared with the explicit joint method. In this approach, the properties of failure criterion (i.e. cohesion and frictional angle) would reduce by a factor until the convergence cannot be met for the slope. That factor would be assumed as a SRF factor. For this model, the SRF for XFEM approach is calculated to be 1.64, while for the explicit method, 1.62 is obtained, which are close. In addition, total displacement for both models is depicted before the failure (SRF=1.6) and as it can be seen, the deformed configurations are very similar.

3.2 *Slope excavation*

In the second example, the excavation of jointed rock slope is considered. The geometry of the example is shown in Figure 6. Two sets of parallel joints are assigned to the rock which are intersecting each other. The rock slope is excavated in 10 stages. At the beginning, the water table is on the ground surface, but during excavation, the water table is lowered at each stage.

The rock is assumed to be over consolidated and modified cam-clay model is assumed as its failure criterion. In the following table, the material properties of rock and joints are given:

The slope is under gravitational loading. After each step of excavation, the stress in all Gauss point is calculated and used to obtain the internal load in the next stage, until the end of excavation. The results of XFEM are compared with traditional explicit joint method. In the following figure, the total displacement is compared at the last stage of excavation between these two methods and it can be seen the results are very close to each other.

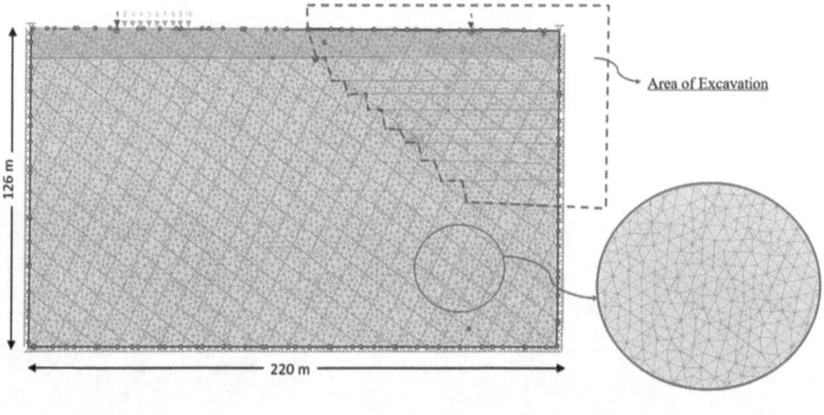

Figure 6. Geometry of the jointed rock in the first stage.

Table 1. Material properties of intact rock and joint.

Unit weight	0.0203 MN/m^3	κ	0.015
Poisson's Ratio	0.3	Over consolidation ratio	2.2
CSL slope	0.67	Joint normal stiffness	500 MPa/m
γ	3.4	Joint tangential stiffness	50 MPa/m

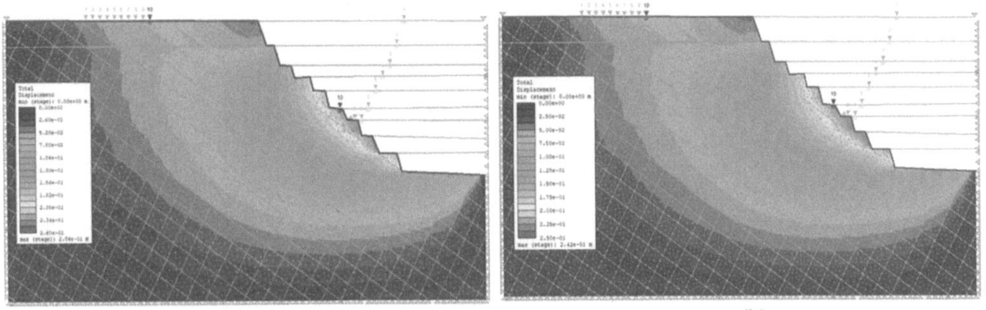

(a) (b)

Figure 7. Total displacement in rock mass at the end of 10th stage (a) XFEM (b) Explicit joint method.

4 CONCLUSION

In this paper, the application of XFEM is explained for a general case where joints are intersecting each other. It has been shown that this method is completely mesh independent and the joints can be placed anywhere in the model regardless of the discretized rock. This benefit of XFEM can eliminate the formation of sliver elements that can happen in traditional approach when multiple joints are placed next to each other.

It has been investigated that the junction functions should be assigned as an additional DOFs, otherwise the results will not be accurate. There are several parameters that can affect the computational cost of XFEM. It has been examined that more intense computation is required in XFEM than the traditional method. However, the mesh independency of XFEM makes it capable of solving similar joint networks with larger mesh sizes, while reducing the cost of calculation.

REFERENCES

[1] M. E. Plesha, 1987, Constitutive models for rock discontinuities with dilatancy and surface degradation, International journal for numerical and analytical methods in geomechanics. 11: 3, 345–362.

[2] R. E. Goodman, R. L. Taylor, T. L. Brekke, 1968, A model for the mechanics of jointed rocks. Journal of soil mechanics and foundation div. 94: 3, 637–659.

[3] T. Belytschko, T. Black, 1999, Elastic crack growth in finite element with minimal remeshing, International journal of numerical method in engineering. 45:5, 601–620.

[4] N. Moës, J. Dolbow, T. Belytschko. 1999, A finite element method for crack growth without remeshing. International journal for numerical method in engineering. 46: 1, 131–150.

[5] N. Moës. T. Belytschko, 2002, Extended finite element method for cohesive crack growth. Engineering fracture mechanics. 69:7, 813–833.

[6] J. Rethore, R. De Borst, A. Abellan. 2007, A two scale approach for fluid flow in fractured porous media. International journal for numerical method in engineering. 71: 7, 780–800.

[7] A. Khoei, M. Vahab, E. Haghighat, S. Moallemi, 2014. A mesh-independent finite element formulation for modeling crack growth in saturated porous media based on enriched-FEM technique. International journal of fracture. 188: 1, 79–108.

[8] D. Deb, K.C. Das, 2010, Extended finite element method for the analysis of discontinuities in rock masses, Geomechanical and Geological engineering. 28: 5, 643–659.

[9] S. Moallemi, J. Curran, T. Yacoub, 2018, On modeling rock slope stability using XFEM, DFNE, 18–1238.

[10] C. Daux, N. Moës, J. Dolbow, N. Sukumar, T. Belytschko, 2000, Arbitrary branched and intersecting cracks with the extended finite element method. International journal for numerical method in engineering. 48: 1741–1760.

[11] T. Belytschko, N. Moës, S. Usui, C. Parimi. 2001, Arbitrary discontinuities in finite elements. International journal for numerical method in engineering. 50: 4, 993–1013.

[12] X. An, G. Fu, G. Ma, 2012, A comparison between the NMM and the XFEM in discontinuity modeling. International journal of computational methods. 9: 2, 1240030–1240058.

Session 18 - Limit equilibrium slope stability analysis

The Evolution of Geotech - 25 Years of Innovation – Hammah et al. (Eds)

Mitigation measures for the protection of working railway lines from landslides: The case study of Altare and Santuario

F. Foria*, R. Giordano & G. Cordua
ETS Srl, Rome, Italy

ABSTRACT: The Italian territory is marked by complex geology, morphology and hydrography. For this reason, hydrogeological instability is a primal issue to phenomena such as landslides and flooding. Within this context, the protection of existing infrastructures plays a significant role along with maintaining the operational continuity and promptly restoring the circulation after a forced traffic interruption.

The railway line Torino-Fossano-Savona is one of the most ancient and a major Italian railway that links the cities of Turin and Savona. In November 2019, the Ligurian area was subjected to a sequence of strong meteorological events that caused landslides and rockfalls. The railway line was interested in a series of landslides which involved ETS Srl (designer) and MICOS Spa (contractor of the works) in the operations of securing the line and the slopes. The line disruption required emergency interventions to restore railway circulation and guarantee the safety of the nearby structures.

1 INTRODUCTION

The Italian territory is marked by complex geology, morphology and hydrography. The Ligurian region is particularly characterized by similar contexts, that lead to high hydrological instability risk and phenomena as landslides and flooding. The railway line Torino-Fossano-Savona is one of the most ancient and a major Italian railway that links the cities of Turin and Savona (Figure 1).

The line has an internal hub at San Giuseppe Cairo towards Savona that splits the line into two different paths: Savona-Torino Via Altare, and Savona-Torino via Ferrania.

In November 2019 a sequence of strong meteorological events caused different landslides and rockfalls involving the two railway lines between San Giuseppe di Cairo and Savona:

- From ch.6+592 to ch.6+663 (Via Altare);
- From ch.13+610 to ch.13+700 (Via Altare);
- From ch. 2+220 to ch. 2+330 (Via Ferrania);
- From ch. 2+070 to 2+130, from ch. 2+160 to 2+200 and from ch. 2+330 to ch. 2+380 (Via Ferrania).

ETS Srl (designer) and MICOS Spa (contractor of the works) were involved in the operations of securing the line and the slopes. For this reason, a systematic approach was fundamental both in the emergency and the design phase.

*Corresponding author
DOI: 10.1201/9781003188339-79

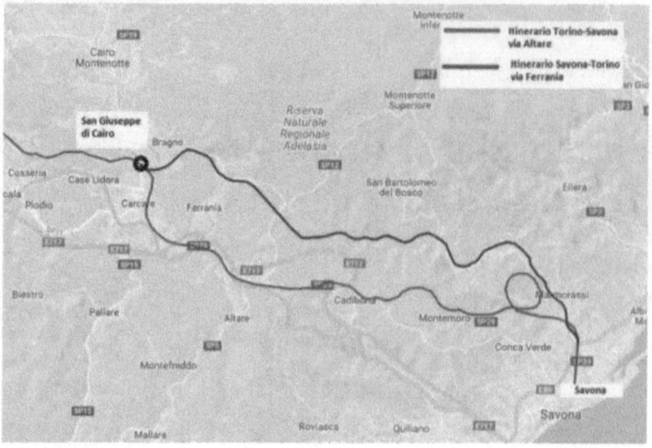

Figure 1. Railway line Savona-San Giuseppe Cairo (Via Altare in purple, Via Ferrania in blue).

This paper is focused more on the main landslide in Via Ferrania, from ch.2+220 to ch. 2 +330, which causes a line disruption that required extensive emergency interventions to restore the railway circulation and to secure the nearby structures.

All these landslides were caused by the topsoil partial saturation, mainly silty, but probably also by the negative effect of the arboreal vegetation in situ.

2 EVENTS OF NOVEMBER 2019

2.1 *Via Altare landslides*

From ch. 6+592 to 6+663 (Via Altare), a shallow landslide near Galleria Santuario portal interested both the debris topsoil and the rocky subsoil. The mechanism did not involve the railway, but clogged a lateral drainage ditch, blocking its outflow. From ch. 13+610 to 13+700 (Via Altare), on the slope above a retaining wall some surface erosion problems arose. The mechanism did not involve structures and railway infrastructures.

Figure 2. (On the left) Landslide from ch. 6+592 to 6+663 (Via Altare)/(On the right) Landslide from ch. 13+610 to 13+700 (Via Altare).

2.2 *Via Ferrania landslides*

From ch. 2+070 to 2+130, a translational landslide on the upper slope of the railway caused large deformation of crib walls, not interfering directly with the railway operation. From ch. 2

+160 to ch. 2+200, a shallow mudflow interested the slope above the retaining wall, interesting also another concrete structure.

Figure 3. (On the left) Landslide from ch. 2+070 to 2+130 (Via Ferrania)/(On the right) Landslide from ch. 2+220 to ch. 2+330.

From ch. 2+220 to ch. 2+330, the main gravitational phenomena occurred. The landslide detachment zone was also adjacent to an existing building, where some anthropic works were involved (drywalls and green area), demolishing a retaining wall at the toe of the slope and directly interesting the railway. The line disruption required emergency interventions to restore railway circulation and guarantee the safety of the nearby structures. The local collapses on the road infrastructure increased the strategic value of the line to ensure the mobility of the area.

Figure 4. Landslide from ch. 2+220 to ch. 2+330 (Via Ferrania).

3 PRELIMINARY ACTIVITY

After the manifestation of the landslides, several on-site inspections were carried out, in order to have a better understanding of the manifested phenomena.

The landslide from ch. 2+270 to km 2+330, due to the line disruption, required some urgency interventions, realized between the end of November 2019 and the beginning of December 2019 were required in order to restore railway circulation (Figure 5). Under these conditions, in December 2019 a GPS topographic survey was carried out, detecting the slope geometry and the landslide contours (Figure 5)

After the removing of the slide mass along the railway, a series of concrete blocks were placed at the toe of the slope, arranged in 4 rows.

Figure 5. (On the left) 3D landslide reconstruction / (On the right) Concrete blocks placed at the slope toe.

In January 2020 a campaign of geognostic surveys was also carried out, achieving a geological, hydrological and geotechnical model of the soils. The campaign was made by the following surveys: 1 vertical borehole drilling at an average depth of 10 m; 1 vertical borehole drilling at an average depth of 15 m; 3 sub-horizontal boreholes drilling at an average depth of 10 m; 14 standard penetration tests (SPT); 5 mechanic laboratory tests on undisturbed soil samples; 23 soil physics laboratory tests on disturbed soil samples; 3 seismic refraction tomography.

Before the beginning of the design process, some other preliminary activities were done. Along these, the integration of the topographic survey with a specific study of the photographic data acquired during the on-site inspections. This study allows the designer more details of the slope geometry, of the landslide and about the exact numbers and positioning of the concrete blocks.

It is important to specify that in Italy, from March 2020, started the Covid-19 emergency and that until May 2020, the entire Italian territory was under a full lockdown regime.

In the meanwhile, the topographic and geognostic survey was undertaken to achieve a better comprehension of the phenomena, along with the processing of the data from the site (Foria, 2019; Foria, 2020). The last part was fundamental to tackle the design and the support to the Client and the Contractor even in lockdown condition induced by COVID-19.

This situation affected also the logistic and the organization of all the activities, making it necessary a distance communication between the technical personal and the construction site staff.

4 DESIGN AND STABILITY ANALYSES

After reaching a sufficient knowledge of the phenomena, in March 2020 the design phase began. The design choices considered the followings aspects: kinematic characteristic of the landslides; geotechnical and stratigraphic properties of the soils; hydraulic properties of the instability mechanism; pre-existing structures in the area; geometry and dimensions of the landslides.

4.1 *Via Altare landslides*

The two minor landslides in Via Altare from ch. 6+592 to 6+663 and from ch. 13+610 to 13 +700 were studied by preliminary inspections on the site, specific geological studies and a geotechnical model based on back-analysis procedures using the collected data.

For the study of the manifested phenomena (for the geotechnical characterization of soils) and the design of the interventions, the software Rocscience Slide2D was used (Figure 6).

Having the gravitation problems geometrical shallow conditions, their resolution was conducted by soil-nailing interventions.

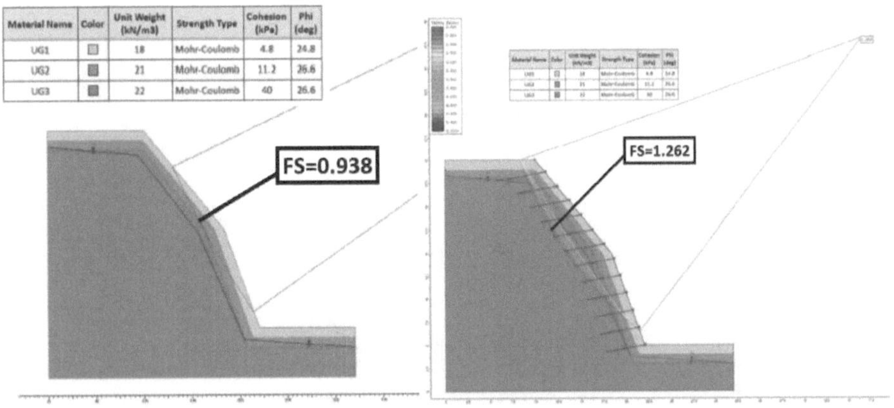

Figure 6. Rocscience Slide 2D analysis outputs.

4.2 *Via ferrania landslides*

For the main landslide from ch. 2+220 to ch. 2+330 (Via Ferrania) the first step was to define the type of failure mechanism that was involved. In order to establish a landslide volume, the following data were put together: the landslides contours coming from the photographic and topographic data; the extension of the tension cracks on the upper, and lateral contours of the landslide (on average between 1 and 1.5 m); the stratigraphy coming from the geognostic boreholes; a correlation between the shear wave velocity Vs outputs coming from the seismic refraction tomography and the results from boreholes stratigraphy in order to assign a maximum velocity limit to the landslide surface.

The type of kinematic was a composite landslide, with a curved form on the upper part of the slope, and linear similar form down. The mechanism is also defined as "progressive" because the failure did not happen simultaneously along its surface.

The slide geometry had a width variable from 10 m to 16 m, with a length of around 40 m, and a depth of 2-3 m from the topographic surface.

The main cause of the landslide is connected to hydraulic problems, due to bad control of the water on the slope.

After the elaboration of all the geotechnical tests coming from geognostic investigations, and the geotechnical characterization of the soils, it was necessary to proceed with a back-analysis procedure in order to characterize also the failure surface.

The geological context is made of topsoil, partially interest by the landslide and made of debris materials from altered rocky sub-soil. Instead, the downer layer has a schistose origin.

These analyses were carried out through Rocscience Slide2D, in order to evaluate the safety factor of the failure surface. All the analysis were simulated using the Morgenstern-Price Method. The sloping topography, along with the stratigraphy of the soil, the failure surface and its tension cracks, were modelled in three different analysis (Figure 7), allowing to define the geotechnical parameters in a failure condition for the landslide (effective cohesion c'=0).

The resulting critical state friction angle was 30°. Table 1 summarizes the characterization with a Mohr-Coulomb failure criterion.

Being the failure surface shallow (average depth around 2-3 m), it was decided a design line composed by intervention as soil-rock nailing systems and underground micropiles bulkhead, in addition to drainage systems, water control systems and the restoration of the walls washed out by the landslide.

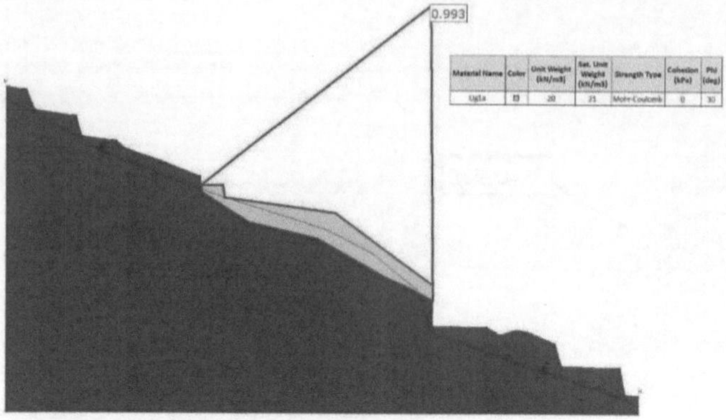

Figure 7. Output example of the Rocscience Slide 2D back-analysis.

Table 1. Characteristic geotechnical parameters for the Units (UG).

	γ		c'			φ'			E'	k	
	γ_n	γ_{sat}	c'$_r$	min	max	φ'$_r$	min	max	loading	min	max
Geotechnical Units	[kN/m^3]		[kPa]			[°]			[MPa]	[m/s]	
UG1a (landslide)	20	21	0	-		30	-		30	$5x10^{-8}$	10^{-6}
UG1b	20	21	-	5		-	32		30	$5x10^{-8}$	10^{-6}
UG2	20	21	-	5	10	-	34	35	40	10^{-7}	10^{-6}
Filling	19	19	-	0		-	35		30	-	

It was also used the Bustamante and Doix method for the geotechnical characterization of the designed micropiles and soil-nails.

These measures were also thought to be in continuation with those already realized between ch. 2+330 and ch. 2+380.

The following measures have been established:

- Consolidation injections and installation of anchor rods for the consolidation of the lower wall still in place;
- Slope clearing, re-profiling and installation of soil-nailing systems;
- Realization of two underground micropiles bulkheads for the slope consolidation;
- Realization of one micropiles bulkhead at the toe of the slope, for the excavation, retaining necessary to the concrete cast of the new lower retaining wall (after the concrete blocks removing). The bulkhead is after solidarized with the concrete cast of the wall;
- Installation of sub-horizontal drains on the wall structure;
- Installation of water control system (water channels connected by water collecting wells). The soil-nailing systems are characterized by the following works: laying of the reinforced 3D geomat; laying of steel mesh reinforcement with steel wire; installation of soil-nails with fixing steel plates arranged in 5 staggered rows.

The project phases and the construction methodology were modelled through commercial software for geotechnical analysis problems.

The global stability analysis of the designed interventions was carried out by Rocscience Side 2D (Figure 8). The final geometries, along with all the interventions were modelled comparing the safety factors obtained with the minimum values required by national technical standards. All the analysis were performed using the Morgenstern-Price method.

Works started in May 2020, requiring during their execution a support design phase, where some intervention required small changes due to executive needs (i.e. modifying the layout of the bulkhead on the slope).

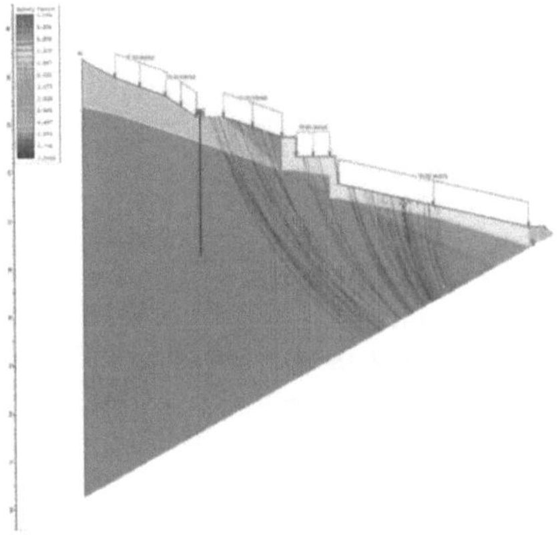

Figure 8. Rocscience Slide 2D design analysis output.

5 CONCLUSIONS

The systematic emergency interventions, survey and design of landslides in a Ligurian area is presented. The approach and design choices were successful. The railway line was disrupted only for a few nights and the final construction works are finished in accordance with the working line scheduling and the strategic needs of the railway line.

ACKNOWLEDGEMENTS

Special thanks to the ETS Team joining this project: Fabio Pulvirenti, Emanuele Miraglia, Luca Terrile and Gabriele Avancini. Thanks for the supervising of Domenico Chiaino and Gabriele Miceli. A further thanks to the contractor MICOS Spa for the support to the design activities and RFI to push in the direction of safe and sustainable design.

REFERENCES

Clouterre, R. 1991. Soil Nailing Recommendations 1991 (English translation by Federal Highway Administration). Report No. FHWA-SA-93-093.

Foria, F. et al. 2019. ARCHITA: an innovative multidimensional mobile mapping system for tunnels and infrastructures. MATEC Web of Conferences. Vol. 295. EDP Sciences.

Foria, F. et al. 2020. Galleria Olmata, from survey to construction: an integrated design approach for the renewal of railway tunnels. WTC 2020.

Lancellotta, R. 2008. Geotechnical engineering. CRC Press.

Morgenstern, N. U. & Price, V. E. 1965. The analysis of the stability of general slip surfaces. Geotechnique, 15(1),79–93.

The Evolution of Geotech - 25 Years of Innovation – Hammah et al. (Eds)

Can 2D cross sections be safely extrapolated?

C.J. MacRobert, T. Mutede & N. de Koker
Department of Civil Engineering, Stellenbosch University

ABSTRACT: Although slope failure is inherently three dimensional, slopes have typically been analysed using 2D limit equilibrium approaches, largely due to computational limitations. This is changing as commercial 3D limit equilibrium computer programs become available. A parametric study was used to shed light on how slope heterogeneity influences safety, and whether it is safe to simply extrapolate (or extrude) a 2D cross-section to carry out a 3D analysis. It was found that critical 3D slip surfaces were dependent on toe conditions in the central zone of the surface and crest conditions in the side zones (of the critical slip surface). In some cases, the critical factor of safety of a 3D heterogeneous slope was lower than the 2D factor of safety computed for the weakest cross-section. Consequently, even the weakest 2D cross section may not reflect the safety of a slope, with this being even more so if a 3D analysis is carried out on an extrapolated 2D section. If 3D approaches are to be used, site investigation work will need to be more extensive than currently utilized for establishing 2D sections.

1 INTRODUCTION

Difficulties in selecting inter-column force functions, combined with convergence challenges in optimization algorithms has hampered the development of three-dimensional (3D) slope stability analyses despite formulations having been proposed several decades ago (Fell et al. 2015). Several assumptions have to be made and a true 'rigorous method' satisfying all equilibrium conditions has yet to be developed. Nevertheless, recent advances have led to commercially available packages such as TSlope, SVSLOPE and Slide3.

For many cases, 3D factors of safety are higher than two-dimensional (2D) factors of safety, although the ratio between the two can vary between 1.0 and 1.4 (Fell et al. 2015). To date practice has relied on 2D analysis, but 3D is increasingly being adopted. It has long been accepted that in back-analysis, neglecting strong 3D effects can lead to an overestimation of strengths, giving this move to 3D analysis credence. However, 3D formulations can have difficulties searching for critical slip surfaces, particularly if dependent on weak layers. Furthermore, a reliance on 3D effects to show a factor of safety is adequate is often considered unwise as the safety of the situation is likely marginal (Fell et al. 2015). Much of this advice is based on extrapolating 2D cross sections and does not necessarily take into account the heterogeneity parallel to the crest of a slope.

The safety of extrapolating 2D cross-sections will be dependent on the degree of heterogeneity inherent in a slope. Investigations of the influence of heterogeneity on the shape of critical failure surfaces (Hicks and Spencer 2010) found that when heterogeneity is small, failure surfaces extend along the entire length of slopes with 2D and 3D factors of safety being similar. These studies further found that with increased heterogeneity, failure surfaces tend to become more localized with a growing contrast between 2D and 3D factors of safety.

DOI: 10.1201/9781003188339-80

Current practice, particularly on tailings storage facilities where embankment walls are long and regular in outer profile, is to carry out in-situ testing on a distinct cross-section from which geotechnical cross-sections are developed, after which stability is analyzed using 2D formulations. When 3D analysis is carried out, it is also often based on this single profile, despite the comparison of multiple cross-sections revealing obvious heterogeneity parallel to the crest. It is unclear how accounting for this heterogeneity would influence calculated factors of safety. This study seeks to investigate how heterogeneity influences the safety of extrapolated 2D cross-sections to determine 3D factors of safety.

2 METHODOLOGY

The investigation was conducted via three main steps, as summarized in Figure 1.

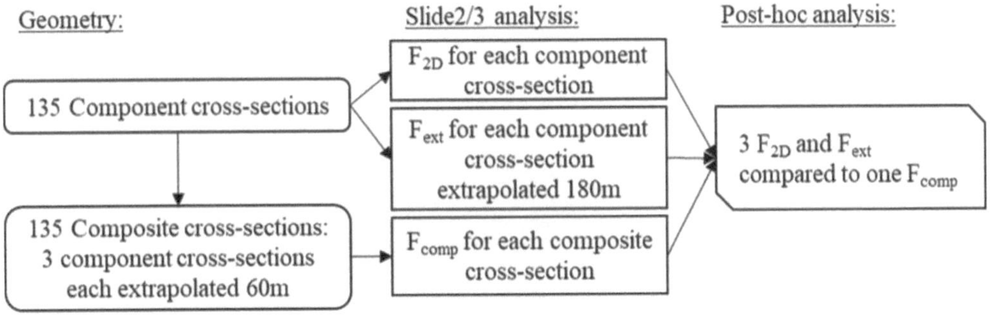

Figure 1. Methodology schema.

2.1 *Geometry*

To investigate the difference between 2D and 3D factors of safety, a parametric study based on a simple cross-section typical of an upstream tailings dam was undertaken (see Figure 2). The profile consisted a fixed outer geometry with adjustable parameters describing the inner geometry. The outer geometry consisted of a 30 m high, 1V:2H slope and base length of 150 m. The inner geometry consisted of two zones, an outer section defined by effective stress parameters (ϕ = 31° and c = 10 kPa) separated by a stepped boundary from an inner section defined with total stress parameters (s_u = 60 kPa). The distance ordinate of each step point (defining six layers) could vary randomly within the given bounds. In total 135 component cross-sections were generated in this initial study.

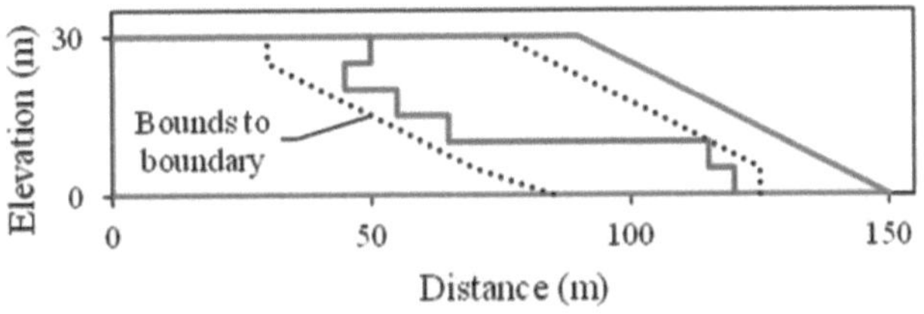

Figure 2. Example of random profile: Blue line is outer boundary and red line is an example of an inner stepped boundary.

2.2 *Slide2/3 analysis*

For each component cross-section a 2D factor of safety (F_{2D}) was calculated using the Spencer limit equilibrium formulation (Spencer 1967). Slide2 was used with non-circular slip surfaces and the auto refine search method (an iterative approach that progressively changes the geometry of a slip surface to converge to the critical surface). Each component cross-section was then extrapolated 180 m, and a 3D factor of safety (F_{ext}) calculated using the Spencer limit equilibrium formulation (the formulation used is an extensions of work done by Huang et al. (2002) and Cheng and Yip (2007)). Slide3 was used with ellipsoidal slip surfaces and the cuckoo search method (a random walk algorithm).

For this initial study, 135 composite cross-sections were formed by joining three component cross-sections each extrapolated 60 m (Figure 3). Care was taken to ensure that no combinations were repeated. No periodic boundaries were set at the lateral ends (i.e. the search zone was limited to the three components). For each of these 135 composite cross-sections a 3D factor of safety (F_{comp}) was determined. F_{comp} is the critical factor of safety taking into consideration heterogeneity parallel to the crest.

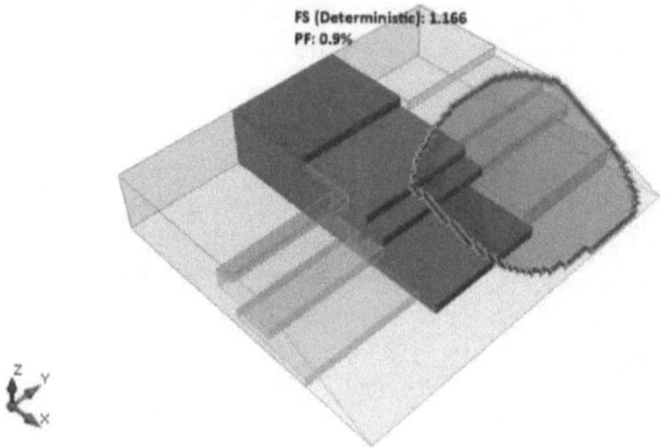

Figure 3. Example of a composite section.

2.3 *Post-hoc analysis*

This analysis resulted in 135 data sets consisting of three F_{2D} values, three F_{ext} values and a single F_{comp} value which could be compared. The heterogeneity in a given composite slope was quantified by calculating the coefficient of variation of the three component factors of safety (i.e. F_{2D} and F_{ext}) making up a composite slope.

3 RESULTS AND DISCUSSIONS

3.1 *F_{ext} compared to F_{comp}*

If parameters had been established for a single cross-section through a slope (as is common practice), we would only be able to determine a single F_{2D} or single F_{ext} value. However, the actual factor of safety for the slope would be closer to F_{comp} as this value incorporates the fact that an adjacent unsampled cross-section may have weaker material. A slope will naturally fail where it is weakest and this section may unfortunately not have been sampled. To accommodate this, state-of-practice dictates that certain factors of safety must be adopted to limit the potentially unsafe situation arising. Consideration is first given to comparing F_{ext} values to F_{comp} values.

It is perhaps unsurprising that in 60% of cases maximum F_{ext} was larger than the corresponding F_{comp} value (i.e. the overall minimum factor of safety for a composite slope taking into account heterogeneity), or that in 53% of cases the median F_{ext} value was larger than F_{comp}. What is surprising was that in 40% of the cases *minimum* F_{ext} was higher than F_{comp} values. This implies that even if we managed to sample the weakest cross-section this may not reflect the weakest condition.

The mean difference between component F_{ext} and F_{comp} values resulting in safe and unsafe scenarios are plotted against the coefficient of variation of the 3 component F_{ext} values in Figure 4. The coefficient of variation (of the three F_{ext} making up a composite slope) is a measure of the heterogeneity inherent to a given composite slope. The higher this value the greater the variation in the conditions between the three sections making up the composite slope. For instance, the composite slope with the highest coefficient of variation (0.22) had F_{ext} varying between 1.03 and 1.60 (smallest F_{comp} was 1.12).

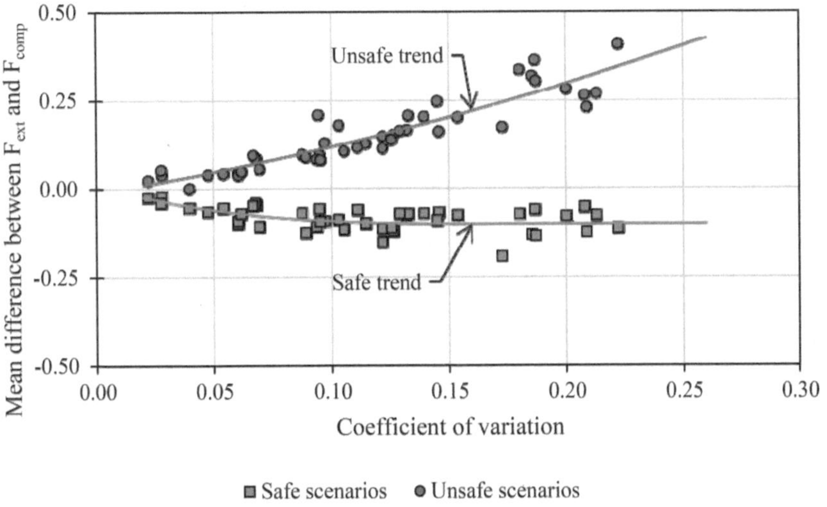

Figure 4. Magnitude of differences as a function of profile variability.

Figure 4 shows that the difference in factor of safety increases much faster with heterogeneity for unsafe scenarios compared to safe scenarios. For unsafe scenarios the mean difference between F_{comp} and F_{ext} ranged from 0.01 to 0.40 (the maximum difference in the current study was 0.48). For safe scenarios, the mean difference increased from -0.02 and remained constant at about -0.10 (the maximum difference in the current study was -0.28).

3.2 *F_{2D} compared to F_{comp}*

Consideration is now given to comparing F_{2D} to F_{comp}, a situation closer to current practice which relies predominantly on determining F_{2D} and ignoring any 3D effects (whether positive or negative). For 30% of the analysis sets, maximum component F_{2D} values were greater than F_{comp}, 25% of median F_{2D} values were greater than F_{comp}, and 19% of minimum F_{2D} values were greater than F_{comp}. Thus, even if the weakest section is sampled and a 2D factor of safety computed there is still a possibility that, within the bounds of the variability set for this problem, a weaker 3D section exists. In the light of contrasts between 2D and 3D factors of safety discussed earlier, this observation seems counterintuitive, warranting further consideration in the following section.

In a similar fashion to Figure 4, the mean difference between F_{2D} and F_{comp} is plotted against coefficient of variability of the three component F_{2D} values in Figure 5. For comparative purposes the trends from Figure 4 are included in Figure 5 as dotted lines. For low heterogeneity (coefficient of variability < 0.07) there were no unsafe F_{2D}. The F_{ext} unsafe trend line formed an upper boundary to mean F_{2D} differences, illustrating the relative safety of carrying out a 2D analysis as opposed to extrapolating a 2D section and carrying out a 3D analysis. The maximum difference between F_{2D} and F_{comp} in the current study was 0.41. With this safety, comes increased conservatism for safe scenarios. For these safe scenarios, the difference in F_{2D} was greater in all cases, being on average -0.15 (a maximum difference of -0.41 was recorded in the current study).

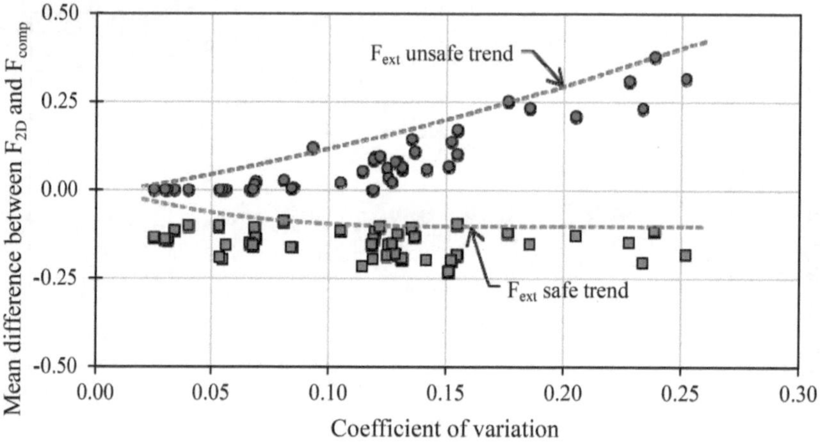

Figure 5. Magnitude of differences as a function of profile variability.

3.3 *Reasons for observed differences*

The observation that in some geometrical combinations the overall factor of safety (F_{comp}) is lower than both the lowest F_{2D} and F_{ext} making up the profile is at first counterintuitive. Consideration of the distributions of heterogeneity reveals that cases with lower F_{2D}/F_{ext} values tended to be associated with profiles with weak zones closer to the toe (e.g. Component cross-section 2 in Figure 6). Even when weak zones were close to the crest, critical slip circles (for F_{2D}/F_{ext}) tended to come out through the toe even if the strong zone was wide in this region (e.g. Component cross-section 3 in Figure 6). Weak 3D combinations arose when the central component had a weak zone close to the toe and side components had weak zones close to the crest. For example, when Component cross-section 2 was sandwiched between Component cross-section 1 and 3, F_{comp} was 1.15 (i.e. lower than all component F_{2D} and F_{ext}). This arises because for 3D stability, the central portion of the slip surface is largely controlled by toe conditions and side portions of the slip surface are controlled by crest conditions.

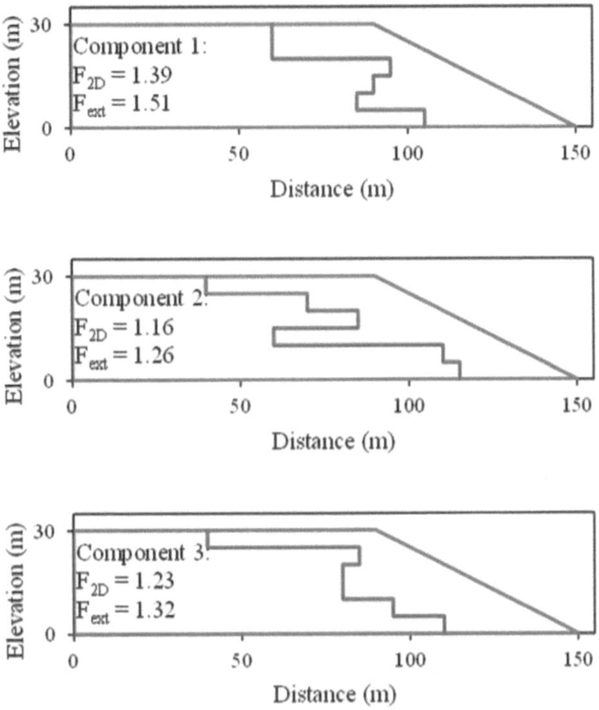

Figure 6. Examples of component cross-sections making up a composite slope.

4 CONCLUSIONS AND RECOMENDATIONS

Using a parametric study in which the inner geometry of a typical upstream tailings dam was varied the following conclusions were made:

- Heterogeneity can result in critical 3D slip surfaces with factors of safety below values computed using weak 2D sections. Critical 3D slip surfaces were found to be dependent on toe conditions in the central zone of the surface and crest conditions in the side zones (of the critical slips surface). Consequently, 2D cross-sections with weak toes conditions can be even weaker in 3D due to weak crest conditions in adjacent zones.
- For 40% of cases the actual minimum 3D factor of safety for a slope (F_{comp}) was lower than the lowest component 3D factor of safety (F_{ext}).
- For 20% of the cases the actual minimum 3D factor of safety for a slope (F_{comp}) was lower than the lowest component 2D factor of safety (F_{2D}).
- The mean unsafe difference between F_{ext} and F_{comp} was higher at all levels of heterogeneity compared to mean differences between F_{2D} and F_{comp}.
- These observations suggest that if insufficient information is available to define heterogeneity parallel to the crest of an embankment it is safer to use 2D analysis over extrapolated 3D analysis. However, this does result in slightly more conservative results if sampling is fortuitously done in a weak layer and heterogeneity provides positive 3D effects (although this would not be known).
- Although this study is in its early stages, design margins of safety should take into consideration that the maximum unsafe difference between F_{ext} and F_{comp}, and between F_{2D} and F_{comp} were 0.48 and 0.41 respectively.

The following recommendations are made for further study:

- As heterogeneity plays an important role, additional cross-sections should be studied. For the case in question, this should include additional layers, wider bounds and narrower distances over which component cross-sections are joined. Other typical embankment geometries should also be tested.
- Only factors of safety are compared in this study, however, by including material variability probabilities of failure can be compared.
- Only the Spencer method was used in this study, and the difference for other limit equilibrium formulations can be explored.

REFERENCES

Cheng, Y., and Yip, C. (2007). "Three-Dimensional asymmetrical slope stability analysis extension of Bishop's, Janbu's, and Morgenstern–Price's techniques." *Journal of geotechnical and geoenvironmental engineering*, 133(12),1544–1555. 10.1061/(ASCE)1090-0241(2007)133:12(1544)

Fell, R., MacGregor, P., Stapledon, D., Bell, G., and Foster, M. (2015). *Geotechnical engineering of dams*, CRC Press, London, UK

Hicks, M. A., and Spencer, W. A. (2010). "Influence of heterogeneity on the reliability and failure of a long 3D slope." *Computers and Geotechnics*, 37(7-8), 948–955. 10.1016/j.compgeo.2010.08.001

Huang, C.-C., Tsai, C.-C., and Chen, Y.-H. (2002). "Generalized method for three-dimensional slope stability analysis." *Journal of geotechnical and geoenvironmental engineering*, 128(10),836–848. 10.1061/(ASCE)1090-0241(2002)128:10(836)

Spencer, E. (1967). "A method of analysis of the stability of embankments assuming parallel inter-slice forces." *Geotechnique*, 17(1),11–26. 10.1680/geot.1967.17.1.11

The Evolution of Geotech - 25 Years of Innovation – Hammah et al. (Eds)

Characterization and stabilization of a mine waste dump-landslide

E. Türk & Ö. Dumanlilar
Akademi Etüt Proje Müh. Müş. Ltd, Ankara, Turkey

S.O. Akbas
Gazi University, Ankara, Turkey

ABSTRACT: Extensive deformations due to a mine waste dump-activated landslide necessitated temporary relocation of about 700 m section of Malkara – Hayrabolu State Highway located in the northwestern Turkey. This paper presents the design actions and measures quickly taken to stabilize this 510 m long, 250 m wide mass movement. Due to rapid mobilization of the landslide mass, it became necessary to estimate the position and properties of the sliding surface mainly through field observations and numerical methods. The analyses indicated that the required safety level under seismic loads could only be reached after an extensive amount of waste dump removal in addition to the construction of at least four rows of drilled shafts. Due to extensive cost of the drilled shaft-based remediation method, a less costly alternative that limits the damage to the highway under seismic loads was proposed.

1 INTRODUCTION

A 700 m section of the Malkara-Hayrabolu State D100 Highway located about 7.5 km north of Malkara county in the northwestern Turkey (Figure 1) had to be temporarily closed due to excessive and progressively developing deformations on and around the pavement observed since January 2020 (Figure 2). Therefore, detailed geotechnical studies including field observations, laboratory studies, and slope stability analyses were carried out to understand the possible causes of the deformations and to determine the triggering factors and reliable remedial measures required to stabilize the highway.

2 FIELD & LABORATORY FINDINGS

The site is located in a valley with a mean elevation of 119 m, and lies mainly on Oligocene aged 1000 m thick Danisment formation that surfaces generally as residual claystone, and characterized also by limestone and sandstone units with occasional lignite layers. Years long mining operations targeting this lignite created waste dumps that extend over large distances with heights of up to 40 m in the east and west of the D100 highway (Figure 3).

During site investigations, clear indications of an about 510 m long, 250 m wide landslide in the west-east direction with a crest at the top of the waste dump and a toe just on the highway was observed as the main cause of the deformations. Signs of the mass movement include cracks up to 1 m in width concentrated on the waste dump berms roughly perpendicular to the direction of the movement, various scarp and minor scarp features that are characterized by sudden elevation differences up to 2 m that are indications of a multi circular failure, and clear heaving on the toe area (Figure 4). Note that the maximum height of the heave material

DOI: 10.1201/9781003188339-81

around the toe was observed to increase from 2.0 m to 4.0 m between two site visits conducted in July and August of 2020, which is a clear demonstration of the activity and high velocity of the landslide.

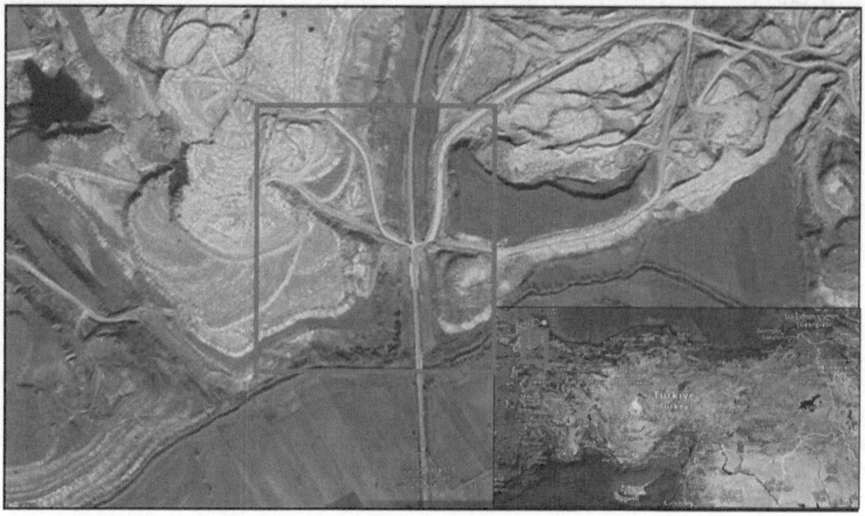

Figure 1. Location and aerial photo of the study area.

Figure 2. Damage observed on the highway.

Figure 3. View of waste dumps and the highway.

Figure 4. Observed signs of the mass movement on the waste dump material.

Based on the observations, a geotechnical investigation program that consists of five borings with a total length of 153 m was conducted. All borings originally included inclinometer setups. However, due to instability issues none of the borings could be drilled on the waste dump, and they had to be relocated and drilled on or near the highway. In addition, even after the relocation, no meaningful inclinometer readings could be taken, because all of the casings were broken before the first reading due to the rapid progression of the mass movement. Therefore, a significant amount of uncertainty was inevitably involved during the characterization and later in the analysis of the landslide.

The data obtained from the boreholes indicate that a medium stiff to hard clay layer of 10 to 15 m in thickness, which is determined to be the weathered product of the underlying rock, with occasional silt layers overlies the parent claystone unit of Danisment formation. The SPT N values measured within the clay layer ranges between 3 and 59, and generally shows a nearly linear increasing trend from surface to the depth of the claystone. A shallow ground water table is measured between depths of 0.9 and 3.5 m.

Laboratory studies were carried out on 68 disturbed, 13 undisturbed soil, and 15 rock core samples. Natural and saturated unit weights, consistency limits, and shear strength parameters were measured on soil samples. The index properties, particle size distribution, and consistency limits were determined according to relevant ASTM standards. Samples were categorized according to USCS (Unified Soil Classification System) (ASTM 2011). The soil samples taken from the study area was classified generally as CH (high plasticity clay), with a few high plasticity silt (MH) samples. The results of the laboratory tests are summarized in Table 1.

Table 1. Summary of laboratory test results.

Property (average)	CH	MH	Clays tone
Plastic limit (PL)	30	36	-
Plasticity index (PI)	35	30	-
Unit weight (γ)(kN/m^3)	18	18	21
Und. shear strength (s_u) (kPa)	85	83	-
Uniaxial strength (q_u) (MPa)	-	-	2.4

As mentioned previously, many uncertainties arose due to time constraints as well as the limited number and the less than ideal location of borings. Therefore, the geometrical characterization of the landslide was based mostly on information gathered from aerial photos, site reconnaissance studies and field investigations supported by laboratory tests. Without any meaningful inclinometer data, the location of the sliding surface was estimated based on a circular shape assumption connecting the toe, the crown and the depths at which the probes were broken in the boreholes (Figure 5). Fortunately, those depths also coincide with the clay-claystone contact at the state highway location.

3 STABILITY ANALYSES

It became obvious that the landslide occurred due mainly to uncontrolled deposition of mine waste dump on east and west of the state highway. Lack of any surface drainage features and the additional water from the broken utility pipes may also have contributed to the instability. Rocscience Inc.'s Slide2 (Rocscience, 2018) was used to conduct limit equilibrium based slope stability analyses. For the high plasticity clay and silt, no drained test result was available. Therefore, the effective stress friction angle (ϕ') values for these fine grained soils were estimated from the PI as recommended by Kenney (1959) and Olson

(1974). Considering the limited data, a conservative approach was taken to determine the effective cohesion (c') by using about 1/20 of s_u. Mass Mohr-Coulomb fit parameters for the claystone were determined using RocLab (Rocscience, 2003) using measured q_u, GSI, and intact modulus. No reliable information was available for the waste dump material. However, it is apparent that it mainly consisits of excavated soils of the area, therefore, parameters equivalent to CH and MH were conservatively adopted. The parameters employed in the analyses are summarized in Table 2.

The residual shear strength parameters of the assumed failure surface were determined using a limit equilibrium back-analysis approach (Figure 6) and are also presented in Table 2, considering that the factor of safety (FS) in the existing geometry without any mitigation is close to unity.

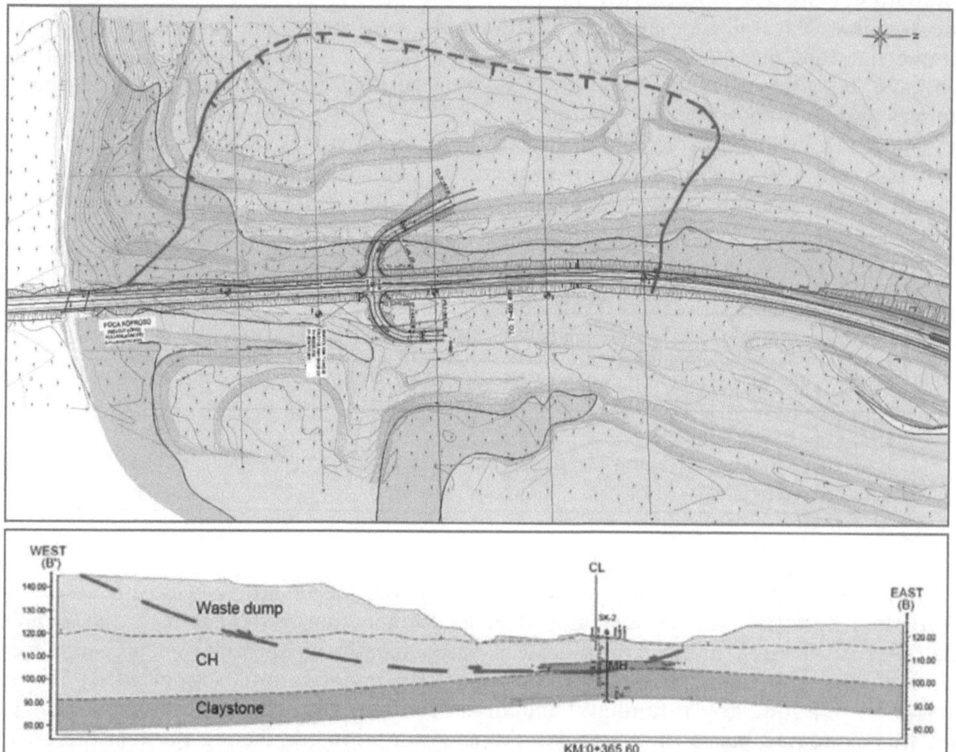

Figure 5. Estimated geometry of the landslide.

Table 2. Geotechnical parameters adopted in the limit equilibrium analyses.

Parameter (average)	CH	MH	Waste dump	Claystone	Failure surface
Unit weight (γ) (kN/m³)	18	18	20	21	18
Effective cohesion (c') (kPa)	5	3	3	53	1*
Effective stress friction angle (ϕ') (°)	28	28	28	16	10*

* Residual values.

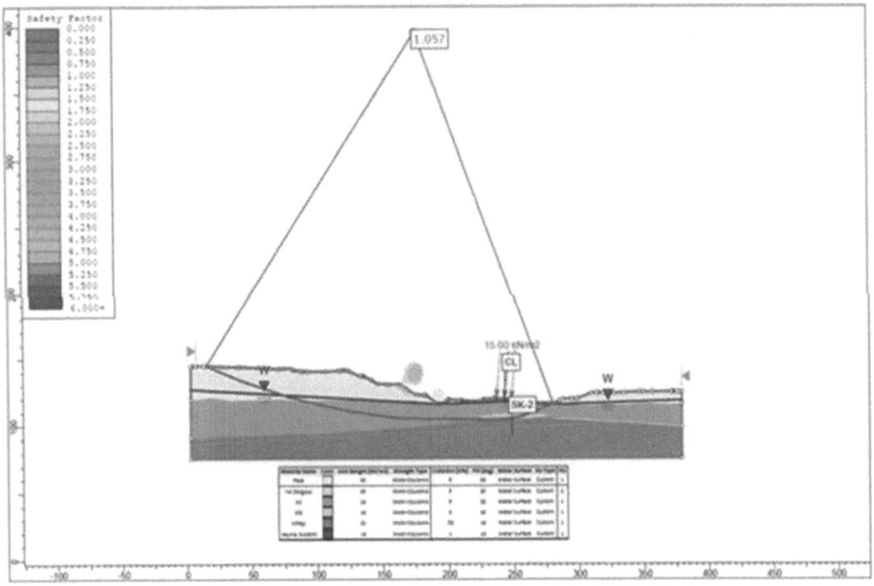

Figure 6. Back analysis of the landslide.

4 STABILIZATION OF THE LANDSLIDE

A parametric limit equilibrium-based slope stability study that involves both removing and slope flattening of waste dump material was conducted. Both circular and non-circular failure surfaces were considered due to the uncertainties that exist due to lack of inclinometer data. The obtained FS values under static loading conditions for various design mitigation options are summarized in Table 3.

Table 3. Results of the parametric study for mitigation options.

	FS	
Alternative	Circular	Non-Circular
130 m removal + 2H:1V*	1.32	2.22
95 m removal + 2H:1V*	1.07	1.31
130 m removal + 3H:1V*	1.72	3.50

* Indicates the target for the remaining waste dump slope after flattening.

As can be seen from Table 3, the first stabilization design alternative, which involves removal of the waste dump material completely for a distance of 130 m as measured from the state highway center line, and flattening the remaining waste dump slope to a target value of 2H:1V meets the target FS value of 1.3 for analyses that are based on a failure surface specified with residual strength parameters as usually adopted by Turkish State Highways practice.

On the other hand, when the analyses are repeated under seismic loading conditions with utilizing a pseudo-static coefficient of 0.18g, which is the half of the peak ground acceleration (PGA) value specified for the site by the Turkish Earthquake Code (AFAD, 2018), the FS reduces well below the target value of 1.1 (Figure 7).

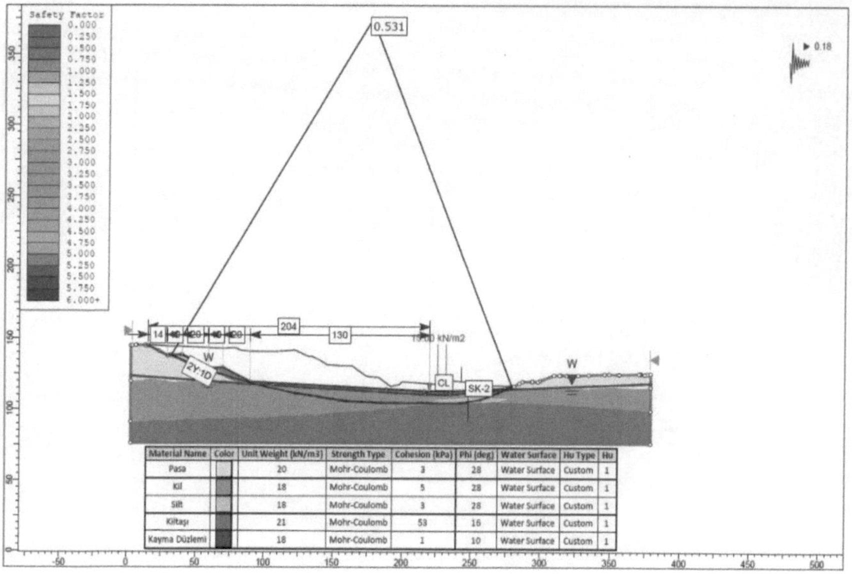

Figure 7. FS under seismic conditions for the selected design alternative.

To overcome this problem, the selected design alternative was modified by the addition of drilled shafts. A series of parametric limit equilibrium analyses coupled with 2D finite element (FE) computations conducted using Rocscience's RS2 (Rocscience Inc. 2014) finite element modelling software indicated that, due mostly to the relatively large amount of mass involved and high water table level, at least 4 rows of 1.2 m diameter 25 m long drilled shafts are required to reach the specified target FS value of 1.1 under seismic loading conditions (Figure 8). For comparison, finite element analyses were repeated using Plaxis 2D (Brinkgreve et al. 2017) and the results that form the basis for drilled shaft design were found to be in good agreement with those that were obtained with RS2 (Table 4).

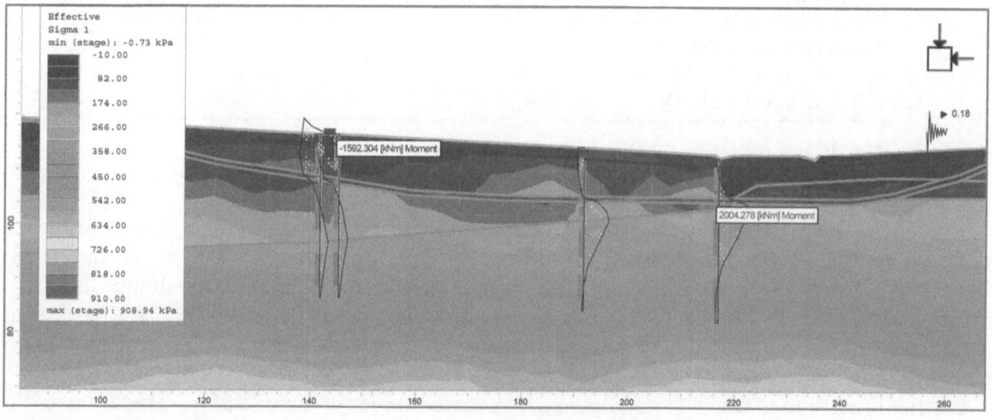

Figure 8. Bending moments on the drilled shafts obtained by RS2 under seismic loads.

Table 4. Comparison of FEM results obtained by RS2 and Plaxis under seismic loading.

	RS2	Plaxis 2D
Max. deformation (m)	1.20	1.18
Axial load (kN/m)*	934	1048
Shear force (kN/m)*	819	1015
Bending moment (kN.m/m)*	2004	2497

* Indicates the maximum value on the drilled shaft.

5 SUMMARY & CONCLUSIONS

A case study of a slope stability problem threatening a major highway was presented. The driving force was the result of uncontrolled dumping of mine waste. Very limited time was available to come up with a practical and applicable solution to continue operating the lifeline infrastructure without any further delay. Field observations, laboratory studies, and slope stability analyses were carried out to understand the possible causes of the deformations and to determine the triggering factors in the absence of any reliable inclinometer readings. Therefore, numerical studies that combine limit equilibrium and finite element methods using Slide2 and RS2 had to be efficiently combined. A comparison of results with FE program Plaxis indicated close agreement with those obtained through RS2. In the end, a solution that involve significant material removal and slope flattening was recommended.

REFERENCES

ASTM 2011. *Standard practice for classification of soils for engineering purposes (Unified Soil Classification System)*. ASTM D2487-11, West Conshohocken, PA

Brinkgreve, R.B.J., Engin, E. & Swolfs, W.M. 2017. *Plaxis 2D manual*. Rotterdam: Balkema.

Disaster and Emergency Management Presidency (AFAD) 2018: *Turkish Seismic Code for Buildings*. Ankara: Prime Ministry of Turkey.

Kenney, T.C. 1959. Discussion. *Jnl. Soil Mechanics and Foundations Division ASCE* 85(SM3): 67–79.

Olson, R.E. 1974. Shearing Strength of Kaolinite, Illite and Montmorillonite. *Jnl. Soil Mechanics and Foundations Division ASCE* 100(GT11): 1215–1229.

Rocscience 2003. *RocLab*. www.rocscience.com. Toronto, Ontario, Canada.

Rocscience 2014. *RS2 Version 9.0 – 2D Finite Element Analysis*. www.rocscience.com. Toronto, Ontario, Canada.

Rocscience 2018. *Slide Version 2018 – 2D Limit Equilibrium Slope Stability Analysis*. www.rocscience.com. Toronto, Ontario, Canada.

The Evolution of Geotech - 25 Years of Innovation – Hammah et al. (Eds)

Case study – Development of a practical design for a landslip site using advanced geotechnical software

D. Egan & A. Zaharescu
Remedy Geotechnics Ltd., Daventry, UK

I. Williams
GeoConsult Ltd., Swansea, UK

ABSTRACT: Several different software packages have been used to develop a practical design for the development of a site located on an active landslide overlooking Portland Harbour, on the cost of Dorset, UK. Three new dwellings are proposed which will require extensive foundation engineering works to stabilise the slope, retain the public highway, and support the new development.

This paper considers the implementation of a range of computer-based geotechnical software packages used to model the geotechnical process at play on the site and to design a practical to install foundation solution.

A combination of embedded retaining wall, slope stability, pile design and finite element software was used to examine the stability of the slope, assess the forces on the structural elements and size the foundation works. This required complex soil-structure interaction modelling to size the individual elements and consider the interaction between them.

1 INTRODUCTION

Three large detached luxury houses are to be built overlooking Portland Harbour at Old Castle Road, Weymouth (Figure 1). Each house will be of two storey construction embedded in the slope with the roof level with Old Castle Road. The site has been affected by landsliding over a long period of time. Between 1999 and 2014 the rear scarp of the landslide has retreated by approximately 7.5m and with a recent episode of slippage in March 2020 to within 3m of Old Castle Road. Portland Harbour was originally a partly sheltered bay located behind the shingle bank of Chesil Beach. In the 19th century major breakwaters were constructed to protect the bay from south-easterly winds.

The general geology of the site is characterised by the Corallian Formation of the Upper Jurassic, West (2012), which appears as the Sandsfoot Grit overlying Sandsfoot Clay members with the limestone of the Clavellata Beds at depth. Figure 1 shows the site setting.

1.1 *Mechanism of landsliding*

Prior to the construction of the harbour the main mechanism of landsliding was wave erosion unloading of the slope's toe creating retrogressive failures progressing inland. The wave regime following harbour construction has been one of low energy, with evidence of accretion of the beach and low rates of toe erosion. Recent slope failure has almost certainly been the result of episodes of high groundwater and flow through the Sandsfoot Grit (exacerbated for the 2020 slip by significant surface water run-off from Old Castle Road). Without intervention

DOI: 10.1201/9781003188339-82

the scarp will progress inland until a stable long-term angle of repose is achieved making Old Castle Road impassable and cutting off access to around 90 properties.

1.2 *Current site development proposals*

The site was acquired by a new landowner in 2020 and the immediate imperative was to secure Old Castle Road. A line of contiguous 450mm diameter bored piles spaced at 600mm centres were installed to the rear of the road kerb line (the kerb piles). The kerb piles were designed for a temporary retention height of up to 3m to facilitate the construction works platform on the main development site.

Figure 1. Site setting.

2 SITE STABILISATION PHILOSOPHY

The overall site development philosophy will:

i) create a stable development area such that the houses can be built to the requirements of the approved planning consent;
ii) improve the long-term stability of the slope such that the houses can be safely supported; and
iii) ensure the proposed house foundation and stabilisation measures are practical and buildable given the prevailing site constraints.

Figure 2 shows the proposed cross section through the middle of the site indicating the key elements of the slope stabilisation and house foundation support system which comprises:

i) kerb piles adjacent to Old Castle Road, implemented to protect the road and allow later excavation in front of the piles to create a working area to construct the permanent works;
ii) an upper retaining wall of 750mm diameter bored piles spaced at 900mm centres to support the ground and work in conjunction with the kerb piles to provide a 7m retained height from the house slab level to the road above, as well as providing support to the ground floor slab and house structure;
iii) bearing piles located under the centre of the house floor slab;
iv) a lower row of more widely spaced 600mm diameter piles supported by an inclined passive drilled and grouted passive tieback;

v) counterfort drains in the slope below the house running down to the beach to control groundwater levels; and
vi) toe loading and slope angle reduction provided by a low height anchored gravity gabion wall and up-filling of ground on the lower slope using free-draining fill.

In developing the final solution, it was essential to ensure the stability of the site could be maintained throughout the construction process.

The proposed construction sequence is to: -

i) install the kerb piles from Old Castle Road to provide temporary support and enable the site below to be developed;
ii) construct the toe retaining structure and up-filling along with counterfort drains within the lower slope;
iii) construct a temporary working platform to allow installation of the lower line of 600mm piles, passive tiebacks and bearing piles;
iv) construct a temporary high level working platform (reusing material from the lower platform) to enable construction of the upper line of 750mm diameter piles; and
v) complete the substructure reinforced concrete works and house construction.

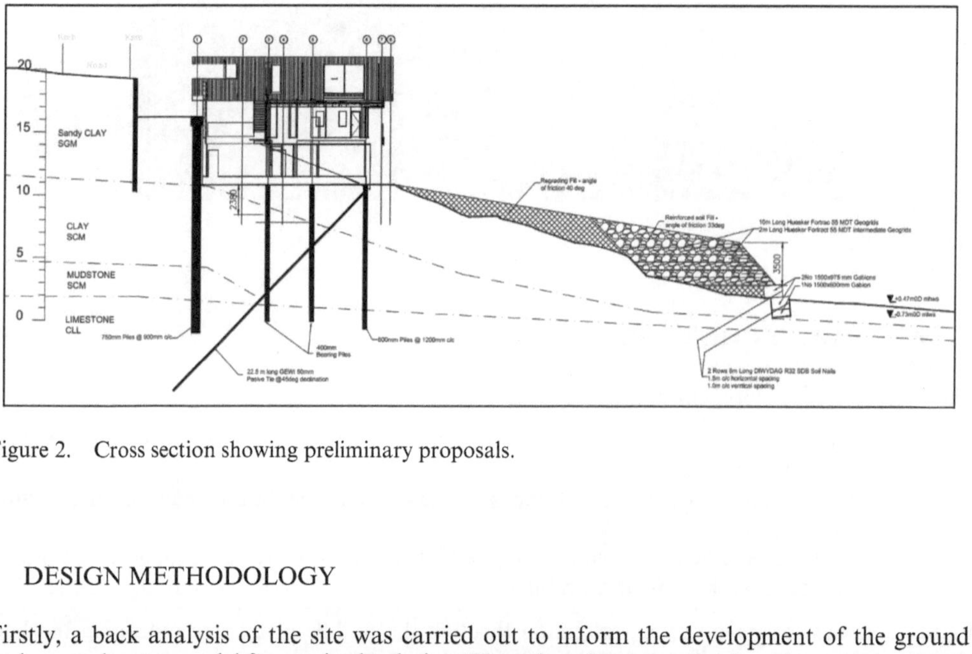

Figure 2. Cross section showing preliminary proposals.

3 DESIGN METHODOLOGY

Firstly, a back analysis of the site was carried out to inform the development of the ground and ground water model for use in the design. Then the scheme was broken down into component elements, and each element sized (the elemental design). Each element was generally sized in isolation from the effects of the other elements. A notable exception being the modelling of the 'push-pull' of the floor slab forces on the upper and lower lines of piles.

On completion of the elemental design all of the elements were brought together in a composite model in Slide2 to represent, as completely as possible, the relevant interactions between the separate elements.

Attempts were also made to model the stabilisation works by means of finite element modelling to replicate the full construction sequence using RS2. This proved to be impracticable not least because implementing cautious shear strength parameters within the FE model led to a simulation that was unstable from the outset. This probably reflects an inherent degree of conservatism in the estimated shear strength parameter values and/or aspects of soil behaviour that could not be fully captured by the Mohr-Coulomb plasticity model adopted for the work. Enhancement of the calibrated shear strength parameters was not considered appropriate or justifiable for an *a priori* design of the stabilisation works. It was felt that the number of uncertainties and consequential assumptions required to progress a finite element model could impair the integrity of the model solution and potentially degrade its reliability for use as the primary design tool. However, RS2 was helpful in providing insight into the general mode of slope failure using the shear strength reduction model (see below).

3.1 *Development of ground model*

The ground model was established from the limited available ground investigation data and back analysis of the slope using limit equilibrium analyses in Slide2 (Figure 3a) and finite element analyses in RS2 using the shear strength reduction method (Figure 3b) to recreate, as best as could be achieved, the mechanism of failure observed at the site.

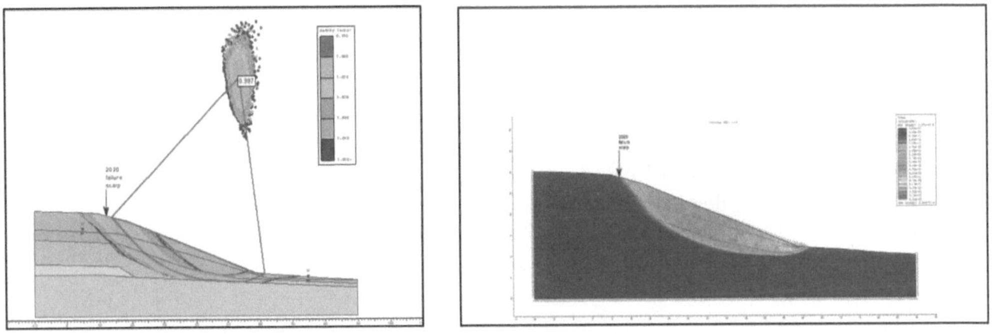

Figure 3. (a) Pre-2020 failure Slide2 back analysis. (b) RS2 back analysis.

Both analyses returned results that are consistent with the mode of failure observed in the field and serve to provide a degree of confidence that the inferred ground model and estimated shear strength parameters are reasonably representative and suitable for design purposes.

3.2 *Kerb piles*

The first, very urgent, task was to design and construct the kerb piles to provide protection to Old Castle Road whilst also having a future role in supporting the ground enabling a temporary works platform to be constructed to allow installation of the upper row of 750mm diameter contiguous piles on the main site.

The elemental design of all the retaining walls on the site was undertaken using the flexible retaining wall programme Wallap run in the subgrade reaction mode. Wallap computes the pile length required for stability (in accordance with the requirements of EC7) together with the bending and shear forces generated in the piles. The kerb pile analysis was straightforward as the ground behind the retaining wall was horizontal (with a traffic surcharge) and the degraded slope in front of the piles was modelled as a sloping berm.

3.3 *Toe regrading and drainage*

It had been decided early in the design process that once the kerb piles were in place the most practical way to open up the site would be by starting work at the toe of the slope. The first elements to be considered in the design were the counterfort drains, toe retaining structure and slope re-grading. The overall stability was confirmed using Slide2 and internal design of the toe retention was by hand calculation and in-house spreadsheets. The critical driver for the regrading works was to achieve a satisfactory factor of safety (FOS) so the temporary works berms could be constructed above for the installation of the bored piles.

3.4 *Upper pile wall*

The focus of the design then moved to the upper 750mm diameter contiguous piled retaining wall. The pile length and wall forces were calculated using Wallap. The main purpose of the 750mm diameter piles was to retain a 7m height from the house floor slabs to the road, working in conjunction with the kerb piles. The interaction of the 450mm diameter kerb and upper 750mm diameter house piles was iteratively modelled in Wallap. In the long-term, propping support will be provided by the floor slab reacting onto the lower line of piles to optimise the design of the upper retaining wall. To verify there was compatibility between the strength and stiffness (and hence forces mobilised) between the upper and lower piles an iterative approach was required.

3.5 *Lower pile line*

The function of the lower line of piles was to support vertical building loads and provide reaction to the lateral forced mobilised in the floor slab due to its propping of the upper piles in the long term. They will also provide some protection to the houses if, in the very long term, some ground loss in the slope below occurred. Passive raking tiebacks were incorporated into the design of the lower piles to help dissipate the lateral loading imparted by the floor slab. The lower line of piles comprised 600mm diameter bored piles at 1.2m spacing, with 200mm diameter passive tiebacks at typically 2m spacing.

3.6 *Bearing piles*

Six 400mm diameter bearing piles will be incorporated into the foundations for each house to help support vertical loading and reduce the span of the floor slab between the upper and lower lines of piles and prevent any appreciable foundation loads being applied to the slope face.

4 COMPOSITE DESIGN

Having completed the elemental designs it was necessary to ensure that the various elements worked together, and the composite design was modelled within Slide2 (Figure 5).

4.1 *Slide2*

Slide2 was used to investigate the overall degree of improvement in slope stability at various stages of the construction process and arising out of the incremental installation of the individual elements. A particular feature of Slide2 is that the resistance provided by piles and passive tension piles can be incorporated within the limit equilibrium analysis. The piles were modelled with a defined shear resistance and the passive tiebacks with a defined pull-out resistance and assumed un-bonded length.

4.1.1 *Modelling pile shear resistance*

Pile shear resistance can be modelled within Slide2 with one of four options: i) importing properties having first pre-processed them in RS Pile; ii) Equivalent Fluid Weight (EFW); iii) Ito & Matsui; and iv) shear resistance input.

For this design method iv) shear resistance was adopted due to its ease of implementation and rational approach, and the fact that the piles were relatively closely spaced rendering the Ito & Matsui approach inappropriate.

4.1.1.1 Slide2 Pile shear resistance input approach

With all the methods the shear resistance provided by a pile appears to be incorporated within an individual slice force calculation and, depending on its magnitude, increases the available stabilising resistance and hence calculated FOS for a particular failure surface geometry. It is vitally important therefore that the shear resistance entered into the programme is representative of that which can actually be mobilised by the piles and this in turn is dependent on several factors including the magnitude of movement between the ground and the pile, the location of the slip surface along the pile length, the degree of embedment of the pile below the slip surface and the structural capacity of the pile. Since a limit equilibrium analysis does not consider ground movement it is necessary, prior to entering Slide 2, to establish the available resistance that a pile can provide for the design situation and limit state being considered. Figure 4 shows the relationship between soil displacement and mobilised shear resistance for a 750mm pile subjected to soil displacement of 10mm to 60mm. To obtain these data, the lateral displacement behaviour of the pile was modelled in Oasys ALP. Thereafter two design situations were investigated within Slide 2D i) the ULS condition where it was assumed that significant ground displacement could take place and hence the shear resistance is governed by the structural capacity of the pile (435kN) and which would be available at a ‘large’ ground displacement, with the overall situation being considered satisfactory at a FOS no less than 1.5, and ii) the SLS Condition where it was assumed that ground displacement would be limited to 10mm and hence a mobilised shear resistance of 191kN would be available, with the overall situation being considered satisfactory at a FOS no less than 1.2. As is can be observed in Figure 5a, in the ULS condition the lowest FOS is 2.376 passing through the piled structure, whilst for the SLS condition the minimum FOS found is 1.950 as shown in Figure 5b.

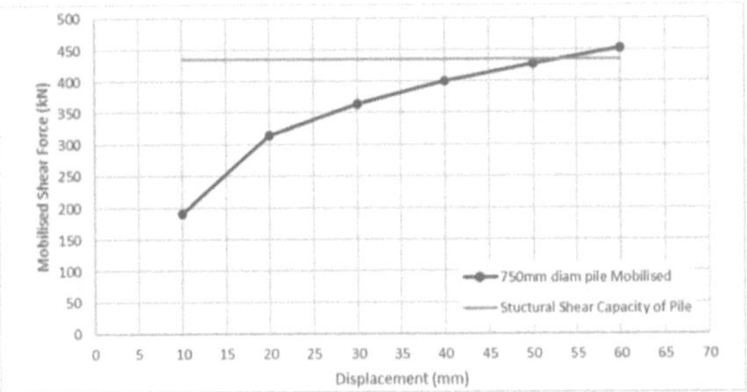

Figure 4. Mobilised shear force and soil displacement relationship for a 750mm pile.

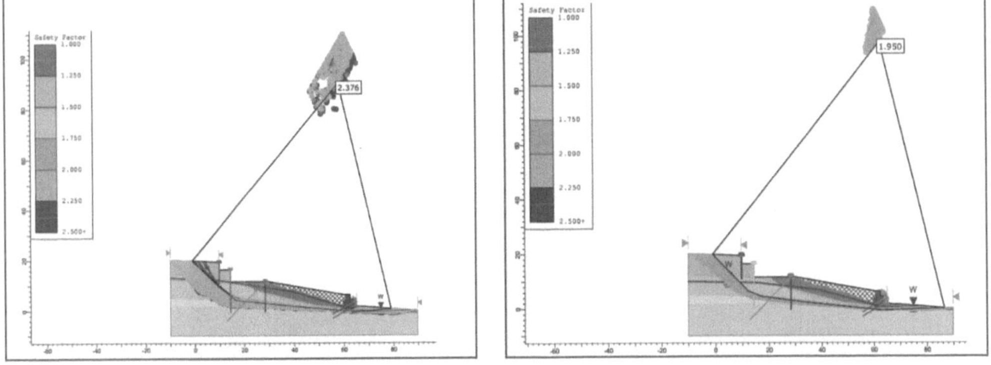

Figure 5. a. Slide2 composite ULS analysis. b. Slide 2 composite SLS analysis.

5 SUMMARY

The development of a complex landslide site required the use of several specialist geotechnical software tools. The design approach was first to design each discrete element of the stabilisation and foundation solution before bringing them all together to undertake a combined design verification. The combined design verification was undertaken using Slide2, which has the benefit of being able to incorporate the effect of shear resistance from stabilising piles and tension elements but has the limitation that it is a limit equilibrium analysis that does not consider ground movement. Attempts were made to model the performance of the stabilisation works using RS2, but this proved to be impracticable due to the unstable state of the failed slope and difficulties in calibrating the FE model to the required degree of confidence for it to be incorporated into the final design process.

REFERENCES

West, I. 2012. Portland Harbour, Dorset. Geologhy pf the West Coast of Southern England. https://wessexcoastgeology.soton.ac.uk/Portland-Harbour.htm. 2012

The Evolution of Geotech - 25 Years of Innovation – Hammah et al. (Eds)
© 2022 copyright the Author(s), ISBN 978-1-032-03645-8
Open Access: www.taylorfrancis.com, CC BY-NC-ND 4.0 license

Author Index